fundamental physical constants

SYMBOL	VALUE	UNITS (mks)
c	2.998×10^8	msec^{-1}
ϵ_0	8.85×10^{-12}	farad m^{-1}
μ_0	$4\pi \times 10^{-7}$	henry m^{-1}
h	6.625×10^{-34}	joule-sec
k	1.38×10^{-23}	joule °K^{-1}
q	1.602×10^{-19}	coul
m_0	9.106×10^{-31}	kg
q/m_0	1.759×10^{11}	coul/kg
kT/q	25.9×10^{-3}	volts
ev	1.602×10^{-19}	joule
N	6.024×10^{23}	molecules per mole

semiconductor electronics

JAMES F. GIBBONS

Department of Electrical Engineering
Stanford University

SEMICONDUCTOR ELECTRONICS

McGRAW-HILL BOOK COMPANY

New York   San Francisco   St. Louis
Toronto   London   Sydney

PREFACE

MOST STUDENTS approach their first course in electronics with a sense of excitement and anticipation, and rightly so. They are aware of the impact electronics has on our society. They are also aware of the promises made by the electronics industry and the graduate schools that exciting careers are open for electronics engineers and scientists. As a result, they are generally a group of very highly motivated students.

The first course in electronics must present to these students some of the concepts and principles of electronics that will be broadly applicable in the remainder of their professional careers while at the same time maintaining their high level of interest and motivation. It is generally agreed that a basic understanding of the operating principles for electronic devices (excluding devices whose behavior cannot be properly understood without recourse to Maxwell's equations) and a basic competence in the analysis and design of circuits using these devices are two important components of the first course, regardless of the field of specialization a student ultimately selects.

It is also a fact that in today's crowded curriculum, the first electronics course may provide the only formal training many students receive in these two areas. The course must therefore provide a working knowledge of this material for students who will later specialize in information theory, control systems, and radio propagation, and even for those students whose major interest is in other branches of engineering or the basic sciences. At the same time, it must be a foundation course for students who elect to specialize in electronic device physics, device design, active network theory, or microelectronics.

My colleagues and I realized that even as late as 1960 our first course in electronics did not meet the needs of this broad range of students adequately, nor did it accurately reflect the needs and the state of modern electronics. The reasons for this seemed clear. The available texts for such a course gave about equal coverage to vacuum tube and semiconductor electronics, whereas the vast majority of electronic circuits and systems that modern engineers and scientists come into contact with, whether as users or designers, employ solid-state components. The available texts also gave considerably more emphasis to the formal analysis of electronic circuits than to flexible principles of circuit design; and they gave only a survey coverage to the physical principles that determine the behavior of electronic devices, particularly semiconductor devices. By contrast, many of the exciting careers in electronics that industry and the graduate schools advertise are for students who have a thorough grasp of the physical principles that govern solid-state device behavior and who have the ability to design electronic circuits using new solid-state components.

It seemed clear to us that it was time for a fresh start. Accordingly, I was given the opportunity of beginning a series of pedagogical experiments aimed at producing an entirely new first course in electronics—not an updated revision of an existing course. The first experiment consisted of a two-quarter sequence of lectures (about 60 one-hour lectures in all), given to a volunteer group of about 20 junior-level students in electrical engineering plus a few physics, chemistry, and physiology majors with a prior interest in electronics. The aim of these lectures was to give the students (1) a firm physical grasp of what holes and electrons are and how they behave in semiconductors; (2) a clear understanding of the effects that occur at a *pn* junction when bias is applied to it; (3) valid physical descriptions (quantitative and qualitative) of how junction diodes and transistors operate; (4) an understanding of the processes by which models are made, for both circuits and devices (and in particular the need for making key approximations); and (5) realistic experience in designing amplifiers, oscillators, and simple switching circuits.

I attempted to keep the center of attention focussed on major effects, all of which are relatively simple. The lectures did not delve deeply into subtle physical phenomena or extreme sophistication in circuit design. At the same time, they were definitely not intended to be a survey in any way. I also wanted to make sure that adequate attention was given to experimental electronics, and that the equal partnership of theory and experiment was emphasized. To accomplish this objective, a number of lecture demonstration experiments were developed to give the students an appreciation of basic experimental realities and some insight into the quality of the agreement between theory and experiment. I also had the students build some simple circuits (a power supply, an audio amplifier, an oscillator, and a multivibrator) as part of their homework. The circuits had to meet certain specifications, and were graded with meters and oscilloscopes whose scales were marked off "A, B, C, D" according to how nearly the circuit met the prescribed specifications.

The course met with an enthusiastic reaction among the volunteer group of students. It therefore seemed reasonable to try it next on a larger group of students. To accomplish this, a limited preliminary edition consisting of some of the lecture notes (particularly those topics that were unavailable in other texts) together with instructions for building some of the demonstrators was published. This preliminary edition was used by electronics teachers at a number of schools across the country and by teachers in the Electrical Engineering Department at Stanford who had little prior experience in teaching semiconductor electronics. It was also used by me and by other teachers as the basis for industrial "in-house" courses for practicing electrical engineers who grew up in the vacuum-tube era, and for self-study by a number of individual students at Stanford (primarily mathematics and physics majors for whom electronics was a hobby). Finally, the preliminary edition was carefully reviewed by teachers who did not use it in class but who were thoroughly familiar with the subject matter.

This extensive testing produced many valuable suggestions that have been incorporated into this book. In particular:

1. A number of new experiments have been included and old ones have been revised to be more effective. Experimental techniques and results are described in detail at appropriate points in the text, using oscilloscope traces and other direct records of experimental performance. This approach gives the student an appreciation of experimental problems and it permits him to make calculations by using experimental data and thus to compare theory and experiment without actually seeing the experiments performed.

2. A large number of numerical examples have been introduced, particularly in the sections on circuit design.

3. All circuit-design examples use data taken directly from transistor data sheets rather than "typical" parameters.

4. The basic ideas necessary to understand the operating principles and applications of vacuum tubes have been included, using the same theory-and-experiment format found elsewhere.

To insure that the basic aims and philosophy were not diluted by the increase in book length that was necessary to incorporate these suggestions, certain key chapters of the final book were made available to a cross-section of students, teachers, and reviewers for their comments. Their reactions suggest that the book will satisfy serious students in electronics, whether they intend to specialize further in this subject or not.

Preparation. For the most part, I have attempted to write the book for students of a certain level of scientific maturity rather than a specific background preparation. The maturity level assumed at the outset is that of a beginning junior in electrical engineering or physics. By this I mean that the students have either had or can quickly learn (from listed references) the elements of Fourier analysis, some basic atomic physics, and some basic circuit theory.

As the subject unfolds, a mastery of mesh and nodal circuit analysis, in both its differential equation and Laplace transform aspects, is required. Electrical engineering majors who take the course at Stanford obtain this material from a concurrent course in network analysis; again references for additional reading are provided at appropriate points for students who are not taking the circuits course. The references have proved to be an acceptable substitute, though a progressive increase in maturity level of the student is required.

Teacher's Manual. The book has been organized so that fundamental ideas are presented at the beginning of each section and elaboration follows. The intention here is to provide for flexible use of the book by those who have specific needs and specific limitations. The entire book can readily be covered in a full academic year (90 lectures). Suggestions for courses of 15, 30, 45, and 60 lectures are given in the Teacher's Manual for *Semiconductor Electronics,* together with solutions to selected problems.

Acknowledgments. It will be evident from earlier statements that I am heavily indebted to many people for their contributions to this book. I particularly want to acknowledge the positive influence of many stimulating discussions I have had with students who have taken the course, both at Stanford and in industry. Their reactions, together with comments and reactions from teachers and students at other universities, have helped to provide the encouragement necessary for carrying out the work.

I am also indebted to my colleagues in the Electrical Engineering Department at Stanford, not only for giving me the opportunity to develop this course, but also for much assistance and encouragement along the way; and to the University administration in general and Professor J. G. Linvill in particular for creating an environment in which I have been encouraged to do research work, graduate teaching, and industrial consulting in semiconductor electronics. Writing an undergraduate text on the subject would have been difficult indeed without this background.

Finally, I am indebted to Professors J. G. Linvill, D. F. Tuttle, and R. J. Smith, and several members of the Semiconductor Electronics Education Committee for many useful discussions that have helped shape the philosophy of the book; to Professor G. L. Pearson and Mr. K. G. Sorenson for assistance in preparing demonstration experiments; to Mrs. M. Cloutier for expert secretarial services; and to the McGraw-Hill Book Company for making the book available in a preliminary edition.

J.F.G.

CONTENTS

17 LUMPED MODELS AND THE HIGH-FREQUENCY BEHAVIOR OF TRANSISTORS

18 HIGH-FREQUENCY CIRCUITS AND TRANSISTOR LIMITATIONS

semiconductor electronics

INTRODUCTION

ELECTRONICS is the rather vague name that is usually applied to a broad field of activities, all of which are in some way concerned with the generation, transmission, and reception of *information*. To begin our studies of this subject, it will be helpful to introduce some basic ideas and briefly describe some typical electronic systems so that we can obtain a somewhat clearer picture of just what electronics is, and how the topics which receive the greatest attention in this book are related to the whole.

The word "information" has been italicized in the definition of electronics because this is a primary characteristic which distinguishes an "electronics" system and electronics engineering from a "power" system and power engineering. Some examples will elucidate this distinction.

Any electric current has two attributes of fundamental importance: it carries *energy* and, in a suitably defined sense, it carries *information*. The sense in which information is carried is that the form and/or presence of an electric current implies that an event has happened.[1]

These two attributes of an electric current are inseparable, though frequently one of them is of primary interest. For example, the fact that a light bulb is on conveys the information that a certain sequence of switches is closed, that the electric power station is operating, and so on. However, the power generation and distribution system is constructed to energize the light bulb, not to answer questions about the positions of various switches. On the other hand, a radio transmission and reception system is designed for the primary purpose of conveying information. The form of the electric current which drives the loudspeaker enables us to identify the sounds that originate in the broadcast studio. We may wish to ensure that the speaker creates sufficient air-pressure variations to make the sound clearly audible, so we shall be concerned with the energy content of the electric current supplied to the speaker. However, the energy content is not the most important attribute of the electric current in this case. We classify a system as being electronic if the *information content* of the currents and voltages in it are of most importance. A power system is one in which the *energy attribute* of the current is the key factor.[2]

We usually refer to a quantity that carries information (a current, a voltage, or a sound-pressure variation, for example) as a *signal*. In an electronic system, we are concerned with transmitting signals from their point of origin to a desired destination, and then presenting them in an appropriate way to an observer. The signal may change its form several times during this process, though finally it will be reconstructed in a way which enables the observer to understand the message.

It is convenient to represent the signal processing that occurs in an electronic system by means of a block diagram; that is, a set of interconnected boxes in which each box performs an assigned signal-processing function. The boxes are simply labelled in the diagram according to the function they perform. We shall use this technique to illustrate some of the types of signal processing that occur in electronic systems common to our everyday experiences. The names of certain of the boxes, which are also called functional blocks or subsystems, will be italicized when they are first mentioned. The components needed to build these (and many other) subsystems, the manner in which the components are connected to achieve the desired subsystem functions, and the overall performance characteristics

---

[1] We shall expand on this definition specifically in Chap. 16 and by examples elsewhere in the text.

[2] Electrical engineering is not cleanly divided by this definition (for example, a power supply in a radio receiver is part of an electronic system in which the energy content is of greatest importance), but it is more adequate for our purposes than other definitions based on light or heavy currents, power flow, the importance of electronic phenomena, and so on.

of the resulting systems and subsystems will be discussed at length in various parts of the book.

Signal processing in some electronic communication systems. As our first example, let us consider the simple telephone system shown in Fig. I · 1a. We shall take the original signal in this system to be the air-pressure variations created by the person speaking. These air-pressure variations are directed at the carbon microphone in the telephone handset, where they produce corresponding fluctuations in the resistance of the microphone. In the simplest system, a dc voltage is applied to the microphone through wires that connect the telephone set to a "central station." The amplitude of the current that circulates in these wires is inversely proportional to the instantaneous resistance of the microphone, and is thus a measure of the instantaneous air pressure at the microphone. The variations in amplitude of this current with time contain the spoken message.

The spoken message can be reproduced by causing a replica of the signal current $i_s(t)$ to flow in an earphone at the receiving end. In the earphone the current flows in a coil of wire with a diaphragm attached to one end of the coil. The coil is placed in a local magnetic field produced by a

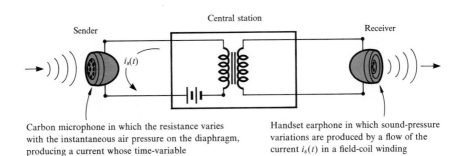

Carbon microphone in which the resistance varies with the instantaneous air pressure on the diaphragm, producing a current whose time-variable part $i_s(t)$ contains the message to be transmitted

Handset earphone in which sound-pressure variations are produced by a flow of the current $i_s(t)$ in a field-coil winding

(a) Partial illustration of a simple telephone system, mod I

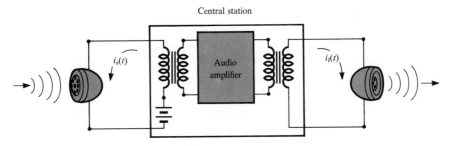

(b) More elaborate system employing an audio amplifier to increase the power level of the signal before sending it on to the receiver

FIG. I · 1   SOME COMMON ELECTRONIC SYSTEMS FOR COMMUNICATION

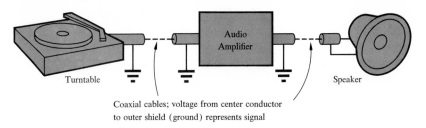

Coaxial cables; voltage from center conductor
to outer shield (ground) represents signal

FIG. I · 2 BLOCK DIAGRAM OF AN AUDIO REPRODUCTION SYSTEM

small bar magnet, so that current flow in the coil produces motion of the diaphragm. The earphone thus reconverts the electrical signal into a sound-pressure signal.

The energy content of the signal current $i_s(t)$ is large enough so that with fresh batteries and a properly selected transformer to couple the earphone and microphone, telephone communication over several tens of miles can be established with the simple system shown in Fig. I · 1a. However, for greater distances, it is necessary to increase the energy content of the signal before sending it on to the receiver. This can be accomplished in an *audio amplifier,* which is located in the central station (see Fig. I · 1b). By using a pair of amplifiers (or other possible techniques), a two-way communication can, of course, be established between the two parties.

A basically similar system to that shown in Fig. I · 1 is the general audio reproduction system shown in Fig. I · 2. This system consists of a loudspeaker, an audio amplifier, and a "program source." If our program source is a record, the original signal is the information contained in the grooves that are made in the record during the recording (or record-reproduction) process. The phonograph cartridge converts this information into a voltage whose amplitude is proportional to the instantaneous sound pressure which we wish the speaker to reproduce. However, the available output power from the phonograph cartridge may be only 10^{-10} watt, whereas we may want to deliver 1 watt of signal power to the voice coil of the loudspeaker to develop a suitable volume of sound. Again, the audio amplifier supplies the required increase in signal power.

We can modify the foregoing concept or change certain details to build public-address systems, "stereo" sound systems, and so on. However, all these systems are built with the assumption that the distance between the sender, or original signal, and the receiver is small; or that wires can conveniently be used to connect the sender and receiver. When communication over long distances is required, such as in radio broadcasting or long-distance telephony, then changes in the system operating philosophy must be made to make it economically feasible.

In a radio broadcast, for instance, we may want a transmitter to provide signals at all points within a radius of a few hundred miles. To do this, we radiate or broadcast electrical signals out into space by forcing a signal current to flow in an antenna. However, for an antenna to radiate energy efficiently, its size must be approximately the same as the free-space wavelength λ of the signal being radiated. For a 1 kcps tone, which is in the middle of the audio-frequency range, a simple loop antenna would have to be about 10^5 m (~ 60 miles) long to meet this requirement!

Shorter antennas can be built by using sophisticated antenna-design techniques. However, partially to avoid the problem of building such an antenna, the information-bearing signals in a radio broadcast are transmitted on a *carrier*. This carrier is a sine wave with a frequency high enough to make efficient radiation possible with an antenna of reasonable size. For example, an AM radio station broadcasting at a carrier frequency of 1 mcps can use an antenna which is roughly 100 m, or about 300 ft high, to achieve efficient radiation. (Again shorter antennas can be built using more sophisticated antenna-design techniques.)

The basic technique used to send and receive information on a carrier was invented by E. H. Armstrong in 1918 and is called the *superheterodyne principle*.[1] A block diagram showing the basic features of an AM radio broadcasting station, which utilizes the superheterodyne principle, is shown in Fig. I·3. We shall take as our original signal the sound-pressure variations created by a person speaking into the microphone at the broadcasting studio. The sound pressure at the microphone as a function of

[1] Some interesting chapters in the history of electronics can be developed from inventions and patents that were filed between 1900 and 1935. Several key inventions during this period belong to Armstrong, including the superheterodyne or frequency-shifting principle (further discussed in Chap. 16); the regeneration principle (1912), in which a vacuum tube was used to produce sine-wave oscillations by purely electronic means; the super-regenerative receiver (1924); and the technique of FM broadcasting and receiving (1934).

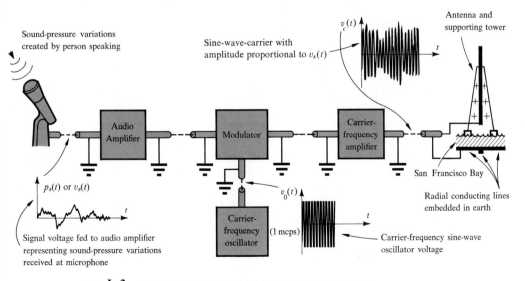

FIG. I·3   **SIGNAL PROCESSING IN AN AM RADIO TRANSMITTER**

time $p_s(t)$ will have a pattern like that shown in Fig. I·3. These sound-pressure variations are converted into an equivalent electrical signal $v_s(t)$ by the microphone and then amplified to a suitable level. They are then fed to one pair of terminals of a subsystem called a *modulator*. A carrier-frequency sine wave generated at the broadcast transmitter in an *electronic oscillator* is fed into another pair of terminals of the modulator.

In the modulator the information-bearing signal and the carrier are multiplied together. The output of the modulator contains several different signals, among them a signal at the carrier frequency *whose amplitude is at every instant proportional to the amplitude of the information-bearing signal* $v_s(t)$ *which was fed into the modulator*. This amplitude-modulated, or AM, signal can be selected and amplified by a *carrier-frequency amplifier,* so that only the amplitude-modulated carrier signal is fed to the antenna and radiated.

The radiated signal is carried through space by means of traveling electric and magnetic fields. These traveling fields will induce time-varying voltages in receiving antennas (a few feet of wire in a simple case), the time variation of the induced voltage being a replica of the radiated signal.

The information-bearing part of the induced voltage is contained in the amplitude variations of the received signal, or the carrier *envelope*. In its most elementary form, the signal processing that takes place in an AM receiver consists in simply detecting the envelope of the received signal.

The subsystem which accomplishes this task is called an *envelope detector*. To a first approximation, an envelope detector is a circuit whose output is equal to the peak positive (or negative) amplitude which the carrier takes on during each cycle. Within this approximation, the output of an envelope detector will appear as shown in Fig. I·4a.

In practice, one of the simplest and most widely used envelope detectors consists of a *detector diode* followed by an appropriately chosen parallel RC network. A simple AM receiver which uses such a detector is shown in Fig. I·4b. Here we employ an antenna (a few feet of wire) to pick up the radiated signal and an LC resonant circuit, or tuning network, to select the desired station. If the tuning network is adjusted to be resonant at the 1 mcps carrier frequency shown in Fig. I·3, then the output signal of the envelope detector will be the $v_d(t)$ shown in Fig. I·4b. Except for a dc component, $v_d(t)$ essentially reproduces the original $p_s(t)$, so we have reproduced an electrical equivalent of the information-bearing waveform.

If we have an efficient receiving antenna and/or we are very close to the broadcast transmitter, $v_d(t)$ may have sufficient energy content to drive a pair of earphones directly. Usually the amplitude of $v_d(t)$ will be too small for this, however, so we follow this detector with an audio amplifier to build up the signal power to the point where it can drive a loudspeaker.

In more sophisticated receivers, we use the superheterodyne principle in the receiver to effect a better rejection of stations broadcasting on nearby carrier frequencies and more sensitivity (that is, better reception in situations where the received signal may have a maximum amplitude on the

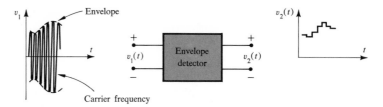

(a) *To a first approximation, the envelope detector is required to produce an output level equal to the peak positive (or negative) of the input during each cycle of the carrier.*

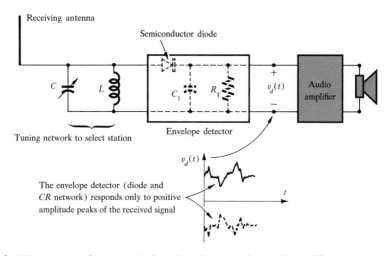

(b) *A simple AM receiver employing a typical envelope detector and an audio amplifier*

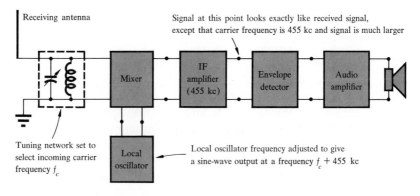

(c) *A superheterodyne AM receiver*

FIG. I · 4 SIGNAL PROCESSING IN AN AM RECEIVER

order of a 1 μv). A block diagram of such a receiver is shown in Fig. I · 4c. Here the incoming signal and a sine wave which is generated in the receiver (local electronic oscillator) are multiplied together in a *mixer*, which is nothing more than a "local modulator." The local oscillator is not adjusted to the carrier frequency, however. Instead, for any position of the tuning capacitor, the local oscillator frequency is 455 kc above the carrier frequency. The local oscillator is tuned at the same time that station tuning is being effected.

The output of the mixer contains, among other things, a signal at a frequency of 455 kc which has the same variations in amplitude with time that the received signal has. In other words, the local oscillator and mixer simply shift the received signal to a new carrier frequency. This new carrier frequency, or *intermediate frequency,* as it is called, is, of course, the same for all stations.

An intermediate frequency, or IF, amplifier is then built to select this signal from the output signal of the mixer and increase its energy content to the point where envelope detection, followed by a small amount of audio amplification, is sufficient to provide an ample volume of sound.

Information coding. Having defined at least broadly the idea of a "signal" and described the signal-processing steps which occur in some rather simple communication systems, let us now consider more carefully the nature of signals in electronic systems in general.

Electronic systems usually employ *transducers* (for example, the microphone and earphone in a telephone system, but more generally, loudspeakers, television camera and picture tubes, etc.) at their input and output terminals, the function of the transducers being to convert the signal from its nonelectrical form to an electrical signal which contains the same information, or vice versa. In the electrical signal, the information is carried in (or represented by) a certain *code;* perhaps the amplitude of the electrical signal, its polarity, or its time dependence carries the required information. There are two types of codes used to carry the information: *analog* and *digital.*

In an analog code, *the instantaneous value of the coded electrical signal is directly proportional to the instantaneous value of the original signal.* In an AM radio broadcast, the amplitude of the carrier-frequency sine wave is made to conform to the instantaneous value of the original signal generated in the broadcast studio. This is one example of an analog code. The spoken information in a long-distance telephone transmission and the picture information in a television broadcast are also broadcasted on an AM carrier. However, there are many other types of analog codes and analog signals. For example, an FM radio station uses an analog code in which the frequency of the transmitted signal varies in accordance with the information being sent. The sound information in a television broadcast is also sent in such a code.

In the digital (or quantized) coding scheme, the signal information is represented by a series of pulses. The number, duration, and/or frequency of the pulses may carry the required information. Frequently the pulses in the code are thought of as having a standard amplitude, though this is unnecessary, since the information being transmitted does not usually depend on the pulse amplitude.

Many familiar electronic systems employ digital-coding schemes in transmitting information. The dot-dash code of telegraphy is one example of a digital code, where *pulse length* is important. In a television system, a pulse code is used to synchronize the electron beam which is swept across the face of picture tube with a similarly sweeping beam in the camera tube at the studio. Here the *repetition frequency* of the pulses carries the synchronizing information. In the telephone system, the number dialed to initiate a call is sent to the central station as an appropriate sequence of pulses. In this case, a *number* of pulses which is proportional to the rotation of the dial is transmitted as the dial returns to its zero position. Numerical information and instructions which are fed to a digital computer are also coded digitally (by the presence of a hole at a particular location in a card, for example).

Digital codes may also be used for transmitting analog information, though some information is lost in the process. For example, the face of a television picture tube can be considered to contain about 200,000 picture elements. The shade of each picture element, on a continuous scale from black to white, is transmitted every $\frac{1}{30}$ sec in a television signal. However, more information is transmitted in such a signal than is needed to identify the picture. The coarse details in a scene can sometimes be identified by painting each picture element either black or white, according to whether the picture element in the real scene is nearer to black than white, or vice versa. Most of the finer details in a typical picture can be obtained by dividing the black-to-white scale into only eight shade intervals, and transmitting one of these possible shades for each picture element. In both these latter cases, a digital code is used to transmit the scene.

Some characteristics of analog and digital systems. The systems required to process these two basic types of signals are quite different in their characteristics. If analog information which is contained in the amplitude of a voltage is to be amplified, then the output voltage of the amplifier must be accurately proportional to the input signal amplitude. Otherwise, information will be lost in this processing step. A circuit in which the output variable is proportional to the input variable is called a *linear transducer,* and circuits which process analog signals must be linear transducers (in so far as the information-bearing aspect of the signal is concerned).

By contrast, in a digital system, the amplitude of the pulse usually contains no information and can become badly distorted in the course of transmission. In a television receiver, for example, the pulses which synchronize the electron beam in the picture tube need not have a shape exactly like that of the originally transmitted pulse to adequately synchronize the sweep.

Since the amplitude of the pulses contains no information in the examples discussed, the circuits used to process these digital signals are frequently *not* linear transducers. Pulse-regeneration circuits may be more appropriate, in which the input pulse simply triggers a circuit which then produces a pulse of standard size and shape. In this way the waveform of a digital signal can be restored when necessary without changing the information content of the signal.

Test signals for analog and digital systems. The difference in the nature of analog and digital systems is also reflected in the types of signals that are used to measure or specify the performance required in the system. The basic test signal for a digital system is a pulse, while the basic test signal for an analog system is a sine wave if the steady-state performance of the system is a design objective or a step function if the transient response is most important. It will perhaps be useful to expand briefly on the use of the sine wave as a test or design signal in an analog system (we shall take up this topic in more detail in Chap. 16). So far we have described the information-bearing signals in an analog system as being a rather random-looking pattern of peaks and troughs (for example, $v_s(t)$ in Fig. I·3). Such a signal, which is typical of the information-bearing signals that a broadcasting system actually handles, is not particularly reminiscent of a sine wave. However, the apparently random-looking signals which analog systems must actually handle can be represented as a sum of sine waves, with appropriate frequencies, amplitudes, and phases. As a result, we frequently specify and test the performance of an analog system in terms of its ability to handle *sine waves* over a certain *range* or *band* of frequencies. The theoretical justification for this step was initiated by the French mathematician, J. B. J. Fourier (1768–1830). He and those who followed him showed that any "reasonable" function of time $f(t)$ can be represented by a sum of sine waves, where by "reasonable" we exclude only certain pathological functions which do not occur in physical systems.

Unfortunately, the exact sine-wave or Fourier representation of a real communication signal always requires an infinite number of sine waves, a large number of them having arbitrarily high frequencies. Of course, a real communication system cannot transmit a sine wave of arbitrarily high frequency, so it cannot reproduce the original signal exactly. However, there will generally be some range of frequencies, or *principal frequency spectrum,* as we shall call it, within which the most important sine waves that are required in the Fourier representation of a given signal will be located.

The principal frequency spectrum of a signal will be determined by the nature of the source that produces the information-bearing signals and/or the characteristics of the observer. For example, speech can be adequately represented by sine waves in the frequency range of about 200 cps to 4 kcps, these limits being established by the processes human beings use when they are speaking. Music requires a frequency band of about 30 to 15 kcps for "high-fidelity" reproduction, the limits in this case arising from the response characteristics of the ear to sound-pressure variations. A television scene may require sine waves with frequencies in the range of

0 to 4 mcps for its reproduction, these limits being established from the characteristics of the television and camera tubes and the eye.

We shall give a more detailed discussion of the relation between a communication signal and its principal frequency spectrum in Chap. 16. We shall see there that signal processing can be very usefully described by observing the changes in the principal frequency spectrum of a signal as it passes through a system. Until then, however, we shall be concerned primarily with the response of devices and circuits to sine waves, taking for granted the fact that a sine-wave signal gives a reasonable basis for designing and testing an analog system.

Objectives and plan of attack.   From the preceding discussion it can be appreciated that the design of a complete electronic system represents a rather complex task. If the system is intended for communication purposes, then we first need to know the principal frequency spectrum of the signals which the system is to handle. We must then decide how this information is to be encoded, how it should be transmitted, how it may be decoded, and so on. Several alternate signal-processing schemes will usually be possible, and a decision must be made among these alternatives based on feasibility, cost, reliability, and so on. We must then decide how to construct the various subsystems, assemble and test them, and then assemble and test the entire system.

If the system is intended for computation, the initial and most important effort will usually be spent in deciding how the logical operations should be performed, what types of memory are required, where they should be located, and so on. Again, we arrive ultimately at a block diagram of the final system, which must then be built with real components.

Because there are a number of separate but interrelated ideas involved in studying and designing electronic systems, it generally proves to be efficient educational policy to split the subject of electronics into smaller parts, which are described by such terms as *systems theory, electronic circuits and devices, electromechanics* (that part of it which deals with information transducers), *radio propagation, computer logic and design,* and so on. A partial organization chart for the general subject of electronics based on these subdivisions is shown in Fig. I · 5. In each subdivision, one studies in so far as possible the techniques and general principles which find applications in a great many electronic systems.

In this book we shall largely follow the established pattern of subject division by considering primarily electronic devices and circuits. Under this heading we consider primarily subsystems in which both the input and output signals are electrical quantities (voltages or currents). In this class of systems, all the signal-processing functions described so far (and many others) can be obtained by properly combining the familiar RLC elements of linear circuit theory with *two new elements:* a *nonlinear element* (a diode) and an *amplifying element* (a transistor or a vacuum tube).

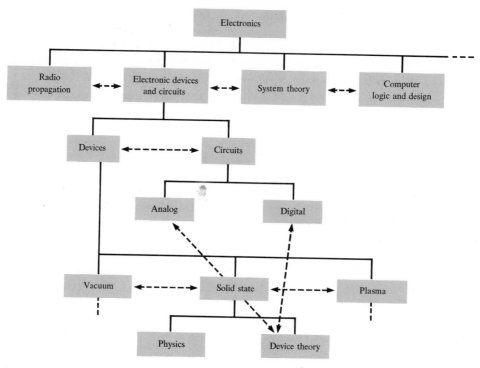

FIG. I·5 A PARTIAL ORGANIZATION CHART FOR ELECTRONICS

In analog systems, nonlinear elements are used to produce the frequency-shifting signal transformations that occur in modulators, mixers, and envelope detectors. In digital systems, nonlinear elements are used for performing logic, steering pulses in desired directions, and so on. Diodes with suitably chosen R, L, and C elements can perform all those functions; or the nonlinear characteristics of amplifying elements can also be employed.

In both analog and digital systems, amplifying elements are used to increase the power level of a signal. They are also used to build electronic oscillators for both analog and digital systems, and with appropriate combinations of RLC elements they provide the frequency-selective amplification required in IF and carrier frequency amplifiers.

The tremendously increased range of possibilities which the addition of diodes and amplifying elements gives to a circuit or systems designer can be fully exploited only after the operating characteristics of these two new components are fully understood. Our study of electronic devices and circuits therefore has two main objectives: (1) to learn how diodes and amplifying elements (primarily semiconductor diodes and transistors) operate, and (2) to learn how to design and build some of the most useful subsystems with them.

Of course, the study of electronic devices and circuits has its own subdivisions, some of which are indicated in the organization chart of

Fig. I·5. As suggested there, the subject can be divided into "device" and "circuit" subheadings. This traditional (but increasingly blurred) division corresponds roughly to the objectives (1) and (2) given in the preceding paragraph. Within the device area, there are further possible subcategories, according to whether the device is a vacuum device or a solid-state device, and whether it is intended primarily for analog or digital applications or both. The device studies also can center on "basic physics" or on device theory and characterization.

The circuits area can also be divided into those circuits which are intended to operate on analog signals and those which process digital signals. There are again subclassifications based on signal frequency, power level, and so on, which have not been included in Fig. I·5. As it is, the figure is sufficiently complete to show that there are many possible paths which can be followed in studying the electronic devices and circuits subdivision of electronics.

Let us now explain the choices we have made, the plan of attack we shall use, and then get to work. As we have said, this book is primarily concerned with the theory and characterization of semiconductor devices (largely *pn*-junction diodes and transistors) and their applications in electronic circuits. More emphasis is given to analog circuits than to digital circuits, more to low- and medium-power applications (less than 10 watts) than to high-power applications, and more to signals with frequency components at and below 100 mcps than to signals with important frequency components above this range. In many cases the principles which apply in one of the undiscussed areas are identical, or at least similar, to principles which will be discussed, though there are always important details which a discussion of principles does not suggest.

A feeling for the limitations imposed by the choices mentioned above can be obtained from Fig. I·6. Here power and frequency are plotted logarithmically on the coordinate axes, and curves are drawn to indicate the regions of this plane in which solid-state devices (very largely semiconductor devices) predominate, where solid-state devices are in the majority, and where vacuum tubes predominate. Most, if not all, electronic circuits which are employed in radio and television receivers, audio reproduction equipment, computers, some mobile communications equipment, instrumentation, and some industrial control applications are or can be made with solid-state devices. When very high power at very high frequencies is required, as in the final stages of various types of transmitters, vacuum tubes predominate. Our primary interest centers on circuits and devices for lower-power applications, so our concentration on semiconductor devices is a justifiable one.

We shall, however, briefly describe the physical operating principles for vacuum tubes which can be used in the applications of interest to us, and develop appropriate circuit models for them. The intention here is to provide examples of some of the principles of space-charge control (which apply in all tubes), to discuss certain specifically important tubes (such as

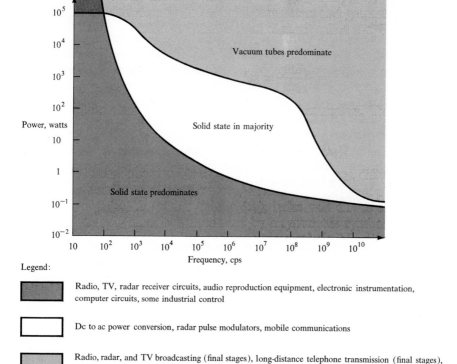

Legend:

Radio, TV, radar receiver circuits, audio reproduction equipment, electronic instrumentation, computer circuits, some industrial control

Dc to ac power conversion, radar pulse modulators, mobile communications

Radio, radar, and TV broadcasting (final stages), long-distance telephone transmission (final stages), mobile communications, some industrial control, some electronic instrumentation

FIG. I · 6 POWER-FREQUENCY PLOT OF AREAS OF PRINCIPLE APPLICATION FOR SOLID-STATE AND VACUUM DEVICES

a television picture tube), and to show the power of various techniques used in making circuit models for active devices. However, actual circuit-design work is carried out exclusively with semiconductor devices. (It can be mentioned in passing that most of the tubes used in that part of Fig. I · 6 marked by "tubes predominate" are considerably more complicated than the simple diode, triode, and pentode structures we shall study.)

The point of view which we shall use in studying our subject is to emphasize the relationship between theoretical models and experimental realities at every stage in the development. Theoretical models, whether they be models for physical processes, models for understanding device behavior, or models for circuit analysis, form an important part of any subject, because without them the understanding of principles and the creative application of principles to practical problems are very difficult.

However, we need to bear in mind that any theoretical model is based on approximations which must be made to obtain a manageable model. The approximations represent a compromise between simplicity and accuracy of representation, and imply a limited range of validity for any model. Sometimes the limits of validity are apparent from the nature of the approximations, though more frequently this will not be the case. In most cases a comparison of theoretical and experimental results is needed to establish the limits of validity of a theoretical model and to give a more realistic feeling for the practical aspects of the subject than numerical examples or problems can do.

One of the results of adopting this philosophy of presentation is that the "physics," "devices," and "circuits" subheadings of the subject tend to be deemphasized as separate entities. This is because the experimental verification of a physical principle (or a combination of principles) always involves building some type of device, and that device invariably can perform interesting circuit functions. We are thus naturally led to study the device properties and applications of thermistors, photoconductors, solar cells, electroluminescent devices, photodiodes, and so on, along the way toward our main goal, which is the development and application of the transistor. The intended result of this free circulation between physics, devices, and circuits is to provide a solid basis for understanding and applying electronic devices, a basis which increasingly involves concrete experiences in all these areas.

1 PROPERTIES AND MODELS OF THE ELECTRON

THE DESIRABLE properties of diodes and amplifying elements are obtained by controlling the flow of electrons in a solid or a vacuum. It is therefore appropriate to begin the study of electronics by examining (1) the basic physical attributes of the electron and (2) the means by which we may predict its motion in fields of force. Taken together, these properties form what we shall call a *model* of the electron.

There are two models of the electron which have been used in the development of the subject of electronics. These are the *classical model* and the *quantum-mechanical model*. They differ primarily in the manner in which electronic motion is predicted. It is known that the classical model is actually incorrect, though it gives a very satisfactory approximation to actual behavior in many cases. The quantum-mechanical model is believed to furnish a very accurate representation of electronic behavior in any physical circumstance.

Electronic motion in vacuum tubes can be satisfactorily studied without reference to the quantum-mechanical model. In semiconductor electronics, both models are used. For studying the electrical characteristics of most semiconductor devices, a modified classical model of the electron is adequate. The quantum-mechanical model is used to understand the basis of the modified classical model and to provide a physical basis for our fundamental ideas about the structure of semiconductors.

In this chapter we shall briefly discuss the physical concepts on which these models are based. We begin by stating those properties of the electron which are common to both models. Next we consider how electronic motion in a vacuum is predicted by the classical model, both to develop some familiarity with this model and to provide a comparison for later work on electronic motion in a semiconductor. We then describe some basic features of the quantum-mechanical model and state the general criterion by which we may decide when this model needs to be used.

1·1 FUNDAMENTAL PROPERTIES OF THE ELECTRON

In all models of atomic structure thus far developed, the electron is considered to be a fundamental particle. It bears the smallest known unit of electrical charge, 1.6019×10^{-19} coul, and it has the smallest known finite mass,[1] 9.1066×10^{-31} kg. Since the electron is a fundamental (that is, indivisible) particle, its charge and mass are fundamental physical constants. They are tabulated on the inside front cover of this book, together with other fundamental constants.

In addition to charge and mass, the classical electron is sometimes assumed to have other mechanical properties. However, these properties do not enter present calculations of electronic motion, and they need not be considered further here.

1·2 THE CLASSICAL MODEL

In addition to its physical attributes, we must specify how electronic motion is to be calculated to complete the definition of our model. *The fundamental assumption of the classical model is that the motion of the electron should be in accordance with Newton's laws.* By using this assumption, we may predict the trajectory or path which the electron should follow in any given force field. As examples, we shall consider the two experiments

---

[1] The mass quoted is the rest mass of the electron. Relativistic changes in mass with velocity are considered in Prob. 1·6.

from which the electronic charge and mass were first determined. The mathematical techniques to be used are the basis of the highly developed subject of electron ballistics. In fact, as a by-product of the discussion, we shall learn the principles by which electron beams are deflected in modern oscilloscopes and television picture tubes.

1·2·1 *Measurement of charge-to-mass ratio.* The first fundamental property of the electron to be determined was its charge-to-mass ratio. The experiment was carried out by Sir J. J. Thomson in 1898. He was later awarded the Nobel prize in physics for this and other work.[1]

Thomson's apparatus is shown schematically in Fig. 1·1. The evacuated glass tube contains a cathode C, and anodes A and B with rectangular slits for collimating the beam. Electrons are produced within the space between C and A by passing the discharge of an induction coil through this region, thus ionizing the rarefied gas remaining in the tube. These electrons are attracted to A by virtue of its positive potential, and some of them pass through the slits in both A and B. They emerge as a small electron beam of rectangular cross section. The electrons in the beam then drift with a uniform velocity through the space between B and a screen S. The screen is coated with a fluorescent material, so when the electrons strike it, a small fluorescent patch P_1 is produced.[2]

Now suppose a potential difference V_y is applied between the deflection plates D and E, E being positive. The spot will then appear at P_2, having been deflected downward by the electrostatic field. The amount of the deflection can be calculated by considering the trajectory of a particular electron in the beam.[3] The calculation is illustrated in Fig. 1·2, which gives the reference coordinates and the important geometrical parameters. The following calculations assume mks units.

[1] The details of this experiment, together with an interesting biographical sketch of Sir J. J. Thomson, may be found in M. H. Shamos, "Great Experiments in Physics," Holt, Rinehart and Winston, Inc., New York, 1959.

[2] The "evacuated tube" still had enough positive ions and surface charge inside the glass to provide a current to the screen and thus prevent the electrons from building up a charge. Modern display tubes, which *are* very highly evacuated, provide other means for accomplishing this.

[3] The density of electrons in the beam is small enough that the effects of mutual repulsion may be neglected.

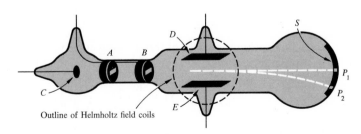

FIG. 1·1   TUBE USED BY SIR J. J. THOMSON FOR DETERMINING q/m FOR AN ELECTRON

Properties and models of the electron   19

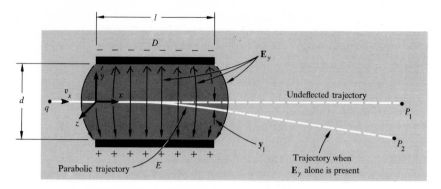

The potential difference of V_y volts produces a constant electric field in the central region between D and E of

$$\mathbf{E}_y = \frac{V_y}{\mathbf{d}} \tag{1·1}$$

where $\mathbf{d}$ is the spacing between D and E in meters.[1] Thus, during the time the electron is within the range $0 < x < l$, it experiences a constant y-directed force of

$$\mathbf{F}_y = -q\mathbf{E}_y = \frac{d(m\mathbf{v}_y)}{dt} \tag{1·2}$$

the last identity being Newton's second law. If the mass of the electron is assumed to be constant, Eq. (1·2) predicts a constant y-directed acceleration of

$$\mathbf{a}_y = \frac{-q\mathbf{E}_y}{m_0} \tag{1·3}$$

where m_0 is the electron mass and q is the magnitude of its charge ($-q$ is therefore the actual charge of the electron).

The x-directed velocity of the electron during this time is unknown, but constant. Hence, if the electron enters the deflection space at $t = 0$, it emerges at

$$t = \frac{l}{v_x} \tag{1·4}$$

Neglecting the effect of the fringing electric fields, the electrons thus follow a parabolic trajectory, being deflected by an amount

$$y_1 = \frac{1}{2}\,\mathbf{a}_y t^2 = -\frac{1}{2}\frac{q}{m}\,\mathbf{E}_y\left(\frac{l}{v_x}\right)^2 \tag{1·5}$$

---

[1] Symbols set in boldface type are vectors. Scalar quantities and the magnitudes of vectors are set in italic type.

upon emerging from the deflection space. Now E_y and l are known, and y_1 can be inferred from $P_1 - P_2$ by geometrical considerations. Therefore, Eq. (1 · 5) provides a means of measuring q/m, if v_x can be determined.

Thomson determined v_x as follows. A pair of coils with diameters equal to the length of the deflection plates were placed in front of and behind the tube so as to produce a reasonably constant magnetic field $\mathbf{B}$ in the deflection space (Helmholtz field coils shown in Fig. 1 · 1).

Now an electron moving with a velocity $\mathbf{v}$ through a magnetic field $\mathbf{B}$ will, in general, experience a force

$$\mathbf{F} = -q(\mathbf{v} \times \mathbf{B}) \qquad (1 \cdot 6a)$$

In this formula, $\mathbf{F}$, $\mathbf{v}$, and $\mathbf{B}$ are vectors, and $\mathbf{v} \times \mathbf{B}$ is the vector product of $\mathbf{v}$ and $\mathbf{B}$. [Note that $\mathbf{F}$ is at right angles to $\mathbf{v}$ (and to $\mathbf{B}$), so it can change only the direction of motion of the electron, not its energy.]

Thus, if the magnetic field in Fig. 1 · 1 is directed *toward* the reader, the electron will experience an upward force of magnitude

$$F_y = qv_xB \qquad (1 \cdot 6b)$$

since $\mathbf{v}$ and $\mathbf{B}$ in Eq. (1 · 6a) are at right angles.

By adjusting the current in the coils, the magnetic force given in Eq. (1 · 6b) can be made to exactly cancel the electric force given in Eq. (1 · 2), and the spot will then remain undeflected in passing through the deflection space (see Fig. 1 · 3). The value of B which achieves this condition is found by equating (1 · 6b) and (1 · 2):

$$v_xB = E_y \qquad (1 \cdot 7)$$

By making the required adjustment and then measuring B and V_y, we obtain

$$v_x = \frac{V_y}{Bd} \qquad (1 \cdot 8)$$

From this, q/m in Eq. (1 · 5) may be determined.

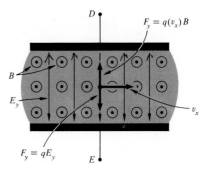

FIG. 1 · 3   THE CANCELLATION OF y FORCES DUE TO E_y AND B

Of course, errors are introduced by fringing electric and magnetic fields, and by the fact that the electrons which enter the deflection space do not all have the same v_x. However, a value of q/m which is within 10 percent of the presently accepted value can be determined in this way.

1·2·2 Cathode-ray tubes. The display tubes used in modern oscilloscopes and television receivers do not differ in principle from the tube used by Thomson to determine q/m. In fact, these display tubes are called *cathode-ray tubes,* since in Thomson's day the electron beam was thought by many physicists to consist of rays emanating from the cathode. We shall digress briefly to describe a tube of the type used in a modern oscilloscope.

A schematic diagram of such a tube is shown in Fig. 1·4. The cathode K is heated to provide a source of electrons for the tube. The positive potentials applied to A_1 and A_2 attract the electrons through the grid G. The grid has a small hole in its center, so that only a pencil-like beam of electrons is transmitted through it. A bias potential V_g is applied between the cathode and grid to permit adjustment of the spot intensity on the screen.

The pencil beam of electrons emerging from the grid passes through holes in A_1 and A_2, the pair forming an electric-field lens by which the electron beam may be accurately focused on the screen. The complete beam-forming assembly shown in Fig. 1·4b is known as an *electron gun.*

The beam of electrons thus produced is then electrostatically deflected by voltages applied to the horizontal- and vertical-deflection plates. A display tube for a television receiver is basically similar, except that magnetic deflection is used.[1]

---

[1] The basic features of these deflection systems provide an interesting application of the use of the classical model and are the subject of Probs. 1·1–1·5.

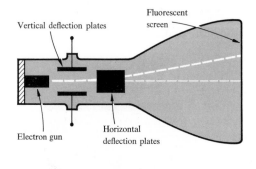

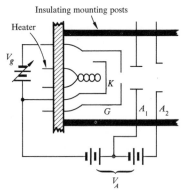

(a) *Schematic diagram of oscilloscope display tube*   (b) *Partial detail of an electron gun with bias voltages*

FIG. 1·4   CATHODE-RAY TUBE

1·2·3 *Measurement of the electronic charge.* We shall continue our development of the classical model by considering the experiment in which the charge on the electron was first accurately measured. The experiment was performed by R. A. Millikan in 1906.[1] The value of electronic charge which he obtained is within 0.4 percent of the presently accepted value. Millikan also was awarded a Nobel prize in physics for this work and other pioneering efforts in the photoelectric effect.

A schematic illustration of Millikan's apparatus is shown in Fig. 1·5. By means of an atomizer, a cloud of fine oil droplets is blown into a dust-free chamber C. One of these droplets is allowed to fall through a pinhole p into the space between the plates of a horizontal air capacitor. The pinhole is then closed by means of an electromagnetically operated cover (not shown). The droplet is observed by means of a telescope.

The droplet will fall under the action of gravity toward the lower plate. The air behaves as a viscous medium in which the droplet falls with constant velocity. Before it reaches the lower plate, an electric field of strength between 3,000 and 8,000 volts/cm is created between the plates, with the upper plate positive. If the droplet has picked up a negative electric charge from its surroundings (ionized by cosmic rays, etc.), it will be pulled toward the upper plate. Before it strikes the upper plate, the field is switched off and the droplet is allowed to fall again. By continued operations of this type the droplet can sometimes be kept within the capacitor plates for periods of an hour.

The basic mathematics for determining q from this experiment is as follows. Suppose the oil droplet has a total charge Q on it. Then it may be made to stand still between the capacitor plates if the electric field is adjusted so that

$$QE_y = Mg \qquad (1·9)$$

[1] The details of this experiment, together with an interesting biographical sketch of Millikan, may be found in M. H. Shamos, *op. cit.*

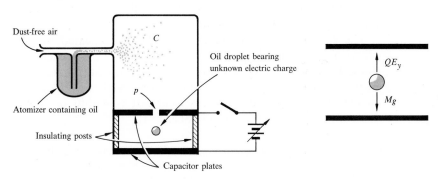

(a) *Schematic illustration of experiment for determining the electronic charge q*

(b) *Forces operating on oil droplet*

FIG. 1·5 MILLIKAN'S OIL-DROP EXPERIMENT

where E_y is the electric field between the plates, M is the mass of the droplet, and g is the magnitude of the acceleration of gravity (see Fig. 1·5).

We can determine Q from Eq. (1·9) if we know M. The mass M is determined as follows. When the field is switched off, the droplet will fall toward the lower plate. The velocity of descent is constant by virtue of the viscosity of the surrounding medium. From Stokes' law we have

$$F_y = Mg = 6\pi\mu r v_d \tag{1·10}$$

where μ is the coefficient of viscosity, r is the radius of the droplet, and v_d is the velocity of descent. Thus the radius of the droplet may be calculated from Eq. (1·10):

$$r = \frac{Mg}{6\pi\mu v_d} \tag{1·11}$$

Assuming the droplet to be spherical, its mass is then

$$M = \tfrac{4}{3}\pi r^3 \rho \tag{1·12}$$

where ρ is the density of the oil. Substituting Eq. (1·11) into Eq. (1·12), we obtain

$$M = \tfrac{4}{3}\pi(\tfrac{3}{2}\mu v_d)^{3/2}\, \rho^{-1/2}\, g^{-3/2} \tag{1·13}$$

Using this in Eq. (1·9) we find

$$Q = \tfrac{4}{3}\pi(\tfrac{3}{2}\mu v_d)^{3/2}\, \frac{g^{-1/2}}{E_y\rho^{1/2}} \tag{1·14}$$

Since all the quantities on the right-hand side are known or measurable, Q can be determined.

Of course, this formula [Eq. (1·14)] needs to be corrected for the bouyancy of air. Furthermore, the validity of using the bulk mass density to determine M for the small drop needs to be checked, and Stokes' law needs to be validated to the accuracy desired. However, these and other questions were successfully resolved by Millikan, and the resulting modified formula gives highly accurate results.

The experimental determination of Q was repeated by Millikan and others many times, always with the result that Q was some (small) integral multiple of a fundamental charge. This fundamental charge is assumed to be the electronic charge q, and the value obtained by Millikan was accepted as the standard value for many years. It is within 0.4 percent of the presently accepted value given in Sec. 1·1.

1·2·4 *Summary of the classical model.* Thus, by assuming the electron to be essentially a point particle which obeys Newton's laws, we are able to determine its mass and charge. These determinations rest on predictions of the trajectory of an electron in certain man-made fields of force. The theory implies that these trajectories can be precisely calculated if the force fields are accurately known. However, the theory is not put to a really stringent experimental test in either of the measurements we have discussed, because a small difference ($\sim 10^{-4}$ cm or less) between the actual end point of a trajectory and its calculated end point would be nearly impossible to measure and would be of no practical significance.

The theory *does* face a crucial test when we attempt to use it to predict electronic motion on a distance scale which is small compared to 10^{-4} cm— and it fails this test completely. Electronic motion in atoms, molecules, and crystals (where the natural unit of distance is the Angstrom, 10^{-8} cm) simply cannot be understood using the classical model.

The quantum-mechanical model of the electron was developed in an effort to unravel the problem of how the small-scale behavior of electrons could be predicted. In the next several paragraphs we shall qualitatively describe some of the basic features of this model, emphasizing those points which will be helpful in our future development of the physical properties of semiconductors.

1·3 THE QUANTUM-MECHANICAL MODEL

During the period of about 1900 to 1930, there was a gradual accumulation of experimental evidence which suggested that particle dynamics on a very small scale could not be understood by simply extrapolating our knowledge of how particles behave on a large scale. In particular, the trajectory of an electron in a hydrogen atom cannot be calculated by simply applying Newton's laws and the laws of classical electricity.

Quantum mechanics was developed in an effort to obtain a consistent description of particle behavior on an atomic scale. The basic theory was presented in 1926–1927 by Schrödinger, Heisenberg, Dirac, and Born, all of whom received Nobel prizes for their work.[1]

The theory predicts a form of electronic motion in atoms and crystals which is completely foreign to our tangible experiences with particle dynamics. At the same time, it verifies the use of Newton's laws for studying large-scale dynamics. The proper development and application of the theory requires considerable familiarity with advanced mathematics and is properly left to a course in quantum mechanics. However, there are certain basic properties of the model which can be simply described and which will be very helpful in interpreting quantum-mechanical results when they arise in later studies.

1·3·1 *The Heisenberg uncertainty principle.*

It should be emphasized at the outset that in the quantum-mechanical model, the electron is still considered to be a fundamental particle with mass m and charge q. The basic difference between the quantum-mechanical model and the classical model is in the exactness with which electronic motion is predicted. This differ-

---

[1] W. Heisenberg in 1932, E. Schrödinger in 1933, P. A. M. Dirac in 1933, and M. Born in 1954.

ence can be most simply discussed with the aid of the *Heisenberg uncertainty principle*.

On a qualitative level, the Heisenberg uncertainty principle deals with the problem of simultaneously specifying or measuring the position and velocity of a given particle.[1] On a more quantitative level, it requires that we use *probability distributions* to specify position and velocity, and then it makes precise statements about these probability distributions.

Like Newton's laws in classical physics, the uncertainty principle may be thought of as a basic assumption or postulate in quantum physics. Its validity rests on the fact that it seems to characterize nature correctly.[2] We shall give several examples to verify this point. However, these examples should not be construed as "proving" the uncertainty principle. They merely demonstrate that the uncertainty principle is in keeping with our present experiences.

We begin by stating the uncertainty principle as it applies to the measurement of position and velocity of a particle. To do this, we suppose that a particle is moving along the x axis with a velocity v_{x0}. Then, according to the uncertainty principle, when we make a measurement to determine the position of the particle, it will "interact" with the measuring apparatus in such a way that its velocity will change. The precise nature of the interaction is not (and cannot be) specified, so the precise magnitude of the velocity change is uncertain. However, a range for the particle's velocity after the position measurement can be specified. In particular, if the position measurement shows that the particle is in the range $x_0 \pm \Delta x$, then its velocity will most probably be in the range $v_{x0} \pm \Delta v_x$, where $[\Delta x]$ and $[\Delta v_x]$ must satisfy the *uncertainty relation*

$$[\Delta x][\Delta v_x] \sim \frac{\hbar}{m} \tag{1 $\cdot$ 20}$$

In this relation $\hbar$ is Planck's constant divided by 2π, and m is the mass of the particle. $\hbar$ has the value 1.054×10^{-34} joule-sec.

To see that this principle does indeed violate our intuition about particle dynamics, we consider the problem of forming a beam with a collimating aperture such as the one shown in Fig. 1 $\cdot$ 6. The aperture defines a range of positions which particles must have to pass through it ($r = 0 \pm r_0$). If a uniform flux of horizontally directed, noninteracting particles (for example, neutrons) were incident on the aperture, then we would expect a pencil-like beam of particles to emerge from it. However, the uncertainty principle suggests that this precise result will not occur. Instead, the particles which pass through the aperture may somehow interact with it, emerging with some nonzero radial velocity. The emerging beam will therefore be conical rather than cylindrical, as our intuition would suggest (see Fig. 1 $\cdot$ 6b).

[1] Other forms of the uncertainty principle which relate angle and angular momentum, energy, and time are developed in Prob. 1 $\cdot$ 11. These are of secondary importance for our present discussion, however.

[2] See P. A. M. Dirac, The Physicist's Picture of Nature, *Scientific American,* vol. 28, no. 5, May, 1963, pp. 45–54, for an interesting discussion of this point.

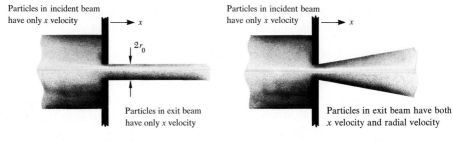

Particles in incident beam
have only x velocity

$2r_0$

Particles in exit beam
have only x velocity

Particles in incident beam
have only x velocity

Particles in exit beam have both
x velocity and radial velocity

(a) *Classical action of a collimating aperture*

(b) *Quantum-mechanical effect of collimating aperture*

FIG. 1·6 **CROSS SECTION OF A CYLINDRICAL COLLIMATING APERTURE WITH CLASSICAL AND QUANTUM-MECHANICAL EXIT BEAM**

We can judge whether this effect will be of practical importance by simply estimating the perturbation which may be introduced into the trajectory of a particle by the aperture interaction.

As an example, let us suppose that the particles in the beam are electrons which have been accelerated through a potential of 1,000 volts before reaching the aperture. Then their x velocity will be

$$v_x = \left(\frac{2qV_A}{m_0}\right)^{1/2} \sim 1.9 \times 10^7 \text{ m/sec}$$

If the diameter of the aperture is 1 mm, the magnitude of their radial velocity after passing through the aperture will be

$$[\Delta v_r] \sim \frac{\hbar}{m_0 r_0} \sim 0.23 \text{ m/sec}$$

The apex angle of the cone is therefore about

$$\theta = \frac{2[\Delta v_r]}{v_x} \sim 2 \times 10^{-8} \text{ rad}$$

This means the electrons in the beam could travel about 60 miles before the aperture interaction would cause the diameter of the beam to double. We can therefore conclude that the uncertainty principle will have a negligible effect on electron-beam dynamics in a 2-ft-long cathode-ray tube.

However, when we reduce the diameter of the aperture in an effort to localize the electrons in the beam much more precisely, the uncertainty principle does produce a noticeable effect. For example, if the diameter of the "collimating aperture" considered earlier were reduced to 1Å, then the conical beam emerging from it would have an apex angle of about 12°.

Of course, this is an impossibly small collimating aperture for a practical electron gun. However, atomic cores in a typical metallic crystal *are* separated by a distance of about 1Å, and a free electron moving in such a crystal will pass through many 1Å "collimating apertures" as it travels through the crystal. In such a case the motion of the electron is completely dominated by the uncertainty principle, and we must look to quantum mechanics to properly understand its behavior.

Limits of validity of the classical model. It is perhaps apparent from the foregoing examples that the uncertainty principle provides us with a very convenient tool for distinguishing those situations in which the classical model is valid from those where quantum mechanics is needed. In general, we can say that when the uncertainty introduced into the trajectory of a particle by the "measuring apparatus" makes a negligible difference in the subsequent classical trajectory, then the classical model (that is, Newton's laws) will give accurate results. Otherwise, quantum mechanics must be used. Using this rule, we can show that

1. The motion of "heavy" particles such as bullets, golf balls, planets, etc., is unaffected by the uncertainty principle (Prob. 1 · 12).

2. The motion of electrons is unaffected by the uncertainty principle as long as the electrons are not localized very accurately. Man-made focusing apertures and deflection systems do not constitute a serious limit on localization, so the classical model is valid for predicting electronic motion in vacuum tubes, high-energy accelerators, etc. (Prob. 1 · 13 and 1 · 14).

3. The motion of electrons on an atomic-distance scale (which includes motion in molecules and crystals) is completely determined by quantum mechanics.

1 · 3 · 2   *The use of probability in the quantum-mechanical model.* In the mathematical theory of quantum mechanics, the uncertainty which arises in small-scale particle trajectories is reflected in the use of *probability distributions* for defining the position and velocity of a particle.[1] To see how this comes about, let us consider the trajectory of an electron in the deflection space shown in Fig. 1 · 7. If the electron enters this deflection space horizontally at $t = 0$, classical calculations indicate that after a time t has elapsed, the y position of the electron should be y_0, and the y velocity should be v_{y_0} (see Fig. 1 · 7).

Now the fact that the electron is in the deflection space at all implies a limitation on its possible range of y positions, $[\Delta y]$. The uncertainty principle then assures us that uncontrollable and unknown changes in the trajectory will result from the *quantum-mechanical interaction* of the electron with the deflection plates (similar to the aperture limitation indicated in Fig. 1 · 6).

This uncertainty is reflected by saying that *there is at best only a certain probability of finding the electron near y_0, and only a certain probability that its velocity will be near v_{y0}.* The position and velocity of a particle are therefore represented by a *position-probability density curve $p_1(y)$* and a

---

[1] The definitions of probability, probability density, and related ideas are developed in R. P. Feynman, "The Feynman Lectures on Physics," vol. 1, Addison-Wesley Publishing Company, Inc., Reading, Mass., 1964, chap. 6, for the reader who is unfamiliar with these ideas. An excellent discussion is also given in C. Sherwin "Introduction to Quantum Mechanics," Holt, Rinehart and Winston, Inc., New York, 1959, pp. 19–25.

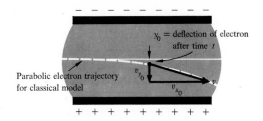

FIG. **1 · 7**   CLASSICAL SPECIFICATION OF SIMULTANEOUS y POSITION AND y VELOCITY

velocity-probability density curve $p_2(v_y)$ in the mathematical theory. Hypothetical curves for the present case are shown in Fig. 1 · 8. We interpret these curves by saying that P_y, the probability that the electron will be found in the region between y and $y + \Delta y$, is

$$P_y = \int_y^{y+\Delta y} p_1(y)\, dy \qquad (1 \cdot 29)$$

Similarly, the probability that an electron will be found with y velocity between v_y and $v_y + \Delta v_y$ is

$$P_{v_y} = \int_{v_y}^{v_y+\Delta v_y} p_2(v_y)\, dv_y \qquad (1 \cdot 30)$$

Equations (1 · 29) and (1 · 30) are interpreted graphically in Fig. 1 · 8.

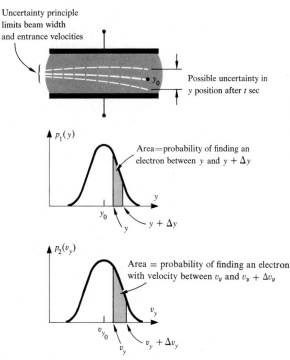

FIG. **1 · 8**   PROBABILITY-DENSITY CURVES USED TO SPECIFY POSITION AND VELOCITY IN QUANTUM-MECHANICAL MODEL

Several comments will help to emphasize the basic importance of these probability density curves:

1. The results of a quantum-mechanical calculation of position or velocity will *always* be in the form of probability density curves [usually $p_1(x,y,z,t)$, $p_2(v_x,v_y,v_z,t)$]. These curves give as much information about position and velocity as quantum mechanics will allow.

2. The use of probability in defining position implies that we cannot follow the precise trajectory of a particular electron in the deflection space. However, we can say that if the experiment is repeated many times under identical conditions, the measured positions of all the electrons after t sec will produce the distribution shown.

3. Rather than calculate a particle trajectory, in quantum mechanics we calculate the trajectory of the position-probability curve. There is a differential equation, called Schrödinger's equation, which can be used to calculate the form and motion of the position-probability curve when the forces that operate on the particles are specified.

One important conclusion which arises from a study of this equation is that the motion of the center of gravity of the position-probability curve obeys Newton's laws. This is the basis of the rule stated in Sec. 1·3·1. However, the form of the position-probability curve may be such that the motion of its center of gravity is hardly representative of the entire curve. For example, the motion of the center of gravity of a beam which has passed through a circular collimating aperture gives no indication that the emerging beam is conical. Again, if the forces which act along one possible particle trajectory are significantly different from those along another, the form of the position-probability curve will spread out rapidly with time, and the motion of the center of gravity will give an insensitive or even misleading indication of actual behavior.

4. The Heisenberg uncertainty principle is interpreted quantitatively by saying that the $p_1(y)$ and $p_2(v_y)$ curves are not independent. Instead, if the "typical width" of the $p_1(y)$ curve is $[\Delta y]$ (see Fig. 1·9), and the "typical width" of the $p_2(v_y)$ curve is $[\Delta v_y]$, then

$$[\Delta y][\Delta v_y] \gtrsim \frac{\hbar}{m} \tag{1·31}$$

Again, we do not know where in the $p_1(y)$ or $p_2(v_y)$ curves to find a given electron. We may only give probabilities. And as we pin the particle down ($[\Delta y] \to 0$) more and more closely, the velocity distribution gets wider and wider (or vice versa, as in Fig. 1·9).

Thus in the quantum-mechanical model, we visualize a position-probability curve moving through space, changing both its general motion and its geometrical form in accordance with the fields that operate on the particles and the obstacles through which they must pass. The trajectory of the position-probability curve tells us as much about particle trajectories as quantum mechanics will allow.

Applications to atomic theory. The success of quantum mechanics in

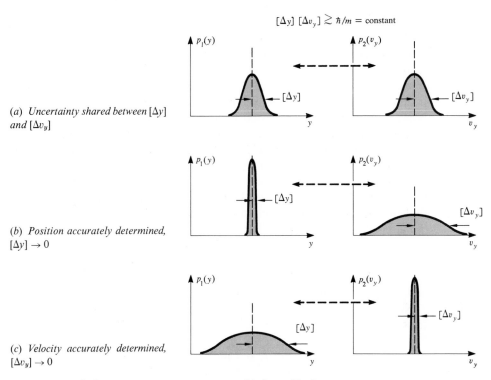

$$[\Delta y]\,[\Delta v_y] \gtrsim \hbar/m = \text{constant}$$

(a) *Uncertainty shared between* $[\Delta y]$ *and* $[\Delta v_y]$

(b) *Position accurately determined,* $[\Delta y] \to 0$

(c) *Velocity accurately determined,* $[\Delta v_y] \to 0$

FIG. 1·9   RECIPROCAL UNCERTAINTIES $[\Delta y]$ AND $[\Delta v_y]$ IN THE HEISENBERG UNCERTAINTY PRINCIPLE; ALL CURVES PROBABILITY-DENSITY CURVES

explaining the electronic structure of atoms was one of the first and most spectacular achievements of the theory. We shall consider this topic in some detail in the next chapter. As a preliminary to that study, however, it is useful to show that even a qualitative understanding of the idea that electronic motion must be described by a position-probability curve is enough to provide an essentially correct picture of the basic size of atoms.

Let us consider the basic problem of explaining the size of atoms. We will consider a hydrogen atom as being roughly representative of atomic size.

If the classical model of the electron were valid on an atomic scale, then we would expect the size of a hydrogen atom to be about the same as the size of its nucleus ($\sim 10^{-12}$ cm in diameter). This conclusion can be reached by the following argument. Suppose we visualize the electron to be circulating around the nucleus in orbits similar to the orbits of the planets circulating around the sun. The centrifugal force in any orbit can be made to exactly balance the coulomb attraction, just as it balances gravitational attraction in the planetary analog. Therefore, the atomic system would appear to be perfectly stable. However, according to the laws of classical electromagnetic theory, the electron will radiate energy, due to the fact that it is always being accelerated. This radiated energy decreases the mechanical energy of the electron in its orbit, so the electron should gradually spiral into the nucleus, finally coming to rest on top of it.

This difficulty can be avoided by the following argument. We know from our previous discussion that at any given instant the best we can do is specify the probability of finding the electron at a given radius from the proton. A possible position-probability density curve is shown in Fig. 1·10. Now, if this curve is compressed toward the nucleus, as in Fig. 1·10b, the most probable radial velocity will increase, the electron's average kinetic energy will increase, and the electron will (on the average) move back out. Opposite tendencies occur when the center of gravity of $p_1(r)$ moves away from the proton. Thus, fluctuations in the $p_1(r)$ curve cause compensating fluctuations in the $p_2(v_r)$ curve. This can be roughly interpreted by saying that the electron will move in and out around the proton, its velocity and radial position continuously changing.

In this motion there will be an equilibrium position-probability distribution and some most probable radius. In order to estimate this radius, we assume that the average radial position is about the same as its uncertainty, and similarly for the velocity (see Fig. 1·10a).

Now, the total energy of the electron in the field of the proton is (classically)

$$E = \frac{1}{2} mv^2 - \frac{q^2}{4\pi\epsilon_0 r} \qquad (1\cdot34)$$

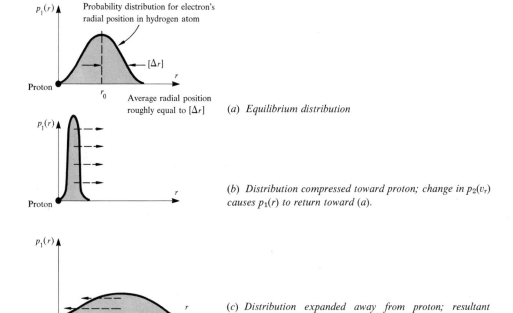

(a) Equilibrium distribution

(b) Distribution compressed toward proton; change in $p_2(v_r)$ causes $p_1(r)$ to return toward (a).

(c) Distribution expanded away from proton; resultant changes in $p_2(v_r)$ cause $p_1(r)$ to return toward (a).

FIG. 1·10   USE OF THE UNCERTAINTY PRINCIPLE TO ESTIMATE ATOMIC SIZE

assuming the zero or reference level of E is when $v = 0$ and $r = \infty$. Furthermore, we have a side condition from the uncertainty relation:

$$(r)(v_r) \sim [\Delta r][\Delta v_r] \sim \frac{\hbar}{m} \qquad (1 \cdot 35)$$

Now, if we assume that the total velocity of the electron may be roughly approximated by its average radial velocity, we may use Eq. (1·35) to rewrite Eq. (1·34) as

$$E \cong \frac{1}{2} m \left(\frac{\hbar}{mr}\right)^2 - \frac{q^2}{4\pi\epsilon_0 r} \qquad (1 \cdot 36)$$

And if we further assume that the most probable radius r_0 is that one which minimizes E, then we find r_0 by setting $dE/dr = 0$:

$$\frac{dE}{dr} = -\frac{\hbar^2}{m}\frac{1}{r_0^3} + \frac{q^2}{4\pi\epsilon_0 r_0^2} = 0 \qquad (1 \cdot 37)$$

$$r_0 = \frac{4\pi\hbar^2\epsilon_0}{mq^2} = 0.528 \times 10^{-10} \text{ m} = 0.528\text{Å}$$

This number is the Bohr radius. Of course, the derivation cannot be trusted for accuracy, but it does suggest that quantum mechanics, via the uncertainty principle, can be used as a basis for "explaining" the size of atoms. We might also interpret this result by saying: the fact that atoms do not collapse to nuclear dimensions is predicted by quantum mechanics but not by the classical model.

1·4 SUMMARY AND PROJECTION

In this chapter we have presented the two models of the electron which we shall need for later work. We have also stated a rule which we can use to decide when the classical model is valid and when the quantum-mechanical model must be used.

The quantum-mechanical model will be most useful in the next chapter where position-probability densities will play a fundamental role in understanding atomic structure and the crystal structure of semiconductors. It will also be useful in Chap. 3 when we develop the modified classical model of an electron, which we use in discussing conduction processes in semiconductor crystals. After Chap. 3, we shall use the modified classical model almost exclusively.

REFERENCES

Harman, W. W.: "Electronic Motion," McGraw-Hill Book Company, New York, 1953, chaps. 1 and 2.

Ryder, J. D.: "Electronic Engineering Principles," 3d ed., Prentice-Hall, Inc., Englewood Cliffs, N.J., 1961, chap. 1.

Feynman, R. P.: "The Feynman Lectures on Physics," vol. 1, Addison-Wesley Publishing Company, Inc., Reading, Mass., 1964, chaps. 37 and 38.

Eisberg, R. M.: "Fundamentals of Modern Physics," John Wiley & Sons, Inc., New York, 1961, chap. 3.

Landé, A.: "From Dualism to Unity in Quantum Mechanics," Cambridge University Press, London, 1960.

PROBLEMS

1·1 This problem deals with some basic features of the electron gun shown in Fig. 1·4. Suppose the electrons are emitted at the cathode with zero velocity. Then show that those which pass through A_2 have an x-directed velocity

$$v_x = \left(\frac{2qV_A}{m}\right)^{1/2}$$

What is v_x for a V_A of 2,000 volts?

1·2 Consider the electrostatic vertical-deflection system shown in Fig. P1·2. Using the results of Prob. 1·1, show that the y deflection is

$$y = \frac{LlV_y}{2dV_A}$$

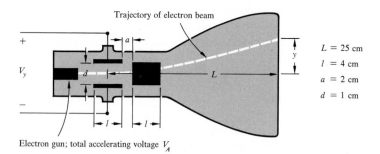

Trajectory of electron beam

$L = 25$ cm
$l = 4$ cm
$a = 2$ cm
$d = 1$ cm

V_y

Electron gun; total accelerating voltage V_A

FIG. P1·2 ELECTROSTATIC DEFLECTION SYSTEM

Compute y for $V_y = 10$ volts, $V_A = 2,000$ volts. The necessary dimensions are given in Fig. P1·2. Compare the force on the electron due to the electric field in the deflection space with that due to gravity.

1·3 Using the dimensions and voltages given in Prob. 1·2, compute the time required for an electron to pass through the vertical-deflection space. If a sine wave of voltage were applied to the deflection plates, at what frequency would the electrons experience no net deflection in passing through the deflection space?

1·4 Assume that a sine-wave signal voltage

$$V_y = 40 \sin (2\pi \times 10^5 t)$$

is applied to the vertical-deflection plates of the tube shown in Fig. P1·2. It is desired to display this voltage on the oscilloscope screen as a function of time. In order to do this, a sweep-voltage waveform of the type shown in Fig. P1·4 is

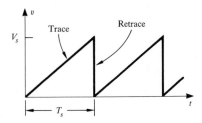

FIG. **P1·4** SWEEP-VOLTAGE WAVEFORM

applied to the horizontal-deflection plates. The beam is swept across the screen at constant velocity during the trace, and returned to its initial position during the retrace so rapidly as to fail to register (sometimes the beam is cut off during the retrace). Calculate V_s and T_s which will trace two cycles of the V_y given above in a horizontal distance of 4 cm. Since controls are provided to adjust the position of the beam, the absolute dc level of the sweep-voltage waveform can be neglected. The horizontal-deflection plates may be assumed to be identical to the vertical-deflection plates.

1·5 It is possible to deflect the beam by use of a magnetic field transverse to the beam direction, the field being obtained from a deflection coil mounted outside the tube (see Fig. P1·5). Assume that the deflection coil produces a uniform magnetic field of B webers/m², over a length of l_m m. Show that the beam deflection at the screen will be approximately

$$ y = \sqrt{\frac{q}{m}} \frac{B}{\sqrt{2V_A}} l_m L $$

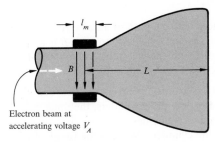

FIG. **P1·5** TOP VIEW OF A MAGNETIC DEFLECTION SYSTEM

where L is the distance from the center of the field to the screen in meters, and the other symbols are as previously defined.

1·6 According to the theory of relativity, the mass of a particle is related to its energy by

$$ E = mc^2 $$

so that $$ dE = c^2 \, dm = F \, ds \qquad (a) $$

where F = force, s = distance. Now, from Newton's second law, we have

$$ F \, ds = \frac{d(mv)}{dt} \, ds = v \, d(mv) \qquad (b) $$

Using these relations, show that

$$\frac{m}{m_0} = \frac{1}{\sqrt{1 - (v/c)^2}} \tag{c}$$

where $m = m_0$ when $v = 0$.

Now, if an electron is accelerated through a potential of V_A volts, its gain in energy is

$$\Delta E = qV_A \tag{d}$$

Using the expression for m as a function of v derived in Eq. (c) above, show that

$$\frac{v}{c} = \sqrt{1 - \frac{1}{(1 + qV/m_0c^2)^2}}$$

Check this formula against the points shown in Fig. P1·6.

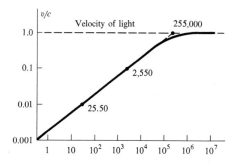

FIG. P1·6   ELECTRON VELOCITY AS A FUNCTION OF ACCELERATING POTENTIAL

1·7 Show that the path of an electron moving with velocity v_0 perpendicular to a uniform magnetic field of strength B is a circle. Find (1) the radius r and (2) the time required for the electron to complete one revolution. (For a related problem, see Prob. 1·10.)

1·8 Two concentric quarter-cylinders are shown in Fig. P1·8. A bias voltage V_0 is applied between the cylinders, the inner cylinder being positive. Electrons with a distribution of velocities are now shot horizontally into the center of the space between the cylinders. What is the range of entrance velocities for those electrons which come out? Electrons which strike either plate may be assumed to be lost.

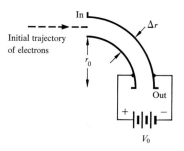

FIG. P1·8   ELECTROSTATIC VELOCITY FILTER

1·9 An electron with an energy in the horizontal direction of 100 ev enters the deflection space shown in Fig. 1·3 in the text. The deflection plates are 2 cm long and separated 2 cm. The lower plate is at ground potential; the upper plate is at +100 volts. Find the magnetic field required to prevent deflection of the electron as it passes through the deflection space. Express your answer for B in webers/m². Neglect fringing fields.

1·10 A simplified cyclotron particle accelerator is shown in Fig. P1·10. The cyclotron works as follows. Two hollow cylinders, called D's, are located opposite each other. A uniform magnetic flux is maintained through each D, and a high-voltage ac source is connected between the D's. A source of charged particles (usually protons) is located near the center of one D. This source emits a group of particles just when the voltage difference between the D's is maximum. The particles are accelerated across the gap to (1), and then deflected in a circular path, due to the magnetic field. They emerge again at (2) after a half-cycle of variation of the voltage. They are again accelerated across the gap, and travel from (2) to (3) with an increased velocity and radius. This process continues until the particles become sufficiently energetic for the required purpose.

Assume the magnetic field in the D's to be fixed at B webers/m² and the particles to be protons. Neglect relativistic mass changes. Develop general formulas for (1) the particle velocity after crossing the gap n times, and (2) the required frequency of the voltage source. State your assumptions clearly, and carefully define any parameters (such as magnetic-field strength) that you need in your analysis. (For a related problem see Prob. 1·7.)

If you are interested in considering this problem further, remove the restriction that the mass of the particle stays constant (see Prob. 1·6).

Within each D the trajectories are semicircles

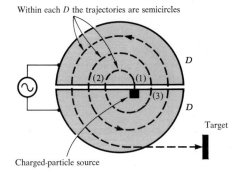

Charged-particle source

FIG. P1·10 TOP VIEW OF A SIMPLIFIED CYCLOTRON

1·11 An electron travels with a velocity v_{x_0} toward a normally closed shutter, as shown in Fig. P1·11. The shutter is opened at time $t = 0$ and then closed Δt

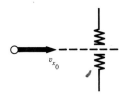

FIG. P1·11 SHUTTER FOR DETERMINING POSITION OF AN ELECTRON

sec later. If the electron makes it through the shutter, what is the perturbation in its x velocity?

Its energy before passing through the aperture was $E = mv_{x_0}^2/2$. The perturbation in v_x also changes E. Show that

$$\Delta E \, \Delta t \geq \hbar$$

This is another form of the Heisenberg uncertainty relation.

1·12 The barrel of a certain rifle has an inside diameter of approximately 0.25 in. Under optimum conditions of aiming and firing, what would be the approximate spread of shots entering a target at 50 ft?

1·13 Using the dimensions for the electrostatic deflection system shown in Fig. P1·2 and assuming that horizontal- and vertical-deflection systems are identical, estimate the size of the smallest spot which can appear on the screen due to the aperture limitation provided by the deflection system.

1·14 The Millikan procedure for measuring q involves measuring the velocity of descent v_d of a small oil droplet in the chamber. Estimate the limit of accuracy in determining q if the measurement of v_d were the only source of error.

1·15 Use Eqs. (1·34) and (1·36) to estimate the minimum energy for a hydrogen atom.

1·16 Classically, an elastic ball can bounce on an ideal horizontal surface with any amount of total energy and any maximum height. Imagine a helium atom ($m \cong 6.6 \times 10^{-27}$ kg) bouncing against gravity, with perfect reflection, on an ideal horizontal surface. Estimate the minimum energy and most probable height when the uncertainty principle is used to limit the possible motion.

2

SOME STRUCTURAL PROPERTIES

OF SEMICONDUCTORS

TRANSISTORS, semiconductor diodes, and other *pn*-junction devices are made from single crystals of semiconductor material, primarily germanium and silicon. There are other semiconductor devices (e.g., photoresistors, thermistors, and certain types of cathodes) which are not usually made from single-crystal material, though they could be and their gross behavior is usually explained as if they were. Thus a study of perfect semiconductor crystals is basic to an understanding of many different types of devices.

In this chapter and the next one, we shall see that it is the structural properties of semiconductor crystals which determine how electricity is conducted in the crystal, and how the crystal will respond to sources of energy such as heat and light. From a study of structural properties, we shall be able to define the basic charge carriers, holes and electrons, and give an elementary description of their properties. We will also show how the number of charge carriers may be controlled by light, temperature, and chemical doping.

2·1 ORBITAL MODELS AND ATOMIC STRUCTURE

Many of the properties of semiconductor crystals can be explained from the orbital models of the combining atoms. These orbital models have their theoretical justification in quantum mechanics, and ample experimental verification in the laws of chemistry and spectroscopy. In this section we shall review the principles which are used in the construction of orbital models of atoms. Our immediate objective is to construct an orbital model of a germanium atom.

In making orbital models, one thinks of atoms as being crudely akin to miniature solar systems, with the nucleus being the central body around which electrons may rotate in allowed patterns. The nucleus itself is composed of protons and neutrons. The proton is a fundamental particle which bears a positive charge equal in magnitude to the charge on the electron, and has a mass which is about 1,800 times that of the electron. The neutron, which is also a fundamental particle, has the same mass as the proton but no charge. The nucleus of an atom is composed of protons and neutrons in the following proportions: if Z is the atomic number and $Y(Y > Z)$ is the atomic weight, then there are Z protons and $Y - Z$ neutrons. There are also Z electrons surrounding this nucleus to make the structure electrically neutral.

The electrons surrounding the nucleus are to be thought of as rotating around it in certain three-dimensional patterns. The detailed mathematical form of these patterns is predicted by quantum mechanics. For many purposes, however, the results of the theory can be adequately summarized by drawing what are called *orbital boundary surfaces*. The basic idea in drawing an orbital boundary surface is to circumscribe the volume of space in which a particular electron will most probably be found. The term "most probably" is significant, since any information which we give about an electron must be of a probabilistic or statistical character, to be in keeping with the fundamental laws of quantum mechanics.

2·1·1 The hydrogen atom: Bohr model. As a simple and very useful illustration of these ideas, let us consider some of the results which arise in the quantum-mechanical treatment of the hydrogen atom ($Z = 1$, $Y = 1$) and compare them with the more familiar Bohr treatment. Here we have only one electron to deal with, and it moves around a single proton, which we may take to be fixed at the origin of coordinates.

In the Bohr treatment of the hydrogen atom,[1] it is postulated that in the steady-state situation, the electron moves around the proton in stable *orbits,* in which the angular momentum J of the electron must be

$$J = n\hbar \tag{2.1}$$

In this formula, n is any positive integer 1, 2, 3, ...

When this postulate is used together with Newton's laws, it is found that only certain orbits are allowed. Each orbit corresponds to a discrete energy level for the atom. The energy levels are given by

$$E_n = -\frac{m_0 q^4}{8\epsilon_0^2 h^2 n^2} \tag{2.2}$$

where the subscript n denotes that n is the only *variable* in Eq. (2.2). The reference level of energy ($E = 0$) corresponds to $n \to \infty$, and describes the situation when the electron and proton are separated from each other by infinite distance, and the electron has zero velocity.

The orbits corresponding to these energies are circular trajectories with radii

$$r_n = \frac{n^2 \epsilon_0 h^2}{\pi m_0 q^2} \tag{2.3}$$

In its lowest energy state ($n = 1$), the atom has an energy

$$E_1 = -\frac{m_0 q^4}{8\epsilon_0^2 h^2}$$

When the atom is in this lowest energy state, the electron moves around the proton in a circle whose radius is

$$r_1 = \frac{\epsilon_0 h^2}{\pi m_0 q^2}$$

r_1 is called the *Bohr radius.*

For all the orbits, the entire motion is in one plane, and there are definite values of potential and kinetic energy associated with each trajectory, or orbit. Figure 2.1 suggests the essential results of this model.

Bohr further postulated that absorption and emission of radiation were accompanied by transitions between two of these stable orbits. A transition from an orbit corresponding to the energy level E_n to the orbit corresponding to E_m involves a photon of frequency

$$\nu_{mn} = \frac{E_m - E_n}{h} \tag{2.4}$$

[1] Bohr's first results on atomic theory were published in 1913, and are considered to be the forerunner of modern quantum theory. In 1922 Bohr was awarded a Nobel prize for his general activities in this field.

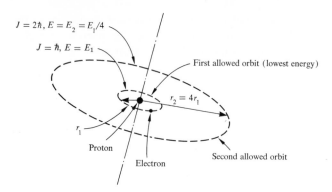

$J = 2\hbar, E = E_2 = E_1/4$

$J = \hbar, E = E_1$

First allowed orbit (lowest energy)

$r_2 = 4r_1$

r_1

Proton

Electron

Second allowed orbit

FIG. 2·1   BOHR MODEL OF THE HYDROGEN ATOM

Conservation of energy requires that the photon be absorbed by the atom if $E_m > E_n$ and be emitted if $E_m < E_n$.

With the aid of this postulate, one can predict the general features of the absorption and emission characteristics of the hydrogen atom, though the fine details of the spectrum cannot be understood, and even the general features of radiation from elements other than hydrogen are difficult to predict. The principal point of interest is that in the Bohr treatment one uses Newton's laws plus cleverly selected postulates to predict the electronic behavior in a hydrogen atom.

2·1·2   *The hydrogen atom: quantum-mechanical model.* Describing the electron as a point particle which must move in one of a series of definite orbital trajectories is, of course, essentially a classical description. In the quantum-mechanical treatment of the hydrogen atom, each of these orbital trajectories is replaced by one or more position-probability functions. Fortunately, each of the possible position-probability functions defines a very limited region of space in which the probability of finding the electron is high. By circumscribing these regions, we can once again obtain a valid and very useful geometrical interpretation of atomic structure.

The procedure for calculating the position-probability functions for an electron moving in a hydrogen atom was developed by Schrödinger and published in 1926. In this classic paper, Schrödinger laid much of the foundation for modern quantum theory. We shall briefly consider the philosophy of his method, because of its bearing on our later work.

Schrödinger postulated that the position-probability functions $p(x,y,z)$ for an electron which is bound to an atomic nucleus could be found by first solving the differential equation

$$\frac{-\hbar^2}{2m} \left(\frac{\partial^2 \psi}{\partial x^2} + \frac{\partial^2 \psi}{\partial y^2} + \frac{\partial^2 \psi}{\partial z^2} \right) + V(x,y,z)\, \psi = E\psi \qquad (2\cdot5)$$

for the function ψ, and then using the rule

$$p(x,y,z) = |\psi(x,y,z)|^2$$

to find the position-probability function $p(x,y,z)$.

Equation $(2 \cdot 5)$ is called Schrödinger's equation.[1] In it,

$\hbar$ = (Planck's constant)$/2\pi$

m = mass of the particle (electron, in this case)

$V(x,y,z)$ = potential energy which the electron has at the point (x,y,z) $(-q^2/4\pi\epsilon_0 r$ for the hydrogen atom problem)

E = total energy of the electron, which is an *unspecified* constant in Eq. $(2 \cdot 5)$

Schrödinger gave some heuristic arguments to explain how he arrived at Eq. $(2 \cdot 5)$, though the success of his method is not related to these arguments. Rather, its success lies in the fact that it correctly predicts the measurable properties of electrons in atomic systems. One should therefore use Eq. $(2 \cdot 5)$ in the same way that Newton's laws are used for studying large-scale particle dynamics—simply assume it is true and see what predictions it makes.

From a purely mathematical standpoint, Eq. $(2 \cdot 5)$ is a completely defined differential equation when V and E are specified. Of these two quantities, V is entirely determined (for the hydrogen-atom problem), but E is an adjustable constant which is at our disposal. We are therefore at liberty to choose an arbitrary value of E (for example, 1 joule), and then attempt to find the corresponding ψ function.

When this is done, an interesting thing happens: while ψ functions can be found for any value of E, the ψ functions which are obtained for many choices of E are such that $|\psi|^2$ cannot be meaningfully interpreted as a probability-density function. For instance, if $|\psi|^2$ is to be a meaningful probability density, then it is necessary that

$$\int\!\!\int\!\!\int_{-\infty}^{\infty} p(x,y,z)dx\ dy\ dz = \int\!\!\int\!\!\int_{-\infty}^{\infty} |\psi(x,y,z)|^2 dx\ dy\ dz = 1$$

to ensure that the electron can be found somewhere. For many choices of E, however, the corresponding solution to Eq. $(2 \cdot 5)$ is such that the integral of $|\psi|^2$ diverges.[2] Thus it happens that there are only certain values of E which give solutions to Schrödinger's equation that can be properly interpreted as probability-density functions.

Schrödinger showed that his equation would yield only acceptable probability-density functions for an electron in a hydrogen atom if the total electron energy were one of the following set of values:[3]

---

[1] There is a more general equation which must be used when the potential-energy function V depends on the time.

[2] An excellent elementary treatment, which shows by simple numerical calculations how $|\psi|^2$ functions diverge for incorrect choices of E, can be found in C. W. Sherwin, "Introduction to Quantum Mechanics," Holt, Rinehart and Winston, Inc., New York, 1959, chap. 3.

[3] The complete solution to the Schrödinger equation for the hydrogen atom can be found in most quantum-theory texts; e.g., R. B. Leighton, "Principles of Modern Physics," McGraw-Hill Book Company, New York, 1959, chap. 5.

$$E_n = \frac{-m_0 q^4}{8\epsilon_0^2 h^2 n^2} \qquad n = 1, 2, 3, \ldots \qquad (2\cdot6)$$

These energy levels are identical to the energy levels which occur in the Bohr treatment, though of course they arise in an entirely different way.

The lowest energy level predicted by Schrödinger is again

$$E_1 = \frac{-m_0 q^4}{8\epsilon_0^2 h^2} = -13.6 \text{ ev}$$

and is only a function of fundamental physical constants.[1] This energy E_1 is called the *ground-state* energy. The reference level of energy ($E = 0$) corresponds to $n \to \infty$, and describes the situation when the electron and proton are separated from each other by infinite distance.

Schrödinger also showed that for each allowed energy level E_n, there is a set of time-independent position-probability density functions. These position-probability densities take the place of orbits in the Bohr theory in the sense that they specify the electronic motion as accurately as quantum theory will permit.

In the ground state of the hydrogen atom, there is only one possible position-probability density function. It is

$$p(r) = \frac{1}{\pi a_0^3} e^{-2r/a_0} \qquad (2\cdot7)$$

where
$$a_0 = \frac{\epsilon_0 h^2}{\pi m_0 q^2} \cong 0.53\text{Å} \qquad (2\cdot8)$$

Like E_1, a_0 is only a function of fundamental physical constants. It is also equal to the Bohr radius in numerical value.

The form of the function given in Eq. $(2\cdot7)$ shows that the motion is three-dimensional and that $p(r)$ is spherically symmetric, instead of planar as the Bohr theory suggests.

From a knowledge of the position-probability density function, we can calculate the *total* probability P of finding the electron inside a sphere of radius r_0:

$$P(r_0) = \int_0^{r_0} p(r)4\pi r^2 \, dr$$
$$= 1 - e^{-2r_0/a_0}\left(1 + \frac{2r_0}{a_0} + \frac{2r_0^2}{a_0^2}\right) \qquad (2\cdot9)$$

The exponential factor ensures that the probability of finding the electron at a radius much larger than a_0 is very small. Note that $P(\infty) = 1$, in agreement with the physical condition that the electron must be found somewhere.

For our purposes, it is useful to have a graphical way of representing these probability functions. There are two procedures in common use:

---

[1] The electron volt is a unit of energy. 1 ev is the energy gained by an electron in falling through a potential difference of 1 volt: 1 ev = 1.6×10^{-19} joule.

1. $p(r)$ is plotted as a function of r.

2. An *orbital boundary surface* is drawn. This is the surface obtained by setting $p(r) = $ const and adjusting the constant so that the probability of finding the electron outside this surface on a random observation is some small, definite number (for example, 0.1 or 0.05). These two possibilities are illustrated in Fig. 2·2.

The orbital boundary surface gives the best "picture" of the hydrogen-atom ground state which quantum mechanics can provide: a proton surrounded by a spherical surface *within which the electron will most probably be found.*

We can further interpret the orbital boundary surface by two comments:

1. The time-average motion of an electron around the proton will roughly "fill in the volume" represented by the boundary surface.

2. If simultaneous measurements of the position of the electron in N independent hydrogen atoms (all in their ground states) are made, 90 percent of the electrons will be found within the boundary surface.

The boundary surface thus gives us a good idea of the geometry of the hydrogen atom in its ground state. Of course, it is possible to excite the hydrogen atom to a higher energy level (by shining light of appropriate wavelength on it, for example). If the atom is brought to the state represented by $n = 2$ in Eq. (2·**6**), then there will be two types of position-probability functions (called *s*- and *p*-type functions) and two types of orbital boundary surfaces. Or, if the atom is in an energy state represented by $n = 3$, there will be three types of orbital boundary surfaces; *s*- and *p*-, as before, and now also a *d*-type orbital.[1]

The orbital boundary surfaces for *p*-type and some *d*-type orbitals are shown in Fig. 2·3. As the figure suggests, there are three independent *p*-type orbitals. There are also five independent *d*-type orbitals.

[1] Position-probability density functions for *s*-, *p*-, and *d*-type orbitals for three different energy levels are given in R. B. Leighton, *op. cit.*, p. 175. A photographic representation of these orbitals is also shown on p. 178.

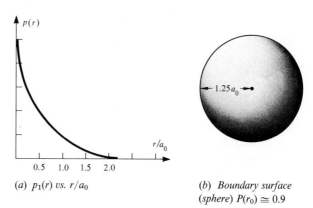

(a) $p_1(r)$ *vs.* r/a_0

(b) *Boundary surface*
(*sphere*) $P(r_0) \cong 0.9$

FIG. 2·2 GRAPHICAL REPRESENTATION OF THE QUANTUM-MECHANICAL ELECTRON ORBITAL
FOR THE HYDROGEN ATOM IN ITS GROUND STATE

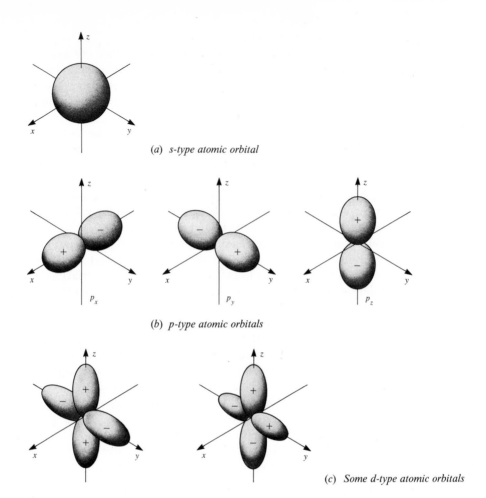

(a) s-type atomic orbital

p_x

p_y

p_z

(b) p-type atomic orbitals

(c) Some d-type atomic orbitals

FIG. **2·3** GEOMETRICAL FORMS FOR s, p, AND SOME d ORBITALS; THE $+$ AND $-$ SIGNS
REFER TO THE SIGN OF ψ, THE GEOMETRICAL SURFACE TO $|\psi|^2$

All the orbitals corresponding to a given value of n have roughly the same radial extent. Orbitals corresponding to higher values of n have larger average radii than orbitals corresponding to lower values of n, however.

Again, we emphasize that these orbital boundary surfaces do not define a "trajectory" in the Newtonian sense of the term. They merely show the regions of space that are most probably occupied by an electron. However, they do give us a sense of the most probable geometry of the hydrogen atom in any given energy state, and this is really what we need for most qualitative purposes.

2·1·3 Generalized atomic orbitals. While the actual electron orbitals for elements other than hydrogen are not precisely s, p, and d types, a sufficiently accurate picture of atomic structure, for many purposes, can be obtained by assuming that they are. These orbitals are then called *atomic orbitals*. They have the following important properties:

1. They are three-dimensional in space.

2. Each orbital has a characteristic geometric shape. *s*-type orbitals are spherically symmetric [though not necessarily as simple as Eq. (2·7) suggests], *p*-type orbitals are similar to dumbbells, etc.

3. Only *s*-type orbitals are allowed in the lowest energy level. Both *s*- and *p*-type orbitals are allowed in the next-to-lowest energy level. Orbitals of the *s*-, *p*-, and *d*-types are allowed in the next energy level, and so forth.

4. Only two electrons can be in any given orbital. These electrons may be thought of as spinning around an axis, and they must be spinning in opposite directions. (This is called the *Pauli exclusion principle*. Since only one electron can be in a given orbital with a given direction of spin, all others are "excluded.")

These results are summarized in Fig. 2·4. In this figure, the first energy level, or shell, can accommodate two electrons in *s*-type orbitals. Their opposite spins are indicated on the diagram. These electrons are called 1*s* electrons, the 1 signifying the energy level and the *s* signifying the type of orbital. The next energy shell (2) has eight possible positions for electrons, though (except for hydrogen) the 2*s* configuration, or cell, is actually at a slightly lower energy than the 2*p* configuration. The 2*s* cell will hold two electrons while the 2*p* cell will hold six. The third shell has two 3*s* electrons, six 3*p* electrons, and ten 3*d* electrons, each at slightly different energy levels. This scheme continues through 4*s*, 4*p*, 4*d*, 4*f*, 5*s*, . . . levels to infinity.[1]

A knowledge of this electronic energy-level scheme, and the geometric

[1] As the figure shows, the 4*s* and 3*d* energy levels are nearly the same, as are 5*s* and 4*d*, and 6*s* and 4*f*. These facts are important in explaining the properties of atoms in the "iron group," for example.

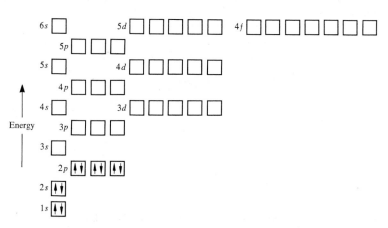

FIG. 2·4 ENERGY LEVELS AND CELLS FOR THE GENERALIZED ATOMIC ORBITALS; EACH CELL WILL HOLD NO MORE THAN TWO ELECTRONS.

arrangement of orbitals implied by it, is sufficient to enable one to construct models from which many of the physical properties of atoms may be inferred. For example, the occurrence of the periodic table, the combining proportions of atoms, a detailed explanation of chemical valence, and the theory of atomic spectroscopy all follow from these simple orbital pictures. We will use the orbital model of a germanium atom to develop these points.

2·1·4 Orbital model of a germanium atom. The element germanium is number 32 in the periodic table. It occurs in several isotopic forms, the most abundant of which has an atomic weight of 74. The nucleus for this isotope is composed, therefore, of 32 protons and 42 neutrons. There are 32 electrons surrounding this nucleus, to make the structure electrically neutral.

The 32 electrons may be considered to be added in the following way. The first two electrons go to the lowest energy level; that is, to the $1s$ energy level. These electrons have opposite spins, as indicated in Fig. 2·5. The next two electrons go into the $2s$ level with opposite spins. The next six electrons go into the $2p$ orbitals. This process is continued through the $4s$ level, after which there are only two electrons left. These two electrons occupy two of the six possible $4p$ levels, leaving four vacant levels.[1]

Occasionally this information is summarized in the form

$$(1s)^2(2s)^2(2p)^6(3s)^2(3p)^6(3d)^{10}(4s)^2(4p)^2$$

The superscripts indicate the number of electrons in each type of orbital. This form is sometimes called the *electronic configuration* of the atom.

2·1·5 Orbital models and atomic spectroscopy. Of course, the configuration just given is only the ground-state configuration for the germanium atom.

[1] When a given type of orbital is not completely filled (such as the $4p$ orbital for germanium), the electrons tend to avoid being in the same space orbit so far as possible. They also tend to have their spins parallel, as in Fig. 2·5.

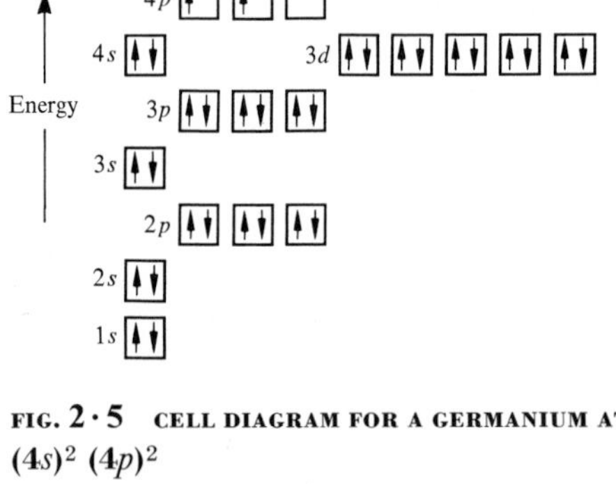

FIG. 2·5 CELL DIAGRAM FOR A GERMANIUM ATOM: $(1s)^2$ $(2s)^2$ $(2p)^6$ $(3s)^2$ $(3p)^6$ $(3d)^{10}$ $(4s)^2$ $(4p)^2$

The 32 electrons can be distributed among the allowed orbitals in other ways, in which case the atom is said to be in an *excited state*.

Absorption. If we suppose that an atom is initially in its ground state, then energy will have to be supplied to it to bring it to an excited state. The minimum energy needed to excite a germanium atom is that required to promote an electron from the 4s state to the 4p state: $E_{4p} - E_{4s}$. This particular excited state may be written as

$$(1s)^2(2s)^2 \cdots (4s)(4p)^3$$

The energy may be supplied by shining a light source which contains photons of frequency

$$\nu = \frac{E_{4p} - E_{4s}}{h} \tag{2 \cdot 10}$$

onto the germanium atom. When this transition occurs, the germanium atom is said to have *absorbed* a photon.

The next possible absorption transition is from a 4p to a 4d state, with an energy difference of $E_{4d} - E_{4p}$, and an associated photon frequency

$$\nu = \frac{E_{4d} - E_{4p}}{h}$$

However, transitions between two arbitrary orbitals (for example, $4s \to 5s$) may not take place. Quantum mechanics provides the means for predicting which transitions can occur. The predictions are based on the geometrical properties of the orbitals.[1]

Emission. There is an inverse process accompanying the absorption process, which is called *emission* of a photon. If a germanium atom is put into an excited state, and then isolated, it will return at some time in the future to its ground state, emitting its excess energy as a photon (or photons). For example, if the atom is in a $(1s)^2(2s)^2 \cdots (4s)(4p)^3$ state, a single photon of frequency given by Eq. (2 · 10) will be emitted when the atom returns to its ground state.

The time at which the excited atom will emit this photon is not exactly known (in fact, it is governed by one of Heisenberg's uncertainty relations). However, if N independent atoms are all known to be in the same excited

---

[1] The existence of "allowed" transitions provides one of the important distinctions between the Bohr model and the quantum-mechanical model. In a hydrogen atom, for example, a transition from a 1s state to a 2s state is not allowed for the mathematical reason that any two s-type orbitals have basically the same spatial symmetry. The difference in spatial symmetry between s- and p-type orbitals is such that a transition from a 1s state to a 2p state is allowed. This distinction accounts in part for the fine structure in the absorption spectrum of hydrogen. Such detail cannot be predicted by the Bohr theory, where there is only one circular orbital for each energy level.

state, at $t = 0$ (and there is no exciting light source), then the number n in this state after t sec will be

$$n = Ne^{-t/\tau} \qquad (2 \cdot 11)$$

τ is called the *average lifetime* of the excited state, and can be calculated when the excited state is specified. The lifetime of an atomic excited state is closely related to the very important concept of the lifetime of excited states in a crystal. We shall meet this concept again in connection with the *recombination* of charge carriers in a semiconductor.

The fact that atoms, molecules, and crystals absorb and emit photons during a transition between two allowed energy levels is the physical basis of the important field of experimental *spectroscopy*. The term "spectroscopy" includes many quite different investigative techniques: optical spectroscopy, microwave spectroscopy, infrared spectroscopy, X-ray spectroscopy, etc. However, in each case, the quantity measured is the frequency or wavelength (and sometimes polarization) of photons which are emitted or absorbed during a transition between two states of the atomic or molecular system being studied. From this information, the characteristics of energy levels (and atomic orbitals) can be inferred and compared with theory. The verification is extremely precise and gratifying.[1]

For general interest, a diagram of the experimentally observed absorption

---

[1] Considerable technical interest also centers on the spectra of atoms, molecules, and crystals because of the possibility of using certain transitions to build devices (called *masers*) which can amplify very-high-frequency signals. For an interesting explanation of the operation of a maser, see two articles by A. L. Schawlow: Optical Masers, *Sci. Am.,* June, 1961, p. 52, and Advances in Optical Masers, *Sci. Am.,* July, 1963, p.34.

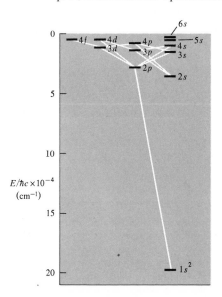

FIG. 2·6 ENERGY DIAGRAM FOR A NEUTRAL HELIUM ATOM, SHOWING SOME OF THE ALLOWED TRANSITIONS; AS SUGGESTED, VERTICAL TRANSITIONS ARE NOT OBSERVED.

"lines" for a neutral helium atom is given in Fig. 2·6. The presence of allowed transitions (for example, $2s \leftrightarrow 3p$) and forbidden ones (for example, $2s \leftrightarrow 3s$) is apparent on this diagram. Similar diagrams for many atoms can be found in the literature of atomic spectroscopy.[1]

2·1·6 *Reduced atomic structures and the periodic table.* The ground-state cell diagram for a germanium atom shown in Fig. 2·5 may be simplified for some purposes by the realization that the electrons in the inner orbits (through $3d$) essentially shield a portion of the nuclear charge. We may thus represent germanium by a "reduced" cell diagram with an effective nuclear charge of $+4$ electronic units. Two of the four electrons which surround this reduced structure are to be thought of as being in s-type orbitals and two in p-type orbitals. The chemical properties of germanium are therefore similar to those of carbon, silicon, and tin, which also possess this same reduced atomic structure (see Prob. 2·4). Similarly, the reduced atomic structures for potassium, sodium, and lithium are all the same, so these elements have similar chemical properties. The equivalence of reduced atomic structures partly accounts for the arrangement of atoms in groups in the periodic table.

2·2 CRYSTAL STRUCTURE

We shall now turn our attention to the subject of crystal structure. In this section we shall see how conductors may be distinguished qualitatively from semiconductors and insulators, and how some of the simpler crystal structures for metals and semiconductors can be explained from the reduced-atomic-orbital models previously described.

When atoms of a given type form a solid, they most frequently take up an orderly three-dimensional arrangement which we call a crystal. For each possible type of crystal, we can identify a fundamental building block,[2] which can be repeated in space to generate as large a crystal as we please.

In principle, the arrangement of atoms in the fundamental building block can be predicted from a knowledge of the reduced atomic structure for the atom. As a practical matter, the positions of the atoms are usually measured by X-ray techniques, and the reduced atomic structures are used to explain the known arrangement and calculate its properties.

2·2·1 *Metallic crystals.* Highly conductive metals such as copper, aluminum, lithium, and silver usually crystallize in close-packed structures where there are many nearest neighbors. For example, the fundamental building block for a lithium crystal is the *body-centered cubic* structure shown in Fig. 2·7.

---

[1] See, for example, H. E. White, "Introduction to Atomic Spectra," McGraw-Hill Book Company, New York, 1934.

[2] When properly defined, the fundamental building block is called a *unit cell.*

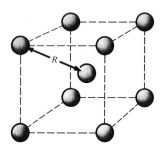

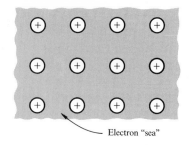

Electron "sea"

(a) Building block for body-centered cubic crystal (example: lithium metal)

(b) Sea of electrons neutralizing ion cores in a two-dimensional representation

FIG. 2·7 A TYPICAL METALLIC CRYSTAL STRUCTURE. BONDING IS ACCOMPLISHED VIA A UNIFORM DISTRIBUTION OF FREE ELECTRONS; THERE ARE NO LOCALIZED BONDS CONNECTING ION CORES.

In such a structure, a given lithium atom has eight nearest neighbors (at distance R in Fig. 2·7) and six next-nearest neighbors (at a distance $2R/\sqrt{3} = 1.15R$).

To explain the properties of these crystals, we regard the individual atoms as being ionized and "floating" in a sea of electrons (see Fig. 2·7). In a lithium crystal, for example, each atom contributes its $2s$ electron to the sea, and becomes singly ionized [$Li^+ = (1s)^2(2s)^0$]. The ionized $2s$ electrons are free to wander throughout the crystal. They are thus considered to be "shared" by all the ion cores in the crystal.

The "binding" in such crystals is due to a balance of two forces: (1) the attraction between the positively charged atomic cores and the electron sea, and (2) the repulsion of the atomic cores by each other. The binding is rather weak, so that considerable movement of the positive ions is possible without great expenditure of energy. This accounts for the general plasticity of metals.

In order for the electron sea to be formed, it is necessary that in the separate atoms there be one or more easily ionized electrons. These electrons remain "free" over a wide temperature range and give the metal its high conductivity. Conduction occurs as a result of the freedom of these electrons to drift slowly through the crystal in response to an applied electric field.

2·2·2 **Semiconductor crystals.** By comparison to the close packing of ions encountered in a metallic crystal, the atoms in a semiconductor crystal have few nearest neighbors. A semiconductor crystal is thus sometimes said to be "very open" (that is, there is a large volume per atom).

Semiconductor crystals are held together by highly localized electronic bonds between neighboring atoms. These bonds are very strong and directional in their character; thus, the crystals are quite hard, have a high melting point, and tend to fracture when stressed.

The two most important semiconductor materials for making transistors and diodes, at present, are germanium and silicon, so we shall study these crystals most extensively. However, the properties to be described are shared by many other semiconductor materials, which will be mentioned later.

It is known from the way X rays are scattered by pure germanium crystals that a given germanium atom is bonded to four nearest-neighbor atoms in the manner shown in Fig. 2·8. In this figure, there is an atom at the center of the box, the four neighboring atoms defining the four vertices of an equilateral pyramid.

Each bond has two electrons in it (with opposite spins), and for this reason it is called an *electron-pair* or *covalent* bond. The bonding action arises from the electrostatic attraction between the electrons in the bond and the atoms at the ends of the bond. The angle defined by any two bonds in Fig. 2·8 is $\simeq 110°$.

2·2·3   *The covalent-bonding orbitals*

To explain the bonding arrangement shown in Fig. 2·8, let us first imagine that the crystal is "blown up" to the point where neighboring atoms are essentially isolated from each other, but still in their correct relative positions. Then the electrons surrounding each atom will be in their normal atomic orbitals. Now, if we let the atoms gradually come together, maintaining their correct relative positions, the outermost, or valence, electrons in each atom will ultimately begin to feel some effect from the neighboring atoms. This will cause a change in the valence-orbital patterns of each atom, to account for the influence of the neighboring atom. (In Schrödinger's equation, there is a change in the potential energy of these outer electrons, due to the neighboring atoms.)

The nature of the change in the valence-orbital patterns depends on the crystal being studied. In a metallic crystal, where there are many nearest neighbors, the valence electrons will experience a nearly uniform attractive force in all directions, due to the neighboring nuclei. The effect of this attraction is to ionize the valence electrons from their parent atoms as they are brought to their final positions. By contrast, the parent atoms in a semiconductor crystal have few nearest neighbors. On account of this, the valence electrons experience highly directed forces from neighboring nuclei. These forces are not sufficient to ionize the valence electrons, but they are strong enough to have a major effect on the geometrical shape of the valence orbitals.

The final result is that in a germanium crystal the two 4s and the two 4p

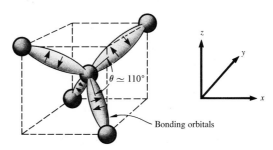

FIG. 2·8   BONDING ORBITALS AND BASIC STRUCTURE OF A GERMANIUM CRYSTAL

electrons of the parent atom reside in a set of "cigar-shaped" orbitals which are directed toward the four nearest neighbors. These orbitals are called the *covalent-bonding orbitals*. One such orbital is shown in Fig. 2·9a. The mathematical theory shows that these orbitals are really mixtures of the 4s and 4p atomic-valence orbitals, the mixing being caused by the fields of the neighboring atoms. The electrons in these bonding orbitals have energy levels which are lower than the atomic 4s energy level, which suggests that the covalent bond is a stable one. Furthermore, all the valence electrons of the parent atom are used in these bonds, so there are none left over to produce conduction. A perfect germanium crystal is therefore an insulator.

Of course, the electrons surrounding the neighboring atoms will be described by similar crystalline orbitals. The bonds connecting two neighboring atoms will therefore appear as shown in Fig. 2·9b. The strength of the bond is determined primarily by the amount that the two orbitals in a given bond overlap. The overlap in crystals of the type we are considering is great, and the bonds are accordingly very strong.

2·2·4 *Geometry of silicon and germanium crystals.*

A large germanium crystal may now be generated by fitting germanium atoms together in the pattern suggested in Fig. 2·8. A section of a crystal so formed is shown in Fig. 2·10. The atoms in the upper-left-hand corner of this figure are accentuated, to indicate the fundamental nature of Fig. 2·8 in the crystal construction.

The cube shown in Fig. 2·10 has the property that it may be repeated in three mutually perpendicular directions to generate as large a crystal as we please.[1] The sides of this cube are, by definition, the *lattice constant* of the solid, designated *a* in Fig. 2·10. The cube has four atoms in its interior, an atom in the center of each of its faces (each of which is shared with an adjoining cube), and eight atoms at the cube vertices (each of which is shared with seven other elementary cubes). Thus, the number of atoms in a crystal is eight times the number of elementary cubes; or, equivalently, there are eight atoms in a cube of size a^3.

X-ray measurements of the lattice constants for germanium and silicon show that the lattice constant *a* equals 5.62×10^{-8} cm for germanium, and 5.42×10^{-8} cm for silicon. Accordingly, germanium contains 4.52×10^{22} atoms/cm³, while silicon contains 5.00×10^{22} atoms/cm³.

[1] In this case, however, the cube is not the unit cell for the crystal.

Reduced atomic core

Covalent bonding orbital

(a) *A typical valence electron orbital in a germanium crystal*

(b) *Atoms bonded together with crystalline valence orbitals*

FIG. 2·9   ELECTRON ORBITALS FOR VALENCE ELECTRONS IN A GERMANIUM CRYSTAL

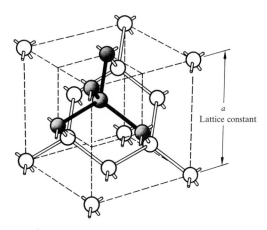

_a_
Lattice constant

FIG. 2·10 FUNDAMENTAL BUILDING BLOCK FOR THE DIAMOND CRYSTAL STRUCTURE

Before leaving the subject of semiconductor crystal structure, we should mention one important point. We speak of forming bonding orbitals from atomic orbitals in order to give a reasonably accurate picture of how the electrons are arranged in the bonds. However, we must realize that all the valence orbitals together form _one continuous three-dimensional chain_ through a perfect crystal. For this reason, we cannot suppose that a given valence electron _must stay_ in a given bond. Rather, it may move about through the crystal, via the system of valence orbitals, _exchanging_ places with other valence electrons as it goes.[1] To be precise, we have to consider both the _system_ of bonding orbitals and all the bonding electrons as being a property of the crystal, not of the local atom. This viewpoint is essential for understanding many of the basic properties of semiconductors, and we shall return to it several times.

2·2·5 _Some mechanical properties of the covalent bond._ The covalent bond indicated schematically in Fig. 2·10 is quite strong, and gives rise to some interesting mechanical properties which are useful for understanding both the electrical properties of semiconductors and the reasons for the adoption of certain fabrication procedures.

The mechanical strength of the bond may be appreciated from the fact that industrial diamonds and silicon carbide are popular materials for making cutting and grinding wheels. A further indication of bond strength is the fact that the melting points of pure carbon, silicon, and germanium in the diamond structure are quite high: 3550°C, 1420°C, and 958°C, respectively. For this reason, silicon-carbide rods are widely used as heating elements for high-temperature furnaces.

A somewhat surprising fact, also indicative of bond strength, is that these materials have quite large thermal conductivities. For example, at room temperature, copper (one of the best heat conductors) has a thermal conductivity of about 4.5 watts/cm-°C, while silicon and germa-

---

[1] No net current flow is involved in these exchanges, of course.

Some structural properties of semiconductors 55

nium have thermal conductivities of 0.84 and 0.63 watts/cm-°C, respectively. And at 20°K, copper has a thermal conductivity of 50 watts/cm-°C, while germanium crystals, which are rich in the atomic weight 74 isotope, have thermal conductivities of 40 watts/cm-°C.

Since carbon, silicon, and germanium are insulators or semi-insulators in the pure diamond-crystal form, they do not have large numbers of free electrons to participate in heat conduction, as metals do. Instead, the thermally induced motion of the lattice atoms being transmitted through the crystal is responsible for heat conduction. This motion may be visualized by picturing each atom as a mass point and the bonds as tightly wound springs. The spring-mass system then couples motion which is thermally induced in one region of the crystal to other neighboring regions.

2·2·6 *Preparation of germanium single crystals*

There are several popular methods for preparing germanium crystals suitable for use in transistor fabrication. One of the most important procedures for preparing transistor-grade germanium or silicon is called the Czochralski pulling technique. A schematic diagram of the apparatus necessary for producing germanium crystals in this manner is shown in Fig. 2·11. The growing technique is as follows.

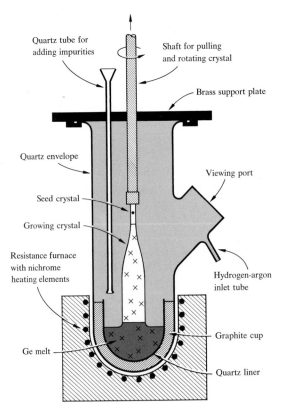

FIG. 2·11 APPARATUS FOR GROWING SINGLE CRYSTALS OF GERMANIUM
(*After W. C. Dunlap, An Introduction to Semiconductors, John Wiley and Sons, 1957*)

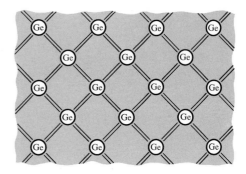

FIG. 2·12   TWO-DIMENSIONAL REPRESENTATION OF A GERMANIUM CRYSTAL AT $T = 0°K$

A charge of polycrystalline germanium is inserted into a quartz crucible, which is in turn mounted in a graphite cup. This assembly is lowered into a resistance furnace in a large quartz envelope.

The pulling head above the furnace jar holds a seed made from a previously prepared single crystal. The seed is dipped into the melted germanium and held at just the proper temperature, so that the seed does not melt or the melted germanium freeze. The pulling head provides a controlled rate of withdrawal of the crystal, and single crystals of germanium grow on the seed continuously. The crystal orientation of the growing crystal will be identical to that of the seed crystal. Provision is usually made for rotating the crystal as it grows, to provide some slight stirring action in the germanium melt. Donor and acceptor atoms[1] may be added to the melt at any time by means of the quartz tube shown in Fig. 2·11. In order to ensure that oxygen and other deleterious impurities do not get grown into the crystal, an atmosphere of hydrogen and argon is admitted into the tube during the growth.

Crystals grown by this process are usually 3–5 in. in length and 0.5–1 in. in diameter.

2·2·7   A two-dimensional representation of a germanium crystal.   While the three-dimensional aspects of the lattice are interesting and useful, it is also convenient to draw a two-dimensional representation of the diamond lattice for some purposes. Our principal use of this model will be in defining the conducting particles, indicating the existence of an ionization energy, and discussing the effects which doping has on the electrical conductivity of the structure.

In Fig. 2·12 we show a two-dimensional representation of the perfect germanium crystal shown in Fig. 2·10. Each valence bond is represented by two lines to indicate that there are two electrons associated with each bond. Figure 2·12 indicates the situation which exists at absolute-zero temperature, where all the lattice atoms are essentially at rest in their proper positions[2] and all the valence bonds are complete. At this temperature, germanium or silicon is a perfect insulator.

In order to make the crystal conduct electricity, we must generate some

---

[1] To be described in Sec. 3·4.

[2] They have some net zero-point motion to satisfy the uncertainty principle.

free carriers in it. There are several ways of doing this: (1) shine electro-magnetic radiation (for example, light) on the crystal and produce carriers by the photoelectric effect; (2) generate free carriers by supplying thermal energy (heat) to the crystal; or (3) dope the crystal with impurities which introduce free carriers. We shall consider these methods in some detail in the next chapter.

REFERENCES

Feynman, R. P.: "The Feynman Lectures on Physics," vol. 1, Addison-Wesley Publishing Company, Inc., Reading, Mass., 1964, chaps. 37 and 38.
Sherwin, C. W.: "Introduction to Quantum Mechanics," Holt, Rinehart and Winston, Inc., New York, 1959, chaps. 2, 3, and 4.
Leighton, R. B.: "Principles of Modern Physics," McGraw-Hill Book Company, New York, 1959, chap. 5.
Eisberg, R. M.: "Fundamentals of Modern Physics," John Wiley & Sons, Inc., New York, 1961, chaps. 4–8.
Adler, et al.: "Introduction to Semiconductor Physics," SEEC Notes, vol. 1, John Wiley & Sons, Inc., New York, 1964, chap. 1.
Wert, C. A. and R. M. Thomson: "Physics of Solids," McGraw-Hill Book Company, New York, 1964, chaps. 2, 7, and 8.

DEMONSTRATIONS

2·1 *Mechanical illustration of crystal growth.* An effective demonstration of the "building up" of a large crystal from the structure shown in Figs. 2·8 and 2·10 can be made as follows. Make eight small cubic blocks from either wood or heavy paper. Color the corners of each block in the manner indicated in Fig. D2·1a. The box may then be thought of as having an atom in its center and four equidistant nearest neighbors.

The eight small boxes may now be put together to form a large cube with the structure shown in Fig. 2·10 (except that now every small box cannot be considered to have an atom in its center, as Fig. 2·10 indicates). The four small boxes forming any given face of the large cube may now be moved to new positions to illustrate the building process. One possible sequence is suggested in Fig. D2·1b.

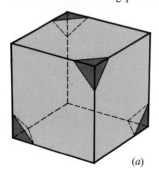

(a)

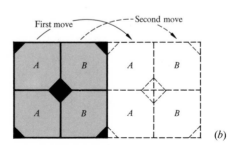

(b)

FIG. D2·1

2·2 *Molecular model of the diamond lattice.* An interesting model of the diamond crystal structure may also be made from small rubber balls and toothpicks. If two different colors of balls are used, the crystal can be constructed so that balls of one color are bonded only to balls of the other color. The features of crystal structure suggested in Prob. 2·8 will then be very apparent, as balls of a given color represent one sublattice. This demonstration is also useful in visualizing semiconductor crystals such as gallium phosphide (to be discussed in Sec. 3·5) and in showing how planes of atoms line up in different orientations of the crystal.

An inexpensive and very useful kit of parts for making this molecular model can be obtained from Edmund Scientific Co., Barrington, New Jersey, Stock No. 30,413. The kit contains 50 sponge rubber balls, 50 wooden rods, a rod sharpener, and directions for making several molecular models.

PROBLEMS

2·1 If a hydrogen atom is in its ground state, what is the probability that the electron will be found at a distance of more than three Bohr radii from the proton?

2·2 Instead of describing the ground state of a hydrogen atom with the position-probability density function $p(r)$ given in Eq. (2·7), we may describe it with a *radial-probability density* given by

$$R(r) = 4\pi r^2 p(r)$$

$R(r)\, dr$ is the probability that the electron will be found in a spherical shell between r and $r + dr$. Plot $R(r)$ for the ground state of a hydrogen atom. At what radius does this curve have its maximum value?

2·3 What will be the appropriate wavelength of the light emitted when a hydrogen atom drops from a $2p$ state to a $1s$ state?

2·4 Write down the ground-state electronic configuration [that is, $(1s)^2(2s)^2\cdots$] for carbon and silicon, and develop the corresponding energy-cell diagrams. Repeat for lithium, potassium, and sodium.

2·5 A singly ionized lithium atom $[(1s)^2(2s)^0]$ may be considered to be a positively charged sphere of radius 3Å. Estimate roughly the number of atoms per cubic centimeter (or free electrons per cubic centimeter) in a perfect crystal of lithium.

2·6 Calculate the mass density ρ of germanium and silicon from the known crystal structure.

2·7 Calculate the distance between next-nearest neighbor atoms in a germanium crystal.

2·8 The fundamental building block for a face-centered cubic (fcc) lattice is a cube with atoms at each corner and in the center of each face. It is sometimes useful to think of the diamond lattice as being two interpenetrating face-centered cubic sublattices, each with lattice constant (a). The exterior atoms in Fig. 2·10 demonstrate one of these sublattices. Verify that the interior atoms also form part of an fcc sublattice. Draw two cubes (without atoms) which interpenetrate each other in the same way as the diamond sublattices. What can you say about the bonding between atoms in a given sublattice? (These features of crystal structure are illustrated by Demonstration 2·2.)

3 ELECTRONS AND HOLES

THERE ARE TWO charge-bearing particles which account for the conduction properties of semiconductors, the *electron* and the *hole*. The presence and properties of both types of carriers are absolutely fundamental to an understanding of how semiconductor devices operate. In order to emphasize their importance, we shall devote this entire chapter to them.

The existence and basic properties of these carriers will be introduced by means of a series of experiments. Simplified theoretical explanations of these experiments will be given, though a really thorough theoretical basis for the properties of these carriers cannot be given without assuming considerable familiarity with Schrödinger's equation and its application to problems in solid-state physics.

3·1   INTERNAL PHOTOELECTRIC EFFECT IN SEMICONDUCTORS

In Fig. 3·1 we show a schematic experimental arrangement for studying the photoelectric effect in semiconductors. A monochromatic beam of light of frequency ν and wavelength λ emerges from the light source. Both the intensity of the beam and the light frequency ν are assumed to be continuously adjustable. A semiconductor crystal of length l is placed in the path of the beam, and a photodetector is placed behind the crystal.

When ν is very low, the light passes through the crystal undisturbed; that is, the crystal is transparent to low-frequency radiation. As ν is increased, however, we find that, at a certain critical frequency ν_c, the crystal begins to *absorb* the incident radiation. The crystal exhibits strong absorption of radiation between this critical frequency ν_c and some upper frequency, which is usually in the X-ray range.

If the light intensity reaching the detector is denoted by $I_d(\nu)$ and the light intensity leaving the light source is $I_s(\nu)$, it is found that

$$I_d(\nu) = I_s(\nu)e^{-\alpha l} \qquad (3 \cdot 1)$$

within the absorption range.[1] In this expression, α is called the *absorption constant,* and l is the length of the crystal in the direction of propagation of the light beam. α is a function of frequency and is usually quoted in the dimensions of centimeter^{-1}.

Experimental curves of α versus photon energy for a germanium crystal are shown in Fig. 3·2.[2] The electromagnetic spectrum is shown in Fig. 3·3 for comparison. In both figures the photon energy is related to the photon frequency ν by the formula $E(ev) = (h\nu/q)$.

Apart from the phenomenon of absorption, two other effects may be observed in this experiment:

1. If side contacts are placed on the crystal specimen, as in Fig. 3·1,

---

[1] It is assumed that reflection losses at the surface of the crystal are negligible, or have been accounted for. Equation (3·1) thus relates only to the internal transmission of light through the crystal.

[2] Actually, the optical transmission experiment indicated in Fig. 3·1a is impractical when $\alpha \geq 10^4$ cm^{-1}, due to the extremely thin samples required for accurate measurements (see Prob. 3·2).

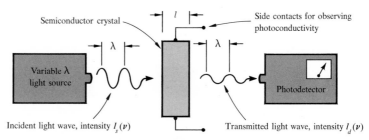

FIG. 3·1   SCHEMATIC ARRANGEMENT FOR OBSERVING OPTICAL AND PHOTOELECTRIC EFFECTS IN SEMICONDUCTORS

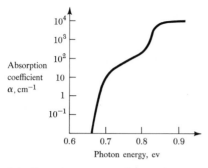 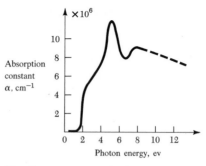

(a) Absorption spectrum of germanium near the absorption edge (optical-transmission experiment)

(b) Absorption spectrum of germanium over wide photon-energy range (optical-reflection experiment)

FIG. 3·2 OPTICAL-ABSORPTION CONSTANT VS. PHOTON ENERGY FOR GERMANIUM

the conductivity of the crystal will be observed to increase when ν is in the absorption range. This phenomenon is called *photoconductivity*.

2. If the photon source is sending in photons of frequency $\nu > \nu_c$ in the absorption range, the crystal may reradiate energy at a frequency ν_c. This phenomenon is called *fluorescence*.

These experimental facts may be explained in two different ways. One of them (an oversimplified, semiclassical explanation) focuses attention on the spatial location and motion of charges in the crystal. This is the simplest way to visualize both the photoconductivity process and the motion of charges in transistors and other semiconductor devices. As a result, this semiclassical model is used extensively in studies of semiconductor electronics and is entirely adequate for these purposes.

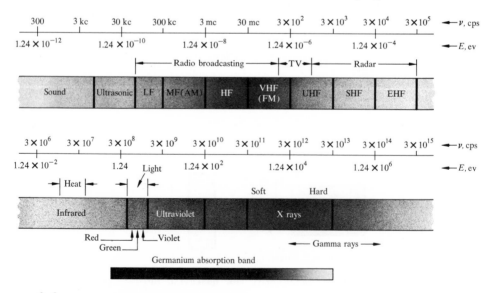

FIG. 3·3 ELECTROMAGNETIC-ENERGY SPECTRUM

However, as with all oversimplified models, the semiclassical model has serious limitations if it is used without a thorough appreciation for its limits of validity. These limits are established by the quantum-mechanical model, in a manner to be briefly described in Sec. $3 \cdot 3$.

$3 \cdot 1 \cdot 1$ *The photoelectric threshold.* Let us first consider the critical frequency ν_c at which absorption begins. The semiclassical explanation of this is as follows. As the frequency of the incident radiation increases, the energy of the incident photons increases, and they finally become sufficiently energetic to *ionize* the covalent bond—in other words, a valence electron can absorb enough energy from an incident photon to leave its valence site and become a *free electron*[1] (that is, free to wander about through the crystal). This process is shown schematically in Fig. $3 \cdot 4$.

Each liberation of a valence electron in the crystal of Fig. $3 \cdot 4$ is accompanied by the creation of a vacancy or imperfection in the valence structure. This imperfection in the valence structure is called a *hole*. It should be apparent that in any pure crystal, such as that shown schematically in Fig. $3 \cdot 4$, the mechanism just described, called *photogeneration,* creates equal numbers of holes and free electrons. The rate at which holes and electrons are created by photogeneration is a function of the parameters of the semiconductor material and the incident photon flux.

The photoelectric effect thus gives us an *ionization energy* for atoms in the crystal, which we denote by $\mathcal{E}_i$. This ionization energy is just the energy which must be given to a valence electron before it can break away from the covalent bond. $\mathcal{E}_i$ will vary somewhat with temperature, usually being a small percentage lower at room temperature than it is at $0°K$. The main features of the photoelectric effect are, however, unaffected by temperature.

A set of values for the room-temperature ionization energy for different

[1] These "free" electrons do not have sufficient energy to escape from the crystal. The excitation process is therefore properly called an internal photoelectric effect. If very highly energetic photons strike the germanium crystal, some electrons will gain enough energy to leave the crystal. This process is called photoelectric emission, and we shall consider it later in more detail.

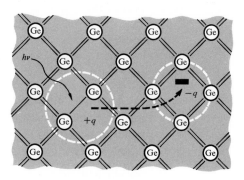

FIG. $3 \cdot 4$ PRODUCTION OF ELECTRONS AND HOLES BY PHOTOGENERATION

TABLE 3·1

| | | Photoelectric threshold | |
Material	E_g, volts	v_c, critical frequency, cps	λ_c, critical wavelength, cm
Germanium	0.67	1.68×10^{14}	1.85×10^{-4} (infrared)
Silicon	1.12	2.69×10^{14}	1.12×10^{-4} (infrared)
Gallium phosphide	2.25	5.46×10^{14}	0.55×10^{-4} (visible)
Carbon (Diamond)	5.20	12.6×10^{14}	0.25×10^{-4} (ultraviolet)

semiconductors is given in Table 3·1. It is customary to quote $\mathcal{E}_i$ in joules. The symbol E_g is used for $\mathcal{E}_i/q$. The units of E_g are volts.

We can observe from Table 3·1 that the photoelectric threshold for germanium (and silicon) lies in the infrared region, while the threshold for gallium phosphide lies in the visible region. The threshold for carbon in the diamond form is in the ultraviolet region.

The fact that the critical frequency for gallium phosphide crystals lies in the visible range makes possible a simple demonstration of this photo-electric effect. Gallium phosphide is a III–V–compound semiconductor and has the largest ionization energy of the materials presently being considered for transistors and diodes. We shall have more to say about its crystal structure later. However, for the present we may utilize the properties of gallium phosphide to clarify some of the points just discussed.

The demonstration apparatus is shown in Fig. 3·5. A 35-mm slide photograph of an optical-diffraction grating is mounted in front of a microscope light to project the spectrum of visible light onto a screen made of translucent paper. A piece of opaque paper with two holes cut in it is then mounted on a wand, and a small piece of gallium phosphide crystal is mounted behind one of the holes. By placing the wand between the light source and the paper screen, the relative light transmission through the open hole and the gallium phosphide can be observed on the screen. The gallium phosphide will appear opaque in the blue and violet part of the spectrum, but will transmit as well as the open hole in the red and yellow part of the spectrum.[1]

The reason for this is that the high-energy part of the incident light (corresponding to the colors green through violet, and the wavelengths 0.5×10^{-4} cm through 0.35×10^{-4} cm) is absorbed by the crystal, causing excitation of electrons from the valence structure to the free electron condition. The green-to-red wavelengths are passed through the crystal with little absorption, however, so we see a generally orange color. For the same reason, we cannot observe any change in the color of a neon lamp when a gallium phosphide crystal is placed in front of it, because the neon lamp is already red.

[1] More detailed instructions and materials for preparing this demonstration are described in Demonstration 3·1 at the end of this chapter.

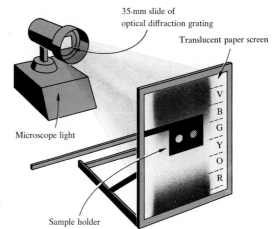

35-mm slide of
optical diffraction grating

Translucent paper screen

V
B
G
Y
O
R

Microscope light

(a) Apparatus for creating light spectrum and
testing gallium phosphide crystal

Sample holder

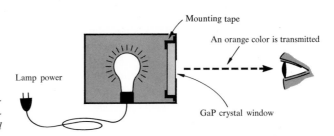

Mounting tape

An orange color is transmitted

Lamp power

(b) Alternate apparatus for demonstrating the photoelectric threshold in a gallium phosphide crystal

GaP crystal window

FIG. 3·5 PHOTOELECTRIC EFFECT IN A GALLIUM PHOSPHIDE CRYSTAL

3·1·2 *Conductivity of the photogenerated carriers.* We have also mentioned
the fact that the conductivity of the crystal will be observed to increase
when light in the absorption range is shined on it. This change in conductivity with light is called photoconductivity and is the basis of modern
electronic devices called photoconductors; that is, resistors whose resistance value depends on the light signal being applied to them.[1]

The phenomenon of photoconductivity can also be demonstrated with
the apparatus shown in Fig. 3·5a. One first applies ohmic contacts to the
gallium phosphide window and then connects these contacts to an electrometer. The electrometer indicates a decrease in resistivity of the gallium
phosphide as it is moved through the spectrum from red to violet.[2]

The explanation of photoconductivity leads us directly to one of the

---

[1] A more detailed description of these devices and their applications is given in Chap. 4.

[2] The details for construction of this demonstration are given in Demonstration 3·2. The
demonstration can also be done with a window of germanium or silicon, in which case the
leads of an ohmmeter can be pressed directly on the sample to measure the resistance
change. A commercial cadmium sulphide photoconductor can also be conveniently used in
this demonstration. The absorption edge for this material lies in the deep red, so its conductivity changes rapidly as the sample is scanned from the yellow part of the spectrum through
the red.

central issues in the theory of semiconductor devices: a definition of the charge carriers and an explanation of their motion. In this first encounter with these carriers, we shall present some workable, but oversimplified, models which are easily remembered. The validity of these models will be discussed in Sec. 3·3.

We shall begin the explanation of photoconductivity by supposing that the electrons which are ionized from the covalent bonds can move through the crystal in much the same manner as electrons do in a metallic crystal. Thus, when an electric field is applied to the crystal, as in Fig. 3·6, these electrons are free to "drift" through the crystal, and they therefore contribute to an increase in conductivity.

However, *only about one-half of the total change in conductivity of a germanium or silicon* crystal can be accounted for by this mechanism. To account for the remainder, we must assume that *the vacancy in the valence structure can also conduct a current.* It is just this fact which makes a semiconductor so significantly different from a typical metal, and gives rise to the possibility of transistor action. We emphasize the basic idea as follows:

The vacancy left in the covalent bond by the liberation of a valence electron behaves as if it were a free, conducting particle with a positive electronic charge of $q(1.6 \times 10^{-19})$ coul and a mass approximating that of the free electron. This "apparent" particle is called a hole.[1]

Let us expand on this idea briefly. First of all, it may be appreciated by inspection of Fig. 3·6 that within the dashed circle surrounding the hole there is a net charge of $+q$ coul, since this volume was neutral before the free electron was liberated and carried off $-q$ coul. Of course, this positive charge is actually associated with an unneutralized atom. However, when we speak of the vacancy as a "particle" (that is, hole) we associate the positive charge with the hole. Thus we say that the *hole* carries a positive charge of q coul.

Furthermore, this hole, or imperfection in the valence structure, can move through the lattice in the following way. When an electric field

---

[1] More thorough discussions will be given in Sec. 3·3.

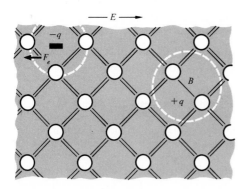

FIG. 3·6   OVERSIMPLIFIED DIAGRAM OF THE MECHANISM OF CONDUCTION FOR ELECTRONS

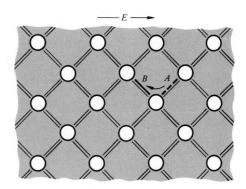

FIG. 3·7 OVERSIMPLIFIED DIAGRAM OF THE MECHANISM OF CONDUCTION FOR HOLES; VALENCE ELECTRON AT A MOVES INTO PREVIOUSLY EMPTY BOND AT B

is applied to the crystal, the otherwise symmetrical paths of the valence electrons are warped in the direction of the field. In Fig. 3·7, for example, a valence electron in position A, which is close to the hole, can now move into it rather easily, leaving a new hole in its original site. Thus, in this simple model the motion of the hole may be visualized as a *transfer of ionization* from one atom to another, carried out by the field-aided motion of the valence electrons. The direction of motion of the hole is the same as the direction of the applied electric field. Finally, note that the motion of the hole is independent of the motion of the electron which was originally liberated to produce it. If we think of the hole and the free electron as two particles, then the particles are *independent,* and give independent contributions to the photoexcited change in conductivity.

3·1·3   The absorption band.   When we accept the proposition that electrons and holes can move in the crystal, we are in a position to understand at least qualitatively why the crystal exhibits continuous absorption of radiation over a *band* of frequencies (rather than at discrete frequencies, as an isolated atom does). For, if the charge carriers can move, they can have kinetic energies. And thus, when ν is increased above $\nu_c = \mathcal{E}_i/h$, the electrons and holes are created with some initial velocities. On this basis we would expect that the crystal will absorb radiation at any frequency $\nu > \nu_c$, unless bounds exist on the amount of kinetic energy which may be given to the carriers.

3·1·4   Fluorescence and recombination of the photogenerated carriers.   Finally, we stated that if the incident photons have a frequency $\nu > \nu_c$ (but are still in the absorption range), the crystal will *emit* radiation at the frequency ν_c. This phenomenon is called *fluorescence* and is analogous to the emission of radiation which occurs when an excited atom returns to its ground state.

To pursue the analogy, we may say that the *crystal* is in an excited state when electron-hole pairs exist within it. It returns to its ground state when electrons and holes *recombine* to perfect the valence structure.

Electrons and holes 67

Now, unlike an isolated atom, the crystal may be excited by any incident radiation in a broad absorption band. In such a situation, the charge carriers will have nonzero initial velocities and will move away through the crystal. However, in this motion they will collide with the lattice. During each collision, there will be energy transferred between the lattice and the carriers. A given electron will sometimes lose and sometimes gain energy. On the average, however, it loses energy until its kinetic energy approaches zero. Then, in a perfect crystal, it can only lose its remaining energy by returning to a broken bond. When the ionization energy is given up, a photon of frequency ν_c is produced.[1] This gives rise to the fluorescence or emission of light at ν_c. Since this process is associated with the recombination of a hole and an electron, it is also given the name *recombination radiation*. It is the opposite of the photogeneration process, by which electron-hole pairs were produced.

The phenomenon of fluorescence can be demonstrated by illuminating a gallium phosphide crystal which is held at dry-ice temperature with an ultraviolet light or "black lamp" (*cf.* Demonstration 3 · 3 at the end of this chapter.) The crystal will be observed to emit a reddish-orange glow. Later, we shall describe a very important class of devices in which the conditions for emitting light are established by electric currents.

3 · 2 THERMAL EXCITATION OF VALENCE ELECTRONS

We have seen in the preceding section that an insulating (or nearly insulating) semiconductor crystal can be made to conduct by shining light on it. The action of the light is to produce free carriers, both electrons and holes, in the solid.

Naturally, other sources of energy can also produce free carriers in the solid. In particular, they may be produced by *thermal* excitation of the semiconductor lattice. The result is that a semiconductor crystal which is an "insulator" near $0°K$ may become a reasonably good conductor at room temperature. (This fact is sometimes used to classify materials roughly into "insulator" or "semiconductor" categories.)

The basic idea in the thermal excitation of carriers is as follows. As the temperature of a perfect crystal is raised from $0°K$, both the lattice atoms and the valence electrons will absorb heat. The lattice atoms will vibrate about their ideal positions, producing a continuous fluctuation of the lengths and angles of the valence bonds. The energy of electrons in these bonds will be a continuously varying function in time.

Occasionally, these fluctuations will give electrons in some of the bonds

[1] The situation just described relies on the assumption that the crystal is perfect. Actual crystals have imperfections (missing atoms, foreign atoms, etc.) which frequently can "trap" electrons and holes and immobilize them very effectively. These traps expedite the recombination of a hole and electron. The nature of most traps is such that during a recombination event the energy $\mathscr{E}_i$ is given to the crystal as heat, rather than in the creation of a photon. In such cases no fluorescence is observed. This is usually the case for germanium and silicon.

sufficient energy to escape from the bond. We may then say that the bond has been *thermally ionized*. This state of affairs is shown in Fig. 3·8. As in the photoionization process, the liberation of each valence electron is accompanied by the creation of a hole, or vacancy, in the valence structure. Thus, the *thermal-generation* mechanism also creates equal numbers of holes and electrons. The rate at which holes and electrons are created by thermal generation is a function of the parameters of the semiconductor material and the temperature.

The thermal-generation rate *per unit volume* is given by the symbol g and is specified to be a function only of temperature for a given material:

Thermal-generation rate = $g \equiv g(T)$ electron-hole pairs/cm³-sec   (3·2)

Two facts concerning this process of thermal generation will now be noted and should be clearly understood (these two comments are applicable to both photogeneration and thermal generation):

1. Whereas at absolute zero (and in the absence of illumination) the crystal is both microscopically and macroscopically space-charge neutral, it is only macroscopically neutral at any finite temperature (or in the presence of illumination in the absorption range, or both).

2. The inverse process of recombination is always acting to return free electrons to the holes in the valence structure, thereby perfecting the valence structure on a local basis.

Thus, there are two opposing *dynamic*[1] mechanisms for the regulation of electron and hole densities in a crystal: generation and recombination. By specifying the actual forms of the generation and recombination functions, we will be able to calculate the equilibrium-carrier densities which exist for a given condition of temperature (and illumination).

3·2·1   *Thermal-generation rate.*   Using the methods of statistical mechanics, it is possible to show that the thermal-generation rate in a semiconductor whose photoionization threshold energy is qE_g joules is

$$g(T) = AT^a e^{-(qE_g)/kT} \qquad (3·3)$$

[1] The meaning of the term *dynamic* will be discussed further in Sec. 3·2·3.

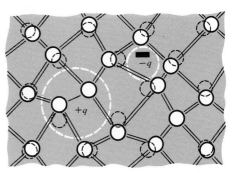

FIG. 3·8   THERMAL-GENERATION MECHANISM; SMALL DASHED CIRCLES SHOW EQUILIBRIUM POSITIONS OF CRYSTAL ATOMS; INSTANTANEOUS ATOMIC POSITIONS ARE SOLID CIRCLES.

where T is the absolute temperature, k is Boltzmann's constant (see inside front cover for value and units), E_g is the photoelectric threshold in volts, q is the magnitude of the electronic charge, and A and a are constants which depend on the semiconductor material.

This result is of sufficient importance in our future studies to warrant some discussion of how it arises.

In order for a valence electron to be promoted to the free-electron state, it has to obtain an energy of at least $\mathcal{E}_i$ from somewhere. In a thermal-ionization event, this energy must come from the thermodynamic inter-actions of the valence electron with its surroundings. We will describe this interaction in two ways, each of which contributes some insight into the process.

We stated earlier that a valence electron was more or less free to move through the lattice via the valence-bonding chain, exchanging places with other valence electrons as it goes. In this motion, it interacts with many other electrons and many lattice atoms. These other electrons and lattice atoms serve as a large "heat bath" at temperature T, with which the valence electron is in thermal contact.

Now, it is shown in statistical mechanics that when a small system (one-valence electron) is in thermodynamic contact with a large one (essen-tially the rest of the crystal)—the large one being in equilibrium at a tem-perature T—the probability per second that the small system can absorb an energy $\mathcal{E}$ from the large system is

$$e^{-\mathcal{E}/kT}$$

If we accept this result on faith, then we would expect that the probability that a given valence electron could gain an energy $\mathcal{E}_i$ from the crystal is proportional to

$$e^{-\mathcal{E}_i/kT}$$

Therefore, we might expect that the number of bonds per cubic centimeter which will become ionized per second, which is $g(T)$, should also be pro-portional to $e^{-\mathcal{E}_i/kT}$.

There are other factors in the detailed analysis which account for the T^a term in Eq. (3·3). However, such factors are usually unimportant, since the principal temperature-dependent factor is the exponential.

An alternate way of visualizing the generation process is based on the theory of black-body radiation. According to this theory, any black body which is in thermal equilibrium at a temperature T contains electro-magnetic radiation at all frequencies from zero to infinity. To employ this idea, let us suppose that the surfaces of a germanium crystal are treated so that the interior of the crystal becomes a black body.[1] Then there will be photons with energies in all ranges from very low to infinity in the crystal. Those photons with an energy of $\mathcal{E}_i$ or greater can produce

---

[1] The fact that the crystal faces are not perfectly reflecting, for photons with less energy than the photoionization energy, is not critical in this argument.

hole-electron pairs by the normal photogeneration process. Thus, the thermal-generation process can also be viewed as a photogeneration process, in which the number of available pair-producing photons is determined by the temperature of the crystal.

In this case, a theoretical derivation of the generation rate can be based on the theory of black-body radiation and the optical absorption properties of the crystal. The result is that the generation rate is once again proportional to

$$e^{-\mathscr{E}_i/kT}$$

Basically, this result arises from the fact that the number of photons per cubic centimeter with an energy of $\mathscr{E}_i$ or greater is itself proportional to $e^{-\mathscr{E}_i/kT}$, as long as $\mathscr{E}_i \gg kT$.

3·2·2 Recombination. We may specify a general form for the recombination rate of carriers by making the following observations. The number of recombination events per unit volume per unit time will certainly be proportional to the average free-electron concentration (or density) n, since these are the electrons available for recombination. The recombination rate will also be proportional to the average hole density p, since a hole is needed for each recombination. The recombination rate for hole-electron pairs must therefore be

$$R = rnp \tag{3·4}$$

where R = recombination rate, pairs/cm³-sec
n = free-electron density, electrons/cm³
p = free-hole density, holes/cm³
r = proportionality constant; function of parameters of semiconductor material and temperature

r may be calculated, in principle, when the recombination mechanisms are specified, though, as a practical matter, r is measured. Equation (3·4) is called a *mass-action* law by analogy with similar chemical processes.[1]

3·2·3 Thermal-equilibrium carrier densities in intrinsic crystals. We have stated the basic condition for thermal equilibrium in so far as carrier densities are concerned: there must be as many electron-hole pairs generated per second as are recombining per second. Specifying the temperature of the crystal is equivalent to specifying the thermal-generation rate. Specifying the electron and hole densities is equivalent to specifying the recombination rate. Therefore, at any given temperature there will exist one particular density of holes and electrons at which the recombination rate and the thermal-generation rate are in balance. This is the thermal-equilibrium density of holes and electrons for the given crystal temperature. These densities satisfy the equation

[1] See Prob. 3·20 for an expanded analogy concerning recombination, generation, and chemical processes.

Thermal-generation rate $= g(T) = rnp =$ recombination rate   (3·5)

Now, for pure, or *intrinsic,* crystals, we know that the electron and hole densities are always equal to each other. Therefore, *in an intrinsic crystal,* we denote n by n_i and p by p_i, and we have

$$n_i = p_i \qquad (3\cdot6)$$

Combining Eqs. (3·3), (3·5), and (3·6), we obtain

$$n_i = \sqrt{g(T)/r} = BT^b e^{-qE_g/2kT} \qquad (3\cdot7)$$

where T is the absolute temperature (°K).[1] An exact analysis shows that $b = \frac{3}{2}$.

This gives the *thermal-equilibrium density* of electrons (and holes) which we expect to find in a *pure* or *intrinsic* semiconductor crystal at an absolute temperature T(°K). The factor B can be evaluated when the semiconductor material is specified, though usually n_i is obtained directly from crystal-resistivity measurements (*cf.* Prob. 3·9). Again, the principal temperature-dependent factor is the exponential.

For germanium at $T = 300°$K, $n_i = 2.5 \times 10^{13}$/cm³, while for silicon at $T = 300°$K, $n_i = 1.5 \times 10^{10}$/cm³. We found earlier that there are about 10^{23} valence bonds/cm³ in a pure germanium crystal, so roughly one out of every 10^{10} of them is ionized. This reflects the basic nature of the exponential-probability function described in Sec. 3·2·1. It is very improbable that a given valence electron will obtain sufficient energy from the thermal field to leave the valence bond.

Finally, we shall make two points about the balance of generation and recombination which emphasize the dynamic nature of these processes:

1. If a crystal at a temperature T is illuminated with a steady light source which produces $g(I)$ pairs per second per unit volume, the electron and hole densities will change to satisfy

$$g(I) + g(T) = rnp \qquad (3\cdot8)$$

when the equilibrium condition is reached. However, the equilibrium condition cannot be established instantaneously. If the light is switched on at $t = 0$, the carrier densities will be below their final equilibrium values for a short time while the new equilibrium condition is established (see Sec. 4·4 for details).

2. Even in a thermal-equilibrium situation with no illumination there will be instantaneous fluctuations of the generation rate and recombination rate. These fluctuations can be indicated schematically in the manner shown in Fig. 3·9. Here, the actual electron density in an intrinsic crystal is plotted as a function of time. The average value is n_i. But during certain intervals of time, $g(T) > R$, so the plot has a positive slope; while during others, $g(T) < R$, so the plot has a negative slope. The fact that $g(T)$ and R vary with time and are not always precisely equal to each other is what is implied when we say $g(T)$ and R are in a *dynamic* balance.

---

[1] The demonstration of this important result will be deferred until the next chapter.

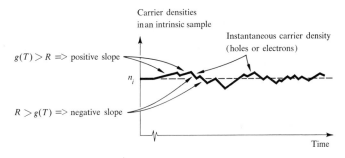

FIG. 3·9   VARIABILITY OF GENERATION AND RECOMBINATION RATES OVER SMALL TIME INTERVALS

If a fixed electric field is applied to such a crystal, the current which flows will be proportional to the instantaneous value of n. The current will therefore have an average value determined by n_i, but it will also exhibit small fluctuations about this average value. These fluctuations are called *noise* (in this case, generation-recombination noise). The amount of noise depends on the exact details of the dynamic processes of generation and recombination.

3·2·4   *Thermionic emission.* In the preceding sections we have seen that valence electrons can be promoted to the free-electron state by giving them an energy $\mathcal{E}_i$. The electrons in a solid may also be emitted into a vacuum if they can obtain enough energy to overcome the forces of attraction (both in the solid and at its surface) which normally keep them in the solid. The required energy may be supplied to the electrons in several forms, of which heat and light are the most common. The corresponding emitters are called *thermionic cathodes* and *photoelectric cathodes,* respectively. Thermionic cathodes are, of course, the source of electrons for most electron tubes.

We may study the emission process with the aid of the thermal-generation probability formula

$$g(T) \propto e^{-\mathcal{E}/kT}$$

To apply this formula, we denote the energy which a free electron needs to escape from the solid by W_F, which is called the *work function* of the solid (metal or semiconductor). By similar arguments to those given above, the probability per second that a free electron can gain enough energy to escape is

$$g(T) \propto e^{-W_F/kT} \qquad (3\cdot9)$$

This emission probability is proportional to the maximum current density J_{max} which can be obtained from a given cathode material, operating at a given absolute temperature T. When the details are worked out, the result is

$$J_{max} = A_e T^2 e^{-W_F/kT} \qquad (3\cdot10)$$

which is called *Richardson's equation*. A_e is a constant which is characteristic of a particular emitter. When a current density of magnitude J_{max} is actually drawn from a cathode, it is said to be operating under *temperature-limited* conditions. We shall return to this matter in Chap. 6.

It is clear that the exponential factor again dominates, so the operating temperature required to make J appreciable is determined by W_F, which, in turn, depends on the attraction for an electron that arises from atoms within the solid and at the "energy barrier" at the solid-vacuum interface. This energy barrier arises from foreign atoms deposited there and from forces associated with incomplete bonding of the surface atoms.

Typical values of W_F range from 1 to 4 volts for metals and certain oxides to 4 to 6 volts for most semiconductors without special surface treatment. For the metal tungsten, $W_F = 4.5$ volts, so operating temperatures of $2500°\text{K}$ are required to produce a J of 1 amp/cm². Tungsten cathodes are usually used in the high-power output tubes for broadcasting stations, primarily because these cathodes can withstand the continuous ion bombardment which occurs in such tubes.

For tubes of lower-power level (cathode-ray tubes, small receiving-type tubes, etc.), ion bombardment of the cathode is less severe, and oxide-coated cathodes are more popular. These cathodes have a $W_F \simeq 1$ volt. They consist of a mixture of barium and strontium oxides, which is sprayed onto a hollow nickel cylinder. Heater wires for this type of cathode are then inserted into the nickel tube. Operating temperatures are around $1000°\text{K}$.

Before leaving the subject of thermionic emission, there are two comments which may be instructive:

1. In a thermionic cathode, J_{max} may be varied by varying T. When this is accurately done, Eq. (3·10) applies over a range of about 6 decades

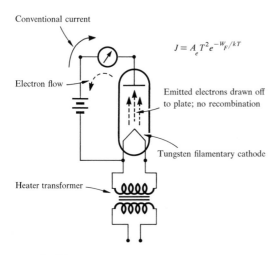

FIG. 3·10 TEMPERATURE-LIMITED EMISSION; THERE IS NO "RECOMBINATION" OF EMITTED CARRIERS, SO $J \propto e^{-W_F/kT}$

(that is, from $J_{max} = 10^{-6}$ amp/cm² to $J_{max} = 1$ amp/cm² for a tungsten cathode). This attests to the wide-range validity of the generation-rate formula (and can be used to provide a demonstration of it.)

2. Equation $(3 \cdot 10)$ for thermionic emission and Eq. $(3 \cdot 7)$ for the thermal-equilibrium carrier density in a semiconductor differ only in that recombination is present in one and not the other. In an intrinsic semiconductor, valence electrons are "emitted" into the free-electron condition at a rate proportional to $e^{-\mathcal{E}_i/kT}$. They recombine at a rate proportional to n_i^2, however, so n_i is proportional to $\sqrt{g(T)}$ or $e^{-\mathcal{E}_i/2kT}$.

In a thermionic emitter, there is no "recombination." Instead, the emitted electrons are drawn off to a positively biased anode as rapidly as they evaporate from the cathode (see Fig. $3 \cdot 10$). The current in the external circuit resupplies electrons to the cathode, of course.

3·3 THE MECHANISM OF CONDUCTION

Previously we have seen that a steady stimulation of a crystal by heat and/or light will produce an average concentration of holes and electrons within the crystal. We have also indicated briefly that these carriers are free to move. To explain photoconductivity, we have given a highly schematic picture of how each particle may conduct a miniature current through the crystal. We now wish to develop the conduction mechanism further and, in the process, set the semiclassical model on a firmer foundation.

3·3·1 *Electrons, holes, and the uncertainty principle.* In order to do this, let us begin by asking whether the localization of electrons and holes implied by Figs. $3 \cdot 2$ and $3 \cdot 4$ is in keeping with quantum-mechanical limits on localization. We shall concentrate first on electrons.

Let us suppose that the free electrons are in thermal equilibrium with the lattice at an absolute temperature T. Then the majority of them will have thermal energies in the range of $0-2\ kT$, with an average energy of kT. (Details of the actual distribution are not important.)

This distribution of energy will correspond to a velocity distribution of roughly

$$0 < v < \sqrt{\frac{4kT}{m}}$$

In other words, the velocity of a given electron will be uncertain by roughly $\sqrt{kT/m}$. But then, from the uncertainty principle we *deduce* that the *position* of such an electron must be uncertain by *at least* an amount

$$\Delta x \simeq \frac{\hbar}{m\Delta v} \cong \frac{\hbar}{\sqrt{kTm}} \qquad (3 \cdot 11)$$

If we set $T = 300°K$, this gives $\Delta x \sim 20 \times 10^{-10}$ m $= 20$Å. Since the fundamental crystal building block shown in Fig. $2 \cdot 10$ is only 5Å on a side, a sphere 20Å in diameter will cover about 100 crystal atoms. Thus, a typical electron in the crystal certainly cannot be localized to the extent implied in Figs. 3·2 and 3·4.

As might be expected, this imprecise localization has a tremendous influence on electron dynamics in a crystal. To appreciate this, we recall that if the fields that operate on a particle vary significantly across the uncertainty width [Δx], then we must resort to quantum mechanics for accurate calculations of particle dynamics (*cf.* Sec. 1·3·3). In this case, we have two types of fields: (1) an externally applied field which usually does not vary significantly in a distance of 20Å and (2) atomic fields which do vary significantly in a distance of 20Å. In fact, the atomic fields are periodic functions of distance in the crystal, with a period of about 5Å. The peak fields are also much stronger than the typical externally applied fields. As a result, the basic motion of electrons is determined by the *atomic* fields and must be calculated by quantum mechanics. The external fields act as only a small perturbation.

Now let us consider the hole. We stated earlier that it could be thought of as essentially a particle. But then, by a similar calculation, it too will have an uncertainty in position of about 20Å. But in this case, the problem caused by the uncertainty in position is perhaps even worse; for we have interpreted hole motion as a local transfer of valence electrons between neighboring bonds which are separated from each other by much less than 20Å. Thus our previous description of hole motion suggests that it occurs entirely within the 20Å sphere. Clearly, the motion of the hole must be calculated by quantum mechanics.

The quantum-mechanical treatment of this problem shows that the concepts of the electron and hole as *classical particles* are valid under certain special circumstances. We shall consider these circumstances briefly in the next section.

3·3·2 Effective mass. One of the most important results of the quantum physics of semiconductors is that *the basic concepts of holes and electrons as essentially classical particles are valid when the externally applied fields are much weaker than those associated with the internal space-periodic electric fields produced by the lattice atoms.* It is then possible to correctly represent the quantum-mechanical motion of a free electron in the combined lattice and external fields as the motion of an imaginary classical electron with a charge q, but with an *effective mass m_e^**. *In a perfect crystal, the imaginary particle responds only to the external fields.*

The basic idea is illustrated schematically in Fig. 3·11. On the left-hand side of this figure is shown the semiconductor lattice with a free electron and an applied electric field. Quantum mechanics must be used to compute electronic motion in this case, taking into proper account both the lattice fields and the applied field. The mass used for the electron in these calculations is its free-space mass.

The effective-mass approximation is illustrated on the right-hand side of

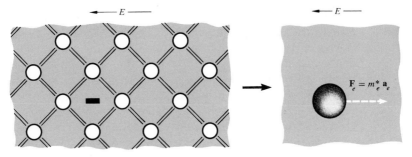

(a) *Actual electron of mass m_0, moving in a perfect lattice with applied external field (with variations in the perfect lattice due to surfaces and atomic jiggling neglected), is equivalent to (b).*

(b) *Crystalline electron with effective mass m_e^* moving in unimpeding medium under influence of applied external field.*

FIG. 3·11   EFFECTIVE-MASS APPROXIMATION FOR ELECTRONS

Fig. 3·11. Here, the (perfect) semiconductor lattice has been erased. The electron has been replaced by a negatively charged ball about 20Å in diameter. The ball has an effective mass m_e^* and responds to external forces in accordance with Newton's laws.[1]

As an example of the idea, if an electric field **E** is applied to a perfect crystal, a free electron within the crystal will experience an acceleration given by

$$m_e^* \mathbf{a}_e = q\mathbf{E} \qquad (3 \cdot 12)$$

where m_e^* is the effective mass of the electron.

Of course, real crystals are not perfect. They have surfaces, where the lattice fields certainly cease to be periodic. The atoms within the crystal are also jiggling around about their equilibrium positions (*cf.* Prob. 3·7), so the internal fields are not perfectly periodic. However, these imperfections can be accounted for by assuming that the classical particle can be accelerated for only a certain period of time t before being deflected. We shall consider this in Sec. 4·1. For the present, we wish to emphasize the following facts.

The major quantum features of the problem are *removed* by adopting the idea that the "free electron" *in the crystal* does not have its free-space mass, but some effective crystalline mass.

The effective mass of a free crystalline electron is positive. Furthermore, the internal electric fields are usually very large compared to applied fields, so the effective-mass concept is quite valid for our needs.

This is perhaps easier to accept than the fact that *holes* are also characterized by an effective mass, and that this effective mass is also *positive*. That is, the quantum-mechanical features of valence-electron motion can

---

[1] The ball is still called a "free electron," however. It would be preferable to invent a new name for it, but the dual usage of the term "free electron" is too deeply rooted to be changed.

be described in classical terms by the concept of a *hole as a classical particle with a positive effective mass and a positive charge.*

The effective-mass approximation for holes can be illustrated with a figure exactly like Fig. 3·10, except that the apparent particle bears a positive charge and has its own effective mass m_p^*. As in the free-electron case, the actual motion of the vacancy in the valence structure via the valence electrons must be calculated by quantum-mechanical methods. However, the hole, which is the classical counterpart of the valence-bond vacancy, responds only to externally applied fields, in accordance with Newton's laws as applied to a perfect crystal.[1]

Thus, in the effective-mass approximation, a perfect crystal becomes a smooth, homogenous medium in which electrons and holes move as classical particles in response to external fields only.

It is important to point out the difference between this idea and the one suggested earlier for hole motion. From our earlier discussion, it would appear that the conduction process for valence electrons must really be looked upon as the field-stimulated motion of a local valence electron into an empty bond. There, the concept of a hole seems to be merely a convenient way to describe the net effect of valence-electron motion.

However, the quantum-mechanical result gives a considerably different view of the matter. If we wish to use the laws of classical physics (that is, Newton's laws) to describe current flow through the valence bonds, we *must* calculate the motion of a hole. The Hall effect (to be described in Sec. 4·3) and the thermoelectric probe (to be described in Sec. 3·4·5) will, in fact, give experimental proof of this. Of course, conduction in the valence structure can also be described in terms of the motion of valence electrons, but then quantum-mechanical calculations are necessary.

3·3·3 *Cyclotron resonance.* Perhaps the most convincing argument one can give for the concept of effective mass is that crystalline effective masses are *measurable.* In fact, several techniques may be used to measure the effective masses of holes and electrons. We shall describe one technique, called cyclotron resonance, which is simple in principle and has the honor of having been used first for this purpose.[2]

To begin with, let us suppose that a classical electron is moving with velocity v_0 in free space. A uniform magnetic field of strength B is now switched on, as in Fig. 3·12a. Then the electron will begin to move in a circular trajectory in a plane perpendicular to the direction of B. The circle will have a radius of

$$r = \frac{m_0 v_0}{qB} \tag{3·13}$$

---

[1] It would be more consistent to always refer to the missing valence bond as a "vacancy" and to the apparent particle as a "hole". However, the terms "hole" and "vacancy" are customarily used interchangeably, so this logical convention will not be adopted.

[2] The details of this experiment may be found in C. Kittel, "Introduction to Solid State Physics," 2d ed., John Wiley & Sons, Inc., New York, 1962, pp. 371ff.

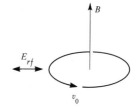

(a) *Circular orbit of electron in free space with velocity v_0 in perpendicular magnetic field B*

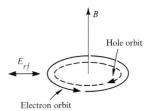

(b) *Orbits of electrons and holes for cyclotron resonance in a crystal*

FIG. 3·12 SCHEMATIC DIAGRAM OF CYCLOTRON-RESONANCE EXPERIMENT

where m_0 is the free-space mass of the electron. The electron will complete one revolution every

$$T = \frac{2\pi m_0}{qB} \quad \text{sec} \tag{3·14}$$

This corresponds to an angular frequency of

$$\omega_c = \frac{qB}{m_0} \tag{3·15}$$

Since v_0 drops out and B is known, a measurement of ω_c will give values for q/m. ω_c is called the *cyclotron frequency*. [The reason for this name can be understood by comparing Eq. (3·15) with the results of Prob. 1·10.]

ω_c can be measured as follows. An rf electric field can be passed through the medium containing the B field and orbiting particles, as shown in Fig. 3·12. The orbiting particles will then absorb energy from this electromagnetic field when its frequency is ω_c, thus serving to locate the cyclotron-resonance frequency. The velocity of the particle (and radius of the orbit) will increase as energy is absorbed, though ω_c stays the same (*cf.* also Prob. 1·10).

The absorption characteristic that accompanies cyclotron resonance has been used in several ways. Its first application was in the explanation of the absorption of radio waves by the ionosphere. We may think of the ionosphere, roughly, as a cloud of free electrons which "shadows" large portions of the earth's surface. This free-electron cloud exists in the magnetic field of the earth, which is between about 0.33 and 0.63 gauss. When radio waves of frequency ω_c are being sent through this medium, the electrons will absorb energy from them. Thus, the propagation range of signals at frequency ω_c will be lower than the range at other frequencies. This effect is demonstrated in Fig. 3·13.

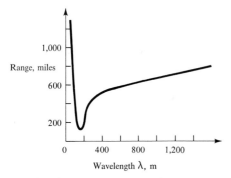

FIG. 3·13 ABSORPTION OF RADIO WAVES IN THE IONOSPHERE DUE TO CYCLOTRON RESONANCE OF FREE ELECTRONS IN THE EARTH'S MAGNETIC FIELD

Cyclotron-resonance absorption can also be used to measure the effective masses of carriers in solids. Equation (3·15) then becomes

$$\omega_c = \frac{\pm qB}{m^*} \tag{3·16}$$

where m^* is the effective mass of the carrier and the $\pm$ sign is included to account for both holes and electrons. Measurement of ω_c will now yield the *effective mass* of the particle. The direction of rotation can also be measured, to tell whether the particle is a hole or an electron (see Fig. 3·12b). A typical plot of cyclotron-resonance absorption for holes in germanium is shown in Fig. 3·14. Some simple calculations with the data given in Figs. 3·13 and 3·14 are the subject of Probs. 3·8–3·10.

Effective masses for holes and electrons range from $\frac{1}{100}$ of the free-space mass to several times the free-space mass of an electron, depending on the material.

It is worthwhile to summarize the past several ideas at this point:

1. There are two types of conducting particles in a semiconductor

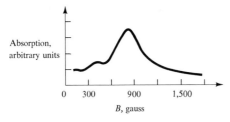

FIG. 3·14 DATA FROM A CYCLOTRON-RESONANCE EXPERIMENT ON HOLES IN GERMANIUM FOR A PARTICULAR CRYSTAL ORIENTATION IN THE MAGNETIC AND RF FIELDS; TEST FREQUENCY = $8,900 \times 10^6$ CPS; MAGNETIC FIELD B VARIED TO ACHIEVE CYCLOTRON-RESONANCE CONDITION

(*From Lax, Zeiger, and Dexter, Physica, vol. 20, p. 818, 1954*)

crystal: electrons and holes. The electron bears a charge of $-q$ coul and the hole bears a charge of $+q$ coul.

2. In a pure crystal, an electron-hole pair is created when a valence electron receives an energy $\mathcal{E}_i$ from its surroundings. This process is correctly visualized as the breaking of a valence bond.

3. Electrons and holes are continuously recombining to perfect the valence structure on a local basis.

4. The steady-state densities of holes and electrons in a crystal are determined by the dynamic balance of the recombination and generation mechanisms. In the particular case of thermal generation only, in a pure crystal,

$$g(T) = rnp = rn_i^2$$

There is a total of $2n_i$ carriers/cm³. These carriers give the crystal its *intrinsic conductivity*.

5. The *motion* of electrons and holes in a crystal is usually studied by means of the effective-mass approximation. In this approximation, the effect of the actual crystal fields is accounted for by assigning crystalline masses or effective masses to the carriers. When these effective masses are used, the carriers respond only to externally applied fields in a perfect crystal.

3·4 DOPED SEMICONDUCTORS

Our discussion has so far been limited to pure (intrinsic) semiconductor materials and the means of producing free carriers in them. In practically all cases of importance, however, selected chemical impurities are added to the parent semiconductor material to provide free carriers for conduction. The semiconductor is then said to have been *doped*.

For germanium and silicon, the usual impurities of interest are the elements in the third or fifth column of the periodic table. Fifth-column elements are called *donor impurities*, since they donate free electrons to the crystal. Third-column elements are called *acceptor impurities*, since they accept electrons from the valence structure and thereby create holes. The common donor impurities are antimony, arsenic, and phosphorus. Common acceptor impurities are aluminum, boron, gallium, and indium. A description of the influence of these impurities on the carrier densities will now be given.

3·4·1 n-*type semiconductors.* An n-type semiconductor is made by substituting elements from column VA of the periodic table into an otherwise pure crystal of germanium or silicon. These impurities may be included in the melt from which the semiconductor crystal is grown (*cf.* Sec. 2·2·6) or added later by alloying or diffusion techniques (*cf.* Sec. 5·1).

The reduced atomic structure for elements in column VA of the periodic table is as follows. There is an atomic core with an effective charge of

$+5q$ coul. This atomic core is surrounded by five valence electrons. Two of them are in s-type orbitals and three are in p-type orbitals (see Prob. 3·11).

When such an atom replaces a germanium atom in a germanium lattice, the three electrons in p-type orbitals and one of the electrons in the s-type orbital will form covalent bonds with the neighboring germanium atoms, as in Sec. 2·2·3. These bonds are very strong, so that *the atomic core of the donor atom is tightly bound into the germanium lattice.*

At absolute-zero temperature, the remaining valence electron will describe a path around its parent nucleus which is very similar to the path of an electron around a hydrogen atom in the ground state. To appreciate this, let us employ the effective-mass approximation to study the behavior of this fifth valence electron. According to this, the crystal itself is thought of as a smooth, unimpeding medium. Within this medium, there are now two charged particles: the partially neutralized impurity atom and the fifth valence electron contributed by the impurity atom.

The impurity atom, which is bound in the lattice, has an effective charge of $+q$ coul (see Fig. 3·15). This is because its effective *nuclear* charge is $+5q$, of which $4q$ is neutralized by the bonding electrons. The bound impurity atom thus plays the part of the proton in the hydrogen atom.

The fifth valence electron moves in the field of this bound charge, though screened from it by the dielectric constant of the germanium crystal. Its motion can be calculated by assigning to this electron the effective mass of a free electron in the crystal and then calculating how this free electron would revolve around the bound positive charge in a medium whose relative dielectric constant is that of the germanium crystal (see Probs. 3·12 and 3·13 for details). The answer is that the electron will be in an s-type hydrogen orbital, with an average radial position

$$r = \frac{h^2 \epsilon_r \epsilon_0}{\pi m^* q^2} = \epsilon_r \frac{m_0}{m^*} r_0 \qquad (3·17)$$

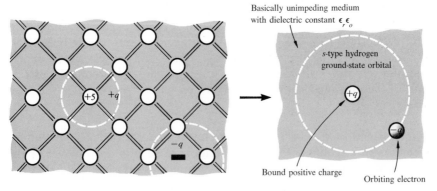

(a) *Donor impurity atom (bound in the lattice) and fifth valence electron*

(b) *Representation of (a) at low temperatures, using effective-mass approximation*

FIG. 3·15   AN n-TYPE GERMANIUM CRYSTAL AT LOW TEMPERATURES

and with a ground-state energy

$$E = \frac{m^* q^4}{8h^2(\epsilon_r \epsilon_0)^2} = \frac{m^*}{m_0}\left(\frac{1}{\epsilon_r}\right)^2 E_0 \qquad (3\cdot 18)$$

where r_0 and E_0 are the average radius and ionization energy of a hydrogen atom in the ground state, ϵ_r is the relative dielectric constant, m^* is the effective mass of a free electron, and m_0 is the free-space electron mass.

If we use $\epsilon_r = 16$ for germanium, and $m^*/m_0 = 0.2$, we find $E \cong 0.01$ ev. Thus, the fifth valence electron can be set free in the crystal by giving it an energy of about 0.01 ev. It can then escape the binding influence of its parent nucleus. (Actual binding energies are listed in Table $3\cdot 1$.)

Now, at room temperature we may suppose that this electron will have an average thermal energy of $kT_{room} = 0.025$ ev. Therefore, the probability that this electron will be free from its parent nucleus at room temperature is high.

As an approximation, we will assume that each donor atom which is added to an otherwise perfect crystal of germanium will, at room temperature, donate one free electron to the crystal.[1] This electron is free to conduct in the same way as any other free electron. The impurity atom itself is bound into the lattice. It cannot move and therefore cannot conduct a current. It does, however, contribute a charge of $+q$ coul to the crystal. The crystal as a whole is still electrically neutral, since there is one mobile electron for every bound positive charge.

3·4·2 *Carrier densities in an* **n-***type semiconductor.* The previous discussion indicates that the addition of donor impurities will increase the electron concentration in the solid. However, when the average free-electron density is increased by doping, the average hole density will decrease. We shall explain why this is so and calculate the new carrier densities in this section.

The argument which we use to calculate the carrier densities in doped crystals is essentially the same as that used for calculating intrinsic carrier densities: the recombination rate and the generation rate are calculated separately and then set equal to determine the equilibrium carrier concentrations.

Recombination rate. The general formula for the recombination rate is

$$R = rnp$$

To evaluate R we need to determine n and p. It will not be necessary to actually know r, as we shall see later.

To evaluate n and p, let us suppose that a crystal of germanium has been doped with N_d donors/cm³. Let the resulting electron concentration be n (per cm³) and the hole concentration be p (per cm³).

---

[1] A precise treatment shows that essentially all donors are ionized at room temperature so long as $N_d < 10^{19}$. For $10^{21} > N_D > 10^{19}$, "recombination" of free electrons with ionized donors produces $n \simeq \sqrt{10^{19}N_d}$. Changes in the bulk properties of the crystal begin to occur at $N_D \sim 10^{21}$. See Probs. $3\cdot 13$ and $3\cdot 22$ for further details.

Now, electrons are produced either by ionization of a valence bond or by ionization of a donor. In either case, each free electron leaves a balancing positive charge in the crystal. The crystal is therefore electrically neutral. We may express the electrical neutrality as

$$\text{Total negative charge} = \text{total positive charge}$$

And if we assume all the donors to be ionized, we can write, on a per-unit-volume basis,

$$n = N_d + p \tag{3·19}$$

Note that some of the positive charge is bound and some is mobile.

In the presence of these carrier densities, the recombination rate must be

$$R = rnp = r(N_d + p)p \tag{3·20}$$

Thermal-generation rate. To complete the calculation of the thermal-equilibrium carrier density, we shall now show that *the thermal-generation rate does not change appreciably as the crystal is doped.* To illustrate this, consider a doping density $N_d = 10^{17}/\text{cm}^3$ in a germanium crystal. This represents roughly one doping atom for every 10^5 germanium atoms. The crystal still appears to be a very pure germanium crystal. Therefore, the energy required to break a valence bond has not changed as a result of doping,[1] and the thermal-generation rate g is the same in doped and undoped crystals.

Now we know, from the discussion given in Sec. 3·2·3, that the average value of $g(T)$ at a given temperature T can be written as

$$g(T) = rn_i^2 \tag{3·21}$$

Carrier densities in an n-type semiconductor. We can now set the generation rate equal to the recombination rate to obtain the thermal-equilibrium carrier densities:

$$rn_i^2 = g(T) = R = r(N_d + p)p \tag{3·22}$$

This equation may be solved to yield

$$p = \frac{-N_d + N_d \sqrt{1 + (4n_i^2/N_d^2)}}{2} \tag{3·23}$$

Equation (3·23) is usually simplified by making the following observation. In a typical doped crystal, $N_d \gg n_i$. Therefore

$$\sqrt{1 + 4n_i^2/N_d^2} \cong 1 + \frac{2n_i^2}{N_d^2} \tag{3·24}$$

(The validity of this approximation is the subject of Prob. 3·14.) When this result is used in Eq. (3·24), we have

$$p = \frac{n_i^2}{N_d} \tag{3·25}$$

[1] This assumption will not be valid when $N_D \gtrsim 10^{20} - 10^{21} \text{ cm}^{-3}$.

The corresponding electron density is

$$n = N_d + \frac{n_i^2}{N_d} \simeq N_d \qquad (3 \cdot 26)$$

Thus, for most doping concentrations, the effect of the donor doping is to fix the electron concentration at $n = N_d \gg n_i$. The recombination mechanism then fixes the equilibrium hole concentration at $p = n_i^2/N_d \ll n_i$. Increasing the electron concentration by doping thus decreases the equilibrium hole concentration (see Prob. 3 · 15).

It is convenient to give special symbols to these thermal-equilibrium carrier densities in doped material. We call the thermal-equilibrium electron density in a donor-doped crystal n_n. The thermal-equilibrium hole concentration is denoted by p_n.

EXAMPLE   If we set $N_d = 10^{17}/cm^3$, we have in a germanium crystal

$$n_n = 10^{17} \text{ electrons/cm}^3$$

$$p_n = \frac{n_i^2}{N_d} = 6.25 \times 10^9 \text{ holes/cm}^3$$

Note that in this example, the hole density is depressed below p_i by a factor of roughly 10^4. The electron concentration has been increased by a corresponding factor. This reflects the nature of the mass-action recombination law, in which the product of p_n and n_n is constant (and equal to n_i^2).

Electrons are now the majority (more numerous) carriers, and holes are the minority (less numerous) carriers. The majority of the current in such a semiconductor will be carried by *negative* charges, so the semiconductor is called an *n*-type semiconductor.

3 · 4 · 3   p-type semiconductors.   A *p*-type semiconductor is made by replacing germanium atoms with elements from column IIIA of the periodic table. Such an element has, in its neutral state, an effective nuclear charge of $+3q$ coul. It is surrounded by three electrons, two in *s*-type orbitals and one in a *p*-type orbital.

When such an atom is used to replace a germanium atom in the germanium lattice, the valence electrons of the impurity atom will fail to complete the normal valence structure. Instead, they form three more-or-less normal covalent bonds with the neighboring germanium atoms. This creates a hole in the lattice, as shown schematically in Fig. 3 · 16a.

However, the motion of valence electrons within the valence-bonding orbitals is such that the hole is not confined to the position indicated in Fig. 3 · 16a. Rather, it can move away to some new position, as shown in Fig. 3 · 16b. When it does so, however, a net negative charge is created in the vicinity of the acceptor atom (and a net positive charge in the vicinity of the hole). The attraction between these charges limits the extent to which the hole can move away.

Once again, the impurity atom itself is tightly bound in the lattice. It contributes a charge of $-q$ coul, since in the crystal it is surrounded

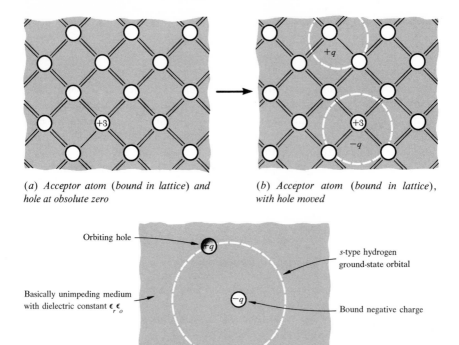

(a) Acceptor atom (bound in lattice) and hole at absolute zero

(b) Acceptor atom (bound in lattice), with hole moved

Orbiting hole

s-type hydrogen ground-state orbital

Basically unimpeding medium with dielectric constant $\epsilon_r \epsilon_o$

Bound negative charge

(c) Effective-mass approximation for hole in field of acceptor atom

FIG. 3 · 16 A p-TYPE SEMICONDUCTOR AT LOW TEMPERATURES

by one more electron than it has in the neutral state. The bound negative charge is neutralized macroscopically by the presence of the hole.

In terms of the effective-mass picture, the hole is a mobile positive charge. At low temperatures in an acceptor-doped crystal, it performs a hydrogenlike orbit around a negative charge which is bound in the lattice (Fig. 3 · 16c.) At room temperature, however, the hole has a high probability of being ionized from the attraction of its parent atom. Thus, at room temperature we shall assume that every acceptor atom in the crystal contributes a free hole for conduction purposes. Actual binding energies for acceptor-atom hole-ground states are listed in Table 3 · 2.

3 · 4 · 4   Carrier densities in a p-type semiconductor. By analogy with the n-type semiconductor, a doping density of N_a acceptor atoms/cm³ will produce a free-electron concentration of

$$n_p = \frac{-N_a + N_a \sqrt{1 + (4n_i{}^2/N_a{}^2)}}{2} \simeq \frac{n_i{}^2}{N_a} \qquad (3 \cdot 27)$$

and a hole concentration of

$$p_p = N_a + n_p \simeq N_a \qquad (3 \cdot 28)$$

The symbol n_p is given to the equilibrium electron concentration in an acceptor-doped crystal. The hole density is denoted by p_p. *Positive* charges (that is, holes) are now the majority carriers, so the resulting material is called a *p*-type semiconductor.

Thus, a given bar of semiconductor material always has two types of carriers in it. By the process of chemical doping, we can select one of these carriers as the majority carrier. We can therefore make two basically different types of material, *n*-type and *p*-type. All semiconductor devices which we shall consider can be synthesized from appropriate combination of these two basic material types (see Probs. 3·18 and 3·19).

For future reference, the important results of this section are summarized in Table 3·2.

3·4·5 Demonstration of p- and n-type conductivity. In samples which are sufficiently heavily doped, there is a simple experiment which can be performed to determine whether a piece of material is *n*- or *p*-type. For our purposes, this experiment can be considered to give an experimental proof of the existence of two types of carriers in a semiconductor.

The experiment consists of locally heating a small spot on a slice of a doped semiconductor crystal (by simply pressing a small soldering iron on the slice, for example) and then measuring the voltage which is developed between this hot spot and a second point in a neighboring region which is cooler. The sign of the potential difference will indicate the conductivity type of the slice.

The theory for this effect can be explained qualitatively by referring to Fig. 3·17. In Fig. 3·17a we show an *n*-type slice of semiconductor in which a high-impedance voltmeter (or a sensitive galvanometer) is connected between a hot point on the slice and a cold point. The electrons under the hot point move away to cooler regions, in an attempt to conduct

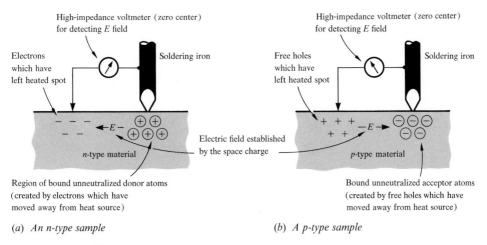

(a) *An n-type sample*

(b) *A p-type sample*

FIG. 3·17   HOT-POINT-PROBE EXPERIMENTS FOR DETERMINING CONDUCTIVITY TYPE

TABLE 3·2

Material type	Mobile negative charge density, electrons/cm³	Bound negative charge density N_a, acceptor atoms/cm³	Mobile positive charge density, holes/cm³	Bound positive charge density N_d, donor atoms/cm³	Relations among quantities 1. $n + N_a = p + N_d$ 2. $np = n_i^2$
Intrinsic	n_i	0	p_i	0	1. $n_i = p_i$
n-type $\left(\text{substitute} \; \text{+5} \; \text{for} \; \text{Ge} \right)$	n_n	0	p_n	N_d	1. $n_n = N_d + p_n \cong N_d$ 2. $p_n \cong n_i^2/N_d$
p-type $\left(\text{substitute} \; \text{+3} \; \text{for} \; \text{Ge} \right)$	n_p	N_a	p_p	0	1. $p_p = N_a + n_p \cong N_a$ 2. $n_p \cong n_i^2/N_a$

Element	Binding energy in hydrogenlike orbit, ev	
	Germanium	Silicon
Acceptor impurities:		
Boron	0.0104	0.045
Aluminum	0.0102	0.057–0.067
Gallium	0.0108	0.065–0.071
Indium	0.0112	0.16
Donor impurities:		
Phosphorus	0.0120	0.045
Arsenic	0.0127	0.049–0.056
Antimony	0.0096	0.039

the heat away from the localized hot spot. However, this sets up a charge unbalance and, consequently, an electric field in a direction which tends to return the electrons to the hot spot. The integral of this electric field is the voltage which the high-impedance voltmeter will measure.

If the slice is p-type instead of n-type, then holes are forced away from the hot spot. The direction of the electric field will be reversed, and the sign of the potential will also change. Thus, the sign of the voltmeter reading gives an indication of the conductivity type (and its magnitude gives a rough indication of the resistivity of the sample). This experiment can be considered to give a proof that both holes and electrons exist, since the sign of the potential changes when the doping atoms change from n-type to p-type.[1]

The change in sign of the potential also shows that our Newtonian intuition gives us a valid picture of the motion of holes, but it does *not* adequately describe the motion of valence electrons. To be perfectly sure of this, let us try to predict the sign of the voltmeter reading in a p-type sample, basing our reasoning directly on the motion of valence electrons. Since we know it is really the valence electrons that move, we might reasonably expect them to leave the hot region. But the transport of negative charge away from the hot region will result in the same sign of voltmeter reading for both p- and n-type samples, which is not the observed result. In fact, the observed result suggests that valence electrons tend to collect under the hot point rather than be driven away from it.

We can resolve this problem only by accepting the fact that our intuition is basically Newtonian, and it frequently does not give us a correct representation of quantum-mechanical behavior. In the present case, our intuition works properly only when we visualize a p-type semiconductor as having an abundance of mobile, positively charged carriers.

3·5   CRYSTALLINE ORBITALS AND ENERGY BANDS

In the previous sections of this chapter, we have developed a semiclassical model for studying the behavior of electrons and holes in a semiconductor. We have referred primarily to quantum mechanics to justify this semiclassical picture. In this section we wish to comment briefly on the relationship between this semiclassical picture and the one which we obtain by using quantum mechanics throughout.

We may profitably begin by considering the brief development of atomic theory given in Chap. 2. There we found that an electron in an atom was to be visualized as being in one of a series of allowed orbital patterns. These orbital patterns arise from a solution to Schrödinger's equation, in which it can be shown that physically acceptable solutions correspond to certain allowed energies. In order to give a quantitative discussion of atomic spectroscopy, chemical valence, the periodic

---

[1] Actually, the hot-point probe is unsatisfactory for indicating the conductivity type of very lightly doped samples. In fact, the probe reads "n type" when the sample is actually intrinsic. Reasons for this will be developed later.

table, etc., we need to know what the allowed energy levels are and what the corresponding orbital functions look like.

A similar situation exists in the quantum theory of solids. In order to make quantitative predictions of electronic behavior in a solid, we must first solve Schrödinger's equation to find out what energy levels and orbital patterns are allowed for electrons in the particular crystal of interest.

The results will again be expressible in the form of an energy-level diagram for electrons in the crystal and a set of corresponding crystalline orbitals. We shall mention briefly several key features of the matter which are illustrated in Fig. 3·18:

1. The energy levels for all the valence electrons are found to lie in an energy *band* (that is, an energy range in which there are many allowed energy levels, adjacent levels being very closely spaced). For germanium, this energy band arises from the atomic 4s- and 4p-energy levels, though the actual energy levels of the crystalline valence orbitals are lower than the corresponding atomic levels.

2. The valence orbitals are only the *ground-state* orbitals for the valence electrons. In addition to these, there are many other unoccupied crystalline orbitals of higher energy. The allowed energy levels for these higher-energy orbitals also fall into bands.

3. The crystal may be excited by promoting an electron from a valence orbital to one of these excited crystalline orbitals. This leaves a vacant energy level in the valence band, which is interpreted as a hole.

4. There is an energy difference between the highest-valence energy level and the lowest energy level for an excited orbital. This energy difference is called the energy gap, and its magnitude is simply the photoionization energy $\mathcal{E}_i$ (*cf.* photo-electric threshold and fluorescence experiments).

5. Electrons can move freely through the crystal in excited orbitals, and holes can move freely through the crystal in valence orbitals (*cf.* Demonstration 3·2).

6. The presence of an absorption band, rather than a set of discrete absorption frequencies, arises from the presence of allowed energy bands which are separated by an energy gap.

A deeper study of semiconductor physics usually begins with a detailed examination of the properties of these crystalline orbitals and energy bands and with the theoretical means by which the distribution of electrons among the possible orbitals and energy levels can be determined.[1]

[1] The interested reader may wish to scan through Part II of W. Shockley, "Electrons and Holes in Semiconductors," P. Van Nostrand Company, Inc., Princeton, N.J., 1950, or R. A. Smith, "Semiconductors." Cambridge University Press, New York, 1961, chaps. 2 and 4.

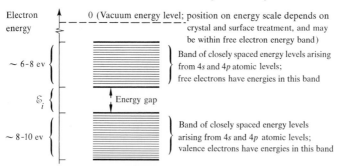

FIG. 3·18 ENERGY-LEVEL DIAGRAM FOR ELECTRONIC ORBITALS IN A TYPICAL SEMI-CONDUCTOR CRYSTAL

In this chapter we have seen that there are two types of current-carrying particles in a semiconductor crystal: electrons and holes. The effective-mass approximation was introduced to enable us to treat them as essentially classical particles.

Several concepts have been introduced to describe the nature of these particles and the processes by which they may be created in a crystal. In most cases, experimental demonstration of these concepts has been described. The most significant concepts and the supporting lines of evidence are as follows:

1. Electrons and holes behave as essentially independent classical particles in a semiconductor crystal. These particles have appropriate crystalline masses, and they are free to move in accordance with Newton's laws in a perfect crystal.

The main experimental evidence for these statements is the cyclotron-resonance experiment. However, the existence of two particles and their freedom to move, even in an imperfect crystal, is also suggested by the thermal-probe experiment. A quantitative explanation of photoconductivity and optical absorption also requires the assumption of two more-or-less free particles:

2. Electrons and holes can be generated by supplying enough energy to a valence bond to ionize it.

The main experimental evidence we have presented for this is the photoelectric-threshold demonstration. The existence of photoelectric and thermionic emission from semiconductors suggests that at least electrons can be generated in this way. Also, semiconductors have an intrinsic electrical conductivity which arises from thermal generation of electron-hole pairs.

3. Electrons and holes are continuously recombining in a semiconductor crystal. The fact that crystals do not simply fall apart under the influence of the carrier-generation mechanisms is adequate proof of this. However, the fluorescence experiment, which is the dual of the photogeneration experiment, also indicates the presence of recombination. We have not yet described in detail the recombination mechanisms which are important in most silicon and germanium crystals.

4. Semiconductors may be doped to increase the number of electrons or holes. The increase in the concentration of one type of carrier automatically involves a decrease in the concentration of the other. We can demonstrate with the thermal-probe experiment that doping produces two conductivity types. We have assumed that, at room temperature, the doping atoms are all ionized (at least for the most typical concentrations of doping atoms). On the basis of this assumption, we have calculated the equilibrium carrier densities in doped materials.

REFERENCES

Shockley, W.: "Electrons and Holes in Semiconductors," D. Van Nostrand Company, Inc., Princeton, N.J., 1950, chap. 1.

Adler, R. B., et al: "Introduction to Semiconductor Physics," SEEC Notes, vol. 1, John Wiley & Sons, Inc., New York, 1964. This volume also has an interesting appendix on laboratory experiments, some of which can be used as alternate demonstrations to those given here and at the end of Chap. 4.

Shive, J. N.: "The Properties, Physics, and Design of Semiconductor Devices," D. Van Nostrand Company, Inc., Princeton, N.J., 1960.

DEMONSTRATIONS

3·1 *Photoelectric threshold.* The photoelectric-threshold demonstrator shown in Fig. 3·4 consists of three main parts: (1) a light source and diffraction grating for producing a color spectrum, (2) a sample and sample holder, and (3) a translucent projection screen. These parts can be assembled as follows:

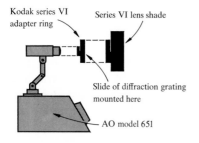

Kodak series VI adapter ring

Series VI lens shade

Slide of diffraction grating mounted here

AO model 651

(*a*) *Light source and diffraction grating*

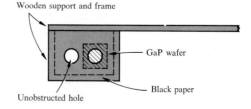

Wooden support and frame

GaP wafer

Unobstructed hole

Black paper

(*b*) *Sample holder*

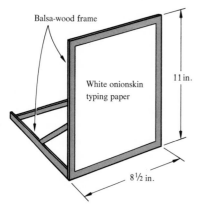

Balsa-wood frame

White onionskin typing paper

11 in.

8½ in.

(*c*) *Projection screen*

FIG. D3·1 **PHOTOELECTRIC-THRESHOLD DEMONSTRATION**

1. A suitable spectrum for the demonstration may be obtained from an American Optical Company model 651 Microscope Light by fixing a transmission-type diffraction across its face. A suitable grating can be obtained from Edmund Scientific Company, Barrington, N.J. The grating (stock number 40,272) has a usable area of about ⅞ in.^{2} and comes mounted in a 2-in.^{2} cardboard holder of the type in which 35-mm slides are customarily mounted.

The grating may be taped directly to the microscope lamp holder or cut down to fit a Kodak series VI photographic adapter ring. The barrel of the microscope lamp is 1^{17}⁄₆₄ in. in diameter, so a 1¼-in. friction-fit adapter ring fits snugly. A Tiffen series VI lens shade screwed into the adapter ring makes a neat assembly.

2. A convenient sample and sample holder for the demonstration is a small wafer of gallium phosphide about ¼ in. on a side and 0.035 in. thick. The material is available from either Monsanto Chemical Company, 800 N. Lindbergh Avenue, St. Louis, Mo., or Semi-Metals, Saxonburg, Pa. It is preferable that the sample be lapped and polished to produce clear surfaces. Cadmium sulphide can also be used for the demonstration.[1]

The gallium phosphide wafer can be mounted on a microscope slide or a piece of cardboard, using epoxy cement. A suggested mounting arrangement is shown in Fig. D3 · 1b. The sample holder shown in Fig. 3 · 5 of the text is mounted on black construction paper and has two 2¼-in.^{2} glass-slide mounts glued to the front and back of the wooden supporting frame, to protect the sample.

3. The projection screen is a piece of 8½ × 11-in. white onionskin typing paper glued to a balsa-wood frame. If the light source is mounted about 2 to 3 ft from the screen, a usable spectrum can be displayed over almost the entire screen.

3 · 2 *Photoconductivity.* The light source and screen described in Demonstration 3 · 1 are useful also for demonstrating photoconductivity. The simplest demonstration is obtained from a commercial cadmium sulphide photoconductor (for example, GE A33 CdS photocell). The resistance of the photoconductor is measured with an ohmmeter as the sample is passed through the spectrum. Photoconductivity in germanium and silicon can be measured by pressing ohmmeter leads against a wafer of the material and turning the microscope light on and off.

Care must be taken to observe the photoconductivity of the gallium phosphide, since it is a very high-resistance material. Satisfactory ohmic contacts can be made by the following procedure. Rinse the sample in trichlorethylene, acetone, and then distilled water. Place the clean specimen on a heating element in an inert atmosphere (nitrogen or argon) with two small globules of tin resting on top of it in the positions where contacts are to be made. A small amount of hydrazine dihydrochloride solution is used as a flux. The material is heated until the tin melts and then allowed to cool. Leads can be attached to the tin "dots" using a low melting point solder (such as an indium base solder). After attaching the leads, the contacts and the area on the crystal around them should be coated with an opaque insulating material such as General Cement Anti-Corona Dope. A clear polystyrene cement applied to the exposed surface reduces surface leakage.

When leads have been attached, the photoconductivity can be measured as the gallium phosphide crystal is passed through the spectrum. However, the intensity of

[1] A polished CdS wafer with ohmic contacts which can be used for both the photoelectric threshold and photoconductivity demonstrations can be obtained from the author at a nominal charge. Address inquiries to the author, Electrical Engineering Dept., Stanford University.

the projected spectrum is low, so a variation of about 200 kohms (compared to a dark resistance of several megohms) is all that can be obtained. To make this variation readily observable, the bridge circuit shown in Fig. D3 · 2 can be used.

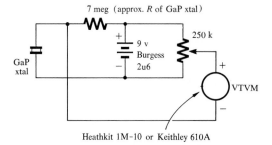

7 meg (approx. *R* of GaP xtal)

FIG. D3 · 2   BRIDGE CIRCUIT FOR DETECTING PHOTOCONDUCTIVITY IN GALLIUM PHOSPHIDE

3 · 3   *Fluorescence.*   To demonstrate fluorescence simply, it is necessary to choose a material in which the radiative-recombination process dominates nonradiative processes, which are always present. One common material for this purpose is zinc sulphide, which may be obtained in wafers from the previously listed suppliers. A simple demonstration consists of a wafer of zinc sulphide taped to a piece of black construction paper. The wafer is illuminated with an ultraviolet lamp, such as the Mineralight SL-3660, available from Scientific Glass Apparatus, Inc. (Bloomfield, N.J., Fullerton, Calif., and Elk Grove Village, Ill.).

To observe fluorescence in gallium phosphide, special fluorescing gallium phosphide must be ordered, and it must be cooled to at least dry-ice temperatures, so that the radiative-recombination process can dominate.

3 · 4   *Conductivity-type tests for semiconductor materials.*   A simple demonstration of the two types of conductivity for semiconductor materials can be made with a small soldering iron and a dc microammeter with a 50-μa full-scale movement (preferably zero center). The resulting apparatus is called a hot-point probe and is described in Sec. 3 · 4 · 5. The experimental arrangement is shown in Fig. D3 · 4. Suitable samples are 0.1 to 1 ohm/cm *p*- and *n*-type germanium or silicon wafers

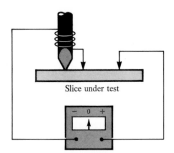

Slice under test

FIG. D3 · 4   SIMPLE HOT-POINT PROBE

Note: To aviod ac pickup, the soldering iron should be turned off while the conductivity-type measurement is being made.

of the approximate dimensions given in Demonstration 3·1. The material can be obtained from the previous suppliers[1] or from Central Scientific Company, 1700 Irving Park Road, Chicago 13, Ill., by ordering their Semiconductor Demonstration Kit, stock number 80394.

PROBLEMS

3·1 Selenium crystals transmit a red light when placed in the apparatus shown in Fig. 3·4b. Estimate E_g for selenium. Express your answer in both joules and electron volts.

3·2 A germanium crystal of thickness l is placed in the apparatus shown in Fig. 3·1. Photons of 0.7 ev energy are shined on this crystal. Neglecting reflection losses, what value of l will give a ratio of transmitted light intensity of 0.1? Repeat for a photon energy of 0.9 ev. For comparison purposes, a typical page of this book is about 0.01 cm thick.

3·3 A beam of monochromatic photons of energy 1 ev is shined on a germanium crystal. The intensity of the beam is 1 mw. Assuming that all the photons are absorbed in the crystal and produce electron-hole pairs, what is the photogeneration rate (that is, number of hole-electron pairs produced per second)? Assume that these hole-electron pairs all recombine radiatively, each recombination event giving off a photon of energy $\mathcal{E}_i = qE_g$. How much of the light energy is converted into heat energy within the crystal?

3·4 The variation of intrinsic carrier density with temperature is of significance in many semiconductor devices. To calculate this variation it is customary to approximate the square of Eq. (3·7) by

$$n_i^2 = B^* e^{-qE_g/kT}$$

Using this approximation, show that n_i^2 is doubled when T increases above room temperature by about 7°C in germanium and 5°C in silicon.

3·5 An oxide-coated thermionic cathode has a work function $W_F \sim 1$ volt. Light in an appropriate frequency range may be shined on such a cathode to cause it to emit electrons. Approximately where does this frequency range begin? What part of the electromagnetic spectrum does this light fall in. Metals generally have work functions on the order of 2 to 5 ev. In what range of the electromagnetic spectrum will these materials begin to act as photocathodes?

3·6 A phototube consists of a large light-sensitive cathode and a small anode in an evacuated tube (see Fig. P3·6). With the biasing arrangement shown, the anode can be either positive or negative with respect to the cathode. Current-voltage characteristics for such a device will appear as shown in Figs. P3·6b and P3·6c. Explain these characteristics.

3·7 How precisely will the uncertainty principle allow us to localize a lattice atom in a germanium crystal at room temperature?

3·8 Estimate the earth's magnetic field from the data given in Fig. 3·12.

---

[1] Semi-Metals, in Saxonburg, Pa., sells an AAPT set consisting of a germanium p-type bar, a germanium n-type bar, and an n-type wafer.

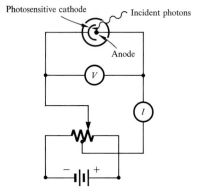

Photosensitive cathode — Incident photons

Anode

V

I

(a) Phototube and biasing arrangement

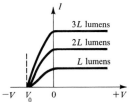

I

$3L$ lumens

$2L$ lumens

L lumens

$-V$ V_0 0 $+V$

(b) I-V plot for various intensities of illumination; illumination frequency f fixed, and $f > W_F/h$

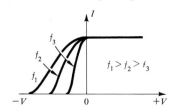

I

f_3

f_2

f_1

$f_1 > f_2 > f_3$

$-V$ 0 $+V$

(c) I-V plot for various frequencies of illumination; illumination intensity held fixed

FIG. P3·6 BASIC PROPERTIES OF A VACUUM PHOTOTUBE

3·9 Estimate the effective mass of holes in germanium from the data given in Fig. 3·13. Similar data for different orientations of the crystal in the applied fields show absorption peaks at $B \sim 700$ gauss and $B \sim 900$ gauss. This means that effective masses are actually somewhat different in different crystalline directions.

3·10 The factor B in Eq. (3·7) can be found from an approximate theoretical analysis to be

$$B^2 = 32\left(\frac{\pi kT}{h^2}\right)^3 (m_e^*)^{3/2}(m_h^*)^{3/2}$$

where m_e^* and m_h^* are the effective masses for free electrons and holes, respectively. Using $m_e^* = 0.2\ m_0$ and the value of m_h^* obtained from Prob. 3·9, calculate B^2. Compare this with the value obtained from Eq. (3·7), using the known room-temperature value of n_i. An accurate room-temperature value of E_g can be calculated from

$$E_g = 0.782 - 3.9 \times 10^{-4}\ T \qquad °K$$

3·11 Phosphorus is a typical element of group V_A of the periodic table. It has an atomic weight of 31 and an atomic number of 15. Work out the ground-state configuration for this atom and its reduced atomic structure. Write down the atomic numbers of the next two donor elements (that is, group V_A). Look up the names of these elements in a periodic table.

3·12 Equations (3·17) and (3·18) are based on the assumption that the bulk value of the dielectric constant can be used to discuss the orbital motion of the fifth-valence electron. To check the validity of this assumption, calculate r_0 in angstroms for a germanium crystal. Use $m^*/m_0 = 0.2$ and $\epsilon_r = 16$. Express your answer for r_0 in terms of the lattice constant for germanium. Does this seem sufficient to justify the use of the bulk value of ϵ_r?

3·13   The calculation of the energy required to ionize a donor atom rests partly on the implicit assumption that the donors are far enough apart to be independent of each other. At what donor concentration would you expect the ground-state hydrogenlike electronic orbitals of the free electrons of two adjacent donors to overlap enough to make this assumption questionable?

3·14   The effective mass of a free electron in silicon is approximately $0.4m_0$. The bulk dielectric constant is $\epsilon_r = 12$. Calculate the approximate ionization energy for a donor atom in the silicon lattice. Compare the value to those given in Table 3·1.

3·15   It is customary to use Eq. (3·25) rather than (3·23) to calculate p_n in an n-type semiconductor. Assuming all donors to be ionized (that is, $N_d \lesssim 10^{14}$), will Eq. (3·25) overestimate or underestimate the value given in Eq. (3·23)? At what value of N_d is there a 10 percent difference between the values of p calculated by these two formulas?

3·16   What percentage of covalent bonds in a pure silicon crystal are broken at a temperature of $300°K$? Repeat for silicon with a donor concentration of $10^{16}/cm^3$. concentration of $10^{16}/cm^3$.

3·17   An n-type semiconductor is said to be *extrinsic* if $N_d \geq 5\ n_i$, *intrinsic* if $N_d \leq 5\ n_i$. Derive a formula for the temperature at which $N_d \sim 5\ n_i$. Calculate the temperature for this condition in germanium and silicon bars doped with $N_d = 10^{16}$ impurities/cm³.

3·18   Derive Eqs. (3·27) and (3·28) from first principles.

3·19   An ingot of germanium is pulled from a melt containing 100 g of germanium and 3.22×10^{-6} g of antimony. Assuming the antimony atoms are uniformly distributed, show that the density of antimony atoms is $8.7 \times 10^{14}/cm^3$. Find p and n for this bar at room temperature. [The average atomic weight of germanium is 72.6, so 72.6 g of germanium contain 6×10^{23} atoms (Avogadro's number). The atomic weight of antimony is 121.7.]

3·20   Suppose the crystal being pulled in Prob. 3·19 is 2 cm in diameter. When the crystal reaches a length of 4 cm, 4×10^{-4} g of gallium is added to the melt, and the remainder of the melt is then grown onto the existing crystal. Discuss carefully the effect of growing both donors and acceptors in the crystal. What will be the conductivity type and doping density of the material grown after the gallium phosphide is added? Calculate p and n for this case, and plot p and n through the entire crystal. You may assume the n and p values to change abruptly at the point of gallium addition, though this assumption will need to be modified later (Chap. 5). The structure just considered is called a "grown" pn junction. The density of germanium is 5.3 g/cm³.

3·21   In this problem we will consider an interesting analogy between semiconductors and electrolytes. The following equation can be written for the electrolytic dissociation of water:

$$H_2O \rightleftarrows H^+ + (OH)^-$$

In the dissociation process, an H_2O molecule must absorb an energy E from its surroundings and then split into an H^+ ion and an $(OH)^-$ ion. This is the ion-generation process. At the same time, H^+ and $(OH)^-$ ions are "recombining" to form H_2O. The recombination rate is proportional to the product of the concentrations of H^+ and $(OH)^-$ ions.

$$[H^+][(OH)^-] = K$$

1. Give an argument to justify this recombination-rate formula.

2. Now consider adding a base such as NaOH to the water. Assuming the NaOH is entirely ionized, write down the equations which you would use to compute the concentrations of Na^+, H^+, and $(OH)^-$ ions.

3. Repeat for an acid such as HCl.

4. Describe what happens when HCl and NaOH are added in equal concentrations.

5. The dissociation of the water molecule into H^+ and $(OH)^-$ ions is analogous to the breaking of a valence bond to produce an electron-hole pair. Write a "chemical equation" which governs this process.

6. Make a table to compare the doping processes in semiconductors with the doping of the water. Include in this comparison the quantitative relations obtained above.

3·22 The binding energy for an electron in a donor-doped germanium crystal is about 0.01 ev (cf. Table 3·2, p. 88). Is it a reasonable approximation to assume that all the donor atoms are ionized at 0°C? At −50°C? Is it a reasonable approximation to assume that all donor atoms in a silicon crystal are ionized at room temperature?

4

CARRIER MOTION IN SEMICONDUCTORS

IN THE LAST chapter it was shown that there are two carriers of electricity in a semiconductor crystal: electrons and holes. The effective-mass approximation was introduced so that these carriers could be treated as classical particles, in so far as their response to applied fields is concerned.

We now wish to study more thoroughly the ballistics of these classical particles in a semiconductor. We shall be concerned with four basic processes: drift, diffusion, motion in perpendicular electric and magnetic fields, and recombination.

The importance of these basic processes can hardly be overemphasized. Taken together with the photoelectric and thermal-generation processes described in the last chapter, they form the physical basis for all the semiconductor devices we shall discuss in this book. Elementary combinations of these processes lead to some simple but useful devices, which we shall consider in this chapter. We shall then be prepared to consider more sophisticated combinations of the basic processes which occur in semi-conductor diodes and transistors.

4·1   CARRIER DRIFT IN AN ELECTRIC FIELD

In a perfectly periodic lattice potential, a free electron responds to a small externally applied electric field by developing an *acceleration* which is proportional to **E**:

$$\mathbf{a} = \frac{-q}{m^*}\, \mathbf{E} \qquad\qquad (4 \cdot 1)$$

The free electron in a perfect crystal therefore behaves exactly like the free electron in a vacuum, except that m_0 is replaced by m^*. This is a statement of the effective-mass approximation.

However, the lattice potential for carriers in a real crystal is not perfectly periodic. In fact, important departures from perfect periodicity arise because

1. The lattice atoms are thermally vibrating about their equilibrium positions.

2. Ionized impurity atoms which have been added to control the carrier densities will disrupt the perfectly periodic lattice potential.

On account of these imperfections, a picture of the actual lattice potential at some instant of time would show some regions where the potential is nearly perfectly periodic and others where there are significant deviations from perfect periodicity.

We may include this situation within the effective-mass approximation by supposing that carriers can be accelerated through the nearly periodic regions in accordance with Eq. (4 · 1), but that they then interact with the nonperiodic potential at the ends of these regions. The interaction will usually be such that the energy gained by the carrier from the field will be transfered to the lattice, and the carrier will be sent off at a new angle unrelated to its angle of incidence. We describe this by saying the carrier "collides with the lattice" (though it really "collides" only with the field produced by a group of lattice atoms that are significantly out of place).

We shall consider the nature of these collisions further in Sec. 4 · 1 · 3. For the moment, we wish to emphasize that because of these collisions, the large-scale effect of an electric field on a free electron in a *real* crystal is to cause it to *drift* through the crystal with a uniform velocity. When the electric field is small enough, the drift velocity is proportional to the electric field:

$$\mathbf{v}_n = -\mu_n \mathbf{E} \qquad\qquad (4 \cdot 2)$$

where $\mathbf{v}_n$ = velocity of the electron

$\quad \mathbf{E}$ = electric field

$\quad \mu_n$ = proportionality constant between v and E, called the *electron mobility*

The minus sign in Eq. (4·2) accounts for the fact that the electron moves in a direction opposite to that of the applied field. The mobility μ_n depends on the type of material, the doping level, and the temperature, in a manner to be described later.[1]

Thus, a perfect crystal is analogous to a vacuum, while a real crystal is analogous to a viscous medium, such as oil, where Stokes' law indicates that a steady force $\mathbf{F}$ acting on a particle produces a constant velocity which is proportional to $\mathbf{F}$. The collisions of the free carrier in the lattice act as a sort of drag on their motion.

To understand more clearly why $\mathbf{v}$ is proportional to $\mathbf{E}$ in real crystals, it is useful to first consider the thermal equilibrium motion of a free electron in an isolated crystal. We shall concentrate on electrons, though the same arguments apply for holes.

Free-electron motion in an isolated crystal. Let us consider an isolated crystal of germanium in thermal equilibrium at a temperature T. We visualize the highly idealized trajectory of a free electron in such a crystal as a series of straight lines between collision points, as shown in Fig. 4·1a. These lines have length $l_1, l_2, \ldots, l_n$, the average of which is the mean free path $\bar{l}$. The paths are also characterized by free times $t_1, t_2, \ldots, t_n$, the average being the *mean free time $\bar{t}$.*

The collisions are considered entirely random in nature; after many collisions the expected displacement of a given particle will be nearly zero.

Between collision points, the electron behaves as a free particle, moving with constant velocity. The exact velocity with which the electron traverses a given path depends on the momentum and energy transfered between it and the lattice during the last collision. These collisions are such that, on the average, the free electron will have an energy of roughly kT on each path. It therefore has an average thermal velocity along each path of

---

[1] Equation (4·2) is valid in germanium for $E \lesssim 1{,}000$ volts/cm, and in silicon for $E \lesssim 5{,}000$ volts/cm. The reason for this is commented on briefly in Sec. 4·1·3.

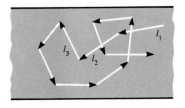

(a) *Trajectory in thermal equilibrium; $E_x = 0$*

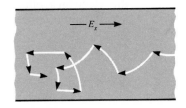

(b) *Trajectory in presence of E_x*

FIG. 4·1 FREE ELECTRON TRAJECTORIES IN A "REAL" CRYSTAL

$$v_T \sim \sqrt{\frac{2kT}{m^*}} \qquad (4\cdot3)$$

where m^* = effective mass of the carrier
k = Boltzmann's constant
T = absolute temperature, $°K$

At room temperature, v_T is between 10^6 and 10^7 cm/sec. Typical room-temperature values of $\bar{l}$ and $\bar{t}$ are 1,000Å and 10^{-12} sec. In terms of our previous argument, $\bar{l}$ is the average linear dimension over which the lattice potential appears to be essentially periodic; $\bar{t}$ is the time required for the electron to move through this region when traveling at the (constant) velocity given by Eq. (4·3).

Drift motion in an electric field. Now, under the influence of an applied electric field $\mathbf{E}_x$, the free paths become curved. In fact, they are segments of parabolas, like the trajectory of a falling body. A particular electron is considered to start each path with the same velocity it would have had in the absence of the applied field. During its free flight, it gains momentum and energy from the electric field, and transfers these to the lattice at the end of its free flight (*cf.* Prob. 4·2). Thus energy is transfered from the electric field to the crystal (which we interpret externally as joule heating), and the electron suffers a net displacement (see Fig. 4·1b).

Under normal conditions, the added displacement due to the field in a free time t_i is small compared to the corresponding free path length l_i. When this is so, the electric field does not disturb the basic thermal motion. The motion in the presence of the field then consists of a slow drift of the equilibrium position, around which the carrier rapidly executes its random motion.

4·1·1 Carrier mobility. A formula for the mobility of an electron can be simply derived on the basis of these fundamental events. To do this, we suppose that the time duration between collisions does not change when an external field is applied. Then, in the presence of a field $\mathbf{E}_x$, the average impulse supplied between collisions will be simply $-q\mathbf{E}_x\bar{t}_e$, where $\bar{t}_e$ is the mean free time between collisions for an electron.

The average momentum lost per collision is just equal to the product of the electron's effective mass m_e^* and its average velocity $\mathbf{v}$. Equating average impulse to average momentum loss then gives

$$m_e^*\mathbf{v} = -q\mathbf{E}_x\bar{t}_e \qquad (4\cdot4)$$

or

$$\mathbf{v} = -\left[\frac{q\bar{t}_e}{m}\right]\mathbf{E} \qquad (4\cdot5)$$

The electron mobility is therefore

$$\mu_n = \frac{q\bar{t}_e}{m_e^*} \qquad (4\cdot6)$$

A similar derivation leads to the formula

$$\mu_p = \frac{q\bar{t}_p}{m_p^*} \tag{4·7}$$

for the mobility of a hole, where $\bar{t}_p$ is the mean free time between collisions for a hole, and m_p^* is the effective mass of the hole.

Thus, the mobilities depend only on the mean free times for a given type of carrier in a given crystal. We shall consider very briefly the major physical processes which determine these mean free times in the next section.

4·1·2 *Variation of mobility with temperature and doping level.* There are two principal scattering mechanisms which determine the mean free time $\bar{t}$ and thereby the mobility. These are *lattice scattering* and *impurity scattering*. A few comments on these processes will give a better perspective of the theory of mobility.

Lattice scattering. This type of scattering arises because the lattice atoms in a real crystal are vibrating about their equilibrium positions (even at absolute zero, according to the uncertainty principle). At any instant, the lattice atoms will be squeezed together in some regions and pulled apart in others, relative to their perfect spacing. Because of this, the lattice potential in the crystal will not be perfectly periodic, and the conditions for the effective-mass approximation will not be met.

However, for temperatures of interest, the vast majority of the lattice atoms do not have enough energy at any given instant to depart very far from their equilibrium positions. Therefore, *major* departures of the lattice potential from its ideal periodic form rarely occur, and then, over only small portions of the crystal.

We may visualize this situation by considering the interior of the crystal to be made up of a series of microscopic crystals within which the lattice atoms are roughly in their equilibrium positions. At the interfaces between the crystals, the lattice atoms are significantly out of position and the potential fails to be even approximately periodic. The effective-mass approximation holds within each subcrystal, but fails at the subcrystal boundaries. Of course, the subcrystal boundaries will be a function of time, due to the random motion of the atoms.

The subcrystal boundaries are the sites at which the electron exchanges energy and momentum with the crystal. As it moves through a boundary, its momentum is randomly redirected. Energy is exchanged because the location of the boundary itself is slowly changing with time.[1]

The lattice-scattering mechanism produces a mean free time $\bar{t}_L$, which decreases with increasing temperature. This is primarily because the

[1] The details of these momentum and energy exchange mechanisms are described in W. Shockley, "Electrons and Holes in Semiconductors," D. Van Nostrand Company, Inc., Princeton, N.J., 1959.

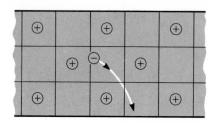

FIG. 4·2 COULOMB SCATTERING OF AN ELECTRON BY A BOUND POSITIVE CHARGE

number of atoms which are not in their equilibrium positions increases as T is increased. The mean free time (and thus the mobility) due to lattice scattering can be shown to vary as T^{-a}, where a is between 1.66 and 2.7, depending on the type of carrier being scattered and the crystal (germanium or silicon).

Impurity scattering. A second variation of the crystal potential from its purely periodic form occurs when ionized donor or acceptor atoms are introduced into the lattice.[1] Thus, moving carriers may also be scattered by impurities. The manner in which a carrier is scattered by a charged scattering center is indicated schematically in Fig. 4·2.

The magnitude of the scattering effect of an ionized impurity depends on the amount of time the carrier spends in the field of the impurity. Thus, the effectiveness of an impurity-scattering center decreases as the velocity of the approaching carrier is increased and therefore is more effective at low temperatures than at high temperatures.

The mean free time for impurity scattering is denoted by $\bar{t}_I$ and can be shown to vary as $T^{3/2}$. Naturally, the number of impurity-scattering events increases as the number of ionized impurities is increased (that is, as the sample is doped more heavily).

Combined impurity and lattice scattering. In an actual crystal the mean free time will be determined by both scattering mechanisms. If the scattering mechanisms are assumed to act independently, the actual mean free time can be approximated by

$$\frac{1}{\bar{t}} = \frac{1}{\bar{t}_L} + \frac{1}{\bar{t}_I} \tag{4·8}$$

We can also define a mobility due to pure lattice scattering μ_L, and one due to pure impurity scattering μ_I [by use of $\bar{t}_L$ or $\bar{t}_I$ in Eqs. (4·6) and (4·7)]. Then, the actual mobility is approximately

$$\frac{1}{\mu} = \frac{1}{\mu_I} + \frac{1}{\mu_L} \tag{4·9}$$

[1] Since the potential introduced by ionized impurities is primarily a function of only spatial coordinates (and not time), impurity scattering produces only a redirection of the momentum of the particle. In such collisions the carrier does not give up the energy which has been added to it by the field. The energy will ultimately be given up during a lattice-scattering event.

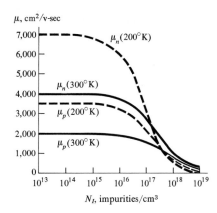

FIG. 4·3   MOBILITY OF HOLES AND ELECTRONS IN GERMANIUM VS. IMPURITY DENSITY

Typical mobility data.   Curves of electron and hole mobility in germanium versus doping level are shown in Fig. 4·3 for two different temperatures. The experimental methods used to obtain these curves will be described later. For each curve, lattice scattering is dominant at low doping densities, so the mobility is constant at the value determined by lattice-scattering phenomena (*cf.* Prob. 4·3). As the doping level is increased, impurity scattering becomes more important, and μ drops.

   Actually, the mobility of a given carrier also depends slightly on whether the crystal is *n* type or *p* type, though this detail is not shown on Fig. 4·3 and is insignificant for our purposes.

   It should be pointed out that the vertical scale (that is, μ scale) in Fig. 4·3 is in square centimeters per volt-second. From its defining equation

$$\mathbf{v} = \mu\mathbf{E} \qquad (4\cdot10)$$

we see that μ is described by length units squared per volt-second. In the mks system, μ therefore should be in square meters per volt-second. However, since typical distances are measured in centimeters, the units adopted for μ are cm²/volt-sec.

4·1·3   *Variation of mobility with electric field.*   To conclude the discussion of mobility, we shall discuss briefly the major limitation in the ideas presented thus far. This has to do with the assumption that the basic random motion of a carrier is not affected by the applied field.

   To discuss the validity of this assumption, let us consider an example. Suppose we apply a field of $E = 100$ volts/cm to a relatively pure germanium crystal. Then, from Eq. (4·2) and Fig. 4·3, a free electron should acquire a drift velocity of 4×10^5 cm/sec. If we assume its mean free path is about 1,000Å and its mean free time is 10^{-12} sec, it should then suffer about 25 randomizing collisions in drifting a distance equal to its mean free path. This many collisions will be adequate to transfer the energy and momentum gained per path to the lattice.

   If E is increased to 1,000 volts/cm, we expect the carrier to suffer only 2 to 3 randomizing collisions in drifting a distance $\bar{l}$. Then our assumptions about

Carrier motion in semiconductors   105

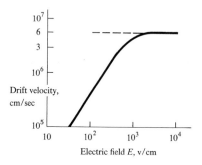

FIG. 4·4   ELECTRON VELOCITY VS. DRIFT FIELD IN A LIGHTLY DOPED *n*-TYPE GERMANIUM CRYSTAL; *v* PROPORTIONAL TO E UP TO $v \sim 3 \times 10^6$ CM/SEC, $E \sim 1,000$ VOLTS/CM

momentum and energy transfer are no longer valid, and our picture requires considerable modification.

The nature of these modifications is beyond our present interests. However, their effect is that the carrier velocity ceases to increase linearly with E, and instead approaches an approximately constant value. A curve demonstrating this phenomenon is shown in Fig. 4·4.

There is thus a limit on the velocity which we can give to carriers in a crystal by subjecting them to an electric field. There is a corresponding limit in the high-frequency performance of semiconductor devices, and a limit on the validity of Ohm's law in a semiconductor bar. For most purposes, however, we will not be concerned with these limits.

4·2   CALCULATION OF THE CONDUCTIVITY OF A SEMICONDUCTOR BAR

As a first application of the mobility concept, we will calculate the conductivity of a bar of semiconductor material. The bar is shown in Fig. 4·5. It is assumed to be of cross section A, and to have n electrons per unit volume and p holes per unit volume. An electric field E is applied in the direction indicated. This causes holes and electrons to move in opposite directions, but because of the negative sign on the charge of the electron, the hole and electron contributions to the total current add.

Now, the number of electrons crossing the plane P in one second is Anv_n. Hence, the electron current density at plane P is

$$j_n = \frac{I_n}{A} = qnv_n = qn\mu_n E_x \qquad (4 \cdot 11)$$

Similarly, the hole current density (evaluated at plane P) is

$$j_p = \frac{I_p}{A} = qpv_p = qp\mu_p E_x \qquad (4 \cdot 12)$$

Therefore, the total current density is

$$j = j_p + j_n = q(n\mu_n + p\mu_p)E_x$$

But by definition, the conduction current and electric field are related through the equation

$$j = \sigma E$$

where σ is the conductivity. Hence

$$\sigma = q(n\mu_n + p\mu_p) \tag{4·13}$$

and the resistivity ρ is

$$\rho = \frac{1}{\sigma} = \frac{1}{q(n\mu_n + p\mu_p)}$$

To obtain some useful numbers, we calculate σ for intrinsic germanium at room temperature. In this case, $\mu_p = 1{,}900$ cm²/volt-sec, $\mu_n = 3{,}900$ cm²/volt-sec, $n_i = p_i = 2.5 \times 10^{13}$ cm⁻³. Therefore

$$\sigma = 1.6 \times 10^{-19}(2.5 \times 10^{13})(5{,}800) = 23.2 \times 10^{-3}(\text{ohm-cm})^{-1}$$

$$\rho = \frac{1}{\sigma} = 43 \text{ ohm-cm} \tag{4·14}$$

When the germanium is doped, n and p will change, as indicated in Chap. 3. As an example, suppose we doped the germanium with donor impurities to the level $N_d = 10^{17}$ cm⁻³. Then, from Fig. 4·3, $\mu_n = 2{,}000$ cm²/volt-sec, and the conductivity of the bar is

$$\sigma = q(\mu_n n_n + \mu_p p_n) \cong q\mu_n n_n = 1.6 \times 10^{-19} \times 2{,}000 \times 10^{17}$$
$$= 32 \text{ (ohm-cm)}^{-1} \tag{4·15}$$

$$\rho = \frac{1}{\sigma} \cong 0.03 \text{ ohm-cm}$$

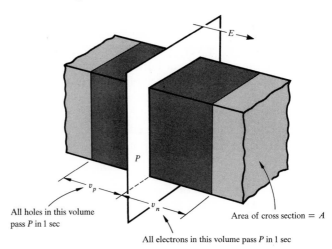

All holes in this volume pass P in 1 sec

Area of cross section $= A$

All electrons in this volume pass P in 1 sec

FIG. 4·5   CALCULATION OF CONDUCTIVITY

Problems $4 \cdot 5$ and $4 \cdot 6$ deal with further aspects of the effect of doping on the conductivity of a semiconductor crystal.

It is instructive to note that the resistivity of the doped bar is more than 1,000 times less than the resistivity calculated for intrinsic material (43 ohm-cm). This comparison indicates that the resistivity of a semiconductor bar may be controlled over quite wide limits by doping. We may also control the resistivity *type* by doping with either donor or acceptor inpurities. These are characteristic properties of the most useful semiconductors and distinguish them from insulators (such as rubber) or good conductors (such as copper). As mentioned in the last chapter, the ease of controlling the amount and type of conductivity is of great significance in both transistor theory and fabrication.

A common procedure by which the conductivity of a semiconductor crystal is measured is described in Prob. $4 \cdot 7$.

$4 \cdot 2 \cdot 1$ Temperature coefficient of resistivity in semiconductor material. Another characteristic which is frequently used for classifying materials as semiconductors is the temperature coefficient of their resistivity. If we return to the calculation of σ for *intrinsic* germanium, we may rewrite Eq. $(4 \cdot 13)$ as

$$\sigma = q(\mu_n + \mu_p)n_i \qquad (4 \cdot 16)$$

Using the formula for n_i given in Chap. 3, we may write

$$\sigma = q(\mu_n + \mu_p)BT^{3/2}e^{-qE_g/2kT} \qquad (4 \cdot 17)$$

Now, in Eq. $(4 \cdot 17)$, μ_n and μ_p are functions of temperature, of the approximate form T^a. Thus, Eq. $(4 \cdot 17)$ can be rewritten as

$$\sigma = \text{const } T^{a*}e^{-qE_g/2kT} \qquad (4 \cdot 18)$$

where $0 \lesssim a* \lesssim 3$. Then, since T^{a*} varies much more slowly with T than the exponential factor, the conductivity *increases* as T increases (or, equivalently, the resistivity decreases with increasing T).

This is a very important property, for, among other things, it leads to the possibility of building *thermistors,* or resistors with negative temperature coefficients. Such elements find applications in several areas of electronics and will be discussed shortly. Before doing so, however, it is instructive to calculate the rate of change of the resistivity with temperature to see how big the effect is.

To do this, we first approximate n_i by

$$n_i = Ce^{-qE_g/2kT} \qquad (4 \cdot 19)$$

in recognition of the fact that the principal contribution to temperature dependence will come from the exponential factor.[1] Then

$$\rho = \frac{1}{\sigma} \cong \frac{e^{+qE_g/2kT}}{q(\mu_n + \mu_p)C} \qquad (4 \cdot 20)$$

---

[1] A less approximate analysis is the subject of Prob. $4 \cdot 8$.

To find out how rapidly ρ varies with T, we calculate the fractional change in ρ for a given fractional change in T. That is

$$\frac{\partial\rho/\rho}{\partial T/T} = \frac{\partial\rho}{\partial T}\frac{T}{\rho} = -\frac{qE_g}{2kT} \qquad (4\cdot21)$$

At room temperature, for germanium,

$$-\frac{qE_g}{2kT} \cong -13.4$$

Hence, a 1 percent change in temperature ($\simeq 3°C$) gives rise to a 13 percent decrease in resistivity. The effect is therefore quite marked.

It should be noted that the size of the effect depends greatly on whether the semiconductor material is intrinsic or not. If the material is heavily extrinsic (say, n-type), then

$$\sigma \simeq q\mu_n n_n$$

The temperature behavior of σ is then determined entirely by that of μ, as in a normal metal. In such cases, the percentage change of σ with changes in T will be small; and, in semiconductors, will have a sign which depends on the doping level (in germanium the sign is $+$ for $N_d < 10^{17}$, $-$ for $N_d > 10^{17}$; see Fig. $4\cdot3$). True thermistor behavior reflects primarily the nature of the thermal-generation mechanism, and therefore is seen only when the thermally generated carrier densities control the conductivity.

Thermistor action can be simply demonstrated by measuring the resistance of a bar of high-resistivity germanium as the temperature of the bar is varied (see Demonstration $4\cdot1$). Because of the dominance of the exponential factor in determining the resistance of the bar, such an experiment qualitatively verifies the formula for n_i derived in the previous chapter.

$4\cdot3$ THERMISTORS

In its simplest form, a thermistor consists of an intrinsic semiconductor bar to which ohmic contacts are made as shown in Fig. $4\cdot6$. Popular materials for commercially available thermistors are high-resistivity germanium, mixtures of nickel and strontium oxides, and certain uranium oxides, with the selection depending on the particular application at hand. The low-frequency v-i characteristic of a thermistor may be understood very simply by the construction given in Fig. $4\cdot7$. Here, a number of straight lines are drawn emanating from the origin, each line representing the resistance of the thermistor near $(V = 0, I = 0)$ at the given temperature. In making such a drawing, we are tacitly assuming that the temperatures indicated with each line are maintained in the semiconductor bar independent of the voltage-current product, which is also being applied to the bar.

Now, a *self-heated* thermistor works as follows. As voltage is applied to the terminals of the thermistor in the manner indicated in Fig. $4\cdot7$, a

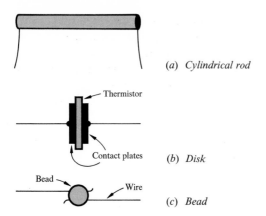

(a) *Cylindrical rod*

Thermistor

Contact plates

(b) *Disk*

Bead

Wire

(c) *Bead*

FIG. 4·6 SEVERAL TYPES OF THERMISTORS

current will flow which is equal to the voltage divided by the room-temperature resistance of the thermistor. Power will be dissipated in the thermistor in the amount $V \cdot I$, but since V and I are small, this power will have only a negligible heating effect. As V is increased, however, the power gets larger and the thermistor begins to heat up. But as the temperature increases, the resistivity decreases, and so, for a given voltage the current must be larger than it would have been without any resistivity decrease. In terms of Fig. 4·7, this means that the characteristic must curve toward one of the resistance lines which apply for higher temperature.

Naturally, the amount of curvature which will be obtained for a given voltage-current product will depend on the thermal environment of the thermistor. For example, if the thermistor is in good thermal contact with a large copper block, which in turn is mounted in the San Francisco Bay, it may be quite difficult to raise the temperature of the thermistor very far. On the other hand, if the thermistor is hanging free in air, or mounted in

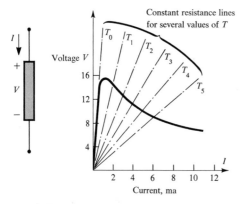

FIG. 4·7 SELF-HEATED THERMISTOR WITH V-I CHARACTERISTIC

an evacuated envelope, its temperature may rise quite appreciably as voltage and current are applied. In the latter case, continued heating will occur and the v-i characteristic of the thermistor will exhibit a differential negative resistance as shown in Fig. 4·7. This differential negative resistance, together with the large resistivity change which is characteristic of thermistors, forms the basis for many important applications of thermistors. We will briefly describe some of these now.[1]

Thermometry. Since the temperature coefficient of resistance for a thermistor is very large (about ten times higher than that of an ordinary wirewound resistor), thermistors can be used with excellent results as the sensing element in conventional resistance thermometers. In most practical cases, the thermistor is used in one arm of a Wheatstone bridge (see Prob. 4·9). Of course, the source voltage and bridge resistances must be adjusted so that the current flowing through the thermistor does not produce heating effects which are comparable to those being detected. Under optimum conditions, such a thermometer can detect a temperature change of about 0.0005°C.

Since thermistor elements can be designed for high-resistance values, lead resistances are ordinarily negligible in comparison with the element resistance. The thermistor can therefore be located in contact with engine blocks, ventilating ducts, etc., while the measuring circuitry is more conveniently located elsewhere. Thermistor contact thermometers are also used in weather balloons, on storage tanks, on buildings, etc., and, in some cases, to measure rates of air flow in pipes, where the cooling effect of the air can be related to the velocity of flow.

Thermistor thermometers may also be used when the object whose temperature is to be measured is inaccessible, in motion, or at too high a temperature for normal contact thermometry. In such cases, heat is gathered from the object and focused onto the thermistor by a suitable set of infrared lenses and mirrors. These thermometers are called thermistor bolometers and are used primarily in military operations as thermal detectors for detecting a hot object in a field of view.

Temperature compensation. Since the temperature coefficient of a thermistor is negative, while that of a metallic resistor is positive, the series connection of a thermistor and a resistor will produce a combination which may have nearly zero temperature coefficient over a limited temperature range (see Prob. 4·11). Figure 4·8 shows graphically a case of this kind of series compensation. With more elaborate combinations of thermistors and wirewound resistors, it is possible to make a composite "resistor" whose value changes by less than 1 percent over a temperature range of about ±30°C around room temperature.

---

[1] A more complete description of thermistor applications will be found in J. N. Shive, "The Properties, Physics, and Design of Semiconductor Devices," D. Van Nostrand Company, Inc., Princeton, N.J., 1960.

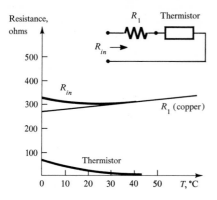

Resistance, ohms

R_1 Thermistor

$R_{in} \rightarrow$

500

400

R_{in}

300

R_1 (copper)

200

100

Thermistor

0   10   20   30   40   50 T, °C

FIG. 4·8 SIMPLE TEMPERATURE-COMPENSATION SCHEME
(*After J. N. Shive, The Properties, Physics and Design of Semiconductor Devices,
D. Van Nostrand Company, Inc., 1960, p. 59*)

Dynamic negative resistance in the thermistor characteristic.[1] It can be seen by inspection of Fig. 4·7 that dV/dI is a negative quantity for the thermistor over part of its operating range. This is an interesting fact and will be discussed briefly to clarify some important issues.

In all the applications mentioned previously, only ac voltages and currents are applied to the thermistor. Therefore, the thermistor operates around the ($V = 0$, $I = 0$) point in the V-I plane, as shown in Fig. 4·9a. In this figure, it is assumed that the ac signals do not affect the temperature of the thermistor. Therefore, the V-I characteristic for the thermistor is just like that for a resistor for these applications (that is, a straight line of positive slope through $V = 0$, $I = 0$), except that the resistance is a sensitive function of temperature: in such cases the thermistor is used as a transducer. Figure 4·9a emphasizes this point.

However, if a dc current I of 5 ma is passed through the device, shown in Fig. 4·9b, the *dc* voltage V will be 10 volts, and dV/dI, the *dynamic* resistance, will be *negative*. This means that if a small ac current is added to the dc current, the alternating component of voltage across the thermistor will be 180° out of phase with the ac current. The *ac* voltage and current are related by

$$v_{ac}(t) = R_{dynamic} i_{ac} = \frac{dV}{dI} i_{ac}(t) \qquad (4·22)$$

and $R_{dynamic}$ is a negative quantity. The operating path for such a signal is shown in Fig. 4·9b.

Frequently, devices with such a negative dynamic resistance are useful for making various types of switching circuits and oscillators. We shall

[1] In this and later sections, the following nomenclature is used to denote quantities which have both dc and incremental (or ac) parts: the total value of a voltage or current is represented by a lower-case letter (with capital subscripts, if appropriate); the dc value is represented by a capital letter (with capital subscripts, if appropriate); and the incremental value is represented by a lower-case letter with lower-case subscripts. Thus $v(t) = V + v_{ac}(t)$.

consider some of these devices and their applications later. For the moment, we wish to emphasize that the thermistor is *not* usually useful for such applications.

The reason for this is that at *each* point on the *V-I* characteristic of Fig. 4·8, the thermistor is supposed to have reached the appropriate steady temperature *T*. Now, heat can only be exchanged relatively slowly between the thermistor and its environment. Therefore, the ac signals will have to be low-frequency signals, to insure that the instantaneous operating point implied by dc plus ac currents falls on the characteristic drawn in Fig. 4·8.

If a high-frequency ac current is applied in series with a bias current of 5 ma, the thermistor of Fig. 4·9c will essentially operate along the positive resistance line labelled T_4. In other words, the dynamic resistance of

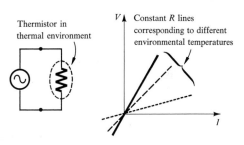

(a) Use of a thermistor as a resistor with large (and incidentally negative) temperature coefficient

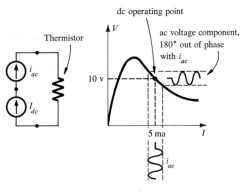

(b) Application of dc and low-frequency ac currents to a thermistor to illustrate negative dynamic resistance

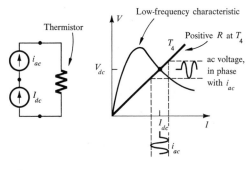

(c) Application of dc and high-frequency ac currents to a thermistor to illustrate positive dynamic resistance at a point where low-frequency dynamic resistance is negative

FIG. 4·9   DIFFERENT TYPES OF THERMISTOR BEHAVIOR

the thermistor is positive at high frequencies, even though the bias point might suggest otherwise.

It is always important to distinguish between the dynamic resistance dV/dI and the dc or static resistance V/I, particularly when dV/dI may be of one sign while V/I is of another. It is also important to understand the conditions which cause dV/dI to be different from V/I, to avoid trying to apply a device incorrectly. For common thermistors, dV/dI is negative only for slow changes in I and V; that is, low-frequency signals. Because of this, the negative dynamic resistance is of use only in low-frequency applications.

4·4 CARRIER MOTION BY DIFFUSION

We began this chapter with a consideration of how carriers drift in a steady electric field. We have seen that it is possible to derive an expression for the mobility of a carrier on the basis of microscopic scattering processes. We now wish to consider a second mechanism by which carriers may move within a solid, *diffusion*. Motion of carriers by diffusion is of fundamental importance in transistors and diodes and needs to be thoroughly understood.

4·4·1 Fundamental ideas. If a puff of smoke of uniform density is released in the center of a closed room, the size of the puff will gradually grow, and the average density of the smoke molecules within the puff will decrease until the smoke is uniformly distributed throughout the room. Similarly, if a bottle of perfume is opened in the room, perfume molecules will evaporate until there is a uniform density of perfume molecules throughout the room.

The mechanism by which the smoke molecules or perfume molecules are propagated through the room is called *diffusion*. The fundamental idea behind the diffusion process is that, on account of their random thermal motion, particles will tend to flow from regions where they are heavily concentrated to neighboring regions of lower concentration. For example, each smoke particle at the boundary of the puff has an equal probability of leaving the boundary or returning to the puff on its next free path. Therefore, roughly half of the smoke molecules on the boundary leave the puff on their next step. There are no smoke molecules beyond the boundary to reenter the puff, so the boundary increases in size. The increase is accompanied by a decrease in density of smoke particles at the boundary, so particles from the next layers within the puff experience a similar situation to those on the boundary; and they then diffuse into the boundary layer. For any particle in the puff, the probability that it will move toward the center of the puff is the same as the probability that it will move toward the boundary layer. However, the density of particles toward the boundary layer is always less than the density toward the center, so there is an average flow from the center toward the boundary layer.

It is possible to show that the rate of flow due to diffusion is proportional to the difference in concentration of the particles at adjacent points. If $C(r,t)$ is the concentration or density of smoke particles at point r, time t, the particle flux J, or rate at which particles *leave* the boundary is proportional to $\partial C/\partial r$:

$$J \propto -\frac{\partial C}{\partial r}$$

The minus sign accounts for the fact that there is a net flow outward when the concentration is decreasing with increasing r.

Both holes and electrons diffuse in a semiconductor crystal. The diffusion always tends to equalize differences in carrier densities at adjacent points. The current flow associated with this diffusion is more important than current flow due to drift in most semiconductor devices.

In the next few sections, the important relations which govern the diffusion of carriers will be developed, and an interesting relation between the drift and diffusion processes will be described.

4 · 4 · 2   *Diffusion as a random-walk problem.* We will begin by describing an idealized, one-dimensional example of a diffusion process, in order to become thoroughly familiar with the basic idea. The example is known as a *random walk* in mathematical statistics, for reasons which will be apparent later.

Suppose a man stands in the middle of a long street. Let his initial position be $x = 0$. He flips a fair coin to decide which way along the street to take his first step. After taking this step, he again flips the fair coin to decide whether his next step shall be in the same direction as the last one, or in the opposite direction, and similarly at the end of each step. For simplicity, we will suppose that the steps are of a uniform size, that each step is completed in a uniform time interval, and that the time needed for flipping the coin can be neglected.

We are now interested in computing the probability that the man will get a certain number of steps away from $x = 0$ after a certain time has elapsed. The computation is readily done with the aid of the simple construction shown in Fig. 4 · 10. At the beginning of his random walk, $t = 0$, the man is at $x = 0$. He takes steps of size l. On his first step, he is equally likely to go right or left. Hence, *before the first step is taken*, his probability of ending up at $x = l$ is ½, and his probability of ending up at $x = -l$ is ½.

Now consider the second step. There are two possibilities. If the man ended up at $x = l$ after the first step, then on the second step, he is equally likely to end up at $x = 2l$ or $x = 0$. On the other hand, if the man ended up at $x = -l$ on the first step, then on the second step he is equally likely to end up at $x = -2l$ or $x = 0$. Thus, after two steps, there are three places where the man could be: $x = -2l$, $x = 0$, $x = 2l$. There are two ways for him to end up at $x = 0$ (first step left, second step right and vice versa) and one way each for him to end up at $x = 2l$, $x = -2l$. If we wish to place a bet, *before the proceedings begin*, that the man will reach

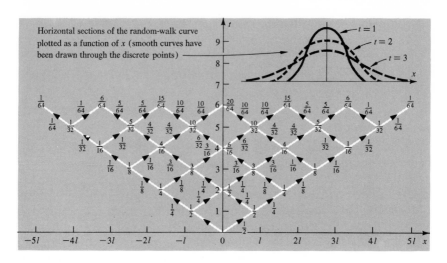

(a) The random-walk construction

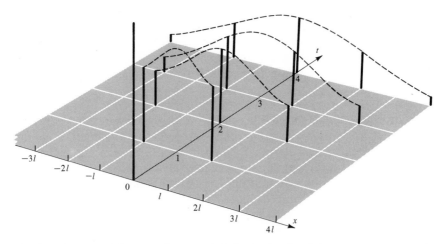

(b) Three-dimensional diagram of (a) in which the height of each line is proportional to the probability of arriving at the point (x,t)

FIG. 4·10   THE RANDOM-WALK PROBLEM

$x = 2l$, we should ask for at least four to one odds, because the probability of his reaching $x = 2l$ is ¼ (see Fig. 4·10).

Before proceeding with more steps, we note that the probability values for the man's position after the second step may readily be derived from his first-step situation by applying a "random-walk" construction to each position. The construction is indicated in Fig. 4·10, and consists of multiplying the probability of reaching $x = l$ by one-half and assigning these numbers (one-fourth, one-fourth) to the lines going to $x = 0$, $x = 2l$. Similarly for $x = -l$. The results are then added to give the total probabilities for reaching $x = -2l, 0, 2l$. Using this idea, the reader can construct

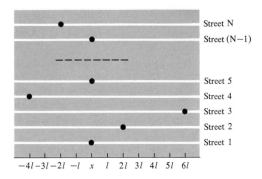

FIG. **4·11**   *N* SIMULTANEOUS RANDOM-WALK PROBLEMS

the remainder of Fig. 4·10*a*. Figure 4·10*b* is a three-dimensional representation of the random-walk problem, in which the height of each line is equal to the probability of reaching the particular point at the designated time.[1]

To make the random walk more analogous to particle diffusion, we may begin with *N* men, situated at identical locations on *N* parallel streets. Each man performs his random walk entirely independently of all others. Then, after six steps, we expect to find $^{20}\!/_{64}$ of the men at $x = 0$, $^{15}\!/_{64}$ of them at $x = \pm 2l$, $^{6}\!/_{64}$ of them at $x = \pm 4l$, and $^{1}\!/_{64}$ of them at $x = \pm 6l$ (see Fig. 4·11). If *N* is small (say, 128), the *actual* distribution of men may be quite different from this. On the other hand, if *N* is 10^{13}, the *relative* porportions of men in the different locations will be insignificantly different from that calculated above.

4·4·3   *The law of diffusion flow.* We now wish to carry the random-walk concept over to the problem of determining the *average* rate at which particles will diffuse across a given surface.

To do this, we have drawn an arbitrary *continuous* curve in Fig. 4·12 to represent an instantaneous, one-dimensional distribution of diffusing particles. We shall assume for simplicity that all particles have the same free-path length *l* and step time t_l, though they have been distributed densely along the *x* axis at random; that is, not just at the positions $x = 0, l, 2l$, etc. The distribution function is intended to be truly continuous.

We now wish to calculate how many particles cross the plane at x_0, from left to right per unit time per unit area. Now, only those particles which are within a distance *l* of the plane at x_0 can cross it on their next free path. If the *density* of particles is denoted by $\dot{n}(x)$, then there are approximately

$$ln(x_0 - l/2)$$

---

[1] Some very useful mathematical features of this probability distribution are the subject of Prob. 4·12.

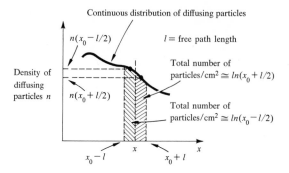

FIG. 4 · 12   DEVELOPMENT OF THE LAW OF DIFFUSIVE FLOW

particles per unit area in a strip of length l to the left of the plane at x_0 (see Fig. 4 · 12) and

$$ln(x_0 + l/2)$$

particles per unit area in the strip on the right. In a step time t_l, the total number M of particles which cross the plane from left to right per unit area, will be

$$M\Big|_{x_0} = \frac{1}{2}\, ln(x_0 - l/2) - \frac{1}{2}\, ln(x_0 + l/2)$$

This equation can be rewritten approximately as

$$M\Big|_{x_0} \cong -\frac{1}{2} l^2 \frac{\partial n}{\partial x}\Big|_{x_0}$$

and since $\partial n/\partial x$ and M are measured at the same plane, we may drop the subscript x_0:

$$M \cong -\frac{1}{2} l^2 \frac{\partial n}{\partial x}$$

Since M particles per unit area cross the plane in a time t_l, the average flow per unit area, or current density, is

$$j \cong -\frac{1}{2} \frac{l^2}{t_l} \frac{\partial n}{\partial x} = -D \frac{\partial n}{\partial x} \qquad (4 \cdot 23)$$

or *the diffusion current (density) is proportional to the slope of the distribution.* The minus sign accounts for the fact that particles will diffuse from regions of high n to regions of lower n.

The quantity D in Eq. (4 · 23) is called the *diffusion constant.* For the present problem, where all particles have the same step size and step time, its value is

$$D = \frac{l^2}{2t_l}$$

Values of D for electrons and holes in semiconductors will be given in Sec. 4·4·4.

The approximation sign can be removed by the artifice of assuming that l and t_l approach zero in such a way that D remains finite. Then, the current density is written exactly as

$$j = -D \frac{\partial n}{\partial x} \tag{4·24}$$

which we shall call the *law of diffusive flow*.

4·4·4 *The diffusion of holes and electrons.* All that is required to take over the results of our previous discussion to the case of electrons and holes is to identify l and t_l, the step size and step time.

Let us concentrate on electrons. We know that the electrons are in thermal equilibrium with the crystal and are therefore to be found flitting about very much as they were in our calculation of mobility. For present purposes, we consider them to be flitting in one dimension.

Of course, the electrons, in their random walk, do not always take steps of a fixed length in a fixed time. Instead, they have a distribution of step sizes (free path) and free times.

When probability distributions exist for the step size y and the step time t, we must calculate D by the formula $D = \frac{1}{2}(\overline{y^2}/\bar{t})$. When this is properly done, it can be shown that $\overline{y^2} = 2(\bar{l})^2$ where $\bar{l}$ is the *mean* free path.[1]

We may then write the diffusion coefficient as

$$D = \frac{(\bar{l})^2}{(\bar{t})}$$

Diffusion constants for holes and electrons in intrinsic germanium and silicon crystals are given in Table 4·1.

TABLE **4·1**

	Diffusion constant for holes, D_p	Diffusion constant for electrons, D_n
Germanium	50 cm²/sec	100 cm²/sec
Silicon	12 cm²/sec	35 cm²/sec

4·5 THE RELATION BETWEEN DRIFT AND DIFFUSION PROCESSES

It is interesting to note that both the mobility and the diffusion constant are determined by the same fundamental events (that is, carrier scattering)

[1] A derivation of a valid distribution function (for either the step size y or the step time t), together with expressions for $(\overline{y^2})$ and $(\overline{t^2})$, may be found in W. Shockley, "Electrons and Holes in Semiconductors," D. Van Nostrand Company, Inc., Princeton, N.J., 1950, pp. 200–204.

and depend on such things as the mean free path and the mean free time. It is therefore not particularly surprising to find that a relationship exists between D and μ. This was first recognized by Einstein, and the equation connecting D and μ is therefore called the Einstein relationship.

4·5·1 *The Einstein relation.* The Einstein relation is

$$\frac{D}{\mu} = \frac{kT}{q} \qquad (4 \cdot 25)$$

where D is the diffusion constant, μ is the mobility, k is Boltzmann's constant, T is the absolute temperature, and q is the magnitude of the electronic charge. To see how this relationship is obtained in our simple one-dimensional problem, we take our expressions for D and μ and form D/μ:

$$\frac{D}{\mu} = \frac{\bar{l}^2}{\bar{t}} \frac{m}{q\bar{t}} = \left(m \frac{\bar{l}^2}{\bar{t}^2} \right) \frac{1}{q} \qquad (4 \cdot 26)$$

Now $\bar{l}$ is the mean free path, and $\bar{t}$ is the mean free time, and their ratio is a velocity. This velocity is, in fact, just the thermal velocity; that is, the velocity which the free electrons have, due to absorbing heat from their environment. We call this velocity v_T. Using this in Eq. $(4 \cdot 26)$ we obtain

$$v_T = \frac{\bar{l}}{\bar{t}}$$

$$\frac{D}{\mu} = m v_T^2 \frac{1}{q} \qquad (4 \cdot 27)$$

Now $m v_T^2 / 2$ is the energy which the electron has as a consequence of being in thermal equilibrium with the lattice, which is at temperature T. It is known from the principle of equipartition of energy that

$$\frac{m v_T^2}{2} = \frac{kT}{2} \qquad (4 \cdot 28)$$

Therefore we have

$$\frac{D}{\mu} = \frac{kT}{q}$$

While we have only shown the relation to be valid for a one-dimensional example, it is also true in three dimensions.

4·5·2 *The Shockley-Haynes experiment.*[1] We are now in a position to understand one of the classic experiments performed with semiconductors. This experiment illustrates very nicely the concepts of drift and diffusion, and allows us to verify that the ratio of D to μ is indeed kT/q. A diagram

---

[1] The original experiment is described in J. R. Haynes and W. Shockley, *Phys. Rev.*, vol. 75, p. 691, 1949.

of the apparatus is shown in Fig. 4·13. It consists of a long semiconductor bar with *ohmic* contacts made to each end. For the particular case shown in Fig. 4·13a, the bar is *n*-type, though the experiment can be done with either an *n*-type or a *p*-type bar. In either case, the ohmic end contacts simply provide a low-resistance contact to the semiconductor which will carry current equally well in either direction.

An adjustable dc voltage source is applied between the ends of the bar, thus creating an electric field in the sample and causing current to flow. This current is basically a *majority-carrier* current.

Emitter and *collector* points are also placed on the bar as indicated in Fig. 4·13. A pulse voltage source is connected between the emitter point and one of the end contacts, while the collector point is connected through a resistor and a battery to the other end contact. An oscilloscope is connected across the collector resistor to observe the current that flows in the collector point.

When the pulse generator delivers a *positive* pulse through the emitter point to the *n*-type bar, two things happen:

1. *A positive pulse appears almost immediately on the oscilloscope screen.* This occurs because the emitter and collector points act partially as taps on a resistor (the semiconductor bar), so some fraction of the applied pulse voltage is fed to the oscilloscope input.

2. *At a later time, a bell-shaped pulse also appears on the oscilloscope screen.* We interpret this fact as follows: When the positive pulse is applied to the emitter point, it causes holes to be injected into the *n*-type bar. This immediately sets up a local positive space charge around the emitter point. But such a local space charge can be maintained for only a very short time in the *n*-type semiconductor because electrons will be drawn to the region to neutralize the injected charge. An equal number of electrons will be supplied to the bar through the ohmic contacts to keep the entire bar in a space-charge neutral condition. All of this action occurs in about 10^{-12} seconds in a 1 Ω-cm bar of material.[1]

Thus, the applied pulse succeeds in creating very quickly a neutral plasma of holes and electrons under the emitter point. The electric field provided by the dc voltage applied between the ends of the bar causes these holes to drift down the bar toward the collector point.[2] The collector battery is connected to provide an attractive field for holes in the vicinity of the collector point, so that as the injected charge passes beneath the collector point, some of the holes will be extracted, providing a signal current in the collector resistor.[3]

[1] The net space charge actually decays exponentially with time, with a time constant τ equal to ϵ/σ where σ is the conductivity of the material and ϵ is its dielectric constant. The quantity σ/ϵ is called the *dielectric relaxation time* of the solid.

[2] The distribution of majority carriers (electrons) in the bar has to change with time to maintain local space-charge neutrality as the minority carriers (holes) drift. However, the majority-carrier density is so large that these changes have a negligible effect on the experiment.

[3] It is necessary to "form" the collector point in order for this action to occur. Forming techniques are described in Demonstration 4·3.

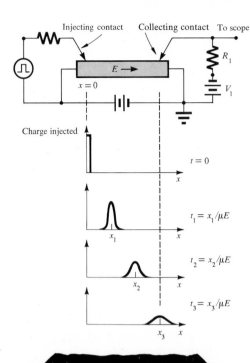

(a) Schematic diagram of experimental apparatus for n-type bar. Experimental data given in Prob. 4·13.

(b) Development of the minority-carrier pulse as it drifts toward the collector point

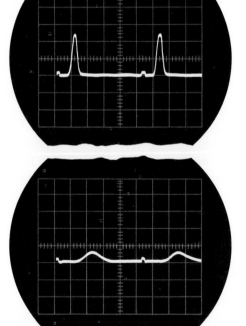

(c) Collector signal obtained in Shockley-Haynes experiment under normal conditions. A p-type bar was used for this experiment, with emitter-to-collector spacing of $\simeq0.5$ cm. Horizontal scale: 20 μs/cm. Vertical scale: -0.5 v/cm. Sweeping field: 7.5 v/cm. Two cycles of the input pulse waveform are shown on the oscilloscope trace.

(d) Collector signal obtained in Shockley-Haynes experiment under normal conditions. A p-type bar was used for this experiment, with emitter-to-collector spacing of $\simeq0.5$ cm. Horizontal scale: 20 μs/cm. Vertical scale: -0.5 v/cm. Sweeping field: 3.75 v/cm. Two cycles of the input pulse waveform are shown on the oscilloscope trace.

FIG. 4·13 SHOCKLEY-HAYNES EXPERIMENT SHOWING SIMULTANEOUS DRIFT AND DIFFUSION OF CARRIERS

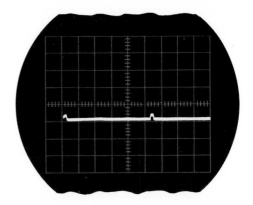

(*e*) *Collector signal obtained in Shockley-Haynes experiment with direction of sweeping field reversed from that used for Fig.* 4 · 13*c*. *All other conditions remain the same.*

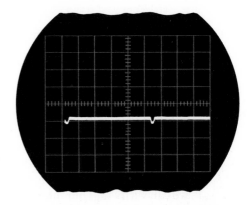

(*f*) *Collector signal obtained in Shockley-Haynes experiment with polarity of input pulse reversed from that used in Fig.* 4 · 13*c*. *All other conditions remain the same.*

The shape of the pulse obtained at the collector can be explained as follows: The initial voltage pulse can be considered to inject essentially a plane of charge at the emitter. This plane of charge is now swept down the bar, diffusing as it goes. The average velocity of the plane of charge is just

$$v = \mu_p E \tag{4·29}$$

If we run along the bar with this velocity, the center of the injected pulse will appear to be stationary, and diffusion of the injected charge about its center will be occurring. Thus, as the injected charge drifts down the bar, it diffuses from being a sharply defined plane into a bell-shaped curve.

The shape of the minority-carrier pulse and its arrival time at the collector can be used to measure μ_p and D_p. To obtain μ_p, we measure the spacing between the emitter and collector points (say L_1) and the time after the initial pulse at which the center of the charge distribution passes beneath the collector point [say t_1. The measurement of t_1 is made directly from Fig. 4 · 13*c* or 4 · 13*d*.] We can then compute the average drift velocity of holes from the formula $v_p = L_1/t_1$. The electric field in the bar can be obtained by measuring the distance between the end contacts (L_2) and the voltage applied between them (V_0). The resulting drift field E is V_0/L_2, and the hole mobility is then

$$\mu_p = v_p/E = L_1 L_2/V_0 t_1$$

We can measure D_p by measuring the spread of the charge distribution as it passes beneath the collector point. If the drift time t_1 is short enough so that the collected pulse appears to be reasonably symmetrical about its center, its form should be $\exp(-x^2/4D_p t_1)$. (See Prob. 4·12.) The actual spatial width of the pulse can be calculated by measuring the time Δt at which the collected pulse is down to $1/e$ of its peak amplitude and then multiplying Δt by the average velocity with which the pulse is being swept down the bar. D_p is then obtained from the formula

$$D_p = (v_p \, \Delta t)^2/4t_1$$
$$= (L_1 \, \Delta t)^2/4t_1{}^3$$

An experimental check of the Einstein relationship is obtained by computing

$$\frac{D_p}{\mu_p} = \frac{L_1}{L_2}\left(\frac{\Delta t}{t_1}\right)^2 \frac{V_0}{4}$$

and showing that the value is approximately 25 mv at room temperature.

Experimental data taken on an n-type bar are given in Prob. 4·13, from which the reader may verify that correct numerical values of μ_p and D_p for lightly doped germanium are obtained from the Shockley-Haynes experiment. However, in addition to values for μ and D, several other things can be learned from the Shockley-Haynes experiment. In particular, we can verify that our interpretation of the bell-shaped pulse as a minority-carrier pulse is correct. To do this, a series of reproduced oscilloscope traces showing additional features of the experiment are given in Figs. 4·13c to f. To supplement the data given in Prob. 4·13, a p-type germanium bar 2 cm long was chosen for these experiments. The spacing between emitter and collector points was approximately 0.5 cm.

In Fig. 4·13c, the normal conditions for the experiment are employed with a sweeping voltage of 15 volts. In Fig. 4·13d, the sweeping voltage is changed to 7.5 volts. This doubles the drift time and increases the pulse width as expected. In this latter case, the leading edge of the minority-carrier pulse is somewhat sharper than the trailing edge because the diffusion time is larger for those carriers that produce the trailing edge.

We can verify that minority carriers are responsible for the bell-shaped pulses by reversing the polarity of the sweeping voltage. This will cause the injected minority carriers to be swept down the bar in a direction opposite to the collector point, so no second pulse should be observed. The oscilloscope trace shown in Fig. 14·3e verifies this point.

We can verify another important feature of the experiment by reversing the polarity of the emitter pulse. No minority-carrier injection then occurs at the emitter point, so no minority-carrier pulse should be observed at the collector. The experimental verification of this point is shown in Fig. 4·13f.[1]

Thus the Shockley-Haynes experiment shows us that it is possible for an emitter point to *inject minority carriers* into a semiconductor and that a fraction of these injected carriers can be collected at a properly made and properly biased collector. This is, in fact, the basic principle of transistor action, and the device shown in Fig. 4·13a is called a filamentary transistor. Modern transistors employ different geometries to obtain high-frequency response, to eliminate the need for a battery to supply the sweeping field, and to collect essentially all of the minority carriers injected by the emitter. However, the basic principle is still the same—minority carriers are injected into a bar at an emitter junction and collected in the attractive field created at a properly biased collector junction.

4·5·3 Summary of drift and diffusion processes. Before studying carrier ballistics in semiconductors further, it may be appropriate to summarize the main points in the preceding sections of this chapter:

1. Because carriers collide frequently with lattice imperfections, they develop a *drift velocity* in response to an applied electric field:

$$\mathbf{v}_p = \mu_p \mathbf{E} \qquad \mathbf{v}_n = -\mu_n \mathbf{E} \tag{4·30}$$

The resulting *drift-current densities* in a semiconductor with n electrons per unit volume and p holes per unit volume are

$$\mathbf{j}_p = q\mu_p p\mathbf{E} \qquad \mathbf{j}_n = q\mu_n n\mathbf{E} \tag{4·31}$$

2. On account of their random thermal motion, both types of carriers may also diffuse in a semiconductor bar. The diffusion process does not require an electric field. In diffusive flow, the particle flux is proportional to the density gradient of the diffusing particles. The resulting *diffusion-current densities* are (in one dimension)

$$j_p = -qD_p\frac{\partial p}{\partial x} \qquad j_n = qD_n\frac{\partial n}{\partial x} \tag{4·32}$$

where p is the hole density and n is the electron density at the plane where j is evaluated. The factors of q are added to convert particle-flux density to current density. The minus sign is absent in the formula for j_n because of the sign on the electronic charge (that is, q in Eqs. (4·32) is a positive constant, 1.6×10^{-19} coul).

3. D and μ are both large-scale manifestations of carrier-collision processes. They may be calculated when the mean free path and mean free time are known, and their ratio is

$$\frac{D_p}{\mu_p} = \frac{D_n}{\mu_n} = \frac{kT}{q} \tag{4·33}$$

---

[1] The space-charge conditions set up near the emitter when the emitter pulse voltage is reversed will be described more fully in Chap. 5.

4·6 MOTION IN PERPENDICULAR ELECTRIC
AND MAGNETIC FIELDS: THE HALL EFFECT

In addition to drift motion in an applied electric field, the motion of free carriers in combinations of electric and magnetic fields is important and has many interesting applications. Perhaps the most basic phenomena in this class is the Hall effect, though the cyclotron resonance mentioned earlier and other phenomena such as magnetoresistance have also played an important role in the development of semiconductor fundamentals. We shall concentrate on the Hall effect, since it is the most widely encountered of these phenomena in practical electronics.

4·6·1 *Mathematics for the Hall effect*. We shall consider the semiconductor bar as shown in Fig. 4·14, which is assumed to be doped with donor impurities, making the conductivity n-type. The bar has an electric field of strength E_x applied in the x direction, and a magnetic field of strength B_z applied in the z direction.

The electric field E_x produces a carrier velocity

$$\mathbf{v}_n = -\mu_n \mathbf{E}_x \tag{4·34}$$

which is perpendicular to the direction of the magnetic field. There is therefore a y-directed force of magnitude

$$F_y = |-q\mathbf{v}_n \times \mathbf{B}| = q\mu_n E_x B_z \tag{4·35}$$

on each electron in the bar. The formula for F_y is correct when B is in webers per square meters and v is in meters per second. Therefore μ must be in square meters per volt-second in this special case.

All of the electrons in the bar move upward in response to this force, leaving an unneutralized layer of donors (that is, bound positive charge) on the bottom of the bar. Thus, a net negative charge is developed on top of the bar, and a net positive charge appears on the bottom. An electric field E_y is developed between these charges which opposes the effect of the magnetic force F_y.

The vertical motion of the mobile charges quickly establishes a steady-state condition in which the electric and magnetic vertical forces balance:

$$|qE_y| = |F_y| \tag{4·36}$$

The trajectories of the electrons are then horizontal once again. The steady-state value of E_y is

$$E_y = \mu_n B_z E_x \tag{4·37}$$

On account of E_y, a voltage will be developed which is positive at the bottom of the bar and negative at the top. The magnitude of this voltage will be

$$V_H = W E_y = (W\mu_n B_z)E_x \tag{4·38}$$

where W is the width of the bar (see Fig. 4·14). The dependence of V_H on μ suggests that the Hall effect will be very small in metals, where μ is very small.

In some applications of the Hall effect to be described later, it is convenient to rewrite Eq. (4·38) in the form

$$V_H = I_x B_z \frac{R_H}{t} \qquad (4\cdot39)$$

R_H is called the Hall constant, and its value is

$$R_H = \frac{1}{qn} \qquad (4\cdot40)$$

where q is the electronic charge and n is the free-electron density. I_x is the current flowing through the bar, t its thickness, and B_z the magnetic-field strength.

To get an idea of the magnitude of this effect, we evaluate V_H for an n-type germanium bar in which $N_d < 10^{17}$, so $\mu_n \approx 3{,}900$ cm²/volt-sec. We assume $W = 0.01$ in., $B_z = 10^3$ gauss $= 0.1$ weber/m², $E_x = 5$ volts/cm. Then, converting all quantities to mks units, we have

$$V_H = (2.5 \times 10^{-4})(3{,}900 \times 10^{-4}) \times 10^{-1} \times 500 \cong 5 \text{ mv}$$

The effect is therefore quite sizable.

An interesting phenomenon now occurs if a p-type bar is substituted for the n-type bar: the polarity of the Hall voltage *reverses*. We may readily explain this by noting that now positive holes are driven to the top surface instead of electrons, and a layer of bound-negative charge (acceptor atoms) is uncovered on the bottom of the bar.

The fact that the Hall voltage changes sign when the bar is changed from n-type to p-type is an important fact indeed, for it supports our concept of the hole as a positively charged classical particle in a semiconductor. If we were to assume that the motion of a hole could be equally well represented

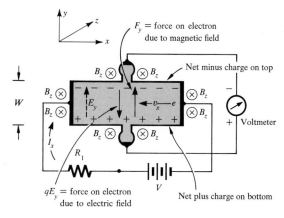

FIG. 4·14   THE HALL EFFECT IN n-TYPE MATERIAL

by the field-stimulated motion of a classical valence electron, then we would get the same sign for the Hall voltage of both n- and p-type bars. This experiment thus supports the position that the concept of the hole as a charge carrier must be adopted if we wish to employ classical physics in the derivation of our formulas.

As a practical matter, the sign of the Hall voltage can be used to determine the conductivity type of semiconductor material which is being investigated. We shall consider this briefly in the next section.

4·6·2 *Applications of the Hall effect in determining electrical properties of semiconductor crystals.* The Hall effect is a very important tool for studying some of the basic electrical properties of semiconductor materials. By combining measurements of the Hall effect and conductivity of a sample of material, one can deduce the carrier densities, mobilities, and various other quantities of interest. We shall consider how this may be done in this section.

Let us first reconsider Eq. (4·39), in which the Hall voltage V_H is written as

$$V_H = \frac{R_H I_x B_z}{t}$$

Now I_x, B_z, and t can be specified by the experimenter, and the corresponding V_H can be measured. Then from Eq. (4·39), R_H can be determined:

$$R_H = \frac{V_H t}{I_x B_z} \qquad (4·41)$$

But, according to Eq. (4·40), R_H depends only on q and the carrier density n in an extrinsic n-type sample. Thus, the carrier density in the sample can be determined:[1]

$$n = \frac{1}{q R_H} \qquad (4·42)$$

By making Hall voltage measurements over a range of temperatures, n can be obtained as a function of temperature. This permits one to plot the concentration of ionized donor atoms as a function of temperature. From such a plot it is possible to determine both the donor concentration and the binding energy of the donor center.

By combining Hall measurements with conductivity measurements, the carrier mobility can also be obtained. Again, dealing with an extrinsic n-type sample, we have

$$\sigma_n = q \mu_n n \qquad (4·43)$$

and therefore
$$\mu_n = \sigma_n R_H \qquad (4·44)$$

with a similar result for the mobility μ_p in a p-type semiconductor. Authentic numerical data for evaluating n and μ_n by this method are given in Prob. 4·17.

Again, by measuring σ and R_H as functions of temperature, the temperature dependence of μ can be established. Also, at a given temperature, measurements of

---

[1] The derivation above is somewhat oversimplified, because it assumes that *all* electrons *always* have their average drift velocity. When the statistical distribution of drift velocities is properly included, one finds

$$n = \frac{3\pi}{8} \frac{1}{q R_H}$$

σ_n and R_H in samples of differing impurity content will give μ_n as a function of the doping concentration. This is the usual way in which curves such as those shown in Fig. 4·3 are obtained.

Precise experimental measurements of σ and R_H are usually made on a sample such as the one shown in Fig. 4·15. This sample is cut from a bar of the material under investigation with an ultrasonic drill. The side arms are used to make large-area contacts for measuring σ and V_H. This is essential for minimizing various problems which occur when point contacts are used.

Of course, a large magnet with precisely known magnetic field is used in place of the small hand magnet in experimental work. However, the hand magnet has enough magnetic field (about 500 gauss) to cause an appreciable Hall voltage on the particular sample shown and can be used in the simple demonstration of the Hall effect shown in Fig. 4·16 (a less sophisticated demonstration is suggested in Demonstration 4·4).

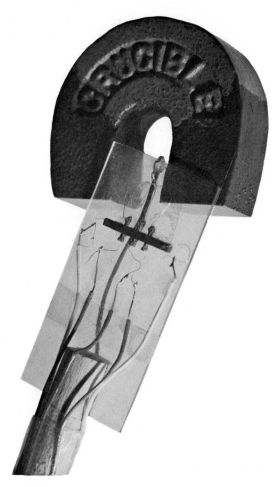

FIG. 4·15 CLOSE-UP VIEW OF HALL SAMPLE AND MAGNET

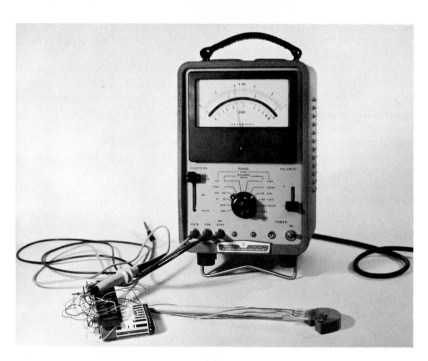

FIG. 4·16 A DEMONSTRATION OF THE HALL EFFECT

4·6·3 *Applications of the Hall effect in electronics.* Applications of the Hall effect in practical electronics center around two different aspects of the phenomenon.

1. For a fixed I_x, the Hall voltage V_H is proportional to the magnetic field B_z (or the B_z component of a generally oriented B). Thus, the Hall effect can be used as a magnetic-field meter. With some care in circuitry and the selection of appropriate semiconductor materials, magnetic fields on the order of a fraction of a gauss may readily be measured. Hall-effect flux meters find important applications in exploring the magnetic fields around machinery, large magnets, and in geophysical work where fluctuations in the earth's magnetic field are caused by oil deposits and other items of interest beneath the surface of the earth. The Hall effect has also been used as the basis of a clip-on ammeter.

2. If both I_x and B are controlled by input variables, then the output Hall voltage is proportional to the product of the two input variables. In this way, the Hall effect is used as a linear *multiplier*. Several different applications of the multiplying feature of the Hall effect are given in Table 4·2.

4·7 RECOMBINATION

Before proceeding to the study of *pn* junctions and junction devices such as transistors and diodes, we need to study the concept of recombination

TABLE 4·2   APPLICATIONS OF THE HALL EFFECT AS A MULTIPLIER

Application	Input 1 (current I_x)	Input 2 (magnetic field B)	Input 3 (mechanical displacement of element)	Hall voltage represents
Multiplier	Variable voltage or current	Variable B achieved by varying current through field winding	None	Input 1 $\times$ Input 2
Wattmeter	Load voltage, ac or dc	Load current, ac or dc	None	Real power or volt amps
Varmeter	Load current, ac	Load voltage, ac	None	Imaginary power
Resolver or rotary position transducer	Constant amplitude, ac or dc	Permanent magnet	Rotary	$f(\theta)$, where θ is the angle between the flux B and control current
Magnetic tape or drum reader	Constant amplitude, ac or dc	Magnetic field from information stored on tape	Rectilinear or rotary	Signal on tape or drum
Wave analyzer	Signal to be analyzed	Sweep signal generator	None	Proportional direct current at each harmonic
Variable attenuator	Input signal	Variable control current	None	Attenuated signal

in some detail. The idea of recombination was introduced in Chap. 3 as a balancing process for carrier generation. It is thus not directly responsible for carrier motion. However, the recombination process always affects the transportation of carriers from one place to another, and it is appropriate to consider it in this chapter.

4·7·1   *Fundamental ideas.* It will be recalled from Chap. 3 that there are two processes which regulate the free carrier densities in a uniform semiconductor bar: generation and recombination. A recombination event occurs when a free electron drops into a hole (or vacant valence bond) to complete the valence structure on a local basis. Recombination thus reduces the number of free electrons and holes. Other processes such as thermal- or photogeneration produce hole-electron pairs, and in a steady-state condition the average generation and recombination rates balance.

Now, in most physical situations of interest to us, the recombination rate per unit volume (that is, the number of hole-electron pairs which recombine per second per unit volume) can be written as

$$R = rnp \tag{4·45}$$

where r is a number which can be determined when the physical processes responsible for the recombination are specified. n and p are the instantaneous densities of electrons and holes, which we shall assume to be uniform throughout the bar for the present.

It was also shown that the average thermal-generation rate can be expressed in terms of the intrinsic carrier density by

$$g(T) = rn_i{}^2 \tag{4·46}$$

so that recombination and thermal generation are in dynamic balance when

$$R = g \quad \text{or} \quad np = n_i{}^2 \tag{4·47}$$

Let us now consider a particular, uniform semiconductor bar in which the steady-state electron and hole densities are n_0 and p_0, respectively. Then

$$n_0 p_0 = n_i{}^2 \tag{4·48}$$

Now, assume that a short flash of light falls on the entire semiconductor bar. The light will generate holes and electrons in equal concentrations. Immediately after the flash, the total electron and hole concentrations will be

$$n = n_0 + n_v \quad p = p_0 + p_v \tag{4·49}$$

where n_v and p_v are the excess concentrations created by the light flash (assumed to be uniformly distributed through the bar).

Now, the recombination rate immediately after the light flash is

$$R = rnp = r(n_0 + n_v)(p_0 + p_v) \tag{4·50}$$

while the thermal-generation rate is still

$$g = rn_i^2 \qquad (4 \cdot 51)$$

Under these circumstances, $R > g$, and there will be a *net recombination rate U*. The general form for U is

$$U = R - g = r(np - n_i^2) \qquad (4 \cdot 52)$$

Immediately after the flash, U will have the value

$$U = r[(n_0 + n_v)(p_0 + p_v) - n_i^2]$$

The term "net recombination rate" is used to emphasize that both the generation and the recombination processes are still active, though temporarily out of balance.

Because there is a net recombination rate, the values of n and p will decay with time. The decay will continue until n and p have been reduced to n_0 and p_0, at which point recombination and thermal generation will again be in balance.

4·7·2 Mathematics of simple recombination processes. In order to study the recombination process mathematically, we observe that the rate of loss of holes by recombination is the same as the rate of loss of electrons by recombination, so that

$$U = -\frac{dp}{dt} = -\frac{dn}{dt} \qquad (4 \cdot 53)$$

Using Eq. (4 · 53) with Eq. (4 · 52), we find[1]

$$-\frac{dp}{dt} = -\frac{dn}{dt} = r(np - n_i^2) \qquad (4 \cdot 54)$$

Equation (4 · 54) is actually the general equation describing the balance between generation and recombination, in terms of the parameter r (as yet unspecified), the instantaneous values of n and p, and the thermal-equilibrium density n_i. When $np > n_i^2$ (after the flash of light, for example), then Eq. (4 · 54) indicates that p and n decrease with time until $np = n_i^2$, at which point generation and recombination are in balance. If by some means the np product is depressed below the value n_i^2, there will be a *net generation rate* of carriers. That is, dp/dt and dn/dt will be positive, so n and p will increase with time until the np product reaches n_i^2.

Recombination in n-type material. Generally, we are interested in the net recombination rate in material which is basically either *n*-type or *p*-type.

---

[1] Photogeneration can be included in Eq. (4 · 54) by simply expanding the equation to read

$$-\frac{dp}{dt} = r(np - n_i^2) - g_v$$

where g_v represents the photogeneration rate.

If we consider n-type material for the moment, then the thermal-equilibrium densities n_0 and p_0 used in the previous argument become

$$n_0 = n_n \qquad p_0 = p_n \qquad (4 \cdot 55)$$

Now, a flash of light will create holes and electrons in equal numbers ($p_\nu = n_\nu$), and recombination removes holes and electrons in equal numbers. Therefore, since the semiconductor bar was space-charge neutral before the light flash, it will also be space-charge neutral at any time after the flash. Thus, even though n and p are functions of time after the flash, we still have space-charge neutrality at any instant:

$$n(t) = N_d + p(t) \qquad (4 \cdot 56)$$

Therefore, the general recombination formula becomes

$$-\frac{dp}{dt} = r[(N_d + p)p - n_i^2] \qquad (4 \cdot 57)$$

Equation (4·57) is a nonlinear, ordinary differential equation describing how p varies with t after the light flash. We seek its solution under the conditions that $p(t)$ has an initial value

$$p(0) = p_n + p_\nu \qquad (4 \cdot 58)$$

The equation itself ensures that the final value $p(\infty)$ will be p_n.

The solution of Eq. (4·57), which applies for most cases of interest, is found as follows. First, suppose the light flash creates only a small number of carrier pairs compared to N_d. Then Eq. (4·57) may be simplified to

$$-\frac{dp}{dt} = rN_d(p - p_n) \qquad (4 \cdot 59a)$$

We then define a minority-carrier *lifetime* by the formula

$$\frac{1}{\tau_p} = rN_d$$

so that Eq. (4·59a) may be rewritten as

$$\frac{dp}{dt} = -\frac{p - p_n}{\tau_p} \qquad (4 \cdot 59b)$$

The solution to Eq. (4·59b) is

$$p = p_n + Ae^{-t/\tau_p} \qquad (4 \cdot 60)$$

To evaluate A, we note that at $t = 0$,

$$p = p_n + p_\nu$$

so that

$$A = p_\nu \qquad p = p_n + p_\nu e^{-t/\tau_p} \qquad (4 \cdot 61)$$

Alternately, Eq. (4·61) may be recast as

$$p - p_n = p_\nu e^{-t/\tau_p} \qquad (4 \cdot 62)$$

This equation may be interpreted by saying that the *excess* carrier population dies away to zero in an exponential fashion. The decay process is characterized by the parameter τ_p, the hole lifetime. The physical processes which determine τ_p will be described in Sec. $4 \cdot 7 \cdot 3$.

Recombination in p-type material. An entirely similar argument may be used to study the recombination of electrons in *p*-type material. The linearized recombination law becomes

$$-\frac{dn}{dt} = \frac{n - n_p}{\tau_n}$$

so long as $n \ll N_a$. The electron lifetime in *p*-type material, τ_n, is defined by

$$\tau_n = \frac{1}{rN_a}$$

Then an excess density of electrons created by a momentary light flash will decay with time according to

$$n = n_n + n_\nu e^{-t/\tau_n}$$

Discussion of the analysis. There is one significant point about the recombination phenomenon which tends to get lost in the mathematical analysis. The assumptions which are made to linearize and solve the basic Eq. $(4 \cdot 57)$ focus attention on the minority carriers. It is assumed that the majority-carrier concentrations do not change significantly when the equilibrium between R and g is disturbed. This ultimately leads us to characterize the recombination process by a *minority*-carrier lifetime. Of course, every recombination involves a majority and a minority carrier, so it would perhaps be more correct to specify a "recombination lifetime" for the type of material being used.

$4 \cdot 7 \cdot 3$ *Physical basis of the recombination process.* The recombination of electrons and holes may take place in several different ways. In a given piece of semiconductor material, there will usually be more than one process by which recombination occurs, though one process may frequently be the dominant one.

In some respects, the simplest process to visualize is the "direct-recombination" process, by which an electron drops directly into a hole. This process is merely the inverse of the photoelectric effect described in Chap. 3. When a direct recombination occurs, a photon of energy $\mathcal{E}_i$ is given off. This photon may be reabsorbed in the crystal or emitted. The direct-recombination process is important in certain III-V-compound semiconductors, such as gallium arsenide. It is of less importance in germanium and silicon. Experimental evidence for direct recombination will be given later when *pn*-junction electroluminescence is discussed in Chap. 6.

In germanium and silicon, recombination occurs primarily at so-called "recombination centers." A recombination center may be thought of as a defect in the crystal which can trap a moving electron or hole and then bind it tightly, so that, when its recombination partner comes along, the recombination can be effected. Several different types of defects behave in this way. For example, small traces of gold or oxygen or places where lattice atoms are missing can act as recombination centers.

When recombination occurs through a recombination center, a photon of energy $\mathscr{E}_i$ is usually not generated, even though the recombination process is ultimately accompanied by the liberation of $\mathscr{E}_i$ joules of energy. As an electron moves into the field of a gold atom in the silicon lattice, for example, it will "spiral" in toward the atom, giving up energy to the lattice in the form of heat.[1] It will give up an energy of nearly $\mathscr{E}_i/2$ before it reaches its ground state in the trap. When a hole is trapped to complete the recombination process, its excess energy will also be transferred to the crystal as heat, so that the entire ionization energy $\mathscr{E}_i$ is given up as heat during the recombination process.

There are theories from which one may obtain an expression for the net recombination rate U when the recombination mechanism is specified. For gold atoms in the silicon lattice, for example, U may be written approximately as

$$U = \frac{np - n_i^2}{\tau_p n + \tau_n p} \qquad (4 \cdot 63a)$$

in which

$$\tau_p = \frac{2.5 \times 10^7}{N_{Au}} \qquad (4 \cdot 63b)$$

and

$$\tau_n = \frac{1.6 \times 10^7}{N_{Au}} \qquad (4 \cdot 63c)$$

where N_{Au} is the density of gold atoms. In heavily doped n-type material, $n \cong N_d \gg p$, so U reduces to

$$U = \frac{p - p_n}{\tau_p} \qquad (4 \cdot 64)$$

in which τ_p is determined by the number of gold atoms in the silicon lattice. A similar simplification can be made for p-type material.

Similar forms of U can be derived for oxygen impurities, missing host atoms in the lattice, etc. It is usually these imperfections which determine the recombination lifetime in germanium and silicon crystals.

Very good crystals of germanium may have lifetimes on the order of 1 to 10 msec. Typical lifetimes in silicon crystals are on the order of 10–50 μsec. Lifetimes of 1 msec are required in certain types of devices (for example, power rectifiers), and lifetimes of 1 nanosec are required in others (for example, computer diodes). The value of lifetime which one would like to have depends on the device being considered and the application to which it is to be put, as we shall see more clearly later.

It is therefore of considerable importance to be able to control the recombination lifetime. As a practical matter, present methods of lifetime control consist in using the best obtainable crystal and then doping it with atoms which act as recombination centers to reduce the lifetime to the desired value.

Recombination centers may also be produced in semiconductor materials by

---

[1] The gold atom may be considered to have many excited orbitals through which the electron drops to reach its ground state in the trap.

irradiating them with high-energy particles. Since the lifetime is always decreased by these processes, the characteristics of a given semiconductor device will change drastically upon radiation of high-energy particles. This fact is of great concern to engineers attempting to design systems using semiconductor components for applications in nuclear instrumentation and deep space probes.

4·7·4 *Measurement of lifetime in semiconductor crystals.* One of the most popular methods of measuring lifetime in a semiconductor crystal is based on the creation of a nonequilibrium density of hole-electron pairs by means of a light flash. The apparatus is shown diagrammatically in Fig. 4·17 and described in detail in Demonstration 4·5. A rectangular bar of the semiconductor material in which we wish to measure the minority-carrier lifetime is suitably polished and ohmic contacts are placed on its ends. The sample is placed in a circuit such as the one shown in Fig. 4·17. A light source which emits periodic bursts of light is then shined on the filament, and the voltage across the sampling resistor R_s is observed with an oscilloscope.

The effect of the light burst is to create excess carriers in the bar, thus increasing the conductivity of the bar and reducing the voltage drop across it. The excess-carrier densities decay according to an $\exp(-t/\tau)$ law, resulting in an exponential rise of the voltage across the semiconductor and an exponential decay of the voltage across R_s. The time constant for this exponential can readily be measured and is equal to the recombination lifetime in the semiconductor sample.

A reproduction of an actual oscilloscope trace obtained with the apparatus in Fig. 4·17 is shown in Fig. 4·18. The light source for this experiment was a General Radio Strobotac emitting 100 pulses of light per second. Each pulse is about 10^{-8} sec in duration. The horizontal scale on the oscilloscope trace is 10 μsec/cm; the vertical scale is 50 mv/cm (which is unimportant). On the basis of Fig. 4·18, we can estimate the recombination lifetime in this particular sample to be 15 μsec.

This procedure for measuring lifetime becomes poor when the lifetime

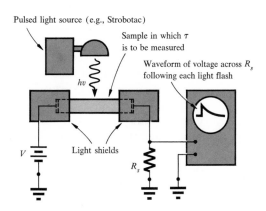

FIG. 4·17   APPARATUS FOR MEASUREMENT OF LIFETIME IN A SEMICONDUCTOR BAR

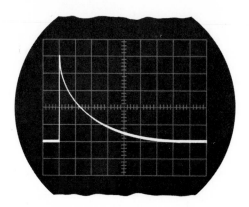

FIG. 4·18   OSCILLOSCOPE TRACE OBTAINED FROM APPARATUS SHOWN IN FIG. 4·17;
HORIZONTAL SCALE: 10 μs/cm

is very short and/or the doping density of the sample is large, for then the change in conductivity of the sample is difficult to detect. Other methods of measuring lifetime have been devised for these cases.

4·8 PHOTOCONDUCTORS

The fact that the carrier densities in a semiconductor bar are affected by light (of the correct wavelength) is the basis of an important class of devices called photoconductors. In fact, the apparatus shown in Fig. 4·17 for measuring lifetime in a semiconductor bar contains, in effect, a photoconductor. If the pulsed light source used in that experiment were replaced by a steady light source, then the resistance of the semiconductor sample would be decreased as long as the light source were operating.

To make these statements more quantitative, consider an n-type semiconductor bar in which the electron density is

$$n_n = N_d + p \cong N_d \qquad (4·65)$$

The hole density in thermal equilibrium is p_n, and the recombination lifetime for carriers in the n-type bar is τ_p. The bar is considered to be in thermal equilibrium, so the net thermal-generation rate is exactly balanced by the recombination rate.

Suppose now that a steady light in the absorption range for the particular semiconductor is shined on the bar (see Fig. 4·19). The light will generate carrier pairs in the bar, increasing the density of electrons and holes above their thermal-equilibrium values. The recombination rate will also be increased, and a new equilibrium condition will be reached in which the sum of thermal generation and photogeneration is balanced by recombination.

We may calculate the increased density of electrons and holes in the presence of the light source as follows. Suppose the light generates g_ν electron-hole pairs per second per unit volume. Then, when the carrier densities have reached equilibrium, we must have

$$g_\nu + g(T) = R = rnp \tag{4 \cdot 66}$$

This may be rewritten as

$$g_\nu = r(np - n_i^2) \tag{4 \cdot 67}$$

and if we utilize the assumption that the bar is n-type

$$g_\nu = rn_n(p - p_n) = \frac{p - p_n}{\tau_p} \tag{4 \cdot 68}$$

Therefore, in the presence of the light, the new equilibrium carrier densities are

$$p = p_n + g_\nu\tau_p \qquad n = p + n_n = n_n + g_\nu\tau_p \tag{4 \cdot 69}$$

Now, the conductivity of the semiconductor bar is

$$\sigma = q(\mu_n n + \mu_p p) \tag{4 \cdot 70}$$

This formula can be written to emphasize the change in conductivity due to the light, as

$$\sigma = q(\mu_n n_n + \mu_p p_n) + q(\mu_n + \mu_p)g_\nu\tau_p = \sigma_0 + \Delta\sigma_\nu \tag{4 \cdot 71}$$

If the initial conductivity σ_0 is low, then the change in conductivity caused by the light may be quite appreciable. The materials problem associated with photoconductors is therefore one of finding a material which can be readily fabricated into many different device geometries, has a high intrinsic resistivity, and is sensitive to light in the desired range.

Common commercial photoconductors are made from cadmium sulphide, lead sulphide, and selenium. Lead sulphide and cadmium sulphide photoconductors are usually made by evaporating the material onto an insulating substrate and then applying electrodes on the ends, as shown in Fig. 4·20a. Selenium photocells are made by winding two fine nickel wires around an insulating card and then painting a thin layer of molten selenium onto one side of the insulator, as in Fig. 4·20b. The selenium is then allowed to cool properly, so that the wires are embedded in a semiconducting selenium

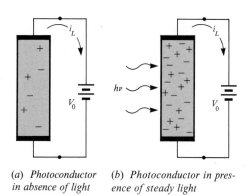

(a) Photoconductor in absence of light   (b) Photoconductor in presence of steady light

FIG. 4·19   A SIMPLE PHOTOCONDUCTOR

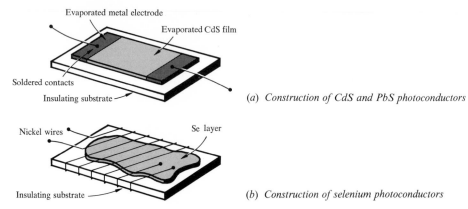

(a) Construction of CdS and PbS photoconductors

(b) Construction of selenium photoconductors

FIG. 4·20 PHYSICAL CONSTRUCTION OF SOME SIMPLE PHOTOCELLS

layer, making ohmic contact with it. Light falling anywhere on the selenium layer causes a local increase of conductivity and, therefore, of the conductance measured between the wires.

Selenium cells of this type can be made to have a dark resistance on the order of 10^6 ohms. The resistance may be decreased to 5 percent of its dark value upon exposure to direct sunlight. Commercial cadmium sulphide photocells may also have dark resistances of 10^6 ohms and operating resistances of 10^3 ohms when placed at a distance of 6 inches from a small flashlight. Photoconductivity is therefore a sizable effect, and it has a number of important industrial applications.

4·9 SUMMARY

In this chapter we have seen that holes and electrons which are in thermal equilibrium in a semiconductor lattice execute a random thermal motion. This motion is visualized as the high-speed traversal of a series of short, straight paths between collision points. The random motion leads to the following effects:

1. If an electric field is applied to the crystal, the holes will drift in the direction of the field (electrons opposite to it), still executing their rapid, random thermal motion about their drifting average position. The drift velocities are

$$v_p = \mu_p E \qquad v_n = -\mu_n E$$

where E is the electric field, and μ_p and μ_n are the hole and electron mobilities. The resulting drift-current densities in a semiconductor with n electrons per unit volume and p holes per unit volume are

$$j_p = q\mu_p p E \qquad j_n = q\mu_n n E$$

The total drift-current density is

$$j = j_p + j_n = q(\mu_p p + \mu_n n)E$$

2. The conductivity of a semiconductor crystal is defined by

$$\sigma = \frac{j}{E}$$

and can be expressed in terms of p and n as

$$\sigma = q(\mu_p p + \mu_n n)$$

The conductivity of a semiconductor crystal increases rapidly with temperature when the thermal-generation mechanism is an important source of carriers. This leads to the possibility of thermistor action.

3. On account of their random thermal motion, both types of carriers also diffuse in a semiconductor bar. The diffusion process does not require an electric field. In diffusion flow, the particle flux is proportional to the density gradient of the diffusing particles. The resulting diffusion-current densities are, in one dimension,

$$j_p = -qD_p \frac{\partial p}{\partial x} \qquad j_n = qD_n \frac{\partial n}{\partial x}$$

where D_p and D_n are the hole and electron diffusion constants.

4. D and μ are both large-scale manifestations of carrier-collision processes. They may be calculated when the mean free path and mean free time are known, and their ratio is

$$\frac{D_p}{\mu_p} = \frac{D_n}{\mu_n} = \frac{kT}{q}$$

We have described experiments which verify these theoretical ideas and allow us to measure diffusion constants and mobilities.

5. We have also seen that the motion of holes and electrons in a crystal is sensitive to the presence of a magnetic field. One manifestation of this behavior is the Hall effect. The Hall effect is useful as a tool for physical investigations and as a transducer for electronic control and measurement.

6. In addition to their dynamics of motion, holes and electrons in a semiconductor crystal are continuously being generated and recombining. The net recombination rate U can be expressed as

$$U = R - g$$

where R is the recombination rate per unit volume and g is the generation rate per unit volume. Usually, R can be written as

$$R = rnp$$

and the thermal generation rate can be written as

$$g(T) = rn_i^2$$

so that the net recombination rate becomes

$$U = r(np - n_i^2) - g_e$$

where g_e is the generation rate due to external influences such as light, particle bombardment, and so on.

In all cases

$$U = -\frac{dp}{dt} = -\frac{dn}{dt}$$

Furthermore, in an n-type sample, where $n \simeq n_n$, we write

$$U = rn_n(p - p_n) - g_e$$

We define a minority-carrier lifetime

$$\tau_p = \frac{1}{rn_n}$$

and speak of a minority-carrier net recombination rate

$$-\frac{dp}{dt} = \frac{p - p_n}{\tau_p} - g_e$$

Similarly, in p-type material

$$-\frac{dn}{dt} = \frac{n - n_p}{\tau_n} - g_e$$

In both cases, we assume that the majority-carrier densities are not seriously affected by fluctuations in the minority-carrier density.

The dependence of the equilibrium carrier densities on the external generation rate g_e leads to one experimental method of measuring τ_p and to the possibility of making photoconductors and radiation detectors.

REFERENCES

Adler, R. B., et al: "Introduction to Semiconductor Physics," SEEC vol. 1, John Wiley and Sons, Inc., New York, 1964, chap. 1 and Laboratory Experiments Appendix.

Valdez, L. B.: "The Physical Theory of Transistors," McGraw-Hill Book Company, New York, 1961, chaps. 8 and 9.

Shockley, W.: "Electrons and Holes in Semiconductors," D. Van Nostrand Company, Inc., Princeton, N.J., 1950, chaps. 1, 3, and 11.

Bube, R. H.: "Photoconductivity of Solids," John Wiley and Sons, Inc., New York, 1960.

Shive, J. N.: "The Properties, Physics and Design of Semiconductor Devices," D. Van Nostrand Company, Inc., Princeton, N.J., 1960.

DEMONSTRATIONS

4·1 *Conductivity of nearly intrinsic germanium vs. temperature.* By making conductivity measurements on a nearly intrinsic sample of germanium, thermistor

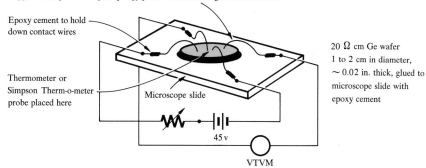

Copper wires, prebent to give springy pressure contact on germanium surface

Epoxy cement to hold down contact wires

Thermometer or Simpson Therm-o-meter probe placed here

Microscope slide

20 Ω cm Ge wafer 1 to 2 cm in diameter, ~ 0.02 in. thick, glued to microscope slide with epoxy cement

45 v

VTVM

FIG. D4·1 CONDUCTIVITY OF NEARLY INTRINSIC GERMANIUM VS. TEMPERATURE.

Notes: 1. *The soldering iron or other heat source should be placed near but not on the germanium wafer. The underside of the microscope slide can be heated.*
2. *Phosphor-bronze contact wires are somewhat better for the contacts. The contact points should be cut with a pair of diagonal cutters and etched or filed to produce a well-defined contact. All four contacts should be in line.*

action can be demonstrated (the negative temperature coefficient of ρ), and the energy required to create hole-electron pairs can be estimated. Suitable apparatus for the demonstration is shown in Fig. D4·1. A wafer of high-resistivity germanium (20 ohm-cm, or higher at room temperature) is used as the sample. The sample can be mounted on a microscope slide with epoxy cement or a casein glue and pressure contacts made as suggested in Fig. D4·1. A constant current of approximately 1 ma is passed through the outer two probes and a voltage difference is measured across the inner two probes. The sample is then heated with a soldering iron to a temperature of about 50°C and allowed to cool. The sample temperature can be conveniently monitored with a Simpson THERM-O-METER with its temperature-measuring thermocouple contacting the sample at a point near the voltage probes. If the voltage is measured at a series of temperatures and plotted as a function of temperature on semilog paper, E_g can be estimated.

The sample resistivity can be determined with reasonable accuracy if the contact points shown in Fig. D4·1 are uniformly spaced. For this case, the resistivity of the sample is

$$\rho \simeq \frac{\pi}{\ln 2}\frac{V}{I} W$$

4·2 *Thermistors.* A good kit of commercial thermistors for demonstration purposes is the Fenwall G200 Experimental Kit. It may be obtained from a local supplier or from McKnight and Company, 1436 El Camino Real, Menlo Park, Calif. The differential negative resistance of a "fast" thermistor can be displayed with a Tektronix type 575 Transistor Curve Tracer or the simpler scheme shown in Fig. D4·2. A Wheatstone-bridge thermometer also provides an interesting demonstration.

4·3 *Shockley-Haynes experiment, demonstrating carrier drift and diffusion.* The most important demonstration of physical principles which can be performed in this course is the Shockley-Haynes experiment. This experiment illustrates diffusion and drift phenomena very nicely and also forecasts the importance of narrow base

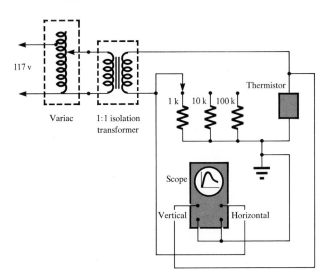

FIG. D4·2   A SIMPLE CIRCUIT FOR DISPLAYING THE *V-I* **CHARACTERISTIC OF A THERMISTOR**

width in transistors. The preparation of this demonstration takes time, but it is worth it.[1]

A diagram of the experimental arrangement is shown in Fig. D4·3, together with a reproduction of the output waveforms which are obtained when the experiment is working properly. The pulse source is an HP 212A, or equivalent, pulser, with the output pulse adjusted to about 0.5 μsec. The sweeping field should be variable between 2 and 10 volts/cm.

The germanium bar for the experiment must have a long surface lifetime so that the injected pulse has appreciable amplitude when it arrives at the collector. A detailed set of preparation instructions is given below. The germanium wafer should have a bulk lifetime on the order of 100 μsec to begin with. A suitable wafer is about 1 to 2 cm in diameter and 0.02 to 0.1 in. thick, and can be obtained from Monsanto, Semi-Metals, or Cenco (Semiconductor Demonstration Kit number 80394).

The wafer is prepared as follows:

1. Polish one side of the wafer on a wet cloth or a wet glass plate with polishing alumina such as Linde A lapping powder (a fine emery paper is a possible substitute). After polishing, rinse the sample in distilled water and dry on a clean filter paper, always handling the sample with a clean pair of tweezers.

2. Coat the polished rod with polystyrene cement, except for about 0.5 cm on each end.

3. Roughen the uncoated ends of the rod with steel wool.

4. Make contacts on the end surfaces. This may be done in either of two ways. The most satisfactory way is to electroplate gold or rhodium onto the roughened ends. However, a solder contact can be applied. To do this, a coreless solder (for example, American Smelting and Refining Company NEATPAK) and an appropriate soldering flux should be used. The soldering flux can be made by mixing two parts zinc chloride with one part ammonium chloride and diluting with water. Several

---

[1] Details for constructing the Shockley-Haynes apparatus are also given in W. J. Leivo, "Teaching Transistor Physics," *Am. J. Phys.*, vol. 22, Dec., 1954.

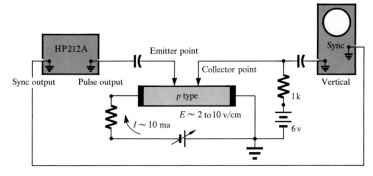

(a) *Experimental arrangement*

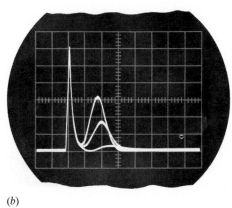

(b)

FIG. D4·3 **THE SHOCKLEY-HAYNES EXPERIMENT**

drops of the flux can be applied to the sample with an eyedropper. The solder is then melted on the tip of a clean, well-tinned soldering iron and applied to the roughened surface. The iron should be applied long enough to insure that the germanium becomes well heated.

5. Remove polystyrene on the middle of the sample with toluene.

At this point there are two possible ways to proceed. A sample can be prepared with a minimum of further treatment; though its surface properties will not be stable and repreparation will be necessary, unless the experiment is done immediately. The minimum further treatment consists in etching the sample in boiling 3 percent hydrogen peroxide or C.P.4 (directions given below) for 2½ to 3 min. The wafer is then ready for contact application.

Or, a surface preparation which has excellent long-term stability (at least five years on a well-prepared sample) can be made as follows:[1]

6. Cover the electrodes made in step 4 with polystyrene cement.

7. Etch the rod in C.P.4 for 2½ to 3 min. It is convenient to attach the sample to a platinum wire with polystyrene cement for this operation. C.P.4 is a standard germanium-etching solution, which is prepared by mixing five parts nitric acid, three parts hydroflouric acid, and three parts acetic acid.

[1] A prepared sample of this type with ohmic contacts attached to its ends can also be obtained from the author for a nominal charge. Address inquiries to the author, Electrical Engineering Department, Stanford University, Stanford, California.

8. Rinse the sample in distilled water and dry on clean filter paper.

9. Hang platinum wire and the sample in toluene to loosen the sample. Remove the sample and rinse it thoroughly in toluene, making sure it is very clean.

10. Prepare a saturated antimony oxychloride suspension by placing a few crystals of antimony chloride in distilled water.

11. Place a negative electrode of tantalum or platinum in the suspension and connect it to the negative terminal of a 1½-volt cell. Connect the positive terminal to a pair of tweezers, holding the sample by an electrode. Lower the sample into the suspension until all of the etched surface has been covered. Allow the sample to remain in the suspension for 5 min.

12. Remove the sample from the suspension, rinse thoroughly with distilled water, and dry on clean filter paper.

13. Place the sample in hot synthetic ceresin wax and allow the sample to come to the temperature of the molten wax.

14. Remove the sample from the wax, and while the sample is kept warm over a hotplate, remove the excess wax with a clean filter paper. When the sample cools, it is ready for experimentation.

Four contacts are required for the experiment. A mounting arrangement like that shown in Fig. D4·1 is suitable. The two outer contacts (either copper or phosphor-bronze wires) are brought down on the prepared electrodes. The two inner contacts are the emitter and collector points. They must break through the wax and press solidly against the sample. These points should be a few millimeters apart and approximately in line with the end contacts.

It is necessary that the points make rectifying contacts to the semiconductor. This can be checked with an ohmmeter. A forward-to-backward resistance ratio of 100 or more for each point is desirable, though the experiment will work with less.

The diode with the highest resistance ratio is selected as the collector. It may work adequately without further adjustment, though if a noisy signal results the collector contact must be "formed." This is done by charging a 1-μfarad capacitor to 100 volts and then discharging it through the diode, making sure that the direction of positive current flow in the capacitor discharge is the same as the direction of positive current flow when the ohmmeter reads low resistance.

4·4 *Hall effect.* Demonstration of the Hall effect is important in that it provides direct experimental confirmation that two types of mobile carriers exist. The apparatus is not complicated to construct. Using either pressure contacts of the type

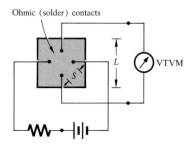

Ohmic (solder) contacts

Probe spacing $S \simeq 0.3$ cm

Sample thickness $\cong 0.01$ to 0.1 in.

Sample length $L \sim 0.5$ to 1 cm

FIG. D4·4   HALL EFFECT SAMPLE CONSTRUCTION

described in Fig. D4·1 or solder contacts of the type described in Demonstration 4·3, place four ohmic contacts on a p- and an n-type sample, in the positions shown in Fig. D4·4. The sample can be mounted on a glass slide with carnauba wax, epoxy cement, or Elmer's Glue. Two of the points are used as current contacts and two as Hall voltage probes. The change of sign of the Hall voltage with a change of conductivity type can be easily demonstrated with this apparatus. Quantitative measurements of Hall voltage and conductivity can also be made with this setup.[1]

4·5 *Carrier recombination and lifetime.* Carrier recombination and lifetime can be demonstrated using the sample prepared for the conductivity measurements (Fig. D4·1) or that prepared for the Hall effect experiment. The experiment consists of exciting the sample with a burst of light and measuring the photoconductivity decay with time.

The experimental setup is shown in Figs. 4·17 and 4·18 in the text. The battery is a 6-volt or other convenient size. The resistor R in Fig. 4·18 is made roughly equal to the resistance of the sample. A convenient light source is the General Radio Strobotac type 1513-A.

Only the two end contacts are used in the experiment. To avoid contact effects, these contacts should be shielded from the light burst. An oscilloscopic trace of a typical waveform from this experiment is given in Fig. 4·18 in the text.

4·6 *Particle diffusion.* A very simple demonstration of particle diffusion is obtained by lighting a small incense stick and placing it in the corner of a room. The time when people in various parts of the room notice the scent is readily observable.

4·7 *Random walk.* A simple visual demonstration of the random walk can be made by mounting match sticks or long cocktail picks which have been painted

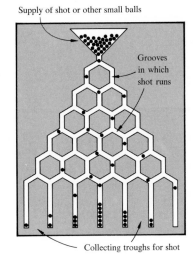

Supply of shot or other small balls

Grooves in which shot runs

Collecting troughs for shot

FIG. **D4·7** PROBABILITY DEMONSTRATION

black in a piece of polyfoam, in the manner suggested in Fig. 4·10b. A string can be glued over the tops of the picks, as suggested in Fig. 4·10b, to approximate the

[1] Buehler, Shockley, Pearson, Hall measurements using corbinolike current sources in thin circular disks, *Applied Physics Letters,* vol. 5, no. 11, Dec. 1, 1964, pp. 228–229.

continuous distribution. A HEXSTAT PROBABILITY DEMONSTRATOR can also be used to provide a simple demonstration of the random walk. The demonstrator is available from Lansford Press, 2516 Lansford Avenue, San Jose 95125, Calif. It consists of a board with grooves, made in the manner shown in Fig. D4·7, and a supply of small steel balls, which can be rolled from the top of the board to the bottom.

PROBLEMS

4·1   Consider a single free electron in the isolated bar of germanium shown in Fig. 4·1. This electron is performing a random motion about an equilibrium position. Between each collision point it conducts a miniature current through the crystal. Suppose each path has a length of 1,000Å, and that the thermal velocity of the electron is $v_T = \sqrt{2kT/m^*} = 10^7$ cm/sec. Make a rough sketch of the voltage which should appear across the terminals due to this single electron. How does the maximum amplitude of this voltage depend on T? When there are many electrons in the crystal, their different directions of motion and separate currents at any instant will tend to cancel, though for short-time intervals the cancellation will not be perfect. Thus, ac voltages develop in materials due to the thermal motion of the carriers. Such voltages are called noise voltages.

4·2   A bar of intrinsic germanium 2 cm long has a 10-volt battery connected across it, as in Fig. P4·2. How long does it take an electron to drift the length of the bar? How much energy does the electron deliver to the lattice during this transit? Sketch the drift time as a function of battery voltage for $10v < V < 20,000$ (a pulse generator is substituted for the battery and the sample geometry is changed considerably in the higher ranges of V).

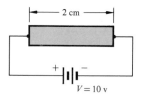

FIG. **P4·2**

4·3   Estimate the values of τ_I for electrons at $T = 200°K$ and $T = 300°K$ from the data given in Fig. 4·3. On the assumption that $\tau_I + AT^a$, estimate a from this data.

4·4   A metallic conductor has on the order of 10^{23} free electrons/cm³ and a typical resistivity of 10^{-5} to 10^{-6} ohm-cm. Estimate the mobility of free electrons in metallic conductors from this data. What is the corresponding value of mean free time, assuming $m_e^* = m_0$?

4·5   Assuming the mobilities of electrons and holes to be independent of impurity concentration, derive an expression for the minimum conductivity σ_{min} of a germanium crystal, achievable by choice of doping conditions. What is the ratio of σ_{min} to the intrinsic conductivity σ_i? Under what conditions would $\sigma_{min} = \sigma_i$?

4·6   In Prob. 3·18, Chap. 3, it was found that 3.22×10^{-6} g of antimony distributed uniformly in a 100-g crystal of germanium produced 8.7×10^{14} atoms of antimony/cm³. Assuming that these donors are all ionized, calculate the conductivity and resistivity of the crystal. What would be the resistance of a rectangular specimen of this crystal of cross section 1 mm² and length 2 cm?

4·7 The basic apparatus for measuring the resistivity of a semiconductor crystal is shown in Fig. P4·7. The apparatus is called a four-point probe. The four points are pressed down on a thin slice of the crystal whose resistivity is to be measured.

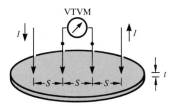

FIG. P4·7

A voltmeter is connected across the two inner probes and a current I is supplied to the sample and removed from it via the outer probes. The resistivity of the slice is then calculated from

$$\rho = \frac{V}{I} \frac{2\pi t}{\ln 2}$$

The formula assumes that the slice is thin, so that the current is uniformly distributed across the thickness t (at least at the point where it flows underneath the voltage probes). Derive this formula for ρ.

4·8 The purpose of this problem is to study the temperature dependence of the various factors affecting the conductivity of an intrinsic germanium crystal. To do so, it is convenient to take the logarithm of Eq. (4·17):

$$\ln \sigma = \text{const} + \ln (\mu_n + \mu_p) + \ln T^{3/2} - \frac{qE_g}{kT}$$

Then $\Delta(\ln \sigma) = \Delta\sigma/\sigma$ is the percentage change in σ brought about by a change of ΔT in T. On the basis of this, discuss $\Delta\sigma/\sigma$ versus ΔT for metals, heavily doped semiconductors, and intrinsic semiconductors.

4·9 A Wheatstone bridge with a germanium-flake thermistor in one arm is shown in Fig. P4·9.

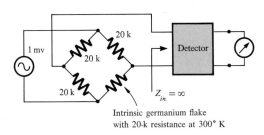

FIG. P4·9

1. How much voltage is fed to the detector for a 1°C change in temperature at the thermistor?

2. Obtain the Thévenin equivalent generator, as seen by the detector. Verify your answer to (1).

3. Assume now that the detector has an input resistance of 200 ohms. Plot the current into the detector vs. T from 0 to 40°C. List your assumptions clearly.

4·10 A thermistor T_h is embedded in the block of an internal-combustion engine to measure its temperature. The circuit is shown in Fig. P4·9, with each bridge resistor being 3,000 ohms. The germanium flake is replaced by a thermistor which has a resistance of 3,000 ohms at 27°C. Assuming that the thermistor is in good thermal contact with the engine block, draw a calibration curve for the instrument: engine-block temperature T vs. detector input voltage, up to 200°C. You may assume that only the exponential factor need be considered in determining the variation of resistance of T_h with temperature, and that $E_g = 1.2$ volts.

4·11 Thermistors and wire-wound resistors are frequently used in series to produce a combination resistor with zero-temperature coefficient (at some specified temperature). The simplest arrangement is shown in Fig. P4·11. Suppose R_1 is a

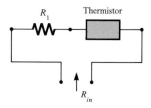

FIG. P4·11

wire-wound resistor with 600-ohm resistance at room temperature. Its resistance changes with T at the rate of 1 percent per °C. Assuming the thermistor is made from intrinsic germanium, what resistance should it have to insure that R_{in} has zero temperature at room temperature?

4·12 Let us consider some useful properties of the random-walk problem. If the step size l and step time t_l in a random walk are decreased in such a way that

$$D = \frac{l^2}{2t_l}$$

remains constant, then the diagram plotted in Fig. 4·10a will become a surface. The equation of this surface is

$$\frac{A}{\sqrt{t}} e^{-x^2/4Dt}$$

where A is a constant. It is interesting to consider how accurately this "continuous" solution represents the discrete solution shown in Fig. 4·10a. To study this, set $A/\sqrt{t} = 20/64$ and compare the continuous solution with the one given in Fig. 4·10a for $t = 6t_l$. Repeat for $t = 4t_l$ with $A/\sqrt{t} = 6/16$, and $t = 2t_l$ with $A/\sqrt{t} = \frac{1}{2}$. What do you conclude from these calculations?

4·13 The data indicated in Fig. P4·13 were obtained from the Shockley-Haynes apparatus described in Demonstration 4·3.

Calculate D, μ, and D/μ from this data. The output pulse may be assumed to be of the form

$$A e^{-x^2/4Dt}$$

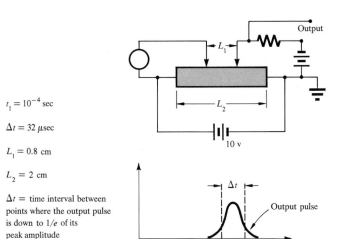

$t_1 = 10^{-4}$ sec

$\Delta t = 32\ \mu$sec

$L_1 = 0.8$ cm

$L_2 = 2$ cm

$\Delta t =$ time interval between points where the output pulse is down to $1/e$ of its peak amplitude

Input pulse of holes injected at $t = 0$

FIG. P4 · 13

4 · 14   The central capacitor in the RC network shown in Fig. P4 · 14 is charged to 1 volt and then switched into the network. Plot the voltage distribution vs. x for

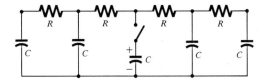

FIG. P4 · 14

$t = 0$ and $RC = 0.1, 0.2, 0.3,$ and 0.4. You may assume that the currents flowing in the resistors are constant throughout each time interval at the values they had at its beginning. Is charge properly conserved in your calculations? Explain the relation of this problem to diffusion.

4 · 15   Using the data given in Fig. 4 · 3, sketch D_p and D_n versus the ionized impurity concentration N_I for germanium at room temperature.

4 · 16   Suppose that $n(x)$ is an arbitrary density distribution of particles along the x axis. The particles move along the x axis by diffusion only. Therefore, the number which enter the small strip of width Δx per unit time is $-D\dfrac{\partial n}{\partial x}\Big|_{x_0}$; the number which leave per unit time is $-D\dfrac{\partial n}{\partial x}\Big|_{x_0+\Delta x}$. Furthermore, the increase in total population of the region in time Δt can be written in terms of the density as

$$\Delta x\left(\frac{\partial n}{\partial t}\Big|_{x_0+\Delta x/2}\right)\Delta t$$

From these facts show that n obeys the diffusion equation

$$\frac{\partial n}{\partial t} = D\frac{\partial^2 n}{\partial x^2}.$$

Carrier motion in semiconductors   *151*

4·17 The following data are taken from G. L. Pearson and J. Bardeen, Electrical properties of pure silicon and silicon alloys containing boron and phosphorus, *Phys. Rev.,* vol. 75, 1949, pp. 865ff. A boron-doped sample has $\sigma = 0.8$ ohm/cm and $\log_{10} R_H = 1.5$ (R_H measured in cubic centimeter per coulomb) at $T = 167°$K. Calculate μ_p and p for this sample at $T = 167°$K.

4·18 Use the data given in Fig. 4·18 to evaluate the lifetime of the sample shown there.

4·19 Assume that a flash of light falls uniformly on the specimen given in Prob. 4·2, creating 10^{13} carrier pairs. The sample may be assumed to be thin enough that the photogenerated carriers are distributed uniformly throughout the specimen.

1. What is the resistance of the specimen just after the flash?
2. The specimen has a hole lifetime of 10 μsec. Plot the resistance of the specimen as a function of time after the flash, assuming the recombination formula

$$U = \frac{p - p_n}{\tau_p}$$

3. Evaluate U from

$$U = r(np - n_i^2)$$

immediately after the flash and compare with the value estimated by the simpler formula. What effect would this have on the plot obtained in (2)?

4·20 A chopped light is focused on a photoconductor as shown in Fig. P4·20. The dark resistance of the photoconductor is 10 kohms. Its resistance under steady-state illumination by the light source is 1 kohm. The chopper allows light to strike

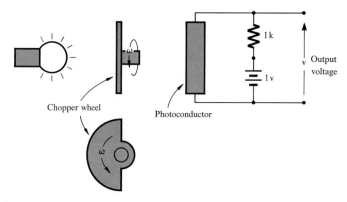

the photoconductor for 1 msec and then the light is shut off for 1 msec. Plot the output voltage of the circuit shown and dimension it as accurately as you can. The minority-carrier lifetime in the photoconductor is 1 μsec.

5

pn JUNCTIONS

A *pn* JUNCTION is a metallurgical boundary in a semiconductor crystal in which holes are the majority carriers on one side of the boundary and electrons are the majority carriers on the other. The *pn* junction is of great importance because it is, in effect, the "control element" for most semiconductor devices. It is therefore of fundamental importance in the operation of transistors and diodes, and we must thoroughly understand its properties before we can move on to a study of these devices.

The control property arises from the fact that the carrier densities (in particular, the *minority* carrier densities) on each side of the junction are regulated by the voltage which is applied across the junction. This control property, which we call the "law of the junction," is the principal subject of this chapter.

153

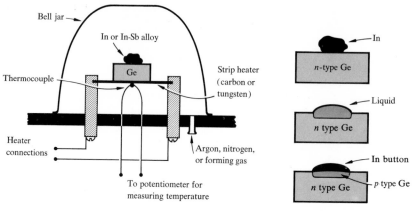

(a) *Experimental arrangement for preparing an alloyed junction*

(b) *Steps in preparation*

FIG. 5·1   PREPARATION OF A *pn* JUNCTION BY THE ALLOYING TECHNIQUE

5·1   SOME METHODS OF FABRICATING *pn* JUNCTIONS

Before beginning our study of junction properties, it will be helpful to consider briefly two of the methods which are commonly used in fabricating *pn* junctions. This will then provide some idea of the accuracy of the idealizations which we will make in developing the mathematics of a *pn* junction.

5·1·1   *Alloyed junctions.*   One common method for making *pn* junctions is called *alloying*. A crude schematic indication of the process is shown in Fig. 5·1. Here a small pellet of indium is placed on an *n*-type germanium slab. The system is then heated to a temperature of about 500°C, where the indium and some of the germanium melt to form a small puddle of molten germanium-indium mixture (Fig. 5·1b). The temperature is then lowered, and the puddle begins to solidify.[1] Under proper conditions the initial portion of the recrystallized material will be a single crystal with the same crystal orientation as the parent germanium slab, but with indium atoms substituted in the germanium lattice. (The alloying process is, in fact, very closely related to the procedure described in Sec. 2·2·6 for growing single crystals from a melt.) Usually, the density of indium atoms obtained in this way is about 0.01 to 0.1 percent of the density of germanium atoms, which is sufficient to overcompensate the original *n*-type material and to make it heavily *p* type.

As the recrystallization proceeds, the remaining molten mixture becomes increasingly rich in indium. When all the germanium has been redeposited,

[1] A physical metallurgist's explanation of the alloying process is usually given in terms of a eutectic diagram. An introduction to these diagrams is given in A. G. Guy, "Physical Metallurgy for Engineers," Addison-Wesley, 1962, pp. 102–109.

the remaining material appears as an indium button, which is frozen onto the outer surface of the recrystallized portion (Fig. 5 · 1d). This button is a good mechanical and ohmic contact and serves as a suitable base for soldering on leads.

The pn metallurgical junction. Near the beginning of the recrystallized region, the germanium will change from n type to p type. The nature of this change is approximately represented in Fig. 5 · 2a, where the net doping density is plotted as a function of distance into the germanium slab, along the center line of the recrystallized region. There is some surface in the recrystallized region at which the net doping density is zero. This surface is called the *metallurgical junction.*

By selecting appropriate orientations of the crystal and by careful control of the alloying process, it is possible to make junctions which change from p type to n type in a distance of 100 to 200Å. Since this distance is small compared to other critical distances to be described later, the impurity distribution in such a junction may frequently be considered to change abruptly, from N_a to N_d, as shown in Fig. 5 · 2b. Such a junction is called a *step junction.*

5 · 1 · 2 Diffused junctions. A second very useful method of preparing pn junctions is by *solid-state diffusion.* In the diffusion process, impurities of one conductivity type are made to diffuse at elevated temperature into a semiconductor body of opposite conductivity type. This produces a skin of converted material with a pn junction between this skin and the parent body beneath.

One of the popular methods of making diffused pn junctions is described schematically in Fig. 5 · 3. Part a of this figure shows a quartz boat with a grooved platform across the bottom. A charge of the impurity element to be diffused is placed beneath the platform (usually in the form of an oxide). Wafers of the parent semiconductor material which have been cut,

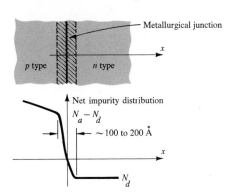

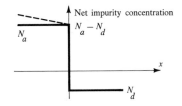

(a) *Approximate impurity distribution in an alloyed junction*

(b) *Idealized impurity distribution for an alloy junction (step-junction approximation)*

FIG. 5 · 2 **IMPURITY DISTRIBUTION IN AN ALLOYED pn JUNCTION**

lapped, and etched to the desired size and shape are then mounted in the grooved platform, and a quartz plate is put on top of the boat.

The boat is then inserted into a diffusion furnace, as shown in Fig. 5·3b, where the temperature is accurately known and controlled. Typical temperatures in the diffusion furnace are between 800°C and 1200°C, depending on the desired nature of the junction.

At these temperatures, many of the impurity atoms which have been introduced into the boat will be in a gaseous phase, and a high concentration of doping atoms will surround the semiconductor wafers (see Fig. 5·3c). The temperature is also such that the lattice atoms of the parent semiconductor will be highly excited, so that many of them will leave their lattice sites and take up new positions in the crystal and on its surface. At the same time, the impurity atoms diffuse into the semiconductor material (by virtue of the existing concentration gradient of impurity atoms) and there occupy the lattice vacancies created by the wandering atoms of the host lattice. During the process, a flow of very pure inert gas is maintained in the tube to keep out unwanted impurities (though the stream may sometimes carry the impurity which is to be diffused into the semiconductor).

The mathematical aspects of this doping process are described by the *diffusion equation* (developed in Prob 4·16). For the present problem, the concentration C of the diffusing material may be considered to be constant in the medium surrounding the semiconductor. A solution of the diffusion equation which fits this boundary condition is then

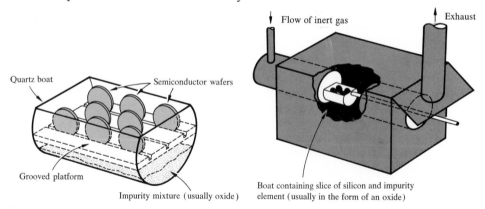

(a) *Boat containing impurity material and parent semiconductor wafers*

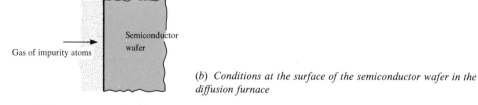

(b) *Conditions at the surface of the semiconductor wafer in the diffusion furnace*

FIG. **5·3** **PREPARATION OF A** *pn* **JUNCTION BY SOLID-STATE DIFFUSION**

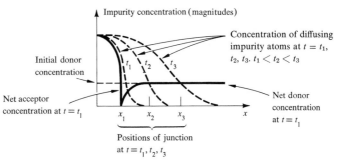

FIG. **5·4** **IMPURITY CONCENTRATION IN A DIFFUSED JUNCTION**

$$C = C_0\left[1 - \frac{1}{(\pi Dt)^{1/2}} \int_0^{x/2(Dt)^{1/2}} e^{-y^2/4Dt} \, dy\right]$$

In this formula, C_0 is the surface concentration of the diffusing species, D is a diffusion constant which describes the random motion of the diffusing atom in the semiconductor lattice and t is the duration of the heat treatment. The diffusion constant D is a sensitive function of the temperature, and it is therefore necessary to control the temperature quite accurately.

It should be emphasized that when the heat treatment is over (that is, when the boat is removed from the furnace), all the atoms in the semiconductor (host and impurities) are frozen into the positions which they occupied at that time. The impurity atoms are not free to wander through the semiconductor at room temperature; they are fixed in the positions they obtained in the metallurgical treatment.

Figure 5·4 shows a sketch of the diffused impurity concentration profile in an n-type semiconductor. If the impurity were boron, for example, with a surface concentration exceeding the donor concentration N_d in the n-type semiconductor, then the surface layers of the semiconductor will have an excess of acceptors over donors and accordingly will be p type to a depth given by the intersection of the acceptor and donor profiles. It is apparent from an inspection of Fig. 5·4 that by diffusion technology we may produce a junction in which the doping changes quite abruptly from p type to n type (that is, step junctions, with N_d and N_a arbitrary), or we may produce a junction in which the doping changes only gradually from p type to n type. In the latter case, we refer to the junction as being a *graded junction*. The external properties of a graded junction will be different from those of a step junction in a manner to be made clear later.

5·2 QUALITATIVE DESCRIPTION OF A *pn* JUNCTION
IN THERMAL EQUILIBRIUM

We now proceed to describe the state of affairs which exists at and around a *pn* step junction in thermal equilibrium. Figure 5·5 shows

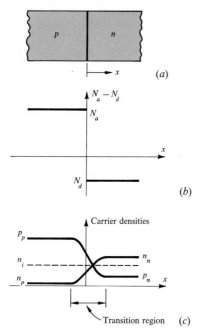

FIG. 5·5 IDEALIZED IMPURITY DISTRIBUTION IN A *pn* JUNCTION AND RESULTING CARRIER DENSITIES

a one-dimensional model of a *pn* junction, a step distribution of impurities (with N_a and N_d arbitrary), and the resulting carrier densities. Of course, real junctions are not one-dimensional nor do they have impurity distributions which change abruptly. However, the principal features of junction action can be developed with this model and later modified to suit actual conditions.

We begin by observing that, at distances sufficiently far from the metallurgical junction, the carrier densities have the same values they would have in uniform material (that is, without a *pn* junction). In the *n*-type body, the electron density will be n_n and the hole density p_n; and in the *p*-type body, the hole density will be p_p and the electron density n_p. Furthermore

$$n_p p_p = n_n p_n = n_i^2$$

Since $p_p \neq p_n$ and $n_p \neq n_n$, there will be a region around the metallurgical junction, called the *transition region*, in which the hole density will change smoothly from p_p to p_n, and the electron density will change smoothly from n_n to n_p (see Fig. 5·5c).

We may satisfy ourselves that this is so by the following heuristic argument. We shall suppose that an impenetrable barrier is established at the metallurgical junction. The hole density is p_p, right up to the metallurgical junction, after which it drops abruptly to p_n. Now, when the barrier is removed, the difference in concentration of holes on the two sides of the junction will produce a diffusion of holes from the *p* side to the *n* side (similar arguments apply for electrons which diffuse from the n side to the p side).

If the diffusing carriers were uncharged, the diffusive motion would continue until a uniform concentration of holes and electrons existed throughout the entire semiconductor bar. However, such a condition cannot occur when the diffusing particles are charged. To appreciate this, we observe that when a hole moves from the p side to the n side, it carries a positive charge with it, and it leaves a *bound* negative charge (an unneutralized acceptor atom in the lattice) behind. Similarly, when an electron moves from the n side to the p side, it carries a negative charge with it and leaves a *bound* positive charge (unneutralized donor atom in the lattice) behind. In this way, the diffusion of holes and electrons produces two small regions around the metallurgical junction, in each of which there is a net space charge.[1] On the whole, the material is still neutral, but there is a dipole-charge layer surrounding the metallurgical junction. The extent of the dipole layer is the same as the transition region, so the transition region is sometimes called the *space-charge region* or *space-charge layer*.

The presence of a space-charge layer implies (via Gauss' law) that an electric field will be developed in the transition region. The direction of this electric field is such that it tends to push holes back into the p-type material and electrons back into n-type material. In this way the diffusion process sets up by its own operation a counteracting electric field. For an unbiased junction—that is, a junction not connected to a source of power —one can conclude that an equilibrium condition will be established in which the tendency for carriers of a given type to diffuse in one direction is exactly balanced by their tendency to drift in the other. A graphical representation of the equilibrium situation is given in Fig. 5·6, which shows sketches of the net charge density and the corresponding electrostatic field (more accurate drawings are given in Fig. 5·7).

5·3 CONTACT POTENTIAL AT A pn JUNCTION

Associated with this internal electrostatic field there will be an electrostatic potential, for which we use the symbol ϕ. This electrostatic potential is just the *contact potential* which always develops when two dissimilar materials are joined together. ϕ is also called the *built-in voltage* at the pn junction. By definition, of course,

$$\phi = -\int_{l_1}^{l_2} \mathbf{E} \cdot \mathbf{dx} \qquad (5 \cdot 1)$$

where l_1 and l_2 and sign conventions for ϕ and $\mathbf{E}(x)$ are defined in Fig. 5·6.

We shall show in this section that the magnitude of the contact poten-

---

[1] The relative importance of mobile and bound carriers in this space charge layer is incorrectly given by this argument, since recombination of the mobile carriers is neglected (see Sec. 5·4·1).

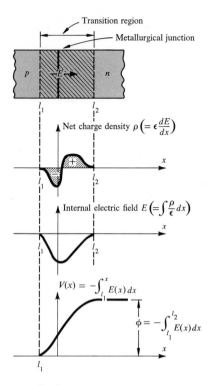

(a) pn junction with reference direction for E

(b) Dipole-space-charge layer. Negative charge consists of electrons and unneutralized acceptor atoms. Positive charge consists of holes and unneutralized donor atoms.

(c) Electric field produced by (b)

(d) Electrostatic potential with respect to left-hand terminal

FIG. 5·6 NET CHARGE DENSITY, ELECTRIC FIELD, AND ELECTROSTATIC POTENTIAL DEVELOPED IN THE TRANSITION REGION AT A STEP *pn* JUNCTION; THE CENTER OF THE DIPOLE-SPACE-CHARGE LAYER IS NOT NECESSARILY AT THE METALLURGICAL JUNCTION.

tial can be written as

$$|\phi| = \frac{kT}{q}\ln\frac{p_p}{p_n} = \frac{kT}{q}\ln\frac{n_n}{n_p} = \frac{kT}{q}\ln\frac{N_aN_d}{n_i^2} \tag{5·2}$$

However, before doing this, we should emphasize that this contact potential cannot be measured by connecting a voltmeter across the junction. This is because contact potentials develop at every junction between dissimilar materials (in particular between the voltmeter leads and the semiconductor) and the sum of these contact potentials around a closed loop must be zero (as long as all the junctions are at the same temperature)—otherwise power could be generated simply by joining wires together in a closed loop.

5·3·1   The balance of drift and diffusion.   The derivation of Eq. (5·2) is based on an idea implied in the qualitative discussion of conditions in the transition region given in Sec. 5·2. There, it was suggested that in thermal equilibrium, the carrier-density and electric-field curves have to be such that the tendency for carriers of a given type to diffuse in one direction is exactly cancelled by their tendency to drift in the opposite direction. This must be true at each point within the transition region.

The contact potential can be obtained from this basic idea in the following steps:

1. We first prove formally that the drift and diffusion tendencies for holes and electrons must cancel *independently*. To do this, we observe that there can be no total current flowing across a junction to which no source of power is attached—the device is not a generator:

$$j = j_n + j_p = 0 \qquad (5 \cdot 3)$$

Equation $(5 \cdot 3)$ can be satisfied either by setting both j_n and j_p equal to zero or by setting j_n equal to $-j_p$. The latter alternative requires that both holes and electrons be passing across a given plane in the same direction with identical velocities. This latter condition is certainly not an equilibrium situation, since there is a net transportation of particles involved. Hence we must require

$$j_n = j_p = 0 \qquad (5 \cdot 4)$$

in thermal equilibrium.

2. Now, j_n and j_p each have two components: a drift component and a diffusion component. Thus, in thermal equilibrium

$$j_p = q \left[\mu_p p(x) E(x) - D_p \frac{\partial p}{\partial x} \right] = 0 \qquad (5 \cdot 5)$$

$$j_n = q \left[\mu_n n(x) E(x) + D_n \frac{\partial n}{\partial x} \right] = 0 \qquad (5 \cdot 6)$$

where the positive direction of E is assumed to be the same as positive x. Thus drift and diffusion tendencies for holes and electrons balance separately.

Now, by definition,

$$E(x) = - \frac{\partial V}{\partial x} \qquad (5 \cdot 7)$$

so we may write Eq. $(5 \cdot 5)$ as

$$- \mu_p p \frac{dV}{dx} - D_p \frac{dp}{dx} = 0 \qquad (5 \cdot 8)$$

where p and V are functions of x. Equation $(5 \cdot 8)$ may be separated as follows:

$$- \frac{\mu_p}{D_p} \frac{dV}{dx} = \frac{1}{p} \frac{dp}{dx} \qquad (5 \cdot 9)$$

We now use Eq. $(5 \cdot 9)$ to obtain p as a function of V by integrating both sides:

$$\frac{d}{dx} \left(- \frac{\mu_p}{D_p} V \right) = \frac{d}{dx} (\ln p - \ln C) \qquad (5 \cdot 10)$$

where C is a constant. Thus, we finally obtain

$$\ln \frac{p}{C} = - \frac{\mu_p}{D_p} (V_p - V_C) \qquad (5 \cdot 11)$$

Using the Einstein relationship,

$$\frac{\mu_p}{D_p} = \frac{q}{kT} \tag{5·12}$$

we have

$$\ln \frac{p}{C} = -\frac{q}{kT}(V_p - V_C) \tag{5·12}$$

In Eq. (5·12), $V_p - V_C$ is the potential difference between the plane where the hole density is p and the plane where it is C. If we let p be the hole density at the left-hand end of the sample, p_p, and let C be the hole density on the right-hand end, p_n, we have

$$\ln \frac{p_p}{p_n} = -\frac{q(V_p - V_n)}{kT} \tag{5·13}$$

We now call ϕ the potential *difference* between the p and n regions, being positive on the n-type side (so $V_p - V_n = -\phi$). Then

$$\ln \frac{p_p}{p_n} = \frac{q\phi}{kT} \tag{5·14}$$

Equation (5·14) is the relationship we seek, for it tells us the (contact) potential difference between the p and n sides—with n positive with respect to p (a similar result for graded junctions is studied in Prob. 5·3).

On physical grounds, it is apparent that ϕ must be independent of whether Eq. (5·5) or Eq. (5·6) is used to obtain it. We prove this by direct manipulation:

$$\mu_n n E = -D_n \frac{dn}{dx} \Rightarrow \frac{\mu_n}{D_n} \frac{dV}{dx} = \frac{1}{n} \frac{dn}{dx} \Rightarrow \ln \frac{n}{B} = \frac{q}{kT}(V_n - V_B) \tag{5·15}$$

and in particular,

$$\ln \frac{n_n}{n_p} = \frac{q(V_n - V_p)}{kT} = \frac{q\phi}{kT} \tag{5·16}$$

To show that Eq. (5·16) and Eq. (5·14) are equivalent, we observe that

$$n_n = N_d \qquad p_p = N_a \qquad n_p = \frac{n_i^2}{N_a} \qquad p_n = \frac{n_i^2}{N_d}$$

so

$$\phi = \frac{kT}{q} \ln \frac{p_p}{p_n} = \frac{kT}{q} \ln \frac{n_n}{n_p} = \frac{kT}{q} \ln \frac{N_a N_d}{n_i^2} \tag{5·17}$$

EXAMPLE. An example will help give a feeling for the typical order of magnitude of ϕ. Suppose we have a pn step junction in germanium in which the resistivity on the p side is 10^{-3} ohm-cm and the resistivity on the n side is 1 ohm-cm. Then

$$1/\rho_p = \sigma_p = q\mu_p p_p \Rightarrow p_p \cong 3.3 \times 10^{18}/\text{cm}^3$$

in the p-type material, and

$$1/\rho_n = \sigma_n = q\mu_n n_n \Rightarrow n_n \cong 1.6 \times 10^{15}/\text{cm}^3$$

in the *n*-type material. (The temptation to carry calculations to more than two significant figures should be resisted since empirical data is rarely that accurate.)
Using the form

$$\phi = \frac{kT}{q} \ln \frac{N_a N_d}{n_i{}^2} = \frac{kT}{q} \ln \frac{p_p n_n}{n_i{}^2}$$

we find
$$\phi = 0.41 \text{ volt}$$

Contact potentials for typical silicon junctions are on the order of 0.7 volt (because $n_i{}^2$ is lower by a factor of nearly 10^6; see Prob. 5 · 1).

5·4   PROPERTIES OF THE UNBIASED TRANSITION REGION

So far we have seen that a dipole-charge layer develops in the transition region of an unbiased *pn* junction. We have associated an electric field E and a contact potential ϕ with this space-charge layer. However, there are several other properties of the transition region which we need to know in order to thoroughly understand junction behavior in practical devices. We shall develop these properties briefly in this section, emphasizing the nature of the arguments, from which the necessary mathematics will follow.

5·4·1   *Zero-bias width of the transition region.*   One of the most useful properties of the transition region in our future work is its width. The transition region extends a distance l_1 into *p*-type material and a distance l_2 into *n*-type material (see Fig. 5 · 7). We shall calculate these widths in this section.

Let us observe first that *the arguments for calculating ϕ presented in Sec. 5 · 3 were carried to completion without ever specifying E(x). This is because of the fundamentally thermodynamic nature of ϕ.*

Of course, we may also use the methods of *electrostatics* to obtain a relation between ϕ and the space charge that exists in the transition region, and it is in this way that we shall determine l_1 and l_2. For convenience, we will make some specific assumptions about carrier-density variations within the transition region to obtain a simple form for the space charge there. The accuracy of the values of l_1 and l_2 will be affected by these assumptions, even though the method of calculating l_1 and l_2 is itself perfectly general.

The basic ideas involved in calculating the width of the transition region are as follows:

1. From a knowledge of the doping densities and carrier densities, we can develop an expression for the net charge density $\rho(x)$ in the transition region.[1] The simplest and most useful approximation is that *the contribution of holes and electrons to the space charge is negligible,* so that $\rho(x)$ is only a

---

[1] It is customary to use the symbol ρ for both the resistivity of a material and for the net charge density in any medium.

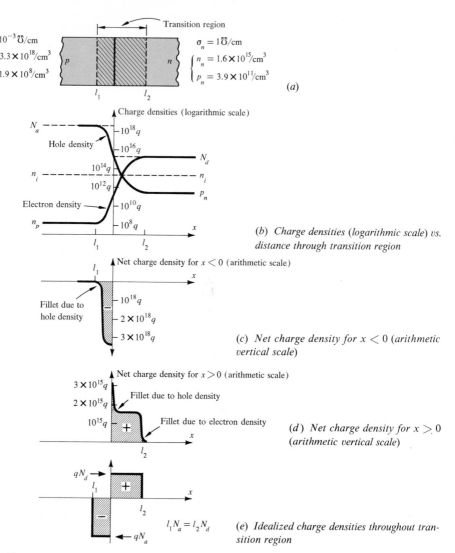

$\sigma_p = 10^{-3}\,\mho/cm$
$\begin{cases} p_p = 3.3 \times 10^{18}/cm^3 \\ n_p = 1.9 \times 10^{8}/cm^3 \end{cases}$

Transition region

$\sigma_n = 1\,\mho/cm$
$\begin{cases} n_n = 1.6 \times 10^{15}/cm^3 \\ p_n = 3.9 \times 10^{11}/cm^3 \end{cases}$

(a)

(b) Charge densities (logarithmic scale) vs. distance through transition region

(c) Net charge density for $x < 0$ (arithmetic vertical scale)

(d) Net charge density for $x > 0$ (arithmetic vertical scale)

(e) Idealized charge densities throughout transition region

FIG. 5·7   CHARGE DENSITIES AND APPROXIMATIONS TO THEM USED FOR CALCULATING $E(x)$ AND ϕ

function of N_a and N_d:

$$\rho(x) = f(N_a, N_d) \qquad (5 \cdot 18)$$

2. Using Gauss' law, we can calculate $E(x)$ from $\rho(x)$:

$$\epsilon \frac{dE}{dx} = \rho(x) \qquad (5 \cdot 19)$$

ϵ is the dielectric constant for the semiconductor crystal. In the simplest case, $E(x)$ can be calculated from a "Gauss pillbox" by inspection.

3. Using

$$\phi = \int_{l_1}^{l_2} -E(x)\,dx \qquad (5 \cdot 20)$$

we can compute ϕ from E. Now ϕ will be a function of N_a, N_d, l_1, and l_2. However, we already know from an entirely independent argument that

$$\phi = \frac{kT}{q} \ln \frac{p_p}{p_n} \qquad (5 \cdot 21)$$

and this will enable us to calculate the lengths l_1 and l_2 by setting the ϕ found in Eq. (5·20) to that found by the previous thermodynamic arguments.

Charge density determination. Figure 5·7 shows a simple approximation for the space-charge density $\rho(x)$. In order to be specific, the data used in a preceding example for calculating ϕ have been assumed ($\sigma_p = 10^{-3}$ mho/cm, $\sigma_n = 1$ mho/cm).

The charge densities associated with both the mobile carriers and the bound doping agents are plotted in Fig. 5·7b. The vertical scale on this figure is *logarithmic* in order to compress the variation into a plottable interval. Now, as we proceed from p type to n type, we see the hole density dropping and the electron density increasing. This is viewed against a background of N_a acceptors per volume on the p side of the metallurgical junction and N_d donors per volume on the n side.

A careful study of Fig. 5·7b reveals three important points:

1. The $p(x)$ and $n(x)$ curves are such that

$$p(x)n(x) = n_i^2 \qquad (5 \cdot 22)$$

for all x. This is because the junction is in thermal equilibrium, with zero bias applied, so generation and recombination must balance. (The np product will *not* be n_i^2 if the junction is not in thermal equilibrium and/or there is bias applied to it.)

2. The excess of holes in the interval $0 < x < l_2$ is *less* than the deficiency of holes in the interval $l_1 < x < 0$. In a previous heuristic argument intended to show that a transition region would be formed by diffusion of carriers, we suggested that each hole which leaves the p-type side carries a positive charge with it into the n-type side, and leaves a negative charge behind (in the form of an unneutralized acceptor atom). We should, then, expect the excess of holes in the $0 < x < l_2$ interval to equal the deficiency of holes in the $l_1 < x < 0$ interval. However, this is not a thermal-equilibrium condition; that is, when the impenetrable barrier of the previous argument is removed, holes will indeed diffuse from the p material to the n material, but they will also recombine with electrons in the n material, in a tendency toward thermal equilibrium. Thus, the mobile positive charge in $0 < x < l_2$ need not be equal to the *bound* negative charge in $l_1 < x < 0$.

3. Over most of the transition region in p-type material, the actual charge density is N_a. This is because the electron contribution is insignificant, and the hole contribution is only significant when the density of holes is greater than about $0.1N_a$. To emphasize this, the net charge density on the p-type side of the metallurgical junction (that is, for $x < 0$) is replotted on an *arithmetic* vertical scale in Fig. 5·7c.

A similar situation holds for the net charge density on the n-side of the metallurgical junction. Over most of this region, the actual densities of holes and electrons are small compared to the underlying donor doping density. There are exaggerated "fillets" on the charge curves contributed by the mobile carriers at the extreme ends of the space-charge layer, but the charge-density profile still looks basically rectangular (see Fig. 5·7d).

It is customary to idealize these two curves into rectangular charge blocks for use in computing $E(x)$. The idealizations are shown in Fig. 5·7e. There must be the same total charge in each block (so that E goes to zero at l_1 and l_2); thus

$$l_1 N_a = l_2 N_d$$

Of course, the actual width of the space-charge layer will be somewhat greater than calculations based on this idealization will suggest. However, the error is usually small, and there is no loss of principle involved in making the approximation.

Determination of $E(x)$ from $\rho(x)$. The determination of $E(x)$ from $\rho(x)$ is a straightforward application of Gauss' law. We have

$$\epsilon \frac{dE}{dx} = \rho(x) \tag{5·23}$$

Since $\rho(x)$ is constant in each region, E must be a straight line in each region. The resulting $E(x)$ is

$$E(x) = -E_{max}\left[1 + \frac{x}{l_1}\right] \qquad 0 > x > l_1$$

$$E(x) = -E_{max}\left[1 - \frac{x}{l_2}\right] \qquad 0 < x < l_2 \tag{5·24}$$

$$E_{max} = \frac{q l_1 N_a}{\epsilon} = \frac{q l_2 N_d}{\epsilon}$$

$E(x)$ is plotted in Fig. 5·8b.

Determination of ϕ from $E(x)$. We finally determine ϕ from $E(x)$ by

$$\phi = -\int_{l_1}^{l_2} E(x)\, dx = \frac{E_{max}(l_1 + l_2)}{2} \tag{5·25}$$

The integral is just the area under the $E(x)$ triangle shown in Fig. 5·8b. ϕ may be divided into V_{b0} and V_{0a}, where V_{b0} is the electrostatic potential rise between a point "b" deep in p-type material and $x = 0$; and V_{0a} is the potential rise from $x = 0$ to a point "a" deep in n-type material. Then

$$V_{b0} = \frac{1}{2} E_{max} l_1 \qquad V_{0a} = \frac{1}{2} E_{max} l_2 \tag{5·26}$$

These features appear in Fig. 5·8c.

Determination of l_1 and l_2 from ϕ. To determine l_1 and l_2 from ϕ, we first use

$$l_1 N_a = l_2 N_d \tag{5·27}$$

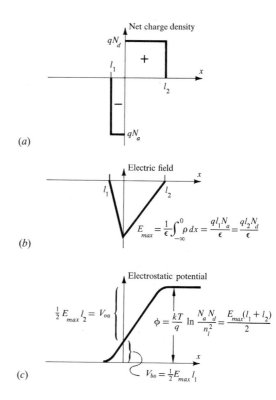

FIG. 5·8   DETERMINATION OF E AND ϕ FROM ρ

to rewrite Eq. (5·25) as

ELECTROSTATIC ARGUMENT
$$\phi = \frac{qN_a}{2\epsilon} l_1^2 \left(1 + \frac{N_a}{N_d}\right) \qquad (5·28)$$

But we also know that

THERMODYNAMIC ARGUMENT
$$\phi = \frac{kT}{q}\ln\frac{N_a N_d}{n_i^2} \qquad (5·29)$$

so
$$l_1 = \left[\left(\frac{kT}{q}\ln\frac{N_a N_d}{n_i^2}\right)\left(\frac{2\epsilon}{qN_a}\frac{N_d}{N_d + N_a}\right)\right]^{1/2} \qquad (5·30)$$

and then
$$l_2 = \left[\left(\frac{kT}{q}\ln\frac{N_a N_d}{n_i^2}\right)\left(\frac{2\epsilon}{qN_d}\frac{N_a}{N_d + N_a}\right)\right]^{1/2} \qquad (5·31)$$

EXAMPLE To continue with our previous example, we may use

$$N_a = 3.3 \times 10^{18}/\text{cm}^3 \qquad N_d = 1.6 \times 10^{15}/\text{cm}^3$$

Then

$$l_1 = 3.3 \times 10^{-8} \text{ cm} \qquad l_2 = 6.8 \times 10^{-5} \text{ cm}$$

which suggests that essentially all of the transition region is in n-type material.

We may also estimate E_{max}, V_{b0} and V_{0a} from these results. Since $l_1 \ll l_2$, we have (using $\phi = 0.4$ volt from the previous example)

$$E_{max} \cong \frac{2\phi}{l_2} \sim 1.2 \times 10^4 \text{ volts/cm} \qquad V_{b0} \sim 0.2 \text{ mv} \qquad V_{0a} \sim 0.4 \text{ volt}$$

DISCUSSION OF ASSUMPTIONS. In order to calculate l_1 and l_2 simply, we assumed that the charge density in the transition region could be represented by two idealized charge blocks. To obtain this charge distribution, we neglected the "fillets" on the edges of the charge-density curves in Figs. $5 \cdot 7c$ and $5 \cdot 7d$, and we assumed that the doping density changed abruptly from N_a to N_d. We now wish to check whether these assumptions are valid for the calculated values of l_1 and l_2.

It can be rather simply shown that the fillets on the outer edges of the charge-density curves given in Figs. $5 \cdot 7c$ and $5 \cdot 7d$ each account for about $2kT/q$, or 50 mv of the total contact potential. To see this, we recall from Eq. $(5 \cdot 12)$ that the variation of p across the space-charge layer may be written as

$$p(x) = p_p \exp \left\{ -\frac{q[V(x) - V_p]}{kT} \right\}$$

Therefore, when we proceed from $x = l_1$ into the transition region to the point where $[V(x) - V_p]$ is $2kT/q$,

$$p(x) = p_p e^{-2} \sim 0.14 p_p$$

If this distance is appreciably less than l_1, then the value of l_1 calculated by the method above will give a satisfactory estimate of the actual space-charge penetration in p-type material.

For the present case, since V_{b0} is only 0.2 mv, we know that the calculated value of l_1 gives a highly erroneous estimate of the actual length of the transition region in p-type material.

On the other hand, the calculated value of l_2 is quite accurate. Usually the total width of the transition region will be accurately known because $\phi \gg 4kT/q$, but the individual dimensions l_1 and l_2 can be trusted only when the potential drop in these regions is large compared to kT/q.

Regarding the validity of the step-junction approximation, the discussion of junction-fabrication techniques given in Sec. $5 \cdot 1$ indicates that in practical junctions there is at least a 100- to 200-Å width around the metallurgical junction in which the net doping density $N_a - N_d$ is not constant. The step-junction approximation for such a doping profile is valid when $l_1 + l_2 \gg 100$ to 200Å, as in the present case. The step-junction approximation is not adequate for a precise calculation of l_1 in the present example, nor for calculating either l_1 or l_2 for graded junctions (in which the doping changes gradually from n type to p type over the whole transition region). The calculation of l_1 and l_2 for such cases is considered in Prob. $5 \cdot 3$.

$5 \cdot 4 \cdot 2$ *The balance between drift and diffusion again.* The procedure used in Sec. $5 \cdot 3$ for calculating ϕ was based on the physical idea that in thermal equilibrium, drift and diffusion currents for each type of carrier were in balance. Mathematically this idea is expressed by the equation

$$j_p = -qD_p \frac{\partial p}{\partial x} + q\mu_p p(x)E(x) = 0 \qquad (5 \cdot 32)$$

for holes, with a similar equation for electrons.

When a bias is applied to the junction, the balance between drift and diffusion is upset, and j_p assumes a non-zero value. As we shall see later, a forward bias causes the diffusion current to exceed the oppositely directed drift current, and a net current flows from the p-type material to the n-type material. A reverse bias upsets the drift-diffusion balance in the opposite direction, with the result that a net current flows from n-type material to p-type material.

However, the difference between the two terms in Eq. (5 · 32) required to produce typical levels of current flow is very small compared to the magnitude of either term. To see that this is so, let us estimate the magnitudes of the drift and diffusion components given in Eq. (5 · 32), using the values of l_1 and l_2 given in the preceding section.

The diffusion tendency for holes may be approximated as follows. The hole density changes from a value of p_p to a value of p_n in a distance of l_1 and l_2. Therefore, the average hole-density gradient across the transition region is

$$\frac{\partial p}{\partial x} = -\frac{p_p - p_n}{l_1 + l_2} \sim -\frac{p_p}{l_2} \sim -5 \times 10^{22}/\text{cm}^4 \qquad (5 \cdot 33a)$$

The numbers are taken from the preceding example.

Our formula for the diffusion current density in the presence of a density gradient then indicates that

$$j_p = -qD_p \frac{\partial p}{\partial x} \sim 1.6 \times 10^{-19} \times 50 \times 5 \times 10^{22} \text{ amp/cm}^2$$

$$= 400,000 \text{ amp/cm}^2 \qquad (5 \cdot 33b)$$

which is an enormous current density. However, we need to recognize that the length of the transition region is only about seven times the mean-free path of a carrier, so the average value of D_p cannot be used with confidence. But even when statistical fluctuations in D_p and various other refinements are taken into account, the drift-diffusion balance still occurs at a very large average current density.

When we wish to pass a current through the pn junction, which is small compared to the large drift and diffusion tendencies, *the thermodynamic equilibrium between drift and diffusion within the transition region will still be essentially maintained.* The carrier-density distributions need to change only slightly to accommodate typical (or even large) forward currents. This idea will be very useful to us in the next section where we study the effects of applying a bias to a pn junction.

5·5 THE APPLICATION OF BIAS TO pn JUNCTIONS

So far our discussion of the transition region has been limited to the case of thermal equilibrium, zero-applied bias. We now wish to generalize this discussion to include the case of applied bias. We shall show that if a bias voltage v_a is applied across the pn junction with polarity opposite to that of ϕ, then *at every point in the transition region the product of the electron density n and the hole density p is*

$$np = n_i{}^2 e^{qv_a/kT} \tag{5·34}$$

This result will be called the *law of the junction,* and will be of fundamental importance in our studies of semiconductor diodes and transistors.

5·5·1 *Ohmic contacts.*
In order to apply a bias to a junction, we must first add contacts to it. This is done schematically in Fig. 5·9, where a pn junction is shown with two contact surfaces. The contacting surfaces will form junctions with the semiconductor material and contact potentials will develop across these junctions. Unlike the contact potential at a pn junction, however, the contact potential at a metal-semiconductor junction may have either polarity, depending on the metal and the conductivity type of the semiconductor. This leads to the possibility of producing either

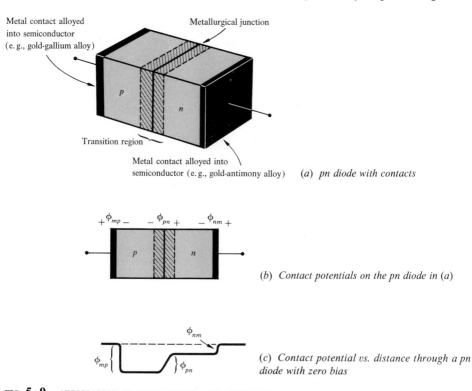

Metal contact alloyed into semiconductor (e. g., gold-gallium alloy)

Metallurgical junction

p

n

Transition region

Metal contact alloyed into semiconductor (e. g., gold-antimony alloy)   (a) *pn diode with contacts*

$+\overset{\phi_{mp}}{}-\quad -\overset{\phi_{pn}}{}+\quad -\overset{\phi_{nm}}{}+$

p n

(b) *Contact potentials on the pn diode in (a)*

ϕ_{nm}

ϕ_{mp} ϕ_{pn}

(c) *Contact potential vs. distance through a pn diode with zero bias*

FIG. 5·9 APPLICATION OF CONTACTS TO A pn JUNCTION

rectifying contacts, like a *pn* junction, or *nonrectifying* contacts between a metal and a semiconductor.

To confine the important junction effects to the semiconductor *pn* junction, it is necessary to choose contacting metals which produce non-rectifying contacts to the semiconductor and/or to prepare contacts which, even though rectifying, will pass very large currents with only a very small voltage drop across the metal-semiconductor junction. In either case, the contacts are called *ohmic contacts*.

The latter alternative is the most commonly adopted technique. A popular ohmic contact to *n*-type material is made by using a gold-antimony mixture for the contact and alloying this material into the *n*-type body. In this way we produce an essentially continuous transition from the relatively lightly doped *n*-type semiconductor body to a layer which is essentially metallic. Electrons may then readily flow from the metal into the semiconductor.

Contacts to *p*-type material may be made with a gold-gallium alloy, the objective again being to dope the semiconductor so heavily that it becomes a gallium-gold metal, to which external contact is readily made. Since the external wires carry current by electron flow, the "ohmic" contact to *p*-type material must be capable of readily exchanging holes for electrons. The mechanism for this is provided by the fact that, in the crystalline disruption inherent in the semiconductor-to-metal transition, there are many recombination centers, particularly when gold is included in the contact material. These recombination centers provide the sites at which the exchange of holes for electrons proceeds.

The contact potentials which develop at the metal-semiconductor surfaces just described are shown in Fig. $5 \cdot 9c$. As the figure indicates, there is no net contact potential between the gold metal on one surface and the gold metal on the other; that is, the sum of the contact potentials at all the junctions is zero:

$$\phi_{mp} = \phi_{pn} + \phi_{nm} \qquad (5 \cdot 35)$$

5·5·2 The law of the junction. Let us now consider Fig. $5 \cdot 10$, in which a voltage v_a is applied to the *pn* diode. We assume that the amount of this voltage which is absorbed across the contact surfaces is negligible, so that all of the applied voltage will be dropped in either the *p* body, then *n* body, or across the transition region. In order to concentrate on essentials, we will assume at the outset that the currents which will flow when a bias is applied are small enough to neglect voltage drops in the *p* and *n* bodies of the diode. Then, the entire magnitude of externally applied voltage appears across the transition region. *This means that the actual potential drop across the transition region must now be* $\phi - v_a$ *instead of* ϕ.

In order for this to occur, the transition region shrinks a bit on both sides. l_1, l_2, and E_{max} must be reduced sufficiently so that

$$-\int_{l_1}^{l_2} E(x)\, dx = \phi - v_a \qquad (5 \cdot 36)$$

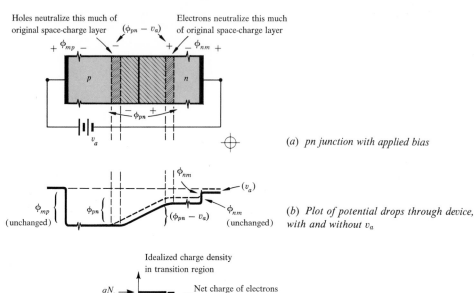

Holes neutralize this much of original space-charge layer

$(\phi_{pn} - v_a)$

Electrons neutralize this much of original space-charge layer

$+\ \phi_{mp}\ -$

$-\ \phi_{nm}\ +$

p n

$-\ \phi_{pn}\ +$

v_a

(a) pn junction with applied bias

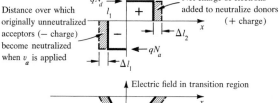

ϕ_{nm}

(v_a)

ϕ_{mp} ϕ_{pn}

(unchanged)

$(\phi_{pn} - v_a)$

ϕ_{nm}

(unchanged)

(b) Plot of potential drops through device, with and without v_a

Idealized charge density in transition region

qN_d

l_1

$+$

Net charge of electrons added to neutralize donors (+ charge)

Distance over which originally unneutralized acceptors (− charge) become neutralized when v_a is applied

$-$

Δl_2

x

$-qN_a$

Δl_1

Electric field in transition region

x

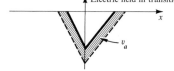

v_a

(c) Charge density and electric field in transition region. Shaded areas show effects of forward voltage on transition region.

FIG. 5·10 CHANGES IN THE TRANSITION REGION UPON APPLICATION OF A FORWARD BIAS

As is suggested in Fig. 5·10, holes must move in on the left to neutralize the previously unneutralized acceptors over a distance Δl_1, and electrons must move into the transition region on the right to neutralize previously unneutralized donors over a distance Δl_2. For the idealized charge densities shown in Fig. 5·10c, it is possible to show that

$$\Delta l_1 \propto \sqrt{v_a} \qquad \Delta l_2 \propto \sqrt{v_a} \qquad \Delta E_{max} \propto \sqrt{v_a}$$

Now, when the transition region shrinks, the density gradient for both holes and electrons within the transition region increases. At the same time, the electric field at every point decreases. Thus, the balance between drift and diffusion has been upset in favor of diffusion. However, if the net current density passed through the transition region is small compared to the drift-diffusion current balance, then the densities must adjust so that thermal-equilibrium conditions are nearly established again.

If we designate the hole density at the edge of the transition region on the p side (that is, $x = l_1$) by p_{tp}, and the hole density at the edge of the transition region on the n side (that is, $x = l_2$) by p_{tn}, we must then have

$$p_{tn} = p_{tp} \exp\left[\frac{-q(\phi - v_a)}{kT}\right] \tag{5.37}$$

Equation (5·37) is the exact analog of the thermal-equilibrium relationship

$$p_n = p_p \exp\left(-\frac{q\phi}{kT}\right) \tag{5.38}$$

except that now the voltage drop across the transition region is $\phi - v_a$ and the carrier densities are no longer p_p and p_n. In fact, when we set $v_a = 0$, Eq. (5·38) reduces to Eq. (5·37), because then $p_{tn} = p_n$ and $p_{tp} = p_p$.

In a similar way, the electron density n_{tp} at $x = l_1$ is related to the electron density n_{tn} at $x = l_2$ by

$$n_{tp} = n_{tn} \exp\left[-\frac{q(\phi - v_a)}{kT}\right] \tag{5.39}$$

Now, using Eqs. (5·39) and (5·37), we may verify that

$$p_{tp}n_{tp} = p_{tn}n_{tn} \tag{5.40}$$

That is, the product of electron and hole densities is the same at each edge of the transition region. In fact, it is possible to show that the product of electron and hole densities is constant throughout the transition region, not just at the edges (see Prob. 5·4).

Now, electrical neutrality at the edges of the space-charge layer requires

$$p_{tp} = N_a + n_{tp} \qquad n_{tn} = N_d + p_{tn} \tag{5.41}$$

If we assume $N_a \gg n_{tp}$, then

$$p_{tp} = p_p$$

so that, in Eq. (5·37), we have

$$p_{tn} = p_p e^{-q\phi/kT} e^{qv_a/kT} = p_n e^{qv_a/kT} \tag{5.42}$$

If we also assume that $N_d \gg p_{tn}$, then

$$n_{tn} = n_n$$

so that

$$n_{tn}p_{tn} = n_n p_n e^{qv_a/kT} = n_i^2 e^{qv_a/kT} \tag{5.43}$$

which is the law of the junction.

By a similar development, or the direct use of Eq. (5·43), it is possible to show that

$$n_{tp} = n_p e^{qv_a/kT} \tag{5.44}$$

An idealized sketch indicating the variation of n and p through the transition region is shown in Fig. 5·11 for the simple case where $N_a = N_d$.

Though the above arguments were introduced for the case of a small forward bias, the same arguments hold when reverse bias is applied. In a reverse-bias condition, the transition region is widened to accommodate

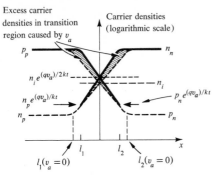

Excess carrier densities in transition region caused by v_a

Carrier densities (logarithmic scale)

p_p

n_n

$n_i e^{(qv_a)/2kt}$

n_i

$n_p e^{(qv_a)/kt}$

$p_n e^{(qv_a)/kt}$

n_p

p_n

x

l_1

l_2

$l_1(v_a = 0)$

$l_2(v_a = 0)$

FIG. 5·11 CARRIER DENSITIES IN THE TRANSITION REGION WITH FORWARD BIAS v_a APPLIED. THE JUNCTION IS ASSUMED TO BE SYMMETRICAL FOR SIMPLICITY IN REPRESENTATION.

the applied voltage, and the balance between drift and diffusion is upset in favor of the drift current. Formally, the results are the same as those for a forward bias, if it is recognized that now v_a is a negative number. Equation (5·43), then, suggests that the np product is decreased below n_i^2 in the transition region and at its edges.

The observations made above have revealed that the balance between drift and diffusion is essentially maintained in spite of typical current flow in the junction. From this idea, we have shown that *the application of bias to a junction regulates the density of carriers at the edges of the transition region.*

We shall see in the next chapter that *the currents that flow in a pn junction diode can be simply related to the carrier densities which exist at the edges of the transition region.* Thus, the relationship of voltage to current in a *pn* diode is an indirect one:

Applied voltage $\Rightarrow$

excess carrier densities in transition region $\Rightarrow$ current flow

The equation is also reversible:

Current flow $\Rightarrow$

excess carrier densities $\Rightarrow$ voltage across the transition region

5·6 JUNCTION CAPACITANCE

Thus far, potential drops across the transition region have been associated with a space-charge layer which extends on both sides of the metallurgical junction. In the idealized-charge-block model, the magnitude of the total charge on either side of the metallurgical junction may be written

$$Q = ql_1 N_a A = ql_2 N_d A \tag{5·45}$$

where A is the area of the junction, and the other symbols are as previously defined. The potential drop associated with given values of l_1 and l_2

is found from

$$\phi - v_a = -\int_{l_1}^{l_2} E(x)\,dx = \frac{q}{2\epsilon}(N_a l_1^2 + N_d l_2^2) \qquad (5\cdot46)$$

Now, a change in v_a requires a readjustment of l_1 and l_2 to the proper values. This is accomplished by the essentially instantaneous movement of majority carriers into (or out of) the transition region (see Fig. 5 · 12).

However, a change in l_1 and l_2 implies a change in the charge associated with the dipole layer through Eq. (5 · 45). Thus, this charge must be moved into or out of the transition region to accommodate a change in junction voltage. And this means we must associate a *capacitance* with the transition region. The capacitance will be a nonlinear function of voltage, so it is necessary to consider carefully how it should be defined.

Figure 5 · 13 has been drawn to aid in this discussion. A generalized charge-voltage curve is drawn and a bias point (Q_0, V_0) is shown on the curve. If a small change ΔV is made in the voltage, a charge ΔQ will have to be supplied. The magnitude of ΔQ will be approximately

$$\Delta Q = (\text{slope of } QV \text{ curve})\, \Delta V \qquad (5\cdot47)$$

The associated capacitance is

$$C = \frac{\Delta Q}{\Delta V} = \text{slope of } QV \text{ curve} \qquad (5\cdot48)$$

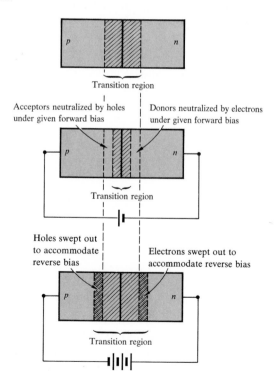

FIG. 5 · 12   THE TRANSITION REGION UNDER VARIOUS BIAS CONDITIONS

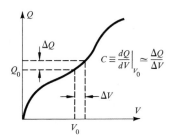

Because this C is defined only for small changes in V, it is called a *small-signal capacitance*.

If the QV plot were a straight line emanating from the origin, then $\Delta Q/\Delta V$ and Q/V would be equal. Then

$$C = \frac{Q}{V} \qquad (5 \cdot 49)$$

However, Eq. (5·49) is a special case of the more general Eq. (5·48).

To apply this idea to a junction, we shall first determine the Q-V relationship and then find dQ/dV.

The charge on one "plate" of the capacitor (that is, from $x = 0$ to $x = l_1$) is

$$Q = -qAN_a l_1 = -qA\,|N_d l_2| \qquad (5 \cdot 50)$$

If we assume $N_a \gg N_d$ for simplicity, then l_2 may be written as

$$l_2 = \sqrt{\frac{2\epsilon(\phi - v_a)}{qN_d}} \qquad (5 \cdot 51)$$

Hence
$$Q = -A\,\sqrt{2\epsilon qN_d}\,\sqrt{(\phi - v_a)} \qquad (5 \cdot 52)$$

and
$$C = \frac{dQ}{dV} = \frac{A\,\sqrt{2\epsilon qN_d}}{2\,\sqrt{\phi - v_a}} \qquad (5 \cdot 53)$$

Equation (5·53) can be recast in two useful ways:

$$C = \frac{C(0)}{\sqrt{1 - \dfrac{v_a}{\phi}}} \qquad (5 \cdot 54)$$

$$C = \frac{\epsilon A}{l_2} \qquad (5 \cdot 55)$$

Equation (5·54) expresses C at any given applied voltage v_a in terms of the zero-bias capacitance $C(0)$ and the built-in potential ϕ. Equation (5·55) shows that the capacitance can be calculated by the same formula as a parallel plate capacitor of area A spacing l_2.[1] In these terms, we can say that the spacing between the plates of the capacitor is a function of voltage, so C is also a function of voltage.

In either case, the junction capacitance increases with forward bias and decreases with reverse bias on the diode. As $v_a \to \phi$, the spacing of the capacitor "plates" approaches zero, so $C \to \infty$.

As a practical matter, the junction capacitance dominates the high-frequency impedance of a pn junction when it is operated in a reverse-bias condition (and below its reverse breakdown voltage.) Other phenomena to be described in the next chapter dominate the impedance in the forward-bias region.

EXAMPLE. For the junction parameters considered earlier

$$\sigma_p = 10^{-3} \text{ mho/cm} \qquad \sigma_n = 1 \text{ mho/cm}$$

we found that

$$l_1 = 3.3 \times 10^{-8} \text{ cm} \qquad l_2 = 6.8 \times 10^{-5} \text{ cm}$$

Thus, if the area of the junction were 10^{-3} cm², its zero-bias small-signal capacitance would be

$$C = \frac{\epsilon A}{l_2} = 27 \text{ pfarads}$$

(1 pfarad $= 10^{-12}$ farad). Since the built-in voltage is 0.4 volt for this junction, a reverse bias of 10 volts will change the capacitance to

$$C(-10 \; v) = \frac{C(o)}{\sqrt{1 + |v|/\phi}} = \frac{27}{\sqrt{26}} \sim 5.4 \text{ pfarads}$$

5·6·1 Applications of the voltage-variable capacitance characteristic of a pn junction. The fact that a reverse-biased pn junction possesses a voltage-variable capacitance of useful size for electronic circuits leads to several interesting applications. One of the simplest of these is voltage tuning of an LC resonant circuit. The basic circuit diagram for this arrangement is shown in Fig. 5·14. The tank circuit is simply shunted by the diode. Z is an RF-isolation resistor (or choke) and C_2 is a dc-blocking capacitor. The reverse bias across the diode is varied by adjusting R_1. The resulting changes in capacitance of the diode change the resonant frequency of the tank circuit. This basic scheme of tuning may be used to frequency modulate an oscillator or automatically control the frequency of a local oscillator in a radio receiver.

The voltage-variable capacitance of a pn junction also finds applications

---

[1] When N_a and N_d are comparable, the mathematical manipulations are somewhat more complicated, but it is still possible to show that the small-signal capacitance of the junction is

$$C = \frac{\epsilon A}{l_1 + l_2}$$

This and some other interesting results on junction capacitance are the subject of Probs. 5·6–5·10.

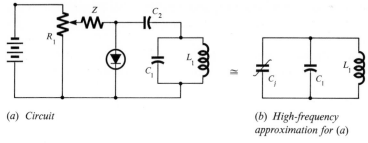

(a) Circuit

(b) High-frequency
approximation for (a)

FIG. 5·14   USE OF JUNCTION CAPACITANCE FOR TUNING A TANK CIRCUIT

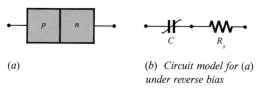

(a)

(b) Circuit model for (a)
under reverse bias

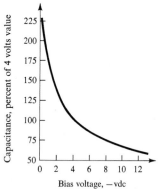

Bias voltage, −vdc

(c) Capacity vs. bias voltage (all types)

FIG. 5·15   CHARACTERISTICS OF A COMMERCIAL VARACTOR DIODE
[Permission of Pacific Semiconductors, Inc.]

ELECTRICAL SPECIFICATIONS

Varicap type number	Capacitance @ −4 volts ($\mu\mu f$)	Maximum operating voltage (volts)	Series resistance		Typical Q @ 50 mc
			Maximum (ohms)	Typical (ohms)	
V20	20	−20	18	8.5	18.7
V27	27	−20	14	7.5	15.7
V33	33	−20	12	6.6	14.6
V39	39	−20	10	5.4	15.1
V47	47	−20	7.5	4.4	15.4
V56	56	−15	7	4.2	13.5

in voltage-controlled filter networks, self-balancing bridge circuits, and in parametric amplifiers and harmonic generators.

Since "voltage-variable capacitor" is a rather unwieldly name, diodes which are made for this purpose are usually called *varactors*. Typical manufacturer's data on a commercially available varactor is given in Fig. 5·15. The maximum operating voltage for these devices (and *pn* junctions in general) will be taken up in the next chapter.

5·7 SUMMARY

The ideas in this chapter which are most important for our future study may be summarized as follows.

There is a region around a *pn* metallurgical junction in which a dipole layer of space charge exists. This region is called the transition region or space-charge layer. For a *step junction,*

1. The width of the space-charge layer is determined by the applied voltage through the formulas

$$l_1 = \left[\frac{2\epsilon(\phi - v_a)N_d}{qN_a(N_a + N_d)} \right]^{1/2} \tag{5·56}$$

$$l_2 = \left[\frac{2\epsilon(\phi - v_a)N_a}{qN_d(N_a + N_d)} \right]^{1/2} \tag{5·57}$$

where ϕ is the contact potential, and v_a is the applied voltage (positive when v_a is of opposite polarity to ϕ).

2. The contact potential ϕ depends only on the doping densities N_a and N_d, through the formula

$$\phi = \frac{kT}{q} \ln \frac{N_a N_d}{n_i{}^2} \tag{5·58}$$

3. The product of electron and hole densities at any point within the transition region is

$$np = n_i{}^2 e^{qv_a/kT} \tag{5·59}$$

At $x = l_2$, this formula gives

$$p = p_n e^{qv_a/kT} \tag{5·60}$$

so long as $p \ll N_d$. At $x = l_1$,

$$n = n_p e^{qv_a/kT} \tag{5·61}$$

so long as $n \ll N_a$. Equation (5·59) is called the law of the junction. Equations (5·60) and (5·61) are the most frequently used forms of it.

4. There is a nonlinear, voltage-dependent junction capacitance associated with the space-charge layer. The capacitance is

$$C = \frac{\epsilon A}{l_1 + l_2} \tag{5·62}$$

where l_1 and l_2 are voltage dependent [Eqs. (5·56) and (5·57)]. C may also be written as

$$C = \frac{C(o)}{\sqrt{1 - v_a/\phi}} \qquad (5 \cdot 63)$$

for the step junction.

REFERENCES

Shockley, W.: "Electrons and Holes in Semiconductors," D. Van Nostrand Company, Inc., Princeton, N.J., 1950, chap. 12.

Linvill, J. G., and J. F. Gibbons: "Transistors and Active Circuits," McGraw-Hill, Book Company, New York, 1961, chap. 3.

Gray, P. E., D. DeWitt, A. R. Boothroyd, and J. F. Gibbons: "Physical Electronics and Circuit Models of Transistors," SEEC Notes, vol. 2, John Wiley & Sons, Inc., New York, 1964, chap. 2.

Middlebrook, R. D.: "An Introduction to Junction Transistor Theory," John Wiley & Sons, Inc., New York, 1956, chap. 6.

Ramey, R. L.: "Physical Electronics," Wadsworth Publishing Company, Belmont, Calif., 1961, chap. 6.

Moll, J. L.: "Physics of Semiconductors," McGraw-Hill Book Company, New York, 1964, chap. 7.

DEMONSTRATION

pn-JUNCTION ACTION Most of the demonstrations of *pn*-junction action in semiconductors are related to diode performance and thus can be better understood after the material of the next chapter is digested. One useful demonstration of junction action *per se,* however, is as follows. Measure the capacitance-versus-voltage characteristic of several types of diodes in the circuit arrangement shown in the figure. Relate the measurements to the fabrication process (step junction, linearly graded junction, etc.). Estimate the built-in potential for one of the diodes from these measurements (see Probs. 5·7 and 5·10).

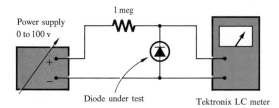

Power supply 0 to 100 v — 1 meg — Diode under test — Tektronix LC meter

FIG. D5·1

PROBLEMS

5·1 A *pn* step junction is made with $N_a = N_d = 10^4 n_i$, in material for which $n_i = 10^{13}/\text{cm}^3$, $\epsilon = 10^{-12}$ farad/cm. Assuming a rectangular-charge distribution, determine ϕ, the length of the transition region, and the maximum electric field in the transition region (at zero bias). Repeat using $n_i = 10^{10}/\text{cm}^3$ (to approximate a silicon junction.)

5·2 Show that if the potential drop at a point x in the transition region of a step junction is measured with respect to the point $x = l_1$, then the hole and electron densities at any point x can be written

$$p = p_p e^{-qv_x/kT} \tag{a}$$
$$n = n_p e^{qv_x/kT} \tag{b}$$

Using these results, show that v_x must satisfy the following equation:

$$\frac{d^2 v_x}{dx^2} = + \frac{q}{\epsilon}(- p_p e^{-qv_x/kT} + n_p e^{qv_x/kT} + N_d - N_a) \tag{c}$$

Equation (c) is a nonlinear equation, the solution of which will yield the precise form of $v(x)$. The use of Eqs. (a) and (b) will then yield the exact carrier-density profiles in the transition region.

5·3 *pn* junctions made by the diffusion process can have doping profiles like those shown in Fig. P5·3. The net doping density $d(x)$ in the vicinity of the metallurgical junction can be considered to vary linearly with x, as shown. The transition region extends an equal distance l on both sides of the metallurgical junction.

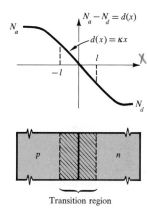

FIG. P5·3

Assuming $K = 10^{22}/\text{cm}^4$,

1. Plot the net doping density as a function of x from $x = 0$ to $x = 5 \times 10^{-5}$ cm. Use a logarithmic scale for the doping density.
2. On the assumption that the charge contributed by mobile carriers in the transition region (including the edges) can be neglected, sketch the electric field in the transition region and its integral, the electrostatic potential. Obtain formulas for $E(x)$ and ϕ, for arbitrary values of K and l.
3. Show that the contact potential can be written as

$$\phi = \frac{kT}{q} \ln \frac{K^2 l^2}{n_i^2}$$

4. Plot the ϕ's obtained in (2) and (3) as a function of l, assuming $K = 10^{22}/\text{cm}^4$.

Determine the length of the transition region at zero bias. Assume $n_i = 10^{13}/\text{cm}^3$, $\epsilon = 10^{-12}$ farad/cm for simplicity.

5. Discuss the assumptions made in (2) with the use of these results.
(Note: The capacitance of this junction is the subject of Prob. 5 · 10.)

5·4 It is suggested in the text that drift and diffusion tendencies balance each other in the transition region, even when a bias voltage is applied. Using this assumption, show that the hole densities at two arbitrary points within the transition region are related by

$$\ln\frac{p(x = x_1)}{p(x = x_2)} = -\frac{q}{kT}(v_{x_1} - v_{x_2})$$

Derive a corresponding relation for the electron densities at the same two points. Then, show that the product of electron and hole densities is constant throughout the transition region, even in the presence of a bias voltage (so long as the currents due to drift and diffusion tendencies are large compared to the current being passed through the junction.)

5·5 An abrupt junction exists between p and n material in germanium. The acceptor doping level and the donor doping level are equal, each being $10^{17}/\text{cm}^3$. The entire semiconductor bar is illuminated with the result that the hole density in the neutral n material rises to $10^{13}/\text{cm}^3$. No significant current is permitted to flow through the junction.

1. What will a voltmeter read when connected across the junction with its positive terminal on the p side?
2. What is the electron density in neutral p material?
3. What is the length of the transition region when the light is on?

5·6 Show that the small-signal capacitance of a pn step junction can be written as

$$C = \frac{\epsilon A}{l_1 + l_2} = \frac{C(0)}{\sqrt{1 - v_a/\phi}}$$

for arbitrary N_a and N_d (assuming the charge of mobile change within the space-charge layer may be neglected).

5·7 Show that a plot of $1/C^2$ vs. v is a straight line for a pn step junction. Also show that $1/C^2 = 0$ when $v_a = \phi$. Now, using the curve given in Fig. 5 · 15c, plot $1/C^2$ as a function of v and estimate the built-in voltage ϕ for this junction.

5·8 The network shown in Fig. P5 · 8 determines the frequency of the oscillator to which it is connected. Plot this frequency as a function of the position of the control C from 0 at the bottom to 100 at the top. The effect of the oscillator voltage on the junction capacitance can be neglected.

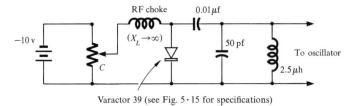

Varactor 39 (see Fig. 5·15 for specifications)

FIG. P5·8

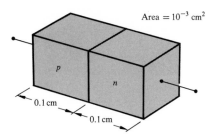

Area = 10^{-3} cm²

p

n

0.1 cm

0.1 cm

Resistivities: $\rho(p$ side$) = 10^{-3}$ Ω-cm

$\rho(n$ side$) = 0.1$ Ω-cm

FIG. **P5·9**

5·9 A germanium diode and its physical parameters are shown in Fig. P5·9. Calculate
1. The built-in voltage ϕ developed across the transition region.
2. The length of the transition region in n- and p-type material.
3. The capacitance of the junction at zero bias and a reverse bias of 10 volts.
4. The body resistance in series with this capacitance.
5. The "cut-off" frequency for this capacitance at zero bias and 10 volts reverse bias. (The cut-off frequency is the frequency at which the device exhibits a capacitive reactance which is equal to the bulk-series resistance. Below this frequency, the impedance of the device is essentially ~~capacitive~~, as long as the reverse-bias condition is maintained.)   *RESISTIVE*

5·10 Show that the capacitance of a linearly graded pn junction (that is, $N_a - N_d = KX$, as in Fig. P5·3) can be written as

$$C = 0.436 \, (qK)^{1/3}\epsilon^{2/3}A(\phi - v)^{-1/3}$$

(neglecting the space charge of mobile carriers in the transition region). Evaluate C for the junction parameters given in Prob. 5·3 (use $A = 10^{-3}$ cm²). Show that the dimensions of the equation are farads. Show that a plot of $1/C^3$ vs. v is a straight line, with one point being $(1/C^3 = 0, \phi = v)$.

6 DIODES AND RELATED ONE-JUNCTION DEVICES

IN THE PRECEDING two chapters we have studied the principles of carrier motion in semiconductors and the properties of *pn* junctions. Using these basic principles, we can analyze the behavior of most semiconductor devices.

Our primary objective in this chapter is to show how the low-frequency properties of a *pn*-junction diode arise from these basic principles. This is a subject of interest in its own right because of the many applications junction diodes have in electronics. However, the ideas to be developed in this chapter are of much more importance than diode applications would indicate; they are, in fact, basic to an understanding of all junction devices.

In particular, all of the ideas needed to understand the behavior of transistors will be presented in this chapter. It is therefore important to understand these ideas thoroughly.

6·1 JUNCTION-DIODE CHARACTERISTICS

It is useful to begin our discussion of *pn*-junction diodes with a brief consideration of the experimental characteristics which they exhibit. For this purpose a typical dc or low-frequency voltage-current characteristic is drawn in Fig. 6·1. The figure has not been drawn to scale, to emphasize important features of the curve. The diode symbol and reference directions for voltage and current are also defined on this figure.

When the diode is biased so that current flows in the direction of the arrow, the diode is said to be in the forward bias or forward conduction range (first quadrant of Fig. 6·1). When the bias causes current to flow in a direction opposite to that of the arrow, the diode is in the reverse bias range (third quadrant of Fig. 6·1).

The experimental characteristics for most junction diodes show that there is a range of forward current and voltage for which the functional relationship between the diode current i and the voltage v is of the form

$$i = I_0 \left(e^{qv/kT} - 1 \right) \tag{6·1}$$

This is a theoretically predictable relationship. It arises from the influence of the junction on the v-i characteristic and usually applies over a rather wide current range (in some cases from 10^{-10} amp to 10^{-2} amp or more).

Equation (6·1) fails to represent the experimental v-i curve when the current becomes high enough to produce a significant voltage drop in the series resistance provided by the contacts and the body of the diode. As a

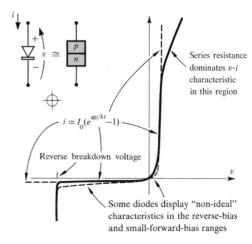

Series resistance dominates v-i characteristic in this region

$i = I_0 (e^{qv/kt} - 1)$

Reverse breakdown voltage

Some diodes display "non-ideal" characteristics in the reverse-bias and small-forward-bias ranges

FIG. 6·1 LOW FREQUENCY v-i CHARACTERISTIC OF A *pn*-JUNCTION DIODE practical matter, the current level at which most of the applied voltage is lost in ohmic drops may be anywhere from 1 ma to 100 amps, depending on the diode design.

Equation (6·1) may also fail to represent the experimental *v-i* curve at very low values of current, though this is a phenomenon which can best be described later.

Equation (6·1) also correctly predicts the reverse-bias *v-i* relationship for many diodes until the diode "breaks down." The reverse breakdown voltage may be anywhere from a few volts to perhaps 1,000 volts, depending on the diode design. Furthermore, this breakdown is not (or *need not* be) a destructive one, and in fact some diodes are purposely made to operate in this region. We shall study in this chapter the physical mechanism for this breakdown phenomenon.

As was the case for low forward biases, Eq. (6·1) may also fail to represent the actual reverse characteristic in some instances, though again this is a phenomenon which will be postponed until later.

From a circuit standpoint, the diode is an interesting element because of the abrupt *nonlinearity* which it exhibits in the region near $i = 0$, $v = 0$. Because of this inherent nonlinearity, the diode finds many important applications in the field of power and signal switching. In these applications, the diode is considered to be ON under forward-bias conditions and OFF under reverse-bias-conditions. Its ON and OFF resistances and other parameters of interest can be determined directly from its *v-i* characteristic.

However, there are certain questions of interest in some switching applications which cannot be answered directly from the *v-i* curve. For example, in a computer application we may be very interested in knowing how long it takes the diode to switch from a given point on its forward conduction curve to a given point on its reverse characteristic; or what type of actuating signal should be used to minimize this switching time. And since time is not a variable in Fig. 6·1, the *v-i* plot does not directly answer questions about the transient performance of the diode.

In order to discuss simply these various aspects of diode performance, we shall develop the theory for a *pn* diode as follows. The *v-i* characteristic, which is a dc or low-frequency characterization of the diode, will be developed in this chapter. In Chap. 7 we shall consider some applications for which the *v-i* characteristic gives an adequate portrayal of the device. Then, in Chap. 8 we shall return to switching problems and a study of the transient behavior of junction diodes.

6·2 DEVELOPMENT OF A MODEL FOR ANALYSIS

The discussion given in Sec. 6·1 indicates that many junction diodes will exhibit a current-voltage law of the form

$$i = I_0(e^{qv/kT} - 1)$$

from reverse breakdown to a forward current where series resistance in the diode becomes important. In this section we wish to develop a model for analysis, which will allow us to concentrate on the physical processes which produce this v-i relationship. A brief inspection of the most common forms for actual diodes will be helpful for this purpose.

It has been indicated in previous chapters that there are basically three ways of preparing pn junctions: They can be introduced into a crystal as it is grown from the melt (Sec. 2·6 and Prob. 6·2); they can be prepared by alloying (Sec. 5·1·1); or they can be prepared by diffusion (Sec. 5·1·2). Sketches of representative diodes which result from these processes are shown in Fig. 6·2.

Now, when a bias voltage v_a is applied to the terminals of any of these diodes, only a portion of it will be developed across the transition region, the remainder being lost in IR drops in the diode body and at the contacts.

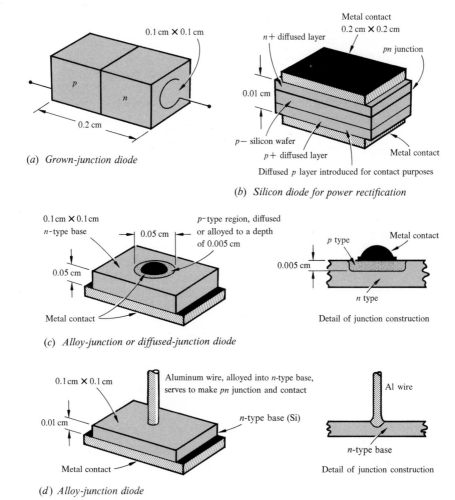

(a) Grown-junction diode

(b) Silicon diode for power rectification

(c) Alloy-junction or diffused-junction diode

(d) Alloy-junction diode

FIG. 6·2 REPRESENTATIVE FORMS OF PRACTICAL JUNCTION DIODES

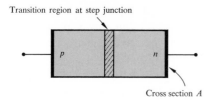

Transition region at step junction

Cross section A

FIG. 6·3   IDEALIZED SEMICONDUCTOR DIODE

As we have seen, the v-i characteristic of a practical diode will exhibit these ohmic effects. But since we are principally interested in the nonohmic relation between junction voltage drop and current, we begin by neglecting resistances in series with the junction, in order to concentrate on the junction action. The effects of series resistance can be added later quite simply.

With this approximation, the bias voltage all appears across the transition region, where its principal effect is to *create and maintain excess carrier densities* (that is, $np > n_i^2$) in the transition region *and* in the neutral n- and p-type bodies near the transition region. The amount of current flow is then determined by the rate at which these excess carriers recombine in the transition region, in the neutral p- and n-type bodies, and on those portions of the surface which are near the junction.

As a practical matter, all of these sites for recombination are important for understanding the overall behavior of an actual diode. However, over most of the forward-bias region, volume recombination in the neutral p- and n-type bodies is the dominant process, so we further simplify an actual diode by neglecting recombination on the surfaces and in the transition region. With this simplification, the current flow in all of the diodes shown in Fig. 6·2 becomes essentially one-dimensional, except for the diode in Fig. 6·2d, where the radial geometry may affect the current and carrier densities, if the junction radius is sufficiently small.

Again, in order to concentrate on the salient physical features of the performance, it is reasonable to set the geometrical differences between diodes aside, and consider a diode in which the current and carrier densities can vary in one dimension only. As a second step, we must consider the changes which need to be made in the analysis to account for recombination in the transition region and on the surfaces, and for nonplanar geometry.

The idealized diode which we wish to analyze is shown in Fig. 6·3. It is planar with a cross section A, it exhibits volume recombination only, and it has no series resistance. For simplicity, the pn junction in this diode will be considered to be a step junction, with N_a and N_d arbitrary.

6·3   QUALITATIVE DESCRIPTION OF THE OPERATION OF AN IDEAL DIODE

We may now describe the operation of the pn-diode model shown in Fig. 6·3 in a qualitative way. Suppose we impress a positive voltage v_a on the diode, as shown in Fig. 6·4. Then, according to the law of the junction, the minority carrier densities on each side of the transition region[1] must increase to

$$p(x = 0) = p_n e^{qv_a/kT} \qquad n(y = 0) = n_p e^{qv_a/kT} \qquad (6 \cdot 2)$$

Now, since the bias causes the hole density at $x = 0$ to become greater than p_n, it creates a density gradient which will cause holes to diffuse into the body of the n-type material. This diffusion of holes into the n-type material produces two effects:

1. It tends to decrease the hole density at $x = 0$ (the n-type edge of the transition region). Holes will then flow from the p-type material through the transition region to maintain the hole density at $x = 0$ at its required value, $p = p_n e^{qv_a/kT}$.

2. It provokes an essentially immediate flow of electrons in the n-type body so that space-charge neutrality is essentially maintained at every point in the n-type body.

In this way a forward bias permits holes to flow from neutral p-type material into neutral n-type material. The holes are said to be *injected* into the n-type body. Electrons will also be injected into the p-type body (which is also instantaneously neutralized by an appropriate flow of holes).

The actual amount of current which flows for a given bias depends on how rapidly these injected carriers leave the junction. This, in turn, depends on which minority-carrier flow mechanisms (diffusion and/or drift) are important, and what the lifetime of minority carriers is in the neutral regions of the diode. For the important case in which minority carriers flow

[1] We use x to denote distance into the n-type body from the n-type edge of the transition region, and y to denote distance into the p-type body from the p-type edge of the transition region to achieve symmetry and simplicity in later results (see Fig. $6 \cdot 4$).

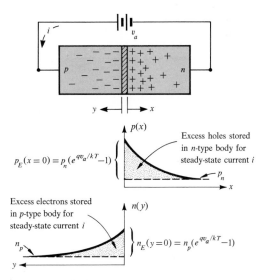

FIG. 6·4 A FORWARD-BIASED DIODE AND MINORITY-CARRIER DISTRIBUTIONS

only by diffusion, it is the *slope* of the hole-density curve at $x = 0$ that determines the rate at which holes diffuse away from the junction, and thus the steady-state hole current which needs to be supplied through the transition region to maintain the excess-hole density at $x = 0$ at its required value.

The process of analysis, therefore, aims at finding a mathematical expression for the excess-hole-density-distribution curve in the n-type body (and the electron density distribution in the p-type body) in the presence of an applied bias, so that the slope of this curve may be evaluated at $x = 0$ to find the hole current injected into the n-type body.

The principal result of the analysis is that the hole-density distribution in the n-type body is given by

$$p(x) = [p_E(x = 0)]e^{-x/L_p} + p_n = [p_n(e^{qv_a/kT} - 1)]e^{-x/L_p} + p_n \quad (6 \cdot 3)$$

where $p_E(x = 0)$ is the *excess* hole density at $x = 0$ and L_p is the *diffusion length*

$$L_p \equiv \sqrt{D_p \tau_p} \quad (6 \cdot 4)$$

This distribution is shown in Fig. $6 \cdot 4$. The decrease in hole density with x reflects the physical fact that as holes diffuse into the n-type body, their density will be reduced by recombination. At distances which are far enough removed from the pn junction ($x \gg L_p$), the hole density is essentially p_n.

The slope of the hole-density profile at $x = 0$ is, from Eq. $(6 \cdot 3)$,

$$\left. \frac{dp}{\partial x} \right|_{x=0} = - \frac{p_n}{L_p} (e^{qv_a/kT} - 1) \quad (6 \cdot 5)$$

from which we obtain

$$i_p(x = 0) \equiv - qAD_p \left. \frac{\partial p}{\partial x} \right|_{x=0} = \frac{qAD_p p_n}{L_p} (e^{qv_a/kT} - 1) \quad (6 \cdot 6)$$

By a similar set of steps the electron current injected into the p-type body will be found to be

$$i_n(y = 0) = \frac{qAD_n n_p}{L_n} (e^{qv_a/kT} - 1) \quad (6 \cdot 7)$$

so that the total current flowing across the junction will be of the form

$$i = I_0(e^{qv_a/kT} - 1)$$

which is the basic current-voltage law for the ideal diode. As we shall show, this law applies for both forward and reverse biases.

$6 \cdot 4$ QUANTITATIVE ANALYSIS OF THE IDEALIZED DIODE

In this section we give a mathematical derivation of the current-voltage relationship for our idealized junction diode. The device model to be analyzed is shown in Fig. $6 \cdot 5$. The following properties of the device are assumed to be known:

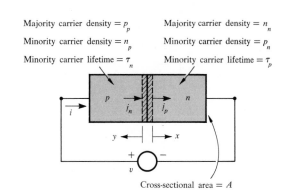

Majority carrier density $= p_p$ Majority carrier density $= n_n$

Minority carrier density $= n_p$ Minority carrier density $= p_n$

Minority carrier lifetime $= \tau_n$ Minority carrier lifetime $= \tau_p$

Cross-sectional area $= A$

FIG. 6·5   AN IDEAL DIODE AND ITS EXCITATION

A Cross-sectional area

p_p,n_n Equilibrium majority-carrier densities, uniform throughout the neutral p and n regions under zero-bias conditions

p_n,n_p Equilibrium minority-carrier densities, uniform throughout the neutral n and p regions under zero-bias conditions

τ_p Lifetime of holes in the n region

τ_n Lifetime of electrons in the p region

As suggested in the preceding section, the flow of current in the diode can be analyzed when the minority-carrier-density curves in the neutral n- and p-type bodies are known. To obtain these curves, we first develop a differential equation (called the continuity equation, for reasons which will shortly be apparent) whose solution will give us the minority-carrier densities as a function of x at each point of the diode.

6·4·1   *The continuity equation for holes in n-type material.*   The mathematical form of the hole-density curve in n-type material is determined by specifying the conditions which this curve must meet at every point. These conditions are developed in terms of the arbitrary hole-density curve $p(x)$ in Fig. 6·6.[1] If we focus attention on the small volume of material between the planes at x and $x + \Delta x$, then we can say that in the steady state, the rate at which holes enter the slab is equal to the rate at which they are being lost within it due to recombination, plus the rate at which they are flowing out across the plane at $x + \Delta x$:

Flow in = net recombination within the volume per unit time + flow out

$$(6\cdot8)$$

This type of equation is called a steady-state continuity equation, since it expresses continuity of hole flow in the volume being considered.[2]

Equation (6·8) leads to a differential equation which the hole-density

---

[1] An alternate but equivalent procedure for specifying the $p(x)$ curve is the subject of Prob. 6·2.

[2] Continuity equations for transient situations will be considered in Chap. 8.

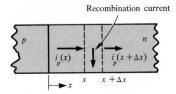

Recombination current

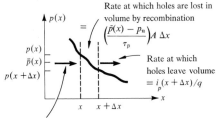

Rate at which holes are lost in volume by recombination

$$= \left(\frac{\bar{p}(x) - p_n}{\tau_p}\right) A \, \Delta x$$

Rate at which holes leave volume $= i_p(x + \Delta x)/q$

Rate at which holes enter volume $= i_p(x)/q$

Rate at which holes enter = rate at which holes are lost by recombination + rate at which holes leave

$$\frac{i_p(x)}{q} \qquad = \qquad \frac{\bar{p}(x) - p_n}{\tau_p} A \, \Delta x \qquad + \qquad \frac{i_p(x + \Delta x)}{q}$$

FIG. 6·6   DEVELOPMENT OF THE STEADY-STATE CONTINUITY EQUATION FOR HOLES IN n-TYPE MATERIAL

curve $p(x)$ must satisfy at every point. To see this, we let $\bar{p}(x)$ be the average hole density in the slab. Then, the average recombination rate within the volume can be written as

$$\frac{\bar{p}(x) - p_n}{\tau_p} A \, \Delta x$$

where the first factor is the average recombination rate per unit volume in the slab, and $A \, \Delta x$ is the volume.

Furthermore, if we let $i_p(x)$ and $i_p(x + \Delta x)$ denote the currents flowing into and out of the slab, respectively (see Fig. 6·8 for definition of signs), then $i_p(x)/q$ and $i_p(x + \Delta x)/q$ represent the hole flows, so that Eq. (6·8) becomes

$$\frac{i_p(x)}{q} = \frac{\bar{p}(x) - p_n}{\tau_p} A \, \Delta x + \frac{i_p(x + \Delta x)}{q} \qquad (6·9)$$

Equation (6·9) can be rearranged to read

$$-\frac{1}{qA} \frac{i_p(x + \Delta x) - i_p(x)}{\Delta x} = \frac{\bar{p}(x) - p_n}{\tau_p} \qquad (6·10)$$

Now, in the limit as $\Delta x \to 0$,

$$\bar{p}(x) \to p(x)$$

$$\frac{i_p(x + \Delta x) - i_p(x)}{\Delta x} \to \frac{di_p}{dx}$$

so Eq. (6 · 10) becomes

$$-\frac{1}{qA}\frac{di_p}{dx} = \frac{p(x) - p_n}{\tau_p} \tag{6 · 11}$$

Equation (6 · 11) is the standard mathematical form of the steady-state continuity equation for holes in an n-type semiconductor. Usually, the factor A is absorbed by writing the equation in terms of the current density j_p:

$$-\frac{1}{q}\frac{dj_p}{dx} = \frac{p(x) - p_n}{\tau_p} \tag{6 · 12}$$

Now, holes can move, in general, by both drift and diffusion, as indicated in the expression for hole current density j_p:

$$j_p = -qD_p\frac{dp}{dx} + q\mu_p p(x)E(x) \tag{6 · 13a}$$

However, recall that we are interested in current levels for which the series resistance of the diode is negligible. Therefore, we shall assume that the n-type body has a sufficient number of majority carriers to permit any current flow required with a negligibly small electric field.[1] Therefore, we may simplify Eq. (6 · 13a) by neglecting the drift term:

$$j_p = -qD_p\frac{dp}{dx} \tag{6 · 13b}$$

This is equivalent to assuming that *holes can move in n-type material by diffusion only.* When Eq. (6 · 13b) is used in Eq. (6 · 12), we have

$$D_p\frac{d^2p}{dx^2} = \frac{p(x) - p_n}{\tau_p} \tag{6 · 14}$$

This is the differential equation which the hole-density curve $p(x)$ must satisfy in the n-type body, assuming that holes move only by diffusion there.

6 · 4 · 2 *The boundary conditions.* In addition to satisfying the differential equation (6 · 14), the hole-density curve must have specific values at the boundaries of the region of interest. In particular, at $x = 0$, the law of the junction requires

$$p(x = 0) = p_n e^{qv/kT} \tag{6 · 15}$$

We also know that the presence of recombination in the n-type body will continuously reduce the excess-hole density, so that at great distances from the junction, there can be no excess density at all. Thus

$$\lim_{x \to \infty} p(x) = p_n \tag{6 · 16}$$

Of course, the n-type body is not infinitely long, but we will see presently

---

[1] Since $p(x) \ll n(x)$, this assumption implies that if the hole current in the n-type body is going to be important at all, it must be because of the diffusion component.

that it only needs to be long compared to a *diffusion length* L_p for the latter condition to apply, where

$$L_p \equiv \sqrt{D_p \tau_p} \tag{6·17}$$

Diodes for which this assumption cannot be made are considered in Prob. 6·3.

6·4·3 *Solution of the steady-state continuity equation.* To obtain the solution of Eq. (6·14) which fits these boundary conditions, we first define the excess-hole density p_E as

$$p_E(x) = p(x) - p_n \tag{6·18}$$

so that Eq. (6·14) becomes

$$D_p \frac{d^2 p_E}{dx^2} = \frac{p_E}{\tau_p} \tag{6·19}$$

The general solution of Eq. (6·19) is

$$p_E = B e^{-x/L_p} + C e^{x/L_p} \tag{6·20}$$

where $L_p = (D_p \tau_p)^{1/2}$. The boundary conditions require

$$p_E(0) = p_n(e^{qv/kT} - 1)$$

$$\lim_{x \to \infty} p_E(x) = 0$$

The second of these conditions can be satisfied in Eq. (6·20) only by setting $C = 0$. Then, the first condition can be applied to yield

$$B = p_E(0) = p_n(e^{qv/kT} - 1) \tag{6·21}$$

Using this in Eq. (6·20), we finally obtain

$$p_E(x) = p_E(0)e^{-x/L_p} = p_n(e^{qv/kT} - 1)e^{-x/L_p} \tag{6·22}$$

The excess-hole density thus decreases exponentially with distance into the *n*-type body, as shown in Fig. 6·7.

Several properties of this curve are shown on Fig. 6·7 and are of interest:

1. The slope of the curve at any point x is

$$-\frac{dp_E(x)}{dx} = \frac{p_E(x)}{L_p} \tag{6·23a}$$

and in particular the slope at $x = 0$ is $-[p_E(0)/L_p]$.

2. The area under the curve from any point x to ∞ is

$$\int_x^\infty p_E(x)\, dx = p_E(x)\, L_p \tag{6·23b}$$

and in particular the total hole charge stored in the *n*-type body is

$$Q_p = qAL_p p_E(0) \tag{6·23c}$$

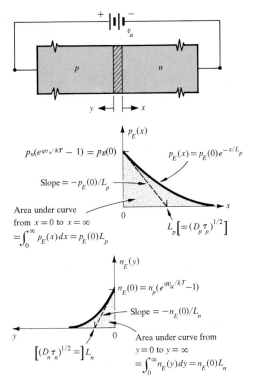

3. The average distance $\langle x \rangle$ that a hole will diffuse before recombining is

$$\langle x \rangle = L_p \qquad (6 \cdot 23d)$$

These properties are the subject of Prob. 6 · 1.

6·4·4 The distribution of electrons in p-type material. By an exactly similar set of steps to those used in the preceding paragraphs, we may find the electron concentration in p-type material.

The steady-state continuity equation for electrons is

$$-\frac{1}{q}\frac{\partial j_n}{\partial y} = \frac{n - n_p}{\tau_n} \qquad y \geq 0 \qquad (6 \cdot 24)$$

where j_n is defined to be positive when conventional current flows to the right. The origin of the distance coordinate y is at the p-type edge of the transition region, and y is positive when measured into the p-type body, as in Figs. 6 · 5 and 6 · 7.

If we assume that electrons can flow only by diffusion in p-type material, then

$$j_n = -qD_n \frac{\partial n}{\partial y} \qquad (6 \cdot 25)$$

When this is substituted into Eq. (6·24) we obtain

$$D_n \frac{d^2 n}{dy^2} = \frac{n - n_p}{\tau_n} \tag{6·26}$$

as the equation which $n(y)$ must satisfy.

The boundary conditions on n are

$$n(y = 0) = n_p e^{qv/kT} \qquad \lim_{y \to \infty} n(y) = n_n$$

The solution to Eq. (6·26) which fits these boundary conditions is

$$n_E(y) = n_E(0)e^{-y/L_n} \tag{6·27}$$

where $n_E(y)$ is the excess electron density in p-type material and

$$n_E(0) = n_p(e^{qv/kT} - 1) \qquad L_n = (D_n \tau_n)^{1/2}$$

Equation (6·27) is also plotted in Fig. 6·7.

6·4·5 The total diode current. So far, our considerations have led to a knowledge of the minority-carrier distributions in the neutral regions of the diode. On the assumption that minority carriers flow only by diffusion, we can now obtain the minority-carrier currents at each point in the neutral regions:

$$i_p(x) = -qAD_p \frac{dp}{dx} = -qAD_p \frac{dp_E}{dx} = \frac{qAD_p p_E(x)}{L_p} \qquad 0 \le x \le \infty \tag{6·28}$$

$$i_n(y) = -qAD_n \frac{dn}{dy} = -qAD_n \frac{dn_E}{dy} = \frac{qAD_n n_E(y)}{L_n} \qquad 0 \le y \le \infty \tag{6·29}$$

Equations (6·28) and (6·29) reveal the very important fact that *the minority-carrier currents are directly proportional to the excess-minority-carrier densities* at each point in the diode body. This proportionality arises from the fact that the diffusion law used to relate current to minority-carrier density in the diode is linear. *All the nonlinearity in the v-i characteristic of an idealized junction diode arises from the nonlinear relation between junction voltage and minority-carrier densities at the edges of the transition region.*

Now, to evaluate the *total* diode current, we must know the sum of the hole and electron currents at a given plane through the diode. We do not actually know this. However, we do know the hole current on one side of the transition region ($x = 0$) and the electron current on the other ($y = 0$). And if we neglect recombination in the transition region, then the hole current entering the transition region at $y = 0$ will be the same as the hole current emerging from it at $x = 0$, and similarly for the electron currents. Therefore, the total current can be computed as

$$i = i_p(x = 0) + i_n(y = 0) = \frac{qAD_p p_E(0)}{L_p} + \frac{qAD_n n_E(0)}{L_n} \tag{6·30}$$

It is customary to define

$$I_{p0} = \frac{qAD_p p_n}{L_p} \tag{6·31a}$$

$$I_{n0} = \frac{qAD_n n_p}{L_n} \tag{6.31b}$$

so that Eq. (6·31b) becomes

$$i = I_{p0}\frac{p_E(0)}{p_n} + I_{n0}\frac{n_E(0)}{n_p} = (I_{p0} + I_{n0})(e^{qv/kT} - 1)$$

$$= I_0(e^{qv/kT} - 1) \tag{6.32}$$

We thus obtain the standard expression for the v-i characteristic of an ideal pn-junction diode. In this solution, I_0 is called the *saturation current* since the reverse current saturates (that is, becomes independent of voltage) at the value I_0 when $v \leq -0.1$ volt.

Several important properties of the ideal v-i characteristic are considered in the problems at the end of the chapter. The temperature dependence of the ideal diode v-i characteristic is the subject of Prob. 6·5. The dynamic resistance of the ideal junction diode is considered in Prob. 6·6.

6·4·6 *Saturation current and the reverse characteristic of the idealized diode.* While the preceding discussion has been carried out with forward-bias conditions in mind, the formal analysis applies equally well for a reverse-bias condition. In this case, the law of the junction indicates that the minority-carrier densities are *depressed* below their equilibrium values in the neutral regions, as shown in Fig. 6·8. There will now be a diffusion

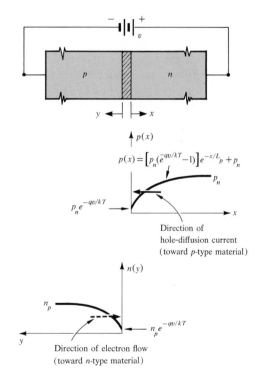

FIG. 6·8 **MINORITY-CARRIER DENSITIES UNDER REVERSE BIAS**

of holes from the neutral n-type material *toward* the pn junction. These holes tend to increase the density of holes at $x = 0$ above the value required by the junction bias, thus causing a hole flow from right to left through the transition region to maintain the hole density at $x = 0$ at its required value.

The hole current flowing through the transition region will again be linearly proportional to the slope of the hole density profile at $x = 0$. However, the law of the junction indicates that after about 0.1 volt of reverse bias is applied, the hole density at $x = 0$ will be about $0.01 p_n$. Further increases in reverse bias can only reduce the hole density at $x = 0$ by 1 percent (that is, to zero), and will therefore have an insignificant effect on the slope of the hole density curve at $x = 0$. Thus, the reverse current is independent of bias voltage after a reverse bias of about 0.1 volt.

6·4·7 The relationship between minority-carrier stored charge and the diode current.

Thus far, our analysis of the junction diode has provided us with a detailed view of the distribution and flow of minority carriers in the neutral diode bodies. The analysis focuses primary attention on the diffusion mechanism as being responsible for current flow, the recombination process entering primarily in the determination of the hole-density profile $p(x)$.

There is an alternate and very useful viewpoint in which the importance of recombination is stressed, however. To develop this viewpoint, let us first recognize that when a diode is conducting a steady current in the forward direction, there will be an excess density of holes stored in a small volume of neutral n-type material near the junction ($\sim 3L_p$ in length). Now these holes have a lifetime τ_p, so on the average the entire population of excess holes in n-type material will recombine every τ_p seconds. In order to maintain a steady-state condition, these holes have to be replaced by holes injected from the p material. Thus, if we let Q_p be the total excess charge of holes stored in the n-type body (see Fig. 6·9), then a charge of Q_p must enter the n-type body every τ_p sec. The average hole current flowing into the n-type body at $x = 0$ is then

$$i_p = \frac{Q_p}{\tau_p} \tag{6·33a}$$

An equal electron current will enter the n-type body on the right to supply the electrons for this recombination, as shown in Fig. 6·9.

Similarly, the total excess population of electrons in the p-type material Q_n will have to be resupplied every τ_n sec, so the electron current entering the p-type material at $y = 0$ will be

$$i_n = \frac{Q_n}{\tau_n} \tag{6·33b}$$

This will also be matched by hole current from the left (see Fig. 6·9). Thus, in the absence of recombination in the transition region, the total

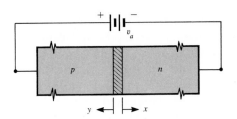

(a) pn diode with forward bias v_a

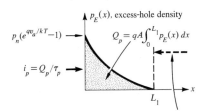

(b) Stored holes in the n-type body have a lifetime τ_p, giving rise to an injected hole current Q_p/τ_p.

$p_n(e^{qv_a/kT}-1) \rightarrow$

$Q_p = qA \int_0^{L_1} p_E(x)\,dx$

$i_p = Q_p/\tau_p \rightarrow$

$p_E(x)$, excess-hole density

L_1

Electron flow in the amount Q_p/τ_p to supply electrons for recombination in n-type body

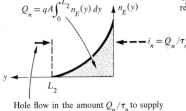

$Q_n = qA \int_0^{L_2} n_E(y)\,dy$

$n_E(y)$

$\leftarrow\!-\,- i_n = Q_n/\tau_n$

(c) Stored electrons in the p-type body have a lifetime τ_n, giving rise to an injected electron current Q_n/τ_n.

L_2

Hole flow in the amount Q_n/τ_n to supply holes for recombination in p-type body

FIG. 6·9   DEPENDENCE OF INJECTED CURRENTS ON TOTAL MINORITY-CARRIER CHARGE STORED IN THE NEUTRAL DIODE BODY

current crossing the junction can be related quite simply to the total excess charge of minority carriers stored in these regions:

$$i = \frac{Q_p}{\tau_p} + \frac{Q_n}{\tau_n} \qquad (6\cdot33c)$$

This fundamental relation can be used to obtain the dc diode current as follows. First, Q_p can be found by integrating the known excess-hole-density function $p_E(x)$:

$$Q_p = qA \int_0^\infty p_E(x)\,dx = qAL_p p_E(0) = qAL_p p_n(e^{qv/kT} - 1) \quad (6\cdot34)$$

Similarly,
$$Q_n = qAL_n n_p(e^{qv/kT} - 1) \qquad (6\cdot35)$$

When these relations are substituted into Eq. (6·33c), we obtain

$$i = I_0(e^{qv/kT} - 1) \qquad (6\cdot36)$$

We shall see later that this method of analysis, called the *stored-charge formulation*, is very useful for studying the transient performance of junction diodes and understanding the internal operation of transistors.

6·4·8   Majority-carrier effects in the junction diode.   Our analysis of the junction diode has to this point been concerned primarily with the distribution and flow of minority carriers. To complete our discussion of the operation of the diode, we shall next consider briefly the role of the majority carriers. In this section we will use our information about total current and minority-carrier densities to deduce the distribution and flow of majority carriers. We do this as follows:

1. Space-charge neutrality has been assumed. Since the minority-carrier distributions are known, the majority-carrier concentration may be found from space-charge neutrality requirements:

$$n(x) = p(x) + N_d, x \geq 0 \qquad p(y) = n(y) + N_a, y \geq 0 \qquad (6 \cdot 37)$$

2. Majority-carrier current at every point is obtained by the requirement of current continuity. The majority-carrier current has both drift and diffusion components. The diffusion component is evaluated from the knowledge of majority-carrier density gradients obtained in step 1. The drift component is then obtained by subtraction.

When the majority carrier densities and currents have been established we will be able to check several assumptions we have made in the analysis.

6·4·9 *An illustrative example.* While a straightforward analysis for majority-carrier effects based on the ideas presented in (1) and (2) above will give general formulas for the majority-carrier currents and densities, it is somewhat simpler to carry out this analysis in an illustrative example. We thus proceed to analyze in detail the germanium diode shown in Fig. 6·10. This diode is typical of grown-junction diodes, and has the parameters which have been used in previous examples.

As starting information for the example, we assume that the resistivities for each region ($\rho_p = 10^{-3}$ ohm-cm, $\rho_n = 1$ ohm-cm), the area of the diode (10^{-2} cm), the lengths of the p- and n-type bodies (0.1 cm each), and the minority carrier lifetimes ($\tau_p = \tau_n = 10 \, \mu\text{sec}$) are known. Note that these data are only material parameters and geometrical specifications, and by choosing different materials and dimensions we could design a diode with different properties.

(a) *Zero-bias calculations.* To review some of the ideas presented in Chap. 5 we first consider the pn junction with zero bias applied. The built-in voltage, or contact potential, between the p- and n-type bodies may be calculated as follows:

$$\phi = \frac{kT}{q} \ln \frac{p_p}{p_n} = 0.41 \text{ volt}$$

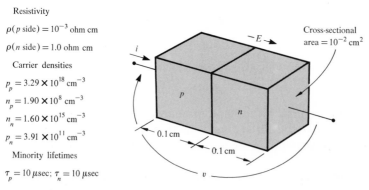

Resistivity

$\rho(p \text{ side}) = 10^{-3}$ ohm cm

$\rho(n \text{ side}) = 1.0$ ohm cm

Carrier densities

$p_p = 3.29 \times 10^{18} \text{ cm}^{-3}$

$n_p = 1.90 \times 10^{8} \text{ cm}^{-3}$

$n_n = 1.60 \times 10^{15} \text{ cm}^{-3}$

$p_n = 3.91 \times 10^{11} \text{ cm}^{-3}$

Minority lifetimes

$\tau_p = 10 \, \mu\text{sec}; \tau_n = 10 \, \mu\text{sec}$

Cross-sectional area $= 10^{-2} \text{ cm}^2$

0.1 cm

0.1 cm

FIG. 6·10 GERMANIUM DIODE AND PARAMETERS

The extent of the transition region into p- and n-type material may now be obtained as follows:

$$\phi = \frac{l_1{}^2 \, qp_p(1 + p_p/n_n)}{2\epsilon} \qquad l_2 n_n = l_1 p_p$$

$$l_2 = 6.81 \times 10^{-5} \text{ cm} \qquad l_1 = 3.31 \times 10^{-8} \text{ cm}$$

We thus see that the extent of the transition region in n-type material is appreciably larger than the extent of the transition region in p-type material (see Fig. 6 · 11a). This is to be expected from the fact that the p-type side is much more heavily doped than the n-type side. If the junction were a linearly graded junction instead of a step junction, the transition region would extend equally into the p- and n-type bodies. Of course, we must remember that the computation just completed is approximate, in that it neglects edge effects and the charge of carriers in the transition region, but it does give a good idea of the extent of the transition region, nonetheless.

From the above evaluation of the width of the transition region, we may now determine the depletion capacitance for zero bias.

$$C(0) = \frac{\epsilon A}{l_1 + l_2} = 208 \text{ pfarads}$$

Similarly, for a reverse bias of 10 volts, the formula for the junction capacitance of the step junction given in Chap. 5 is

$$C(-10) = C(0) \left(\frac{\phi}{\phi + 10} \right)^{1/2} = 42 \text{ pfarads}$$

(b) *Minority-carrier distributions under forward bias.* We now obtain the distribution of carriers when 150 mv dc forward bias is applied to the diode. The diffusion lengths are

$$L_p = (D_p \tau_p)^{1/2} = 2.2 \times 10^{-2} \text{ cm} \qquad L_n = (D_n \tau_n)^{1/2} = 3.21 \times 10^{-2} \text{ cm}$$

The excess hole density at $x = 0$ is

$$p_E(0) = p_n(e^{qv/kT} - 1) = 1.3 \times 10^{14}/\text{cm}^3$$

from which $\qquad p_E(x) = 1.3 \times 10^{14} e^{-x/2.2 \times 10^{-2}}/\text{cm}^3$

which is plotted in Fig. 6 · 11b.

A precisely similar set of calculations yields

$$n_E(y) = 6.2 \times 10^{10} e^{-y/3.21 \times 10^{-2}}/\text{cm}^3$$

which is not visible on the scale of Fig. 6 · 11b. Note that the distance scale makes

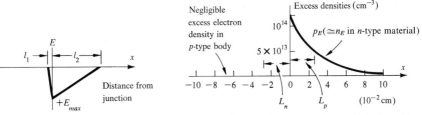

(a) *The transition region* (b) *Distribution of excess carriers under 150 mv bias*

FIG. 6 · 11   PERFORMANCE OF THE GERMANIUM pn-JUNCTION DIODE SHOWN IN FIG. 6 · 10 WITH A FORWARD BIAS OF 150 MV

the transition region look to be of zero width, since it is very small compared to the diffusion lengths.

(c) *Calculation of minority-carrier currents.* The minority-carrier currents may be evaluated by using Eqs. (6·28) and (6·29). The hole current in n material is (see Fig. 6·11c)

$$i_p \text{ (diffusion)} = -AqD_p\frac{dp_E}{dx} = 456 \times 10^{-6}e^{-x/L_p} \text{ amp}$$

and the electron current in p material is

$$i_n \text{ (diffusion)} = -AqD_n\frac{dn_E}{dx} = 0.3 \times 10^{-6}e^{-y/L_n} \text{ amp}$$

The total diode current is therefore

$$i = i_p(x = 0) + i_n(y = 0) = 456.3 \times 10^{-6} \text{ amp}$$

which is essentially the same at the hole current as $x = 0$. This result could have been anticipated by calculating the saturation currents I_{p0} and I_{n0}.

$$I_{p0} = 1.4 \times 10^{-6} \text{ amp} \qquad I_{n0} = 9.7 \times 10^{-10} \text{ amp}$$

The doping at the p material is so high compared to that of the n material, that the current crossing the junction is essentially all hole current. Hence, from this point on we shall neglect electron current in the p material. Furthermore, we note that we can make the injection of one carrier predominate the injection process by control over the doping levels. This will be an important fact for understanding transistor behavior.

Since the diode current must be constant throughout the diode, and its value is 456×10^{-6} amp at the junction, it must be 456×10^{-6} amp everywhere. This is illustrated in Fig. 6·11c. The entire current is supplied by the drift of holes in the p-type material.

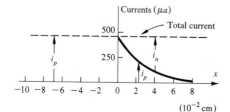

(c) *Currents in n material with* 150 mv *bias*

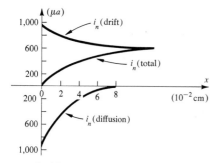

(d) *Distribution of currents under* 150 mv *bias*

FIG. 6·11 PERFORMANCE OF THE GERMANIUM pn-JUNCTION DIODE SHOWN IN FIG. 6·10 WITH A FORWARD BIAS OF 150 MV (CONT'D.)

(*d*) *Calculation of majority-carrier currents in n-type material.* The majority carrier current in *n*-type material is obtained by subtracting the hole current in *n*-type material from the total current. The result is shown in Fig. 6·11*d* and is given by the formula

$$i_n = 456 \times 10^{-6}(1 - e^{-x/L_p}) \text{ amp}$$

This electron current has both drift and diffusion components. To demonstrate this, we observe that from the assumption of space charge neutrality, the electron density must have the form

$$n(x) = n_n + p_E(x) = n_n + p_n(e^{qv/kT} - 1)e^{-x/L_p}$$

The *excess*-electron density in *n*-material is the same as the *excess*-hole density there (see Fig. 6·11*b*).

There is a diffusion current associated with the gradient of electron density whose value is

$$i_n(\text{diffusion in } n \text{ material}) = qAD_n \frac{\partial n}{\partial x} = -940 \times 10^{-6}e^{-x/L_p} \text{ amp}$$

This component of the electron current is plotted in Fig. 6·11*d*.

To obtain the drift component of electron current in *n* material we algebraically subtract the diffusion component from the total electron current. This gives

$$i_n(\text{drift in } n \text{ material}) = 456 \times 10^{-6}(1 + 1.05e^{-x/L_p}) \text{ amp}$$

DISCUSSION OF ASSUMPTIONS We are now in a position to check several of the assumptions which have been made in the analysis. First of all, we observe that an electric field must be present in both the *n*-type and *p*-type bodies to produce the majority carrier drift currents. This electric field is supposed to be sufficiently small so that (1) ohmic voltage drops in the diode body are small, (2) minority carrier drift currents are negligible, and (3), space-charge neutrality is maintained.

Let us consider the *p*-type body first. Since we are neglecting electron flow in *p*-type material, the hole density has the value p_p everywhere. Therefore, the resistivity is constant ($\rho_p = 10^{-3}$ ohm-cm), so that the voltage drop across the *p*-type body is

$$v_p = \frac{\rho_p L}{A} i = 4.56 \ \mu\text{v}$$

This is negligible in comparison to 150 mv, the applied bias, as assumed.

In the more lightly doped *n*-type body a different situation exists. The electric field there may be found from

$$i_n = qA\mu_n \, n(x)E(x)$$

$$E(x) = 0.0456 \frac{1 + 1.05e^{-x/L_p}}{1 + 0.08e^{-x/L_p}}$$

Now to check the previous points:

1. The ohmic drop in *n*-type material is

$$v_n = \int_0^L E(x) \, dx$$

Since $0.045 < E(x) < 0.086$ volt/cm, and the length of the *n*-type body is 0.1 cm, we must have $0.0045 < v_n < 0.0086$ volt. Thus, the voltage drop in the *n*-type body is less than 5 percent of the applied voltage. This is satisfactory, though for somewhat higher currents this voltage drop cannot be neglected.

2. The holes in *n* material also drift in this field as well as diffuse. We may therefore check our assumption that holes flow principally by diffusion. We per-

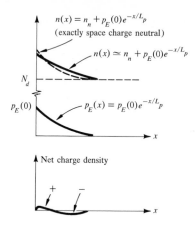

$n(x) = n_n + p_E(0)e^{-x/L_p}$
(exactly space charge neutral)

$n(x) \simeq n_n + p_E(0)e^{-x/L_p}$

N_d

$p_E(0)$

$p_E(x) = p_E(0)e^{-x/L_p}$

Net charge density

FIG. 6·12 EXAGGERATED SKETCH ILLUSTRATING HOW ELECTRON DIFFUSION CREATES A SLIGHT CHARGE UNBALANCE AND A CORRESPONDING ELECTRIC FIELD

form the calculation at $x = 0$, where both the field and the minority-carrier density have their maximum values. Thus

$$i_p(\text{drift in } n \text{ material}) \leq qA\mu_p p_E(0)E(0) = 35 \times 10^{-6} \text{ amp}$$

which is less than 10 percent of the diffusion current, due to holes at $x = 0$. This approximation is again considered to be satisfactory, though this is the worst violation of our assumptions so far.

3. Finally, we check the assumption of space-charge neutrality by employing Poisson's equation in the following way.

$$\rho = \epsilon \frac{dE}{dx} \simeq \frac{\epsilon(0.0456)}{L_p} 1.05 e^{-x/L_p}$$

$$\rho(x = 0) \simeq 3 \times 10^{-12} \text{ coul/cm}^3$$

In order to produce this space charge, we need an electron-density unbalance of $n = \rho/q \simeq 2 \times 10^7$. Thus, the fractional density unbalance is

$$\frac{n}{n_n} = \frac{2 \times 10^7}{1.6 \times 10^{15}} \sim 10^{-8}$$

or one in every 10^8 electrons needs to be removed to produce the required space charge at $x = 0$. The assumption of space-charge neutrality is a very good approximation.

The space-charge unbalance can be thought of as arising in the following way. The creation of an excess hole-density profile $p_E(x)$ in the n-type body will bring about the creation of an essentially equal excess-electron-density profile. Since the $n_E(x)$ curve has a slope at every point, electrons will tend to diffuse down the gradient of this profile (see Fig. 6·12). This then creates the charge unbalance computed in the example. However, it also creates an electric field in a direction to bring these electrons back. In fact, we again have a balance between drift and diffusion tendencies set up. The space-variable part of the electric field in the n-material is sometimes said to be set up to "hold the electrons in place against their density gradient."[1] This electric field is in a direction to aid the flow of holes into the *n*-type body, so the diode current is increased on this account.

RESUME A diagram illustrating the philosophy of the analysis, the self-consistency of our principal assumptions, and the analysis procedures of the last several sections is given below:

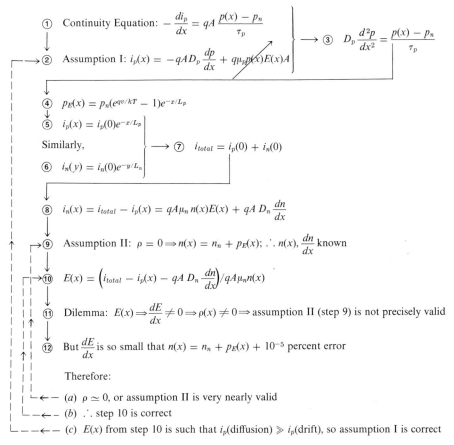

① Continuity Equation: $-\dfrac{di_p}{dx} = qA\,\dfrac{p(x) - p_n}{\tau_p}$

② Assumption I: $i_p(x) = -qAD_p\,\dfrac{dp}{dx} + q\mu_p p(x)E(x)A$

③ $D_p\,\dfrac{d^2p}{dx^2} = \dfrac{p(x) - p_n}{\tau_p}$

④ $p_E(x) = p_n(e^{qv/kT} - 1)e^{-x/L_p}$

⑤ $i_p(x) = i_p(0)e^{-x/L_p}$

Similarly,

⑥ $i_n(y) = i_n(0)e^{-y/L_n}$

⑦ $i_{total} = i_p(0) + i_n(0)$

⑧ $i_n(x) = i_{total} - i_p(x) = qA\mu_n\,n(x)E(x) + qA\,D_n\,\dfrac{dn}{dx}$

⑨ Assumption II: $\rho = 0 \Rightarrow n(x) = n_n + p_E(x);\ \therefore\ n(x),\ \dfrac{dn}{dx}$ known

⑩ $E(x) = \left(i_{total} - i_p(x) - qA\,D_n\,\dfrac{dn}{dx}\right)\Big/qA\mu_n n(x)$

⑪ Dilemma: $E(x) \Rightarrow \dfrac{dE}{dx} \neq 0 \Rightarrow \rho(x) \neq 0 \Rightarrow$ assumption II (step 9) is not precisely valid

⑫ But $\dfrac{dE}{dx}$ is so small that $n(x) = n_n + p_E(x) + 10^{-5}$ percent error

Therefore:

(a) $\rho \simeq 0$, or assumption II is very nearly valid

(b) $\therefore$ step 10 is correct

(c) $E(x)$ from step 10 is such that i_p(diffusion) $\gg i_p$(drift), so assumption I is correct

6·5 REVERSE BREAKDOWN IN *pn* JUNCTIONS

To this point we have been primarily concerned with that part of the *v-i* characteristic of a *pn* diode that is attributable to minority-carrier flow in the neutral diode body. When this condition applies, the diode current *i* is related to *v* by

$$i = I_0(e^{qv/kT} - 1) \tag{6·38}$$

---

[1] This same mechanism is responsible for the "built-in" drift fields which can occur when a semiconductor does not have a uniform density of doping atoms.

Perhaps the most important deviation of real diodes from this ideal behavior occurs under reverse-bias conditions, where Eq. (6·38) predicts that when v is negative, $i = -I_0$ (provided $|v| \gg kT/q$). When we compare this prediction with actual data, we see that it is valid up to some maximum reverse voltage V_B, after which the diode current begins to increase sharply. A typical v-i plot including this effect is given in Fig. 6·1. Naturally, unless the current is limited by the external circuit, appreciable heating will occur when the diode breaks down, and the junction may be permanently damaged. With appropriate heat sinking, however, it is possible to make devices which dissipate watts at their breakdown voltage. These diodes are variously called avalanche diodes, Zener diodes (usually a misnomer as indicated below), or breakdown diodes. They find applications in the fields of voltage regulation, clipping, and for voltage reference purposes. The reverse-breakdown effect is also important in transistor applications where in most cases it establishes a limit on the maximum collector voltage which may be applied to a transistor. Our purpose here is to explain the physical theory for this characteristic briefly. Some of its applications will be indicated in the next chapter.

The physical phenomenon which gives rise to reverse breakdown in most pn junctions is called the avalanche effect. The avalanche effect is a nondestructive carrier-multiplication effect which occurs in a pn junction when it is reverse-biased to a sufficiently large voltage.

Figure 6·13 has been drawn to aid in discussing the avalanche mechanism. This figure shows an expanded version of the space-charge layer, or transition region, which is developed for three different reverse voltages: V_0, $2V_0$, and $4V_0$. A step junction has been assumed, with the doping on the p side much greater than that on the n side. The figure indicates that the width of the space-charge layer and the magnitude of the peak electric field both increase as the square root of the applied voltage so that the area under the triangles is equal, in each case, to the applied voltage. (These conclusions may be obtained mathematically by a simple application of Poisson's equation as is done in Chap. 5.)

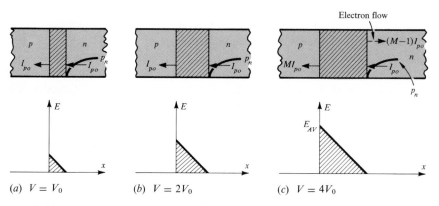

(a) $V = V_0$ (b) $V = 2V_0$ (c) $V = 4V_0$

FIG. 6·13 SPACE-CHARGE LAYER IN A BACK-BIASED pn JUNCTION

The dc current flowing through the transition region in Figs. 6·13a and 6·13b is the saturation current, I_{p0}. I_{p0} is composed of holes which diffuse into the transition region from the n-type body and are then swept across the transition region by the electric field. However, in Fig. 6·13c the hole current emerging from the transition region is greater than that entering it. We denote the emerging hole current by MI_{p0}, where M is called the *avalanche multiplication* factor. In order to preserve current continuity, an electron current of $(M - 1)I_{p0}$ electrons must flow out of the transition region as shown in Fig. 6·13c.

We explain this effect as follows. As the holes which make up the current I_{p0} move through the transition region, they gain energy from the electric field and give it up to the lattice as they collide with it. If the electric field is not too great, the holes can give up all the energy they receive from the field on each collision. However, when the electric field becomes large (as in Fig. 6·13c near the metallurgical junction) the holes gain more energy from the field in a free path than they can give up to the lattice at the end of the free path.[1] Thus, when the electric field is high enough the holes continuously *gain* energy, and after several free paths they may have enough energy to ionize the lattice when they next collide with it. That is, the impinging carrier may give enough energy to the lattice to free a valence electron, thus producing an extra hole and an electron. The products of this ionization (one hole and one electron), together with the original ionizing carrier (which lost almost all of its energy in producing the ionization), are now free to contribute to further carrier multiplication. A pictorial representation of this process for an avalanche multiplication factor of 2 is shown in Fig. 6·14 for a symmetrically doped junction.

6·5·1 *Simplified mathematical treatment of avalanche breakdown.* The carriers which enter the space-charge region and those created within it have a distribution of energies, a distribution of free-path lengths, and a distribution of directions such that not all of the carriers which enter will be successful in creating new carrier pairs. We may describe this situation by saying that there is only some probability P that a given carrier will produce a carrier pair as it travels across the transition region.

Now, let us suppose that N_0 holes enter a transition region in which conditions are such that P percent of them suffer ionizing collisions during their transit across the entire region. Then, since the electron and hole created by an ionizing collision together move the same distance that a hole entering at the edge of the region would, we must also assume that each *pair* created has a P percent chance of creating another pair. Under these conditions the number of holes leaving the region is

$$N_{out} = N_0 + PN_0 + P^2N_0 + \cdots = N_0 \sum_{m=0}^{\infty} P^m \qquad (6·39)$$

The ratio of the number leaving the region to the number which enter is

---

[1] The amount of energy which may be given to the lattice at the end of a free path is limited by quantum-mechanical considerations.

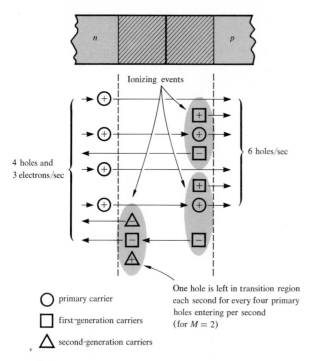

- ○ primary carrier
- ☐ first-generation carriers
- △ second-generation carriers

Ionizing events

4 holes and 3 electrons/sec

6 holes/sec

One hole is left in transition region each second for every four primary holes entering per second (for $M = 2$)

FIG. 6·14 PARTIAL ILLUSTRATION OF THE AVALANCHE MECHANISM

Note: On the average only half of the primary holes produce ionizing collisions for $M = 2$. Similarly only half the first generation electrons produce ionizing collisions, only half the second generation holes produce ionizing collisions, etc.

defined to be the multiplication factor M and may be written as a geometric series

$$M = \frac{N_{out}}{N_0} = \sum_{m=0}^{\infty} P^m = \frac{1}{1 - P} \tag{6·40}$$

in which P is less than one.[1] If $P = 0.1$, M is 1.11; if P is 0.5, M is 2.

Clearly, P must depend in some way on the reverse voltage applied to the pn junction, with the properties that $P = 0$ when $V = 0$, and when $P \to 1$, $V \to V_B$, the avalanche breakdown voltage. Of the many possible functions which satisfy these properties, a reasonable fit for experimental data is obtained from

$$P = \left(\frac{V}{V_B}\right)^n \tag{6·41}$$

[1] In the physical theory of the avalanche effect, one defines an "ionization coefficient" α_i, which is the number of carrier pairs per centimeter of travel of the ionizing carrier. Then, if the width of the space charge layer is denoted by W,

$$P = \int_0^W \alpha_i(E)\, dx$$

with the ionization condition

$$\int_0^W \alpha_i(E)\, dx = 1$$

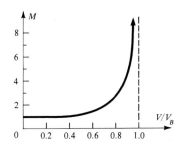

FIG. 6·15 M AS A FUNCTION OF V/V_B FOR $n = 3$

with n as an adjustable index depending on material constants and geometrical factors. Hence, the multiplication factor may be written as

$$M = \frac{1}{1 - (V/V_B)^n} \qquad (6\cdot42)$$

For a silicon pn junction, n is approximately 3; the multiplication factor has been plotted as a function of V/V_B in Fig. 6·15 for this case. It will be observed that M is less than 4 until $V \cong 0.9V_B$. Hence, the multiplication effect appears roughly to have a threshold very near V_B, below which the multiplication is small and above which the multiplication is very large. The reverse current in the diode exhibits this same effect:

$$I = MI_0 \qquad (6\cdot43)$$

6·5·2 *The relation between avalanche breakdown voltage and doping level.*
One of the interesting features of the avalanche process is that the breakdown voltage of a junction can be varied over quite wide limits by simple changes in the manufacturing processes.

To appreciate this, we recall that the peak electric field E_p at a p^+n metallurgical junction varies with both the applied reverse voltage and the doping level of the n material:

$$E_p \sim \sqrt{VN_d}$$

Now the preceding discussion of the avalanche phenomenon may be roughly summarized by saying that avalanche ionization will become important when a sufficiently large peak electric field exists at the metallurgical junction.[1] This, then, suggests that

$$V_B N_d \sim \text{const} \qquad V_B \sim 1/N_d$$

Since E_{AV} increases slightly as N_d increases, V_B is more nearly proportional to $1/(N_d)^{2/3}$ than to $1/N_d$. In any case, it is possible to control the breakdown voltage of a junction diode quite accurately by controlling the

---

[1] In addition to the peak field, the path length of a carrier in the region of the highest field must also enter into the determination of V_B. Since the path length is also related to the doping level, the peak electric field at avalanche breakdown is not exactly constant for all diodes. E_{AV} increases from about 3×10^5 volt/cm to about 6×10^5 volt/cm as the doping level increases from $10^{15}/\text{cm}^3$ to $10^{17}/\text{cm}^3$.

doping level. The avalanche-breakdown voltage for silicon p^+n diodes varies from about 5 volts (for $N_d \sim 10^{18}/\text{cm}^3$) to about 1,000 volts (for $N_d \sim 10^{14}/\text{cm}^3$).

When diodes are made especially for voltage regulation or reference purposes, they are usually called Zener diodes, though the Zener breakdown mechanism, to be described in the next section, has nothing to do with their behavior.

6·5·3 *Zener breakdown.* In order to make diodes with low avalanche breakdown voltages, one needs to use large doping levels in the diode body. However, when the doping levels get too large, another breakdown mechanism, called Zener breakdown, sets in.

Zener breakdown is a direct disruption of valence bonds, which can occur when the electric field strength becomes 10^6 volts/cm or greater. The Zener mechanism does not require the presence of an energetic ionizing carrier, and in fact is responsible for the "dielectric breakdown" of good crystalline insulators such as diamond. While the basic theory for this mechanism is quantum-mechanical in nature, we can appreciate its significance for *pn*-junction operation only by assuming that it will occur when the peak field in the junction reaches about 10^6 volts/cm.

Now, from a previous explanation of the avalanche mechanism, one might assume that a peak field of 10^6 volts/cm is more than enough to cause avalanche breakdown. However, it must be remembered that the basic idea in the avalanche mechanism is that the ionizing carrier must gain enough *energy* to ionize the lattice. If a peak field of 10^6 volts/cm extends over a distance of only 10Å, a carrier moving through it can gain a maximum energy of only 0.1 volt, which would not be enough for an ionizing collision. However, this peak field would still produce Zener breakdown.

Now when one attempts to make a *pn* junction with a reverse breakdown voltage of about 5 volts, the peak fields at the junction will reach and exceed 10^6 volts/cm. The path length in this field will be such that both the Zener mechanism and the avalanche mechanism will contribute to the overall breakdown. And as the reverse breakdown voltage is reduced below 5 volts (by still heavier doping of the *p* and *n* regions), the path length in the high field region is reduced, and then the Zener mechanism will be the predominant cause of the breakdown. This is illustrated in Fig. 6·16.

6·5·4 *Tunnel diodes.* The Zener breakdown mechanism also affects the *forward* characteristic of a *pn* junction when the doping levels in the *p* and *n* materials are high enough. To see this, let us consider a silicon junction with doping levels of $10^{19}/\text{cm}^3$ on both sides. The contact potential is then very nearly one volt. At the same time, the zero-bias width of the space-charge layer is less than 100Å, or 10^{-6} cm. As a result, the peak field in the transition region is 10^6 volts/cm with *no* applied bias. Such a "diode" appears to be "broken down" even in the forward direction, and exhibits a low resistance to voltage of either polarity, as shown in Fig. 6·16a.

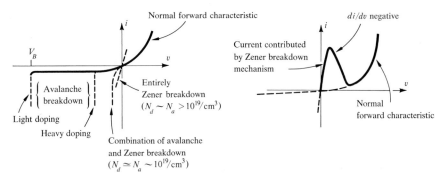

(a) Illustrating the change of reverse breakdown mechanism from avalanche to Zener as doping level increases

(b) Effect of Zener breakdown on forward characteristic in very heavily doped diodes. Diodes of this type are called tunnel diodes, after the quantum mechanical process responsible for Zener breakdown.

FIG. 6·16   EFFECT OF THE ZENER BREAKDOWN MECHANISM ON FORWARD AND REVERSE CHARACTERISTICS OF A *pn*-JUNCTION DIODE

However, since the peak electric field in the junction decreases as forward bias is applied, a point will be reached at which the electric field is no longer adequate to produce the Zener breakdown. Beyond this point, the current falls with increasing forward bias, as shown in Fig. 6·16b. The current rises again when the normal injection mechanisms become dominant.

This behavior gives rise to a region of negative *dynamic* resistance (dV/dI negative), which is very useful for constructing various types of circuits.

A diode with the characteristic shown in Fig. 6·16b is called a *tunnel diode*, since the quantum-mechanical phenomenon responsible for the Zener breakdown is called the tunnel effect. These diodes are sometimes also called Esaki diodes, after their inventor.

6·6   OTHER PROPERTIES OF REAL DIODES

In addition to the reverse breakdown characteristic, real diodes fail to obey the ideal diode law in other ways. As suggested in Secs. 6·1 and 6·2, actual diodes have series resistance, nonplanar geometry, and sometimes exhibit appreciable recombination on the diode surfaces and in the transition region. In this section we shall briefly indicate how the ideal *v-i* relation may be modified to include some of these effects.

6·6·1   *High forward-current region.*   Actual diodes depart from the ideal *v-i* relation at high currents for basically two reasons:

1. Series resistance in the diode bodies absorb an appreciable amount of the voltage drop between the diode terminals. This type of departure can

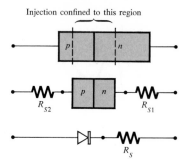

Injection confined to this region

FIG. **6·17**   THE TREATMENT OF SERIES RESISTANCE IN THE DIODE MODEL

be accounted for with the circuit model shown in Fig. 6·17, where the series resistance is considered to be an external element.

2. Low-level injection conditions cease to be valid. It has been repeatedly emphasized that the exponential factor in the ideal current-voltage law arises from the junction relation

$$p_E(0) = p_n (e^{qv/kT} - 1) \qquad (6 \cdot 44)$$

For this relation to be valid, the actual injected hole density must be small compared to N_d:$p(x = 0) \ll N_d$. Otherwise, the actual hole density must be found by a simultaneous solution of

$$np = n_i^2 e^{qv/kT} \qquad (6 \cdot 45a)$$

$$n = N_d + p \qquad (6 \cdot 45b)$$

These equations are such that estimating $p(x = 0)$ by means of

$$p = p_n e^{qv/kT}$$

gives less than 10 percent error as long as $p \leq 0.1N_d$ (see Prob. 6·11). However, it is possible to make diodes (for example, power rectifiers) in which the injected hole density will be much larger than N_d at high current levels. This condition is described by the term *high-level injection*. In the high-level injection limit, Eqs. (6·45) yield

$$p(x = 0) = n_i e^{qv/2kT}$$

The current then becomes roughly proportional to $e^{qv/2kT}$, provided the series resistance of the diode will permit such a current level to be passed.

6·6·2   Low forward-current region.   In the low forward-current region ($I_0 < i < 100I_0$) the approximations made to study carrier motion in the diode *body* fit actual diodes quite well. However, there was an assumption made in the development of the diode model which is not valid at low currents and may affect the *v-i* characteristic in an important way— recombination in the transition region.

Net recombination occurs within any volume where the np product is

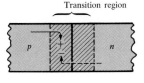

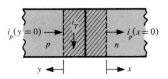

(a) *Illustrating recombination in the transition region*

(b) *Current flow in the transition region including recombination current i_T*

FIG. 6·18 THE EFFECT OF RECOMBINATION IN THE TRANSITION REGION ON THE DIODE CURRENTS

greater than n_i^2. Since the law of the junction indicates that the np product is

$$np = n_i^2 e^{qv/kT} \qquad (6·46)$$

throughout the transition region, there must be net recombination within the transition region when v is positive.

Holes for this recombination flow into the transition region from the left, electrons from the right, as shown in Fig. 6·18a. If we call the current associated with this recombination i_T, then the hole current entering the transition region (shown in Fig. 6·18b) must be

$$i_p(y = 0) = i_T + i_p(x = 0) \qquad (6·47)$$

Some of the hole current entering at $(y = 0)$ is lost by recombination within the transition region, while the rest is injected into the n-type body.

Now, as long as

$$i_p(x = 0) \gg i_T$$

we may neglect i_T, as was done in the ideal diode. To find out when this is an acceptable approximation, we need to know how i_T varies with bias and material parameters.

A study of this problem has been made[1] under the assumption that recombination occurs primarily through traps (for example, crystal imperfections), and that the same trapping mechanism is responsible for recombination in the diode bodies as in the transition region. Then i_T may be written approximately as

$$i_T = \frac{qAW}{\tau_p} n_i \left(e^{qv/2kT} - 1 \right) \qquad (6·48)$$

where W is the width of the transition region. On the basis of this formula, i_T and $i_p(x = 0)$ can be compared (Prob. 6·12). The comparison shows that at zero bias,

$$i_T \gg i_p(x = 0) \qquad \text{for silicon}$$
$$i_T \ll i_p(x = 0) \qquad \text{for germanium}$$

[1] C. T. Sah, R. N. Noyce, and W. Shockley, Carrier generation and recombination in *P-N* junctions and *P-N* junction characteristics, *Proc. IRE*, vol. 45, September, 1957, pp. 1228–1243.

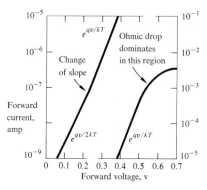

FIG. 6·19 EFFECT OF RECOMBINATION IN THE TRANSITION REGION ON THE i-v CHARACTERISTIC OF A pn-JUNCTION DIODE. FOR CURRENTS BELOW ABOUT 10^{-7} AMP. FOR THIS PARTICULAR DIODE, RECOMBINATION CURRENT SUPPLIED TO THE TRANSITION REGION PREDOMINATES.

[*Data from Sah, Noyce, and Shockley, Proc. IRE, pp. 1228–1234, 1957*]

Thus, i_T may be neglected for germanium diodes operating at room temperature. In silicon diodes i_T dominates the v-i characteristic at low currents. However, since $i_p(x = 0)$ varies as $e^{qv/kT}$, while i_T varies only as $e^{qv/2kT}$, a forward bias will be reached beyond which $i_p(x = 0)$ will again dominate the v-i relation. This behavior is illustrated in Fig. 6·19 (and dotted in on Fig. 6·1).

One final comment on this effect is in order. The derivation of Eq. (6·48) for the ratio of i_T and $i_p(x = 0)$ is based on the assumption that the same recombination mechanism is responsible for recombination in both the diode body and the transition region. In fact, the volume of the transition region is so small that the principal crystal imperfections which cause recombination in the transition region are on its *surface*. By very careful treatment of the diode surfaces, i_T can in fact be reduced, and then the injection current may predominate in silicon diodes even at low current levels.

6·6·3 *Reverse characteristic in the prebreakdown region.* The ideal diode law suggests that after a reverse bias of 0.1 volt is applied, the diode current should remain constant at the value $-I_0$ (until reverse breakdown sets in). This is observed in typical germanium diodes. However, the reverse current may fail to saturate in silicon diodes and instead may increase with reverse bias according to a $V^{+1/2}$ or $V^{+1/3}$ law, depending on whether the junction is a step junction or a linearly graded junction.

This phenomenon again owes its existence to events occurring within the transition region. The application of a reverse bias will cause the product of n and p within the transition region to become much less than n_i^2. In this case, there will be a net *generation* of carriers per unit volume of transition region. These carriers will be swept out of the transition region by the electric field and contribute to the saturation current. And, since the width of the transition region varies as $V^{+1/2}$ for a step junction, its volume does also, and therefore the total contribution of the transition region to the saturation current varies as $V^{+1/2}$.

Of course, in addition to this current there will be the normal saturation current I_0 arising from within the diode body. Whether I_0 is greater than or less than the current generated within the transition region is, again, a question of material parameters and surface treatment. When traps dominate the generation-recombination process in both the body and the transition region, Eq. (6·48) for i_T still applies, but with v negative. In such a case, current generated in the transition region dominates the reverse behavior of a silicon diode but not a germanium diode. The reverse v-i characteristic for a silicon diode will then appear as the dashed characteristic on Fig. 6·1.

6·7   SEMICONDUCTOR DEVICES RELATED TO THE pn-JUNCTION DIODE

In addition to the pn-junction diode, there are several other devices of interest in electronics which employ only a single pn junction. There are also special forms of junction diodes which employ certain features of the behavior which we have not yet emphasized. In this section we shall briefly describe some of these devices. Our objective is to show how the operating principles for a junction diode are used to understand other devices. Applications for some of these devices will be discussed later when other devices with similar characteristics have been presented.

6·7·1   *The field-effect transistor.* One of the earliest solid-state devices to be envisaged as an amplifying element was the field-effect transistor[1] (abbreviated to FET.) The commercial development of FETs has been considerably slower than that of the junction transistor, partly because of technological difficulties concerned with quantity manufacture and partly because the range of possible applications for field-effect transistors is more limited than that of junction transistors. However, there are now available field-effect transistors that have certain distinct advantages over junction transistors, and within their range of applications they are nearly ideal elements.

Figure 6·20a has been drawn to illustrate the operating principle of an FET. While the precise geometry shown in this figure is not now used, it will simplify our discussion of the principles involved.

The principle of operation of this structure is to vary the ohmic resistance between the S and D contacts by varying the voltage applied to the gate G. This is accomplished as follows. A channel is cut in the center of an n-type rod which has ohmic contacts at each end. The end contacts are called the source S, and the drain D. The resistance between S and D is largely controlled by the length L, width W, and thickness t of this channel.

[1] W. Shockley, "Electrons and Holes in Semiconductors," D. Van Nostrand Company, Inc., Princeton, N.J., 1953.

The length and thickness are fixed. However, the width W may be controlled by applying a reverse bias to a p-type layer which has been alloyed or diffused into the structure directly beneath the channel. The p-type layer is called the gate G. When reverse bias is applied between the drain and the gate, a space-charge layer penetrates into the n-type material. This constricts the width W for electron flow from S to D, and therefore increases the ohmic resistance from S to D.

Of course, current flow in the channel will produce a voltage drop along it. As a result, the voltage drop across the space-charge layer, and therefore the width of the channel, will be as shown in Fig. $6 \cdot 20b$. Near the source end of the channel the space-charge layer has a voltage drop of $V_G + V_A$ across it, while near the drain end it has a voltage drop of only V across it (V_A and V_G are defined in Fig. $6 \cdot 20a$).

The actual shape of the channel and the current flowing through it can be calculated when V_A and V_G are specified. The result is that

$$I = G_0 V_p \left[\frac{V_A}{V_p} - \frac{2}{3} \left(\frac{V_A + V_G}{V_p} \right)^{3/2} + \frac{2}{3} \left(\frac{V_G}{V_p} \right)^{3/2} \right] \qquad (6 \cdot 49)$$

where G_0 is the conductance of the channel and V_p is the *pinch-off voltage*, [that is, the value of voltage required for the space-charge layer to com-

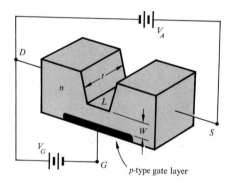

(a) *Construction and biasing of a field effect transistor*

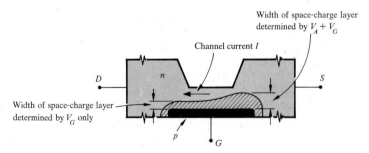

(b) *Modulation of channel width by space-charge-layer control*

FIG. $6 \cdot 20$ CONSTRUCTION AND OPERATING PRINCIPLE OF A FIELD-EFFECT TRANSISTOR

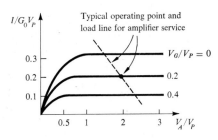

FIG. 6·21 NORMALIZED SOURCE-TO-DRAIN CHARACTERISTIC OF A FIELD EFFECT TRAN-SISTOR WITH NORMALIZED GATE VOLTAGE AS A PARAMETER

pletely penetrate the channel at the source end. Actually, current flow from S to D prevents complete penetration, and the current satures at the value predicted in Eq. (6·49) when $V_A + V_G = V_p$].

From the point of view of the user, the gate lead in this structure is thought of as the control terminal. Thus, the device is characterized for low-frequency circuit applications by a set of curves showing the v-i characteristic for the SD terminals with gate voltage V_G as a parameter. Such a set of curves [obtained from Eq. (6·49)] is shown in Fig. 6·21. They are generally reminiscent of the output characteristics of transistors or vacuum tube pentodes.

To use an FET as an amplifier, biases are applied to produce an operating point in the region where the source-to-drain current is independent of V_A. Then, an input voltage applied to the gate-drain terminals controls the source-drain current.

Interest in these devices arises primarily because the input (that is, gate-drain) circuit has the characteristics of a single reverse-biased diode. An FET therefore requires an exceedingly small dc input current (I_0) and has a very high input impedance.

We shall expand on these points when we return to a consideration of circuit applications in which FETs are useful at a later point. For the present we leave this subject with the observation that the main operating characteristics and the distinctive features of this device can be simply understood from the theory of a junction diode.

6·7·2 *The double-base diode or unijunction transistor.* A second interesting device whose operating principles can be simply understood by an application of junction-diode theory is the unijunction transistor.

The geometrical configuration of a unijunction transistor is shown in Fig. 6·22. It is quite similar to that of the field-effect transistor except that the biases are arranged to utilize the forward-injection characteristic of the *pn* junction. On account of this difference, the unijunction transistor exhibits a negative dynamic resistance over part of its v-i characteristic, making the device useful for certain low- and medium-frequency switching applications.

A qualitative idea of the source of the negative resistance may be obtained from the following argument. Suppose the ends of the *n*-type region are connected to a battery of voltage V_0, and a current source i_C is connected between the control terminal (the p-type material) and the negative end of the n-type bar, as shown in Fig. 6·22. Now, let a current i_C of holes be injected into the n-type bar. These holes will drift into the n-type material which lies below the junction. They may recombine in the n-type body, or they may drift all the way to the ohmic contact and recombine there. In either case, there will be an *increase in conductivity* of the part of the n-type bar that lies below the pn junction. The increase of conductivity will be proportional to the control current i_C.

Now the voltage drop v_2 from a point inside the n-type bar at the height of the pn junction to the common terminal (see Fig. 6·22) will depend on this change in conductivity, or *conductivity modulation* as it is called. If i_C is zero, so there is no conductivity modulation, then

$$v_2 = \frac{V_0 l_1}{l_1 + l_2} = V_{20} \qquad (6 \cdot 50)$$

where l_1 and l_2 are defined on Fig. 6·22. As i_C is increased from zero, v_2 will drop until essentially all of V_0 is dropped across the upper part of the bar.

The voltage from the control terminal to ground is equal to the forward-bias drop across the pn junction (needed to allow the current i_C to flow) plus v_2, and since v_2 starts at V_{20} and drops to essentially zero, v_c must start at $v_j + V_{20}$ and drop to essentially v_j as i_C increases from zero. This characteristic is shown in Fig. 6·23.

The characteristic may be summarized as follows. In the negative i_C range, where $i_C \simeq I_0$, the v-i characteristic between the control terminal and ground appears to be that of a normal pn junction in series with a battery of voltage V_{20}. In the high current, positive i_C range, the heavy conductivity modulation of that part of the n region which is below the pn junction causes the voltage v_2 to become essentially zero, so that the v-i characteristic now appears to be that of a normal pn junction with a moderate series resistance. There is a transition region of negative dynamic resistance between these two extremes.

An approximate mathematical analysis of the i_C-v_2 relation can be

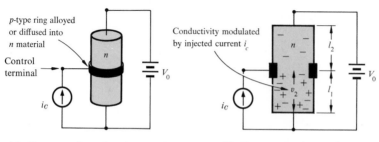

(a) Geometry of a unijunction transistor   (b) Cross-sectional view of (a)

FIG. 6·22 ILLUSTRATING THE GEOMETRY AND OPERATION OF A UNIJUNCTION TRANSISTOR

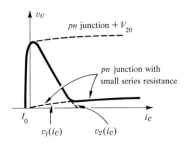

FIG. 6·23 i_C-v_C CHARACTERISTIC AT THE CONTROL TERMINAL OF A DOUBLE-BASE DIODE (OR UNIJUNCTION TRANSISTOR)

obtained from these ideas. The analysis predicts that

$$v_2 = -\frac{\mu_n}{\mu_p}\frac{l_1 l_2}{l_1 + l_2}\frac{\rho_n}{A}i_C + \frac{l_1}{l_1 + l_2}V_0 \qquad (6\cdot51)$$

where l_1 and l_2 are defined in Fig. 6·22, ρ_n is the resistivity of the unmodulated n-type bar, and A is the cross-sectional area of the bar. Equation (6·51) is valid until $v_2 \sim v_j$.

v_2 and v_j, the junction voltage for a given value of i_C, can each be plotted as functions of i_C, as in Fig. 6·23, to determine the actual characteristic. Some features of this v-i characteristic are the subject of Prob. 6·13.

6·7·3 Photodiodes and solar cells. We saw in Chap. 3 that hole-electron pairs can be generated in a semiconductor crystal by shining light in the appropriate wavelength range on it. This phenomenon is responsible for the photoconductivity of semiconductors described in Chaps. 3 and 4. It is also responsible for photovoltaic effects in pn-junction diodes, and leads directly to the possibility of constructing photodiodes and solar cells (large-area photodiodes purposely fabricated to convert light from the sun into electrical power). In this section we shall see how a simple modification of the theory of a junction diode can account for the properties of photodiodes and solar cells.

We begin with a brief consideration of the simplest physical forms which photodiodes and solar cells take. These are shown in Fig. 6·24. The earliest photodiodes were made by simply mounting a grown junction diode between two leads of a "top hat" package, which also has a lens in the top for focusing light on the semiconductor bar. Solar cells utilize a large-area-diffused pn junction. A ring contact is made to the diffused p-type material in order to maximize the area which light can strike. The thickness of the p-type layer is made sufficiently small that nearly all the carrier-producing light can penetrate to the pn junction and beyond.

The v-i characteristic for these photosensitive diodes can be obtained easily from the theory of a junction diode. For simplicity, let us consider that light falls uniformly on a semiconductor bar which contains a pn junction. The effect of the light is to generate g_v hole-electron pairs per second per unit volume. To account for this effect, the continuity equation for holes must be modified to read

$$D_p \frac{d^2p}{dx^2} = \frac{p - p_n}{\tau_p} - g_\nu \qquad (6 \cdot 52)$$

where we have assumed that holes move only by diffusion in the n-type material. Equation $(5 \cdot 52)$ may be interpreted by saying that the net recombination rate per unit volume has to be diminished by g_ν to account for the photogenerated carriers.

Now Equation $(6 \cdot 52)$ can be solved to yield $p_E(x)$, from which the hole current entering the n-type body at $x = 0$ may be deduced. The result is

$$i_p(x = 0) = I_{p0}(e^{qv/kT} - 1) - q(AL_p g_\nu) \qquad (6 \cdot 53)$$

(see Prob. $6 \cdot 14$). Similarly, the electron current entering the p-type body at $y = 0$ is

$$i_n = I_{n0}(e^{qv/kT} - 1) - q(AL_n g_\nu) \qquad (6 \cdot 54)$$

The total current is therefore

$$i = I_0(e^{qv/kT} - 1) - i_\nu \qquad (6 \cdot 55)$$

where $$i_\nu = qAL_n g_\nu + qAL_p g_\nu \qquad (6 \cdot 56)$$

Evidently, the effect of the light on the external diode characteristics can be simply included by measuring the short-circuit diode current in the presence of the light and then including this current in the circuit model shown in Fig. $6 \cdot 25$. From this figure, or Eq. $(6 \cdot 65)$, we can see that the v-i characteristic for a photodiode is identical to that of a normal diode except that it is displaced on the current scale by an amount i_ν, as shown in Fig. $6 \cdot 25$.

Several observations about this characteristic are noteworthy:

1. In most junction-diode photocells, i_ν is almost exactly proportional to the light flux over a wide range of light fluxes.

2. Because I_0 can be a very small number, the ratio of light current to dark current for a given light input is extremely high. In applications

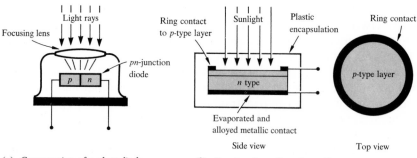

(a) Construction of a photodiode   (b) Construction of a solar cell

FIG. $6 \cdot 24$ SOME pn-JUNCTION PHOTOCELLS

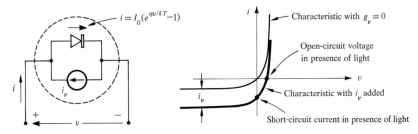

$$i = I_0(e^{qv/kT}-1)$$

Characteristic with $g_\nu = 0$

Open-circuit voltage in presence of light

Characteristic with i_ν added

Short-circuit current in presence of light

FIG. 6·25 i-v CHARACTERISTIC AND CIRCUIT MODEL FOR A PHOTODIODE

where a circuit must discriminate between light ON and light OFF, it is usually this ratio which is important, rather than the actual magnitude of the current increase which the light creates.

3. A solar cell will have a short-circuit current of i_ν and an open-circuit voltage of

$$v_{oc} \cong \frac{kT}{q}\ln\frac{i_\nu}{I_0}$$

However, the power available from this "generator" is not $v_{oc}i_\nu/4$, because the generator has a nonlinear internal impedance. This point is developed further in the next chapter on applications of junction diodes.

6·7·4   Light-emitting diodes and junction lasers.   There are basically two processes by which holes and electrons can recombine in a semiconductor: through traps or by means of a direct recombination event. In both cases, recombination is accompanied by the ultimate liberation of qE_g joules of energy (see Sec. 4·4). However, there is an important difference in the processes. When recombination occurs through traps, the energy qE_g is usually delivered to the crystal as heat, while in a direct recombination event, a photon of energy qE_g is emitted.

Now, our previous studies have shown that all the current that flows in a *pn*-junction diode can be accounted for by recombination somewhere within the structure. When the diode is made of the proper materials, much of this recombination can be of the photon-emitting type, so that when a diode is conducting a forward current it will radiate light. The photon energy for this light will be in a narrow energy range about qE_g. Diodes which do this are called *light-emitting diodes* or *electroluminescent diodes*.

Of course, in most practical semiconductors, both recombination through traps and direct recombination will be present to some extent. The efficiency of these two competing recombination processes depends on the semiconductor material and the operating conditions (principally temperature and injection level).

In silicon and germanium, recombination via traps is the dominant mechanism, so silicon and germanium diodes do not emit a significant amount of recombination light. On the other hand, junction diodes made

Diodes and related one-junction devices   221

from III-V compound semiconductors such as gallium arsenide and gallium phosphide do emit appreciable light, especially when they are operated at low temperatures and high injection levels.

Under most circumstances, these diodes will emit light over a *band* of frequencies roughly centered about the value $\nu_g = qE_g/h$. This is because (1) donor and acceptor doping atoms may participate in the photon-emitting recombination processes, giving rise to some photons with $\nu < \nu_g$, and (2) recombining carriers can have thermal energies on the order of kT, so that some of the emitted photons could have an energy a few kT greater than E_g. Thus, a typical wavelength spectrum of the emitted light appears like that shown in Fig. 6·26a. (The diode is operated at 2°K to maximize the number of radiative recombinations. It will emit light at room temperature, though it is much less efficient.)

The light emitted at any particular frequency in such a spectrum is said to be *incoherent*, by which we mean that the intensity of the light emitted at some future time cannot be predicted from its value at a given time of measurement. Only probabilistic statements can be made. This is because the recombination events occur at random times, so the photons are emitted at random times.

However, under certain special conditions, the light emitted from a junction diode becomes *coherent;* that is, the intensity of the light emitted in a certain direction in space at any future time is predictable from its value at a given time. Mathematically, a coherent light source has an electric-field vector of the form $E(t) = A \sin(2\pi\nu t + \phi)$ where A and ϕ are constant. In the incoherent emission, A and ϕ are random functions of time.

When a junction diode is emitting coherent light, it is called a junction *laser.* The spectrum of the emitted light then changes to that shown in Fig. 6·26b. The reason for this change is that under appropriate circumstances the probability of a given photon-producing recombination event depends on the presence of another photon of that energy near the recombination site. When large numbers of carriers are available for

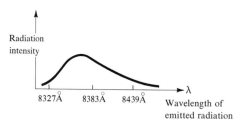

(a) *Emission spectrum from a GaAs diode operated at 2°K. Diode current of* 38 ma

(b) *Emission spectrum of a GaAs laser. Same diode as in (a) except diode current = 50 ma*

FIG. 6·26 CHARACTERISTICS OF LIGHT-EMITTING JUNCTION DIODES AND LASERS
[*After W. E. Howard, et al., IBM Journal, Jan. 1963, p. 74, with permission*]

recombination, and the diode surfaces are treated so as to keep most of the generated photons within the diode body, then these photon-assisted recombination events can predominate. The emission is then called *stimulated emission*. Coherence of the light in a stimulated-emission process is basically due to the fact that a new photon is produced as a given photon travels past a recombination site.

Of course, this description of the light-emitting process is very sketchy, but it is perhaps adequate to suggest that the flow of dc current in a *pn*-junction diode can, under appropriate conditions, produce both coherent and incoherent light. For general illumination purposes, incoherent light is adequate. Coherent light sources are useful in communication systems and various other situations (such as interferometry) where the sine-wave characteristics of the signal are a necessary attribute.

6·8 THE VACUUM DIODE

We conclude this chapter with a brief development of the v-i characteristic of a high-vacuum diode and a discussion of the physical phenomena which produce it. Before doing this, however, some perspective on the importance of this subject is in order.

As a practical matter, semiconductor diodes are preferred over vacuum diodes for the vast majority of practical diode applications. This is because semiconductor diodes do not require heater power, have much less forward-voltage drop for a given current in the forward direction, pass much larger peak currents, have comparable maximum safe operating voltages in the reverse direction, and at least theoretically, have infinite life. A study of the vacuum diode is not undertaken because of its utility in electronic design, but because the operating principles of the vacuum diode are basic to an understanding of most other types of vacuum tubes, just as the understanding of the semiconductor junction diode is basic to other semiconductor devices. For example, the electron gun for any device employing an electron beam (such as a cathode-ray tube, travelling-wave tube, klystron, or electron microscope) employs a vacuum diode. Thus, we study the vacuum diode as an indispensable part of most devices whose operating characteristics arise from the shaping and control of an electron beam in a vacuum.

6·8·1 *Space-charge effects in a vacuum diode.*

Thermionic emission was described in Sec. 3·2·4 by the basic idea that at any given moment some of the free electrons in a cathode can be emitted into a vacuum by borrowing enough energy from their surroundings to escape the forces which hold them in the cathode. Of course, the emitted electrons will have a distribution of energies after emission, and therefore a distribution of outward-directed velocities. The most probable value of this excess energy is $2kT/q$, so the most probable outward energy is almost 0.17 ev for a cathode temperature of $1000°$K.

Now, if a thermionic cathode is sealed into an evacuated envelope and heated to emitting temperature, electrons will be emitted and will ultimately fill the space surrounding the cathode with a negative *space charge*. This space charge will set up an electric field at the cathode surface which tends to repel electrons back into the cathode. Thus, the space charge affects the conditions for emission of other electrons.

The result is that electrons which are emitted with energies of roughly $2kT/q$ or more will be able to successfully penetrate into the space-charge layer, though of course they will be slowed down in the process. Electrons which would have had a small energy after emission will be repelled back to the cathode by the space-charge forces, and as energetic electrons enter the space-charge region, the less energetic electrons in the space-charge cloud will also be forced back to the cathode. A *dynamic* equilibrium will then be set up in which electrons are returning to the cathode at the same rate as they are being emitted.

If a second electrode, called the *anode*, is added in the envelope and externally connected to the cathode with a piece of wire, a small current can be observed in the wire (see Fig. 6·27). This is because the highest-energy electrons emitted can travel completely through the space-charge region and still have sufficient outward-directed velocity to reach the anode. The space-charge region thus acts as a barrier for the flow of electrons from the cathode to the anode. Only the highest-energy electrons can pass this barrier.

If the anode is made slightly positive with respect to the cathode, electrons will be attracted from the space-charge region to the anode. This will reduce the space charge and the repelling electric field at the cathode surface, so that lower-energy electrons can be emitted from the cathode. A new equilibrium will be set up in which electrons with somewhat lower energies than before are passing *through* the space-charge cloud. The maximum energy of electrons remaining in the space-charge region is also reduced. We may summarize this condition by saying that the potential difference between the anode and the cathode has lowered the energy barrier produced by the space-charge cloud.

As the anode voltage is further increased, the collected current increases, and the space charge is reduced. Finally, an anode voltage is reached which is sufficient to draw off all the electrons that can be emitted at the given cathode temperature.

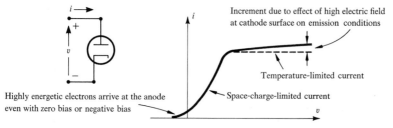

FIG. 6·27   **THE FORWARD i-v CHARACTERISTIC OF A VACUUM DIODE**

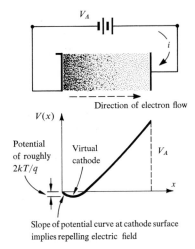

Direction of electron flow

Slope of potential curve at cathode surface
implies repelling electric field

FIG. 6·28   EFFECT OF THE SPACE CHARGE ON THE POTENTIAL DISTRIBUTION IN A ONE-
DIMENSIONAL VACUUM DIODE

When the negative charge in the space between the anode and cathode limits the collected current, the diode is said to be operating under *space-charge-limited conditions*. When the anode voltage is high enough to draw off all the current emitted by the cathode, the diode is said to be operating under *temperature-limited conditions*. The v-i characteristic shown in Fig. 6·27 exhibits these conditions.[1]

The effect of the space charge on the distribution of potential in the space between the anode and cathode is shown in Fig. 6·28. The scale on the figure is highly exaggerated to emphasize the features just described. The repelling electric field at the cathode surface is equal to the slope of the potential energy curve there. The potential minimum point is called the *virtual cathode;* at this position the average electron emitted from the cathode has zero velocity. The depth of the potential minimum is about $2kT/q$ for a zero-bias condition.

Generally, cathodes are operated in the space-charge-limited region since the current is then not a function of the cathode material. Furthermore, random fluctuations in emission are then smoothed out, and the resulting current is less noisy. It is therefore of interest to determine the mathematical form of the v-i characteristic for space-charge-limited conditions. The principal result of the derivation, to be given shortly, is

$$I = BV_A^{3/2} \tag{6·57}$$

where I is the current, V_A is the anode voltage, and B is a constant for a given geometry. B is called the *perveance,* and has the dimensions ampere per volt$^{3/2}$. Typical values of B are usually in the range 10^{-3} to 10^{-6}.

[1] The current in an actual vacuum diode does not saturate at the value suggested by the temperature-limited condition at the cathode. This is because the electric field produced at the cathode by the applied voltage affects the emission conditions. This is called the Schottky effect.

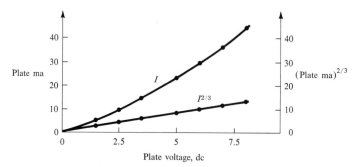

Equation (6·57) is called the *Langmuir-Child* or 3/2-power law. Many
practical vacuum diodes obey this law rather well as long as V_A is greater
than a few tenths of a volt (*cf.* experimental characteristics of a 6AL5
given in Fig. 6·29).

6·8·2   Derivation of the 3/2-power law.   In this section we shall give a
somewhat unconventional derivation of the 3/2-power law for a planar-
diode geometry. The results are of particular interest in an electron gun
such as might be found in a CRT. Furthermore, the method to be employed
is such that it can easily be extended to a cylindrical-diode geometry,
which is the most common form of geometry for other devices we shall be
interested in later.

The derivation of Eq. (6·57) centers on finding the potential distribution
in the space between the *virtual* cathode and the anode when a given current
is flowing. Figure 6·30 has been drawn to illustrate the problem. Current
flow is considered to be one-dimensional only.

We begin by observing that, since the current is due to the motion
of electrons in a vacuum, I can be written as

$$I = A\rho(x)v(x) \qquad (6·58)$$

where $\rho(x)$ is the space-charge density and $v(x)$ is the electron velocity at
the position x.

Now, if we assume that the electrons leave the virtual cathode with
zero velocity, then their velocity at the position x will depend only on the
potential difference they have travelled through in reaching a position x:

$$\tfrac{1}{2}m_0[v(x)]^2 = qV(x) \qquad (6·59)$$

where $V(x)$ is the potential at the position x relative to the potential minimum, and m_0 is the free-space mass of the electron.

Now, since $V(x)$ increases with x, $v(x)$ must also do so, and therefore
$\rho(x)$ must decrease with x to keep I constant. This behavior is sketched in
Fig. 6·30. By using Eqs. (6·58) and (6·59), we can express $\rho(x)$ as

$$\rho(x) = \frac{I}{A}\sqrt{\frac{m_0}{2q}}\,[V(x)]^{-1/2} \qquad (6·60)$$

Now, Gauss' law also gives us an independent relation between $\rho(x)$ and $V(x)$:

$$\frac{d^2V}{dx^2} = -\frac{\rho(x)}{\epsilon} \tag{6·61}$$

When this is combined with Eq. (6·60), we find

$$\frac{d^2V}{dx^2} = -\frac{I}{A\epsilon_0} \sqrt{\frac{m_0}{2q}} [V(x)]^{-1/2} \tag{6·62}$$

as the *nonlinear* differential equation which $V(x)$ must satisfy.

Now, unlike linear differential equations, it is not possible to simply write down *the* general solution to a nonlinear differential equation and then adjust the constants to fit the boundary conditions. Instead, nonlinear differential equations have only special solutions, and the boundary conditions affect the *form* which these solutions must take. In recognition of this, it is sometimes helpful to begin the search for a suitable solution to a nonlinear differential equation by asking what types of functions will satisfy the boundary conditions.

For the present problem, we want a function $V(x)$ which is such that

$$V(0) = 0 \qquad \frac{dV}{dx}(0) = 0 \tag{6·63}$$

so that both the potential and the electric field are zero at the virtual cathode.

Now, any function of the type

$$V(x) = bx^n \qquad n > 1 \tag{6·64}$$

will satisfy these boundary conditions. Of course, to be an acceptable function, $V(x)$ must also satisfy the differential equation. This means that

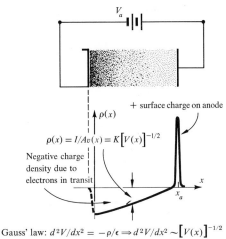

$$\rho(x) = I/Av(x) = K[V(x)]^{-1/2}$$

Negative charge density due to electrons in transit

+ surface charge on anode

Gauss' law: $d^2V/dx^2 = -\rho/\epsilon \Rightarrow d^2V/dx^2 \sim [V(x)]^{-1/2}$

FIG. 6·30   DEVELOPMENT OF THE 3/2 POWER LAW

if only one function of the type shown in Eq. (6·64) is to do the job, then b and n must be such that

$$bn(n-1)x^{n-2} = Kb^{-1/2}x^{-n/2} \tag{6·65}$$

where

$$K = \frac{I}{A\epsilon_0}\sqrt{\frac{m_0}{2q}} \tag{6·66}$$

We must then have

$$n - 2 = -n/2 \qquad bn(n-1) = Kb^{-1/2} \tag{6·67}$$

From these equations we find

$$n = \frac{4}{3} \qquad b = \left(\frac{9K}{4}\right)^{2/3}$$

Thus

$$V(x) = \left(\frac{9K}{4}\right)^{2/3} x^{4/3} \tag{6·68}$$

is an acceptable solution to the differential equation.

Now, at $x = x_a$, the anode spacing, the potential is $V_A + V_T$, where V_T is a small potential on the order of kT/q to account for the fact that the potential at the virtual cathode is not zero. Thus

$$V_A + V_T = \left(\frac{9K}{4}\right)^{2/3} x_a^{4/3} \tag{6·69}$$

We may now substitute Eq. (6·66) for K to obtain

$$\frac{I}{A} = \frac{4}{9}\epsilon_0 \sqrt{\frac{2q}{m_0}} \frac{(V_A + V_T)^{3/2}}{x_a^2} \tag{6·70}$$

Actually, the depth of the potential minimum V_T is a function of the current level. However, since $V_T \leq 0.2$ ev, the relation

$$I = BV_A^{3/2} \tag{6·71}$$

will apply over most of the forward conduction region.

Diodes with cylindrical symmetry. Exactly the same procedure can be used to obtain the v-i characteristic of a diode in which the cathode and anode are coaxial cylinders of length l and radii r_c and r_a, respectively. The differential equation for V is

$$\frac{d}{dr}\left(r\frac{dV}{dr}\right) = \frac{I}{2\pi l\epsilon_0}\sqrt{\frac{m_0}{2q}}[V(r)]^{-1/2}$$

The virtual cathode is assumed to be at $r = r_0$, so we want a solution with

$$V(r_0) = 0 \qquad \frac{dV}{dr}(r_0) = 0$$

A possible function for these boundary conditions is

$$V(r) = b(r - r_0)^n \qquad n > 1$$

Unfortunately, this function will not satisfy the differential equation exactly. However, for values of $r \gg r_0$, it gives an excellent approximation. Then, we can find n and b as before, with the result

$$V(r) = \left(\frac{9I}{8\pi l \epsilon_0} \sqrt{\frac{m_0}{2q}}\right)^{2/3} (r - r_0)^{2/3}$$

Now, at $r = r_a$, the anode radius, $V(r) = V_A$. This gives

$$I = \frac{4}{9} \frac{2\pi l}{\epsilon_0 r_a} \sqrt{\frac{2q}{m_0}} (V_A)^{3/2} = K_1 V_A^{3/2}$$

if $r_a \gg r_0$. The exact solution is

$$I = \frac{K_1 V_A^{3/2}}{\beta^2}$$

where β^2 is a function of r_a/r_0, which may be taken to be unity for $r_a/r_0 \geq 7$. In any case, the current is still proportional to the ³⁄₂ power of the voltage, as in the planar case. The reason for this is that the boundary conditions specify basically the same type of function for both cases.

Another interesting manifestation of the importance of the boundary conditions for this problem is related to the change of the v-i characteristic from space-charge-limited conditions to temperature-limited conditions. Precisely the same differential equation must be satisfied by $V(x)$. However, the electric field is no longer zero at the cathode. The effect of this change of boundary condition is to change the functional form of the V-I relation.

6·9   SUMMARY

The basic results of this chapter may be summarized as follows:
1. The current i that flows in an *ideal pn*-junction diode is related to the applied voltage v by the equation

$$i = I_0(e^{qv/kT} - 1)$$

This relation arises in the following way. The application of voltage to the diode regulates the hole and electron densities in the transition region and at its edges. In particular,

$$p_E(x = 0) = p_n(e^{qv/kT} - 1) \qquad n_E(y = 0) = n_p(e^{qv/kT} - 1)$$

are the excess minority-carrier densities at the edges of the transition region, so long as $p_E \ll N_d$, $n_E \ll N_a$.
2. The diffusive flow of minority carriers into the diode bodies and their subsequent recombination produces the excess minority-carrier distributions

$$p_E(x) = p_E(x = 0)e^{-x/L_p} \qquad n_E(y) = n_E(y = 0)e^{-y/L_n}$$

3. The minority-carrier currents flowing at $x = 0$ and $y = 0$ are found from the law of diffusive flow and are *linearly related to the excess densities at $x = 0$ and $y = 0$:*

$$i_p(x = 0) = \frac{qA \, D_p p_E(x = 0)}{L_p} \qquad i_n(y = 0) = \frac{qA \, D_n n_E(y = 0)}{L_n}$$

By neglecting recombination currents in the transition region, we can obtain the total diode current as

$$i = i_p(x = 0) + i_n(y = 0) = \frac{qA \, D_p p_E(x = 0)}{L_p} + \frac{qA \, D_n n_E(y = 0)}{L_n}$$

$$= I_0(e^{qv/kT} - 1)$$

where
$$I_0 = \frac{qA \, D_p p_n}{L_p} + \frac{qA \, D_n n_p}{L_n}$$

4. The current in a diode can be accounted for entirely by recombination somewhere in the structure. In the ideal case, the steady-state charge of holes residing in the n-type body Q_p can be written as

$$Q_p = qA \int_0^\infty p_E(x) \, dx = qA p_E(0) L_p$$

Since holes have a lifetime of τ_p, this entire charge will have to be resupplied to the n-type body every τ_p sec to maintain equilibrium. Therefore,

$$i_p(x = 0) = \frac{Q_p}{\tau_p} = \frac{qAL_p}{\tau_p} p_E(x = 0) = \frac{qAD_p}{L_p} p_E(x = 0)$$

Similarly,
$$Q_n = qA \int_0^\infty n_E(y) \, dy = qAL_n n_E(0)$$

and
$$i_n = \frac{Q_n}{\tau_n} = \frac{qAL_n}{\tau_n} n_E(0)$$

5. Real semiconductor diodes obey the ideal v-i law reasonably well, though they exhibit reverse breakdown, have series resistance which limits the flow of current in the forward direction, and may have significant recombination or generation of carriers in the transition region.

6. Vacuum diodes exhibit a theoretical v-i characteristic of the form

$$I = B(V_A + V_T)^{3/2}$$

in the range where space-charge-limited conditions apply. B is the diode perveance, V_A is the applied voltage, and V_T is a correction which is added to account for the fact that electrons are emitted from the cathode with finite velocities. Real vacuum diodes obey this characteristic rather well over a substantial portion of the space-charge-limited region.

REFERENCES

Gray, P. E., D. DeWitt, A. R. Boothroyd, and J. F. Gibbons: "Physical Electronics and Circuit Models of Transistors," SEEC Notes, vol. 2, John Wiley & Sons, Inc., New York, 1964, chap. 3.

Linvill, J. G., and J. F. Gibbons: "Transistors and Active Circuits," McGraw-Hill Book Company, New York, 1961, chap. 4.

Valdez, L.: "The Physical Theory of Transistors," McGraw-Hill Book Company, New York, 1961, chap. 11.

Middlebrook, R. D.: "An Introduction to Junction Transistor Theory," John Wiley & Sons, Inc., New York, 1957, chap. 7.

Spangenberg, K. R.: "Fundamentals of Electron Devices," McGraw-Hill Book Company, New York, 1957, chaps. 8 and 9.

Nanavati, R. P.: "An Introduction to Semiconductor Electronics," McGraw-Hill Book Company, New York, 1963, chap. 3.

Moll, J. L.: "Physics of Semiconductors," McGraw-Hill Book Company, New York, 1964, chap. 7.

DEMONSTRATION

pn-DIODE CHARACTERISTICS   The purpose of this demonstration is to display visually the volt-ampere characteristics of a diode. The curve is displayed on a cathode-ray oscilloscope, using the circuit shown. The circuit provides the instantaneous current-voltage relationship by sweeping the diode voltage at a 60-cps rate. A number of commercial curve plotters using refinements of this basic circuit are available, any of which could be used for this demonstration. (The reverse characteristics of the diode should be viewed with high series resistance to avoid overheating.) It is instructive to display and compare the characteristics of a general-purpose diode, a Zener diode, and a photodiode.

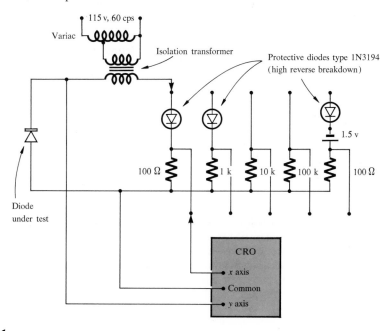

FIG. D6 · 1

PROBLEMS

6 · 1   Equation (6 · 21*a*) indicates that the excess hole density in the *n*-type diode body of a junction diode can be written as

$$p_E(x) = p_E(0)e^{-x/L_p}$$

Show that

$$-\frac{dp_E(x)}{dx} = \frac{p_E(x)}{L_p} \qquad \text{Eq. (6·23a)}$$

$$\int_x^\infty p_E(x)\,dx = p_E(x)L_p \qquad \text{Eq. (6·23b)}$$

Also show that the probability that a hole will recombine in the interval between x and $x + \Delta x$ is

$$\frac{e^{-x/L_p}}{L_p}\,\Delta x$$

Using this equation, show that the average distance $\langle x \rangle$ that a hole will diffuse before recombining is

$$\langle x \rangle = L_p \qquad \text{Eq. (6·23d)}$$

6·2 In the text, the excess hole density $p_E(x)$ in n material is found by solving the steady-state continuity equation. As an alternate procedure for determining $p_E(x)$, we may note that the hole current crossing a plane at x in n-type material has the sole function of resupplying the holes which are lost between x and ∞ due to recombination. Show that this condition leads to the equation

$$-\frac{dp_E(x)}{dx} = \frac{1}{L_p{}^2}\int_x^\infty p_E(x)\,dx$$

which the $p_E(x)$ curve must satisfy. Show also that this equation is equivalent to the steady-state continuity equation.

6·3 The junction current in the p^+n diode shown in Fig. P6·3 consists entirely of holes. The n region from $x = 0$ to $x = W$ has a hole lifetime of τ_p. However, the portion of the n-type body from $x = W$ to $x = L$ has been treated (by sandblasting of gold doping) so that the lifetime for holes in this region is nearly zero. Both regions have an equilibrium hole density of p_n.

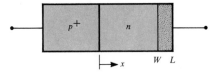

FIG. P6·3

1. Show that the density of holes at $x = W$ is p_n, regardless of the diode current.
2. Determine the v-i characteristic for the diode.
3. Plot $p_E(x)$ for the case $W \ll (D_p\tau_p)^{1/2}$. What fraction of the total current recombines in the body and what fraction in the treated area for this case? (This procedure is actually used to produce diodes for computer applications.)
4. Can the diode current be written in the form

$$i = \frac{Q}{\tau}$$

If so, what are Q and τ?

6·4 Consider the diode (made from a transistor) shown in Fig. P6·4.
1. Assuming the law of the junction applies at $x = W$, what is the hole density there (in terms of v)?
2. Determine the v-i characteristic for this diode. Assume that $W \ll L_p$, the diffusion length for holes in n-type layer.

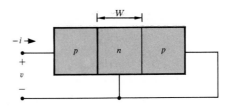

6·5 In this problem we will consider the temperature dependence of the v-i characteristic of an ideal junction diode.

1. A germanium diode has a saturation current of 10^{-6} amp at room temperature. Plot the v-i characteristic for this diode from $i = 10^{-6}$ to 10^{-2} amp at room temperature and at $+30°$C. Use a logarithmic scale for current and a linear scale for voltage (as in Fig. 6·19).

2. If the diode is driven from a constant current source of 1 ma, what is the rate of change of v with T? Use the plot constructed above to show how much v changes for a $30°$C change in temperature. Discuss the possibility of using this effect to make a thermometer.

3. If the diode is driven from a constant voltage source of 0.3 volt (forward bias), what is the rate of change of i with T. Show the total change of i for a $30°$C in T, using the plot constructed in (1).

6·6 *Dynamic resistance of the ideal junction diode.* In some applications of junction diodes we are interested in knowing the resistance which the diode presents to low-frequency sine waves of small amplitude. In these applications we establish a bias point at (I_1, V_1) and then want to know how small changes in voltage Δv are related to small changes in the current Δi. If the signals are small enough, Δv and Δi are proportional. We can then define the resistance of the diode to the small-signal current as

$$r = \frac{\Delta v}{\Delta i} \simeq \frac{dv}{di}$$

r is called the ac, dynamic, or incremental resistance of the diode.

1. Show that

$$\frac{1}{r} = \frac{q}{kT}(I_1 + I_0)$$

where I_1 is the bias current. Evaluate r for a bias current of 1 ma. You may assume $I_0 \ll 1$ ma.

2. Sketch the v-i characteristic for an ideal diode and locate an arbitrary bias point (V_1, I_1) on it. Now, draw two straight lines on this sketch, showing V_1/I_1 and the dynamic resistance. Is V_1/I_1 equal to dv/di? Why?

3. If the voltage applied to the diode is represented by $v_0 + \Delta v$, then the current is

$$i = I_0(e^{q(v_0+\Delta v)/kT} - 1)$$

Use the expansion

$$e^x = 1 + x + \frac{x^2}{2!} + \cdots$$

to rewrite i. Explain the various terms on a sketch of the v-i characteristic. How large can Δv be if we require

$$i = i_0 + r\Delta v \pm 0.1 r\Delta v$$

where r is evaluated at the given bias point.

6·7 Show that in a diode for which $N_a \gg N_d$, the saturation current is equal to the current which would be *generated* in a length L_p of n-type material if the hole density were maintained there at zero.

6·8 Construct a chart which shows how the fundamental ideas presented in all the previous chapters lead to the v-i characteristic of the ideal junction diode.

6·9 A pair of identical diodes are connected back to back (Fig. P6·9), and a voltage of 5 volts is applied to them. The diodes have breakdown voltages of 50 volts. Determine the current in the pair and the voltage drop across each diode. Sketch the v-i characteristic for the diode pair.

6·10 A silicon pn junction has an area of 10^{-2} cm². Its doping levels are $N_a = 10^{19}/\text{cm}^3$, $N_d = 10^{16}/\text{cm}^3$. What will the avalanche breakdown voltage be for this diode? You may assume a peak electric field of 4.5×10^5 volt/cm for the avalanche condition.

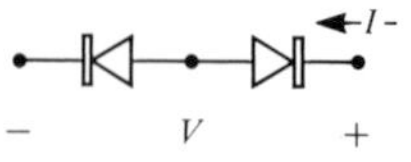

FIG. P6·9

6·11 The hole density at the n-type edge of the transition region should be calculated by a simultaneous solution of Eqs. (6·45a) and (6·45b). Show that the solution of these equations is such that estimating $p(x = 0)$ by

$$p(0) = p_n q^{v/kT}$$

gives less than 10 percent error as long as $p < 0.1N_d$.

6·12 This problem deals with the effect of recombination in the transition region on the v-i characteristic of a pn diode. In order to compare i_T and $i_p(y = 0)$, defined on Fig. 6·19, show from equations given in the text that the ratio of i_T to $i_p(y = 0)$ is

$$\frac{i_T}{i_p} \cong \frac{WN_d}{L_p n_i} e^{-qv/2kT}$$

Evaluate this ratio for the germanium diode shown in Fig. 6·10. How large should v be to ensure $i_T/i_p < 0.1$? Now evaluate i_T/i_p for a silicon diode with the same doping and lifetime as the germanium diode. How large should v be to ensure $i_T/i_p < 0.1$? Compare your prediction with the one shown in Fig. 6·19.

6·13 Typical parameters for a commercially available double-base diode are $\rho_n = 10$ ohm-cm, cross-sectional area of n-type bar $= 0.001$ cm², length of n-type bar $= 0.15$ cm. Plot an approximate i_C-v_C characteristic for this device assuming $V_0 = 12$ volts. The semiconductor material is germanium and the pn junction is situated two-thirds of the way from the ground end of the device.

6·14 Show that the v-i characteristic for a pn-junction photocell may be written as

$$i = I_0(e^{qv/kT} - 1) - i_v$$

by solving the continuity equation (6·52), subject to the appropriate boundary conditions. Also show that i_v can be interpreted as the total current generated by the light within a diffusion length on either side of the junction.

6·15 This problem deals with a comparison of vacuum diode and semiconductor diode characteristics. A 5Y3 vacuum diode intended for power rectification has a perveance of 3×10^{-4} amp/volt$^{3/2}$. Determine the anode voltage necessary for this diode to conduct a current of 100 ma. How much power is dissipated at the anode under this condition?

A 1N488A silicon power rectifier intended for comparable applications has a saturation current of 2.5×10^{-8} amp and at high currents obeys the v-i law

$$I = I_0 e^{qv/1.5kT}$$

(Reasons for the factor 1.5 in the exponent are given in Sec. 6·6·1.) Determine the voltage necessary for this rectifier to conduct a current of 100 ma. How much power is dissipated in the device under this condition?

6·16   The population of the United States is approximately 200 million. Assuming the average age at death is 67 years, and that 200 million is essentially the steady-state population, calculate the average birthrate (that is, how many babies per year; or how many seconds between births on the average, if you are interested). Explain the relationship of this problem to the flow of current in a *pn* diode which has $N_a \gg N_d$.

6·17   Sketch the *i-v* characteristics of a photodiode with i_p as a parameter. Let i_p assume the values $I_0, 2I_0, 3I_0$ for convenience. What applications can you think of for a device with these characteristics?

6·18   A *pn*-junction diode is made in a material for which $E_g = 2.0$ volts. Assuming a forward current of 1 ma flows in the diode, and all the recombination is direct recombination, calculate the light power emitted by the diode.

6·19   Actually, light-emitting recombinations occur infrequently compared to nonradiative recombinations, and some of the light that is emitted is reflected at the surfaces of the semiconductor. As a result, only about 0.1 to 1 percent of the light calculated in Prob. 6·18 is transmitted to an observer. Under these conditions, compare the intensity of the emitted light with that of sunlight at the earth's surface (about 10 watts/ft^2), assuming the diode area is 10^{-2} cm^2. Repeat for moonlight on a clear night (about 5×10^{-5} watts/ft^2).

7

LOW-FREQUENCY APPLICATIONS OF DIODES

IN THE INTRODUCTION to Chap. 6 it was pointed out that the diode is an interesting circuit element because of the inherent nonlinearity of its *v-i* characteristic. In this chapter we will see how various features of this nonlinear *v-i* relation may be exploited to design circuits with performance characteristics which cannot be obtained by *any* combination of the more familiar R, L, and C elements. Throughout the chapter we wish to emphasize that the ideal diode has an equal standing with *R, L,* and *C* elements in the electronic designer's kit of ideal elements.

7·1 IDEAL DIODES AND ELECTRONIC CIRCUIT THEORY

As we mentioned in the introduction, electronic circuits are circuits which utilize electronic devices (such as diodes, transistors, vacuum tubes, cathode-ray tubes, etc.) to perform some desired operation on an electrical signal. Electronic circuit theory is the mathematical study of these circuits.

Electronic circuit theory is different from the more familiar *RLC* circuit theory primarily because electronic circuits are *nonlinear* and/or *nonreciprocal*. The nonlinearity in electronic circuits arises primarily from the nonlinearity associated with the *v-i* characteristic of a diode. The most important type of nonreciprocity arises from the fact that ideal amplifying elements transmit signals in one direction only.

In order to account for these basic effects in the analysis and design of electronic circuits, it is necessary to add two elements to the familiar *RLC* building blocks of linear-network theory. These elements are the *ideal diode* and the *controlled source*. With these five elements, we can then, in principle, analyze (or in practice, estimate) the performance of any electronic circuit, even though the characteristics of actual devices used in the circuit may have little or no relation to the behavior of these ideal elements.

In this chapter we shall be concerned with the ideal diode and its relation to practical diodes and low-frequency diode circuits. We will consider diode transient performance in the next chapter and return to the development of the controlled source at a later point.

7·1·1   Characteristics of an ideal diode.   The ideal diode (not the ideal junction diode) is a self-actuated polarity-sensitive switch. The circuit symbol for the ideal diode[1] and reference directions for *v* and *i* are given in Fig. 7·1*a*; its *v-i* characteristic is shown in Fig. 7·1*b*. As the *v-i* characteristic indicates, the ideal diode behaves like a short circuit in the forward direction and an open circuit in the reverse direction.

The *v-i* characteristic of the ideal diode is said to be *piecewise linear,* because the ideal-diode curve is linear on each separate piece of its characteristic. In fact, for $i > 0$, it behaves like a resistor with $R = 0$; and for $v < 0$, it behaves like a resistor with $R = \infty$. The nonlinear nature of the ideal diode only becomes evident when the circuit operation drives the diode across the *break point* in its *v-i* characteristic.

When we approximate the *v-i* characteristic of a real diode or other device with a piecewise-linear one, we are tacitly assuming that for the appli-

---

[1] We shall use a solid diode symbol to denote an ideal diode and an outline symbol to indicate a semiconductor diode.

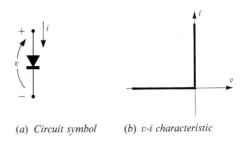

(a) *Circuit symbol*   (b) *v-i characteristic*

FIG. **7·1**   THE IDEAL DIODE

cation we are envisaging, the important aspects of device behavior are entirely concerned with where the *break points* in the characteristic lie, and not at all with the fine-grained nature of the nonlinearities which exist on the separate pieces.

7·1·2 *Relation between the ideal junction diode and the ideal diode.* The best approximation that we have to the ideal diode is the ideal junction diode studied in the last chapter. There we saw that the *v-i* characteristic of the ideal junction diode was of the form

$$i = I_0(e^{qv/kT} - 1) \tag{7·1}$$

To make a piecewise-linear approximation to this characteristic, we observe that

1. In the forward-bias direction, the voltage drop across the diode will be at most a few tenths of a volt, regardless of the magnitude of the current.

2. The current which the diode will pass in the reverse direction will be, at most, a few microamperes, regardless of the magnitude of the voltage.[1]

If a diode application is such that we may neglect the small reverse current and the few tenths of a volt that are necessary to produce forward current, the *v-i* characteristic of the ideal junction diode can be approximated by that of the ideal diode. This approximation is a valid one for many applications of real *pn*-junction diodes.

If the application is such that the forward-voltage drop cannot be neglected, then we can approximate the junction diode with an ideal diode and a battery, as shown in Fig. 7·2b. And if we need to include some idea of the behavior of the junction diode near the origin and its nonzero reverse current, then we can use the circuit shown in Fig. 7·2c.[2] By a continuation of this process, we can approximate the junction diode to any required degree of accuracy with nothing but the five ideal elements of electronic circuit theory. Using similar techniques, we can also approximate the *v-i* characteristic of a vacuum diode with ideal diodes, resistors, and batteries (see Prob. 7·4).

Of course, there are some applications that depend on the exact nature of the *v-i* characteristic of the ideal junction diode or the ideal vacuum diode. For example, we shall see in Sec. 7·7 that the power output obtainable from a solar cell can be most simply calculated by considering the cell to be an ideal *junction diode* in parallel with a light-dependent (that is, controlled) source. In such a study we need not refer to the ideal diode of electronic circuit theory. Furthermore, one should not lose sight of the fact that we must know what the actual *v-i* characteristic (or transient response) of a device is, and frequently how it arises, before we can build up a

---

[1] We assume for the moment that the ideal junction diode does not have a reverse breakdown voltage, though we could as well define both the ideal diode and the ideal junction diode to have adjustable reverse breakdown voltages, as is done in Sec. 7·2·2.

[2] Procedures for analyzing the *v-i* characteristics of networks such as those shown in Fig. 7·2 are the subject of Probs. 7·1, 7·2, and 7·3.

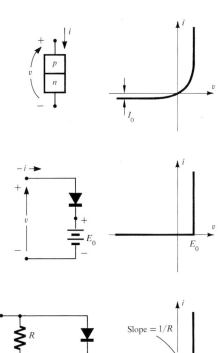

(a) *Ideal pn-junction diode characteristic*

(b) *Piecewise linear approximation to (a) with ideal diode and battery*

(c) *Piecewise linear approximation to (a) with two ideal diodes, resistor, current source, and battery*

FIG. 7·2 **THE IDEAL JUNCTION DIODE AND PIECEWISE LINEAR APPROXIMATIONS TO IT**

model for it from the five ideal elements of electronic circuit theory. Ideal circuit elements do not eliminate the need for understanding the physics of electronic devices. Nevertheless, it is important to begin the study of diode circuits with an appreciation of the fundamental position which the ideal diode has in electronic circuit theory.

7·2 RESISTIVE DIODE CIRCUITS

Circuits which contain only energy sources, resistances and diodes are called resistive diode circuits. Such circuits find important applications in their own right, and in addition they form a basis for understanding more complicated diode circuits which include capacitance and inductance. We shall consider several examples of resistive diode circuits in this section. In each example, the reader should note carefully how the nonlinearity of the diode produces circuit characteristics that could not be obtained otherwise.

7·2·1 *Synthesis of nonlinear-resistance curves.* One of the simplest uses of the diode nonlinearity is in the synthesis of nonlinear-resistance curves, that

is, v-i curves with nonconstant slopes. We have already seen in Fig. $7 \cdot 2$ that the v-i characteristic of an ideal junction diode can be synthesized with ideal diodes, resistors and sources. While this is of considerable analytical use, the synthesis of nonlinear-resistance curves with actual diodes also has practical applications.

For example, in the field of analog computation, nonlinear resistances are frequently synthesized to obtain the solution of nonlinear differential equations. To be specific, let us consider how we might make an electric network to solve the equation

$$\frac{dy}{dx} + f(y) = F(x)$$

$$y(0) = y_0 \qquad\qquad (7 \cdot 2)$$

where $f(y)$ is the nonlinear function shown in Fig. $7 \cdot 3a$. To concentrate on essentials, we assume that $f(y)$ can be adequately represented by the piecewise-linear approximation shown in Fig. $7 \cdot 3a$. We can solve this equation with the circuit shown in Fig. $7 \cdot 3b$. In this circuit, voltage across the capacitor is analogous to y, time is analogous to x, and Kirchoff's current law at the $v_c(t)$ node is analogous to Eq. $(7 \cdot 2)$. The diode-resistor-battery network synthesizes $i = f(v)[\sim f(y)]$. In Fig. $7 \cdot 3$, we are assuming that the required current source $F(t)[\sim$driving function $F(x)]$ is available, though if it is other than a simple function its synthesis may also have to be carried out with diodes, resistors, and amplifiers. The equation is solved by establishing an initial voltage on C corresponding to $y(0)$, then operating S_1 and S_2, and recording the development of v_c with time.

More general types of nonlinear-resistance curves may, of course, be required by other problems. These curves can be synthesized with nothing more than diodes, resistors, batteries, and current generators, as long as the slope of the nonlinear v-i characteristic being synthesized is positive

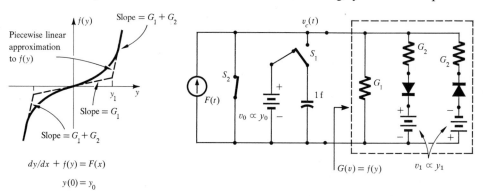

(a) Nonlinear ordinary differential equation

(b) Solution of Eq. $(7 \cdot 2)$ by electrical analog

FIG. $7 \cdot 3$ THE USE OF NONLINEAR CONDUCTANCE IN THE SOLUTION OF A NONLINEAR ORDINARY DIFFERENTIAL EQUATION

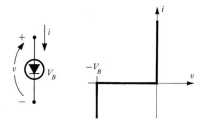

FIG. 7·4   *v-i* CHARACTERISTIC OF AN IDEAL BREAKDOWN DIODE

(or zero) everywhere. The nature of the synthesis procedure for these cases is the subject of Prob. 7·1–7·3.

7·2·2   The ideal breakdown diode and its use in synthesizing nonlinear impedances.   The synthesis of a nonlinear-resistance characteristic which has break points out in the *v-i* plane will always involve the use of sources (such as the batteries shown in Fig. 7·3). As a practical matter, one may not wish to use batteries in an actual synthesis if they can be avoided. In such cases it is useful to note that a semiconductor junction diode with an avalanche breakdown voltage of V_B gives an approximate realization of an ideal diode and a battery in series (with the polarity of the battery opposing the direction of forward current in the ideal diode). We can, in fact, define an *ideal breakdown diode* as an ideal diode with a reverse breakdown voltage which can be specified by the designer.

The circuit symbol and *v-i* characteristic for an ideal breakdown diode are shown in Fig. 7·4. The use of this element frequently enables one to perform a practical synthesis of a nonlinear-impedance characteristic without the use of batteries (see Prob. 7·5).

7·2·3   Diode-logic circuits in digital computers.   Resistive diode circuits also find an interesting application in performing logical operations in a digital computer. Here the ON-OFF character of the diode nonlinearity is used to register whether a condition is met or not.

Some examples of typical diode-logic circuits for a digital computer are shown in Fig. 7·5. In the AND circuit shown in Fig. 7·5*a*, an output of 1 volt is registered whenever voltages are present simultaneously at A and B. In the OR circuit shown in Fig. 7·5*b*, an output of 1 volt is registered whenever *C* or *D* is present. In both circuits the output potential will be UP (or high) when the correct condition is met, and DOWN (or low) when the condition is not met.

An example of how these basic logic circuits can be combined to perform more complicated logical operations is shown in Fig. 7·5*c*. Here an output is obtained whenever *A* and *B* or *C* and *D* are present. We shall study other applications of diode-logic networks in later chapters.

7·2·4   The resistive half-wave rectifier.   One of the most common uses of diodes is for converting the alternating voltage available from the 60-cycle

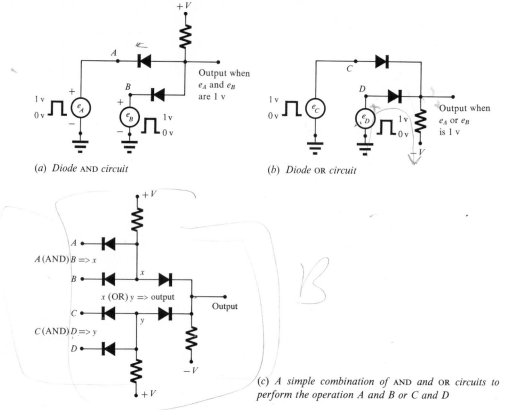

(a) *Diode* AND *circuit*

Output when
e_A and e_B
are 1 v

1 v
0 v

(b) *Diode* OR *circuit*

Output when
e_A or e_B
is 1 v

A (AND) $B \Rightarrow x$

x (OR) $y \Rightarrow$ output

C (AND) $D \Rightarrow y$

Output

(c) *A simple combination of* AND *and* OR *circuits to perform the operation A and B or C and D*

FIG. 7·5 THE USE OF DIODES TO PERFORM LOGIC IN A DIGITAL COMPUTER

power lines to a dc voltage which is suitable for powering electronic equipment. A simple circuit which illustrates the basic idea is the resistive half-wave rectifier shown in Fig. 7·6. In this circuit a voltage source e_s supplies a sinusoidal voltage to the circuit. The source has an internal resistance R_S, which we may also use to account for diode series resistance in a real diode. A resistance R_L acts as the load.

The circuit operates as follows. When the supply voltage e_s is positive, the diode is a short circuit. The supply voltage is then resistively divided between R_S and R_L, so that the load voltage $e_L(t)$ is

$$e_L(t) = e_s(t) \frac{R_L}{R_S + R_L} \qquad e_s(t) > 0 \qquad (7·3)$$

When $e_s(t)$ is negative, the diode is an open circuit, so

$$e_L(t) = 0 \qquad e_s(t) < 0 \qquad (7·4)$$

A compact way of presenting the information contained in these equations is in terms of the *voltage-transfer curve* shown in Fig. 7·6d. Here e_L is plotted against e_S. The figure shows that e_L is zero when e_S is negative.

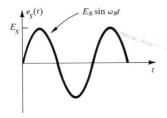

(a) *Input ac voltage*

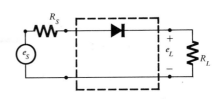

(b) *Power converter with source and load*

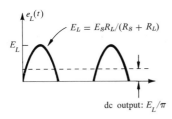

(c) *Output voltage with dc component*

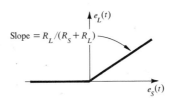

(d) *Voltage transfer curve*

FIG. 7·6 THE RESISTIVE HALF-WAVE RECTIFIER AS A POWER CONVERTER CIRCUIT

When e_S is positive, the voltage-transfer curve is a straight line with slope $R_L/(R_S + R_L)$. The piecewise linearity of the voltage-transfer curve indicates the presence of the ideal diode in the circuit.

The load-voltage waveform for the resistive half-wave rectifier is shown in Fig. 7·6c. It consists of half-sine-wave pulses and is *periodic* in time. It therefore be represented by a Fourier series (that is, a series of sine waves of appropriate amplitude and frequency).[1] The series is

$$e_L(t) = \frac{E_L}{\pi} + \frac{E_L}{2} \cos \omega_s t + \frac{2E_L}{3\pi} \cos 2\omega_s t + \cdots \qquad (7·5)$$

where E_L is the peak amplitude of the rectified wave. (The time origin has been shifted to coincide with the maximum value of e_L for mathematical simplicity in finding the Fourier series coefficients). The first three terms of this Fourier series are shown in Fig. 7·7 to indicate how quickly the terms of the series converge to approximate the actual load voltage.

Equation (7·5) shows that the Fourier expansion of the output voltage $e_L(t)$ contains a dc component plus components at ω_s, $2\omega_s$, ... However, the input voltage is a pure sine wave of frequency ω_s. Thus *frequencies are present in the output which are not present in the driving waveform*. This is, in fact, a general characteristic of nonlinear circuits, and of sufficient importance that we shall digress to comment on it briefly.

---

[1] For readers who are unfamiliar with this topic, an introduction to Fourier series can be found in L. A. Manning, "Electrical Circuits," McGraw-Hill Book Company, New York, 1966, Chap. 20; or H. H. Skilling, "Electrical Engineering Circuits," 2d ed., John Wiley & Sons, New York, 1966, Chap. 14.

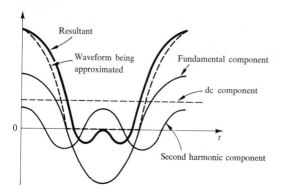

Resultant

Waveform being
approximated

Fundamental component

dc component

0

t

Second harmonic component

FIG. 7·7 THE REPRESENTATION OF A HALF-WAVE RECTIFIED COSINE WAVE BY ITS
FOURIER SERIES

7·3 THE BEHAVIOR OF PERIODIC SIGNALS
IN NONLINEAR NETWORKS

One of the characteristic differences between a nonlinear circuit and a
linear one arises when we consider how these circuits respond when they
are driven with signals which are periodic functions of time, in particular
with sine waves. From the point of view of linear circuit theory, we may
summarize the importance of sine waves in the following two statements:

1. They have the *unique* property of preserving their special variation
with time when they are transmitted through a linear circuit. The waveform of any other function (for example, a square wave or a triangular wave)
cannot in general be passed through a linear network without changes in
its shape.

2. Any practical periodic function of time can be synthesized from an
appropriate combination of sine waves with the correct relative amplitudes
and phases (the Fourier theorem).

These two statements lead to the following procedure for determining
the response of a linear circuit to a periodic driving voltage. We first
resolve the periodic driving signal into its Fourier components, and then
determine how each of these components is passed through the circuit.
The separate components may be added together at the point where we
wish to know the voltage or current waveform.

There will, in general, be changes in the amplitude and phase of each
Fourier component as it passes through a linear circuit, so the actual
waveform of voltage or current at any point in the circuit may bear little
resemblance to the driving waveform. However, any current or voltage
waveform in the linear circuit can only have Fourier components at the
frequencies which are present in the driving waveform.

In contrast to this behavior, nonlinear circuits *characteristically* produce
Fourier components at frequencies which are *not* present in the driving
waveform. And since nonlinearity in electronic circuits arises primarily in

the operation of the ideal diode, we can expect these new frequencies to appear when a diode is driven through its break point by a periodic signal in the course of operation of a circuit.

We can turn this effect to advantage in several ways. For example, dc power is required to operate most electronic devices. However, the most commonly available source of power is in the form of a 60-cps sine wave. By an appropriate use of diodes, transformers, capacitors and resistors we can build a *power supply* to convert this 60-cps voltage into a useable dc voltage. Such a power supply is clearly a nonlinear circuit or system, since it is driven with a 60-cps voltage and its output is a dc voltage. This feature is suggested in Fig. 7·8a.

Again, speech signals lie in a frequency range of 20 to 20,000 cps, which cannot be readily transmitted over long distances. However, signals in a higher-frequency range can. Therefore, in the radio transmission of speech, the Fourier spectrum of the speech waveform is shifted to a higher frequency which is more suitable for transmitting. For example, to transmit a 2-kcps audio tone, a radio station operating at a station frequency of 500 kcps will actually transmit three sine waves at the frequencies 500 kcps, 502 kcps, and 498 kcps (that is, the carrier frequency f_c and two sidebands at $f_c \pm f_{audio}$).

The piece of equipment which achieves this frequency-shifting task is called a *modulator*. As suggested in Fig. 7·8b, it is supplied with the 2-kcps tone and the carrier signal ($f_c = 500$ kcps) at its inputs; its output contains frequencies which are not present in the input signals (for example, 500 kcps + 2 kcps = 502 kcps). A modulator is therefore a nonlinear

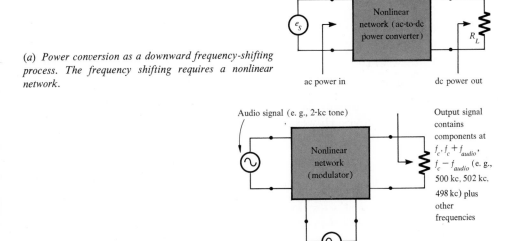

(a) *Power conversion as a downward frequency-shifting process. The frequency shifting requires a nonlinear network.*

ac power in   dc power out

Audio signal (e. g., 2-kc tone)

Output signal contains components at f_c, $f_c + f_{audio}$, $f_c - f_{audio}$ (e. g., 500 kc, 502 kc, 498 kc) plus other frequencies

(b) *Modulation is a frequency-shifting process and requires the use of a nonlinear network.*

Carrier signal, frequency f_c (e. g., 500 kc)

FIG. 7·8   CIRCUITS WHICH REQUIRE NONLINEARITY TO ACHIEVE THEIR OBJECTIVES

system.[1] It can, in fact, be made with nothing more than diodes. However, a modulator cannot be made with any combination of only ideal R, L, or C elements.

There are many other examples of electronic systems and subsystems where nonlinear operations are required to achieve certain objectives. We shall study some examples of these systems in the remainder of this chapter. In all cases the nonlinearity of the diode provides the basis for the circuit characteristics.

7·4 AC-TO-DC POWER CONVERSION

In Sec. 7·2·4, we saw that the resistive half-wave rectifier exhibits the basic possibility of converting ac power to dc power. However, the output voltage of this circuit has a large time-varying component and is hardly useable as a dc voltage for operating electronic devices.

In order to remove this defect, we usually employ some type of filter between the ideal rectifier and the load. Usually, the filter has a very significant effect on the waveforms in the circuit, so it is necessary to analyze each rectifier-filter combination separately. However, each design is aimed at the same basic objective: we want to produce a smooth dc voltage from the ac input voltage.

From the point of view of *load requirements,* we may judge the relative merits of various circuit arrangements by constructing an equivalent circuit for the output of the power supply.[2] As is suggested in Fig. 7·9, the equivalent circuit for the output will contain a dc voltage E_{dc}, an ac voltage $e_r(t)$ (also called the *ripple* voltage), and a source resistance R_S.

Usually, the *design requirements* for a power supply are also stated in terms of the required equivalent circuit. It is customary to state the no-load dc voltage, the full-load dc voltage, the full-load dc current and the peak-to-peak ripple which is acceptable. The nature of the applications for the power supply will determine just how small $e_r(t)$ needs to be. The difference between no-load and full-load voltage divided by the full-load current gives the maximum acceptable value for R_S.

There will usually be more than one possible circuit to achieve given design objectives. However, the various circuits will place different requirements on the diodes and other components. These requirements will be

---

[1] Actually, modulators can be constructed from linear *time-varying* elements, of which a butterfly capacitor driven by a motor is an example. In practice, time-varying elements are usually obtained by driving an element, such as a reverse-biased semiconductor diode, which has a nonlinear C vs. V law, with a time-varying voltage.

[2] The concept of an equivalent circuit for the output of a nonlinear network is only valid in a limited sense and must be used with care. In the present case, we mean that for resistive loads that are not too small, the output terminals of the power supply can be visualized in the manner shown in Fig. 7·8. The parameters of the equivalent circuit shown in this figure cannot be measured in the same way that the parameters of the equivalent circuit for a linear network would be measured, as the analysis to be given will show.

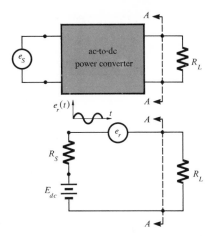

directly related to the cost, reliability and simplicity of a given circuit. Therefore, in addition to load requirements a realistic design must account for these other factors as well.

In this section we will examine several of the more common circuit arrangements for ac-to-dc power conversion. We will give a rather complete analysis of the half-wave rectifier with a capacitor filter, since this circuit can be most simply used to illustrate ideas. Several variations of this circuit will then be presented to show how the design objectives can be met in cases where the simple circuit is not adequate.

7·4·1 *Half-wave rectifier with capacitor filter.* One of the simplest and most widely used circuits for ac-to-dc power conversion is made by connecting a large capacitor across the output of the resistive half-wave rectifier. The circuit is shown in Fig. 7·10. Voltage and current waveforms at various points in this circuit are drawn to aid in discussing the circuit's operating characteristics.[1]

The basic operation of this circuit can be understood as follows. We suppose that C initially has no charge. At $t = 0$ the ac source (with zero source resistance) is switched ON and delivers a sine wave of voltage $e_S(t)$ to the circuit. During the first quarter-cycle of operation, the diode is conducting, and the capacitor charges to the peak ac voltage E_S. As the input voltage falls below E_S, the diode becomes reverse-biased, and the capacitor C discharges through the load resistance in the normal exponential manner. The diode remains reverse-biased for $(T - \Delta t)$ sec (see Fig. 7·10), when the source voltage and the voltage across the filter capacitor are equal. From this point on to the peak voltage, the diode is conducting. During this conducting interval, the capacitor is recharged to the peak voltage E_S.

---

[1] Convenient circuits for demonstrating waveforms in this and other power supplies are given in Demonstrations 7·1 and 7·2.

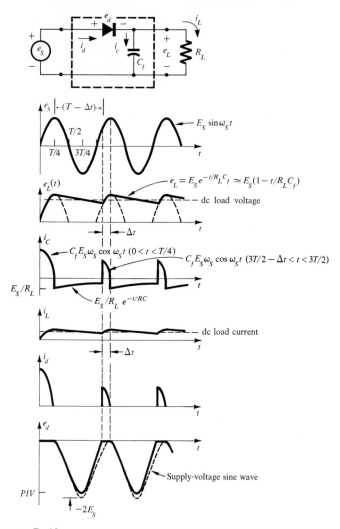

FIG. 7·10 HALF-WAVE RECTIFIER WITH CAPACITOR FILTER AND WAVEFORMS ILLUS-
TRATING ITS OPERATION

Examination of the load voltage reveals that the circuit can deliver a nearly smooth dc voltage to the load R_L, as long as the load current is small enough that the filter capacitor does not discharge too much between the recharging pulses. The waveforms shown in Fig. 7·10 are drawn for this approximation.

It is convenient to perform the quantitative analysis of this circuit with the dc load current i_L as the variable, rather than the value of the load resistance. The analysis then will directly yield the components of the output equivalent circuit.

The load voltage consists of a dc component and a small ripple component. The ripple voltage $e_r(t)$ results from the fact that the load current i_L

is drawn from the filter capacitor between the recharging pulses. The time interval between these pulses is approximately T, the period of the sine-wave supply voltage. If we assume i_L is approximately constant, then the total charge lost by the capacitor is

$$\Delta Q \cong i_L T \qquad (7 \cdot 6)$$

The corresponding change in capacitor voltage is

$$\Delta v_C = \frac{\Delta Q}{C_f} = \frac{i_L T}{C_f} \qquad (7 \cdot 7)$$

The peak-to-peak value of the ripple voltage is

$$\Delta v_C = 2E_R \qquad (7 \cdot 8)$$

Using Eq. (7 · 7) and assuming $e_r(t)$ to be a sawtooth wave, the dc output voltage is found to be

$$E_{dc} = E_s - \frac{\Delta v_C}{2} = E_s - \frac{i_L T}{2C_f} \qquad (7 \cdot 9)$$

We may now begin to construct the equivalent circuit for the output terminals. Equation (7 · 9) indicates that the open-circuit dc voltage is E_s and the series resistance is

$$R_s = -\frac{dE_{dc}}{di_L} = \frac{T}{2C_f} \qquad (7 \cdot 10)$$

where the minus sign accounts for the fact that the dc output voltage will decrease as i_L increases. We may also write R_s in the useful form

$$R_s = \pi X_c \qquad (7 \cdot 11)$$

where X_c is the reactance of C_f at the supply-voltage frequency. These equations for R_s, of course, neglect the series resistance of the diode and the sine-wave supply voltage, and rely on the approximation that the capacitor discharge is adequately represented by a linear approximation to $e^{(-t/R_L C_f)}$. However, these conditions are met in most instances where this simple circuit is useful.

The ripple voltage at the output is a sawtooth voltage whose peak-to-peak amplitude depends on the load current i_L [see Eq. (7 · 8)]. We must therefore represent the ripple voltage by an ac source $e_r(t)$, whose output is a sawtooth, with a peak amplitude[1] of

$$E_R = \frac{T}{2C_f} i_L \qquad (7 \cdot 12)$$

Thus, from the point of view of the load, the output of the power supply may be represented by the equivalent circuit shown in Fig. 7 · 11. Note

---

[1] Actually, the entire amplitude of e_r appears at the output terminals so a large capacitor should be added in parallel with R_s. This correction is usually unimportant and will be neglected for simplicity.

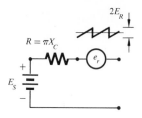

$R = \pi X_C$

$2E_R$

e_r

E_s

that the components in this circuit depend only on C_f, E_S and f_s, the supply frequency. Usually f_s is fixed, so we have only E_S and C_f to use as adjustable parameters in attempting to achieve design objectives with this simple circuit. This lack of flexibility limits this circuit to applications where the full-load dc current is small and a rather sizable ripple voltage is acceptable. These points can be most easily brought out in an example.

EXAMPLE   Let us suppose that we want to design a power supply with an open-circuit output voltage of 25 volts dc. The full-load current required is 5 ma, and the full-load voltage must be at least 23 volts. The peak-to-peak ripple voltage can be at most 5 percent of the dc full-load output voltage. The ac power source available to us delivers a 117-volt rms 60-cps sine wave of voltage. With the addition of a Zener diode, such a power supply would be adequate for operating a small transistor radio. (The advantages obtained by the use of a Zener diode will be considered in Sec. 7·4·6).

These design objectives lead to the following values for the equivalent circuit:

$$E_s = 25 \text{ volts}$$
$$R_s \leq \frac{25 - 23}{5} \times 10^3 = 400 \text{ ohms}$$
$$2E_R \leq 1.15 \text{ volts at } i_L = 5 \text{ ma}$$

To meet these objectives, we can select a stepdown transformer to deliver 25 volts peak voltage when operated from the ac line. We shall neglect its winding resistance for the moment. The circuit is shown in Fig. 7 · 12.

The previous equations for R_s and E_R yield the following values for C_f:

$$2\pi i_L X_c = 2E_R \Longrightarrow C_f = 72.5 \text{ } \mu\text{farads}$$
$$\pi X_c = R_s \Longrightarrow C_f = 20 \text{ } \mu\text{farads}$$

These equations show that the ripple requirement determines the value of C_f. The circuit is shown in Fig. 7·12b.

The shortcomings of this circuit are apparent when we consider the parameter values that are necessary if we wish to tighten the specifications. For example, the ripple-voltage output of this power supply is so large that it would cause an objectionable hum in the loudspeaker of a transistor radio if it were used directly as the power source for the radio.

If we require the power supply to have only 0.1 volts peak-to-peak ripple at $i_L = 5$ ma, then the filter capacitor needs to be increased by a factor of about 10 (to ~750 μfarads). And if we want the full-load current to be 50 ma instead of

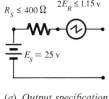

$R_S \leq 400\,\Omega$ $2E_R \leq 1.15$ v

$E_S = 25$ v

(a) Output specifications
of power converter

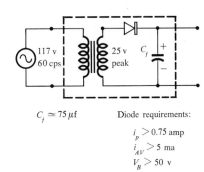

117 v
60 cps

25 v
peak

C_f

$C_f \simeq 75\,\mu f$

Diode requirements:

$i_p > 0.75$ amp
$i_{AV} > 5$ ma
$V_B > 50$ v

(b) Half-wave rectifier designed to meet
specifications given in (a)

FIG. 7·12   HALF-WAVE RECTIFIER DESIGNED TO MEET GIVEN OUTPUT SPECIFICATIONS.
TRANSFORMER PARAMETERS AND DIODE SERIES RESISTANCE NEGLECTED

5 ma, then C_f is again multiplied by a factor of 10. Thus, changes in the requirements, which are actually quite modest, result in impractically large values of C_f. On this basis we may correctly conclude that the simple diode-capacitor network for power conversion is, at best, useful for applications where the full-load current is small (usually a few milliamperes or less) and a per-unit ripple of a few percent can be tolerated.

Diode requirements: peak current.   The need for increasing C_f to meet tighter specifications also puts undue burdens on the diode and the transformer. We can appreciate the nature of these problems by considering the capacitor and load-current waveforms shown in Fig. 7 · 10.

As the discussion of the operating cycle for this circuit suggests, current flows through the diode in pulses. The height and width of these pulses are determined by the output specifications on the power supply. And since peak current and its duration are parameters which diode manufacturers specify, we must have some knowledge of these quantities to select an appropriate diode.

The waveforms and formulas given in Fig. 7 · 10 enable us to easily determine the diode peak current. However, before we use these formulas it will be helpful to give an approximate analysis to emphasize the principal feature of the problem.

The capacitor waveform, of course, must show zero average current when the averaging time is taken to be sufficiently long. In the steady state this means that the charge supplied to the capacitor in each positive pulse must equal the charge delivered to the load during the remainder of the cycle. For approximate purposes, let us suppose that the capacitor-charging current is a series of rectangular pulses of height $\hat{i}_c$ and width Δt. Then the foregoing remarks suggest that

$$\hat{i}_c\,\Delta t \simeq i_L T \qquad \hat{i}_c \sim \frac{i_L T}{\Delta t} \qquad\qquad (7 \cdot 13)$$

Thus

1. $\hat{i}_C$ depends inversely on Δt.
2. $\hat{i}_C$ is many times larger than i_L in a practical design.
3. The attempt to reduce E_R involves decreasing Δt, and therefore requires the use of a diode with a higher peak-current capability.

We may now proceed to estimate the peak diode current as follows. The capacitor-charging component may be written exactly as

$$i_c = C_f \omega_s E_S \cos \omega_s t \qquad (7 \cdot 14a)$$

(see Fig. $7 \cdot 10$). $E_S \cos \omega_s t$ can be estimated as follows. The angle $\omega_s t$ at which conduction begins is such that

$$E_S \sin \omega_s t = E_S - 2E_R$$

Therefore

$$\sin \omega_s t = 1 - \frac{2E_R}{E_S} \qquad (7 \cdot 15a)$$

so that the value of $\cos \omega_s t$ at which conduction begins is

$$\cos \omega_s t = \sqrt{1 - \sin^2 \omega_s t} = \sqrt{1 - \left(1 - \frac{2E_R}{E_S}\right)^2} \simeq \sqrt{\frac{4E_R}{E_S}} \qquad (7 \cdot 15b)$$

Using Eq. $(7 \cdot 14a)$, the peak capacitor current can be written as

$$\hat{i}_c \cong C_f \omega_s \sqrt{4E_R E_S} = \frac{\sqrt{4E_R E_S}}{X_c} \qquad (7 \cdot 14b)$$

We also know that X_c is determined by the allowable ripple

$$\pi i_L X_c = E_R$$

so Eq. $(7 \cdot 14b)$ can be further transformed to read

$$\hat{i}_C = \pi i_L \sqrt{\frac{4E_S}{E_R}} \qquad (7 \cdot 14c)$$

Equation $(7 \cdot 14c)$ gives the peak capacitor current in terms of the design objectives. The peak diode current i_p is approximately

$$i_p \cong \hat{i}_C + i_L = i_L \left(2\pi \sqrt{\frac{E_S}{E_R}} + 1\right) \qquad (7 \cdot 16)$$

For the example considered earlier, Eq. $(7 \cdot 16)$ indicates that the diode needs to be capable of passing a peak current of

$$i_p = 5(2\pi \sqrt{43.5} + 1) \text{ ma} = 212 \text{ ma}$$

in order for the circuit to deliver an average current of 5 ma. As a practical matter, we would select a diode with at least 0.5-amp peak current capability to provide some safety factor.

However, a reduction of E_R to 0.1 volts and an increase of i_L to 50 ma requires an i_p of about 7 amp. This escalation of i_p again suggests that a simple diode-capacitor network is unsuitable for power supplies where low E_R and high i_L are required.

Diode requirements: average current.   Of course, the time duration of the diode current pulses is also important, since this is one of the factors that determines the diode heating, and the heating is ultimately responsible for diode failures (and thus diode ratings). On this basis we should expect manufacturers to specify the time duration of the peak current for their diodes. This is in fact done in certain cases. However, it is more common to specify the average current which the diode can pass. For our case, this is equivalent to specifying Δt. While this is somewhat indirect, there are other circuits to be considered presently, where the average current is a more suitable rating than Δt.

The combined diode-current requirements for our example are therefore a peak-current capability of at least 0.22 amp and an average-current capability of at least 5 ma. A satisfactory diode for this application would have a peak-current rating of at least 0.5 to 1 amp and an average-current rating of at least 10 to 20 ma.

Diode requirements: reverse breakdown voltage.   In addition to diode-current specifications, real diodes are also rated in terms of their reverse breakdown voltages. The diode voltage curve given in Fig. 7·10 shows that a *peak inverse voltage* (PIV) of essentially $2E_S$ will be applied to the diode in the simple diode-capacitor power supply. For the power supply specifications given in our running example, this means the diode will experience a peak inverse voltage of 50 volts. We would then select a diode of about 75 to 100 volts reverse breakdown voltage.

Transformer requirements.   Selecting a transformer for a power supply would be a very involved problem if leakage inductance, magnetizing inductance, winding resistances, etc., all had to be specified. Generally, however, transformers are specified in terms of the circuit in which they are to be used and the power-supply output requirements for which they are intended, so that the transformer selection is a relatively straightforward problem. For the present example, we would select a transformer which is intended to be used in a half-wave rectifier circuit to deliver a dc output of 25 volts at 5 ma. The transformer design will account for the various features required to obtain the output.

There is one interesting feature in this transformer which deserves comment, however. This relates to the winding resistances in the primary and secondary, plus any source resistance in the ac power lines which is reflected into the secondary. In the half-wave rectifier, the diode peak current must flow in the transformer secondary. For the present example, this current is only about 0.2 amp. However, the transformer output voltage is still supposed to be nearly 25 volts under this circumstance. Since we allow the output voltage to drop to 23 volts at full load, we could perhaps allow the transformer output voltage to be 24.8 volts when the 0.2-amp peak current is flowing. This means the total source resistance (neglecting diode drop) should be about

$$R_{source} = \frac{25 - 24.8}{0.2} \sim 1 \text{ ohm}$$

While this is not a difficult requirement to meet, an inexpensive transformer for the present supply would probably not satisfy it. Furthermore, as the power supply output requirements are tightened (more i_L, less E_R, etc.), i_p increases and the source and transformer resistances become even more important.

To summarize, a half-wave rectifier circuit with capacitor filter is satisfactory for load currents of a few milliamperes and reasonably large per-unit ripple. It is an unsuitable circuit for applications where the load current is high and/or the required per-unit ripple is very low.

Most of the other common ac-to-dc power-conversion circuits are modifications of this basic half-wave rectifier circuit, which are aimed at decreasing the size of R_s and $e_r(t)$ and/or reducing the requirements which the circuit places on the diodes and transformers. Several examples will indicate some of the more common procedures which are employed.

7·4·2 Full-wave rectifier with capacitor filter. A simple and very practical means of simplifying the requirements on the components for given output specifications is to use a full-wave rectifier, shown in Fig. 7·13. Here, a center-tapped transformer and two diodes are used to charge the filter capacitor. The polarities on the transformer are arranged so that when e_S is positive, diode D_1 provides current to the filter capacitor (and load),

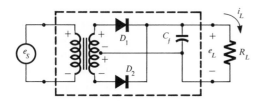

(a) Full-wave rectifier circuit

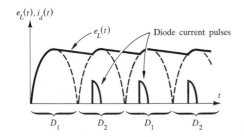

(b) Load voltage and diode current waveforms

FIG. 7·13   THE FULL-WAVE RECTIFIER CIRCUIT AND WAVEFORMS ILLUSTRATING ITS OPERATION

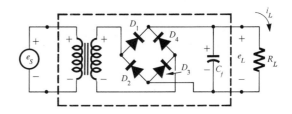

FIG. 7·14 A DIODE BRIDGE FOR OBTAINING FULL-WAVE RECTIFICATION

while when e_s is negative, the diode D_2 supplies the load and capacitor currents. The capacitor thus receives two charging pulses in each cycle of input voltage, instead of one.

For given values R_s and E_R, C_f can be halved (by comparison to the half-wave circuit). The current requirements in the diodes are also reduced. However, the peak inverse voltage applied to the diodes is still the same. The details of this situation are considered in Prob. 7·11.

Since the full-wave rectifier only involves one more diode and a slightly more complicated transformer, it is a simple and inexpensive way to improve the characteristics of the power supply and is more frequently used than the half-wave circuit.

7·4·3 Full-wave rectifier with diode bridge. When the output capabilties of a full-wave rectifier with capacitor filter are adequate for a given application but it is desired to reduce the peak-inverse-voltage requirements on the diodes, the diode bridge circuit shown in Fig. 7·14 can be used. This circuit also eliminates the need for a center-tapped transformer.

The circuit operates as follows. When e_s has the polarity shown in Fig. 7·14, diodes D_1 and D_3 supply the pulses of current to the filter capacitor. When e_s has the opposite polarity, D_2 and D_4 supply the charging pulses.

The diode-current requirements for a given load specification are the same in this circuit as for the full-wave rectifier discussed in Sec. 7·4·2. However, the diodes need to withstand only half as much peak inverse voltage in the bridge-rectifier arrangement. This is an important consideration in power supplies where high dc-output voltages are required; it is here that the bridge rectifier finds one of its principal applications. It is also possible to operate a diode bridge directly from the ac line to give a full-wave rectified output (without the use of transformers).

7·4·4 Full-wave rectifier with an LC filter. When inductance is added to the filter in a power supply, we obtain another adjustable element to use in achieving design objectives and adjusting diode requirements. The advantages which accrue are quite significant, as is suggested by the fact that the LC filtering arrangement to be described is widely used when the load currents required are between 50 ma and about 1 amp. However, these advantages must be balanced against the increased cost and weight of the power supply.

The full-wave rectifier with a choke input filter is shown in Fig. 7·15. The dc output characteristic of this circuit is also plotted in Fig. 7·15 and can be understood as follows. When the load current is zero, the filter capacitor charges up to the peak value of e_S, as in the other full-wave rectifier circuits. As the load current is increased, the capacitor will discharge and the dc-output voltage will drop rapidly. In terms of the equivalent circuit, the output resistance R_s of this power supply is quite high at low load currents. However, after a certain critical value of load current is reached, the source resistance falls abruptly to a very small value.

To see how this arises, let us first consider how the circuit operation changes as L is increased from zero, assuming the dc-load current remains constant. At $L = 0$, the circuit is the same as the full-wave rectifier described in Sec. 7·4·2, where charge is supplied to C_f through the diodes in short, high-amplitude current pulses. The total charge in each pulse divided by half the period of the source frequency must be equal to the required dc-load current. However, when L is increased to a finite value, the

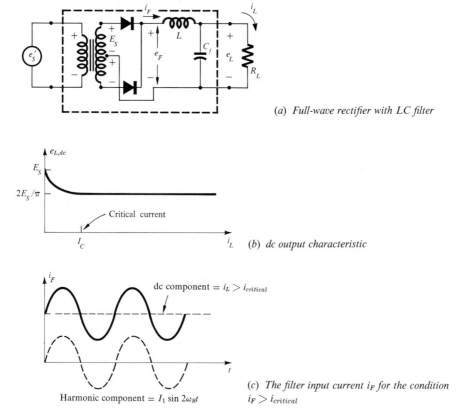

(a) Full-wave rectifier with LC filter

(b) dc output characteristic

(c) The filter input current i_F for the condition $i_F > i_{critical}$

FIG. 7·15 FULL-WAVE RECTIFIER WITH CHOKE INPUT FILTER AND ITS OPERATING CHARACTERISTICS capacitor-charging current will increase less abruptly than before, and will flow over longer periods of time. The result is that each diode conducts for a longer period of time and passes a smaller peak current to maintain the same average current. As L is increased further, a point is reached where the voltage supplied to the filter through the transformer and diodes is not sufficient to reduce the current through L to zero at any point in the cycle.

A similar action occurs for a given L as i_L is increased from zero. When the average current through L is increased to a sufficiently large value, the (rectified) sinusoidal variations in voltage supplied from the transformer are absorbed entirely across L in an attempt to change the current flowing in it. Some changes occur, but the current is never reduced to zero, so the diodes can never become reverse-biased. As a result, the voltage applied to the filter is a full-wave rectified sine wave like that shown dotted in Fig. $7 \cdot 13b$. The dc voltage in such a waveform is $2E_S/\pi$, which appears directly across the load (to the extent that L and C_f are ideal elements). This accounts for the fact that beyond a certain critical load current, the dc-load voltage does not change, as shown in Fig. $7 \cdot 15b$.

The minimum value of the load current for which the diodes conduct during a full half-cycle can be found as follows. The Fourier series for the voltage input to the filter is

$$e_F(t) = E_s\left(\frac{2}{\pi} - \frac{4}{3\pi}\cos 2\omega_s t - \frac{4}{15}\pi \cos 4\omega_s t + \cdots\right) \qquad (7\cdot17)$$

when the time origin is properly selected. From this series we see that the input voltage to the filter consists of a dc component of magnitude $(2/\pi)E_S$, and a set of sinusoidal components at frequencies that are even multiples of the frequency ω_s.

The current supplied to the filter i_F will then consist of a dc component plus a series of sinusoidal components. If we assume that the values of L and C will be such that only the first sine-wave component contributes appreciably to i_F, then the i_F vs t waveform will appear as shown in Fig. $7 \cdot 15c$.

Assuming C_f presents a small impedance compared to L at the radian frequency $2\omega_s$, the current associated with the sine-wave component will have a peak amplitude of approximately

$$I_1 = \frac{4E_S}{3\pi(2\omega_s L)} = \frac{2E_S}{3\pi\omega_s L} \qquad (7\cdot18)$$

which is independent of the load current. When the magnitude of the load current is equal to I_1, then the current to the filter will just drop to zero at its minimum value. Thus, a load current of at least

$$i_L = \frac{2E_S}{3\pi\omega_s L} = i_{critical} \qquad (7\cdot19)$$

must flow to ensure that one or another of the diodes is always conducting.

The advantages of this condition are appreciable. First of all, the equivalent circuit will ideally exhibit zero R_s. This is because the inductor

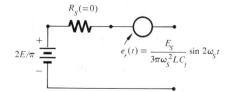

ideally presents zero impedance and the capacitor infinite impedance to the flow of dc current. (The effects of nonideal elements are considered in Prob. 7·12.) Therefore the dc load voltage will be $2E_s/\pi$, independent of the load current; that is, $R_s = 0$. Second, the ripple can be made small without placing undue burdens on the circuit components.

We may calculate the output ripple voltage as follows. The peak amplitude of the sinusoidal-current component is

$$I_1 = \frac{2E_s}{3\pi\omega_s L} \tag{7·20}$$

so the ripple voltage appearing across the load terminals has a peak value of

$$E_R = I_1 X_c = \frac{2E_s}{6\pi\omega_s^2 L C_f} \tag{7·21}$$

assuming C_f has a much lower impedance than the load impedance at the radian frequency $2\omega_s$. The per-unit ripple is therefore

$$\frac{E_R}{E_{dc}} = \frac{1}{6\omega_s^2 L C_f} \tag{7·22}$$

For good filtering, L_f and C_f should be chosen to make $6\omega_s^2 L C_f \gg 1$. Notice that the ripple voltage is independent of the load current, which is not the case for the simple capacitor filter. These facts are summarized in the equivalent output circuit of Fig. 7·16.

EXAMPLE   To compare this circuit with our previous example, let us consider the values of L and C_f required to give a critical current of 5 ma and a ripple voltage of 1.15 volts for a dc output voltage of 25 volts. First of all, we observe that the dc output of 25 volts means that the voltage between one end of the transformer secondary and the center tap must be

$$\frac{2E_s}{\pi} = 25 \text{ v} \Rightarrow E_s = 39.2 \text{ volts}$$

In order to insure that the power supply exhibits zero R_s for an i_L of 5 ma, the previous equations indicate that

$$3\omega_s L \geq \frac{25 \text{ volts}}{5 \text{ ma}} = 5 \text{ kohms} \Rightarrow L \geq 4.4 \text{ henries}$$

Thus a choke of at least 5 henries inductance, capable of passing at least 10 ma, would be used.

To provide a peak-to-peak ripple of $2E_R = 1.15$ volts, we must then select a filter capacitor such that

$$3\omega_s{}^2LC_f = \frac{E_{dc}}{2E_R} \sim 22$$

Using an L of 5 henries then requires a C of approximately 12 μfarad. If we use a C_f of 72 μfarad as before, the per-unit ripple would be reduced by a factor of 6, or the actual ripple voltage would be about 0.2 volt.

There are several features in this example which may be profitably compared with the previous examples.

1. First of all, the average R_s for the LC filtering arrangement for current levels below the critical current is

$$R_s = \frac{E_s - (2/\pi)E_s}{5 \text{ ma}} \sim 3 \text{ kohm}$$

Hence, we should never operate the supply in this range. There are two ways of avoiding this. One is to connect a *bleeder resistance,* in parallel with C_f, as a part of the filter (not the load). If the bleeder resistance has the value E_{dc}/i_{crit}, then the load will always face a zero R_s power supply. For the present example the bleeder would have the value 5 kohm.

This is a valid option when load currents which are large compared to the bleeder current are to be drawn (for example, $i_L \geq 25$ ma). Otherwise, too much power is lost in the bleeder compared to the power delivered to the load. Thus, for low-load currents the simpler full-wave rectifier with capacitor filter is more desirable.

2. The increased flexibility in meeting design objectives with the LC filter is quite remarkable. The dc parameters of the required output determine the transformer secondary voltage, and the size of L. This leaves C as a free parameter to adjust the size of the output ripple, a very convenient state of affairs.

Diode requirements. In addition to the relaxation of requirements on C_f, the LC filtering arrangement also relaxes the diode requirements. It is clear from the i_F waveform given in Fig. 7·15c that the peak diode current can be approximated by

$$i_{peak} = i_L + I_1 = i_L + i_{crit} \tag{7·23}$$

For the present example, the peak current would only be 5 ma more than the load current. Thus, diode peak currents are not a significant problem; and furthermore, requirements on the transformer-winding resistances are also relaxed. It is apparent that the diode average-current rating will be the only significant current parameter for selecting a diode.

A moment's study of the circuit will reveal that since one of the diodes is always ON, the voltage across the other diode will simply be the total voltage across the transformer secondary. The peak inverse voltage will therefore be $2E_S$, or π times the dc output voltage. For a given output voltage, the LC filter requires a higher PIV than any other common circuit.

In summary, the full-wave rectifier with LC filtering offers significant advantages over the simple capacitor filter when dc load currents of 25 ma or more are to be drawn. The circuit offers more flexibility in achieving output objectives, and relaxes the current requirements on the diodes and transformer for a given load current. The peak inverse voltage required in the diodes is higher in the LC circuit, but if this is a problem a diode bridge arrangement can be used in place of two diodes and a center-tapped transformer.

7·4·5 *Full-wave rectifier with CLC filter.* The principal problems with the LC filtering arrangement are, (1) it gives a rather inefficient use of the transformer voltage (only 65 percent of the secondary voltage is fed to the load) and (2) it requires the use of heavy inductors. When the first of these problems is the most important, a CLC filtering arrangement such as that shown in Fig. 7·17 is sometimes used. In comparison with the LC filter, the CLC filter gives higher output voltage and lower $e_r(t)$ for a given supply voltage. However, the output resistance of the CLC circuit is not as good as that of the LC filter. As a result, the CLC filter is primarily used in applications where a reasonably high voltage output is required at a current level of 50–200 ma, but with exceedingly small allowable E_R. The diode current will again flow in pulses so that the peak diode current may again require consideration. Transformer winding resistances will also assume some importance in the design. Detailed design considerations appear in handbooks.

7·4·6 *Regulated power supplies: Zener-diode regulator.* When better low-current regulation than can be supplied with an LC filter is required, or when the size and weight associated with inductors is objectionable, an electronic voltage regulator is used.

The simplest type of electronic regulator uses a Zener diode to fix the

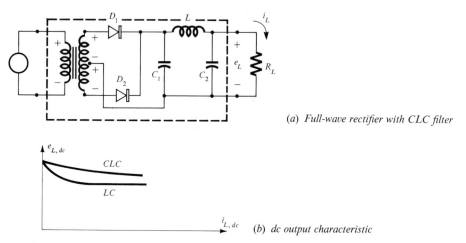

(a) *Full-wave rectifier with CLC filter*

(b) *dc output characteristic*

FIG. 7·17   **FULL-WAVE RECTIFIER WITH CLC FILTER**

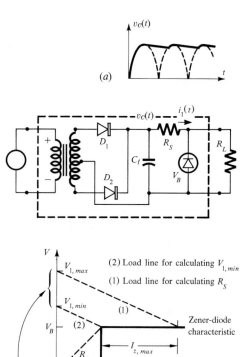

(a)

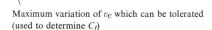

(b) *The action and design of a simple Zener-diode voltage regulator*

(2) Load line for calculating $V_{1,min}$

(1) Load line for calculating R_S

Zener-diode characteristic

Maximum variation of v_C which can be tolerated (used to determine C_f)

FIG. 7·18   A SIMPLE ZENER-DIODE VOLTAGE REGULATOR FOR GIVEN LOAD CURRENT

output voltage at a value equal to the breakdown voltage of the Zener diode. A regulating circuit of this type is shown in Fig. 7·18. The full-wave rectifier supplies pulses of charge to the filter capacitor each half-cycle. Between pulses, the load (consisting of R_s, R_L, and the Zener diode) draws current from the capacitor. The capacitor voltage $v_C(t)$ thus has the waveform shown in Fig. 7·18a.

Ideally, the regulating action of the Zener diode arises as follows. As long as v_C is large enough, the diode will be broken down, leaving a fixed voltage drop across it equal to V_B. Variations in $v_C(t)$ then produce variations in the current $i_1(t)$, which are absorbed by the Zener diode. A constant current equal to V_B/R_L is passed on to the load.

This action is illustrated graphically in Fig. 7·18b. The vertical axis for this figure is voltage, and the horizontal axis is current. The Zener-diode characteristic has been shifted to the right by an amount $I_L = V_B/R_L$, so that the current scale reads i_1 directly. As long as the variations in capacitor voltage are within the limits

$$V_{1,max} > v_C > V_{1,min}$$

the current through the diode satisfies

$$0 < i < I_{z,max}$$

and the circuit regulates.

In designing a Zener-diode regulator, one normally selects the Zener diode first (to get an appropriate output voltage), and then a transformer whose peak output voltage is perhaps 50 percent greater than the Zener voltage. When the transformer and Zener diode are given, both V_B and $V_{1,max}$ are known. R_S can then be determined from the load line [(1) in Fig. 7·18] which passes through $i_1 = I_L + I_{z,max}$, $V = V_{1,max}$. The minimum value of V_1 is determined by using this value of R_S, and recognizing that

$$\frac{V_{1,min} - V_B}{R_S} = \frac{V_B}{R_L}$$

[load line (2) in Fig. 7·18]. When these values have been determined, a value for C_f may be calculated which will produce the expected variations in V_1 (see Prob. 7·13).

When C_f is known, the charging-current waveform can be obtained and diode-current requirements studied. The peak inverse voltage is twice the maximum transformer-secondary voltage, or about three times the dc output voltage.

As a practical matter, a Zener diode intended for power conversion will have a body resistance between several ohms and several tens of ohms. As a result, the Zener-diode characteristic shown in Fig. 7·18 should have a slight positive slope in the regulating region. The changes necessary to account for this aspect of the behavior are the subject of Prob. 7·14.

Thus far, the design has been carried out in terms of fixed i_L. Variations in i_L may also be accommodated at the expense of variations in v_C. There

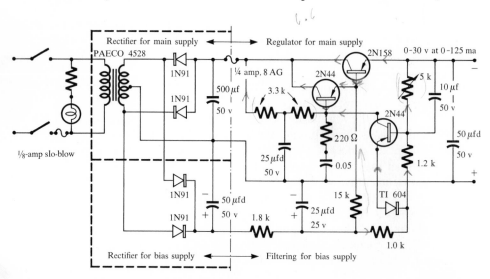

FIG. 7·19 ELECTRONICALLY REGULATED POWER SUPPLY

are many particular situations that are not difficult to calculate, though they are sometimes tedious. In addition to typical data, Zener-diode manufacturers provide slide rules, charts, etc., to help us circumvent the tedium.

7·4·7 *Regulated power supplies for laboratory use.* When the ultimate in regulation is required, a much more elaborate electronic voltage regulator is used. The circuit diagram for an electronically regulated laboratory power supply is shown in Fig. 7·19. Since we have not yet studied transistor amplifiers, we are not yet equipped to understand how this circuit works. However, we are in a position to see that the rather crude rectifier-filter combination shown in Fig. 7·19 can be quite nicely regulated with such a regulator. The particular supply shown in the figure will supply any voltage from zero to 30 volts, at any current from zero to 125 milliamperes. The peak-to-peak output ripple voltage is less than 10 mv.

7·5   DIODE CLAMPING CIRCUITS AND AC VOLTMETERS

In addition to its use in power-conversion circuits, the basic half-wave rectifier circuit can be used for clamping voltages at various levels and in constructing ac voltmeters.

A diode clamping circuit is shown in Fig. 7·20. The circuit is identical to the half-wave rectifier except that the positions of the filter capacitor and the diode are interchanged, and there is no load shown. In this circuit, the capacitor charges to the positive peak value of E_s, as indicated in Fig. 7·20. The output voltage, which is taken across the diode, is

$$e_L(t) = e_s(t) - E_S \qquad (7·24)$$

The waveform of the output voltage is shown in Fig. 7·20b. As this figure shows, the output voltage can only rise to zero volts. For this reason the circuit is said to *clamp* the positive peak of $e_s(t)$ at zero volts.

It is also possible to clamp the negative peak of a signal waveform with the circuit shown in Fig. 7·20c. The explanation of the operation of this circuit is similar to that given above and will not be repeated.

Diode clamping circuits have a number of useful applications. For example, it is necessary that signal voltages appearing at various points in a television or radar receiver have their maximum values clamped at certain levels. Diode clamping circuits are used for this purpose.

As another interesting example, a diode clamping circuit and a lightly loaded half-wave rectifier may be combined to make a peak-to-peak ac voltmeter. The circuit is shown in Fig. 7·21. Diode D_1 forces the voltage appearing at point A to have a minimum value of zero. The maximum value is equal to the peak-to-peak amplitude of the input signal. C_2 then charges through D_2 to the positive peak of the signal appearing at point A; that is, the peak-to-peak value of the input voltage.

If a high-resistance dc microammeter is added as a load on C_2, the meter indication will be proportional to the peak-to-peak value of the

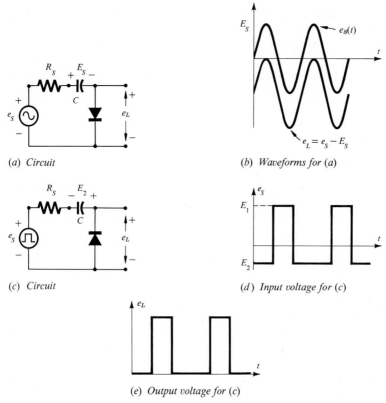

(a) Circuit

(b) Waveforms for (a)

(c) Circuit

(d) Input voltage for (c)

(e) Output voltage for (c)

FIG. 7·20 DIODE CLAMPING CIRCUITS

signal voltage. Usually, such meters will have their scales calibrated to read the rms value of the input voltage, on the assumption that it is a sine wave. When nonsinusoidal voltages are applied, the instrument reads the peak-to-peak value divided by $2\sqrt{2}$.

Voltage doublers. When a sine wave of voltage is applied to the circuit of Fig. 7·21, the no-load output voltage is a dc voltage whose value is twice

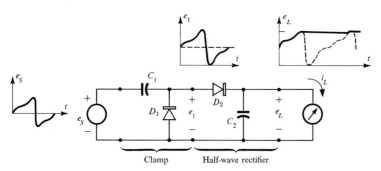

FIG. 7·21 THE COMBINATION OF A DIODE CLAMP AND A HALF-WAVE RECTIFIER TO PRODUCE A PEAK-TO-PEAK VOLTMETER

the peak value of the input wave. The circuit is therefore called a voltage *doubler,* and is frequently used as a power converter to obtain a high-output dc voltage when only very small load currents need to be supplied.

7·6 PRODUCTION AND DETECTION OF AMPLITUDE-MODULATED WAVES

We mentioned in the Introduction and in Sec. 7·3 that, in radio broadcasting, signals in the audio frequency range are transmitted on a high-frequency carrier, which is easier to broadcast than the audio signal and can be separated from other signals transmitted at other carrier frequencies. One of the simplest schemes for accomplishing this utilizes *amplitude modulation.*[1] In this scheme the *amplitude* of the carrier frequency wave is made to conform to the instantaneous value of the audio signal.

The circuit which accomplishes this is called a *modulator.* There are many types of modulator circuits, each having properties of interest for specific applications. For present purposes, we wish to consider only one type of modulator, to illustrate how diode nonlinearity may be used in these applications.

A so-called *diode bridge modulator* is shown in Fig. 7·22. To explain the operation of this circuit, we suppose that the carrier signal at frequency f_c is applied across the bridge in one direction, and a battery and series R_S across it in the other. When the carrier signal is positive, the diodes are conducting. The bridge then presents a low impedance compared to R_S, so very little voltage is fed to the load resistor $R_L(\gg R_S)$. During the negative half-cycle of the carrier voltage, the diodes in the bridge are reverse-biased, so most of the battery voltage appears across R_L. The resulting waveshape at the load is a square wave with a minimum value of zero.

The Fourier series for this square wave has frequency components at dc, f_c, $2f_c$, ... The component at f_c can be written as

$$e_C(t) = A_C \cos \omega_C t \qquad (7 \cdot 25)$$

when the time origin is chosen properly. In this formula, A_C is the amplitude of the component and ω_C is the radian frequency.

Now, suppose a low-frequency sine-wave voltage source is placed in series with the battery. Then the amplitude of the square wave shown in Fig. 7·22 will vary slowly with time:

$$e_L = \frac{(E_0 + v_s \sin \omega_s t)R_L}{R_s + R_L} = \frac{E_0 R_L}{R_L + R_s}(1 + m_s \sin \omega_s t) \qquad (7 \cdot 26)$$

where m_s is the ratio of v_s to E_0.

Since the amplitudes of the Fourier components of the square wave are proportional to the square-wave amplitude, we expect the amplitude of each Fourier component to vary. The component at f_c can now be written as

$$e_C(t) = A_C(1 + m_s \sin \omega_s t) \cos \omega_C t \qquad (7 \cdot 27)$$

---

[1] The amplitude-modulation process will be considered further in Chap. 16.

Low-frequency applications of diodes 265

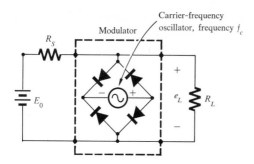

(a) Diode bridge modulator

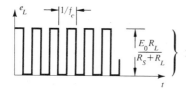

$$\left.\begin{array}{c}E_0 R_L \\ \overline{R_S + R_L}\end{array}\right\}$$ Fourier component at f_c: $e_c(t) = A_C \cos \omega_C t$

(b) Load waveform for (a)

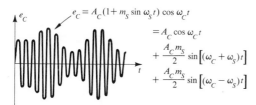

$$e_C = A_C(1 + m_S \sin \omega_S t) \cos \omega_C t$$

$$= A_C \cos \omega_C t$$

$$+ \frac{A_C m_S}{2} \sin \left[(\omega_C + \omega_S)t\right]$$

$$+ \frac{A_C m_S}{2} \sin \left[(\omega_C - \omega_S)t\right]$$

(c) Fourier component at f_c with amplitude variation at frequency ω_s. Actually Fourier representation of $e_c(t)$ requires three frequencies:

$$\omega_c, \quad \omega_c + \omega_s, \quad \omega_c - \omega_s.$$

FIG. 7·22   A DIODE BRIDGE MODULATOR AND WAVEFORMS

Such a waveform is shown in Fig. 7·22c. It is described by saying that the amplitude of the carrier has been *modulated* by the voltage $v_s \sin \omega_s t$.

Using trigonometric identities, we can rewrite Eq. (7·27) as

$$e_C(t) = A_C \cos \omega_C t + \frac{A_C m_S}{2} \sin (\omega_C - \omega_S)t + \frac{A_C m_S}{2} \sin (\omega_C + \omega_S)t \quad (7·28)$$

In Eq. (7·28) we see that the *time-variable Fourier amplitude* given in Eq. (7·27) has been resolved into three new Fourier components with *constant amplitude*. The frequencies of these components are ω_C, $\omega_C + \omega_S$, and $\omega_C - \omega_S$.

The voltage represented by Eq. (7·27) or (7·28) is typical of the signals which are broadcast by an AM (amplitude modulation) radio station. The amplitude of the carrier varies in accordance with the information being broadcast. In the simple case of a sine-wave tone, the amplitude-modulated carrier can be decomposed into a Fourier series consisting of constant-amplitude signals at the carrier frequency f_C and two *side-band* frequencies $(f_C - f_s)$, $(f_C + f_s)$. The side-band frequencies really contain the information being sent.

We will return to a further consideration of modulating systems and their properties in Chap. 16. For the moment we re-emphasize that the modulator is a frequency-shifting, and therefore nonlinear, system. A simple modulator can be made with a diode bridge. Such a modulator is not used for commercial radio purposes, though it is extensively used in carrier telephone systems.

7·6·1 *Detection of an amplitude-modulated wave.* A radio receiver which is tuned to the frequency f_C will receive a signal voltage of the form given by Eq. (7·28), but a very small amplitude (perhaps only a few microvolts peak amplitude). This signal must be amplified in the receiver to the point where a usefully large audio signal can be recovered from it. The process by which the audio signal is recovered is called *envelope detection*. A simple type of envelope detector consists of nothing more than a half-wave rectifier with a capacitor filter and a resistive load (Fig. 7·23a).

For the purpose of detecting the audio information, we want to select a $C_f R_L$ combination which will allow the half-wave rectifier to follow the peak amplitude as nearly as possible (see Fig. 7·23). If the time constant is too large, the filter will respond only to the peaks of the modulated wave, as shown in Fig. 7·23c. If the time constant is too short, very little audio voltage will be obtained. The best value of $C_f R_L$ represents a compromise between these alternatives. In Prob. 7·17 it is shown that a reasonable value for the $C_f R_L$ product is

$$R_L C_f \sim \frac{\sqrt{1 - m_s{}^2}}{m_s \omega_s}$$

where m_s is the modulation index and ω_s is the audio frequency. A maximum value of m_s equal to 80 percent is customarily employed, and the maximum

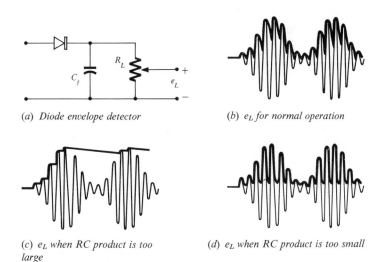

(a) Diode envelope detector

(b) e_L for normal operation

(c) e_L when RC product is too large

(d) e_L when RC product is too small

FIG. **7·23** DIODE ENVELOPE DETECTOR AND WAVEFORMS FOR VARIOUS OPERATING CONDITIONS value of ω_s is usually about $2\pi \times 5 \times 10^3$ cps for a typical AM broadcast. These values suggest

$$R_L C_f \sim 24 \times 10^{-6} \text{ sec}$$

Actual values of R_L and C_f in a typical AM radio receiver are 50 kohms and 500 $\mu\mu$farads, respectively, so the actual product of R_L and C_f is 25×10^{-6} sec.

7·7 pn-JUNCTION SOLAR BATTERIES

The final application of diodes which we shall consider in this chapter will be in solar cells. For this application we need to represent the pn-junction diode by its ideal v-i law (with photogeneration included). The piecewise-linear characterization of the diode which we have used in the previous parts of this chapter is not adequate for the present application.

A schematic physical diagram of a typical solar cell is shown in Fig. 6·24. The cell has a diameter of about one inch, so as to intercept an appreciable amount of sunlight. Under illumination conditions which are typical of a California summer day, an average solar cell will develop about 0.4 volt open-circuit voltage. The corresponding short-circuit current will be on the order of 5 ma. The effective value of the saturation current I_0 is therefore 5×10^{-10} amp.

In a typical application of solar cells for power generation, several cells will be connected in series to increase the effective voltage of the supply. Since there is not a great deal of power available, we are interested in terminating the cell in a load which will absorb maximum power from it. We may readily calculate the optimum terminating impedance as follows.

From the results given in Sec. 6·7·3, the v-i relation for the cell is

$$i = i_\nu - I_0(e^{qv/kT} - 1) \tag{7·29}$$

This gives the output characteristic of the solar generator. Now, the power delivered by this generator to a resistive load can, in general, be written as

$$P_L = v_L i_L \tag{7·30}$$

At the optimum value of v_L (or i_L) we have

$$\frac{dP_L}{dv_L} = i_L + v_L \frac{di_L}{dv_L} = 0 \tag{7·31}$$

or

$$\frac{v_L}{i_L} = -\frac{dv_L}{di_L} \tag{7·32}$$

Equation (7·32) shows that at the optimum operating point the negative of the slope of the generator characteristic equals the slope of the line drawn from the origin to the optimum point. The construction is indicated in Fig. 7·24. For our case, the optimum values of v and i are

$$v = 0.288 \text{ volt} \qquad i = 4.5 \text{ ma}$$

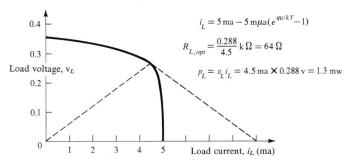

FIG. 7·24 CALCULATION OF OPTIMUM OPERATING POINT FOR SOLAR CELL

7·8 SUMMARY

Diodes are useful circuit elements because their v-i characteristic exhibits an abrupt nonlinearity near $i = 0$, $v = 0$. On account of this nonlinearity they behave basically as polarity-sensitive switches, and the ideal diode is simply defined as the ideal polarity-sensitive switch.

Ideal diodes play a prominent role in electronic circuit theory, and they are useful in making approximate analyses of real diodes and real electronic circuits. Some of the applications of diodes in electronics, such as in non-linear-resistance synthesis, logic circuits, power conversion, modulators, and detectors, were considered in this chapter. We will meet others later, all relating to the need for or use of nonlinearity in processing an electronic signal.

REFERENCES

Zimmerman, H. J., and S. J. Mason: "Electronic Circuit Theory," John Wiley & Sons, Inc., New York, 1962. Chapters 1–4 give an excellent account of the application of piecewise-linear techniques in analyzing diode circuits.

Angelo, E. J.: "Electronic Circuits," 2d ed., McGraw-Hill Book Company, New York, 1958, chap. 2.

Linvill, J. G., and J. F. Gibbons: "Transistors and Active Circuits," McGraw-Hill Book Company, New York, 1961, chap. 15.

Terman, F. E.: "Radio Engineer's Handbook," McGraw-Hill Book Company, New York, 1943.

DEMONSTRATIONS

7·1 FLEXIBLE POWER SUPPLY DEMONSTRATOR. The principles of power supply operation are easily demonstrated by observing waveforms at various points in the circuit shown below. The switches allow the supply to be operated either full- or half-wave and with different combinations of load and filter components. The pot, which provides the load, allows for currents above and below "critical" as deter-

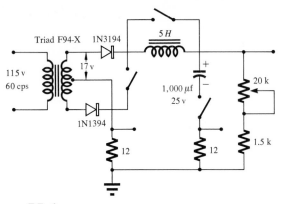

mined by the choke. This critical current condition can be readily observed by means of the 12-ohm resistance in series with the center tap.

7·2 ZENER-DIODE REGULATOR. The principles of operation of a Zener-diode regulator are easily demonstrated by observing voltage levels and waveforms in the circuit shown below. The variac is to reduce the effective line voltage to the point where the regulating action ceases.

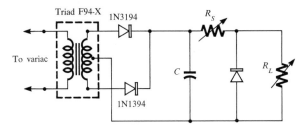

FIG. **D7·2**

If Prob. 7·14 is assigned (using an available Zener diode), student's values can be substituted for R_S, R_L, and C and the regulating action checked.

7·3 MODULATION AND DETECTION WITH DIODES. The use of diodes to produce and detect modulated waves can be simply demonstrated with the circuit shown.

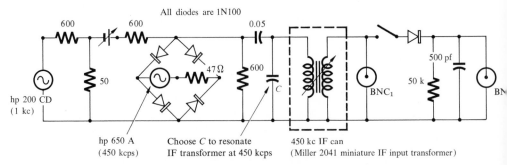

FIG. **D7·3** MODULATION AND DETECTION WITH JUNCTION DIODES

A 450-kcps carrier is selected to conform to the intermediate frequency in an AM radio receiver. The amplitude-modulated wave can be observed at BNC_1; the detected wave, at BNC_2. The modulation index can be varied by changing the ratio of the 1-kc signal voltage to the dc voltage.

EXPERIMENTAL PROJECTS

7·1 Construct a power supply to meet the following specifications:

Output voltage: at least 20 volts, dc
Output current range: 10–40 ma
Ripple factor: 10 percent maximum
Total filter capacitance: not to exceed 40 μfarads

Two diodes (1N2069) and a transformer (Triad F-90X) are required. The rest is up to you.
Diode specifications:

$I_0 = 10$ μa
$PIV = 200$ volts
$I_F =$ minimum at 1.2 volts $= 0.5$ amp
$I_{av} = 750$ ma at 25°C
$\quad\; = 500$ ma at 100°C
$I_{max} = 6$ amp (repetitive pulse; 25°C)

Transformer specifications: furnished with the transformer

7·2 Using a nominal 15-volt Zener diode, add a voltage regulator to the power supply constructed in the previous project. The regulated supply should be capable of delivering about 25 ma for future purposes. Be sure that the Zener diode has a power-dissipation rating adequate for these requirements.

PROBLEMS

7·1 The *v-i* characteristic for a nonlinear two-terminal device in which dv/di is positive (or zero) everywhere can be synthesized with the aid of ideal diodes, resistors, batteries, and current sources. Both synthesis and analysis can usually be performed by combining elementary, one-diode networks in a ladder structure to build up a complicated *v-i* curve.

1. Sketch by inspection as many of the *v-i* characteristics of the networks shown in Fig. P7·1 as you can. All diodes are assumed to be ideal. Note that the *v-i* curve for each network has only one break point, because there is only one diode. The circle labelled I_0 is a current generator which delivers a current of I_0 in the direction of the arrow into any load which is connected to its terminals.

2. When the *v-i* characteristic cannot readily be sketched by inspection, the following method is helpful. Sketch the *v-i* characteristic of each separate element of the network on one set of coordinates and then add them together graphically in accordance with their circuit connections. For example, in Fig. 7·1a, plot separate *v-i* characteristics for the battery, resistor, and diode on one set of coordinates. Now, graphically add these characteristics together, obeying the constraint that the current is the same in all the elements. Construct the *v-i* characteristics for the networks shown in Figs. P7·1d and P7·1f in this manner (and any others that you

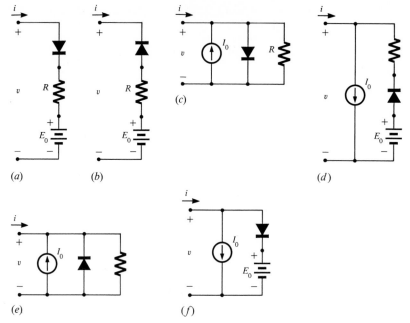

FIG. P7·1

could not sketch by inspection). See also Zimmerman, H. J., and S. J. Mason, "Electronic Circuit Theory," John Wiley & Sons, 1962, chap. 3 for other techniques for sketching v-i characteristics of diode networks.

7·2 1. Sketch the v-i characteristic of each branch of the network shown in Fig. 7·2c (in the text) on one set of coordinates and show how they add together to give the v-i characteristic. Sketch the v-i characteristic which results when these networks are put in series.

2. Modify the network of Fig. 7·2c to include the possibility of avalanche breakdown at a voltage V_B.

7·3 1. Suppose there exists a nonlinear resistor whose terminal characteristic is

$$i = (v + v^4) \text{ ma}$$

where v is in volts. Sketch the v-i characteristic for this device, then synthesize a network of diodes, resistors, batteries, and constant current sources that will have approximately the same characteristic. The range from -1 to $+1$ volt is to be covered, using four ideal diodes.

2. A 1-volt battery and a 1-kohm resistor are now connected in series with the nonlinear resistor. What current flows in the piecewise linear model?

3. A sine-wave voltage source with a peak amplitude of one volt is applied to the nonlinear resistor. Sketch a complete cycle of the current which flows in the piecewise-linear model, using an accurate current scale.

7·4 1. A certain vacuum diode exhibits the v-i relation $i = 10^{-3} v^{1.5}$. Devise a piecewise-linear model which will match the slope of this curve at $v = 0$ volt, 1 volt, and 3 volts. For $v < 0$, $i = 0$. Plot the v-i characteristic of the diode and that of the piecewise-linear approximation on the same set of coordinates and estimate the maximum error in current for $v < 4$ volts.

2. The diode exhibits a temperature-limited saturation current of 50 ma. Include this in the piecewise-linear representation.

7·5 Sketch the v-i characteristic of the circuit shown in Fig. P7·5. The diodes are ideal breakdown diodes with breakdown voltages of V_B volts. How would you use these elements to approximate $f(y)$ in the solution of Eq. (7·2)?

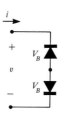

FIG. P7·5

7·6 Sketch the transfer characteristic for the resistive half-wave rectifier when a semiconductor diode is used. The semiconductor diode is adequately represented by the piecewise-linear model shown in the text in Fig. 7·2b. Repeat for the model of Fig. 7·2c.

7·7 Sketch the transfer characteristic of the networks shown in Fig. P7·7, assuming the diodes to be ideal. Indicate clearly the slopes of all lines on these curves. Repeat using semiconductor diodes represented by the model shown in Fig. 7·2b.

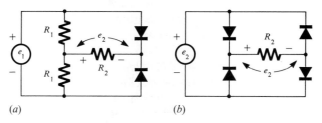

(a)   (b)

FIG. P7·7

7·8 An automobile battery charger is shown in Fig. P7·8. Assuming the diode is ideal and the battery is a 6-volt type, sketch the load-current waveform indicating the important details accurately. What value of R_S should be used to limit the peak charging current to 5 amp? When this value of R_S is used, how much charge is delivered to the battery in one hour? Estimate the peak current for a semiconductor diode with a forward saturation voltage of 1 volt. Repeat for a vacuum diode with $i = 10^{-3} v^{1.5}$.

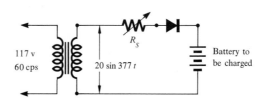

117 v
60 cps
20 sin 377 t
R_S
Battery to be charged

FIG. P7·8

7·9 Find the first three terms of the Fourier series representation for a half-wave rectified cosine wave and a full-wave rectified cosine wave. Using these series, develop equivalent output circuits for the resistive half-wave rectifier and the resistive full-wave rectifier. In each case the source resistance is R_S.

7·10 The half-wave peak rectifier shown in Fig. P7·10a is supplied with the square wave of voltage shown in P7·10b. Determine the parameters for the equivalent output network for this circuit (R_L is to be considered as the load for the network). Let $T = 0.5$ msec.

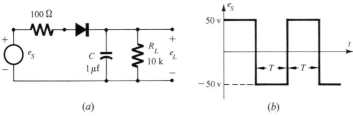

(a)   (b)

FIG. **P7·10**

7·11 1. Develop general formulas for (a) the component values in the equivalent output circuit and (b) the diode requirements for a full-wave rectifier circuit with a single-filter capacitor C_f. State your assumptions and approximations carefully.

2. Design a full-wave rectifier with a single-capacitor filter to produce a power supply with the following output characteristics:

E_{dc} (no load) $= 30$ volts
E_{dc} (full load) $= 25$ volts
Full-load current $= 10$ ma
Peak-to-peak ripple voltage $= 0.5$ volts

The design may be considered complete when C_f and the diode specifications (peak current, average current, and reverse breakdown voltage) are given.

7·12 The supply voltage for the circuit of Fig. P7·12 is $e_s = 500 \sin 377t$; the circuit parameters are $r_d = 400$ ohms, $R_c = 100$ ohms, $L = 10$ henries, and $C = 16$ μfarads.

1. What critical current is needed to achieve low dc output resistance in this power supply?

2. Find the dc load voltage for a load current of two times the critical current. Repeat for four times the critical current and sketch the dc output characteristic. Make any approximations that appear reasonable.

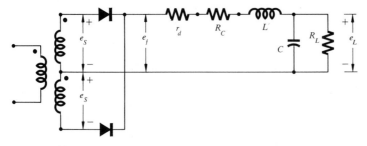

FIG. **P7·12**

3. Find the 120-cps component of voltage across the load for a dc load current of two times the critical current and four times the critical current. Compare your result with the ripple-voltage formula found in the text and comment briefly on any differences which exist.

7·13 In an aircraft, a small motor-generator set is sometimes used to convert a 28-volt dc battery voltage to an ac voltage at a frequency of 400 cps. A power supply such as that shown in Fig. P7·12 is then used to obtain a higher dc voltage (a few hundred volts). What effect does this have on the critical current for an ideal LC filtering arrangement?

7·14 A regulated power supply with an output of 20 volts which is capable of supplying a constant 20-ma load is desired. A Zener diode with $V_B = 20$ volts, $I_{Z,max} = 40$ ma is to be used with a center-tapped transformer whose peak output voltage (to the center tap) is 30 volts. The circuit is shown in the text, Fig. 7·18.

1. Select a series resistance R_S which utilizes the maximum capability of the diode.
2. Develop and solve a differential equation for the voltage across C. Assuming the discharge period is $T/2$, find the value of C which gives the minimum allowable value of v_c.
3. The Zener diode has a body resistance of 10 ohms. Thus, in place of looking like a battery of V_B volts when broken down, it looks as follows:

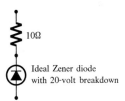

FIG. P7·14

Include this effect in part (1) of this problem.

7·15 Frequently, a Zener diode is used to provide a fixed output voltage to a variable-load resistance. In many such cases a fixed dc supply is available. Suppose we have a 40-volt dc supply and the Zener diode given in part (3) of the previous problem. The anticipated load variations fall in the range

$$1 \text{ kohm} < R_L < 10 \text{ kohm}$$

Select R_S in the circuit shown below and plot your operating path on the Zener diode characteristic.

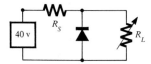

FIG. P7·15

What are the minimum values of R_L and R_S which the circuit can accommodate?

7·16 A peak-to-peak voltmeter is calibrated to read rms voltage for a sine-wave input. What does it read when it is attached to a voltage of $(15 + 5 \sin 377t)$ volts?

7·17 A simple ac voltmeter can be made by attaching a dc microammeter to the circuit shown in Fig. P7·17. The loading of the diode-clamping circuit can be assumed to be negligible. Draw a calibration curve so that the meter will read the rms value of sine-wave input voltages. Disregarding the inertia of the meter move-

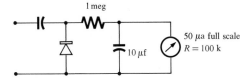

FIG. P7·17

ment, how much fluctuation would you expect when the meter reads at the half-scale position?

7·18 An amplitude-modulated voltage has the form

$$e(t) = A(1 + m_s \sin \omega_s t) \cos \omega_c t$$

To recover the information by an envelope detector such as the one shown in Fig. 7·23a, it is necessary that the voltage across the filter capacitor should fall more rapidly than the modulation envelope. Show that this condition leads to an RC product of

$$RC \sim \frac{\sqrt{1 - m_s^2}}{m_s \omega_s}$$

to avoid distortion. (Consider adjacent positive peaks of the carrier in the region where the modulation envelope has a negative slope.)

7·19 The circuit shown in Fig. P7·19 is called a phase-sensitive detector. Sketch the output of this circuit as a function of ϕ when

$$v_r = V_r \cos \omega t \qquad v_s = V_s \cos (\omega t + \phi)$$

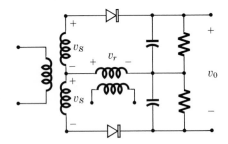

FIG. P7·19

8 TRANSIENT AND HIGH FREQUENCY PERFORMANCE OF pn-JUNCTION DIODES

IN THE LAST TWO chapters we have been concentrating on the low-frequency theory and applications of diodes. It was shown in Chap. 6 that the dc v-i relation for the ideal junction diode was of the form

$$i = I_0(e^{qv/kT} - 1) \qquad (8 \cdot 1a)$$

In Chap. 7, when it was necessary to consider the pn-junction diode specifically, we assumed that the v-i relation, Eq. $(8 \cdot 1a)$, could be generalized to

$$i(t) = I_0(e^{qv(t)/kT} - 1) \qquad (8 \cdot 1b)$$

Equation $(8 \cdot 1b)$ gives a valid characterization of the pn-junction diode when $v(t)$ and $i(t)$ change slowly enough so that the minority-carrier distributions in the p- and n-type diode bodies are always of the form

$$p_E(x,t) = p_E(0,t)e^{-x/L_p} \qquad n_E(y,t) = n_E(0,t)e^{-y/L_n} \qquad (8 \cdot 2)$$

This condition is met when the junction voltage changes so slowly that the diffusion of minority carriers into the diode bodies is sufficiently rapid to ensure that the dc carrier distributions are maintained. When the junction voltages change rapidly with time, however, the assumption of "quasi-equilibrium" carrier distributions is not met, and the v-i characteristic cannot be represented by Eq. $(8 \cdot 1b)$.

In this chapter we shall see how the ideal junction-diode model needs to be modified to account for rapidly changing junction voltages. As in Chap. 6, the principles of carrier motion which arise in this discussion are of as much significance for understanding high-frequency transistor characteristics as they are in understanding high-frequency performance of diodes. They should therefore be thoroughly understood.

8·1 SWITCHING TRAJECTORIES AND TRANSIENT PERFORMANCE

It is useful to introduce the problem of high-frequency and transient characterization by describing the diode response in a simple transient. Figure $8 \cdot 1$ has been drawn to aid in the discussion. Here, we show an ideal junction diode driven by a current source. We suppose that the current source has been delivering a current of I to the diode for $-\infty < t < 0$. At $t = 0$, the current begins to increase linearly with time $[i = I(1 + Kt)]$, and continues to increase until the diode current is twice the original value. Then the current is maintained at the value $2I$.

The diode voltage $v(t)$ is continuously recorded during this time, so that we can locate the diode current and voltage in the v-i plane at each instant. The results of this hypothetical experiment are shown in Fig. $8 \cdot 1$. When K is very small (Fig. $8 \cdot 1a$), the v-i coordinates for the diode will proceed along the dc v-i characteristic from $i = I$ to $i = 2I$. As K is increased (Fig. $8 \cdot 1b$), the diode voltage will not change as rapidly as the dc relation requires, so the v-i path which the diode follows will not be the same as the dc v-i law. As is suggested in Fig. $8 \cdot 1b$, the actual voltage drop across the diode is less than the dc voltage would be for the same current at any point in the transient. The actual path will, of course, have the same end points as before. As K is increased further (Fig. $8 \cdot 1c$), the lag of the diode voltage will become even more noticeable. In the limit of an abrupt change of i from I to $2I$, the entire voltage change takes place after the current change has been applied (Fig. $8 \cdot 1d$). The paths in the v-i plane which the diode follows in this example are called *switching trajectories*. Time is a parameter on each trajectory, being zero at the starting point and infinity at the end point of the transient.

If the current is decreased from $2I$ to I at a later time (when a steady-state condition has been established with the current $2I$ flowing), then

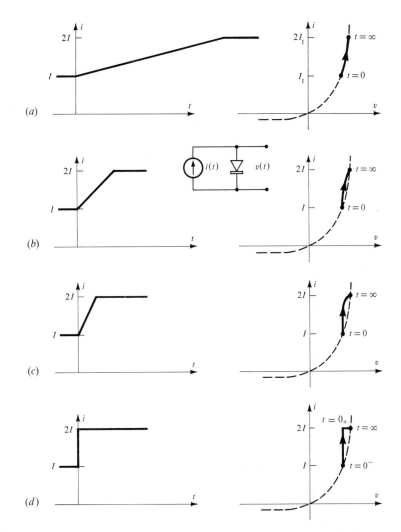

FIG. 8·1 SWITCHING TRAJECTORIES IN A JUNCTION DIODE FOR A SIMPLE TRANSIENT

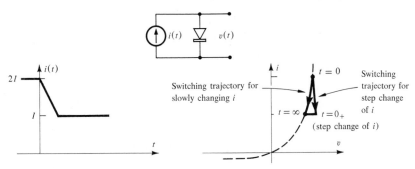

FIG. 8·2 SWITCHING TRAJECTORIES IN A JUNCTION DIODE FOR A DOWNWARD STEP IN FORWARD CURRENT

Transient and high frequency performance of pn-junction diodes 279

again the diode voltage will not follow the diode current. The two extreme switching trajectories corresponding to very slow and step changes in $i(t)$ are shown in Fig. 8·2. High-frequency sine-wave currents will also lead to operating paths which do not follow the dc characteristic, but rather form ellipse-like paths around it.

Thus, the high-frequency and transient behavior of a junction diode can be visualized in terms of an operating path on the v-i plane. In this chapter we shall see how our previous ideas about diode behavior can be expanded to account for transient behavior, and we will compute the transient and high-frequency operating paths for some practical cases.

8·2 PHYSICAL PHENOMENA OCCURRING

IN A SIMPLE SWITCHING TRANSIENT

To introduce the new physical concepts needed to characterize the transient performance of the diode, let us reconsider the transient shown in Fig. 8·1d, this time focusing attention on the internal phenomena which occur during the transient.

In this transient, the current is changed abruptly from I to $2I$ at $t = 0$. We wish to calculate how the voltage across the diode changes with time. The approach which we shall use to do this is as follows. The law of the junction tells us how the diode voltage is related to the minority-carrier densities at the edges of the transition region. This law is valid on an essentially instantaneous basis, so that we may write

$$p(t)|_{x=0} = p_n e^{qv(t)/kT} \qquad n(t)|_{y=0} = n_p e^{qv(t)/kT} \qquad (8 \cdot 3)$$

Equations (8.3) suggest that if we can calculate how the minority-carrier densities at the edges of the transition region respond to the transient excitation, we can calculate how the junction voltage varies with time. The minority-carrier distributions therefore provide the link between diode current and junction voltage drop, even in transient problems. We thus focus attention on how the minority-carrier charge distributions change with time when the diode current is abruptly doubled.

8·2·1 *Qualitative discussion of minority-carrier effects during the transient.* To concentrate on essentials, let us suppose the diode is a p^+n structure; the p-type side is heavily doped, so only the hole distribution in n-type material needs to be considered. We shall assume that holes can flow only by diffusion in the n-type body. We will also assume that current flow is one-dimensional in the diode, and that recombination only occurs within the diode body. The recombination lifetime will be denoted by τ_p, the diode area by A.

We know from the dc analysis given in Chap. 6 that for $t < 0$, the excess hole density in the neutral n-type body is of the form

$$p_E(x) = p_E(0)e^{-x/L_p} \qquad (8 \cdot 4)$$

where x is measured from the edge of the transition region in n-type material and $L_p = (D_p \tau_p)^{1/2}$. This distribution is plotted in Fig. $8 \cdot 3c$.

Furthermore, for $t < 0$, the gradient of the hole density at $x = 0$ is

$$\left. \frac{dp_E}{dx} \right|_{x=0} = - \frac{p_E(0)}{L_p} \tag{8 \cdot 5}$$

and the hole current injected into the n-type diode body at $x = 0$ is accordingly

$$i_p(x = 0) = I = - qAD_p \frac{dp_E}{dx} = \frac{qAD_p p_E(0)}{L_p} \tag{8 \cdot 6}$$

The excess-hole charge stored in the n-type body for $t < 0$ is

$$Q_p = qA \int_0^\infty p_E(x) \, dx = qAL_p p_E(0) \tag{8 \cdot 7a}$$

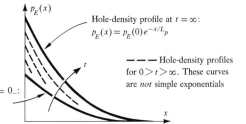

(a) Diode and excitation

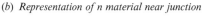

(b) Representation of n material near junction

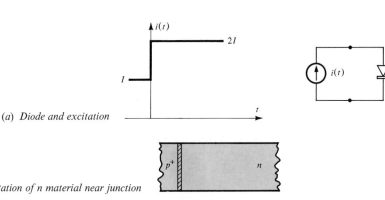

(c) Excess-hole-density profiles at selected times during the transient

Hole-density profile at $t = \infty$:
$p_E(x) = p_E(0)e^{-x/L_p}$

— — — Hole-density profiles for $0 > t > \infty$. These curves are *not* simple exponentials

Hole-density profile at $t = 0_-$:
$p_E(x) = p_E(0)e^{-x/L_p}$

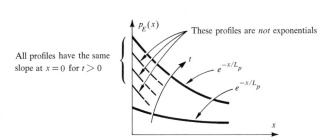

(d) Illustrating the constant slope of each profile at $x = 0$ during the transient

All profiles have the same slope at $x = 0$ for $t > 0$

These profiles are *not* exponentials

e^{-x/L_p}

e^{-x/L_p}

FIG. $8 \cdot 3$ THE TIME DEVELOPMENT OF THE EXCESS-HOLE-DENSITY PROFILES DURING A SWITCHING TRANSIENT

and the current can also be written in terms of Q as

$$I = \frac{Q_p}{\tau_p} \qquad (8 \cdot 7b)$$

These are the dc relations which apply for $t < 0$.

Now, consider what happens at $t = 0$. The current doubles. To accommodate this change, the *slope* of the hole-density curve at $x = 0$ must instantaneously change to *twice* the value it had at $t = 0_-$. There will be no change of the hole charge stored in the diode body, however, since

$$\Delta Q = \int_{0_-}^{0_+} i \, dt = 0 \qquad (8 \cdot 8)$$

Thus, the doubling of the terminal current requires

$$\left. \frac{dp_E}{dx} \right|_{x=0, t=0+} = 2 \left. \frac{dp_E}{dx} \right|_{x=0, t=0-} \qquad (8 \cdot 9a)$$

while charge conservation requires

$$p_E(x = 0, t = 0_+) = p_E(x = 0, t = 0_-) \qquad (8 \cdot 9b)$$

As time proceeds in this transient, the increased hole injection will first manifest itself as an increase in the excess-hole density near $x = 0$, as shown in Fig. $8 \cdot 3c$. Later on, the extra holes will have diffused into the diode body, thus increasing the charge stored in regions which are remote from the junction. The diffusion process takes time, however, so the build-up of charge in the diode body will always lag the build-up of charge near the junction.

During the build-up of charge, the total diode current cannot be accounted for by recombination. However, an equilibrium condition will ultimately be established in which the entire current injected into the diode bodies *can* once again be accounted for by recombination. Then the excess-hole-density profile will again be of the form e^{-x/L_p}. And, since the diode current for $t > 0$ is twice as large as it was for $t < 0$, the final distribution will be twice as large as the original distribution at every point. In particular,

$$p_E(x = 0, t \to \infty) = 2 \, p_E(x = 0, t = 0_+) \qquad (8 \cdot 10)$$

These facts are shown in Fig. $8 \cdot 3$.

$8 \cdot 2 \cdot 2$ *Charge continuity during the transient.*

On the basis of the discussion given above, we can see that at any time in the interval $0 < t < \infty$, more charge is being injected into the n-type body than is necessary to supply the needs of recombination at that moment. This extra charge is reflected in the general build up of p_E in the diode body.

We can phrase this mathematically as follows: Let $Q_p(t)$ be the total excess-hole charge stored in the n-type diode body at any instant during the transient. Then the recombination current at that instant is

$$i_{rec} = \frac{Q_p(t)}{\tau_p} \qquad (8 \cdot 11)$$

If the hole current being injected into the diode body at the time t is equal to the current required by Eq. (8·11), then the diode is in a steady-state condition. If not, then Q_p must be changing with time. If we denote the rate of change of $Q_p(t)$ by $dQ_p(t)/dt$, then the total current being injected into the n-type diode body must be

$$i_p(t) = \frac{Q_p(t)}{\tau_p} + \frac{dQ_p(t)}{dt} \qquad (8 \cdot 12)$$

Equation (8·12) is nothing more than a statement of charge conservation. It is therefore rigorously true, even though we have not specified the exact distribution of excess charge $p_E(x,t)$. In fact, we shall see later that in general it is difficult to find the exact distribution of $p_E(x,t)$. However, when it can be found, then Q_p can be found from

$$Q_p(t) = qA \int_0^\infty p_E(x,t)\, dx \qquad (8 \cdot 13)$$

Equation (8·12) can then be evaluated and must, of course, be found to be true at all times.

To summarize, when the diode current is increased abruptly from I to $2I$, a transient will ensue in which the excess-hole-density distribution changes from one dc profile to another dc profile. During this transient the $p_E(x,t)$ curve will satisfy the following conditions:

1. $p_E(x = 0, t = 0_-) = p_E(x = 0, t = 0_+)$
2. $p_E(x = 0, t \to \infty) = 2p_E(x = 0, t = 0_+)$

3. $\dfrac{\partial p_E}{\partial x}(x = 0, 0_+ < t < \infty) = 2\dfrac{\partial p_E}{\partial x}(x = 0, t = 0_-)$

4. $i_p = \dfrac{dQ_p(t)}{dt} + \dfrac{Q_p(t)}{\tau_p}$

where
$$Q_p(t) = qA \int_0^\infty p_E(x,t)\, dx$$

The fact that the junction voltage does not change instantaneously when the current step is applied is reflected in condition (1) above. Conditions (1) and (2), show, in fact, that for the particular transient under consideration

$$v(t = 0_+) = v(t = 0_-) \qquad v(t \to \infty) = v(t = 0_+) + \frac{kT}{q}\ln 2$$

Thus, we have established several conditions which the $p_E(x,t)$ and $v(t)$ curves must satisfy during the transient and obtained some valuable pictures of the physical phenomena occurring during the switching process. However, we do not yet know how $p_E(x,t)$ actually develops in time; and, in particular, we do not know how $p_E(0,t)$ varies, so we cannot yet calculate how the junction voltage varies with time. We must therefore consider how $p_E(x,t)$ can be determined.

In order to find $p_E(x,t)$, we have to formulate and solve a *time-dependent continuity equation*. As was true for the dc continuity equation obtained in Chap. 6, the time-dependent continuity equation can be obtained by simply specifying the conditions which the $p_E(x,t)$ curve must satisfy at every point in space and time. Unfortunately, the continuity equation turns out to be a partial differential equation. Advanced mathematical techniques are needed to solve it in even simple cases. As a result, several approximate techniques have been developed which focus attention on the simpler physical aspects of the transient and can be used to estimate the behavior of $p_E(x,t)$ in simple situations. We shall consider the simplest one of these in this section. The treatment to be described is called the *stored-charge approximation*, and gives a first-order correction to the dc behavior.

To develop the basic idea, let us reconsider the transient described in Fig. 8·1b. Here $i(t)$ varies rapidly enough that $v(t)$ does not follow the dc characteristic. However, the deviation is not great, so we can suppose that the minority-carrier distributions are not much different from their dc distributions for the given diode current. We therefore *assume* that in such cases *the actual excess-hole-density distribution at any time during the transient can be adequately approximated by an appropriately chosen steady-state distribution.* That is, we assume that at any time during the transient, the x distribution will be of the form e^{-x/L_p}. Time variations are then reflected *only* in the variation of the excess-hole density at $x = 0$. Thus, for slow transients, we may write

$$p_E(x,t) = p_E(0,t)e^{-x/L_p} \tag{8·14}$$

Such a distribution will apply when the current changes only a small amount during the time required for holes to diffuse a distance of 2 or $3L_p$.

Now, in such cases, the instantaneous value of the total stored charge can be quite simply found:

$$Q_p(t) = qA \int_0^\infty p_E(x,t)\,dx = qAL_p p_E(0,t) \tag{8·15}$$

The charge-conservation condition

$$i_p(t) = \frac{Q_p(t)}{\tau_p} + \frac{dQ_p}{dt} \tag{8·16a}$$

then becomes
$$i_p(t) = \frac{qAL_p}{\tau_p}\left[p_E(0,t) + \tau_p\frac{dp_E(0,t)}{dt}\right] \tag{8·16b}$$

Equation (8·16) is a simple first-order differential equation giving $p_E(0,t)$, which we need to know to calculate $v(t)$, in terms of the forcing function $i_p(t)$, which we do know. Thus we may solve (8·16b) directly to find $p_E(0,t)$ [and from it $v(t)$].

Before doing this, however, it is useful to see how Eq. (8·16b) gives the first-order correction to the dc solution. If the dp_E/dt term were very small compared to $p_E(0, t)$, then Eq. $(8 \cdot 16b)$ would reduce to

$$i_p(t) = \frac{qAL_p}{\tau_p} p_E(0,t) \qquad (8 \cdot 17a)$$

This *is* the dc and very low frequency solution. To make this clear, we can apply the law of the junction

$$p_E(0,t) = p_n(e^{qv(t)/kT} - 1)$$

to obtain $\qquad\qquad i_p(t) = I_0(e^{qv(t)/kT} - 1) \qquad (8 \cdot 17b)$

We would expect this solution to apply when

$$\tau_p \frac{dp_E(0,t)}{dt} \leq 0.1 \, p_E(0,t) \qquad (8 \cdot 18a)$$

or $\qquad\qquad \dfrac{dv}{dt} \leq \dfrac{0.1}{\tau_p} \dfrac{kT}{q} \qquad (8 \cdot 18b)$

Thus, when the rate of change of voltage is less than one-tenth the thermal voltage (kT/q) divided by the carrier lifetime, the switching trajectory will essentially follow the dc v-i characteristic.

When dp_E/dt is larger than the value suggested in Eq. $(8 \cdot 18a)$, its presence cannot be neglected. Then we no longer have an instantaneous proportionality between $i(t)$ and $p_E(0,t)$. p_E will be less than the dc relation would suggest, and the junction voltage will be less than the dc v-i relation requires.

To see this clearly, we apply Eq. $(8 \cdot 16)$ to obtain the switching trajectory for a *step change* in the diode current. To be sure, the excess-hole-density profile in such a transient will *not* be a succession of steady-state values. Nonetheless, the accuracy of the model is quite surprising.

Let the step change in current be applied at $t = 0$. For $t < 0$, the diode current will be denoted by I. For $t > 0$, the diode current will be $I + \Delta I$. We leave I and ΔI unspecified for generality. Then Eq. $(8 \cdot 16a)$ gives

$$i = \frac{Q_p}{\tau_p} + \frac{dQ_p}{dt} \qquad (8 \cdot 19)$$

Now, for $t < 0$, the diode is in a dc condition with

$$I = \frac{Q_p}{\tau_p} \Longrightarrow Q_p(0) = I\tau_p \qquad (8 \cdot 20a)$$

For $t \to \infty$, the diode is again in a dc condition with

$$I + \Delta I = \frac{Q_p}{\tau_p} \Longrightarrow Q_p(\infty) = (I + \Delta I)\tau_p \qquad (8 \cdot 20b)$$

Then the general solution to Eq. $(8 \cdot 19)$ is

$$Q_p(t) = A(t) + Be^{-t/\tau_p} \qquad (8 \cdot 21)$$

When Eqs. $(8 \cdot 20a)$ and $(8 \cdot 20b)$ are used, we may evaluate $A(t)$ (it is a constant) and B. The result is

$$Q_p(t) = I\tau_p + \Delta I\tau_p(1 - e^{-t/\tau_p}) \qquad (8 \cdot 22a)$$

Hence,
$$p_E(0,t) = \frac{\tau_p}{qAL_p}\left[I + \Delta I(1 - e^{-t/\tau_p})\right] \qquad (8 \cdot 22b)$$

It is useful to compare this result with the exact result which is obtained by actually solving the partial differential equation of continuity for $p_E(x,t)$ (to be developed in Sec. $8 \cdot 4$). To consider a specific case, we let $I = 0$ and plot $p_E(0,t)/p_E(0,\infty)$. The plot is shown in Fig. $8 \cdot 4$.

As the plot shows, the simple model gives p_E within a factor of 2 of the exact result for $(t/\tau_p) > 0.25$. Of course, for times less than this, the error is increasingly large. However, this would be expected from our previous qualitative discussion, where we saw that the excess density near $x = 0$ increases more rapidly than the excess density elsewhere.

The plot also shows that the stored charge analysis will yield a conservative approximation to the actual switching trajectory. That is, the actual switching transient will be completed before the transient predicted by the quasi-equilibrium approximation. This is a useful point to remember, since an upper limit on the switching time is adequate for many practical problems.

To compute the junction voltage during the switching transient, we use Eq. $(8 \cdot 22b)$ and the law of the junction. There results

$$v(t) = \frac{kT}{q}\ln\left[\frac{p_E(0,t)}{p_n} + 1\right] \qquad (8 \cdot 23a)$$

The stored-charge model then yields

$$v(t) = \frac{kT}{q}\ln\left[\frac{I}{I_0} + \frac{\Delta I}{I_0}(1 - e^{-t/\tau_p}) + 1\right] \qquad (8 \cdot 23b)$$

where we have used the relation

$$I_0 = \frac{qAD_pp_n}{L_p}$$

to simplify the final expression. The variation of $v(t)$ and the switching

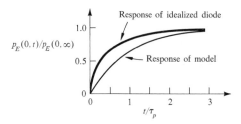

$p_E(0,t)/p_E(0,\infty)$

Response of idealized diode

Response of model

t/τ_p

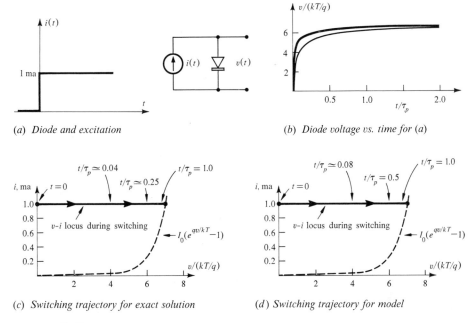

(a) Diode and excitation

(b) Diode voltage vs. time for (a)

(c) Switching trajectory for exact solution

(d) Switching trajectory for model

FIG. 8·5   SWITCHING TRAJECTORIES FOR A SIMPLE TRANSIENT

trajectories for the "exact" solution and the model are plotted in Fig. 8·5 for the case $I = 0$, $\Delta I = 1$ ma, $I_0 = 1$ μa. According to the stored-charge model, the transient will be essentially completed in about $3\tau_p$ sec. The more exact solution reaches the same point in about τ_p sec.

8·3·1   Reverse recovery time in junction diodes.   A transient phenomenon of great practical importance occurs when we attempt to switch a junction diode from forward conduction to reverse bias. We find experimentally that when a diode has been conducting a steady current in the forward direction, then it can pass a reverse current which is appreciably larger than its reverse saturation current I_0 for a short period of time, during which the junction voltage remains *positive*. We call this period of time the *storage-delay time,* because it is caused by the stored minority carriers in the diode body.

Two possible switching trajectories which exhibit this effect are shown in Fig. 8·6. In both cases, the exciting source appears to be a current source for the period of time that the diode remains forward-biased. After the storage delay, however, the junction voltage will begin to rise rapidly. The true current supply will then drive the diode into its reverse-breakdown condition, while the voltage source with series resistance will deliver less current as the junction voltage rises, so the switching trajectory will rapidly return to the dc v-i characteristic at a current of I_0.

The physical mechanism for this effect can be traced once again to the sluggish response of the minority-carrier densities to changes in the diode

Transient and high frequency performance of pn-junction diodes   287

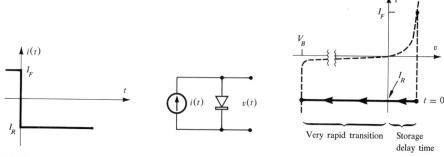

(a) *Reverse-recovery switching trajectory for current-source drive*

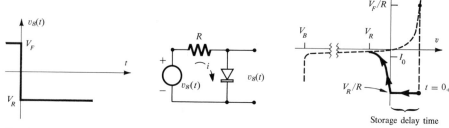

(b) *Reverse-recovery switching trajectory for voltage source and series R*

FIG. 8·6 SWITCHING TRAJECTORIES FOR REVERSE-RECOVERY TRANSIENT

current. The basic idea is developed in Fig. 8·7. At $t = 0$, the hole density has the familiar e^{-x/L_p} distribution. When a reverse current is applied, the *slope* of the hole-density distribution at $x = 0$ will change instantaneously from $-I_F/qAD_p$ to I_R/qAD_p, where I_F is the forward current being passed for $t < 0$ and I_R is the reverse current being passed for $0 < t < t_s$. The storage phase of the transient is completed when $p_E(0,t)$ reaches zero [so $v(t) = 0$]. The storage-delay time is called t_s.

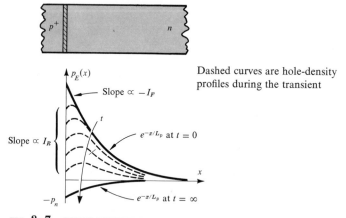

Dashed curves are hole-density
profiles during the transient

FIG. 8·7 EXCESS-DENSITY PROFILES DURING A REVERSE-RECOVERY TRANSIENT

The reverse current consists of holes being drawn across the transition region from the n-type body near $x = 0$ into the p-type body. Ideally, there is no change in $p_E(0,t)$ at $t = 0$, and hence, no change in the junction voltage. However, as time proceeds, the excess-hole density near $x = 0$ will drop, as suggested in Fig. 8·7. This produces a diffusion gradient which allows holes at greater distances from the junction to diffuse toward the junction, an effect which tends to maintain the excess-hole density at $x = 0$ at a positive value. In this way, the charge stored in regions which are remote from the junction have a significant effect on the storage-delay time.

To calculate the storage-delay time for the exact case, as we said, involves mathematical techniques which are beyond our present interest (for instance, the solution of a partial differential equation). However, we can use the simple stored-charge analysis to give a conservative estimate of the storage-delay time.

To do this, we suppose that the excess-hole density at any time can be approximated by a distribution of the form

$$p_E(x,t) = p_E(0,t)e^{-x/L_p} \qquad (8 \cdot 24)$$

Then $p_E(0,t)$ during the transient is still given by Eq. (8·22b):

$$p_E(0,t) = \frac{\tau_p}{qAL_p}[I + \Delta I(1 - e^{-t/\tau_p})] \qquad (8 \cdot 25)$$

The storage delay is completed when $p_E(0,t)$ reaches zero, for then $v(t) = 0$.

For the present case, we may substitute

$$I = I_F \qquad \Delta I = -(I_F + I_R) \qquad (8 \cdot 26)$$

in Eq. (8·25) to yield

$$\frac{qAL_p}{\tau_p}p_E(0,t) = I_F - (I_F + I_R)(1 - e^{-t/\tau_p}) \qquad (8 \cdot 27)$$

The value of t at which $p_E(0,t) = 0$ is then

$$t_s = \tau_p \ln\left(1 + \frac{I_F}{I_R}\right) \qquad (8 \cdot 28)$$

The storage time predicted by an exact analysis of the idealized diode is

$$t_s = \tau_p\left(erfc\ \frac{I_F}{I_F + I_R}\right)^2 \qquad (8 \cdot 29)$$

where $erfc$ means complementary error function, and is a tabulated function.

Again, it is useful to compare these results to appreciate the nature of the stored-charge approximation. This is done in Fig. 8·8 where the normalized storage-delay time t_s/τ_p is plotted as a function of I_R/I_F. The quasi-equilibrium approximation gives a conservative estimate of t_s, with the error increasing to large values when $|I_R| \gg |I_F|$, where e^{-x/L_p} becomes a very poor approximation to the actual transient distribution.

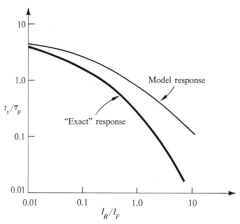

8·3·2 *High-frequency sine-wave excitation: forward bias.* The fact that
switching trajectories do not in general follow the dc v-i characteristic
implies that high-frequency sine waves will also fail to do so. Instead, the
instantaneous diode voltage and current will trace out an operating path
like that shown in Fig. 8 · 9, when a high-frequency sine-wave current is
applied together with a dc bias current.

The nature of this operating path is such that the diode can be represented
by the small-signal circuit model shown in Fig. 8 · 9. In this model, C_j is
the capacitance of the transition region, r_e is the dynamic resistance
of the diode (or slope of the dc v-i characteristic at the operating point),
and C_D is called the *diffusion capacitance*. C_D provides a circuit representation for the effects of minority-carrier storage in the diode body.

To see this, let us consider the first-order correction to the low-frequency
theory which is provided by the stored-charge approach. Suppose the
diode is driven with a small sine-wave current $i_e(t)$, which is superposed on
a dc bias current I. Then the quasi-equilibrium hole distribution in the
diode will vary with time in the manner shown in Fig. 8 · 9d. The figure
shows that minority-carrier charge must be added when the current
increases and removed when the current decreases, which is a capacitive
effect.

The magnitude of the capacitance can be simply estimated as follows.
The diode current $i(t)$, the excess hole density at $x = 0$, and the junction
voltage will all be of the form

$$i_p(t) = I + i_e(t)$$
$$p_E(0,t) = P_E + p_e(t) \qquad (8 \cdot 30)$$
$$v(t) = V_0 + v_e(t)$$

in which capital letters represent dc quantities and lower-case letters represent ac quantities. Now the basic stored-charge relation gives

$$I + i_e(t) = \frac{qAL_p}{\tau_p}\left[P_E + p_e(t) + \tau_p\frac{dp_e(t)}{dt}\right] \qquad (8\cdot31a)$$

or
$$I = \frac{qAL_pP_E}{\tau_p} \qquad (8\cdot31b)$$

$$i_e(t) = \frac{qAL_p}{\tau_p}\left[p_e(t) + \tau_p\frac{dp_e(t)}{dt}\right] \qquad (8\cdot31c)$$

Also, the law of the junction requires

$$p_E(0,t) = p_n(e^{qv(t)/kT} - 1) \qquad (8\cdot32a)$$

which may be manipulated to yield

$$P_E = p_n(e^{qV_0/kT} - 1) \qquad (8\cdot32b)$$

$$p_e(t) = P_E\frac{qv_e(t)}{kT} \qquad (8\cdot32c)$$

if we assume that $P_E \gg p_n$ and $|v_e(t)| \ll kT/q$. Then Eq. $(8\cdot32c)$ in Eq. $(8\cdot31c)$ gives, after some manipulation

$$i_e(t) = \frac{qI}{kT}\left(v_e + \tau_p\frac{dv_e}{dt}\right) \qquad (8\cdot33)$$

Equation $(8\cdot33)$ thus suggests that the diode may be represented by a conductance

$$g_e = \frac{1}{r_e} = \frac{qI}{kT} \qquad (8\cdot34a)$$

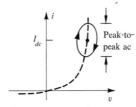

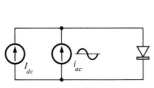
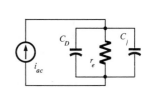

(a) Diode and high-frequency excitation

(b) Operating v-i locus

(c) Small signal circuit model

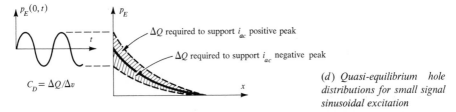

ΔQ required to support i_{ac} positive peak

ΔQ required to support i_{ac} negative peak

$C_D = \Delta Q/\Delta v$

(d) Quasi-equilibrium hole distributions for small signal sinusoidal excitation

FIG. 8·9 HIGH-FREQUENCY SINE-WAVE EXCITATION OF A JUNCTION DIODE

Transient and high frequency performance of pn-junction diodes   291

and a capacitance

$$C_D = \frac{qI}{kT}\tau_p \qquad (8\cdot34b)$$

in parallel. r_e is just the dynamic resistance of the diode at the given operating point (see Prob. $6\cdot6$). This value of C_D may be conveniently remembered by noting that

$$C_D r_e = \tau_p \qquad (8\cdot34c)$$

so the time constant of C_D and r_e is τ_p, independent of the dc current level.

Of course, this is a first-order approximation. But it does predict an elliptical operating path for small-signal high-frequency excitation of the diode, and is adequate for most purposes.

In summary, the stored-charge analysis provides a simple way of accounting for the basic ideas involved in switching transients and high-frequency behavior of junction diodes. It is based simply on (1) the charge-continuity relation

$$i_p = \frac{Q_p}{\tau_p} + \frac{dQ_p}{dt}$$

which is true for the diode and (2) the *assumption* that the excess carrier distributions always have the form e^{-x/L_p}. The theory is simply visualized and gives the first-order correction to the dc theory. For these reasons, it is very useful for estimating the switching performance of diodes, and is similarly quite useful in a first-order description of high-frequency and transient behavior of transistors. We shall see how these ideas apply to transistors at a later point.

However, there are cases where we would like to estimate the performance of a semiconductor diode or transistor more closely than the first-order correction will allow. For such cases we have to either formulate and solve the time-dependent continuity equation for $p_E(x,t)$, or devise a scheme for making successively better approximations to $p_E(x,t)$. Both approaches to the problem lend valuable insights into the transient behavior of the diode and deserve consideration. Since the approximate procedure to be described rests on an understanding of the time dependent continuity equation, we shall develop this first.

$8\cdot4$ THE TIME-DEPENDENT CONTINUITY EQUATION

As was true for the steady-state continuity equation obtained in Chap. 6, the time-dependent continuity equation can be obtained by simply specifying the conditions which the $p_E(x,t)$ curve must satisfy at every point in space and time. Figure $8\cdot10$ has been drawn to illustrate the development. Here we show a small volume of the n-type body bounded by planes located at

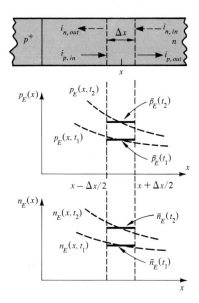

FIG. 8 · 10   THE DEVELOPMENT OF THE TIME-DEPENDENT CONTINUITY EQUATION

$x - \Delta x/2$ and $x + \Delta x/2$. We assume the diode to be in a transient condition so that the carrier densities within the slice are changing with time. Curves of $p_E(x)$ are shown for two different times t_1 and t_2, the difference $t_2 - t_1$ being denoted by Δt.

Now, holes which enter this volume in a time interval Δt either (1) flow out, (2) recombine, or (3) are stored. Of course, the same is true for electrons; but we shall again assume that space-charge neutrality always applies, so majority-carrier phenomena can be deduced when the minority-carrier behavior has been established.

In terms of the slice shown in Fig. 8 · 10, the simple accounting for holes suggested above leads to the equation

$$i_p\big|_{x-(\Delta x/2)}\Delta t = i_p\big|_{x+(\Delta x/2)}\Delta t + \left(\frac{qA\,\Delta x}{\tau_p}\bar{p}_E\right)\Delta t + qA\,\Delta x\,\Delta\bar{p}_E \qquad (8\cdot 35a)$$

where $\bar{p}_E$ is the *average* excess-hole density in the volume. $\Delta\bar{p}_E$ is then the *change* in the average excess-hole density in the volume in the time interval Δt (see Fig. 8 · 10).

If we rearrange the terms, divide by Δt, and take the limit as Δt approaches zero, we find

$$i_p\left(x - \frac{\Delta x}{2}\right) - i_p\left(x + \frac{\Delta x}{2}\right) = \frac{qA\,\Delta x\bar{p}_E}{\tau_p} + qA\,\Delta x\frac{d\bar{p}_E}{dt} \qquad (8\cdot 35b)$$

Now, since $qA\,\Delta x$ is the volume of the slice, the excess-hole charge q_p stored in the volume element is simply

$$q_p = qA\,\Delta x\bar{p}_E$$

Equation $(8 \cdot 35b)$ can be written as

$$-\Delta i_p = i_{pin} - i_{pout} = \frac{q_p}{\tau_p} + \frac{dq_p}{dt}$$

Evidently the first term in Eq. $(8 \cdot 35b)$ represents the average recombination current in the volume element and the second term represents the average storage current.[1]

To obtain the partial differential equation of continuity from Eq. $(8 \cdot 35b)$, we divide both sides by $qA \, \Delta x$ and then take the limit as $\Delta x \to 0$. This yields

$$-\frac{1}{q} \frac{\partial j_p}{\partial x} = \frac{p_E}{\tau_p} + \frac{\partial p_E}{\partial t} \qquad (8 \cdot 35c)$$

where the partial derivatives indicate that p_E is a function of both x and t. Equation $(8 \cdot 35c)$ is identical to the dc continuity equation obtained in Chap. 6, except that Eq. $(8 \cdot 35c)$ contains a term $\partial p_E/\partial t$ to account for the fact that currents must flow whenever there are *changes* in the excess-hole density (or, equivalently, changes in the excess-hole density must accompany changes in the current).

Since we are assuming that holes flow only by diffusion in n-type material, we can write

$$j_p = -qD_p \frac{\partial p_E}{\partial x} \qquad (8 \cdot 36)$$

If we substitute Eq. $(8 \cdot 36)$ into Eq. $(8 \cdot 35c)$ we obtain

$$D_p \frac{\partial^2 p_E}{\partial x^2} = \frac{p_E}{\tau_p} + \frac{\partial p_E}{\partial t} \qquad (8 \cdot 37)$$

Equation $(8 \cdot 37)$ is the *diffusion equation* (*cf.* Prob. $4 \cdot 16$) with a term p_E/τ_p added to account for recombination. It plays exactly the same role in predicting the transient behavior of a diode as does its dc counterpart in predicting the dc characteristics. However, the solution to Eq. $(8 \cdot 37)$ is not so readily found as the solution to the corresponding dc continuity equation. To appreciate this, we first remark that the general solution to a partial differential equation will usually be an infinite series of functions:

$$p_E(x,t) = \sum_{n}^{\infty} A_n f_n(x,t) \qquad (8 \cdot 38)$$

The functions will not, in general, be simple, and $p_E(x,t)$ can not usually be expressed in a closed form.

The boundary conditions will specify the A_n, though this is more complicated than the simpler dc case. To illustrate this point, we show in Fig. $8 \cdot 11$ a schematic picture of the solution surface for Eq. $(8 \cdot 37)$, for a transient in which the diode current is abruptly switched from I to $2I$. At $t = 0$,

---

[1] By integrating both sides of the equation for Δi_p between the limits $x = 0$ and $x = \infty$ and setting $i_p(x \to \infty) = 0$, we obtain Eq. $(8 \cdot 12)$, upon which the stored-charge approach is based.

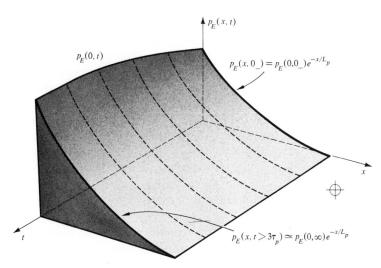

FIG. 8·11 SOLUTION SURFACE FOR PARTIAL DIFFERENTIAL EQUATION GOVERNING BEHAVIOR OF $p_E(x,t)$

the surface reduces to the curve $p_E(0,0_-)e^{-x/L_p}$ for all $x > 0$. In terms of Eq. (8·38), this means

$$p_E(x,0) = \sum_n^\infty A_n f_n(x,0) = p_E(0,0_-)e^{-x/L_p}$$

Also, for all $t > 0$, the partial derivative of $p_E(x,t)$ with respect to x evaluated at $x = 0$ must have its final value:

$$\frac{\partial p_E}{\partial x}(0,t) = \sum_n^\infty A_n \frac{\partial f_n}{\partial x}(0,t) = \text{const}$$

When the functions f_n and the constants A_n are properly selected to satisfy Eq. (8·37) and meet these conditions, the entire solution surface can be plotted. In particular, the $p_E(0,t)$ curve will be known precisely, so we can find the junction voltage by using the law of the junction:

$$v(t) = \frac{kT}{q} \ln \left[\frac{p_E(0,t)}{p_n} + 1 \right]$$

Unfortunately, while the solution surface is an interesting way of visualizing the problem, its determination involves advanced mathematical techniques. Furthermore, it contains far more information than we can use or need, and it gives much greater accuracy than is perhaps warranted by the nature of the idealizations which we have made in arriving at Eq. (8·37).

The question thus arises whether approximations can be made to preserve the main features of the physics but simplify the determination of $p_E(0,t)$, which is what we really want to know. In particular, we are interested in eliminating the need for solving partial differential equations.

Transient and high frequency performance of pn-junction diodes 295

One such approximation is, of course, provided by the stored-charge analysis. In this case, the solution surface is assumed to have the form e^{-x/L_p} for all t. However, it is possible to make considerably better approximations to the ideal behavior than this first-order correction allows.

These higher-order approximations lead to a series of diode models which are called *lumped models*, for reasons that will shortly be apparent. They are based on the idea that an approximate calculation of $p_E(x,t)$ at a few well chosen points will give adequate information about the actual variation of p_E. Calculations with the lumped model can then be visualized as providing a stepwise approximation to the solution surface, as shown in Fig. 8·12. While we do not plot such a surface in working an actual problem, it does help to illustrate the nature of the approximation we are making.

From a mathematical point of view, the principal simplification associated with the lumping approximation is that the partial differential equation (8·37) is replaced by a set of ordinary differential equations. For example, four quite simple, ordinary differential equations need to be solved to produce the solution surface shown in Fig. 8·12. In practical problems an adequate approximation can frequently be made with one or two simple differential equations. Thus the mathematical simplification of the problem is enormous.

The lumping technique therefore provides a way of using relatively simple mathematical tools to obtain an approximate solution to a complicated problem. The ideas to be introduced in developing these models deserve careful attention, inasmuch as the same techniques are applicable in many other physical situations where a meaningful but approximate treatment of a distributed system is adequate.

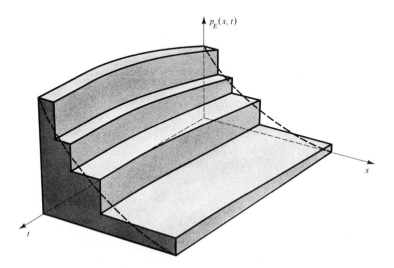

FIG. 8·12 SOLUTION SURFACE FOR A LUMPED VERSION OF THE CONTINUITY EQUATION

8·5 LUMPED MODELS FOR JUNCTION DIODES[1]

To maintain maximum contact with the physical ideas already established, we shall develop the lumped model for a junction diode by first constructing a lumped model for the continuity equation and then considering how junction effects may be added. Other more general methods of developing these models are described in the references at the end of this chapter.

To begin the development, it is desirable to consider carefully the point at which partial differentiation became necessary in the development of the continuity equation. This occurred in obtaining Eq. (8·35c) from Eq. (8·35b). Equation (8·35b) is an *ordinary* differential equation, but it involves the size of the slice through Δx. When we divide by Δx and take the limit, we eliminate the dependence of the continuity equation on Δx, but we introduce partial differentiation in the process. Thus, to avoid the partial differential equation, we shall leave Δx finite and *assume that the average excess-hole density $\bar{p}_E$ gives an adequate description of the excess-hole density throughout the slice.* We then refer to the slice as a *lump* and use $\bar{p}_E$ to provide a complete description of recombination and storage processes within the lump. The lumping approximation will yield accurate results if the width of the lump Δx is small enough so that the actual hole density at each point in the lump does not differ appreciably from the average density. For the present development we shall leave the exact size of Δx unspecified, though when we come to actual problems, we will have to choose the number and size of the lumps in some manner.

We now observe that, since Δx is fixed, the only variable on the right-hand side of Eq. (8·35b) is $\bar{p}_E$. To emphasize this, we rewrite Eq. (8·35b) in the form

$$i_p\big|_{x-(\Delta x/2)} - i_p\big|_{x+(\Delta x/2)} = H_c\bar{p}_E + S_p\frac{d\bar{p}_E}{dt} \qquad (8·39)$$

where

$$H_c \equiv \frac{qA\,\Delta x}{\tau_p} \qquad S_p \equiv qA\,\Delta x \qquad (8·40)$$

Note that H_c and S_p involve only geometrical factors and material parameters. Note also that Eq. (8·39) stresses the linear dependence of recombination current on $\bar{p}_E$ and the linear dependence of storage current on $d\bar{p}_E/dt$.

A schematic representation of the minority-carrier behavior in the lump can be obtained by properly defining symbols and flow laws to represent the recombination process and the storage mechanism. We shall present the necessary ideas in Secs. 8·5·1 and 8·5·2.

---

[1] An edited version of this section appears in chap. 6 of Gray, DeWitt, Boothroyd, and Gibbons, "Physical Electronics and Circuit Models of Transistors," SEEC vol. 2, John Wiley & Sons, Inc., New York, 1964.

8·5·1 *Definition and properties of a combinance.* The recombination current in the volume $A \Delta x$ is represented in Eq. $(8 \cdot 39)$ by the term

$$i_{rec} = H_c \bar{p}_E \qquad (8 \cdot 41)$$

According to our assumptions, there is a uniform distribution of holes in the volume and a uniform distribution of electrons as well. Equation $(8 \cdot 41)$ represents the rate at which holes and electrons recombine when $\bar{p}_E$ is the average hole density and $\bar{n}$ $(= n_n + \bar{p}_E)$ is the average electron density in the lump.

These facts may be represented diagrammatically in the manner shown in Fig. $8 \cdot 13$. In Fig. $8 \cdot 13d$ the *excess*-hole density $\bar{p}_E$ is represented as a *node variable,* as is the electron density, $\bar{n}$. This does not mean, of course, that $\bar{p}_E$ and $\bar{n}$ are separated in space. This technique is used only to provide a means of accounting for these carrier densities. The carriers themselves are uniformly distributed throughout the lump.

Between the two nodes there is drawn an element which is called a

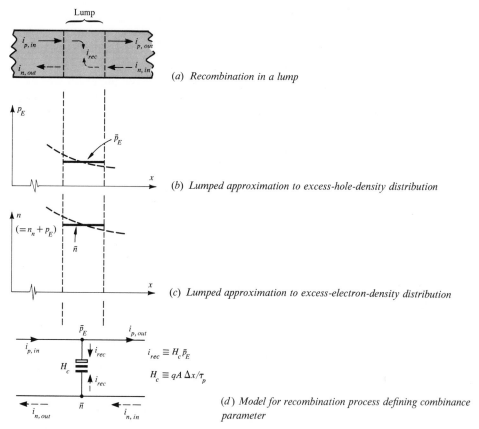

(a) *Recombination in a lump*

(b) *Lumped approximation to excess-hole-density distribution*

(c) *Lumped approximation to excess-electron-density distribution*

$$i_{rec} \equiv H_c \bar{p}_E$$

$$H_c \equiv q A \Delta x / \tau_p$$

(d) *Model for recombination process defining combinance parameter*

FIG. $8 \cdot 13$ LUMPED REPRESENTATION OF THE RECOMBINATION PROCESS, DEFINING THE COMBINANCE ELEMENT

combinance. It is given the symbol H_c and accounts for recombination current within the lump. The *value* of the combinance is, in the present discussion

$$H_c = \frac{qA\,\Delta x}{\tau_p} \tag{8.42}$$

which has the dimensions amperes-cubic centimeters.

The current flowing in the element is defined to be

$$i = H_c \bar{p}_E \tag{8.43}$$

It does not depend on $\bar{n}$, because we are assuming that recombination is proportional to the excess-minority-carrier density only. To emphasize this a box is drawn near the $\bar{p}_E$ node to indicate that only the value of $\bar{p}_E$ is to be used to calculate i_{rec}.

Notice that the physics of the recombination process requires the current entering H_c from the $\bar{p}_E$ node to be *hole current*, while that entering the lower end of H_c is an *electron current*. The value of the electron current is determined by $\bar{p}_E$, and is independent of $\bar{n}$.

8·5·2 *Definition and properties of a storance.* The second term in Eq. (8·39) represents a *hole-storage current*. It was derived from an earlier term, $S_p\,\Delta\bar{p}_E$, which was the *change* in the total charge of holes stored in the lump during the time interval Δt.

Now, the assumption of instantaneous space-charge neutrality requires that any change in $\bar{p}_E$ be balanced by an identical change in $\bar{n}_E$. Thus, if $\Delta\bar{p}_E$ holes are stored in the lump over a time interval Δt, then $\Delta\bar{n}(=\Delta\bar{p}_E)$ electrons are also stored. As in the case of recombination, both holes and electrons are involved in the storage process.

Hole- and electron-storage phenomena can be represented schematically in the manner shown in Fig. 8·14. Once again, $\bar{p}_E$ and $\bar{n}$ are the node variables. Now an element called a *storance* (because it stores holes) is drawn between the two nodes. The storance is given the symbol S_p. It has the *value*

$$S_p = qA\,\Delta x \tag{8.44}$$

and the dimensions coulombs-cubic centimeters. The current flowing into the storance is by definition

$$i_{storage} = S_p \frac{d\bar{p}_E}{dt} \tag{8.45}$$

and the charge of holes stored in it at any time is

$$Q_p = S_p \bar{p}_E \tag{8.46}$$

Once again, a box is used near the $\bar{p}_E$ node to emphasize that storage phenomena are determined entirely by variations in $\bar{p}_E$ (at least in the space-charge neutral approximation).

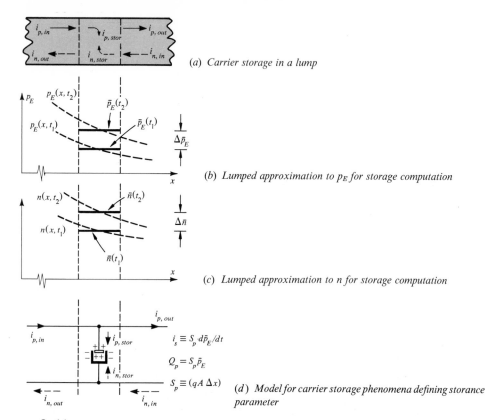

(a) Carrier storage in a lump

(b) Lumped approximation to p_E for storage computation

(c) Lumped approximation to n for storage computation

(d) Model for carrier storage phenomena defining storance parameter

FIG. $8 \cdot 14$ LUMPED REPRESENTATION OF THE STORAGE PROCESS, DEFINING STORANCE ELEMENT

$8 \cdot 5 \cdot 3$ *Schematic representation of Eq. ($8 \cdot 39$).*   We may now combine the results of the last two sections to give a schematic representation of Eq. ($8 \cdot 39$). This is done in Fig. $8 \cdot 15$. The flow laws for H_c and S_p are such that the minority-carrier continuity Eq. ($8 \cdot 39$) is merely Kirchhoff's current law at the $\bar{p}_E$ node.

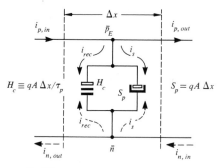

FIG. $8 \cdot 15$ LUMPED REPRESENTATION OF CONTINUITY EQUATION EMPLOYING COMBINANCE AND STORANCE ELEMENTS

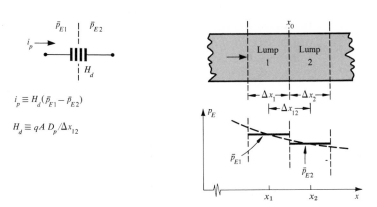

Electron currents flowing into and out of the lump have also been included in Fig. 8·16. They are shown dotted to represent the actual direction of motion of electrons. Kirchhoff's law at the lower node is then the *majority*-carrier continuity equation.

8·5·4 *Lumped representation of flow by diffusion processes.* Carrier-diffusion phenomena enter the lumped model as the means by which carriers are transported from one lump to the next. The diffusion process therefore provides a means of relating $i_p|_{x-(\Delta x/2)}$ and $i_p|_{x+\Delta x}$ to $\bar{p}_E$ and the excess-hole densities in the two adjacent lumps.[1] Figure 8·16 has been prepared to facilitate discussion. The actual hole-concentration profile at a given time t is dotted on this figure. The lumped approximation to it is drawn in solid.

The actual current flowing in across the left face of lump 2 is

$$i_p|_{x_0} = -qAD_p \frac{\partial p_E}{\partial x}\bigg|_{x_0} \tag{8·47}$$

Since we are representing phenomena within each lump by average values of p_E, we must approximate Eq. (8·47) by

$$i_p|_{x_0} = -qAD_p \frac{\bar{p}_E(x_2) - \bar{p}_E(x_1)}{\Delta x_{21}} \tag{8·48}$$

where Δx_{21} is the distance between the centers of the two lumps. We regard Δx_{21} as fixed, and rewrite Eq. (8·48) as

$$i_p|_{x_0} = H_d[\bar{p}_E(x_1) - \bar{p}_E(x_2)] \tag{8·49}$$

---

[1] Of course, holes can move by drift (independent of diffusion), and this process can also be represented in the lumped model. However, we are interested in low-level injection processes in homogeneous semiconductors, so we shall neglect the drift motion.

In Eq. (8·49) the only variables are $\bar{p}_E(x_1)$ and $\bar{p}_E(x_2)$. H_d is a coefficient which depends only on material parameters and geometrical factors.

We are thus led to define a third element called a *diffusance*. Diffusance is given the symbol H_d. Its *value* is

$$H_d = \frac{qAD_p}{\Delta x_{21}} \tag{8·50}$$

and its dimensions are amperes-cubic centimeters. The current flowing from *left to right* in H_d is defined by

$$i_d = H_d[\bar{p}_E(x_1) - \bar{p}_E(x_2)] \tag{8·51}$$

Note that in contrast to recombination and storage, diffusion involves only one carrier type. Furthermore, diffusion is a *bilateral* phenomenon. That is, if $\bar{p}_E(x_2)$ is greater than $\bar{p}_E(x_1)$, holes will diffuse from right to left. Hence, H_d must be represented by a symmetric symbol, whereas H_c and S_p are represented by nonsymmetric symbols. No distinctive marks are made on the terminals of the diffusance element.

8·5·5 *Complete lumped model for hole flow in the n-type junction diode body.* With the aid of the diffusance element, we can draw a lumped model to represent all the hole-flow processes in the n-type body. This is shown in Fig. 8·17. Diffusance elements are simply used to tie adjacent nodes together.

Of course, electrons also diffuse *and* drift between lumps, so there should be elements in the electron line to account for these transport processes. However, for any problem where the terminal behavior can be obtained by studying minority-carrier processes only, this is not necessary. Majority-carrier phenomena are determined in the junction diode from current continuity and space-charge neutrality requirements. Furthermore, the electron densities are nearly the same in adjacent lumps, so we may for simplicity neglect the actual electron-transport processes and connect adjacent electron nodes together by a wire. This will then limit the lumped model to low-level injection conditions in a homogeneous extrinsic semiconductor.[1]

8·5·6 / *A lumped representation of the transition region.* In the preceding sections we have formulated a lumped representation for minority-carrier transport in a neutral semiconductor bar. We shall now develop a lumped model for the pn junction and show how it may be connected to the lumped model just formulated, to build up a model for the entire diode.

There are two ways in which this may be done. The most general procedure is to further develop the lumped model so that it may be extended right through the transition region. To do this, we have to devise

[1] Less restricted lumped models have been developed which characterize semiconductors without limitations on homogeneity, doping level, or injection level. See J. G. Linvill and J. F. Gibbons, "Transistors and Active Circuits," McGraw-Hill Book Company, New York, 1961, chap. 4.

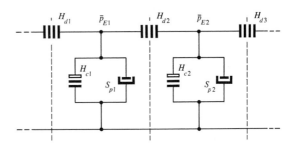

FIG. 8·17 LUMPED MODEL FOR THE CONTINUITY EQUATION IN A HOMOGENEOUS n-TYPE SEMICONDUCTOR, ASSUMING CARRIER FLOW BY DIFFUSION ONLY

a lumped representation for the drift flow of both types of carriers and provide a lumped representation of Gauss' law (via appropriately connected interlump electric capacitances). When this is done, a complete lumped model can be constructed for the entire diode. Separate parts of the diode such as the neutral p region, the neutral n region, and the transition region are then distinguished only by the fact that assumptions made in the analysis allow us to neglect some of the elements of the lumped model in these separate regions.

This procedure is satisfying in the sense that we start with a diagrammatic model containing all the physical processes of carrier flow and recombination and then specialize it for individual problems. However, such a development is also very lengthy, and the principal end results can be easily understood without reference to the entire argument. For our purposes, it will suffice simply to describe the properties of the lumped model of the transition region.

The necessary ideas are illustrated in Fig. 8·18. We recall that, by assumption, the voltage applied to the diode appears across the entire transition region. The effects of this bias voltage are

1. To change the width of the space-charge layer to accommodate the applied voltage v_a

2. To establish minority-carrier densities on each side of the transition region:

$$p_E(x = 0) = p_n(e^{qv_a/kT} - 1) \qquad n_E(y = 0) = n_p(e^{qv_a/kT} - 1) \quad (8·52)$$

The first effect may be accounted for as charge supplied to the junction capacitance C_j. The charging current is a majority-carrier current on each side of the junction.

The second effect serves to establish the minority-carrier densities which are required to produce minority-carrier currents in the diode bodies.

We may represent these effects in the lumped model suggested in Fig. 8·18. A box is drawn around the transition region. A capacitance C_j is connected from the hole-current line on the p side to the electron-current line on the n side. The current which flows in this element is

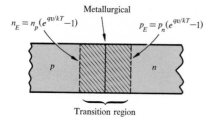

Metallurgical

$n_E = n_p(e^{qv/kT}-1)$ $p_E = p_n(e^{qv/kT}-1)$

p n

Transition region

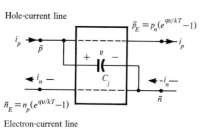

Hole-current line

$\bar{p}_E = p_n(e^{qv/kT}-1)$

i_p $\bar{p}$ i_p

$+$ v $-$

$i_n -$ $-i_n-$

C_j

$\bar{n}_E = n_p(e^{qv/kT}-1)$ $\bar{n}$

Electron-current line

FIG. 8·18   LUMPED REPRESENTATION OF THE TRANSITION REGION AT A *pn* JUNCTION

$$i_{C_j} = \frac{d}{dt}(C_j v) \tag{8·53}$$

where v is the applied voltage. As suggested in Fig. 8·18, this current is a majority current, and *the junction voltage drop appears across C_j*.

The injection effect is represented by simply defining the excess minority-carrier densities at the appropriate nodes by means of the law of the junction.

This model for the junction can then be tied directly to the lumped model for the diode body, as illustrated in Fig. 8·19. To do this we must, of course, set the average excess-hole density in the lump nearest to the space-charge layer equal to the edge density. The width of the first lump should then be small enough that the error between the actual average density and the edge density is not too large.

Figure 8·19 also suggests that interesting contact effects are present, particularly at the *p*-type contact where electrons must be exchanged for holes. The simple model shown does not describe these effects explicitly,

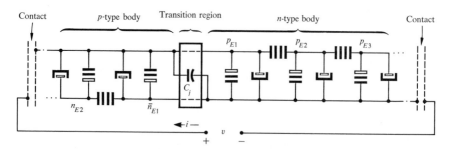

FIG. 8·19   LUMPED MODEL OF A JUNCTION DIODE FOR LOW-LEVEL INJECTION CONDITIONS

though they could be included in a properly developed lumped model. For practical purposes, however, we always assume that current flows freely through the contacts, the necessary carrier conversions being made without appreciable voltage drop. Thus, the lumped model is usually drawn without reference to these contact phenomena (*cf.* Figs. 8 · 20*c* and 8 · 21*b*.)

Finally, it should be remarked that as long as minority-carrier injection is small enough that the majority-carrier density remains essentially undisturbed, the voltage applied to the diode terminals can be thought of as being transmitted directly to the junction capacitance C_j via the majority-carrier lines.[1]

To conclude our discussion of the lumped representation of a *pn*-junction diode, we now wish to show how the lumped model is used in circuit calculations and how it may be used to improve the accuracy of the stored-charge analysis. Before doing this, however, it is useful to point out several features of the physical electronics of junction diodes that are highlighted in the lumped model for the junction diode:

1. All the diode current can be accounted for by recombination and storage in the diode bodies and capacitive current supplied to C_j.

2. The relationships between carrier concentrations and currents are *entirely linear*. The nonlinearity arises only because of the law of the junction.

3. In the *p*- and *n*-type diode bodies at distances which are far removed from the junction, the current flow is entirely a majority carrier process.

8 · 6 USE OF THE LUMPED MODEL TO STUDY A SIMPLE TRANSIENT

To use the lumped model for predicting diode transient behavior, it is first necessary to decide on the number of lumps to be used and to choose their sizes appropriately. This will then specify the magnitudes of the lumped elements.

Of course, the error between the predictions of the lumped model and the results of an exact analysis of the distributed system can be made arbitrarily small by using a large number of very small lumps. However, the labor required to analyze the lumped model increases rapidly as the number of lumps is increased. Thus, we must make a compromise between accuracy of representation and ease of computation. In making this choice we usually favor ease of computation, since, if highly accurate results are desired, the distributed model should normally be used. Usually, how-

---

[1] Such a representation is not strictly accurate, since the variable at the nodes on the majority-carrier current lines is majority-carrier density, not voltage. This problem is resolved in the general lumped model, where each lump is characterized by two carrier densities and a voltage measured with respect to an end terminal. See reference 1, Chap. 4.

ever, device variability is such that a calculation accurate to more than 10 to 20 percent is unnecessary.

With these considerations in mind we shall turn to the problem of specifying a lumped model from which we may calculate the storage-delay time of a pn-junction diode. To do this, we consider the dc excess hole-density distribution shown in Fig. 8·20. It is clear from this sketch that the principal effects of hole injection occur in a distance of $2L_p$ or $3L_p$ from the junction. Thus the lumps can be confined to this region.

8·6·1 One-lump model.

The simplest lumped approximation which we can make to the steady-state distribution is a one-lump approximation. Since the lumped model requires the use of edge density as the density variable in the first lump, it is convenient to use L_p as the length of the lump. Then the steady-state diode current and excess-hole storage will be correctly represented. On this basis

$$H_c = \frac{qAL_p}{\tau_p} \qquad S_p = qAL_p \qquad (8·54)$$

The corresponding lumped model is shown in Fig. 8·20. The hole current which flows into the n-type body in this model is

$$i_p = H_c p_E + S_p \frac{dp_E}{dt} \qquad (8·55)$$

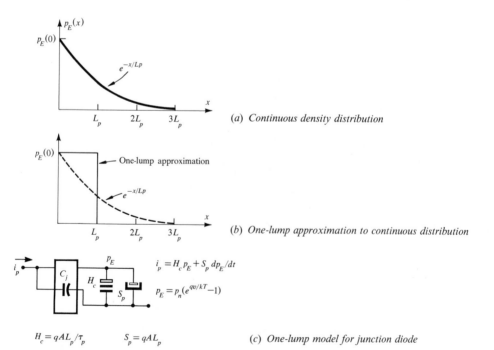

(a) Continuous density distribution

(b) One-lump approximation to continuous distribution

$$i_p = H_c p_E + S_p \, dp_E/dt$$

$$p_E = p_n(e^{qv/kT} - 1)$$

$$H_c = qAL_p/\tau_p \qquad S_p = qAL_p$$

(c) One-lump model for junction diode

FIG. 8·20 DEVELOPMENT OF A ONE-LUMP MODEL FOR THE pn JUNCTION

When the values of H_c and S_p are substituted into this expression, we find

$$i_p = \frac{Q_p}{\tau_p} + \frac{dQ_p}{dt} \qquad (8 \cdot 56)$$

so that the one-lump model is exactly equivalent to the stored-charge model described earlier. Thus it gives the first order correction to the dc behavior. In fact, the dc behavior is determined by H_c only. Adding S_p gives the first-order correction. Adding another lump gives the second-order correction.

Since we have, in essence, already examined the behavior of the one-lump model in the stored-charge analysis, we will not repeat it here. However, a comment on the utility of the lumped model is in order.

In using the lumped model for calculations, we have a schematic representation of the physical device before us instead of a set of equations. This is quite useful in writing equations and for answering many questions about the connection between electrical and physical behavior. As an example, let us consider the high-frequency admittance which the device presents at its terminals. The lumped model shows us immediately that this admittance must consist of C_j in parallel with admittance components due to minority-carrier effects (H_c and S_p).

To determine these minority-carrier admittance components, we simply recognize that the law of the junction provides a transformation between excess-carrier densities and junction voltage drop. As before, if we let

$$p_E = P_E + p_e(t) \qquad v = V_0 + v_e(t) \qquad (8 \cdot 57)$$

then $\qquad P_E = p_n(e^{qV_0/kT} - 1) \qquad p_e(t) = P_E \frac{q v_e(t)}{kT} \qquad (8 \cdot 58)$

The small-signal current flow through H_c is then

$$i_{rec} = H_c p_e(t) = H_c P_E \frac{q}{kT} v_e = g_e v_e \qquad (8 \cdot 59)$$

The small-signal current in S_p is

$$i_S = S_p \frac{dp_e}{dt} = S_p P_E \frac{q}{kT} \frac{dv_e}{dt} = C_D \frac{dv_e}{dt} \qquad (8 \cdot 60)$$

Thus, the voltage-density transformation simply transforms H_c into a conductance and S_p into a capacitance. Since

$$I = H_c P_E \qquad \text{and} \qquad \frac{S_p}{H_c} = \tau_p$$

we again have

$$r_e = \frac{1}{g_e} = \frac{kT}{qI} \qquad C_D r_e = \tau_p$$

In any case, the law of the junction provides the relation between the elements of the lumped model and the elements of the corresponding circuit model.

8·6·2 *A two-lump model.* Since we have already examined the behavior of the one-lump model in the stored-charge analysis, we will proceed to the two-lump model to see how the transient behavior is approximated with it.

It will be recalled that the basic problem with the stored-charge approach for fast transients is that it provides a very inaccurate representation of the density variations near the junction. It is these density variations which react most rapidly to the current change. A two-lump model provides the simplest way of separately accounting for density variations near the junction and density variations at remote points.

The two lumps can, of course, be chosen in an infinite number of ways. Fortunately, as long as we use reasonable judgment in their selection, the results are not strongly dependent on the choice. We therefore choose the first lump to be $L_p/2$ wide, the second one $3L_p/2$ wide. The distance between the centers of the two lumps is then L_p. The corresponding two-lump model is shown in Fig. 8·21. The parameter values are

$$H_{c1} = \frac{qAL_p}{2\tau_p} \qquad S_{p1} = \frac{qAL_p}{2}$$

$$H_{c2} = \frac{3qAL_p}{2\tau_p} \qquad S_{p2} = \frac{3qAL_p}{2} \qquad (8·61)$$

$$H_d = \frac{qAD_p}{L_p} = \frac{qAL_p}{\tau_p}$$

The differential equations which govern this model are obtained by writing Kirchhoff's current law at the two hole nodes:

$$i_p = (H_{c1} + H_d)p_{E1} + S_{p1}\frac{dp_{E1}}{dt} - H_d p_{E2}$$

$$0 = -H_d p_{E1} + (H_d + H_{c2})p_{E2} + S_{p2}\frac{dp_{E2}}{dt} \qquad (8·62)$$

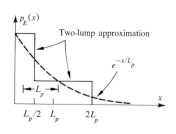

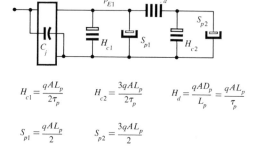

(a) *dc excess density distribution and a possible two-lump approximation to it*

(b) *Two-lump model corresponding to lump selection given in (a)*

FIG. 8·21 DEVELOPMENT OF A TWO-LUMP MODEL FOR THE *pn* JUNCTION DIODE

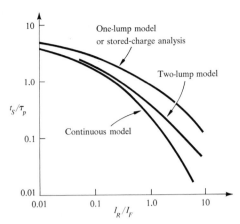

Eqs. (8·62) are two coupled linear differential equations for p_{E_1} and p_{E_2}, which may be solved for $p_{E_1}(t)$ and $p_{E_2}(t)$ when the transient excitation is specified. For the storage-delay problem, we find

$$p_{E_1}(t) = -0.45\frac{I_R}{H} + 0.25\frac{(I_F + I_R)}{H}e^{-t/\tau_p}$$
$$+ 0.2\left(\frac{I_F + I_R}{H}\right)e^{-3.7t/\tau_p} \qquad (8·63)$$

during the storage-delay phase. The storage-delay time is the time at which $p_{E_1}(t) = 0$, which is given by

$$0 = -0.45I_R + 0.25(I_F + I_R)e^{-t_s/\tau_p} + 0.2(I_F + I_R)e^{-3.7t_s/\tau_p}$$

The storage-delay time predicted by this equation is plotted in Fig. 8·22, where it is compared with the results of the "exact" solution. The approximation is considerably better than the one-lump (or stored-charge) analysis gives, though it is still in error for large values of I_R/I_F.

8·7 SUMMARY

In a steady-state situation, current flow in a *pn*-junction diode is associated with the presence of stored minority-carrier charge in the diode body. If the current flowing in the diode is changing with time, there will be changes in the stored minority-carrier charge with time. If the current changes are slow enough, the minority-carrier charge can keep in step with them. The diode current and voltage will then be related by

$$i(t) = I_0(e^{qv(t)/kT} - 1)$$

This law will apply when $dp_E(0,t)/dt \lesssim 0.1\ p_E(0,t)/\tau$, where τ is the minority-carrier lifetime in the diode body.

For rates of change of $dp_E(0,t)/dt$ greater than this, the changes in the minority-carrier charge distribution which occur must be determined as a function of time. The law of the junction is used to relate $p_E(0,t)$ to $v_E(t)$, so that again minority-carrier dynamics provide the link between diode current and voltage.

The theoretical behavior of the minority-carrier density in the diode is obtained by solving a partial differential equation expressing minority-carrier continuity. This equation has been solved for some specific cases of interest, though the general solution is not particularly helpful, and the physical picture tends to be obscured.

As a result, two approximate techniques for studying minority-carrier dynamics in transient situations have been devised. One of these, called the stored-charge formulation or stored-charge approximation, is a quasi-equilibrium solution, in which the minority-carrier distribution is assumed to always maintain its exponential form in distance, but the excess minority-carrier density at the edge of the transition region is allowed to vary with time. This is a useful technique for estimating transient behavior in many cases. Its greatest advantage is simplicity.

Less approximate techniques are necessary if important details in the transient are due to the delayed effects of minority carriers stored in the diode body when a switching transient is initiated. Lumped models provide a simple method for dealing with this problem. If lump sizes are chosen with care, the lumped model can provide a surprisingly good approximation to the behavior of an actual diode. The accuracy of approximation can always be improved by choosing more lumps, though the algebra becomes complicated.

REFERENCES

Linvill, J. G., and J. F. Gibbons: "Transistors and Active Circuits," McGraw-Hill Book Company, New York, 1961, chap. 4.
Linvill, J. G.: "Models of Transistors and Diodes," McGraw-Hill Book Company, New York, 1963.
Gray, P. E., D. DeWitt, A. R. Boothroyd, and J. F. Gibbons: "Physical Electronics and Circuit Models of Transistors," SEEC Notes, vol. 2, John Wiley & Sons, Inc., New York, 1964, chap. 6.
Kingston, R. H.: Switching Times in Junction Diodes and Junction Transistors, *Proc. IRE,* vol. 42, pp. 829–834, 1954.

DEMONSTRATIONS

8·1 DIODE RECOVERY TIME. The recovery time of a diode is interesting and instructive to observe. The circuit shown below can be used to demonstrate the principles of diode recovery. To simplify the response requirements for the scope, a rectifier diode (such as 1N91) can be used. R_f and R_r can be used to adjust the for-

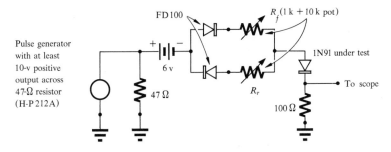

FIG. D8·1

ward and reverse currents of the diode under test (over a limited range). The diode current is measured across a 100-ohm resistor and can be viewed directly on an oscilloscope.

8·2 SMALL SIGNAL IMPEDANCE OF A DIODE. The circuit shown below can be used to measure the dynamic impedance of a low-frequency diode as a function of frequency. The ac voltage at A is proportional to ac current flowing in the diode, and the voltage at B is the diode voltage. The ratio v_B/v_A shows the phase angle of the diode impedance. The generator should supply about 10 mv across the transformer secondary (to keep ac current in the test circuit small). The frequency should be variable up to about 500 kcps. (Many signal generators already have a transformer output, and the circuit could then be simplified by eliminating the transformer and R_s). In addition to impedance measurements, this demonstration provides an estimate of the lifetime in the diode. This result can be correlated with the recovery-time measurement made in demonstration 8 · 1.

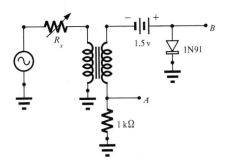

FIG. D8·2

PROBLEMS

8·1 The RC circuit shown in Fig. P8·1 is driven from a sine-wave voltage source with a peak amplitude of 1 volt. $\omega_s = 1$ rad/sec. Sketch and dimension the

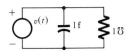

FIG. P8·1

v-i operating locus for this circuit, assuming steady-state conditions have been established. Show the operating point for a few selected times in one complete cycle of the input voltage. What happens to the plot when $G \to 0$, C remaining fixed? $C \to 0$, $G = 1$?

8·2 Sketch the switching trajectory during the storage phase of a reverse-recovery transient in a *pn*-junction diode, using the stored-charge analysis to estimate $p_E(0,t)$ and hence $v_j(t)$. Use $I_R = I_F = 1$ ma, $\tau_p = 0.1$ μsec. Indicate values of the time parameter at several points on the trajectory. The diode may be assumed to have a saturation current of 10^{-8} amp. What is the storage-delay time for this example?

8·3 The diode circuit shown in Fig. P8·3 is driven by current pulses of (*a*) small and (*b*) large size. In working this problem be sure to point out the differences in the physical phenomena which occur in the two cases. Also explain and justify any approximations you may make.

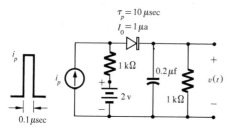

FIG. P8·3

1. Sketch and dimension the response $v(t)$ for a current pulse of height 20 ma.
2. Repeat for a current pulse of height 2 amp.

8·4 The current waveform shown in Fig. P8·4 is applied to a *pn*-junction diode in which $N_a \gg N_d$. The lifetime of holes in *n* material is τ_p and the diode saturation current is I_0.
1. Calculate $p_E(0,t)$ for $t > \tau_p$.
2. What is dp_E/dt for $t = \tau_p$?

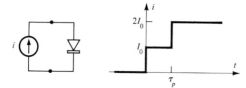

FIG. P8·4

8·5 Chaps. 1–6 and 8 are intended to provide a systematic development of semiconductor physical electronics and its applications to a *pn* diode. Summarize each chapter in a short well-worded paragraph of not more than 100 words. Construct a chart showing the relation of these chapters to each other.

8·6 Calculate the storage-delay time predicted by a one-lump model of a *pn*-junction diode without reference to the stored-charge analysis (that is, use only the lumped model). As boundary conditions use p_E at $t = 0$, dp_E/dt at $t = 0$. Notice that the storance element will not allow p_E to change instantaneously.

8·7 p^+n diode has a reverse saturation current of I_0 amperes. Using a one-lump model, evaluate p_E for reverse junction bias and determine H_c as accurately as you can. If the hole lifetime in n-type material is 10^{-6} sec, determine S_p as accurately as you can. Will the lack of precise determination of these quantities affect the calculation of junction voltage in terms of junction current? Why?

8·8 A p^+n-junction diode is exposed to light with the result that i_v ma of current flows when the diode terminals are shorted. This can be represented in a one-lump model in the manner shown in Fig. P8·8.

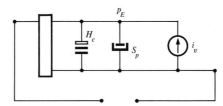

FIG. P8·8

1. Derive the dc v-i relation for the photodiode.
2. Using $i_v = 1$ ma, $I_0 = 1$ μa, $\tau_p = 0.1$ μsec, calculate the junction voltage $v(t)$, assuming the light is switched ON at $t = 0$. Plot the results in the form of a switching trajectory. You may neglect the current required to charge C_j.
3. How much ac voltage develops across the terminals of this diode when a sine-wave current source of peak amplitude 1 μa is attached to the external terminals? The frequency of the sine wave is 2 mcps. The light has been ON for some time (again neglect C_j).

8·9 The parameters of the two-lump model used in Sec. 8·6·2 were chosen somewhat arbitrarily. Evaluate the dc current-voltage relation for this two-lump model and discuss the error which is incurred by its use in the calculation of switching transients.

8·10 Diodes which are intended for computer switching applications are frequently made in the form shown in Fig. P8·10. The width of the n-type region W is very small compared to a diffusion length L_p, and the n-type contact is treated to make the hole lifetime there essentially zero. As a result, it is impossible to maintain an excess density of holes at the contact. Under these circumstances the dc excess hole-density distribution in the n-type body can be approximated by a straight line from p_E at $x = 0$ to zero at $x = W$ (see Fig. P8·10).

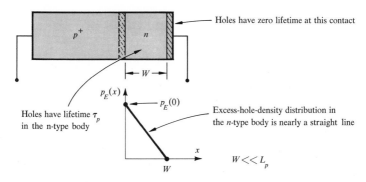

FIG. P8·10

1. Construct a two-lump model for such a diode. Let the lumps be of equal width (each $W/2$) and assume the excess density in the lump nearest the n-type contact to be always zero.

2. Evaluate the parameters on the assumption that the hole lifetime in the n-type body is 10 μs and the width W of the n-type body is 10^{-3} cm. The semiconductor material is silicon.

3. Derive the dc v-i relation for this diode. What fraction of the injected holes recombine in the n-type body and what fraction at the contact to the n-type material? Use the values of τ_p and W given in (1).

4. Convert the current-hole-density equation for the lumped model into a current-vs.-stored-charge equation and interpret the result.

5. Construct a small-signal electrical model for this diode, assuming a fixed-dc bias current. Is the diffusion capacitance greater or less than it would be for a diode made of the same materials but with $W \gg L_p$?

6. Obtain a general formula for the storage-delay time of this diode model. Leave I_F and I_R unspecified, but assume the hole lifetime is so large that recombination may be neglected during the switching transient. Plot your results in normalized form (that is, $t_s D_p/W^2$ vs. I_R/I_F).

7. Read A. S. Grove and C. T. Sah, Simple analytical approximations to the switching times in narrow base diodes, *Solid State Electronics,* vol. 7, no. 1, pp. 107–110, January, 1964. Compare their results with those obtained in part (6) of the previous problem.

9 OPERATING PRINCIPLES FOR JUNCTION TRANSISTORS

IN THE INTRODUCTION to this book, it was stated that the subject of
electronic devices is organizationally a subcategory of electronic circuits
and devices; and that electronic circuits and devices is, in turn, only
a subcategory of the larger subject which we call electronics. However, the
organizational chart is somewhat misleading, because the importance of
the whole subject rests on the fact that we can make devices, such as transistors and vacuum tubes, which can amplify signals. The design of amplifiers and amplifying devices occupies a central position in the entire field

of electronics, and in fact the history of the subject can, in a large measure, be constructed from patents filed on the invention, development, and application of new and better amplifying devices.

From this point onward, we shall be almost entirely occupied with the theory and application of amplifying devices, principally transistors. We shall find that this study provides a common focal point for many subject areas of electronics. For example, the design of an electronic amplifier in which transistors are used as the active elements frequently involves an extensive interplay between circuit theory and physical electronics. The amplifier specifications themselves may arise from the relation between the information-bearing capacity of a signal and its frequency spectrum, the problems associated with long-distance communication, or the nature of the components which are used in data processing or automatic control systems.

A really thorough study of amplifying devices and their applications, therefore, requires both practical and theoretical experience in all the other areas of electronics. While we can only touch on many of these areas in our treatment of amplifiers and amplifying devices, we shall attempt to suggest the interplay of these disciplines as the subject unfolds.

In this chapter we shall develop the basic operating principles of junction transistors. In the next two chapters we will see how simple amplifiers can be designed from graphical characteristics of active devices, and then we will show how the circuit models which are used to represent active elements in electronic circuits are related to the physical theory. In Chaps. 12–15 we will study the basic ideas involved in using active devices in amplifier and switching applications. After a brief consideration of communication systems in Chap. 16, we will then proceed to a study of the high-frequency effects and limitations in transistors and some applications where these limitations are important.

9·1 QUALITATIVE DESCRIPTION OF THE DIFFUSION MODEL OF A TRANSISTOR

The principal purpose of this chapter is to develop the operating principles of *pn*-junction transistors. To do this, we shall first give a qualitative description of how the transistor works. This will be followed by a discussion of some of the more common methods of transistor fabrication, the attempt here being to provide the basis for approximations which we shall make later in analyzing real transistor structures.

A *pnp* transistor (Fig. 9·1) is essentially two *pn*-junction diodes *coupled together* through a common, very narrow *base region*.[1] Usually, one of the *pn* junctions is called the "emitter junction" and one the "collector junction."

---

[1] The operating principles of *npn* transistors are entirely analogous to those for the *pnp* type, with the roles of holes and electrons simply interchanged everywhere.

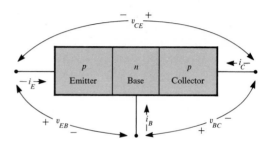

However, one should not attach too much significance to these names, because in many applications the "collector junction" may be acting as the emitter and the "emitter junction" as the collector.

Because the transistor is a three-lead device, three currents and three voltages may be defined at its terminals. These are defined with subscripts and reference positive directions in Fig. 9·1. Of course, all the voltages and currents defined on the figure are not independent. In particular, Kirchhoff's current law requires

$$i_E + i_B + i_C = 0$$

so that only two of the terminal currents can be specified independently. Similarly, the loop voltages must add to zero

$$v_{EB} + v_{BC} + v_{CE} = 0$$

so that only two of the terminal-pair voltages are independent. Thus, only two terminal-pair voltages and two currents are required to characterize the terminal behavior of the transistor.

For operation of the transistor as an amplifier, the emitter junction is forward-biased and the collector junction is reverse-biased. For the *pnp* transistor this means v_{EB} and i_E will be positive, while v_{CB} and i_C will be negative. Under these conditions,

1. *The collector current will be nearly equal to the emitter current in magnitude.*

2. *Both currents will be controlled by the emitter-base voltage.*

3. *Both currents will be nearly independent of the collector-base voltage.*

The emitter-base voltage therefore controls the flow of power in the collector circuit. This control action is the basis of amplification in the transistor.

The mechanism of control can be understood by focusing attention on the distribution and flow of *minority carriers* in the base region. As in the junction diode, this distribution depends on the geometry of the transistor, the flow and recombination processes for minority carriers in the base, and the boundary conditions on minority-carrier densities imposed by the junctions. To explain the significance of these factors, we consider the series of idealized models of a *pnp* transistor shown in Fig. 9·2. Minority-

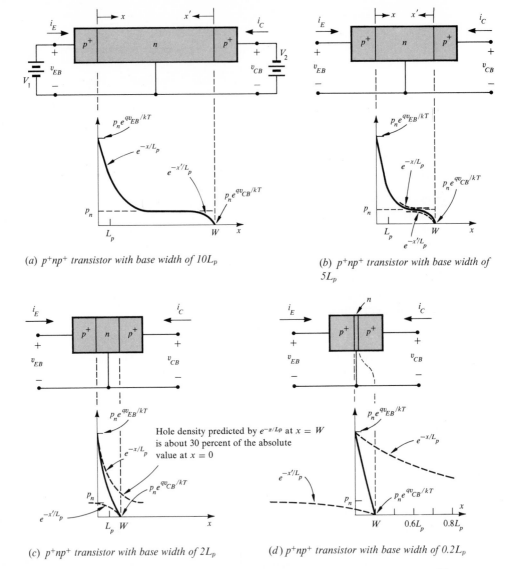

(a) p^+np^+ transistor with base width of $10L_p$

(b) p^+np^+ transistor with base width of $5L_p$

Hole density predicted by e^{-x/L_p} at $x = W$ is about 30 percent of the absolute value at $x = 0$

(c) p^+np^+ transistor with base width of $2L_p$

(d) p^+np^+ transistor with base width of $0.2L_p$

FIG. $9 \cdot 2$ THE DEPENDENCE OF THE MINORITY-CARRIER DISTRIBUTION ON THE BASE REGION OF A TRANSISTOR ON W_o AND THE JUNCTION BOUNDARY CONDITION

carrier current flow is assumed to be entirely one-dimensional in this model and minority carriers are assumed to flow by diffusion only. The collector and emitter regions are assumed to be heavily doped in comparison to the base, so that the minority-carrier processes of interest are confined to the base region.

The interplay of base width and junction conditions in determining the minority-carrier distribution in the base is exhibited in Figs. $9 \cdot 2a$–$9 \cdot 2d$. In Fig. $9 \cdot 2a$, the transistor is shown with a base width $W = 10L_p$, where

L_p is the diffusion length for holes in the base region. The emitter-base junction is forward-biased by the battery V_1, while the collector-base junction is reverse-biased by the battery V_2. Because the base is so wide, the hole-density profiles near the emitter and collector are essentially the same as they would be for normal p^+n diodes. In particular, near the emitter the excess-hole-density profile is of the form e^{-x/L_p}. The *absolute* hole density at $x = 0$, denoted by p'_E, is

$$p'_E = p_n e^{qv_{EB}/kT} = p_n e^{qV_1/kT} \qquad (9\cdot1)$$

as required by the law of the junction.

Near the collector the excess-hole-density profile is again of the form e^{-x'/L_p}, where x' is measured into the base from the n-type edge of the collector-base transition region (see Fig. $9\cdot2a$). The law of the junction gives an *absolute* hole density at $x = W$ of

$$p'_C = p_n e^{qv_{CB}/kT} = p_n e^{-qV_2/kT} \ll p_n \qquad (9\cdot2)$$

since we assume $V_2 \gg kT/q$.

The currents flowing across the emitter and collector junctions are found by evaluating the slope of the hole-density profile at $x = 0$ and $x = W$. The emitter current is simply

$$i_E = -qAD_p \frac{\partial p}{\partial x}\bigg|_{x=0} = \frac{qAD_p p_n}{L_p}(e^{qV_1/kT} - 1) \qquad (9\cdot3)$$

and the collector current is

$$i_C = +qAD_p \frac{\partial p}{\partial x}\bigg|_{x=W}$$

$$= \frac{+qAD_p p_n}{L_p}(e^{-qV_2/kT} - 1)$$

$$\cong \frac{-qAD_p p_n}{L_p} \qquad (9\cdot4)$$

The positive sign in the first form of Eq. $(9\cdot4)$ is required because of the reference direction used to define positive i_C.

Equations $(9\cdot3)$ and $(9\cdot4)$ are the same as the equations for the forward and reverse currents of a p^+n-junction diode. The collector and emitter currents are essentially independent of each other; or equivalently, there is negligible coupling between the emitter and collector diodes for such a wide base region.

Now consider Fig. $9\cdot2b$. Here, the transistor has a base width of $5L_p$. The bias voltages V_1 and V_2 are unchanged, so the law of the junction requires that the hole densities at the edges of the base region be the same as before. Now, however, the excess density profile is not exactly exponential in character at the edge of the base. In particular, the slope of the hole-density profile at both $x = 0$ and $x = W$ is somewhat greater than before, so the emitter and collector currents are larger.

Operating principles for junction transistors   *319*

Physically, the increase in collector current arises because a small fraction of the holes which are injected into the base from the emitter can now diffuse into the collector-base transition region before they recombine. Inasmuch as the collector-base junction is reverse-biased, these holes encounter a high electric field in the collector-base transition region, so they are rapidly swept into the collector body, thus contributing to the collector current. The "vacuum cleaner" action of the electric field within the reverse-biased transition region ensures that the hole density at the edge of the transition region will be maintained at a value much below p_n during the collection process.[1]

The increase of collector and emitter currents with decreasing base width is shown more clearly in Fig. 9·2c, where the transistor has a base width of only $2L_p$. Here, the deviation of the actual hole-density profile from the two exponential distributions is quite apparent, particularly in the vicinity of $x = W$. An appreciable number of the holes which are injected at the emitter can now diffuse into the collector-base junction, so that the collector current is now determined primarily by emitter injection.

The final case, shown in Fig. 9·2d, approximately represents the state of affairs which exists in a real transistor. The base width is $0.2L_p$. The hole-density profile is nearly a straight line between the values

$$p'_E = p_n e^{qv_{EB}/kT} \quad \text{and} \quad p'_C = p_n e^{qv_{CB}/kT} \simeq 0$$

Since the hole-density profile is nearly straight, it has almost the same slope at $x = 0$ and $x = W$. This means that the collector current and the emitter current are nearly equal. Furthermore, since the hole density at $x = W$ is not sensitive to the value of V_2, as long as V_2 is large compared to kT/q, both currents are essentially independent of V_2. Therefore, for an expenditure of $V_1 i_E$ watts in the emitter-base circuit, $V_2 i_C \simeq V_2 i_E$ watts can be taken from the collector-base battery. Since $|V_2| \gg |V_1|$, this represents a large potential power gain.

9·1·1 *The role of recombination and space-charge neutrality in transistor operation.* The preceding discussion has been centered entirely on the distribution and flow of minority carriers in the transistor base region. As for the *pn*-junction diode, the basic terminal characteristics of the transistor can be derived entirely from an analysis of minority-carrier flow in neutral regions of the transistor. However, the fundamental role of recombination and its relationship to the base current of a transistor is not adequately emphasized in such a discussion. It is useful to describe this relationship before proceeding to a more exact characterization of the transistor.

Let us again consider the idealized transistor structures shown in Fig. 9·2. In all cases, space-charge neutrality will exist at every point of

---

[1] The applicability of the law of the junction in predicting the exact value of p at $x = W$ is discussed later.

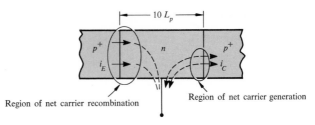

FIG. 9·3   **BASE CURRENT FLOW IN A** p^+np^+ **TRANSISTOR WITH A VERY WIDE BASE**

the base region. The electron density at any plane x will be determined from the excess hole density $p^*(x)$ by the equation[1]

$$n(x) = n_n + p^*(x) \tag{9·5}$$

The electrons required to establish this space-charge neutrality come from the batteries via the base lead.

Furthermore, at every point in the base region where excess electron and hole densities exist, there must be net recombination. Similarly, a deficiency of holes and electrons at any point implies net generation. For example, in the wide-base transistor shown in Fig. 9·2a, electrons flow into the base and then drift to the left to recombine with the excess holes which are injected by the emitter. At the same time, hole-electron pairs are generated in the region near the collector junction where $p(x) < p_n$ (or $p^*(x) < 0$). These holes diffuse into the collector-transition region and form part of the collector current. The generated electrons flow *out* of the base lead to preserve space-charge neutrality. Thus, the base current for the transistor shown in Fig. 9·2a has two oppositely directed components (see Fig. 9·3.)

The *net* base current can be simply found as follows. If $p^*(x)$ is the excess-hole density at the plane x, then the *total excess-hole density* in the base is

$$q_P = qA \int_0^W p^*(x)\, dx \tag{9·6}$$

Geometrically, q_P is the area under the *excess*-hole-density curve in the base region. Since holes have a lifetime τ_p in the base layer, all of these holes will recombine every τ_p sec. They must of course be replenished by injection from the emitter to maintain a steady-state condition.

The electrons required for this recombination enter the base through the base lead. Therefore, the net base current is simply

$$i_B = -\frac{q_P}{\tau_p} \tag{9·7}$$

This is a fundamental relation, from which we may determine the base current for dc excitation of the transistor. We shall use it repeatedly in interpreting and developing the electrical characteristics of transistors.

---

[1] $p^*(x)$ is used to represent the excess-hole density in the base because p_E is reserved to denote the excess density at the emitter.

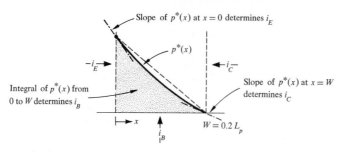

FIG. 9·4 THE MINORITY-CARRIER DISTRIBUTION IN A NARROW-BASE TRANSISTOR. THE
EFFECT OF BASE CURRENT IS TO PRODUCE A SLIGHT CURVATURE OF THE DENSITY DISTRI-
BUTION, SO $|i_E| > |i_C|$.

Of course, Eqs. (9·6) and (9·7) apply for any of the idealized transistors
shown in Fig. 9·2. In the most realistic case (Fig. 9·2d), the existence of
a base current will result in a slight curvature of the hole-density profile,
so that the emitter and collector currents will not be exactly equal. A highly
exaggerated sketch of this effect is shown in Fig. 9·4.

9·1·2 *Estimates of the terminal currents for a reverse-biased collector junction.*
We can use the fact that the hole-density profile is essentially a straight line
in the base region to make some instructive estimates of the terminal
currents in the transistor shown in Fig. 9·2d. The estimates will be based
on the assumption that sufficient reverse bias is applied to the collector-
base junction to make p_C' (the *absolute* hole density at the collector edge of
the base region) essentially equal to zero. This assumption will cause our
results to be slightly in error, but the basic ideas will be correct (and the
results adequate for many practical situations).

Base current. To estimate the base current we use the formulas developed
in Sec. 9·1·1 for the specific case where the hole-density profile in the
base region is a straight line.

From Fig. 9·2d, we find that the total minority-carrier charge stored in
the base region is $qAWp_E'/2$, which is just the area under the hole-density
triangle.[1] The *excess* charge of holes is therefore

$$q_P = \frac{qAWp_E'}{2} - qAWp_n \qquad (9·8)$$

The base current is therefore

$$i_B = -\frac{q_P}{\tau_p} = -\frac{qAWp_E'}{2\tau_p} + \frac{qAWp_n}{\tau_p} \qquad (9·9a)$$

For a typical forward-bias condition on the emitter-base junction, however, $p_E' >>> p_n$, so we shall estimate i_B to be

[1] The fact that the hole-density profile is not exactly straight makes little difference in
estimating q_P, because we only want the total *area* under the curve.

$$i_B \simeq -\frac{qAWp_E'}{2\tau_p} \tag{9 \cdot 9b}$$

Collector and emitter currents. To find the collector and emitter currents we need to know the slope of the hole-density profile dp/dx at both $x = 0$ and $x = W$. If we stick by our assumption that the hole-density profile is a straight line, we will land in the predicament that $i_C = i_E$ and $i_B = 0$. To avoid this, we shall assume that i_C can be calculated from the average slope of the profile, and then we can use Kirchhoff's law to obtain a consistent estimate of i_E[1]:

$$i_E = -(i_B + i_C).$$

Thus

$$i_C = qAD_p \frac{dp}{dx}\bigg|_{x=W} \simeq -qAD_p \frac{p_E'}{W} \tag{9 \cdot 10a}$$

and

$$i_E = \frac{qAD_p p_E'}{W} + \frac{qAWp_E'}{2\tau_p} \tag{9 \cdot 10b}$$

To summarize, the preceding equations and discussion lead to several important conclusions about the behavior of transistors:

1. When the collector-base junction is reverse-biased by more than a few tenths of a volt, p_C/p_n is very small. Then all the terminal currents are essentially determined by p_E and are essentially independent of v_{CB}.

2. The emitter-base voltage v_{EB} determines p_E and therefore the terminal currents.

3. The collector current under the conditions described in Fig. $9 \cdot 2d$ is *much* larger than the reverse current of an isolated pn junction (by a factor of about $(L_p e^{qv_{EB}/kT}/W)$. The emitter current under the same conditions is only a factor of $L_p/W(\sim 5)$ larger than it is for an isolated pn junction with the same forward bias.

4. The *ratios* of the terminal currents are independent of the value of p_E. One ratio which is of particular significance in studying both the physical theory and the circuit properties of transistors is i_C/i_E. This ratio, called the transistor *alpha* (α), will be temporarily defined as

$$\frac{i_C}{i_E}\bigg|_{\substack{collector-junction \\ reverse-biased}} = \alpha = \frac{1}{1 + (W^2/2L_p^2)}$$

When we perform a more correct analysis we will find that i_C can be written as

$$-i_C = \alpha i_E + I_{co}$$

---

[1] There is no *a priori* reason for using the average slope of the profile to estimate i_C rather than i_E (or for using any of a number of other estimates in which part of i_B is associated with i_E and part with i_C). All of these possibilities give nearly the same answers, for the physical reason that i_B is small. The procedure used here has the additional advantage that it is the simplest one which estimates the ratios i_C/i_E and i_C/i_B most correctly.

where I_{CO} is the saturation current of the collector-base diode. We can interpret this equation by saying that the collector collects as part of its reverse current a fraction α of the current injected at the emitter.

5. The power supplied to the transistor in order to produce the emitter and collector currents comes from the emitter battery. It is

$$P_1 = V_1 i_E$$

This power is dissipated in the emitter-base transition region. However, V_1 is small (~ 1 volt, or less), so there is no appreciable heating effect on this account.

The power which flows from the collector battery into the collector circuit is

$$P_2 = V_2 i_C$$

But, since the transistor makes $i_C \simeq i_E$, P_2 can be written as

$$P_2 \cong V_2 i_E$$

Thus the power which flows from the collector battery is controlled by the emitter current. From a circuit standpoint, this power-controlling action is the feature of greatest importance in the transistor.

In the simple circuit shown in Fig. $9 \cdot 2$, all the power leaving the collector battery is dissipated in the collector-base transition region, and only heats up the transistor. However, an appropriate load can be added in series with the collector battery and then the power flowing to the load will be controlled by the current flowing in the low-power emitter circuit.

$9 \cdot 1 \cdot 3$ **Electrical characteristics of an idealized transistor: common-base connection.** The electrical characteristics of a device having a single pair of terminals can be visualized as a single curve in the $v\text{-}i$ plane. A corresponding description of the electrical characteristics at one pair of terminals of a three-terminal device requires either a three-dimensional representation or a *family* of curves in the $v\text{-}i$ plane. For example, the collector current in a transistor is known when the values of v_{EB} and v_{CB} are specified. This means that i_C can be plotted as a surface over the $v_{EB} - v_{CB}$ plane. Alternately, i_C can be visualized as a surface over the $i_E - v_{CB}$ plane, since there is a unique relation between i_E and v_{EB} for a given v_{CB}. These possibilities are illustrated in the sketches of Fig. $9 \cdot 5$ for the idealized transistor just discussed. These figures simply reflect the facts that (1) i_C is supposed to be independent of v_{CB} for sufficient reverse bias on the collector-base junction, and (2) i_C is supposed to be proportional to either i_E or $e^{qv_{EB}/kT}$.

In order to represent either surface in a way which is convenient for calculations, we plot the i_C-v_{CB} relation for selected values of the other variable. The planar character of the i_C vs. (v_{CB}, i_E) surface leads to the simplest and most commonly used representation shown in Fig. $9 \cdot 5d$. Each curve in this figure is called a *collector characteristic*. The family of collector characteristics is a "sideways" contour map of the surface $i_C = f(v_{CB}, i_E)$.

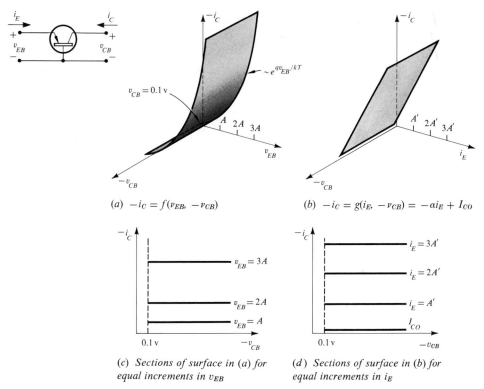

(a) $-i_C = f(v_{EB}, -v_{CB})$

(b) $-i_C = g(i_E, -v_{CB}) = -\alpha i_E + I_{CO}$

(c) Sections of surface in (a) for equal increments in v_{EB}

(d) Sections of surface in (b) for equal increments in i_E

FIG. 9·5 ELECTRICAL CHARACTERISTICS OF AN IDEALIZED *pnp* TRANSISTOR: COMMON-BASE CONNECTION

For the idealized transistor described earlier, we expect each collector characteristic to be perfectly horizontal, as long as $|v_{CB}| \geq 0.1$ volt. We also expect the characteristics to be equally spaced on the i_C axis, if equal increments in i_E are chosen.

The experimental characteristics of a real transistor are amazingly close to those shown in Fig. 9·5d—so close in fact that a single, highly idealized transistor model can be used to represent nearly all transistors for most purposes. However, if one looks at real transistor characteristics in sufficient detail, he will find that:

1. For any given value of i_E, the i_C-v_{CB} curves are not perfectly horizontal. They exhibit a very small slope, so i_C is not completely independent of the reverse-bias voltage v_{CB} in a real transistor (see Fig. 9·6). The reasons for this effect are described in Sec. 9·6·1.

2. For very small values of i_E, the output characteristics are more "crowded" than they are at higher values of i_E (for example, a 0.1-μa increment in i_E will produce a smaller increment in i_C, in the vicinity of $i_C = 0$, than it will in the vicinity of $i_C = 1$ ma). The output characteristics will also show some "crowding" at high values of i_E. A highly exaggerated sketch of these effects is shown in Fig. 9·6. The reasons for these effects are described in Secs. 9·6·3 and 9·6·4.

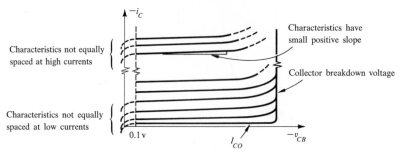

3. At high enough values of v_{CB}, the collector junction exhibits a breakdown effect and i_C becomes very large and independent of i_E. This is an avalanche breakdown similar to the one observed on a simple pn-junction diode (*cf.* Sec. 9·6·2).

Input characteristics. The input characteristics of a transistor may also be visualized as a surface in a three-dimensional space. i_E is the independent variable for this surface; v_{EB} and v_{CB} are the dependent variables.

Ideally, the i_E-v_{EB} relation is independent of v_{CB}. From Eq. (9·10b) we expect the input characteristic to be simply an ideal pn-junction diode characteristic, as shown in Fig. 9·5a. Again, this expectation is generally verified for real transistors, though a detailed examination of the input characteristics of a real transistor reveals that:

1. The emitter-base diode exhibits the same departures from the ideal which a normal pn-junction diode does (it has series resistance for example).

2. The i_E-v_{EB} relation is not completely independent of v_{CB}. Instead, the input characteristics will look somewhat like those shown in Fig. 9·7b for a real transistor (*cf.* Sec. 9·6·1).

However, the outstanding feature of the input characteristics of a real transistor is that they are strikingly similar to the characteristics of a suitably defined ideal transistor. The similarity is maintained even as one

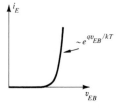

(a) *Ideal*

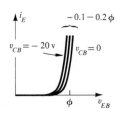

(b) *Typical input characteristics show some influence of v_{CB} on the $i_E - v_{EB}$ relation*

ϕ is approximately the built-in voltage or contact potential at the emitter-base junction (*cf.* Sec. 5·3)

FIG. 9·7 COMMON-BASE INPUT CHARACTERISTICS OF A TRANSISTOR

measures the characteristics of transistors with widely different internal construction. Therefore, when we set up and analyze an ideal transistor model in Sec. 9·3, we do so with the expectation that the results will be of *quantitative* significance for many types of transistors. In fact, the results are sufficiently valid that manufacturers do not supply characteristic curves for many transistor types.

9·2 METHODS OF FABRICATING JUNCTION TRANSISTORS

Before proceeding to a more thorough study of the electrical behavior of transistors, it is helpful to give a brief discussion of practical fabrication techniques. This will enable us to justify several approximations to be made in later analyses.

It is apparent from the preceding discussions that a transistor should be made with a thin-base layer so that transport of minority carriers across it can be accomplished efficiently. The most successful and widely used methods of fabricating transistors are therefore those which produce thin-base layers, while still allowing the emitter and collector bodies to be conveniently formed and to have leads attached to them. The two basic methods of producing junctions, discussed in Chap. 5 (alloying and diffusion), lead to several interesting methods for making transistors.

9·2·1 Alloy transistors.

A very practical method of making general-purpose low-frequency transistors is to alloy the emitter and collector dopants into a semiconductor wafer of opposite conductivity type. The general structure for an alloy transistor is shown schematically in Fig. 9·8. As in the case of a diode, small pellets of dopant are placed on each side of the semicon-

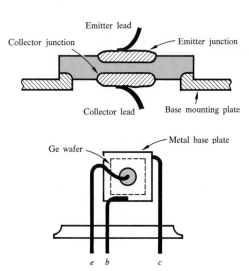

FIG. 9·8 A TYPICAL ALLOY TRANSISTOR

ductor slab and alloyed into the slab to make the emitter and collector junctions. If the crystallographic axes are oriented properly with the faces of the slab, it is possible to obtain an alloy junction which has an almost flat bottom. When the emitter and collector regions have been alloyed, lead wires may be soldered onto the two metal buttons produced by the alloying process to provide an ohmic contact to the emitter and collector bodies. A generally similar process is used to attach an ohmic contact to the base layer, thus completing the transistor structure. In current practice, the parent wafer is in the form of a thin square which is alloyed to a base plate. The base plate has a hole to accommodate the alloyed collector button.

In a typical germanium-alloy transistor, the semiconductor wafer will be about 0.1 in. square and 0.005 in. thick. The alloyed buttons will have diameters of 0.015 to 0.03 in. and depths of about 0.002 in. The final base layer is then about 0.001 in. wide.

9·2·2   Surface-barrier transistors.   A second type of alloy transistor, which is very useful for low-level, high-frequency applications, is a surface-barrier transistor. In this case, one mounts a semiconductor wafer of about 0.003 in. thickness between two coaxial jets of etching solution (Fig. 9·9). The etching reduces the thickness of the central region of the wafer. When the thickness of the wafer at its center has been reduced to about 0.002 in., metal contacts are electroplated into the etched regions, using the same jets as electrolytic vehicles. The wafer is then mounted on a structure making base contact to the wafer with emitter and collector leads contacting the electroplated buttons.

9·2·3   Diffused-base transistors.   The solid-state diffusion process for making *pn* junctions discussed in Chap. 5 is also of considerable importance in making transistor structures. One type of transistor, in which both emitter and collector junctions are made by diffusion, is shown in Fig. 9·10a. To make this transistor, one starts with an *n*-type slab of semiconductor and diffuses a *p*-type impurity into the top surface. Following this, an *n*-type impurity is diffused into the same face with higher surface concentration, but to less depth. Under proper control, the successive diffu-

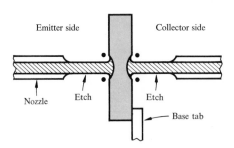

FIG. 9·9   THE SURFACE-BARRIER TRANSISTOR

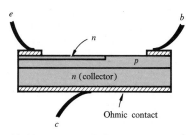

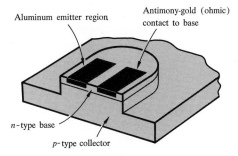

(a) *Transistor in which emitter and base are made by diffusion (double diffused)*

(b) *Transistor with diffused base and alloyed emitter*

FIG. 9·10   THE CONSTRUCTION OF TRANSISTORS USING IMPURITY DIFFUSION TECHNIQUES

sions can be made to yield *pn* junctions only a small fraction of a mil apart, parallel to each other, and independent of the parallelism of the faces of the parent wafer.

It is also possible to combine diffusion and alloying techniques in making transistors such as the one shown in Fig. 9·10b. In this example, a *p*-type slab of semiconductor is used as the collector. An *n*-type doping agent such as phosphorus is diffused into the collector until an *n*-type base layer on the order of 10^{-4} cm thick is produced. Rectangular metal deposits are then laid down on this *n*-type base through suitable masking apertures. One of these deposits is of aluminum, for example, and the other is a gold-antimony alloy. These metals are then alloyed into the *n*-type base, the aluminum producing a *p*-type emitter while the antimony-gold produces an ohmic contact. In this way, a *p*-type emitter region is produced under the aluminum deposit and an ohmic contact to the *n*-type base layer is provided by the gold-antimony. Contacts are now made to the bottom of the original slab and to both metal deposits. A *pnp* structure is thus obtained in which the effective base layer thickness between emitter and collector can be less than 0.5×10^{-4} cm.

To complete the structure, the area around the deposited contact is masked with a pinpoint of wax, and the surface elsewhere is etched away below the collector-junction level. This leaves emitter and base electrodes sitting on a mesa above the collector body.

9·3   DEVELOPMENT AND ANALYSIS
OF THE INTRINSIC TRANSISTOR MODEL

For most purposes in transistor electronics, an adequate description of the electrical behavior of any of the transistor structures just described can be obtained from a rather simple, highly idealized transistor model. In this model, which we call the *intrinsic transistor,* we focus attention directly on the basic features of transistor operation, avoiding extra details which arise from particular fabrication techniques or unusual operating conditions. In the next two sections we shall develop the intrinsic transistor model and explore its operation thoroughly. In Sec. 9·6 we shall comment briefly on how well real transistor structures are approximated by the intrinsic transistor model.

9·3·1 *The intrinsic pnp transistor.* The intrinsic *pnp* transistor is a completely symmetrical structure of uniform cross section A (see Fig. 9·11). The base layer of this structure has width W and a uniform doping level throughout. The base doping density is assumed to be N_d/cm^3; the thermal-equilibrium carrier densities in the base are n_n electrons/cm^3 and p_n holes/cm^3. The emitter and collector bodies are heavily doped in comparison to the base, so minority-carrier injection is confined to the base region. The structure is therefore most properly described as a p^+np^+ transistor.

We assume that holes flow in the base region by diffusion only, and that carrier recombination in the base can be adequately described by the minority-carrier lifetime τ_p. Furthermore, we assume there is negligible recombination or generation of carriers in the emitter-base and collector-base transition regions.

9·3·2 *Electrical characterization of the intrinsic transistor.* The electrical performance of the intrinsic transistor is completely described by a set of two simple equations. Both large- and small-signal performance follow from these equations, so it is important to have a firm grasp of them.

It is useful to describe the nature of the characterizing equations before

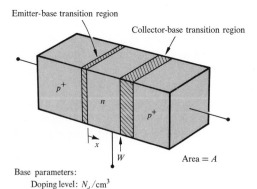

Base parameters:
Doping level: N_d/cm^3
Equilibrium electron density: $n_n(=N_d)$
Equilibrium hole density: $p_n(=n_i^2/N_d)$
Hole lifetime: τ_p

Model properties:
1. Surface currents neglected
2. Emitter and collector bodies heavily doped, so minority carrier flow in them is negligible
3. No recombination or generation in the transition region

FIG. **9·11** THE INTRINSIC TRANSISTOR

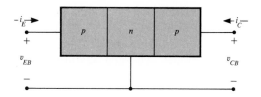

analyzing the intrinsic transistor model. To do this, we first observe that the transistor is a two-port device. That is, it has an input port and an output port, one terminal of the device being common. Two currents and two voltages are required to specify the performance of this two-port box. Two of these variables are selected as *independent variables,* the remaining two being *dependent variables.*

For the connection shown in Fig. 9·12, called the *common-base* connection, we usually express the currents i_E and i_C as functions of the two junction voltages v_{EB} and v_{CB}:

$$i_E = f_1(v_{EB}, v_{CB}) \qquad i_C = f_2(v_{EB}, v_{CB}) \tag{9·11}$$

Our past experience has taught us, however, that junction voltages are only indirectly related to junction currents. The junction voltages control the excess-carrier densities at the edges of the base region, while the excess-hole-density profile in the base region actually determines the junction currents. Therefore, in the analysis we seek to find a set of equations of the form

$$i_E = g_1(p_E, p_C) \qquad i_C = g_2(p_E, p_C) \tag{9·12}$$

where p_E and p_C are the excess-hole densities at $x = 0$ and $x = W$. Then the currents can be expressed in terms of the voltages by using the law of the junction:

$$p_E = p_n(e^{qv_{EB}/kT} - 1) \qquad p_C = p_n(e^{qv_{CB}/kT} - 1) \tag{9·13}$$

The most important result of the analysis will be that, *when i_E and i_C are expressed in terms of p_E and p_C, the equations will be linear:*

$$i_E = a_{11}p_E + a_{12}p_C \qquad i_C = a_{21}p_E + a_{22}p_C \tag{9·14}$$

This fundamental result arises because all of the physical processes which relate currents to minority-carrier densities are linear.

The use of Eqs. (9·13) in (9·14) will then result in

$$i_E = A_{11}(e^{qv_{EB}/kT} - 1) + A_{12}(e^{qv_{CB}/kT} - 1)$$

$$\tag{9·15}$$

$$i_C = A_{21}(e^{qv_{EB}/kT} - 1) + A_{22}(e^{qv_{CB}/kT} - 1)$$

where $A_{ij} = p_n a_{ij}$. These equations, plus Kirchhoff's laws

$$i_E + i_B + i_C = 0 \qquad v_{EB} + v_{BC} + v_{CE} = 0 \tag{9·16}$$

are enough to completely describe the dc performance of the intrinsic transistor.

Equations $(9 \cdot 15)$ are clearly reminiscent of the *v-i* relations for two diodes. The new feature is the presence of *coupling* between the diodes, manifested by nonzero values for A_{12} and A_{21}.

9·3·3 dc minority-carrier distribution in the base region of the intrinsic transistor. To establish the relation between junction currents and minority-carrier densities at the edges of the base region, we must first solve the dc continuity equation to obtain the excess-minority-carrier-density profile in the base region. The physical situation pertinent to the $p^{+}np^{+}$ transistor is shown in Fig. 9·13.

The continuity equation. From Fig. 9·13, we find that hole continuity within the small slab of width Δx requires

$$-\frac{1}{q}\frac{dj_p}{dx} = \frac{p - p_n}{\tau_p} \tag{9·17}$$

in the limit $\Delta x \to 0$. The assumption that holes move only by diffusion in the neutral-base region gives

$$j_p = -qD_p\frac{\partial p}{\partial x} \qquad 0 \le x \le W \tag{9·18}$$

so the differential equation which the hole-density profile must satisfy in the base region is

$$D_p\frac{d^2p}{dx^2} = \frac{p - p_n}{\tau_p} \tag{9·19}$$

If we define the *excess* density

$$p^* = p - p_n$$

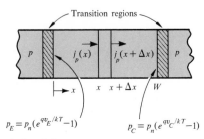

Continuity relation:

Hole flow in = hole flow out + net recombination current

$$\Rightarrow j_p(x) = j_p(x + \Delta x) + [q(p - p_n)/\tau_p]\Delta x$$

$$\Rightarrow -\frac{1}{q}\frac{dj_p}{dx} = \frac{p - p_n}{\tau_p}$$

FIG. 9·13 EXPANDED VIEW OF THE BASE REGION

332 Semiconductor electronics

as the variable of interest, then the general solution of Eq. (9 · 19) is

$$p^*(x) = Ce^{-x/L_p} + De^{x/L_p} \qquad (9 \cdot 20)$$

where $L_p \equiv (D_p \tau_p)^{1/2}$. The constants C and D are evaluated from the boundary conditions.

The boundary conditions. The law of the junction applied at the emitter and collector junctions gives the values which $p^*(x)$ must have at $x = 0$ and $x = W$. When the injection levels are low, so that the injected minority-carrier density remains small compared to the majority-carrier density, we have

$$p^*(x = 0) = p_n(e^{qv_{EB}/kT} - 1) \qquad p^*(x = W) = p_n(e^{qv_{CB}/kT} - 1) \quad (9 \cdot 21)$$

For our purposes, *these boundary conditions will be used for all possible combinations of v_{EB} and v_{CB}.*[1]

Solution of the dc continuity equation. By using Eq. (9 · 21) in Eq. 9 · 20, we can express C and D in terms of $p^*(x = 0)$ and $p^*(x = W)$. To simplify things, we call

$$p^*(x = 0) = p_E \qquad p^*(x = W) = p_C \qquad (9 \cdot 22)$$

Then we find $\quad C = \dfrac{p_E e^{W/L_p} - p_C}{e^{W/L_p} - e^{-W/L_p}} \qquad D = \dfrac{p_C - p_E e^{-W/L_p}}{e^{W/L_p} - e^{-W/L_p}} \qquad (9 \cdot 23)$

The final distribution is shown in Fig. 9 · 14, where positive values of both p_E and p_C are assumed for generality (that is, both v_{EB} and v_{CB} are forward-bias voltages).

Actually, the general values of C and D are generally not used except for proving certain theoretical results. We usually simplify the general solution by observing that because W/L_p is small compared to unity, the exponentials can be simply expanded:

$$p^*(x) = Ce^{-x/L_p} + De^{x/L_p} \qquad 0 \le x \le W$$

$$\simeq (C + D) + \frac{x}{L_p}(D - C) + \text{small terms} \qquad (9 \cdot 24)$$

---

[1] When a large reverse bias is applied to a collector junction, the *absolute* hole density at the junction may be much larger than $p_n e^{-qV_2/kT}$, though it will still be extremely small compared to p_n, so the excess-hole density will still be $-p_n$, as Eqs. (9 · 21) suggest. The reason for this is that in other than low-current low-collector-voltage applications of transistors, the collection of emitted current in the reverse-biased collector-base transition region actually violates the condition on which the junction law is based. However, this proves to be of secondary importance, for when V_{CB} is large and negative, the sweeping action of the electric field within the transition region actually ensures that the hole density at $x = W$ will be small compared to p_n, even though the absolute value of $p^*(x = W)$ may be greater than the junction law suggests, especially at high current levels. Therefore, the junction law gives a valid boundary condition for small collected current, while the actual boundary condition for larger collected currents is not measurably different from the one given by the junction law. This subject is fully discussed in R. D. Middlebrook, Conditions at a reverse-biased *pn* junction in the presence of collected current, *Solid State Electronics,* Nov.–Dec., 1963, p. 555.

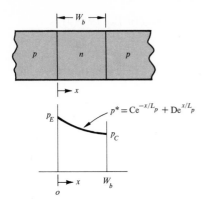

FIG. 9·14 THE SOLUTION OF THE MINORITY-CARRIER CONTINUITY EQUATION IN THE BASE OF A TRANSISTOR

In other words, *the dc minority-carrier distribution is essentially a straight line between the values p_E and p_C.*[1] This fact is worth remembering, because most of the operating characteristics of transistors can be related to it.

9·3·4 *Minority-carrier current flow in the intrinsic transistor.* The emitter and collector junction currents can be evaluated by utilizing the law of diffusive flow:

$$i_E = -qAD_p \frac{\partial p^*}{\partial x}\bigg|_{x=0} \qquad i_C = qAD_p \frac{\partial p^*}{\partial x}\bigg|_{x=W} \qquad (9·25)$$

When the exact form of p^* in Eq. (9·20) is used, we obtain

$$i_E = \frac{qAD_p}{L_p}(C - D) \qquad i_C = \frac{qAD_p}{L_p}(De^{W/L_p} - Ce^{-W/L_p}) \qquad (9·26)$$

When Eqs. (9·23) are used for C and D there results

$$i_E = a_{11}p_E + a_{12}p_C \qquad i_C = a_{21}p_E + a_{22}p_C \qquad (9·27)$$

The values of the a_{ij} are

$$a_{11} = a_{22} = \frac{qAD_p}{L_p}\coth\frac{W}{L_p} \qquad a_{12} = a_{21} = -\frac{qAD_p}{L_p}\operatorname{csch}\frac{W}{L_p} \qquad (9·28)$$

Since $W/L_p \ll 1$, it is customary to make the approximations

$$\coth\frac{W}{L_p} \cong \frac{L_p}{W} + \frac{W}{2L_p} \qquad \operatorname{csch}\frac{W}{L_p} \cong \frac{L_p}{W}$$

to obtain

$$a_{11} = a_{22} = \frac{qAD_p}{W} + \frac{qAW}{2\tau_p} \qquad a_{12} = a_{21} = -\frac{qAD_p}{W} \qquad (9·29)$$

---

[1] The "small terms" in Eq. (9·24) actually account for the base current (see Prob. 9·16).

334 Semiconductor electronics

The accuracy of these approximations is considered in Prob. 9 · 2.

For either Eqs. (9 · 28) or (9 · 29), the a_{ij} depend only on the transistor geometry and material parameters, and are therefore fixed for a given device. The currents can *only* be controlled by controlling p_E and p_C. We shall interpret Eqs. (9 · 27) more fully at a later point.

9 · 3 · 5 Base current in the intrinsic transistor. When i_E and i_C are known, then i_B can always be found from Kirchhoff's current law for the transistor as a whole:

$$i_E + i_B + i_C = 0.$$

However, the physical processes which produce i_B are not emphasized in such a formal step.

As remarked in Sec. 9 · 1 · 1, base current flows whenever there is a net excess or deficiency of minority carriers in the base. If the excess density is represented by $p^*(x)$, then

$$i_B = \frac{-1}{\tau_p} \int_0^W qAp^*(x)\,dx = -\frac{q_P}{\tau_p} \tag{9 · 30}$$

where q_P is the total excess-minority-carrier charge stored in the base (Prob. 9 · 3).

These equations are very important; they show that under dc operating conditions the base current fixes the total minority-carrier charge in the base layer. Geometrically, i_B fixes the area under the $p^*(x)$ curve.

Now, since the minority-carrier density profile is essentially a straight line, we have

$$q_P = qAW\frac{(p_E + p_C)}{2} \qquad i_B = \frac{qAW}{2\tau_p}(p_E + p_C) \tag{9 · 31}$$

Equation (9 · 31) expresses i_B in terms which can be simply used to interpret the dc behavior of the intrinsic transistor.[1]

Résumé. We have now derived the equations which describe the terminal currents of the intrinsic transistor in terms of the excess-minority-carrier densities at the edges of the base region. To summarize, these equations are

$$i_E = a_{11}p_E + a_{12}p_C \cong \left(\frac{qAD_p}{W} + \frac{qAW}{2\tau_p}\right)p_E - \frac{qAD_p}{W}p_C \tag{9 · 32a}$$

$$i_C = a_{21}p_E + a_{22}p_C \cong -\frac{qAD_p}{W}p_E + \left(\frac{qAD_p}{W} + \frac{qAW}{2\tau_p}\right)p_C \tag{9 · 32b}$$

$$i_B = -\frac{q_P}{\tau_p} \cong -\frac{qAW}{2\tau_p}(p_E + p_C) \tag{9 · 32c}$$

---

[1] In Eqs. (9 · 31) i_b is approximated using a straight-line distribution for $p^*(x)$. However, there is no i_B for a strict straight-line distribution. This contradiction is discussed in Prob. 9 · 16.

where
$$p_E = p_n(e^{qv_{EB}/kT} - 1) \qquad (9 \cdot 33a)$$

$$p_C = p_n(e^{qv_{CB}/kT} - 1) \qquad (9 \cdot 33b)$$

Equations $(9 \cdot 32)$ and $(9 \cdot 33)$ embody the essentials of the physical operation of the intrinsic transistor. Its electrical behavior is usually described by

$$i_E = A_{11}(e^{qv_{EB}/kT} - 1) + A_{12}(e^{qv_{CB}/kT} - 1) \qquad (9 \cdot 34a)$$

$$i_C = A_{21}(e^{qv_{EB}/kT} - 1) + A_{22}(e^{qv_{CB}/kT} - 1) \qquad (9 \cdot 34b)$$

to focus attention on terminal currents and voltages.

In either case, two important points should be underlined:

1. The equations expressing i_E, i_C and i_B in terms of p_E and p_C are *linear*. Nonlinearity arises in the intrinsic transistor solely through the law of the junction. As a result of the linearity of Eqs. $(9 \cdot 32)$, it is frequently convenient to employ *superposition* techniques in calculating currents and interpreting performance.

2. The intrinsic transistor is a perfectly symmetrical structure; that is, if it is turned end-for-end, so the terminal marked "collector" is used as the emitter and vice versa, then no change will be observed in its circuit behavior.

This symmetry is reflected in the perfect symmetry of Eqs. $(9 \cdot 32)$. Here, $a_{11} = a_{22}$, $a_{12} = a_{21}$. There are only two independent parameters in Eqs. $(9 \cdot 32)$: a_{11} and a_{12}. The result of this is that there are only two independent "measurements" that need to be made to completely specify the dc behavior of the intrinsic transistor.

9·4 INTERPRETATION OF INTRINSIC TRANSISTOR OPERATION

Equations $(9 \cdot 32)$ and $(9 \cdot 33)$ are the principal mathematical results of this chapter. To appreciate their significance, we shall give several common interpretations of them which emphasize different aspects of transistor behavior.

9·4·1 The Ebers-Moll equations for the intrinsic transistor.
In a classic paper, Ebers and Moll showed that Eqs. $(9 \cdot 34)$ apply to nearly all real transistors over a rather wide range of operating conditions. They also gave a simple interpretation of the constants A_{ij}, in which the "coupled-diode" nature of the transistor is emphasized. We shall develop the coupled-diode interpretation for the intrinsic transistor in this section.

To do this, we shall consider the following series of measurements (*cf.*, Prob. $9 \cdot 4$):

1. Connect the collector and base leads together and then measure the v-i characteristic of the emitter-base diode. The theoretical result is found from Eqs. $(9 \cdot 32a)$, $(9 \cdot 33a)$, and $(9 \cdot 33b)$ [or Eq. $(9 \cdot 34a)$].

$$v_{CB} = 0 \Rightarrow p_C = 0$$
$$\therefore i_E = a_{11}p_E = a_{11}p_n(e^{qv_{EB}/kT} - 1)$$
$$= I_{ES}(e^{qv_{EB}/kT} - 1) \qquad (9 \cdot 35)$$

where I_{ES} ($=A_{11}$) is the saturation current for the emitter-base diode when the collector-base junction is shorted.

Thus the external behavior of the emitter-base diode is identical to that of a normal pn-junction diode. The theory gives the value of I_{ES} in terms of the transistor geometry and material parameters, just as for the normal pn diode. I_{ES} can be computed when these parameters are known, or it can be measured externally. In practice, measurement of I_{ES} is carried out by fitting Eq. (9·35) to the observed forward characteristic of the diode.[1]

The internal behavior of the diode is quite different from the normal diode, however. In particular, *most* of the current injected at the emitter returns via the collector lead rather than the base lead. This transfer of current from emitter to collector circuit is, of course, the essential aspect of the transistor behavior. To emphasize the relationship between emitter and collector currents, we define a *short-circuit current-transfer* parameter as follows:

$$\alpha_{F,i} \equiv -\left. \frac{i_C}{i_E} \right|_{v_{CB}=0} = -\frac{a_{21}}{a_{11}} = -\frac{A_{21}}{A_{11}} \qquad (9 \cdot 36)$$

$\alpha_{F,i}$ is simply the ratio of the short-circuit collector current to the injected emitter current for the diode connection envisaged. The subscript F refers to *forward* because the transistor is being operated in the "forward" direction (that is, the emitter is emitting and the collector collecting). The subscript i refers to the intrinsic transistor.[2]

$\alpha_{F,i}$ can be either measured or computed. To compute it we use Eqs. (9·32a) and (9·32b):

$$\alpha_{F,i} = -\frac{a_{21}}{a_{11}} = \operatorname{sech} \frac{w}{L_p} \cong \frac{1}{1 + W^2/2L_p{}^2} \qquad (9 \cdot 37)$$

$\alpha_{F,i}$ will be very close to unity, since $W/L_p \ll 1$.

In terms of $\alpha_{F,i}$, the collector current can be written simply as

$$i_C = -\alpha_{F,i}i_E = -\alpha_{F,i}I_{ES}(e^{qv_{EB}/kT} - 1) \qquad (9 \cdot 38)$$

The base current is then

$$i_B = -(1 - \alpha_{F,i})I_{ES}(e^{qv_{EB}/kT} - 1) \qquad (9 \cdot 39)$$

Since the dc base current is entirely a recombination current, we conclude that a fraction $(1 - \alpha_{F,i})$ of the injected holes recombine, while $\alpha_{F,i}$ are collected (for the diode connection envisaged).

[1] Reasons for not measuring I_{ES} with reverse bias are suggested in arguments of Sec. 6·6·3.

[2] $\alpha_{F,i}$ is also called the base-transport factor, since it is a measure of the efficiency of transporting holes across the base.

2. The second type of measurement we should make is to connect the emitter and base leads together and then measure the v-i characteristic of the collector-base diode. This will lead to

$$i_C = a_{22}p_C = I_{CS}(e^{qv_{CB}/kT} - 1) \tag{9.40}$$

$$\alpha_{R,i} \equiv \left. \frac{-i_E}{i_C} \right|_{v_{EB}=0} = -\frac{a_{12}}{a_{22}} = -\frac{A_{12}}{A_{22}} \tag{9.41a}$$

$$i_E = -\alpha_{R,i}I_{CS}(e^{qv_{CB}/kT} - 1) \tag{9.41b}$$

Since the intrinsic transistor is perfectly symmetrical, we have

$$I_{CS} = I_{ES} \qquad \alpha_{F,i} = \alpha_{R,i} \tag{9.42}$$

Most real transistors are not symmetrical, however, so separate measurement of I_{CS} and α_R will have to be made.

The measurements or computation of I_{ES}, etc., enable us to rewrite Eqs. (9.32a) and (9.32b) [or (9.34a) and (9.34b)] in the form

$$i_E = I_{ES}(e^{qv_{EB}/kT} - 1) - \alpha_R I_{CS}(e^{qv_{CB}/kT} - 1) \tag{9.43a}$$

$$i_C = -\alpha_F I_{ES}(e^{qv_{EB}/kT} - 1) + I_{CS}(e^{qv_{CB}/kT} - 1) \tag{9.43b}$$

These equations, together with

$$i_B + i_C + i_E = 0 \tag{9.44}$$

give a complete description of the terminal behavior of the transistor. To emphasize that the equations are really linear in the hole-density variable, we can also write

$$i_E = I_{ES}\frac{p_E}{p_n} - \alpha_R I_{CS}\frac{p_C}{p_n} \tag{9.45a}$$

$$i_C = -\alpha_F I_{ES}\frac{p_E}{p_n} + I_{CS}\frac{p_C}{p_n} \tag{9.45b}$$

Equations (9.42), while derived only for the intrinsic transistor, apply to essentially all junction transistors when the complete set of saturation currents and alphas are measured. They are called the *Ebers-Moll equations.* We shall see later that *in all cases*

$$\alpha_R I_{CS} = \alpha_F I_{ES} \tag{9.46}$$

even though

$$\alpha_F \neq \alpha_R \qquad I_{CS} \neq I_{ES}$$

for nearly all real transistors.

9·4·2 Regions of operation of the intrinsic transistor. Of course, the basic equations (9·32) for internal operation of the transistor are valid for any possible combination of the junction voltages v_{EB} and v_{CB}. Similarly, the Ebers-Moll equations, which are essentially Eqs. (9·32a) and (9·32b) with the parameters determined by external measurement, are valid for any dc

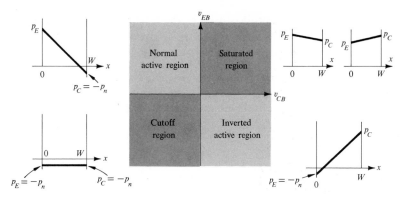

FIG. 9·15 REGIONS OF OPERATIONS FOR A *pnp* TRANSISTOR AND APPROXIMATE MINORITY-CARRIER DISTRIBUTIONS IN THE BASE REGION. P_E AND P_C ARE EXCESS DENSITIES IN THESE FIGURES.

operating condition. However, such generality is frequently not required for studying particular circuits, where the operating range of v_{EB} or v_{CB} is restricted by circuit conditions. Rather, for most circuit applications, it is convenient to define four basic operating regions, corresponding to the four possible combinations of junction voltages. These are

1. *Normal active.* In this region the emitter-base junction is forward-biased and the collector-base junction is reverse-biased ($v_{EB} > 0$ and $v_{CB} < 0$ for a *pnp* transistor).

2. *Inverted active.* In this region, the collector-base junction is forward-biased and the emitter-base junction is reverse-biased ($v_{EB} < 0$, $v_{CB} > 0$ for a *pnp* transistor).

3. *Saturated.* In this region both junctions are forward-biased ($v_{EB} > 0$, $v_{CB} > 0$ for a *pnp* transistor).

4. *Cutoff.* In this region both junctions are reverse-biased ($v_{EB} < 0$, $v_{CB} < 0$ for a *pnp* transistor).

The four regions of operation can be simply visualized on a set of coordinates where v_{EB} is the ordinate and v_{CB} the abscissa. This is shown in Fig. 9·15. The approximate minority-carrier distribution in the base region is also shown for each case. It is always a straight line between the two end points p_E and p_C.

When a transistor is used as an amplifier, operating conditions will usually be confined to the normal-active region. In pulse and digital switching applications, the transistor is considered to be ON when it is in the saturated region and OFF when it is in the cut-off region. It will be driven between these two conditions in the course of circuit operation.

9·4·3 *Intrinsic dc v-i characteristics in the normal-active region.* It is frequently convenient to use special forms of the general equations when operation is confined to a particular region of the v_{EB}-v_{CB} plane. The most commonly encountered equations describe the *output* or *collector characteristics* of the transistor when it is used in the normal-active region. Two types of output characteristics are of most interest. These are the *common-base* and *common-emitter* characteristics.

Intrinsic common-base output characteristics. The common-base output characteristic is a plot of i_C versus v_{CB} with i_E as a parameter. The equation for this characteristic can be obtained quite simply from Eqs. $(9 \cdot 45a)$ and $(9 \cdot 45b)$:

$$i_E = I_{ES}\frac{p_E}{p_n} - \alpha_R I_{CS}\frac{p_C}{p_n} \tag{9·47a}$$

$$i_C = -\alpha_F I_{ES}\frac{p_E}{p_n} + I_{CS}\frac{p_C}{p_n} \tag{9·47b}$$

If we use Eq. $(9 \cdot 47a)$ to solve for p_E/p_n and then substitute the result into Eq. $(9 \cdot 47b)$, we find

$$i_C = -\alpha_F i_E + (1 - \alpha_F \alpha_R)I_{CS}\frac{p_C}{p_n} \tag{9·48}$$

Equation $(9 \cdot 48)$ is usually simplified by writing

$$I_{CO} = (1 - \alpha_F \alpha_R)I_{CS} \tag{9·49}$$

I_{CO} is the saturation current for the collector-base diode when the emitter-base diode is left open (see Prob. $9 \cdot 4$). Then

$$i_C = -\alpha_F i_E + I_{CO}(e^{qv_{CB}/kT} - 1) \tag{9·50}$$

The curves represented by Eq. $(9 \cdot 50)$ may be plotted very simply. First, we suppose that $i_E = 0$. Then the i_C-v_C plot is simply the v-i plot for a *pn* diode with a saturation current of I_{CO}. (See Fig. $9 \cdot 16$). Now, if an emitter current I_{E_1} is made to flow in the emitter lead, the collector current for any given collector-base voltage will simply be larger than before by an amount $\alpha_F I_{E_1}$. The output characteristics of an intrinsic transistor in the common-base configuration can thus be plotted with the aid of a rubber stamp which has a *pn*-diode characteristic plotted on it.

Real transistors sometimes exhibit characteristics which differ markedly from Eq. $(9 \cdot 50)$. In many cases, however, Eq. $(9 \cdot 50)$ gives a sufficiently

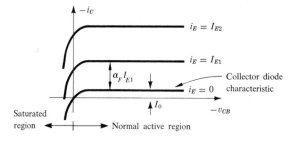

FIG. $9 \cdot 16$ COLLECTOR FAMILY FOR GROUNDED-BASE CONFIGURATION

accurate approximation to actual behavior that only the values of α_F and I_{CO} will be given by the manufacturer.

Equation $(9 \cdot 50)$ may be specialized to the normal-active region by observing that over most of this region, $v_{CB} \ll kT/q$, so the exponential term can be neglected to yield

$$i_C = -\alpha_{Fi_E} - I_{CO} \qquad (9 \cdot 51)$$

Equation $(9 \cdot 51)$ simply shows that the collector junction collects α_{Fi_E} as part of its reverse current, so long as v_{CB} is large and negative (that is, $p_C/p_n \ll 1$).

Common-base input characteristics for the intrinsic transistor. The input characteristics of a transistor in the common-base orientation are plots of i_E vs. v_{EB} with v_{CB} held constant. For the intrinsic transistor in the normal-active mode, Eq. $(9 \cdot 47a)$ reduces to

$$i_E = I_{ES}\frac{p_E}{p_n} + \alpha_R I_{CS} = I_{ES}(e^{qv_{EB}/kT} - 1) + \alpha_R I_{CS}$$

This equation shows that the i_E-v_{EB} relation is basically that of an ideal junction diode, except that i_E is not zero when v_{EB} is zero, due to the coupling between the emitter and collector diodes.

Most transistors obey this general characteristic sufficiently well that the input characteristic is not usually given by the manufacturer. However, there are some effects to be described in Sec. $9 \cdot 6 \cdot 1$ which cause modifications of this ideal relation.

Intrinsic common-emitter output characteristics. For reasons that will be apparent later, the most common connection of the transistor for amplifier service is the common-emitter connection. In this connection the voltages used to describe the transistor performance are measured with the emitter terminal as the common point. The base-to-emitter voltage is denoted by v_{BE} and the collector-to-emitter voltage by v_{CE}. The emitter-base junction voltage is therefore $-v_{BE}$, while the collector-base junction voltage is $v_{CB} = v_{CE} - v_{BE}$.

The output characteristic for this connection is a plot of i_C vs. v_{CE} with i_B as a parameter. From Eqs. $(9 \cdot 45)$ and Kirchhoff's law

$$i_B + i_E + i_C = 0 \qquad (9 \cdot 52)$$

and we find

$$i_C = \frac{\alpha_F}{1 - \alpha_F} i_B + \frac{1 - \alpha_F \alpha_R}{1 - \alpha_F} I_{CS} \frac{p_C}{p_n}$$

$$= \beta_F i_B + \frac{I_{CO}}{1 - \alpha_F}(e^{qv_{CB}/kT} - 1)$$

$$= \beta_F i_B + \frac{I_{CO}}{1 - \alpha_F}(e^{qv_{EB}/kT}e^{qv_{CE}/kT} - 1) \qquad (9 \cdot 53)$$

This equation may also be rather simply plotted, though more care is necessary than before, since one of the defining voltages is not a junction voltage drop. Over most of the normal-active region

$$|v_{CE}| \gg |v_{EB}|$$

however, so Eq. (9·53) can be simplified to

$$i_C = \beta_{FI}i_B - \frac{I_{CO}}{1 - \alpha_F} \tag{9·54}$$

This equation shows that i_C has the value $-I_{CO}/(1 - \alpha_F)$ when i_B is zero.

Typical plots of i_C vs. v_{CE} with i_B as a parameter are shown in Fig. 9·17. From these plots or the preceding analysis, it is apparent that the transistor exhibits *current gain* when operated in the common-emitter configuration: that is, a change in the driving current Δi_B produces a much larger change in collector current Δi_C:

$$\frac{\Delta i_C}{\Delta i_B}\bigg|_{v_{CE}=\text{const}} = \beta = \frac{\alpha_F}{1 - \alpha_F} \sim \frac{2L_p{}^2}{W^2} \tag{9·55}$$

For the curves shown in Fig. 9·17, the β is evidently about 100, since a Δi_B of 0.01 ma produces a Δi_C of approximately 1 ma. β is called the *common-emitter current gain.*

In anticipation of a later need, we point out here that the i_C-v_{CE} characteristics do not all pass through the point ($i_C = 0$, $v_{CE} = 0$), as Fig. 9·17 suggests. Only the $i_B = 0$ curve actually passes through the ($i_C = 0$, $v_{CE} = 0$) point. If the transistor were perfectly symmetrical, the remainder of the curves would pass through $i_C = i_B/2$ when $v_{CE} = 0$. However, the i_B values are so small that this behavior does not appear on the curves shown.

Common-emitter input characteristics for the intrinsic transistor. The input characteristics for the common-emitter configuration are plots of i_B vs. v_{BE} with v_{CE} held constant. By appropriate use of Eqs. (9·47) we can write the

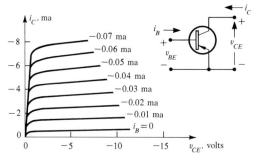

FIG. 9·17 COMMON-EMITTER CHARACTERISTICS FOR A TYPICAL TRANSISTOR WITH i_B AS A PARAMETER

ideal relation for a *pnp* transistor as

$$i_B = -I_{ES}(1 - \alpha_F)\frac{p_E}{p_n} - I_{CS}(1 - \alpha_R)\frac{p_C}{p_n}$$

Again, for the normal-active mode the ideal i_B-v_{BE} relation is essentially that of an ideal junction diode. The input characteristic is not usually shown by a transistor manufacturer, except at high currents, where the intrinsic transistor theory is invalid.

9·4·4 Charge-control interpretation of intrinsic transistor behavior. In our previous discussions of transistor behavior, we have described the base current as arising from the recombination of minority carriers as they diffuse across the base region. This means that we have implicity considered base current to be an "effect" rather than a "cause."

In the common-emitter orientation of the transistor, however, the base current is used to control the collector current. This means that the base current should be thought of as *causing* the flow of emitter and collector currents in the transistor.

The *charge-control* interpretation of transistor behavior focuses attention on the mechanism by which base control is achieved. The charge-control theory gives both an interesting interpretation of the amplifying mechanism and quantitative relationships between the base current and the emitter and collector currents. The quantitative relationships lead to a very convenient geometrical interpretation of the carrier distributions within the transistor, which accompany any low-frequency operating condition. We shall develop these aspects of the charge-control model in this section.

Charge-control interpretation of amplifying mechanism. The charge-control interpretation of the amplifying mechanism is as follows. Let us suppose that a current i_B is being extracted from the base of a *pnp* transistor (that is, electrons are flowing into the base, as shown in Fig. 9·18). The collector junction is assumed to be reverse-biased in this discussion. Let us now follow the history of one of the electrons that makes up the base current i_B. As this electron enters the base region, a hole will be injected into the base from the emitter to maintain space-charge neutrality. Suppose this hole can diffuse across the base and be collected in a time τ_t, which is short compared to the hole lifetime τ_p in the base region. Then, as it does so and is collected, the emitter must inject another hole into the base to maintain space-charge neutrality there (see Fig. 9·18b). Since the recombination lifetime in the base is τ_p, this process will be repeated τ_p/τ_t times (on the average) before the electron recombines with a hole to stop it.

From this argument, the portion of the collector current which arises from the base current i_B is simply

$$i_C = \frac{\tau_p}{\tau_t} i_B \tag{9·56}$$

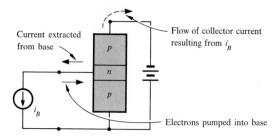

Current extracted from base

Flow of collector current resulting from i_B

i_B

Electrons pumped into base

(a) *pnp transistor with electrons being supplied to the base by an external source*

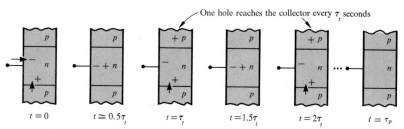

One hole reaches the collector every τ_t seconds

$t = 0$ $t \cong 0.5\tau_t$ $t = \tau_t$ $t = 1.5\tau_t$ $t = 2\tau_t$ $t = \tau_p$

(b) *A time sequence of idealized pictures illustrating the charge-control interpretation of transistor current gain. One hole reaches the collector every τ_t seconds until $t + \tau_p$, when the electron and an injected hole recombine.*

FIG. 9·18   THE CHARGE-CONTROL INTERPRETATION OF TRANSISTOR ACTION

Comparing this with Eq. (9·54), we find

$$\beta = \frac{2D_p\tau_p}{W^2} = \frac{\tau_p}{\tau_t} \qquad (9\cdot57)$$

Evidently, the minority-carrier base-transit time is simply

$$\tau_t = \frac{W^2}{2D_p} \qquad (9\cdot58)$$

(a result which can also be obtained directly from the theory of diffusion).

Relation between minority-carrier charge in the base and the base current.
The minority-carrier densities which are set up in the base in response to the base current flow can be simply obtained as follows. We know that to maintain a steady base current of i_B there must be an excess density of both majority and minority carriers in the base:

$$q_P = -i_B\tau_p$$

Thus, we visualize i_B as being responsible for setting up an excess base charge.

Now, we know from our previous work that the distribution of minority carriers is essentially a straight line across the base. Knowing q_P is equivalent to knowing the area under the distribution curve:

$$q_P = qA \int_0^w p^*(x)\, dx \cong qAW \frac{p_E + p_C}{2}$$

For the particular case under discussion, where the collector junction is reverse-biased (so that $p_C = -p_n$), knowing q_P is equivalent to knowing p_E (and therefore v_{EB}). Hence, if the collector junction is reverse-biased,

$$i_B \rightarrow q_P \rightarrow p_E \rightarrow v_{EB}$$

Charge-control formulation of Eqs. (9·32). We can generalize this idea to include the situation where both junction biases are arbitrary. To do this, we rewrite the fundamental equations (9·32) for the intrinsic transistor as follows:

$$i_E = \left(\frac{qAD_p}{W} + \frac{qAW}{2\tau_p} \right) p_E - \frac{qAD_p}{W} p_C \qquad (9 \cdot 59a)$$

$$i_C = - \frac{qAD_p}{W} p_E + \left(\frac{qAD_p}{W} + \frac{qAW}{2\tau_p} \right) p_C \qquad (9 \cdot 59b)$$

$$i_B = - \frac{qAW}{2\tau_p} (p_E + p_C) \qquad (9 \cdot 59c)$$

Now, because i_E, i_C, and i_B depend *linearly* on p_E and p_C, we can find i_E, i_B, and i_C for any prescribed values of p_E and p_C by superposition techniques: for example,

$$i_E(p_E = B, p_C = C) = i_E(p_E = B, p_C = 0) + i_E(p_E = 0, p_C = C) \quad (9 \cdot 60)$$

This is equivalent to visualizing the actual minority-carrier distribution in the base as being made up of the two special components, shown in Fig. 9·19a, and computing the terminal currents for each case separately.

For the distribution shown in Fig. 9·19b, the total minority-carrier charge stored in the base is

$$q_F = \frac{qAWp_E}{2} \qquad (9 \cdot 61)$$

This excess-minority-carrier charge recombines and is replaced every τ_p sec. This requires a base current of

$$i_B = - \frac{q_F}{\tau_p} \qquad (9 \cdot 62a)$$

From Eq. (9·59b), the collector current is

$$i_C = - \frac{q_F}{\tau_t} \qquad (9 \cdot 62b)$$

and the emitter current is

$$i_E = q_F \left(\frac{1}{\tau_t} + \frac{1}{\tau_p} \right) \qquad (9 \cdot 62c)$$

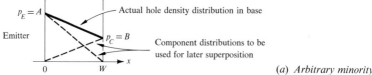

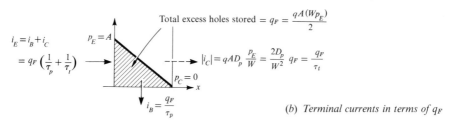

(a) *Arbitrary minority-carrier density profile*

$$\text{Total excess holes stored} = q_F = \frac{qA(Wp_E)}{2}$$

$i_E = i_B + i_C$

$= q_F \left(\dfrac{1}{\tau_p} + \dfrac{1}{\tau_t}\right)$

$|i_C| = qAD_p \dfrac{p_E}{W} = \dfrac{2D_p}{W^2} q_F = \dfrac{q_F}{\tau_t}$

$i_B = \dfrac{q_F}{\tau_p}$

(b) *Terminal currents in terms of q_F*

$|i_E| = qAD_p \dfrac{p_C}{W}$

$= \dfrac{2D_p}{W^2} q_R = \dfrac{q_R}{\tau_t}$

$$\text{Total excess holes stored} = q_R = \frac{qA(Wp_C)}{2}$$

$i_C = i_B + i_E = q_R \left(\dfrac{1}{\tau_p} + \dfrac{1}{\tau_t}\right)$

$i_B = \dfrac{q_R}{\tau_p}$

(c) *Terminal currents in terms of q_R*

FIG. 9 · 19   THE CHARGE-CONTROL FORMULATION OF TRANSISTOR BEHAVIOR

where
$$\tau_t = \frac{W^2}{2D_p} \tag{9·63}$$

as before.

Equation (9 · 62b) can be obtained either from the charge-control description of the amplifying mechanism or by direct calculation from Fig. 9 · 19b, assuming that the average slope of the hole-density profile should be used to calculate the current which diffuses to the collector.

To complete the charge-control formulation, we define a charge component q_R by means of

$$q_R = \frac{qAWp_C}{2} \tag{9·64}$$

This is the excess-minority-carrier charge stored in the base when $p_E = 0$ and p_C is arbitrary. The charge distribution is shown in Fig. 9 · 19c, and leads to the equations

$$i_B = -\frac{q_R}{\tau_p} \qquad i_E = -\frac{q_R}{\tau_t} \qquad i_C = q_R \left(\frac{1}{\tau_p} + \frac{1}{\tau_t}\right) \tag{9·65}$$

Now, since an arbitrary excitation can be built up from the proper combinations of p_E and p_C, or q_F and q_R, the complete equations for the dc terminal currents of a transistor can be simply written as

$$i_E = q_F \left(\frac{1}{\tau_p} + \frac{1}{\tau_t}\right) - q_R \left(\frac{1}{\tau_t}\right) \tag{9·66a}$$

$$i_C = -q_F \left(\frac{1}{\tau_t}\right) + q_R \left(\frac{1}{\tau_p} + \frac{1}{\tau_t}\right) \tag{9·66b}$$

$$i_B = -\left[\frac{q_F}{\tau_p} + \frac{q_R}{\tau_p}\right] \tag{9·66c}$$

which are the charge-control equations for an intrinsic *pnp* transistor. We shall find very important applications for these equations in studying low-frequency transistor circuits and in extending low-frequency concepts to a study of the medium-frequency behavior of transistors.

9·5 FREQUENCY CHARACTERISTICS OF THE INTRINSIC TRANSISTOR

The low-frequency theory of the intrinsic transistor just presented can be used to understand the basic amplifying mechanism in the transistor, to explain and interpret its graphical characteristics, and to study many important transistor applications. These applications fall into two categories.

1. The study of dc circuits such as biasing networks and dc amplifiers, and the determination of the properties of a transistor switch in its ON and OFF conditions

2. The study of transistor amplifiers and oscillators in which the frequencies of operation are limited by passive circuit elements to values for which the transistor merely acts as an amplifying element without frequency-dependent properties

To extend the study of transistor applications beyond these limits, it is necessary to modify the low-frequency theory to account for the frequency-dependent properties of the transistor. The necessary modifications are of two types:

1. Changes in the excess minority-carrier density in the base must accompany changes in the terminal currents, and the dynamics of this process need to be included in the model.

2. The effects of several different parasitic elements (primarily capacitances and resistances) need to be included in the equations and models which are used to study the transistor's behavior.

The relative importance of these factors depends on the transistor and the application. For frequencies which are low compared to the reciprocal of the transit time of minority carriers across the base region

$$f < (2D_p/2\pi W^2)$$

a first-order correction to the low-frequency theory of the intrinsic transistor and the addition of one parasitic capacitance is usually adequate. In some cases (notably transistors intended for use at and below about

10 mcps), these corrections are enough to understand the circuit behavior of a transistor over most of its usable frequency range.

In transistors which are intended for very-high-frequency applications, a more sophisticated account of carrier dynamics is required and careful attention must be given to several parasitic elements.

In this section we will extend the low-frequency theory far enough to understand the behavior of transistors at frequencies up to about $0.1 \times (2D_p/W^2)$. The argument to be used is based on the stored-charge formulation of transistor behavior and will be limited to the case where the collector-base junction is always reverse-biased (that is, the normal-active mode). We will develop a more general model in Chap. 17, where we consider the behavior of transistors at very high frequencies.

9·5·1 *The general relation between base current and base charge.* Basically, the stored-charge approach to transistor theory rests on exploiting the concept of space-charge neutrality in the base region: at every instant, the number of majority carriers entering the base region must equal the number of minority carriers entering. This idea may be used to develop a simple, but perfectly general, equation relating the base current to the stored-base charge.

Let us suppose that the minority-carrier charge stored in the base at a given instant is $q_P(t)$, and that the base current at the same instant is $i_B(t)$. If q_P and i_B satisfy the relation

$$-q_P(t) = i_B(t)\tau_p$$

then a steady-state condition prevails: majority carriers (electrons, in this case) are entering the base region at exactly the rate required to replace those that are recombining with holes. If i_B deviates from the value which just maintains the dc steady state, then q_P will have to change to accommodate the new value of i_B. If i_B is larger than $q_P(t)/\tau_p$, then over a short period of time more majority carriers will enter the base region than are needed to maintain the original steady state. To preserve space-charge neutrality, the minority-carrier charge q_P must increase with time. The rate of increase of q_P is simply the difference between $i_B(t)$ and $q_P(t)/\tau_p$. Similarly, if $i_B(t) < q_P/\tau_p$, then q_P will decrease with time.

We can include both these possibilities in the following equation:

$$-i_B = \frac{q_P(t)}{\tau_p} + \frac{dq_P(t)}{dt} \tag{9·67}$$

In this equation the base current is thought of as serving two functions: it supplies the majority carriers necessary for recombination and it supplies the majority carriers necessary to change the stored charge from one value to another.[1]

---

[1] We could be somewhat more correct by writing Eq. (9·67) in terms of the excess-majority-carrier charge in the base (say, q_m), and then using space-charge neutrality to show that $q_m = q_P$.

Note carefully that the argument used to establish Eq. (9·67) does *not* limit its validity to low-frequency conditions. This equation is always valid and will be just as useful in studying high-frequency transistors as it is for our present purposes.

9·5·2 *First-order correction to dc transistor theory: normal-active mode.* We may use Eq. (9·67) as the basis for deriving a first-order correction to the dc theory of transistor behavior presented earlier. To do this, let us assume that the base current changes so slowly that *the hole-density profile in the base remains essentially straight at all times.*

Since the emitter and collector currents are proportional to the instantaneous slope of the hole-density distribution at the emitter and collector edges of the base, the assumption of a straight-line hole-density profile is equivalent to saying that the instantaneous base current is small compared to the instantaneous collector current. Or, if we divide the base current into a dc part and a small signal part, the basic assumption is met when the small signal or incremental current gain between the base and collector terminals is large. It need not be as large as the dc β, but it needs to be large compared to unity (say at least 10) for the approximation of a straight-line hole-density profile to be valid.

When this assumption is met, the dynamics of transistor behavior can be visualized by simple geometry. In the normal-active mode, where the collector junction is reverse-biased, the straight-line hole-density profile must always pass through an absolute density of zero at $x = W$ [or $p^*(x) = -p_n$ at $x = W$]. A sine-wave variation of i_B with time will then cause the hole-density curve to vary around this pivot point in the manner shown in Fig. 9·20 (more or less like half of a seesaw).

To obtain the terminal currents for this *quasi-static* approximation, we note that in general $q_P(t)$ must be written as

$$q_P = q_F(t) + q_R(t) \tag{9·68}$$

where q_F and q_R are the forward- and reverse-charge components defined in Sec. 9·4·4. But for the normal-active mode, q_R is a (small) constant:

$$q_R = -qAWp_n/2$$

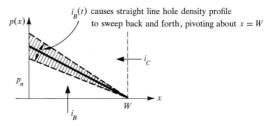

FIG. **9·20** QUASI-STATIC APPROXIMATION OF MINORITY-CARRIER PROFILE IN NORMAL ACTIVE MODE so changes in q_P with time are reflected entirely in q_F. Thus, i_B can be written

$$-i_B = \frac{q_F + q_R}{\tau_p} + \frac{dq_F}{dt} \tag{9.69a}$$

The assumption that q_P always corresponds to a straight-line distribution allows us to write

$$i_C = -\frac{q_F}{\tau_t} + q_R \left(\frac{1}{\tau_p} + \frac{1}{\tau_t} \right) \tag{9.69b}$$

i_E can be found from these equations by Kirchhoff's law.

Since we are assuming that both i_B and q_R are known, Eq. (9.69a) can be used to find $q_F(t)$. $i_C(t)$ is then found from Eq. (9.69b). Furthermore, from the geometry of a triangle we have

$$q_F(t) = \frac{qAWp_E(t)}{2}$$

The law of the junction can then be used to obtain $v_{EB}(t)$:

$$v_{EB}(t) = \frac{kT}{q} \ln \left(\frac{2q_F(t)}{qAW} + 1 \right) \tag{9.69c}$$

Thus, in Eqs. (9.69) we have a complete characterization of the quasi-static terminal behavior of the transistor in the normal-active mode. That is, we can solve for $i_C(t)$ and $v_{EB}(t)$ in terms of $i_B(t)$.

9·5·3 Small-signal model obtained from quasi-static model.
In many applications of transistors an operating point is established (that is, some desirable dc values of I_B, I_C, and V_{CE} are chosen) and a small-signal voltage or current is applied between the base and emitter terminals (with the emitter common). A simple example is the amplifier circuit shown in Fig. 9·21. One of the things we should like to know about this amplifier is how much output signal voltage it provides for a given input signal voltage. In discussing questions of this nature, it is desirable to have a *small-signal circuit model* for the transistor to use in circuit analysis.

We shall give a careful development of both large- and small-signal circuit models for transistors in Chap. 11. However, it is useful to antici-

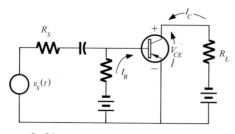

FIG. 9·21   A SIMPLE TRANSISTOR AMPLIFIER

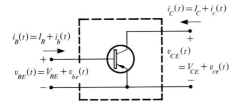

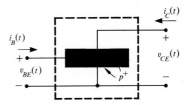

(a) Transistor with total currents and voltages as variables

(b) Idealized physical embodiment of (a)

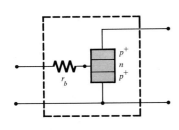

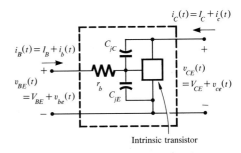

Intrinsic transistor

(c) Model for (b) using intrinsic transistor plus ohmic base resistance

(d) Model for (c) in which junction capacitances have been incorporated in external circuit elements

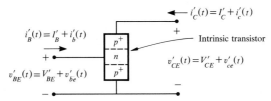

(e) Intrinsic transistor represented by three-terminal box with currents and voltages defined

FIG. 9·22   THE DEVELOPMENT OF A SMALL-SIGNAL CIRCUIT MODEL FOR A TRANSISTOR

pate a part of that discussion here to show how the physical theory just presented leads to a simple and very useful small-signal circuit model. To do this, we shall represent the variables in our theory as follows:

$$y_A(t) = Y_A + y_a(t)$$

The total value of the variable $y_A(t)$ is split up into a dc part Y_A and a time-varying part $y_a(t)$. Capital subscripts are used to denote total or dc quantities; lower-case subscripts denote the time-varying part.

Figure 9·22 has been prepared to illustrate our development of a small-signal model. In Fig. 9·22a we show a transistor with currents and voltages marked on it in accordance with the above notation. In Fig. 9·22b, a physical embodiment of the transistor is shown based on an idealization of an alloy or surface-barrier transistor. In Fig. 9·22c, ohmic resistance between the base lead and the active portion of the transistor is represented as a lumped parasitic base resistance r_b (cf. Sec. 9·6·4 for a more complete discussion). In Fig. 9·22d, the junction capacitances associated with the collector-base and emitter-base transition regions are also removed from the transistor structure and incorporated into a circuit external to the intrinsic transistor. Finally, Fig. 9·22e shows the intrinsic transistor with currents and voltages defined on it. The currents and voltages defined here are marked with primes to indicate that they are *not* the same as the actual currents and voltages at the transistor terminals.

Using the small-signal variables given in Fig. 9·22e, the equations relating the small-signal currents and base charge are, from Eqs. (9·69),

$$i'_b = -\frac{q_f}{\tau_p} - \frac{dq_f}{dt} \qquad (9 \cdot 70a)$$

$$i'_c = -\frac{q_f}{\tau_t} \qquad (9 \cdot 70)$$

The relation between q_F and the base-to-emitter voltage is given by Eq. (9·69c):

$$v'_{EB}(t) = -v'_{BE}(t) = \frac{kT}{q} \ln \left[\frac{2q_F(t)}{qAW} + 1 \right] \qquad (9 \cdot 71a)$$

If we use

$$q_F(t) = Q_F + q_f(t)$$

and suppose that the forward bias on the emitter is such that the "1" can be neglected, then Eq. (9·71a) can be expanded to give

$$-v'_{be}(t) = \frac{kT}{q} \frac{q_f(t)}{Q_F} \qquad (9 \cdot 71b)$$

The expansion is valid as long as $|v'_{be}(t)|$ is always less than about 5 mv.

If Eq. (9·71b) is used in Eq. (9·70a), there results

$$i'_b = A_1 v_{be} + A_2 \frac{dv_{be}}{dt} \qquad (9 \cdot 72a)$$

$$i'_c = A_3 v_{be} \qquad (9 \cdot 72b)$$

in which

$$A_1 = \frac{q}{kT} \frac{Q_F}{\tau_p} = \frac{qI'_B}{kT} \qquad (9 \cdot 72c)$$

$$A_2 = \frac{q}{kT} Q_F = \left(\frac{q}{kT} \frac{Q_F}{\tau_p} \right) \tau_p = A_1 \tau_p \qquad (9 \cdot 72d)$$

$$A_3 = \frac{q}{kT} \frac{Q_F}{\tau_t} = \frac{qI_C}{kT} \qquad (9 \cdot 72e)$$

Now, to make a small-signal circuit model for the transistor *means* to replace the box of Fig. 9·22e with a network which exhibits the same electrical characteristic at its terminals as the box does. For the present case, the network shown in Fig. 9·23 is required. It has an input conductance of A_1 mhos, an input capacitance of A_2 farads, and a current generator at its output, of value $A_3 v_{be}$.

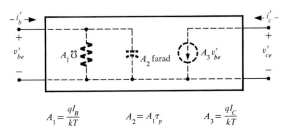

$$A_1 = \frac{qI_B}{kT} \qquad A_2 = A_1\tau_p \qquad A_3 = \frac{qI_C}{kT}$$

In Fig. 9·24 the transistor is represented by the intrinsic transistor model of Fig. 9·23, with the emitter and collector junction capacitances and the base resistance r_b added. The terminal currents and voltages are now marked with the same symbols as those of the real transistor, though this is, of course, still an approximation. In addition, the intrinsic transistor elements are relabelled r_π, C_π, and g_m to conform to standard terminology.

There are several important features of the model shown in Fig. 9·24 which we wish to point out:

1. The current generator expresses the basic amplifying property in the transistor. Its output current is controlled by the base-to-emitter junction voltage. The constant g_m is known when the bias current is specified [Eq. (9·72e)].

2. The elements r_π and C_π of the intrinsic transistor model arise from recombination and charge storage in the base region (q_F/τ_p and dq_F/dt). The time constant of this parallel rC circuit is τ_p, the minority-carrier lifetime.

The value of r_π is specified when the dc operating base current is specified [Eq. (9·72c)]. The value of τ_p is not specified by a transistor manufacturer, so C_π cannot be computed directly. However, the transistor data sheet *will* show the value of $1/(2\pi\tau_t)$, where τ_t is the transit time of carriers from the emitter to the collector. This value is quoted as f_T.[1] Furthermore, the transistor β will also be given, and $\beta = \tau_p/\tau_t$. Hence, τ_p can be estimated from data that is given on the data sheet:

$$\tau_p = \beta\tau_t = \frac{\beta}{2\pi f_T}$$

τ_p and r_π are then used to find C_π.

3. The value of C_{jE} will not be given, though it is usually small compared to the intrinsic transistor capacitance C_π, so we need not be concerned with it at this point.

4. The base resistance r_b will be given on the transistor data sheet if the transistor is intended for high-frequency service (10 mcps and above); otherwise it can be estimated to be in the range 50 ohms $< r_b <$ 200 ohms.

---

[1] Sometimes f_α will be quoted instead of f_T. We will see in Chap. 17 that f_α and f_T are not equal, though they are sufficiently near equal that we may use f_α in place of f_T for computing model parameters.

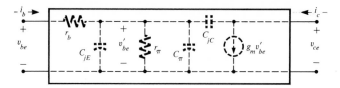

FIG. 9·24 COMMON-EMITTER SMALL-SIGNAL CIRCUIT MODEL INCLUDING PARASITIC JUNCTION CAPACITANCES AND BASE RESISTANCE

5. The model is valid for $|v'_{be}|$ less than about 5 mv. In terms of i_b, this means $i'_b/I'_B < \frac{1}{5}$.

We shall see in later chapters that the model shown in Fig. 9·24 gives a reasonably accurate approximation to the small-signal behavior of most transistors over their useful frequency range.

EXAMPLE To conclude this section, we give a numerical example to show how the circuit parameters of Fig. 9·24 are found from the transistor data sheet. A 2N324 data sheet gives the following typical parameters, measured at a bias point of $V_{CE} = -6$ volts, $I_E = 1$ ma: $\beta = 100, f_T = 4.0$ mcps, $C_c = 18$ pfarads, r_b unspecified. To find r_π, C_π, and g_m, we note that

$$I_B \simeq \frac{I_E}{\beta + 1} \simeq 10^{-5} \text{ amp}$$

This gives

$$r_\pi = \frac{kT}{qI_B} = \frac{kT}{qI_E}(\beta + 1) = 2.5 \times 10^3 \text{ ohms}$$

$$C_\pi = \frac{\tau_p}{r_\pi} = \frac{\beta}{2\pi f_T r_\pi} \simeq 1600 \text{ pfarads}$$

$$g_m = \frac{qI_c}{kT} \simeq \frac{qI_E}{kT} = 40 \times 10^{-3} \text{ mho}$$

9·6 REAL TRANSISTORS

There are several important factors which cause the dc behavior of real transistors to be different from that of the intrinsic transistor model. These factors are as follows:

1. The base width W of real transistors is not constant; instead it varies with the collector-base voltage v_{CB}. Therefore, those intrinsic transistor parameters (for example, α_F, β_F), which depend on W will also depend on v_{CB}.

2. Avalanche multiplication may occur in the collector-base transition region when the collector-base junction has sufficient reverse bias on it.

3. All the emitter current is not injected into the base. This is described by saying that the *emitter efficiency* γ is not unity. γ deviates from unity because there is recombination in the emitter-base transition region and there is minority-carrier injection from the base into the emitter.

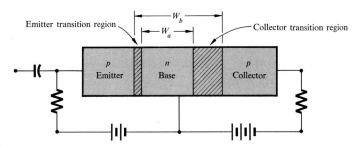

FIG. 9·25 BASE-WIDTH MODULATION DUE TO THE COLLECTOR VOLTAGE

4. The flow of base current causes ohmic voltage drops between the active area of the transistor and the base contact. These effects are usually represented by inserting a base resistance r_b between the base contact and the active area of the transistor. r_b will be a function of base current and collector voltage, in a manner to be described later.

5. There are structural asymmetries in real transistors (for example, the collector and emitter areas are usually not equal).

In this section we shall consider briefly the effects of these factors on the terminal characteristics of a real transistor.

9·6·1 *Base-width modulation.* It was shown in Sec. 5 that the width of the space-charge region surrounding a *pn* junction is a function of the voltage applied to the junction. This phenomenon is responsible for the voltage-variable capacitance of a *pn* junction diode and leads to similar voltage-dependent junction capacitances which have been included in the transistor model shown in Fig. 9·24.

In addition, the voltage dependence of the junction width affects the actual base width of a transistor, as shown in Fig. 9·25. At $v_{CB} = 0$, the base width of the transistor shown there is W_b. As the reverse bias on the collector junction is increased, the space-charge layer must widen to accommodate this increased voltage drop, so the actual width of the base W_a is decreased[1]

$$W_a = W_b - W_1(v_{CB}) \qquad (9·73)$$

Since the base width is a function of v_{CB}, then those parameters of a transistor which depend on W_b must also depend on v_{CB}.

The importance of base-width modulation on the dc characteristics of a transistor can be summarized by the following:

1. It causes both the input and output characteristics for a transistor to be more complicated than those of the intrinsic transistor.

2. In some cases base-width modulation sets an upper limit on the voltage which can be applied to the collector-base junction.

---

[1] The mathematical form of $W_1(v_{CB})$ can be established by the methods of Sec. 9·5. See also Prob. 9·11.

Output characteristics. The effect of base-width modulation on the dc characteristics of a transistor can be simply understood by referring to the output characteristics for the transistor in the common-base connection. According to Eqs. (9 · 50), the collector current in the common-base configuration, normal-active mode, can be written as

$$i_C = -\alpha_F i_E + I_{CO}(e^{qv_{CB}/kT} - 1) \qquad (9 \cdot 74)$$

where, from Eq. (9 · 37),

$$\alpha_F = \frac{1}{1 + W^2/2L_p^2} \qquad (9 \cdot 74b)$$

This equation is derived under the assumption that the base width W is constant. If this were true, then the slope of each curve in the i_C-v_{CB} family would be essentially zero when $|v_{CB}| \gg kT/q$; that is, each curve in the family would be perfectly horizontal over most of the normal-active region.

However, in a real transistor, α_F depends on the base width W, and through this on the collector-base voltage v_{CB}. Since W is reduced, α_F gets larger as the collector-junction reverse bias increases; or a greater fraction of the emitter current is collected for large reverse bias than for small reverse bias. This effect produces a positive slope on the curves in the i_C-v_{CB} family. This effect is shown, highly exaggerated, in Fig. 9 · 26a. Since α_F is nearly equal to unity without base-width modulation, the effect is not really too noticeable on actual transistor characteristics. However, it does lead to the output conductance which real transistors exhibit. Circuit models for real transistors (to be developed in the next chapter) must include this effect.

The effect is much more pronounced in the common-emitter family (i_C vs. V_{CE}, i_B as parameter). Here the equation for i_C in the normal-active region is [from Eq. (9 · 54)]

$$i_C = \beta_F i_B + \frac{I_{CO}}{1 - \alpha_F} \qquad (9 \cdot 75a)$$

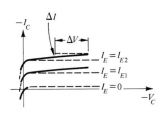

(a) *The effect of base-width modulation on the collector family for common-base operation*

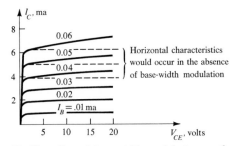

(b) *The effect of base-width modulation on the common-emitter output characteristics of a 2N1611*

FIG. **9 · 26** EFFECTS OF BASE-WIDTH MODULATION ON COLLECTOR CHARACTERISTICS

where
$$\beta_F = \frac{2D\tau}{W^2} \qquad (9 \cdot 75b)$$

D being the diffusion constant for minority carriers in the base, τ their lifetime, and W the base width. Variations in W affect β_F very significantly, so the common-emitter characteristics show appreciable slope, due to the base-width modulation effect. This is shown in Fig. $9 \cdot 26b$ for a 2N1611 (silicon *npn* transistor).

Input characteristics. Base-width modulation is also responsible for the dependence of the input characteristics on the collector-base voltage in a real transistor. The emitter current can be written as

$$i_E = I_{ES}(e^{qv_{EB}/kT} - 1) + \alpha_R I_{CS}$$

In a typical forward-bias condition, $\alpha_R I_{CS}$ is negligible, so the i_E-v_{EB} relation is that of an ideal diode. However, I_{ES} depends inversely on the base width W. Hence, the apparent saturation current of the emitter-base diode depends on the collector-base voltage. It is this fact which causes input characteristics of a real transistor to look like a series of diode curves, when v_{CB} is used as the parameter (see Fig. $9 \cdot 7b$, for example).

Punch-through. In certain transistors (for example, 2N1302) where the emitter and collector are alloyed onto a very lightly doped base layer, base-width modulation can cause the actual base width to become *zero*. When this happens the collector is said to have "punched-through" to the emitter. From Eq. $(9 \cdot 73)$, the collector voltage V_{pt} which must be applied for punch-through must satisfy

$$W_0 = W_1(V_{pt})$$

Calculations of V_{pt} on an actual structure are the subject of Prob. $9 \cdot 11$.

When a punch-through voltage exists, this voltage is also the maximum voltage which may be applied to the collector-base junction. Usually the avalanche breakdown voltage for the collector-base junction is less than the transistor's punch-through voltage, however, in which case the avalanche voltage sets the limit on maximum collector voltage.

$9 \cdot 6 \cdot 2$ *Avalanche multiplication at the collector-base junction.* In Sec. $6 \cdot 5$, it was pointed out that when a sufficient reverse bias is applied to a *pn* junction, carriers moving through the transition region can gain enough energy to ionize the lattice, thus generating new electron-hole pairs.

In a p^+n junction, the primary carriers which are responsible for the avalanche are the holes which make up the saturation current of the junction I_0. These holes diffuse into the transition region, and are then swept across it, creating new carrier pairs as they go. The ratio of the hole current emerging from the transition region to that which enters it is called the *avalanche multiplication factor M. M* can usually be satisfactorily represented by the formula

$$M = \frac{1}{1 - (v/V_B)^n} \tag{9.76}$$

where v is the junction voltage drop (reverse bias), V_B is the avalanche breakdown voltage for the junction, and n is an exponent that depends on the type of semiconductor material and the primary ionizing carrier. Because n is rather large (3 to 7) M is essentially unity until v gets quite close to V_B.

In a p^+np^+ transistor, the conditions required for avalanche can be set up at the collector-base junction by applying a large enough v_{CB}. The effect on the output characteristics will simply be to multiply the hole current entering the collector junction by M. For example, in a common-base connection of a p^+np^+ transistor, the hole current which enters the collector transition region would be

$$i_C' = -\alpha_{FI}i_E - I_{CO} \tag{9.77a}$$

The collector current would then be

$$i_C = Mi_C' = -M(\alpha_{FI}i_E + I_{CO}) \tag{9.77b}$$

As shown in Fig. 9·27, i_C approaches ∞ when v_{CB} approaches the collector avalanche breakdown voltage. This breakdown voltage is usually denoted by manufacturers as BV_{CBO} (meaning *b*reakdown *v*oltage for the *c*ollector-*b*ase junction when the emitter is *o*pen circuited).

The electrons created in the avalanche flow back into the base, and therefore contribute to the real internal base current. This leads to a very interesting effect in the *common-emitter* output characteristics.

Let us suppose that the external base current is supplied by a current source i_B, and that the hole current entering the collector transition region is i_C'. Then the actual number of electrons entering the base per unit time, which we call i_{Beff}, is

$$i_{Beff} = i_B + (M - 1)i_C' \tag{9.78}$$

The primary collector current i_C' is, from Eq. (9·54),

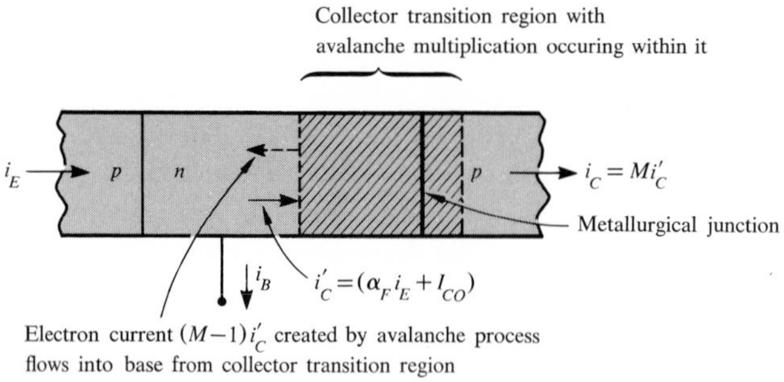

Collector transition region with
avalanche multiplication occuring within it

$i_C = Mi_C'$

Metallurgical junction

$i_C' = (\alpha_F i_E + I_{CO})$

Electron current $(M-1)i_C'$ created by avalanche process
flows into base from collector transition region

FIG. 9·27   TERMINAL CURRENTS IN A TRANSISTOR WHEN AVALANCHE MULTIPLICATION
EXISTS IN THE COLLECTOR-TRANSITION REGION

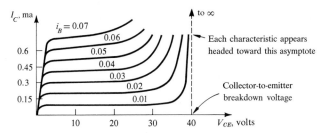

FIG. 9·28 THE EFFECT OF AVALANCHE BREAKDOWN ON THE COMMON-EMITTER OUTPUT CHARACTERISTICS OF A **2N1611**

$$i'_C = \beta_F i_{B_{eff}} + \frac{I_{CO}}{1 - \alpha_F} = \beta_F[i_B + (M - 1)i'_C] + \frac{I_{CO}}{1 - \alpha_F} \quad (9 \cdot 79a)$$

Therefore,
$$i'_C = \frac{\beta_F i_B + I_{CO}/(1 - \alpha_F)}{1 - \beta_F(M - 1)} \quad (9 \cdot 79b)$$

The actual collector current is, of course,

$$i_C = Mi'_C$$

Now, i'_C (or i_C) $\to \infty$ when $\beta_F(M - 1) \to 1$. This is shown in the curves of Fig. 9·28, and since β_F may be a large number, M need not be very much greater than 1 for i'_C to increase rapidly.

The voltage at which $\beta_F(M - 1)$ equals unity is called the collector-to-emitter sustaining voltage and is given the symbol BV_{CEO}. It is usually 40–50 percent of BV_{CBO} (for example, on a 2N1611, the typical value of BV_{CBO} suggested by the manufacturer is 110 volts. The typical value of BV_{CEO} is 40 volts). When a transistor is operated in the common-emitter connection, the maximum collector-to-ground voltage which may be applied is BV_{CEO}.

9·6·3 Emitter efficiency in real transistors. In formulating the intrinsic transistor model, we assumed that recombination in the emitter-base transition region (or the collector-base transition region if it is forward-biased) could be neglected, and that the emitter (and collector) are so heavily doped in comparison to the base that minority-carrier injection from the base into the emitter could be neglected. These assumptions are usually met in most transistors for typical operating conditions, but deviations do occur, and some special devices *depend* on these deviations for their operating characteristics.

In studying this problem it is customary to define the *emitter efficiency* γ as the ratio of the minority-carrier current actually injected into the base to the total emitter current. If we consider a *pnp* transistor, and denote the hole current actually injected into the base by i_{Eb}, then

$$\gamma \equiv \frac{i_{Eb}}{i_E}\bigg|_{v_{CB}=0} \quad (9 \cdot 80)$$

Operating principles for junction transistors 359

The defects which cause γ to be less than unity are shown schematically in Fig. 9·29. There is an electron current i_{nE} injected into the emitter (Prob. 9·12). These electrons recombine with holes in the emitter body (as in a normal diode). This increases the terminal-emitter current without increasing the injected hole current i_{Eb} (or the collector current i_C). Similarly, there is recombination in the transition region and on its surfaces, as described in Sec. 6·6·2. The holes for this recombination also form a part of the terminal-emitter current which does not result in collector current. If this current is denoted by i_T (see Fig. 9·29) then

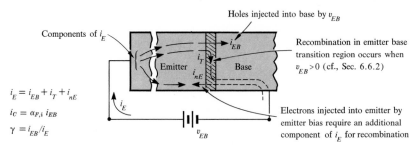

$i_E = i_{EB} + i_T + i_{nE}$

$i_C = \alpha_{F,i}\, i_{EB}$

$\gamma = i_{EB}/i_E$

(a) The defects which produce nonunity emitter efficiency

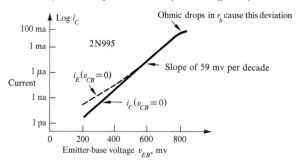

(b) Wide-range measurements of i_E and i_C vs. v_{EB}. Note significant discrepancy between i_E and i_C at low currents

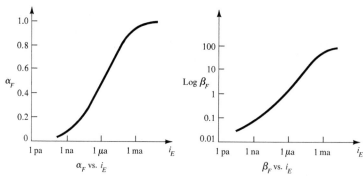

(c) The dependence of α_F and β_F on i_C

FIG. 9·29 CAUSES OF NONUNITY EMITTER EFFICIENCY AND ITS EFFECT ON TRANSISTOR TERMINAL PROPERTIES

$$\gamma = \frac{i_{Eb}}{i_{nE} + i_T + i_{Eb}} \tag{9.81}$$

Since normal transistor action involves only the component i_{Eb}, the previous equations for i_C in terms of i_E and i_B need to be modified. This is usually done as follows. The internal alpha ($\alpha_{F,i}$ of the intrinsic transistor) is defined to be

$$\alpha_{F,i} = \left.\frac{i_C}{i_{Eb}}\right|_{v_{CB}=0} \tag{9.82}$$

Then the real transistor has an α which is defined by

$$\alpha_F = \left.\frac{i_C}{i_E}\right|_{v_{CB}=0} \tag{9.83}$$

so the real α_F is

$$\alpha_F = \gamma\alpha_{F,i} \tag{9.83}$$

Serious deviations of γ from unity are usually associated with the recombination of carriers in the transition region, a phenomenon which is noticeable principally in silicon transistors at low emitter current. As indicated in Sec. 6·6·2, i_T usually increases with v_{EB} according to the law $e^{qv_{EB}/nkT}$, where $n \sim 2$, while i_{EB} varies as $e^{qv_{EB}/kT}$. Therefore, as v_{EB}, or the overall emitter current i_E, increases, i_{EB} will eventually predominate.

These features are brought out in the plots of i_E, i_B, and i_C versus v_{EB} shown in Fig. 9·29b for a 2N995 transistor. To obtain these curves, the collector and base leads are shorted together, so $v_{CB} = 0$. The ratio of i_C to i_E is α_F, and the ratio of i_C to i_B is β_F. α_F and β_F are plotted in Fig. 9·29c to illustrate their dependence on emitter current. Incidentally, the extremely wide range over which log i_C is proportional to v_{EB} shows how very accurate the junction law is on a real transistor.

9·6·4 Base resistance. The flow of base current from the base contact up to and in the active area of the transistor produces voltage drops which have two main effects:

1. They complicate the calculation of actual junction voltages.
2. They cause nonuniform emission across the area of the emitter.

These two effects are described schematically in Fig. 9·30. If voltages v_{EB} and v_{CB} are applied between the external terminals of the transistor, then the junction voltages will be, to a first approximation,

$$v'_{EB} = v_{EB} - i_B R_b \qquad v'_{CB} = v_{CB} - i_B R_b$$

where R_b accounts for the resistance between the base contact and the active area of the transistor.

Since i_B depends on $e^{qv'_{EB}/kT}$, it is not a simple matter to include the effects of R_b on the actual junction voltages. For dc calculations, i_B is a small number, as is R_b, so R_b is usually neglected in making bias calculations or in low-frequency large-signal calculations.

A more serious problem occurs because of the self-biasing effect which the laterally-flowing base current has on the emitter junction. As indicated in Fig. 9·30b, base current must flow laterally through the thin (and frequently lightly doped) active base width to supply recombination current to the central portions of the active region. This base-current flow causes a lateral voltage drop in the base. The polarity of this drop is such that the central portions of the emitter junction have less forward bias than the outer portions. Therefore the outer portions of the emitter carry more of the emitter current than the central portions. This phenomenon is called "emission crowding."

Emission crowding is a serious matter at high-current levels, where the actual current densities at the edges of the emitter may reach damagingly high values. At lower current levels ($I_B \lesssim 0.1$ ma), emission crowding is

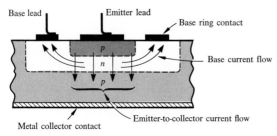

(a) Cross section of a planar pnp transistor

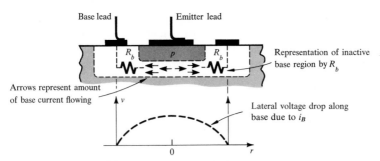

(b) Current paths and lateral voltage drops in the base region

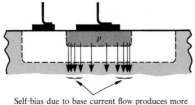

Self-bias due to base current flow produces more emission at the emitter edges than in its center

(c) The emission crowding which occurs from self-bias due to i_B

FIG. 9·30   BASE RESISTANCE AND ITS EFFECTS ON TRANSISTOR OPERATION

362   Semiconductor electronics

not so serious, and the lateral base current flow within the active region can again be approximately accounted for by adding a current-dependent base resistance[1] (see Prob. 9·13).

For most purposes in our studies, it will be sufficient to lump all these effects into a single base resistance r_b, which we add in series with the base lead of an intrinsic transistor to approximate a real transistor.

9·6·5 *Structural asymmetries and their consequences.* The intrinsic transistor has the interesting property that it is completely symmetrical about a plane passed through the center of the base ($x = W/2$). This high degree of symmetry leads to many interesting circuit properties and simplifies circuit thinking and analysis significantly. Of course, most real transistors are not symmetrical. The asymmetries may be of considerable significance in some cases, though on the whole they are not of first-order importance.

The principal asymmetries which one finds in modern transistors are
1. The collector body may be lightly doped in comparison to the base.
2. The collector area is normally much greater than the emitter area.

The effects of these asymmetries on the terminal characteristics of a real transistor are described in Fig. 9·31. The current versus minority-carrier-density relationships are still linear, of course, so we may still use superposition techniques to determine the terminal currents in terms of the minority-carrier densities.

In Fig. 9·31a, we show the approximate flow pattern for holes from the emitter to the collector for the bias condition $v_{EB} > 0$, $v_{CB} = 0$. The analysis of the intrinsic transistor shows that, so long as the base width W is small compared to L_p,

$$i_E = \left(\frac{qA_E D_p}{W} + \frac{qA_E W}{2\tau_p} \right) p_E \qquad (9\cdot84a)$$

$$i_C = - \frac{qA_E D_p}{W} p_E \qquad (9\cdot84)$$

where A_E is the emitter area and p_E is the excess minority-carrier density at the emitter edge of the base region.

On the basis of these equations we define

$$\alpha_F = \frac{i_C}{i_E}\bigg|_{v_{CB}=0} = \frac{1}{1 + W^2/2L_p^2} \qquad (9\cdot85a)$$

$$I_{ES} = \left(\frac{qA_E D_p}{W} + \frac{qA_E W}{2\tau_p} \right) p_n \qquad (9\cdot85b)$$

---

[1] An interesting illustrative calculation of emission crowding and the effective base resistance in the active region of a typical transistor is carried out in P. E. Gray, D. DeWitt, A. R. Boothroyd, and J. F. Gibbons, "Physical Electronics and Circuit Models of Transistors," SEEC, vol. 2, John Wiley & Sons, Inc., New York, 1964, pp. 155–162.

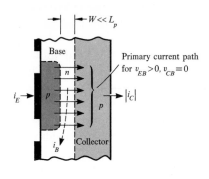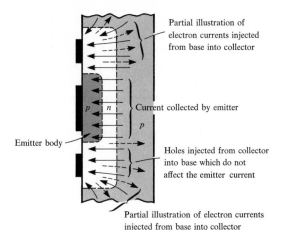

Partial illustration of electron currents injected from base into collector

$W \ll L_p$

Base

Primary current path for $v_{EB} > 0$, $v_{CB} = 0$

i_E p n p $|i_C|$

i_B Collector

Emitter body

p n Current collected by emitter

p

Holes injected from collector into base which do not affect the emitter current

Partial illustration of electron currents injected from base into collector

(a) Flow pattern for holes with $v_{EB} > 0$, $v_{CB} = 0$

(b) Hole and electron currents for the condition $v_{EB} = 0$, $v_{CB} > 0$. The illustration includes electron injection into a collector body which is assumed to be lightly doped in comparison to the base.

FIG. 9·31 THE EFFECT OF STRUCTURAL ASYMMETRIES ON CURRENT PATHS IN A TRANSISTOR

We may then rewrite Eqs. (9·77) as

$$i_E = I_{ES} \frac{p_E}{p_n} \qquad (9 \cdot 86a)$$

$$i_C = -\alpha_F I_{ES} \frac{p_E}{p_n} \qquad (9 \cdot 86b)$$

If we now set $v_{EB} = 0$, $v_{CB} > 0$, two things occur. First, the fact that the collector may be lightly doped in comparison to the base means that a significant number of electrons will be injected into the collector. This means that for a given v_{CB}, the collector current will be greater than it would be if the collector body were heavily doped; or, equivalently, the collector saturation current will be larger for the lightly doped collector body. Second, because the area of the emitter A_E is less than the collector area A_C, many of the holes injected into the base will recombine on the base surfaces opposite to the collector. These effects are partially illustrated in Fig. 9·31b.

Now, we may express the collector current for this bias condition approximately as follows:

$$i_C = \left(\frac{qA_E D_p}{W} + \frac{qA_E W}{2\tau_p} \right) p_C + \left[\frac{q(A_C - A_E)D_p}{L} + \frac{q(A_C - A_E)L}{2\tau_p} \right] p_C$$
$$+ \frac{qA_C D_n}{L_n} n_C \qquad (9 \cdot 87a)$$

In this equation the first two terms are the normal "intrinsic transistor" contributions which arise from applying a forward bias to the collector base junction. The next two terms account for the holes which are injected into the base that cannot be collected by the emitter; the first term in this bracket accounts for holes which diffuse to the base surface opposite the collector before recombining, while the second term accounts for recombination of excess minority carriers in that volume of the base which is outside the "emitter shadow." The final term in Eq. $(9 \cdot 87a)$ accounts for electron injection from the base into the collector. The collector body is assumed to be sufficiently long that this current can be represented by the normal diode expression.

Now, in Eq. $(9 \cdot 87a)$ both n_C and p_C depend on the junction voltage v_{CB}. In particular,

$$p_C = p_n(e^{qv_{CB}/kT} - 1) \qquad n_C = n_{pc}(e^{qv_{CB}/kT} - 1)$$

where p_n is the thermal-equilibrium hole density in the base and n_{pc} is the thermal-equilibrium electron density in the collector body. From these equations we can write

$$n_C = \frac{n_{pc}}{p_n} p_C$$

We can then use this result to rewrite Eq. $(9 \cdot 87a)$. After some manipulation we find that i_C can be expressed as

$$i_C = I_{CS}\frac{p_C}{p_n} \qquad (9 \cdot 87b)$$

Now, the emitter current for $v_{CB} > 0$, $v_{EB} = 0$ is

$$i_E = -\frac{qA_ED_p}{W}p_C \qquad (9 \cdot 87c)$$

We can therefore define a reverse α, α_R, to be

$$\alpha_R \equiv -\frac{i_E}{i_C}\bigg|_{v_{EB}=0} \qquad (9 \cdot 88)$$

Then we can write, for $v_{CB} > 0$, $v_{EB} = 0$,

$$i_C = I_{CS}\frac{p_C}{p_n} \qquad i_E = -\alpha_R I_{CS}\frac{p_C}{p_n}$$

Utilizing the superposition idea, we may now write the equations for i_C and i_E for unspecified values of v_{EB} and v_{CB} as

$$i_E = I_{ES}\left(\frac{p_E}{p_n}\right) - \alpha_R I_{CS}\left(\frac{p_C}{p_n}\right) \qquad i_C = -\alpha_F I_{ES}\left(\frac{p_E}{p_n}\right) + I_{CS}\left(\frac{p_C}{p_n}\right) \qquad (9 \cdot 89)$$

These equations are formally identical to those obtained for the intrinsic transistor except that now

$$I_{CS} \neq I_{ES} \qquad \alpha_R \neq \alpha_F$$

However,

$$\alpha_F I_{ES} = \frac{qA_E D_p p_n}{W} = \alpha_R I_{CS}$$

so *the coupling coefficients between the two "diodes" are independent of the structural asymmetries.*

This conclusion holds even for an exact analysis, so we may completely determine the dc performance of a transistor by making three measurements: I_{ES}, I_{CS}, and one of the alphas (recombination in forward-biased transition regions has to be included separately, since its voltage dependence does not follow the ideal junction law).

9·7 SUMMARY

In this chapter we have presented the basic principles which are responsible for transistor action. We have seen that a reverse-biased collector junction placed sufficiently close to a forward-biased emitter junction can collect•most of the emitted current. This idea is expressed mathematically in the equation

$$i_C = -\alpha_F i_E - I_{CO}$$

More generally, the collector-base and emitter-base junctions in a transistor can be viewed as two coupled diodes. The emitter and collector currents are expressed as

$$i_E = I_{ES}\frac{p_E}{p_n} - \alpha_R I_{CS}\frac{p_C}{p_n}$$

$$i_C = -\alpha_F I_{ES}\frac{p_E}{p_n} + I_{CS}\frac{p_C}{p_n}$$

i_E and i_C depend linearly on the excess densities at the emitter and collector edges of the base region, p_E and p_C, and on the factors α_F, α_R, I_{ES}, and I_{CS}. These latter factors can be measured, obtained from a transistor data sheet, or computed from fabrication data. All four parameters are not given because the relation

$$\alpha_F I_{ES} = \alpha_R I_{CS}$$

must be true for all transistors.

The excess densities p_E and p_C are related to the junction voltage drops through the law of the junction

$$\frac{p_E}{p_n} = (e^{qv_{EB}/kT} - 1) \qquad \frac{p_C}{p_n} = (e^{qv_{CB}/kT} - 1)$$

These relations, together with the previous equations for i_E and i_C give a complete representation of the dc behavior of a transistor.

The base current in a transistor is usefully viewed as setting up the majority-carrier component of the base charge. In the usual connection of the transistor, the common-emitter configuration, signals are applied to the transistor in the form of a base-current or a base-emitter voltage. A small-signal model for this configuration has been derived (Fig. 9·24) from a consideration of time-dependent base-charging currents. The model exhibits the basic amplification property of the transistor through a dependent current generator at its output. It also exhibits the basic frequency characteristics of the transistor through the presence of the $r_\pi C_\pi$ network. The model gives a reasonably accurate representation of the transistor up to frequencies of about 0.1 to 0.5 $[1/(2\pi\tau_t)]$, where τ_t is the transit time for minority carriers across the base region. This frequency limitation arises from the fact that we assumed in the derivation of the model that the hole-density profile was a straight line across the base.

REFERENCES

Linvill, J. G., and J. F. Gibbons: "Transistors and Active Circuits," McGraw-Hill Book Company, New York, 1961, chaps. 4 and 8.

Gray, DeWitt, Boothroyd, and Gibbons: "Physical Electronics and Circuit Models of Transistors," SEEC vol. 2, John Wiley & Sons, Inc., New York, 1964, chaps. 7–10.

Searle, Boothroyd, Angelo, Gray, and Pederson: "Elementary Circuit Properties of Transistors," SEEC vol. 3, John Wiley & Sons, Inc., New York, chaps. 1–4.

Moll, J. L.: "Physics of Semiconductors," McGraw-Hill Book Company, New York, 1964, chap. 8.

Nanavati, R.: "Introduction to Semiconductor Electronics," McGraw-Hill Book Company, New York, 1963, chap. 4.

Valdez, L.: "The Physical Theory of Transistors," McGraw-Hill Book Company, New York, 1961.

Middlebrook, R. D.: "An Introduction to Junction Transistor Theory," John Wiley & Sons, Inc., New York, 1957, chaps. 8, 9, and 10.

DEMONSTRATION

COLLECTOR FAMILY FOR COMMON-BASE AND COMMON-EMITTER OPERATION. Small, portable curve tracers such as the Cubic Model 504 curve tracer require only a scope in addition to the curve tracer for visual display of transistor characteristics. If a curve tracer is not available, one can be made quite simply in the manner shown below.

The finite output conductance (due to base width modulation) on an alloy transistor can be compared to the nearly horizontal characteristics obtained in a mesa transistor (or other transistor where the collector body is more light doped than the base). The difference between common-emitter and common-base collector characteristics may also be readily demonstrated with this set up.

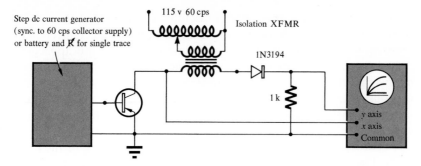

Step dc current generator
(sync. to 60 cps collector supply)
or battery and R for single trace

115 v 60 cps

Isolation XFMR

1N3194

1 k

y axis
x axis
Common

FIG. D9 · 2

PROBLEMS

9·1 Using the exact formula for the collector current of an intrinsic transistor [Eq. (9·27), with Eqs. (9·28) used to define a_{21} and a_{22}], calculate i_C for $W = 10L_p$, $5L_p$, L_p and $0.1L_p$, assuming $p_E = $ const and $p_C = -p_n$. Compare each value with the value of i_C for $W = \infty$ by taking the ratio $i_C(W = L_p)/i_C$ ($W = \infty$). Explain these results.

9·2 A symmetrical germanium transistor has a base width of 10^{-3} cm, and an emitter area of 10^{-3} cm^2. The lifetime of minority carriers in the base is 1 μsec, and the base doping density is $N_d = 10^{15}/$cm^3. The hole-diffusion constant $D_p = 50$ cm$^2/$sec. Treating this transistor as an intrinsic transistor,

1. Calculate the exact parameters a_{11}, a_{12}, a_{21}, and a_{22} given in Eqs. (9·28).
2. Calculate the a_{ij} by the approximate expressions given in Eqs. (9·29) and compare them with the exact values just calculated.

9·3 Using the exact excess-hole distribution $p^*(x)$ given in Eq. (9·20) [with C and D as defined in Eqs. (9·23)], evaluate

$$i_B = \frac{1}{\tau_p} \int_0^W q_P(x) \, dx$$

for arbitrary values of p_E and p_C. Compare the value so obtained with the one which you find by using Eqs. (9·27) and Kirchhoff's current law.

9·4 Draw circuit diagrams indicating how you would measure the parameters which appear in the Ebers-Moll model. The transistor may be assumed to be intrinsic.

9·5 In obtaining the collector characteristics of the common-base connection, it was convenient to define

$$I_{CO} = (1 - \alpha_F\alpha_R)I_{CS}$$

1. Show that this is the current which flows across the collector-base junction when $v_{CB} \ll 0$ and $i_E = 0$ (that is, when the emitter-base junction is left open).
2. Derive a formula for the voltage that appears between the emitter-base terminals when this measurement is made.
3. Sketch the minority-carrier-density distribution which you expect when this measurement is made.

9·6 A diode can be formed from a transistor in the manner shown in Fig. P9·6.

1. Sketch the approximate hole-density distribution across the base of this transistor when v is a small positive voltage.

2. Obtain a formula for the v-i characteristic of this diode, and compare it with the v-i characteristic of the ideal junction diode

$$i = \frac{qAD_p p_n}{L_p}(e^{qv/kT} - 1)$$

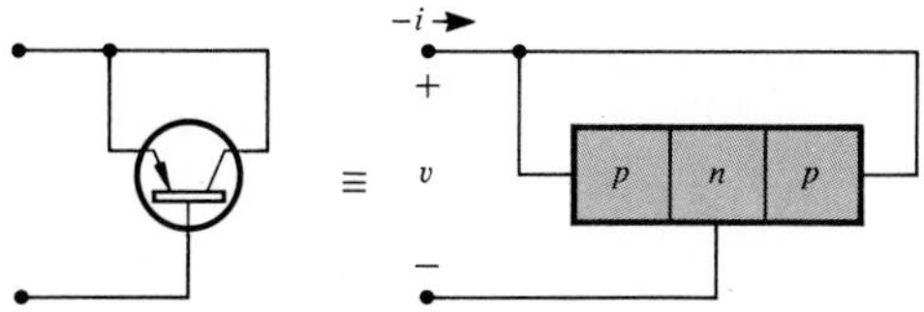

FIG. **P9·6**

9·7 A symmetrical transistor has $\alpha = 0.98$, $I_{CO} = -2$ μa. What are the junction voltages when $i_E = 2$ ma, $i_B = -1$ ma? Draw a hole-density profile in the base for this condition to explain these results.

9·8 Draw the minority-carrier charge distribution for the condition $p_E = 0$, $p_C = -p_n$ and calculate the terminal currents from the charge control formulation. Assume $\tau_t = 2 \times 10^{-9}$ sec and $\tau_p = 1$ μsec.

9·9 A symmetrical silicon-alloy transistor has a base doping density of 10^{16} donors/cm³. The base width is 2.5×10^{-3} cm, and the minority-carrier lifetime in the base is 0.1 μsec. Calculate the parameters of the intrinsic transistor small-signal circuit model, common-emitter connection, for a bias current of $I_E = 2$ ma. The collector-base bias voltage is low enough that the entire width of the base region may be used to compute β.

9·10 Show that the low-frequency small-signal current gain i_c/i_b of the transistor model shown in Fig. 9·24 is $g_m r_\pi$. By substituting values for g_m and r_π, show that $g_m r_\pi = \beta$.

9·11 1. Suppose the collector of a *pnp* transistor has a doping density $N_a \gg N_d$ for the base. Calculate the width of the collector-transition region when a reverse bias v_{CB} is applied to the collector-base junction. Assume $|v_{CB}| \gg \phi$, the built-in voltage of the junction.

2. The "punch-through" voltage V_{pt} for a transistor is the collector voltage which causes the collector-transition region to extend all the way to the emitter. In a 2N1302, V_{pt} is about 25 volts and the base doping density is about $10^{15}/cm³$. Calculate the zero-bias base width.

9·12 The emitter efficiency is defined to be

$$\gamma = \left.\frac{i_{EB}}{i_E}\right|_{v_{CB}=0}$$

where i_{EB} is the current injected into the base by the emitter and i_E is the total current. Let us assume that recombination in the transition region is negligible. Then the emitter junction current has two components: the electron current injected into the emitter i_{nE} and i_{EB}. Assuming the emitter is many diffusion lengths

Operating principles for junction transistors **369**

long for the injected electrons, write down an expression for i_{nE}. Using this and the exact expression for i_{EB} [Eq. (9·27)], find γ. Discuss what can be done to make γ approach unity.

9·13 The base current at which emission crowding will become important can be roughly estimated by assuming that when the lateral voltage drop is on the order of kT/q, emission crowding is just becoming significant. Calculate the base current which produces this condition for a transistor which has a lateral resistance of 200 ohms from the edge of the active base region to its center.

9·14 An idealized germanium-alloy transistor is shown in Fig. P9·14. The base doping density is $10^{15}/cm^3$ and the lifetime for holes in the base is $\tau_p = 10$ μsec. The emitter efficiencies for both the collector and the emitter may be assumed to be unity.

1. Calculate I_{ES}, α_F, α_R, and I_{CS} (approximately) for this transistor.
2. Show that $\alpha_F I_{ES} = \alpha_R I_{CS}$.

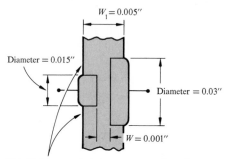

$W_1 = 0.005''$

Diameter $= 0.015''$

Diameter $= 0.03''$

$W = 0.001''$

Hole lifetime on surfaces is zero, so $p = p_n$ here always

FIG. **P9·14**

9·15 A symmetrical transistor with $I_{ES} = 2$ μa and $\alpha = 0.95$ is connected in the circuit shown in Fig. P9·15. Calculate the current which flows in this circuit and the junction voltage drops.

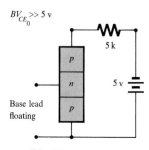

$BV_{CE_0} \gg 5$ v

5 k

p

n

5 v

Base lead
floating

p

FIG. **P9·15**

9·16 1. In the discussion relating to Eq. (9·24), it was indicated that the "small terms" in this equation account for the base current. Using the expansion

$$e^x = 1 + x + \frac{x^2}{2}$$

rewrite Eq. (9·24), including terms in x^2, and evaluate the base current in terms of C and D.

2. An alternate procedure was given in Sec. 9·3·5 for evaluating the base current. Here i_B is obtained from

$$q_P = qA \int_0^W p^*(x)\, dx$$

q_P can be approximated from the straight line distribution for $p^*(x)$ giving Eqs. (9·31). Compare Eq. (9·31) with the equation for i_B obtained in part (1) of this problem. Using the expansion suggested in part (1) for e^x, estimate the error in base current which arises from assuming that the $p^*(x)$ profile is a straight line.

9·17   A 2N1302 has a punch-through voltage of 25 volts. Its β at low voltage is 20. Estimate the minority-carrier lifetime in the base region of this transistor.

9·18   Compute C_c, C_π, r_π, and g_m for the transistor given in Prob. 9·14 for a reverse-bias on the collector-base junction of 6 volts and an emitter current of 1.0 ma. Be sure to allow for the effect of base-width modulation in determining the operating base width.

10 INTRODUCTION TO AMPLIFIERS

TRANSISTORS AND VACUUM tubes are useful circuit elements because, under suitable conditions, they can deliver more *signal power* to a load than they absorb at their inputs. Because of this property, they are called *active* circuit elements. The term "active" distinguishes them from *passive* networks (*RLC*), which cannot deliver more signal power to a load than they absorb from a source. Of course, transistors and vacuum tubes are not active elements at all frequencies, but only up to some maximum frequency. At sufficiently high frequencies, transistors and vacuum tubes behave like ordinary passive networks (that is, three-terminal *RLC* networks).

The fact that active elements can deliver more signal power than they absorb leads us to call them amplifiers and, in some cases, to characterize their behavior by quoting *current gains, voltage gains, and power gains.* We do not mean by this that they violate the law of energy conservation, of course. The signal power delivered to the load comes from a battery or other power supply. The active element only converts some of the dc power available from this supply into ac signal power and makes this signal power available at its terminals.

It is useful to begin our study of the applications of active devices with some simple graphical considerations which highlight these aspects of their behavior. To concentrate on essentials, we shall consider only examples in which the active nature of the component is most apparent. In particular, we shall leave the question of frequency response for a later time, when we have a thorough familiarity with the basic low-frequency properties of active devices.

Our studies in this chapter will be divided as follows. We shall begin by studying the operating characteristics of a basic common-emitter transistor amplifier. We shall then develop briefly the physical theory of triode and pentode vacuum tubes, consider some simple amplifiers which can be made from these devices, and conclude with a perspective view of the contents of the remaining chapters in the book.

10·1 GRAPHICAL CHARACTERISTICS AND LOAD-LINE ANALYSIS OF A COMMON-EMITTER AMPLIFIER

One way of studying the basic signal-amplifying properties of an active device is with the aid of a set of graphical *v-i* curves which characterize the low-frequency electrical behavior of the device at its terminals. In this section we shall use a graphical method to study characteristics of a simple common-emitter amplifier. For this purpose a simple common-emitter amplifier circuit is shown in Fig. 10·1, together with a set of common-emitter output characteristics for a typical *npn* silicon diffused-base transistor.

10·1·1 *Load-line determination.* To study the signal-transmission properties of this circuit, we first determine the values of v_{CE} and i_C which are permitted in the collector circuit.[1] We do this by observing that a permissible pair of values (v_{CE}, i_C) must:

---

[1] The following notation will be used to specify quantities which have both dc and ac components. The total value of a variable is expressed as a lower-case letter with an upper-case subscript: for example, v_{CE}; the dc value of a variable is expressed as an upper-case letter with upper-case subscript: V_{CE}; and the ac value of a variable is expressed with a lower-case letter and subscript, v_{ce}. In the event that v_{ce} is a sine wave, its rms amplitude is expressed by a capital letter with lower-case subscript, V_{ce}. Thus, if v_{CE} contains both a dc and a sinusoidal component,

$$v_{CE}(t) = V_{CE} + v_{ce}(t) = V_{CE} + \sqrt{2}\, V_{ce} \sin \omega t$$

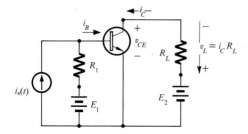

(a) Simple common-emitter transistor amplifier

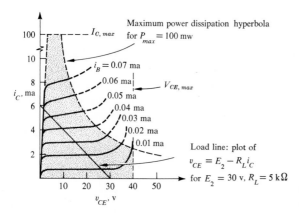

(b) Common-emitter output character-istics for an npn transistor

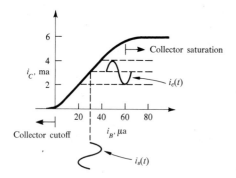

(c) Current-transfer characteristic for (a) assuming no signal-current loss in base biasing network

FIG. 10·1   A SIMPLE COMMON-EMITTER AMPLIFIER

1. Satisfy the *load equation*

$$v_{CE} = E_2 - R_L i_C \qquad (10·1)$$

2. Be compatible with the output characteristics of the transistor

These two conditions are met by constructing a *load line* on the output characteristics of the transistor, as is done in Fig. 10·1b. The load line is simply a plot of Eq. (10·1) for given values of E_2 and R_L. A load line for $E_2 = 30$ volts, $R_L = 5$ kohms, is shown on Fig. 10·1b.

The selection of actual values for E_2 and R_L is usually determined by two factors: (1) the intended uses of the amplifier and (2) the transistor ratings. The interplay of these factors can be understood from the following remarks.

The transistor ratings determine the permissible area of the $v_{CE} - i_C$ plane in which a load line may be located. The transistor manufacturer will specify a maximum collector current, a maximum collector-to-emitter voltage, and a maximum power-dissipation rating. For the transistor shown in Fig. $10 \cdot 1$, these are $I_{C, max} = 100$ ma, $V_{CE, max} = 40$ volts, $P_{max} = 100$ mw. These numbers lead to the following set of inequalities:

$$v_{CE} < V_{CE, max} \qquad i_C < I_{C, max} \qquad v_{CE}i_C < P_{max}$$

These inequalities are all met simultaneously inside the shaded area shown in Fig. $10 \cdot 1b$. The shaded area is the area of the $v_{CE} - i_C$ plane in which the load line may lie. In general, it is good practice to reduce the maximum ratings by about 10 percent and then redefine the permissible area accordingly.

The nature of the application specifies the position of the load line within the permissible area. If it is desired to deliver maximum undistorted power to the load, then E_2 will be as near to the rated maximum collector-to-emitter voltage as the distortion requirements will permit. R_2 will be selected so that either the maximum collector current or the maximum power-dissipation rating is not exceeded. For a small-signal application in which maximum power *gain* is required, R_L will be determined by other conditions, to be described in later chapters, and E_2 will be selected to minimize power losses. R_L for such cases is usually larger than the R_L for maximum power output.

10·1·2 Signal-transfer characteristics. When a load line has been selected (for example, $E_2 = 30$ volts, $R_L = 5$ kohms), the signal-transfer properties of the circuit can be developed. For this particular circuit, the signal-transfer characteristics are given by the *current-transfer curve* (i_C vs. i_B) shown in Fig. $10 \cdot 1c$. This curve may be constructed by simply assuming various values for i_B and reading the resulting value of i_C.

10·1·3 Selection of an operating point. The current-transfer curve (or the load line from which it was derived) can be used to study the signal-transmission properties of the circuit. To do this, one first selects an *operating point* on the load line (or on the current-transfer curve) which reflects the nature of the signals which the amplifier is to handle.

If $i_S(t)$ is a pulse of unknown amplitude, then we would select a bias point on the low current end of the transfer curve. If the maximum amplitude of the pulse were known, and proportionality between the output current $i_C(t)$ and the signal current $i_S(t)$ were important, the operating point would be placed in a range where the output characteristics are equally spaced and parallel, or where the transfer curve is most nearly linear.

If $i_S(t)$ is a wave where essentially equal positive and negative amplitudes are expected, then an operating point near the center of the load line may be chosen, though this will again depend on the expected signal amplitudes.

Once the operating point is chosen, a biasing network will be designed to obtain it. We shall consider the design of biasing networks in detail in Chap. 12, where we will see that the biasing network shown in Fig. $10 \cdot 1$ is not actually a good one. However, it introduces a minimum of confusion into the present line of reasoning, so we shall merely assume that E_1 and R_1 are selected to provide the base-bias current which is required for the envisaged application. For a sine-wave signal current $i_s(t)$, this would mean $I_B \approx 0.03$ ma. If we use $E_1 = E_2$ (same battery), then $R_1 \sim 1$ megohm (neglecting the emitter-base voltage drop, which is small compared to 30 volts).

10·1·4 *Current gain.* If the signal input current were a sine wave with peak value of 10 μa, and the signal-current loss in the base biasing network is neglected, the collector signal current would be essentially a sine wave with 1-ma peak amplitude. The word "essentially" is used because no transfer characteristic is perfectly linear, so a sine-wave input current will not produce a sine-wave output current. The output signal will always be *distorted,* which means that a Fourier analysis of $i_c(t)$ will reveal the presence of harmonics in the $i_c(t)$ waveform.

The harmonic content in $i_c(t)$ can be calculated by performing a graphical Fourier analysis on the $i_c(t)$ waveform which results from a given $i_s(t)$. However, as long as the peak amplitude of $i_s(t)$ is 20 μa or less, distortion in the output waveform will not be serious. In such cases we may define the *signal-current gain* to be the ratio of $i_c(t)$ to $i_s(t)$:

$$A_i = \frac{i_c(t)}{i_s(t)} \tag{10 · 2}$$

A_i is simply the slope of the current-transfer curve. For our example, $A_i = 100$. Usually the word "signal" is dropped, and A_i is referred to as the current gain.

It will be apparent that as long as the excursions of $i_s(t)$ do not drive the transistor out of the region where the current-transfer curve for the circuit is sensibly linear, the current gain will not depend on the waveform of $i_s(t)$; that is, the current gain for a pulse will be the same as that for a sine wave, neglecting transient phenomena. When larger signals are used (particularly sine waves), the definition of current gain must be used with care.

10·1·5 *Voltage gain.* The basic nature of the transistor amplifier is expressed in its current-transfer property. However, transistors are also used in situations where the signal source cannot be represented as a current source. Instead, the signal may come from a voltage source which has some internal impedance R_s. In such cases, the source will usually be coupled to the transistor through a blocking capacitor, as shown in Fig. $10 \cdot 2$, to pre-

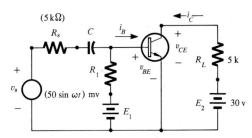

vent dc currents from flowing in the source. As an example, we shall suppose that the signal source is a 50-mv source with a 5-kohm internal impedance. To determine the output current of this amplifier, we need to determine the small-signal base current which the voltage source produces. The current-transfer characteristic can then be used, as before, to obtain the output current.

To determine the small-signal base current, we must first study the input characteristics of the transistor. They consist of a set of i_B-v_{BE} curves plotted with v_{CE} as a parameter. Typical input characteristics for the silicon *npn* transistor of Fig. 10·1 are shown in Fig. 10·3. It is evident from these curves that the i_B-v_{BE} relation is essentially independent of v_{CE} for $v_{CE} > 0.5$ volt. Since this condition will be valid for our amplifier, we only need to consider the one input characteristic which applies for $v_{CE} > 0.5$ volt.

The biasing network will produce a dc base bias current of 30 μa as before, the dc base-to-emitter voltage being approximately 0.55 volt. The signal source will cause i_b and v_{be} to vary about this bias point. To study these variations, the input characteristic has been expanded in the neighborhood of the bias point and presented in Fig. 10·4. Both the total and the incremental values of the input variables are plotted.

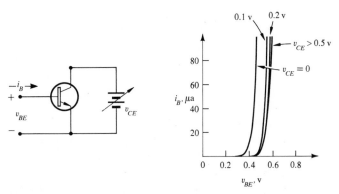

FIG. **10·3**   INPUT CHARACTERISTICS OF AN *npn* SILICON DIFFUSED BASE TRANSISTOR IN THE COMMON-EMITTER CONFIGURATION

Introduction to amplifiers   *377*

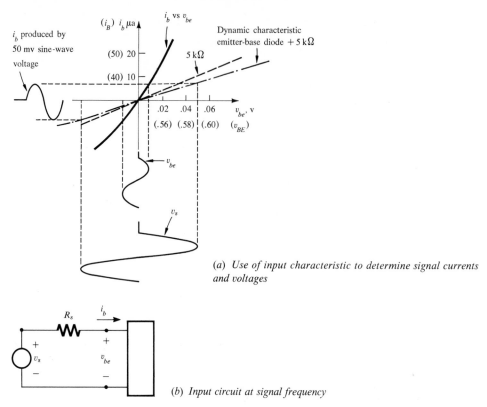

i_b produced by 50 mv sine-wave voltage

i_b vs v_{be}

Dynamic characteristic emitter-base diode $+5\,\mathrm{k}\Omega$

$(i_B)\ i_b\ \mu\mathrm{a}$

$(50)\ 20$

$(40)\ 10$

$5\,\mathrm{k}\Omega$

$5\,\mathrm{k}\Omega$

.02　.04　.06　　v_{be}, v

(.56)　(.58)　(.60)　(v_{BE})

v_{be}

v_s

(a) Use of input characteristic to determine signal currents and voltages

R_s

i_b

$+$

v_s

$+$

v_{be}

$-$

$-$

(b) Input circuit at signal frequency

FIG. 10·4　EXPANDED INPUT CHARACTERISTIC FOR COMPUTING INPUT CIRCUIT PROPERTIES

To estimate the signal current which the source produces, we neglect the signal losses in the bias resistor and assume that the reactance of the capacitor at the signal frequency is very small compared to that of R_s or the input impedance of the transistor. Under these conditions, the signal source faces the simplified circuit shown in Fig. 10·4b.

The base-signal current is determined by a graphical solution of Kirchhoff's laws for the circuit of Fig. 10·4b. The signal voltage v_s is related to i_b through the equation

$$v_s = i_b R_s + v_{be} \qquad (10\cdot3)$$

The v-i characteristic of the 5-kohm resistor and that of the base-emitter diode are added together graphically in Fig. 10·4a to produce the so-called *dynamic v-i characteristic* which the source faces. From this dynamic characteristic, we establish that the base-signal current is approximately a sine wave with 7.5-μa peak value. The word "approximately" is used to emphasize the fact that because the input characteristic of the transistor is nonlinear, the sine-wave voltage source will *not* produce a pure sine-wave base-signal current.

This is not too noticeable in the base-current waveform because, as

Fig. $10 \cdot 4a$ implies, the base current is determined primarily by the 5 kohm resistor. However, the nonlinearity is more apparent in the v_{be} curve, where the peak-positive swing is 8 mv compared with a peak-negative swing of about 12 mv.

We may now proceed to find the output current quite simply. Using the current gain developed before, we have simply

$$i_c(t) = A_i i_b(t) = 100 \times 7.5 \sin \omega t \; \mu a$$

The collector-signal current is therefore a sine wave with peak amplitude of 0.75 ma.

We can also define a *voltage gain* A_v for this circuit in several ways. If we adopt as our definition the ratio of the peak signal voltage developed across the load resistor to the peak voltage of the signal source, we have

$$A_v = \frac{0.75 \text{ ma} \times 5 \text{ kohm}}{50 \text{ mv}} = 75$$

The voltage gain would, of course, be much higher if we used v_{be} as the reference for defining A_v. We shall defer until Chap. 13 questions of which voltage should be used as a reference.

$10 \cdot 1 \cdot 6$ *Power relations in the amplifier output.* The relations governing power flow in the output circuit can be developed as follows. We return to the original 10-μa base-signal current source, for simplicity in the numerical calculations. Then the instantaneous collector current is

$$i_c(t) = I_C + i_c(t) = (3 + 1 \sin \omega t) \text{ ma} \qquad (10 \cdot 4)$$

which is plotted in Fig. $10 \cdot 5a$. From the load line of Fig. $10 \cdot 1b$ we also find

$$v_{CE}(t) = V_{CE} + v_{CE}(t) = (15 - 5 \sin \omega t) \text{ volts} \qquad (10 \cdot 5)$$

$v_{CE}(t)$ is plotted in Fig. $10 \cdot 5b$.

Now, the power leaving the collector battery at each instant is

$$p_B(t) = E_2 i_c(t) \qquad (10 \cdot 6)$$

Since E_2 is constant, the time-average power drawn from the battery, which we denote by $\overline{p_B(t)}$, is

$$\overline{p_B(t)} = E_2 I_C$$

$\overline{p_B(t)}$ is 90 mw for our running example.

The power absorbed by the load at each instant is

$$p_L(t) = R_L i_c^2(t) = R_L[I_C + i_c(t)]^2 \qquad (10 \cdot 7a)$$

Again, assuming $i_c(t)$ is a sine wave, the average power absorbed by the load is

$$\overline{p_L(t)} = R_L(I_C^2 + I_c^2) = P_L + P_l \qquad (10 \cdot 7b)$$

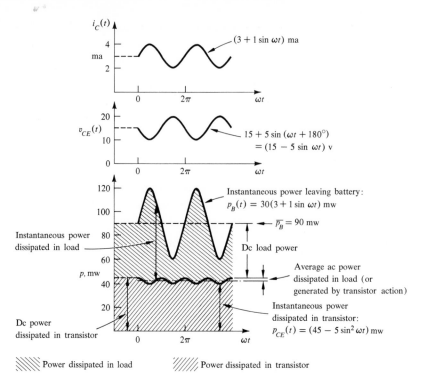

In the absence of an input signal, P_L is 45 mw for our example. The 10-μa
sine-wave signal current causes $\overline{p_L(t)}$ to increase by 2.5 mw:

$$P_l = (I_c)^2 R_L = (10^{-3}/\sqrt{2})^2 \times 5 \text{ kohm} = 2.5 \text{ mw}$$

The power dissipated in the transistor at each instant is

$$p_{CE}(t) = v_{CE}(t)i_c(t) \tag{10·8a}$$

Using Eq. (10·1) to express v_{CE} in terms of i_C and E_2, we may rewrite
Eq. (10·7a) as

$$p_{CE}(t) = [E_2 - R_L i_c(t)]i_c(t) \tag{10·8b}$$

Using Eqs. (10·6) and (10·7a), Eq. (10·8b) may be recast in the form

$$p_{CE}(t) + p_L(t) = p_B(t) \tag{10·9a}$$

Equation (10·9a) simply states that at every instant the power leaving the
battery $p_B(t)$ is equal to the power dissipated in the load, plus that dissipated in the transistor, or, energy is conserved. This is, of course, also true
for the average power:

$$\overline{p_{CE}(t)} + \overline{p_L(t)} = \overline{p_B(t)} \tag{10·9b}$$

For our running example, $\overline{p_{CE}(t)}$ is 45 mw, in the absence of an input signal, and it *decreases* to 42.5 mw when the 10-μa sine-wave drive is applied. These facts are shown in Fig. 10·5c. Note carefully that the average power dissipated in the transistor has its maximum value under no-signal conditions. When an input signal is applied, $\overline{p_L(t)}$ increases and $\overline{p_{CE}(t)}$ decreases, the total remaining constant.

It is possible, of course, to calculate the power dissipated in the transistor directly from the expressions for $v_{CE}(t)$ and $i_{CE}(t)$. We have

$$v_{CE}(t) = V_{CE} + \sqrt{2}V_{ce} \sin(\omega t + \phi) \qquad i_C(t) = I_C + \sqrt{2}I_c \sin \omega t$$

so

$$\overline{p_{CE}(t)} = \frac{1}{2\pi} \int_0^{2\pi} v_{CE}(\omega t)\, i_C(\omega t)\, d(\omega t)$$

$$= V_{CE}I_{CE} + \frac{V_{ce}I_c}{\pi} \int_0^{2\pi} \sin \omega t \, \sin(\omega t + \phi)\, d(\omega t)$$

$$= P_{CE} + P_{ce} \qquad\qquad (10 \cdot 10)$$

Since in our example V_{ce} and i_c are 180° out of phase, $\phi = 180°$, and

$$P_{ce} = -V_{ce}I_c$$

The energy-conservation equation (10·9b) can now be rewritten as

$$\underbrace{P_{CE} + P_{ce}}_{\substack{\text{Transistor-}\\\text{power}\\\text{dissipation}}} + \underbrace{P_L + P_l}_{\substack{\text{Load-power}\\\text{dissipation}}} = P_B$$

Using values from our example

$$45 \text{ mw} - 2.5 \text{ mw} + 45 \text{ mw} + 2.5 \text{ mw} = 90 \text{ mw}$$

Since P_{ce} and P_l are equal and of opposite sign, we may think of the transistor as generating the signal power which is dissipated in the load. Or more correctly, we may visualize the transistor as being a power converter, converting dc power from the battery into the signal power which is dissipated in the load.

In circuit models to be developed later, we shall represent this signal-power-generating capability by using current sources in the transistor circuit models. It must be remembered that this will not imply a violation of the energy-conservation condition.

10·1·7 Power input. How much signal power must be supplied to the transistor at its input to produce the signal power dissipated in the load? To answer this question we again refer to the input characteristics.

We evaluate the input-signal-voltage swing v_{be} by using the input characteristic in the manner shown in Fig. 10·6. The nonlinearity of the input characteristic is indicated by the nonsinusoidal appearance of v_{be}. In principle, this means that we must analyze v_{be}, using the Fourier series, to obtain the signal-power input (which occurs only at the fundamental,

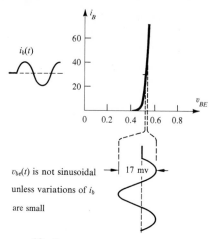

FIG. **10·6** DETERMINATION OF $v_{BE}(t)$ FROM THE INPUT CHARACTERISTIC AND INPUT BASE CURRENT

since $i_s(t)$ has only the fundamental in it). For an estimate, however, we neglect this detail and assume $v_{be}(t)$ to be a sine wave with a peak-to-peak amplitude of 0.017 volts. The average signal-power input is then

$$P_{be} \simeq I_{be}V_{be} \simeq \frac{10 \ \mu a}{\sqrt{2}} \frac{17 \ mv}{2\sqrt{2}} = 42.5 \times 10^{-9} \text{ watts}$$

10·1·8 *Power gain.* The transistor therefore absorbs about 40×10^{-9} watts from the signal source and delivers about 2.5×10^{-3} watts of signal power to a 5-kohm load resistor. The ratio of the signal power output to the signal power input is called the *signal-power gain G*. The transistor gives a signal-power gain of approximately

$$G = \frac{P_l}{P_{be}} \simeq 6 \times 10^4$$

in the circuit of Fig. 10·1.

The ability of a transistor (or vacuum tube) to produce a signal-power gain greater than unity is its most distinctive feature. Because of it, the transistor is called an *active* device.

10·1·9 *Transducer gain.* The preceding calculations show that the transistor can provide a large power gain when it is driven from a current source. However, will the same be true when it is driven from the voltage source, which has considerable internal impedance? The answer is yes, but the reasons are not entirely obvious. After all, most of the power which leaves the signal-voltage source in the circuit of Fig. 10·2 is dissipated in the 5-kohm source resistance. Only a small amount of it is actually delivered to the transistor, so, even though the transistor may have a large power gain, we still cannot be sure that the signal power actually delivered to the load is greater with the transistor in the circuit than it could be without it.

To study this question, let us first calculate the signal power which we could produce in the load without the transistor. The circuit is shown in Fig. 10·7 and consists simply of the source with its internal impedance, and the load. Because the source and load happen to be matched, the signal power that would be delivered to the load in this circuit is

$$P_1 = \frac{|V_s|^2}{4R_s} = 125 \times 10^{-9} \text{ watts}$$

When the transistor is inserted between the source and load, the signal power developed in the load becomes

$$P_1 = |I_c|^2 R_L = \left(\frac{0.75 \times 10^{-3}}{\sqrt{2}}\right)^2 \times 5 \text{ kohm} = 1.4 \times 10^{-3} \text{ watts}$$

The ratio of the actual power supplied to the load to the maximum power which could be supplied by simply matching the source and load is called the *transducer gain* G_T. For this example, the transducer gain is

$$G_T = \frac{1.4 \times 10^{-3}}{1.25 \times 10^{-7}} \simeq 10^4$$

The transducer gain G_T is less than the power gain G because the input of the transistor does not match the source properly.

In general, G_T gives a useful measure of performance because it tells how much "better" the transistor is than a simple passive network which could be used to match the source and load. When $G_T > 1$, some benefit accrues from using the transistor.

Of course, when the input signals are small enough, the input characteristic of the transistor appears to be linear. We can then define a *small-signal input resistance* (slope of the i_b-v_{be} curve), and match the source to the transistor at its input. Then G_T and G are equal. Otherwise, $G_T < G$.

One further word about power gains is necessary. In certain applications, we may be interested in delivering maximum signal current to a load, or obtaining a maximum signal-voltage swing across it, rather than obtaining maximum power gain. There are thus measures of performance other than

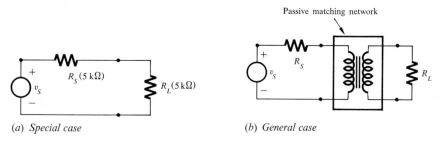

(a) Special case   (b) General case

FIG. 10·7   DIRECT CONNECTION OF SOURCE TO LOAD. SINCE $R_S = R_L$, SOURCE AND LOAD ARE MATCHED. IF $R_S \neq R_L$, A TRANSFORMER OR OTHER "MATCHING NETWORK" CAN BE INSERTED BETWEEN R_S AND R_L.

power gain to be considered. The utility of the power-gain criteria are that they suggest how the device is different from a combination of passive components, and of course, they are useful in situations where power output is the primary concern.

SUMMARY OF PROCEDURE   The procedure which we follow, then, in making a graphical analysis (or design) of an amplifier circuit may be summarized as follows:

1. Using manufacturer's ratings, mark off the permissible area for load lines on the output characteristics of the device.

2. Draw a load line within this area which reflects the amplifier's intended applications (for example, maximum power gain, maximum power output, or others).

3. Construct the transfer curve relating input and output.

4. Locate an operating point on this transfer curve (or on the load line) which reflects the nature of the signals to be processed.

5. Design a biasing network to obtain this operating point.

6. Using the input characteristic, compute the input signal from a graphical solution of Kirchhoff's laws for the input circuit.

7. Compute current, voltage, and power outputs and gains.

Various aspects of the procedure are the subject of Probs. $10 \cdot 1 - 10 \cdot 6$.

$10 \cdot 2$ THEORY OF OPERATION OF VACUUM TUBES

Of course, a graphical analysis of a vacuum-tube amplifier can be carried out along the same lines as were just presented for transistor amplifiers. It is not necessary to understand the physical mechanisms of operation to perform such an analysis, though, of course, one's understanding of the circuit behavior is always improved by a study of the operating principles of the device being used as the amplifier. In this section we shall briefly consider the operating principles of the vacuum triode, leading to a graphical analysis of vacuum-triode amplifiers.

$10 \cdot 2 \cdot 1$ *Operating principles.*   Perhaps the birthdate of modern electronics is best placed in 1907, the year that DeForest invented the vacuum triode. He found that current flow in a vacuum diode could be controlled by inserting a grid of fine wires between the plate and cathode. The control action is most apparent when the grid is negative with respect to the cathode.

To understand the basic control mechanism, we consider the highly idealized structure shown in Fig. $10 \cdot 8$ (actual triode construction will be considered in Sec. $10 \cdot 2 \cdot 3$). Reference directions for voltage and current are drawn on this figure, and the circuit symbol for the triode is also shown. The plate and cathode are plane, parallel electrodes of area A, spaced a distance d apart.

Between the electrodes, and parallel to them, there is a very fine wire mesh, spaced a distance a from the cathode. The mesh is called the grid.

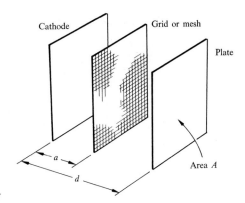

(a) *Idealized plane electrode triode*

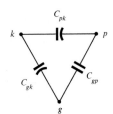

(b) *Circuit model for triode with cold cathode*

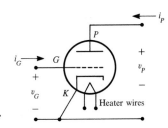

(c) *Circuit symbol for triode*

FIG. 10·8 IDEALIZED TRIODE AND CIRCUIT MODEL WITH CATHODE COLD

When the cathode is unheated, the tube appears to be simply a three-electrode vacuum capacitor. A possible circuit model for such an electrode system is shown in Fig. 10·8b.[1]

Now, if a voltage v_P is applied between the plate and cathode (plate positive) and a voltage v_G is applied between the mesh and cathode (mesh positive), a charge Q_k will appear on the cold cathode surface. The magnitude of Q_k is given by

$$Q_k = C_{gk}v_G + C_{pk}v_P \qquad (10\cdot11a)$$

---

[1] Note that in this model the total capacitance from plate to cathode, when the mesh is left floating, is

$$C_T = C_{pk} + \frac{C_{gk}C_{gp}}{C_{gk} + C_{gp}}$$

which is greater than C_{pk}. If the mesh is very thin, then $C_T = \epsilon_0 A/d$. C_{pk} is less than this, due to the screening action of the mesh. In fact, if the mesh were replaced by a foil, C_T would still be equal to $\epsilon_0 A/d$, but C_{pk} would be zero.

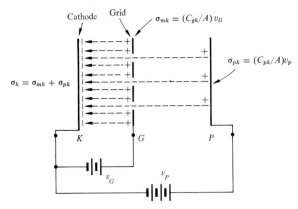

FIG. 10 · 9   IDEALIZED ELECTRIC-FIELD PATTERN IN A PLANE-PARALLEL TRIODE, EXCLUD-
ING PLATE-TO-GRID FIELDS. CATHODE CHARGE DENSITY DEFINED IN TERMS OF APPROPRIATE
INTERELECTRODE CAPACITANCES.

The charge per unit area on the cathode σ_k is

$$\frac{Q_k}{A} = \sigma_k = \frac{C_{gk}}{A} v_G + \frac{C_{pk}}{A} v_P \tag{10 · 11b}$$

From Gauss' law, the electric-field strength at the cathode surface is then

$$E_k = \frac{\sigma_k}{\epsilon_0} \tag{10 · 12}$$

If we define a charge density on the mesh by $\sigma_{mk} = (C_{gk}/A)v_G$ and a charge density on the plate by $\sigma_{pk} = (C_{pk}/A)v_P$, then σ_{mk}/ϵ_0 of the lines of force (or E field), which terminate on the cathode, originate on the grid, while σ_{pk}/ϵ_0 of the cathode E field originates on the plate (see Fig. 10 · 9).[1]

Because field lines from the plate can penetrate the open spaces of the mesh, the potential at the plane of the mesh is not equal to v_G everywhere. Instead, the *average* potential at the plane of the mesh is a function of both v_G and v_P. We can determine an equivalent potential v_{EQ} at the mesh, by using Eq. (10 · 11):

$$Q_k = C_{gk}(v_G + \frac{C_{pk}}{C_{gk}} v_P) = C_{gk}v_{EQ} \tag{10 · 13}$$

Equation (10 · 13) indicates that the electric field at the cold cathode in the presence of the applied voltages v_G and v_P is the same as it would be if v_P were reduced to zero and a potential of v_{EQ} were applied to the mesh alone.

Now, when the cathode is heated, electrons will be emitted and will form a space-charge cloud near the real cathode surface. The electric-field lines originating on the mesh and plate will now terminate on electrons in this space-charge cloud. However, under typical operating conditions, the

---

[1] σ_{mk} and σ_{pk} are not the total charge densities appearing on the grid and plate because the effects of C_{gp} have been omitted. However σ_k is correctly calculated.

electron density in the cloud will be far greater than is necessary to terminate these lines of force. Then the electric field and, hence, emission conditions at the cathode are determined by the space charge of electrons, as in the space-charge-limited diode.

Of course, the electric field produced by the plate and mesh will draw the outermost electrons away from the cloud, as in a normal vacuum diode, giving rise to a current in the tube. If we assume that there is negligible space charge in the region between the mesh and plate, then the electric-field lines which originate on the plate still terminate on charge within the mesh-cathode region. Then the concept of an equivalent mesh potential v_{EQ} is still valid, and we can approximate the magnitude of the cathode current i_K by assuming it is the same that would flow in a diode in which the plate and cathode were separated by a, the mesh-to-cathode spacing, and a voltage v_{EQ} were applied to the diode:

$$i_K = B(v_{EQ} + v_T)^{3/2} \qquad (10 \cdot 14a)$$

where B is the perveance of the equivalent diode (defined in Sec. 6·8). As in the vacuum diode, v_T is a correction for the potential due to the electronic space charge near the cathode. v_T is on the order of 0.1 to 0.2 volt and is normally neglected. We then arrive at the usual approximation

$$i_K = B\left(v_G + \frac{C_{pk}}{C_{gk}} v_P\right)^{3/2} = B\left(v_G + \frac{v_P}{\mu}\right)^{3/2} \qquad (10 \cdot 14b)$$

expressing the cathode current of a triode in terms of the potentials applied to the mesh and plate and the interelectrode capacitances. The ratio of C_{gk}/C_{pk} is denoted by μ, called the *amplification factor*. It measures the relative effectiveness of v_G and v_P in determining the plate current.

10·2·2 Regions of operation of a vacuum triode. As with transistors, various operating regions for the triode can be defined on a plane in which v_G and v_P are the coordinates. Such a plot is shown in Fig. 10·10. The cutoff line on this figure is simply a plot of the equation

$$v_G + \frac{v_P}{\mu} = 0 \qquad (10 \cdot 15)$$

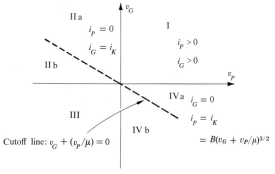

FIG. **10·10** REGIONS OF OPERATION OF A TRIODE (CONSTANT μ).

The cathode current is ideally given by

$$i_K = B\left(v_G + \frac{v_P}{\mu}\right)^{3/2}$$

everywhere. The division of the cathode current between the grid and the plate is, in general, very difficult to calculate. However, from a study of the electron paths in the triode it can be shown that in the first quadrant of Fig. 10·10,

$$\frac{i_P}{i_G} \cong K_1 \left(\frac{v_P}{v_G}\right)^{1/2} \qquad v_P > v_G$$

$$\frac{i_P}{i_G} \cong K_2 \left(\frac{v_P}{v_G}\right)^{2} \qquad v_P < v_G$$

The IIa quadrant is not of much practical interest, because the grid current which flows in this region cannot be effectively controlled by the plate and may also produce enough heat to permanently damage the grid structure.

The control action in the triode is most apparent in quadrant IVa, where v_G is negative but v_P/μ is positive and greater than v_G. Here, we assume that no current flows to the grid; the cathode current and plate current are then equal. Thus

$$i_P = B\left(v_G + \frac{v_P}{\mu}\right)^{3/2} \qquad v_G < 0, \quad \frac{v_P}{\mu} + v_G > 0 \qquad (10 \cdot 16)$$

We shall consider the experimental validity of this formula after a few remarks about actual triode construction.

10·2·3 *Triode construction.* The construction of a triode, of course, depends on its intended uses. Small tubes intended for relatively low power operation (such as would be used in radio and television receivers, audio systems, etc.) are usually constructed in the manner shown in Fig. 10·11a. The cathode is usually an oxide-coated type, the cathode material being simply sprayed onto a hollow nickel sleeve. The grid is a thin wire wound around two supporting posts. The grid is usually elliptical in cross section, as is the plate. Even though the geometry is not planar and the grid is not a mesh, the operating principles are the same as those just described. In fact, we shall see in the next section that some low-power triodes (with less than a few watts maximum plate dissipation) behave almost exactly according to the previous theory.

Triodes intended for higher-power applications provide means for cooling the plate, which may have to dissipate a considerable amount of heat. One possible construction of such a tube is shown in Fig. 10·11b. Here the cathode material is sprayed onto the closed end of a thin nickel sleeve, or a tungsten cathode is used. The grid structure is usually a set of parallel wires which are mounted on a grid plate with a hole in its center.

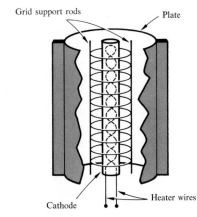

Grid support rods

Plate

(a) *Low-power triode. Typical size: 0.5 to 1 in. diameter, 1 to 2 in. long*

Heater wires

Cathode

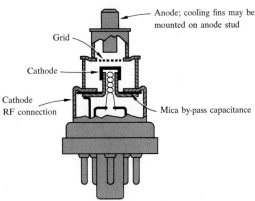

Anode; cooling fins may be mounted on anode stud

Grid

Cathode

Cathode RF connection

Mica by-pass capacitance

(b) *High-power triode construction. For very high-power construction, the position of the plate and cathode is reversed.*

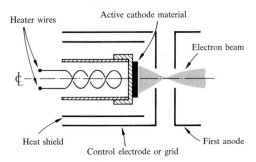

Heater wires

Active cathode material

Electron beam

₵

Heat shield

Control electrode or grid

First anode

(c) *Triode section of a CRT electron gun*

FIG. 10·11   SOME PRACTICAL FORMS OF TRIODE CONSTRUCTION

Triodes are also used in the formation of electron beams for very-high-frequency tubes and for cathode-ray tubes. The construction of the triode section of a cathode-ray-tube electron gun is indicated in Fig. 10·11c. The cathode is usually oxide-coated, the material being sprayed onto the closed end of a thin nickel tube. The grid is a concentric tube with a hole through which electric fields from the plate (or first anode) can penetrate. The first anode also has a hole in it to allow most of the beam to pass through into the focusing and deflection regions.

Introduction to amplifiers   389

10·2·4 *Experimental characteristics of low-power triodes.* The operating characteristics of triodes are usually presented as a set of curves of i_P vs. v_P with v_G as a parameter. If operation in the first quadrant of Fig. 10·10 is permissible, i_G will also be plotted vs. v_P with v_G as a parameter (that is, on same coordinates as i_P curves). Typical curves are shown in Fig. 10·12. The i_P curves are called the *plate family* or *plate characteristics*. The i_G curves are called the grid family.

The usual operating condition is with v_G negative, so i_P and i_K are essentially equal. Under these conditions, the theory presented earlier suggests that the triode plate characteristics should ideally be displaced diode curves, each curve reflecting the ⅔-power law. To check this, we may simply plot $(i_P)^{2/3}$ vs. v_P with v_G as a parameter, as in Fig. 10·13. As is shown on this figure, the triode of Fig. 10·12 behaves "theoretically" over most of the v-i plane. For $v_G \geq 0$, variation from the theory arises because there is appreciable space charge in the grid-plate region (which was neglected in our analysis).

There is also a deviation of the plate current from the ideal ⅔-power behavior at low currents, which arises as follows. If the relation

$$i_P = B\left(v_G + \frac{v_P}{\mu}\right)^{3/2}$$

were obeyed exactly, then the plate current would be zero when

$$v_G + \frac{v_P}{\mu} \leq 0$$

Now, there will be inevitable nonuniformities in the grid wire spacing, so the electric field pattern along the cold cathode surface will be not perfectly uniform. The result is that some parts of the cathode contribute more to the total plate current than others under a given bias condition. We may visualize this by considering a real triode to be made up of several sub-

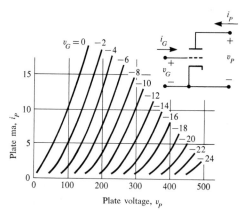

FIG. 10·12 PLATE CHARACTERISTICS FOR A 6SN7 TRIODE. CHARACTERISTICS ARE IN TERMS OF TOTAL VALUES OF v_P, i_P, v_G.

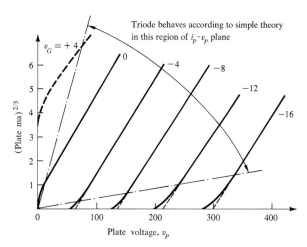

FIG. 10·13 PLATE CHARACTERISTICS OF THE 6SN7 TRIODE REPLOTTED WITH $(i_P)^{2/3}$ VS. v_P

triodes operating in parallel, each with its own value of μ. The plate-current cutoff condition for each subtriode will be different. To cut off the plate current in the real triode, one must apply sufficient bias to cut off the subtriode section which has the lowest value of μ.

In some tubes the grid wires are purposely spaced nonuniformly to obtain a very gradual cutoff of plate current. Such tubes are called remote-cutoff, or variable-μ, tubes and can be used to control the gain of a vacuum-tube amplifier by adjusting grid-bias voltage.

10·2·5 Graphical analysis of a vacuum-triode amplifier. Graphical analysis of a vacuum-triode amplifier can be carried out along the lines that were presented in the summary to Sec. 10·1. We will present a partial analysis in this section to indicate the procedure once again. Prob. 10·8 and 10·9 present other aspects of the graphical analysis of vacuum-triode circuits.

The plate characteristics of one section of a 12AX7 triode are shown in Fig. 10·14. The ratings of the tube are

Maximum plate voltage	400 volts
Maximum plate dissipation	1 watt
Maximum average cathode current	10 ma

These ratings determine the safe operating region shown in Fig. 10·14. A load line is constructed within this region for the arbitrary values $E_B = 300$ volts, $R_L = 100$ kohms. Two transfer characteristics are read from this load line: i_P vs. v_G and v_P vs. v_G. These transfer characteristics are shown in Fig. 10·14c. On the assumption that a sine-wave input voltage will be used, a bias point at $V_G = -2.0$ volts is selected. The corresponding values of I_P and V_P are 0.8 ma and 220 volts, respectively. The bias is obtained with a grid battery for this example. Other more satisfactory biasing schemes will be discussed in Chap. 12.

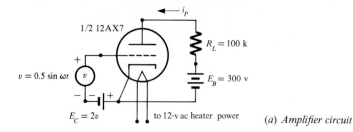

1/2 12AX7

$v = 0.5 \sin \omega t$

$R_L = 100$ k

$E_B = 300$ v

$E_C = 2v$

to 12-v ac heater power

(a) Amplifier circuit

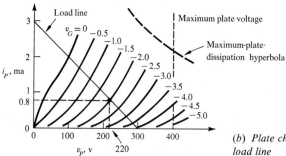

Load line

$v_G = 0$ −0.5
−1.0
−1.5
−2.0
−2.5
−3.0
−3.5
−4.0
−4.5
−5.0

Maximum plate voltage

Maximum-plate-
dissipation hyperbola

i_p, ma

v_p, v

220

(b) Plate characteristics of the 12AX7 with assumed load line

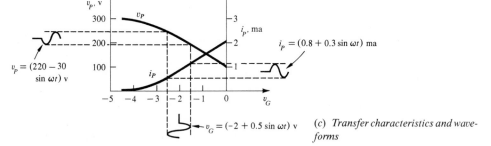

v_p, v

v_P

i_p, ma

$i_p = (0.8 + 0.3 \sin \omega t)$ ma

$v_P = (220 - 30 \sin \omega t)$ v

i_P

v_G

$v_G = (-2 + 0.5 \sin \omega t)$ v

(c) Transfer characteristics and wave-forms

FIG. 10·14 GRAPHICAL ANALYSIS OF A VACUUM TRIODE AMPLIFIER

If a sine-wave signal voltage with a peak amplitude of 0.5 volt is applied at the input, the plate current will appear as shown in Fig. 10·14c. The plate signal current will not be exactly sinusoidal, because 0.5 volt is a sufficiently large grid swing to produce some distortion in this circuit. Neglecting this detail, however, we obtain a plate current of

$$i_P = (0.8 + 0.3 \sin \omega t) \text{ ma}$$

The corresponding plate voltage is

$$v_P = (220 - 30 \sin \omega t) \text{ volts}$$

The dc power dissipated in the load is

$$P_L = 80 \text{ volts} \times 0.8 \text{ ma} = 64 \text{ mw}$$

and that dissipated in the tube is

$$P_P = 220 \text{ volts} \times 0.8 \text{ ma} = 176 \text{ mw}$$

When the grid signal is applied, the average signal power supplied to the load is

$$P_1 = |I_p|^2 R_L = 4.5 \text{ mw}$$

The average power supplied to the load then increases to 68.5 mw, and the average power dissipated in the tube decreases to 171.5 mw.

One of the most striking properties of the tube is apparent when we consider the signal input power required to produce the signal output. Since the grid-cathode diode is essentially an open circuit for reverse biases, only a minute amount of input power is required to supply the output power. The power gain of the amplifier is therefore *very* high. Unfortunately, when amplifier stages are cascaded, high-power gain is not too meaningful for either transistors or vacuum tubes. All stages except the final stage of a vacuum-tube amplifier appear to be signal-voltage amplifiers; and all stages except the final stage of a transistor amplifier appear to be signal-current amplifiers. These remarks will be expanded upon in Chap. 13.

10·3   OPERATING PRINCIPLES OF PENTODE TUBES

The most outstanding defect of the triode as a low-power amplifying device is that the capacitance between its grid and plate is large. This means that at high frequencies there is considerable coupling between the input (grid-cathode) circuit and the output (plate-cathode) circuit. We shall see later that this capacitance has an undesirable effect on both the amplitude and phase of high-frequency signals that are being amplified by the tube.

To avoid this problem, pentode tubes are usually used for low-power, high-frequency amplification purposes. The pentode can be viewed as a triode with two additional grids inserted between the control grid and the plate. The function of these two grids can be described briefly as follows.

First, let us suppose that one additional grid is inserted between the control grid and the plate and connected to ground potential, as in Fig. 10·15a. Then the capacitive coupling between the plate and the grid will be reduced. However, the capacitive coupling between the plate and cathode will be even more drastically reduced, so the plate voltage produces relatively small electric fields at the cathode surface. This is undesirable, since it essentially eliminates the negative-grid control action which the triode has.

A solution to this problem is to operate the additional grid at a positive potential. The added grid then serves essentially as a triode plate with holes in it; or the cathode, control grid, and additional grid, called the

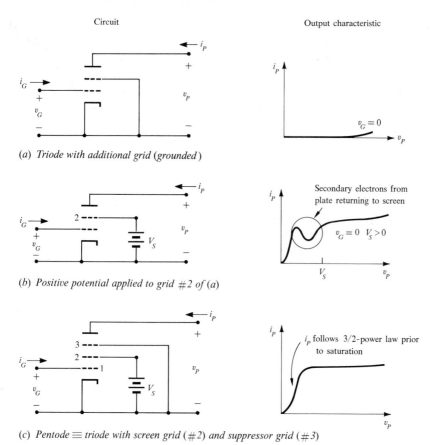

Circuit

Output characteristic

(a) *Triode with additional grid* (*grounded*)

(b) *Positive potential applied to grid #2 of* (a)

Secondary electrons from
plate returning to screen

$v_G = 0 \quad V_S > 0$

(c) *Pentode* ≡ *triode with screen grid* (*#2*) *and suppressor grid* (*#3*)

i_p follows 3/2-power law prior
to saturation

FIG. 10·15   THE EVOLUTION OF THE PENTODE

screen grid, form an electron gun which directs an electron "beam" at the plate. This solution then eliminates the capacitive coupling between the control grid and plate (and the plate and cathode), but plate current still flows.

This new solution also has one new difficulty, however. The impact of an electron arriving at the plate from the electron gun may be sufficient to cause several "secondary electrons" to be emitted from the plate. These "secondary" electrons will return to the plate when the plate is more positive than the screen.[1] Otherwise, they will return to the screen. This causes the undesirable irregularities in the plate characteristics of the tube shown in Fig. 10·15b. The wiggles in the plate current seriously limit the swings of plate voltage which the tube can provide without distortion.

To eliminate this defect, another grid is inserted between the screen and the plate. This grid, called the *suppressor,* is operated at ground potential.

[1] This phenomenon also occurs in the triode, but it is not important, since in normal conditions the plate is always more positive than the grid, and the secondary electrons return to the plate.

Electrons coming from the electron gun are then slowed down to essentially zero velocity by the time they reach the suppressor grid. These electrons form a *virtual cathode* for the plate; the plate and the electron cloud around the suppressor then form a diode. The plate draws its current from this virtual cathode, but because the current available from this virtual cathode is limited to that which the electron gun can supply, the plate current soon saturates, as shown in Fig. 10·15c. Since for any plate voltage the plate is more positive than the suppressor, the difficulty with secondary electrons which are dislodged from the plate is removed.

Of course, the saturation value of the plate current is a function of both the screen-to-cathode voltage and the grid-cathode voltage. However, in typical operation, the screen voltage is held constant, so the plate family for the pentode is plotted with grid voltage as the only running variable.

A plate family for a 6SJ7 is shown in Fig. 10·16. These characteristics are reminiscent of the output characteristics of a transistor, except that the control variable is a voltage, not a current.

10·3·1   Equivalent grid potential. As in the triode, it is once again possible to define an equivalent grid potential. The cathode current may be expressed as

$$i_K = B\left(v_1 + \frac{v_2}{\mu_2} + \frac{v_3}{\mu_3} + \frac{v_P}{\mu_P}\right)^{3/2} \tag{10·17}$$

where the voltages are those of the first, second, and third grids and the plate, respectively, and the μ_i measure the effectiveness of the indicated grid in determining the cathode current (relative to the control grid). The μ_i may be defined as ratios of capacitances, as before.

For the usual operating conditions, v_2 is constant, v_3 is zero, and the electrostatic shielding is such that μ_p is very large ($\sim 10^4$ or more). Therefore, the cathode current can be written as

$$i_K \cong B\left(v_G + \frac{v_S}{\mu_S}\right)^{3/2} \tag{10·18}$$

where v_S is the screen-to-cathode voltage.

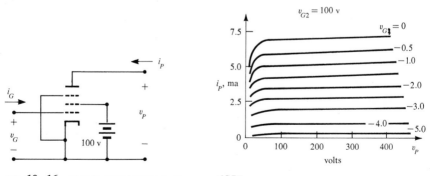

FIG. 10·16   PLATE CHARACTERISTICS FOR A 6SJ7

Of course, since the screen is operated at a positive potential, some of the cathode current will flow to it. The division of the cathode current between the screen and plate depends only on the ratio of the plate and screen voltages

$$\frac{i_P}{i_S} = f\left(\frac{v_P}{v_S}\right) \qquad (10 \cdot 19)$$

but no simple analytical form for f is known.

The pentode is a popular tube because, like the transistor, there is a large region where its plate characteristics are nearly parallel straight lines. This implies that there will be very low distortion of waveforms when signal voltages are being amplified. Problems $10 \cdot 10$ and $10 \cdot 11$ deal with graphical analysis of simple pentode circuits and pentode tube theory.

$10 \cdot 4$ SUMMARY AND PROJECTION

In this chapter we have used graphical techniques to discuss the manner in which active elements such as transistors and vacuum tubes amplify signals. The graphical techniques have the advantage of being very general; they can be applied to study the amplifying characteristics of any active element, without reference to the physical operating mechanisms which produce the graphical characteristics.

On the other hand, graphical analysis is a rather clumsy tool for studying signal transmission when the signals are small. Furthermore, the graphical characteristics do not suggest the presence of frequency limitations in the devices, or the relation between device physics and circuit behavior. The problem of calculating the small-signal characteristics of an amplifier is therefore approached with a *small-signal circuit model* of the device. This model may be developed from the graphical characteristics, or from equations which express the electrical behavior of the device at its terminals in terms of internal operating mechanisms. We shall develop low-frequency small-signal models of transistors and vacuum tubes in Chap. 11.

In Chap. 12, we shall consider the techniques used to obtain a desired operating point, these techniques being based on a combination of the graphical characteristics and small-signal models. In Chap. 13–15, we shall study the applications of active devices in amplifiers, oscillators, and switching circuits.

REFERENCES

Zimmerman, H. J., and S. J. Mason: "Electronic Circuit Theory," John Wiley & Sons, Inc., New York, 1962, chaps. 5, 6, and 7.

Angelo, E. J.: "Electronic Circuits," McGraw-Hill Book Company, New York, 2nd ed., 1964, chaps, 5, 8, and 10.

Spangenberg, K. R.: "Fundamentals of Electron Devices," McGraw-Hill Book Company, New York, 1957, chaps. 10 and 11.

Ramey, R. L.: "Physical Electronics," Wadsworth Publishing Company, Belmont, Calif., 1961, chaps. 6 and 10.

Spangenberg, K. R.: "Vacuum Tubes," McGraw-Hill Book Company, New York, 1948.

DEMONSTRATION

AMPLIFIER DESIGN   The basic results obtained from the graphical analysis of an amplifier can be conveniently demonstrated with the circuit shown in the figure. It is particularly useful to show the effects of bias-point selection, nonlinearity in the current-transfer characteristic, and nonlinearity in the base-emitter input circuit on the circuit waveforms.

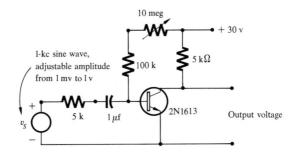

FIG. D10·1   AMPLIFIER FOR DEMONSTRATION OF GRAPHICAL ANALYSIS TECHNIQUES

PROBLEMS

10·1   Let us assume that the transistor amplifier shown in Fig. 10·1b (in the text) is part of a battery-operated receiver. The battery E_2 is a 22.5-volt dry cell. Assuming $R_L = 5$ kohms,

1. Plot the load line on the output characteristics.

2. Sketch and dimension the current-transfer characteristic.

3. Determine R_1 to obtain a bias point in the middle of the transfer characteristic (assuming $E_1 = 22.5$ volts).

4. Determine the signal-output current in the output and the output power for a signal-input current $i_s(t) = 10 \sin \omega t$ μa.

5. Estimate the power gain for this amplifier stage.

6. As the battery ages, it develops some internal resistance. Suppose that after 6 months the battery has an internal resistance of 2 kohms. Repeat parts (1) to (5) for this condition and discuss your results.

10·2   The output characteristics of many transistors can be approximated by saying that they are equally spaced parallel lines with $i_C = \beta i_B$. Suppose $\beta = 50$ for a given transistor.

1. Sketch and dimension the approximate output characteristics of this transistor up to $i_c = 10$ ma, $v_{CE} = 40$ volts.

2. Locate the permissible load-line area for the conditions: $I_{C, max} = 100$ ma, $V_{CE, max} = 35$ volts, $P_{max} = 100$ mw.

3. Construct load lines for the following conditions: $R_L = 5$ kohms, $E_2 = 10$, 20, and 30 volts (in the circuit of Fig. $10 \cdot 1a$).

4. Determine the maximum sine-wave power output for each of these cases, and the base current required to obtain it (both signal amplitude and bias point).

5. Determine the power dissipation in the transistor for each case in (4).

6. Sketch and dimension the current-transfer curves for each case and discuss your results.

10·3 1. Determine the current-transfer curves for the circuit of Fig. $10 \cdot 1$ for the three different load resistances $R_L = 3$ kohms, 5 kohms and 10 kohms.

2. The input characteristic can be *roughly* approximated by the relation

$$i_B = I_B + \frac{v_{be}}{R}$$

where
$$R = \frac{25 \times 10^{-3} \text{ mv}}{I_B}$$

Using a bias point in the middle of each transfer characteristic and the approximate input resistance, compute the power gain for each case.

10·4 Using the transistor characteristics shown in Fig. $10 \cdot 1b$, determine the load lines which should be used with $E_2 = 30$ volts and 40 volts to give maximum power output consistent with the maximum power-dissipation limitation. What is the maximum power output for each case and the base current needed to produce it?

10·5 The amplifier of Fig. $10 \cdot 2$ is driven from a sine-wave voltage source of 10 mv peak voltage. The source impedance is 2 kohms.

1. Determine the base signal current i_b on the assumption that $I_B = 30$ μa.

2. Obtain an approximate value for the input resistance v_{be}/i_b of the transistor for small signals.

3. Assuming the sine-wave signal source has a frequency of 1 KC, select a value of C which will insure that $X_c \le 0.1(R_s + R_{in})$ where R_{in} is the value obtained in part (2).

4. Compute the input-signal current, output-signal current, power gain, and transducer gain for this amplifier.

10·6 1. In the circuit of Fig. $10 \cdot 2$ (in the text), compute and plot the amplitude of the input signal current as a function of the frequency f of the sine-wave voltage source, assuming $C = 0.5$ μfarad. Use $f = 1$ kc, 500 cps, 100 cps, 50 cps, 10 cps for your calculations. The voltage source maintains its output at 10 mv at all frequencies. For ease in plotting, use logarithmic scales for both $|i_b|$ and f.

2. Compute the power output for each frequency, and explain your results briefly.

10·7 1. Describe the measurements you would make to determine C_{pk}, C_{gk}, and C_{gp} in Fig. $10 \cdot 8b$.

2. Measured values of C_{pk} and C_{gk} for a 6SN7 are: $C_{pk} = 1$ pfarad, $C_{gk} = 2.9$ pfarads. The stated value for μ is 20. Explain the reason why $\mu \ne C_{gk}/C_{pk}$.

10·8 From the plate characteristics for the 6SN7 given in Fig. $10 \cdot 12$, plot the i_p-v_G transfer curve for a load line at $E_b = 300$ volts, $R_L = 30$ kohms. Select an operating point which will give the most linear relation between output signal current and grid-cathode signal voltage. Compute the output signal current and voltage for a grid swing of 4 sin ωt volts about this operating point. What is the signal-voltage amplification in this amplifier?

10·9 A 12AX7 is to be operated at the bias point $V_G = -2$ volts, $V_P = 200$ volts. The plate characteristics are given in Fig. 10·14b:

1. Calculate the plate supply voltage (E_b in Fig. 10·14a) that must be used for the following values of load resistance: 50 kohms, 100 kohms, 200 kohms.

2. Sketch and dimension the i_p-v_G transfer characteristic for each case in (1).

3. Sketch and dimension the output current for each case for a grid signal voltage $v_g = 2 \sin \omega t$ volts.

10·10 Using the pentode characteristics shown in Fig. 10·16, construct the load line for a plate supply voltage of 300 volts and load resistance R_L of 100 kohms. Sketch and dimension the corresponding i_P-v_G transfer characteristic. Repeat for $R_L = 50$ kohms.

10·11 The following rule is postulated for converting a set of pentode plate characteristics from one fixed value of screen voltage to another: If the plate characteristics are known for $V_{screen} = E_1$, and V_{screen} is to be changed to KE_1, then multiply the v_P scale by K, each value of v_G by K, and the plate current scale by $K^{3/2}$.

1. Verify this rule.

2. Use the rule to convert the pentode characteristics shown in Fig. 10·16 to $V_{screen} = 50$ volts. Repeat for $V_{screen} = 150$ volts.

11 CIRCUIT MODELS FOR TRANSISTORS AND VACUUM TUBES

THE THEORETICAL ANALYSIS of an electronic circuit (or any other physical system) is a pencil-and-paper activity which is carried out as follows. Using appropriate symbols to represent each important component, first make a diagram in which the symbols are interconnected in the same way they appear in the circuit to be studied. Sometimes the symbols carry with them mathematical relations which express (to an adequate degree of approximation) the electrical behavior of the components they are representing. Other symbols, such as those for a transistor or vacuum tube, are simply invented to facilitate construction of the diagram. In such cases either graphical characteristics or a *circuit model* is used to represent the device. If a circuit model of the real device is to be used, a second diagram is usually made in which the circuit model of the real device is substituted for its diagrammatic symbol.

In either case, when the electrical behavior of each component of the circuit has been correctly expressed, Kirchhoff's laws are used to analyze the circuit diagram which has been constructed. If the models of all the components are well chosen, the analysis will enable one to understand and predict the behavior of the real system.

For studying most transistor circuits it is customary to employ a *circuit model* to represent the device. A circuit model is a schematic network of R's, L's, C's, ideal diodes and *controlled sources* (to be defined shortly) which are connected so that the electrical characteristics of the circuit model *at its terminals* approximate those of the real device with sufficient accuracy for the purposes of the analysis. The internal construction of the circuit model may have little or no relation to that of its physical counterpart.

Generally, we divide circuit models into two categories: large-signal and small-signal. Large-signal circuit models may be either piecewise-linear models or they may contain the nonlinearities associated with ideal junction diodes or ideal vacuum diodes. These models can be developed either from graphical characteristics or from equations which express the terminal behavior of the device in terms of its physical construction and operating principles. Small-signal models are *linear models* which are valid for small excursions about an operating point. They may be derived from the large-signal models or from equations which express the terminal behavior of the device.

For most of the remainder of this book, we shall be involved in the task of making circuit models for transistors and vacuum tubes and using them to study interesting practical systems. In these studies we shall find that the judicious elimination of unessential elements in a circuit diagram and the careful selection of circuit models for the active devices will be key factors in the successful analysis or design of an electronic circuit.

In this chapter we shall first develop piecewise-linear models for transistors and vacuum tubes and then see how small-signal models follow from them. We shall then reconsider the small-signal modeling process from the "black box" or two-port point of view.

11·1   THE CONTROLLED SOURCE

As was indicated in Chap. 7, five ideal elements are required in the analysis of electronic circuits. These are the ideal R, L, and C elements of linear circuit theory, plus the ideal diode and the controlled source. Controlled sources are used to express a special type of unilateral or one-way *coupling* between two parts of a circuit (or device) which cannot be synthesized from a combination of ideal diodes. For example, the coupling between the emitter-base and collector-base diodes of a transistor, which

Circuit   **Output characteristic**

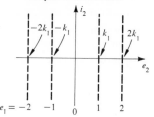

(a) *Input voltage controls output voltage*

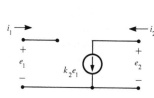
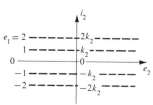

(b) *Input voltage controls output current*

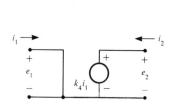
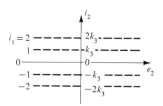

(c) *Input current controls output current*

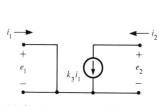
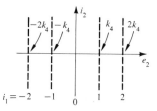

(d) *Input current controls output voltage*

FIG. 11·1   THE THREE-TERMINAL CONTROLLED SOURCE AS A CIRCUIT ELEMENT

is responsible for the amplifying properties of the structure, requires a controlled source for its circuit representation.

For the purposes of electronic circuit analysis, a controlled source is a current or voltage source whose value is determined by the value of a control variable. In most cases the control variable is a voltage or current, and the controlled source and controlling variable usually have one node in common.[1] Four different controlled sources that fit these conditions are shown in Fig. 11·1, along with output characteristics which show the effect of the control variable. Note carefully the reference directions for

[1] Other types of controlled sources will be considered in later chapters.

voltages and currents shown in these figures. The controlled sources shown in Fig. 11·1 are *linear* three-terminal circuit elements. They are also *unilateral,* meaning that a signal applied to the input has an effect on the output, but not conversely. This is in marked contrast to networks of R, L, and C elements, where signals can pass through a three-terminal network in either direction.

Electronic amplifying elements such as transistors and vacuum tubes are not exactly unilateral: they may transmit signals better in one direction than the other, but some transmission in both directions is always possible. Such elements are said to be *nonreciprocal.* Circuit models for nonreciprocal devices may employ two separate controlled sources to account for both forward and reverse transmission, or a single controlled source embedded in an appropriate network of R's, L's, and C's—the passive elements accounting for the reverse-transmission properties. Examples of both types will be given shortly.

11·2   PIECEWISE-LINEAR CIRCUIT MODELS FOR
THE OUTPUT CHARACTERISTICS OF A TRANSISTOR

Using nothing more than resistances, ideal diodes, and controlled sources, it is possible to construct a piecewise-linear circuit model whose terminal characteristics approximate the dc characteristics of a real device to any required degree of accuracy. We shall consider the common-base output characteristics of a transistor, operating in the normal-active region, as an example.

A sketch of the collector characteristics to be approximated is shown in Fig. 11·2a. The main feature of interest in these characteristics is that as long as i_E is positive and v_{CB} is negative, i_C is proportional to i_E and nearly independent of v_{CB}. This suggests that a circuit model of the type shown in Fig. 11·1c could be used with ideal diodes placed appropriately to eliminate the output characteristic of the controlled source in certain quadrants. A simple model which achieves this is shown in Fig. 11·2b, together with its output characteristics. The ideal diode D_C shorts out the controlled-current generator when v_{CB} is positive. The ideal diode D_E will only allow emitter current to flow in one direction. The controlled source is defined to include the saturation current which the real characteristics exhibit.

The circuit model shown in Fig. 11·2b is very useful in problems where a simple representation of the overall collector characteristics of the transistor is adequate. If a somewhat better approximation to actual characteristics is required, however, we can make the following modifications. First, a large-resistance r_c placed in parallel with the collector-base diode will give the output characteristics a constant slope (and thereby represent the effect of base-width modulation on the output characteristics). This is shown in Fig. 11·2c. (The addition of r_c does not account for the influence

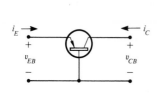
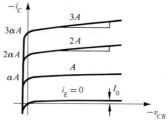

(a) *Transistor with common-base collector family*

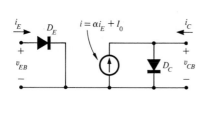
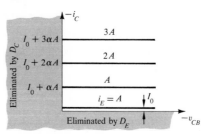

(b) *Circuit model for (a) with two ideal diodes and a controlled source*

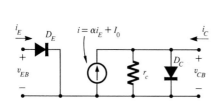
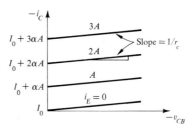

(c) *Addition of r_c to (b) to provide nonzero slope in the output characteristics*

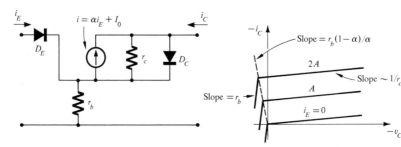

(d) *Addition of r_b to add feedback between collector and emitter*

FIG. 11·2   APPROXIMATING THE COMMON-BASE COLLECTOR FAMILY OF A REAL TRANSISTOR WITH A SERIES OF PIECEWISE LINEAR MODELS

of base-width modulation on the input characteristics, however.) Next, a resistance can be added in the base circuit. This resistive coupling between the input and output provides an additional mechanism by which the emitter current can influence the collector-base circuit, and leads to the output curves given in Fig. 11·2d. Of course, if greater accuracy is required, more elements will have to be used. However, an arbitrarily accurate approximation to the real characteristics can be obtained given sufficient time and patience. Furthermore, one need not know anything about the physical mechanisms which produce the output characteristics of the real device to construct a circuit model for it.

11·3   ELEMENTARY CIRCUIT MODELS FOR TRANSISTORS UTILIZING IDEAL JUNCTION DIODES

Of course, it would be expected that some physical insight into the operating mechanisms of the real device would simplify the construction of circuit models appreciably. As an example of this, we can simply replace the ideal diode D_C in the collector base circuit of Fig. 11 · 2b with an ideal *junction* diode, D_{Cj}; that is, a diode with the current-voltage law

$$i = I_0(e^{qv/kT} - 1) \qquad (11·1)$$

We then remove I_0 from the controlled source and neatly obtain the circuit model shown in Fig. 11 · 3b, whose output characteristic is

$$i_C = -\alpha_F i_E + I_0(e^{qv_{CB}/kT} - 1) \qquad (11·2)$$

which is the *exact* relation obtained for the intrinsic transistor in the preceding chapter. With the addition of a large resistance r_c in parallel with the ideal junction diode, this model provides a representation of transistor output characteristics which is sufficiently accurate that it falls within the typical manufacturing spread of characteristics for a given transistor type over most of the normal-active range (of course, the model is not piecewise-linear any more).

The collector reverse breakdown can also be approximated by simply defining the collector junction diode to have a reverse breakdown voltage BV_{CBO}. This simple expedient will produce the output characteristics shown in Fig. 11 · 3b. The characteristics are physically inaccurate, since only the saturation current I_0 contributes to avalanche multiplication. This can, of course, be remedied by further modeling.

11·3·1   *A circuit model for the Ebers-Moll representation of a transistor.* The circuit model shown in Fig. 11 · 3b can be simply generalized to represent both the forward- and reverse-transmission properties of the intrinsic transistor. We simply replace the ideal diode D_E with an ideal junction diode D_{Ej}; and insert a controlled source in the emitter lead to represent the effect of carrier injection from the collector when the collector-base diode is forward-biased. The control variable is the current flowing in the collector-base diode.

The resulting model is shown in Fig. 11 · 4. It is an exact circuit representation of the Ebers-Moll equations, when the control variables in

Circuit	*Output characteristic*

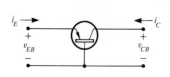
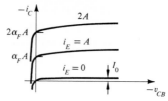

(a) Transistor with common-base collector family

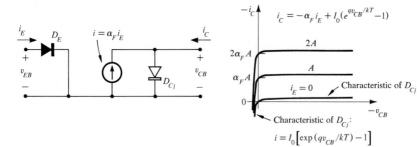

$$i_C = -\alpha_F i_E + I_0(e^{qv_{CB}/kT} - 1)$$

Characteristic of D_{Cj}:
$$i = I_0\left[\exp\left(qv_{CB}/kT\right) - 1\right]$$

(b) Circuit model for (a) using ideal diode in emitter lead and ideal junction diode in collector-base circuit. D_{Cj} exhibits $i = I_0 [exp\ (qv_{CB}/kT) - 1]$.

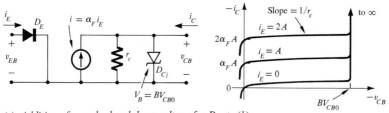

(c) Addition of r_c and a breakdown voltage for D_{Cj} to (b)

FIG. 11·3 THE USE OF IDEAL JUNCTION DIODES IN TRANSISTOR CIRCUIT MODELS

the model are defined correctly (see Fig. 11·4). The model could, of course, be generated directly from these equations without first referring to a graphical representation of the device characteristics.

The model shown in Fig. 11·4 is widely used for calculating the switching properties of a transistor. With the addition of r_b and r_c elements, it is also used to study the small-signal behavior of the transistor.

11·3·2 *Regions of operation of a pnp transistor.* The circuit model of Fig. 11·4 is valid for all possible combinations of junction biases. The four regions of operation of the intrinsic transistor, described in Sec. 9·4·2, correspond simply to the four possible conditions of the diodes shown in Fig. 11·4:

Operating state	D_{Ej}	D_{Cj}
Normal-active	ON	OFF
Saturated	ON	ON
Inverted-active	OFF	ON
Cutoff	OFF	OFF

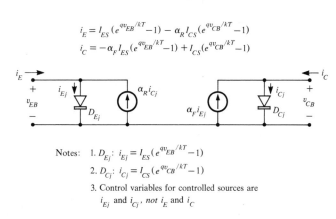

$$i_E = I_{ES}(e^{qv_{EB}/kT} - 1) - \alpha_R I_{CS}(e^{qv_{CB}/kT} - 1)$$

$$i_C = -\alpha_F I_{ES}(e^{qv_{EB}/kT} - 1) + I_{CS}(e^{qv_{CB}/kT} - 1)$$

Notes: 1. D_{Ej}: $i_{Ej} = I_{ES}(e^{qv_{EB}/kT} - 1)$

2. D_{Cj}: $i_{Cj} = I_{CS}(e^{qv_{CB}/kT} - 1)$

3. Control variables for controlled sources are i_{Ej} and i_{Cj}, *not* i_E and i_C

FIG. **11·4** CIRCUIT MODEL OBEYING EBERS-MOLL EQUATIONS FOR THE INTRINSIC TRANSISTOR

11·3·3 *Small-signal model for the normal-active region: common-base connection.* So far, our discussion of circuit models has been based primarily on graphical output characteristics, or sets of equations describing the idealized behavior of the device. The voltages and currents in either representation are *total* quantities. However, when transistors are used in amplifiers, we are frequently interested in knowing how they will transmit small time-varying signals when they are biased at some selected operating point. In studying this type of problem, we usually use either the graphical characteristics or a large-signal model to select an appropriate operating point and then a *small-signal circuit model* to study the behavior of the device around this operating point.

The form of a small-signal circuit model will frequently be similar to the circuit model which applies for total quantities. However, the bias condition will eliminate some of the elements of the complete model from consideration; and the numerical values of the remaining elements need not be the same in the small-signal model as they are in the model based on total quantities. Some examples will help to clarify these points.

Consider first the large-signal models developed in Fig. 11·2. Here ideal diodes are used to approximate the nonlinearity in the terminal characteristics of a transistor. Suppose we are studying an application in which the transistor is to be used in the normal-active mode. Then i_E is positive and v_{CB} is negative, or D_E is ON and D_C is OFF. Then a small-signal model can be generated by simply replacing D_E with a short circuit and removing D_C in any of the models in Fig. 11·2. Furthermore, if the actual bias point in the normal-active region is known, we may choose values of α, r_c and r_b which reflect the local behavior of the graphical characteristics rather than use values for these parameters which give the best fit to the global characteristics of the device.

If the nonlinearities in the transistor characteristics are represented by ideal junction diodes, then we attempt to approximate the small-signal behavior of the diode with a linear resistor. The mathematical basis for this approximation can be appreciated by considering the Taylor series expansion of a nonlinear function. For example, a nonlinear resistor will have a v-i relation of the form

$$i = f(v) \qquad (11 \cdot 3)$$

where i and v are total quantities. An operating point is defined by selecting a value of v, say V_1. Then

$$I_1 = f(V_1) \qquad (11 \cdot 4)$$

or the operating point is $(i = I_1, v = V_1)$. The Taylor series expansion of $f(v)$ about the operating point is simply

$$i = f(V_1) + \frac{df}{dv}\bigg|_{V_1} (v - V_1) + \cdots$$

$$= I_1 + \frac{df}{dv}\bigg|_{V_1} (v - V_1) + \cdots \qquad (11 \cdot 5)$$

This can be rearranged to yield

$$i - I_1 = \frac{df}{dv}\bigg|_{V_1} (v - V_1) \qquad (11 \cdot 6)$$

Now, $i - I_1$ and $v - V_1$ represent the *variations* of i and v around the operating point. If we represent the small-signal variations by $i_s(t)$, $v_s(t)$, then

$$i = I_1 + i_s(t) \qquad v = V_1 + v_s(t) \qquad (11 \cdot 7)$$

From Eq. $(11 \cdot 6)$, $i_s(t)$ and $v_s(t)$ are related by

$$i_s(t) = \frac{df}{dv}\bigg|_{V_1} v_s(t) = \frac{v_s(t)}{r} \qquad (11 \cdot 8)$$

r is called the *incremental, ac,* or *dynamic* resistance of the v-i characteristic at the operating point (I_1, V_1).

If the nonlinear v-i curve is that of an ideal junction diode, then

$$i = I_0(e^{qv/kT} - 1) = f(v)$$

and

$$\frac{df}{dv}\bigg|_{V_1} = \frac{qI_0}{kT} e^{qV_1/kT} = \frac{q(I_1 + I_0)}{kT}$$

where

$$I_1 = f(V_1)$$

Thus, variations in i and v are related to each other by the formula

$$i_s(t) = \frac{v_s(t)}{r_e} \qquad r_e = \frac{kT}{q(I_1 + I_0)}$$

We may carry this over directly to the development of a small-signal model for the intrinsic transistor in the common-base connection, normal-active mode. In this mode the collector-base junction is reverse-biased. Therefore, small variations in v_{CB} produce insignificant small-signal currents in D_{Cj}. Hence, for the purposes of a small-signal model, D_{Cj} and its related controlled source may be neglected (see Fig. 11·5).

The emitter-base junction is forward-biased in the normal-active mode. Let us suppose that the emitter-base junction is passing a steady bias current of I_E. If an additional small-signal current source is added in parallel with the biasing source, the total emitter current will be

$$i_E(t) = I_E + i_e(t)$$

where I_E = dc bias current
$i_e(t)$ = small-signal current

Using our previous results, the emitter-base diode D_{Ej} can be replaced by a resistor r_e of value

$$r_e = \frac{kT}{q(I_E + I_{ES})}$$

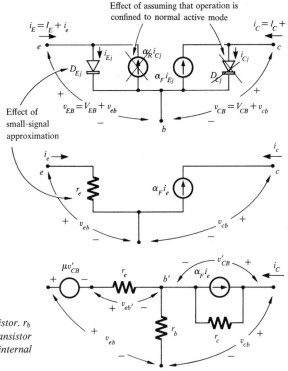

(a) *Large-signal circuit model for Ebers-Moll equations*

(b) *Small-signal model for intrinsic transistor*

(c) *Small-signal T-model for real transistor. r_b and r_c added to approximate real-transistor phenomena. Real base contact is at b; internal base contact at b'.*

where I_{ES} is the saturation current of D_{Ej}. The small-signal emitter current and small-signal emitter-base voltage are then related by

$$i_e = \frac{v_{eb}}{r_e}$$

The resulting model is shown in Fig. 11·5b.

This model can be expanded to include the effects of base-width modulation and base resistance in the manner shown in Fig. 11·5c. Base-resistance effects are approximated with the addition of the resistance r_b. Base-width modulation produces the finite output conductance $1/r_c$. Its effect on the input characteristics is approximated with the controlled-voltage source μv_{CB}. (The mathematical justification of the placement of elements due to base-width modulation is considered in Sec. 11·8.) Usually this latter effect is dropped and the small-signal model then reduces to a T connection of the three resistances r_e, r_b, and r_c, with a current generator αi_e in parallel with r_c. This model is called the T model and is widely used for calculating the small-signal properties of the transistor. We shall employ it for this purpose in Chap. 13.

There is one point about these small-signal models that should be emphasized. The most significant small-signal approximation in them involves replacing D_{Ej} with r_e. This is a critical approximation if we drive the emitter-base terminals with a voltage source, since we must then use r_e to find i_e; it is i_e that we need to know, since i_e is the control variable for the controlled source in the collector circuit. However, if we use a small-signal current source to drive the emitter, the collector current will be proportional to the signal input current, regardless of whether the small-signal approximation in the emitter circuit is valid or not. The emitter-base terminal voltage will be simply a nonlinear function of i_e, which cannot be calculated by using a small-signal resistance r_e. But this may well be an insignificant point. After all, the maximum voltage drop that can occur across the emitter-base diode is small in any case, and it can frequently be neglected in comparison with other voltages in the emitter circuit. Because of this, the simple model shown in Fig. 11·6, which neglects D_E entirely, is frequently used for estimating both large- *and* small-signal circuit performance in the normal-active region.

11·3·4 *Circuit models for the common-emitter connection.* Both large- and small-signal circuit models for the common-emitter connection can be

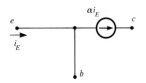

FIG. **11·6** SIMPLE MODEL FOR ESTIMATING CIRCUIT PROPERTIES OF COMMON-BASE CONNECTION IN NORMAL ACTIVE REGION. THE MODEL IS SIMPLY A CONTROLLED SOURCE.

410 *Semiconductor electronics*

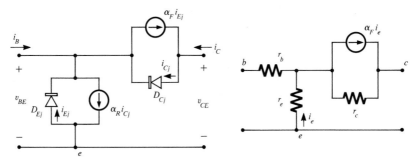

(a) *Possible large-signal model for common-emitter connection, pnp transistor*

(b) *Possible small-signal model for common-emitter connection*

FIG. **11 · 7** MODELS FOR CE CONNECTION DERIVED FROM MODELS FOR CB CONNECTION

obtained by a simple reorientation of the common-base models in the manner shown in Fig. 11 · 7. However, one of the control variables in these models is the emitter current, which is not a particularly convenient situation for common-emitter applications. In the large-signal model this can be remedied quite simply for the normal-active mode. For this case, current flow in D_{Cj} is small, so the controlled source in the emitter lead can be neglected. Kirchhoff's law at the internal node of the model can then be used to express the controlled source in the collector lead of the model in terms of i_B rather than i_{Ej}:

$$\alpha_F i_{Ej} = \frac{\alpha_F}{1 - \alpha_F} i_B = \beta i_B$$

The resulting large-signal models for the common-emitter connection are shown in Fig. 11 · 8 for both *pnp* and *npn* transistors.

For small-signal applications, either the base-emitter voltage or the base current will be known, so we would like to have models in which one of these variables is used to control the current source in the collector. This can be accomplished in the following series of steps. First, in Fig. 11 · 9a

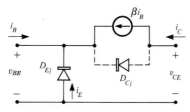

(a) *Large-signal model for CE connection of pnp transistor, normal active mode. D_{Ej} (and D_{Cj}) can be either an ideal diode or ideal-junction diode, depending on the nature of the approximation required.*

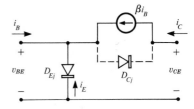

(b) *Large-signal model for CE connection of an npn transistor, normal active mode*

FIG. **11 · 8** LARGE-SIGNAL MODELS FOR TRANSISTORS IN THE COMMON-EMITTER CONNECTION, NORMAL ACTIVE MODE

Circuit models for transistors and vacuum tubes   *411*

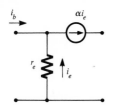

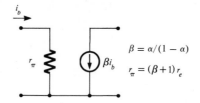

(a) CE small-signal model based on i_e; intrinsic transistor

(b) Equivalent representation based on i_b; intrinsic transistor

$$\beta = \alpha/(1-\alpha)$$

$$r_\pi = (\beta+1)\,r_e$$

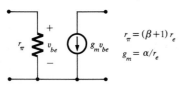

$$r_\pi = (\beta+1)\,r_e$$

$$g_m = \alpha/r_e$$

(c) Representation based on v_{be}; intrinsic transistor

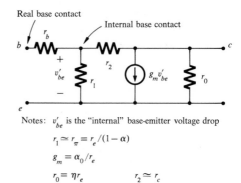

Notes: v'_{be} is the "internal" base-emitter voltage drop

$$r_1 \simeq r_\pi = r_e/(1-\alpha)$$

$$g_m = \alpha_0/r_e$$

$$r_0 = \eta r_e \qquad r_2 \simeq r_c$$

(d) The hybrid-π model for a transistor in the CE connection

FIG. 11·9 SMALL-SIGNAL MODELS FOR THE CE CONNECTION

we show the small-signal T model for the intrinsic transistor (specifically, with no base resistance or base-width modulation effects), oriented so that the emitter terminal is common. The small-signal emitter current can be expressed in terms of the small-signal base current by

$$i_e = -\frac{i_b}{1-\alpha} \qquad (11\cdot9a)$$

The subscript F on α is usually dropped for small-signal models, it being understood that we are operating in the normal-active region. Using Eq. (11·9a), the small-signal emitter-base voltage is

$$v_{be} = -i_e r_e = \frac{r_e}{1-\alpha} i_b = r_\pi i_b \qquad (11\cdot9b)$$

The current flowing in the controlled source is

$$i_c = -\alpha i_e = \frac{\alpha}{1-\alpha} i_b = \beta i_b \qquad (11\cdot9c)$$

We can also express this current in terms of v_{be} by using Eq. (11·9b):

$$i_c = \beta i_b = \frac{\beta}{r_\pi} v_{be} = g_m v_{be} \qquad (11\cdot9d)$$

in which, by appropriate substitutions

$$g_m = \frac{\alpha}{1 - \alpha} \frac{1 - \alpha}{r_e} = \frac{\alpha}{r_e}$$

These transformations among the circuit variables lead to the sequence of models shown in Figs. 11·9b and 11·9c. The model shown in Fig. 11·9c is called the hybrid-π model, for reasons that will be apparent in Sec. 11·6. The hybrid-π model is the most commonly used model for studying common-emitter applications of transistors. The basic hybrid-π model of Fig. 11·9c applies to the intrinsic transistor and is adequate for some purposes. However, in most situations the effects of base-width modulations and base resistance need to be included in this model. A hybrid-π model which is expanded to include these effects is shown in Fig. 11·9d. In this model

$$r_1 \simeq r_\pi = \frac{r_e}{1 - \alpha} \qquad r_2 \simeq r_c \qquad g_m = \frac{\alpha_0}{r_e} \qquad r_0 = \eta r_e$$

where η and r_2 are related to the rate of change of base width with collector voltage. To a reasonable approximation r_2 is essentially equal to the r_c element which appears in the T model, and η is a number which is usually on the order of 10^3. The mathematics necessary for justifying the various parameters in this model will be considered in Sec. 11·8.

11·3·5   *Frequency dependence in hybrid-π models.* One of the principal advantages of the hybrid-π model is that parameters can be easily added to it to account for frequency dependence in the transistor characteristics. In fact, we derived a small-signal model which is valid over most of the usable frequency range of a transistor in Sec. 9·5. This model is shown in Fig. 9·24, and is nothing more than the simple hybrid-π model shown in Fig. 11·9c with two capacitances and a base resistance added.

A more general hybrid-π model is shown in Fig. 11·10. Here the capacitance C_π appearing in parallel with r_π accounts for changes in the charge stored in the transistor base region when the transistor is driven with an ac signal. The capacitance C_c in parallel with r_2 accounts for the collector-base junction capacitance of the transistor. The emitter-base junction capacitance is neglected in comparison with C_π. The expanded model shown in Fig. 11·10 will be used to study transistor amplifiers in Chap. 13.

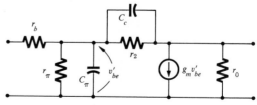

FIG. **11·10**   HYBRID-π MODEL WITH CAPACITANCES TO ACCOUNT FOR FREQUENCY DEPENDENCE

The procedure which we use to construct piecewise-linear circuit models for vacuum triodes is essentially the same as that used for making piecewise-linear models of transistors. We shall limit our considerations to the normal operating region where $i_G = 0$.

By comparing the plate family of a triode with the output characteristics of the controlled sources shown in Fig. 11·1 (or by direct physical reasoning), we see that we need a controlled voltage source in the plate circuit. The control variable for this source is the total grid-cathode voltage v_G.

For a first-order approximation, the output characteristic shown in Fig. 11·1a is suitable, if we eliminate the characteristic in quadrants II, III, and IV of that figure. This may be done by using ideal diodes in the manner shown in Fig. 11·11b. To improve the approximation, we can add a resistance r_p in series with the voltage generator in the "plate lead," a grid resistance r_g, and a diode D_S to ensure that no i_P flows when

Circuit   *Output characteristic*

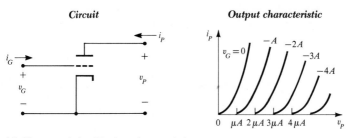

(a) *Vacuum triode with plate characteristics*

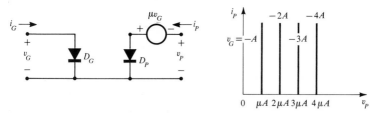

(b) *Circuit model for (a) using two ideal diodes and a controlled source*

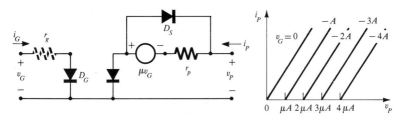

(c) *Addition of r_p to (b) to provide slope to the output characteristics*

FIG. 11·11 **APPROXIMATION OF TRIODE CHARACTERISTICS WITH PIECEWISE LINEAR CIRCUIT MODELS**

$v_G > 0$. The resulting model and its output characteristic are given in Fig. 11·11c. A more accurate representation can, of course, be developed by adding more ideal diodes, resistors, etc., but the representation given in Fig. 11·11c is quite adequate for most purposes. The grid-current characteristics for the models shown in Fig. 11·11c are the subject of Prob. 11.

11·4·1 *Small-signal models for vacuum triodes.* Small-signal models for vacuum triodes are usually drawn on the supposition that the instantaneous grid voltage v_G will always be negative and the instantaneous plate voltage v_P will always be greater than μv_G. As before, the total quantities v_G and v_P may be represented by dc and ac components as follows:

$$v_G = V_G + v_g(t) \qquad v_P = V_P + v_p(t) \tag{11·10}$$

Then the operating point is defined by the values of V_G and V_P, as shown in Fig. 11·12.

The small-signal model is constructed from the large-signal model of Fig. 11·11c by observing that the grid diode D_G is always reverse-biased, so the grid-cathode resistance is essentially infinite. In the plate circuit of the model, the plate diode D_P is always ON, so it need not be shown; the shorting diode D_S is always reverse-biased, so it can be removed.

The value of r_p to be used in the model can be obtained by measuring the slope of the plate characteristic at the known operating point (see Fig. 11·12), or if the triode obeys the ideal theory, r_p may be calculated. The value of r_p is defined by

$$r_p = \left.\frac{dv_P}{di_P}\right|_{v_G=\text{const}} \tag{11·11a}$$

If we use

$$i_P = B\left(v_G + \frac{v_P}{\mu}\right)^{3/2}$$

to describe the i_P-v_P relation, we have

$$\frac{1}{r_p} = \left.\frac{di_P}{dv_P}\right|_{v_G=V_G} = \frac{\tfrac{3}{2}B(V_G + V_P/\mu)^{1/2}}{\mu} = \frac{\tfrac{3}{2}B^{2/3}I_P^{1/3}}{\mu} \tag{11·11b}$$

According to this result, if r_p is measured at one operating point, where the dc plate current is I_{P_1}, then its value at another operating point will simply be

$$r_{p_2} = r_{p_1}\left(\frac{I_{P_1}}{I_{P_2}}\right)^{1/3} \tag{11·12}$$

It is sometimes convenient to make a Thévenin-to-Norton source transformation in the output circuit of the model shown in Fig. 11·12b.

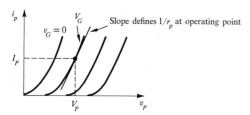

(a) Large-signal triode characteristics with operating point defined

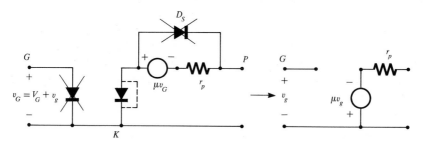

(b) Construction of small-signal model for vacuum triode from piecewise linear model shown in Fig. 10·16c

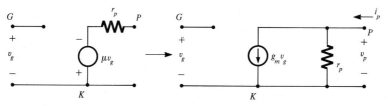

(c) Thévenin-to-Norton source transformation in plate-cathode circuit of a triode

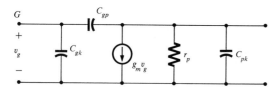

(d) Vacuum tube model with interelectrode capacitances added to account for frequency-dependent properties

FIG. 11·12   TRIODE CHARACTERISTICS AND A TRIODE SMALL-SIGNAL MODEL

The Norton current source has the value $g_m v_g(t)$, where

$$g_m \equiv \frac{\mu}{r_p} \qquad (11\cdot13)$$

Since the i_P-v_P relation has only two parameters in it (B and μ), the three circuit parameters r_p, g_m, and μ are not all independent. The circuit model that results from this transformation is shown in Fig. 11·12c. It is generally similar to the hybrid-π circuit model for a transistor in the CE connection (Fig. 11·9).

The source transformation will be a useful one when r_p is much larger than the values of resistance used in the external plate-cathode circuit. This is not usually true for triodes, but it is for pentodes.

11·4·2 *High-frequency effects in vacuum triodes.* Generally, the cathode-to-plate transit time for an electron in a vacuum tube is so short that the models just derived represent reasonably well the control mechanisms of most low-power triodes. The principal frequency dependence in vacuum-triode characteristics then arises from the interelectrode capacitances in the physical structure. Accordingly, the small-signal models of Fig. 11·12c can be simply modified to obtain those shown in Fig. 11·12d. The models shown in Fig. 11·12d are adequate for representing vacuum triodes intended for applications at frequencies below about 100 mcps, though it is sometimes necessary to add lead inductances in the grid, plate, and cathode leads.

11·5 **PIECEWISE-LINEAR CIRCUIT MODELS EXHIBITING**

THE PENTODE PLATE CHARACTERISTICS

Using the preceding ideas, we can easily construct a piecewise-linear circuit model for the plate characteristics of a pentode. By comparing Figs. 10·16 and 11·1, it is evident that a circuit of the type shown in Fig. 11·1b has the basic property desired, where the grid-cathode voltage is identified as the control variable. The proportionality constant for the controlled source is the transconductance of the model, defined by

$$g_m = \left. \frac{\partial i_P}{\partial v_G} \right|_{v_P = \text{const}}$$

Diodes can be placed in the input and output leads of the circuit model to remove characteristics in the undesired quadrants, shown in Fig. 11·1b. One other addition is necessary, however. The controlled source has zero output when the grid voltage is zero, while the pentode plate current is not zero when the grid voltage is zero. A current generator must therefore be placed in parallel with the controlled source in the plate-cathode lead of the circuit model. This source simply accounts for the plate current which flows due to the influence of the positive screen voltage.

For a better approximation, a resistor r_p is usually added in parallel with the output circuit to account for the slight influence of the plate voltage on plate current in the normal operating range. The resulting model is shown in Fig. 11·13. It must be remembered that since the screen current is not included in this model, it can be used only for studying the influence of grid control on the plate-circuit properties of the pentode. It cannot be used for setting up an operating point, since the screen voltage must be specified, and means for obtaining a relatively constant screen voltage must be provided.

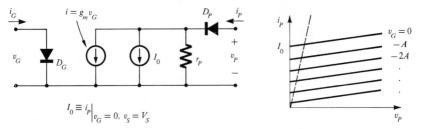

FIG. 11 · 13   PIECEWISE LINEAR MODEL GIVING VALID REPRESENTATION OF PENTODE OUT-
PUT CHARACTERISTICS TO THE RIGHT OF THE DOTTED LINE

11·5·1   Small-signal model for the pentode.   A small-signal model for the pentode can be simply generated from the model shown in Fig. 11 · 13. Diode D_G is eliminated by postulating that the pentode will always be operated in the negative grid region; diode D_P is eliminated by postulating that plate current always flows; and the fixed source I_0 is removed by observing that it only affects dc conditions in the pentode output circuit. The resulting model is identical to the model shown in Fig. 11 · 12d, (utilizing g_m instead of μ). This pentode-circuit model bears a close resemblance to the small-signal model of a transistor operated in the common-emitter connection, a fact which we shall use frequently later.

11·6   THE TWO-PORT NETWORK APPROACH TO SMALL-SIGNAL MODELS

As we have seen in preceding sections, small-signal models can be developed by suitable specialization of large-signal models, the large-signal models usually being of the piecewise-linear form. There is another approach to the construction of small-signal models, however, which does not proceed through the medium of large-signal models. Instead, one visualizes a transistor or a vacuum tube as being simply one specific manifestation of a black box with two pairs of terminals, or two *ports*. He then develops small-signal models directly from small-signal measurements which could be made at the terminals of the box.

Such an approach is very useful for the great generality which it has. We can represent a complete amplifier, a cross-country power-transmission line or a simple resistive voltage divider in the same basic two-port language that we use to describe an active device. We can get a new insight into the properties of networks with the two-port formulation, and we can express network properties (for example, the signal transmission from a source to a load) in terms which are valid for nearly any network. Thus, while we shall be primarily concerned in this chapter with the small-signal representation of transistors and vacuum tubes using the two-port approach, the reader should note the generality of the method for future reference.

11·6·1   Definition of two-port variables.   A two-port network is shown in Fig. 11 · 14. The network is characterized by four variables: two node-pair

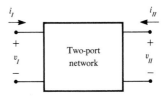

FIG. 11·14   TWO-PORT NETWORK WITH VARIABLES DEFINED

voltages and two currents. Reference-positive directions for these variables are given in Fig. 11 · 14. To study the properties of this network, we should think of two of these variables as being dependent variables and of two as being independent variables. Six different sets of dependent and independent variables are possible:

Dependent	Independent
$i_\mathrm{I}, i_\mathrm{II}$	$v_\mathrm{I}, v_\mathrm{II}$
$v_\mathrm{I}, v_\mathrm{II}$	$i_\mathrm{I}, i_\mathrm{II}$
$i_\mathrm{I}, v_\mathrm{II}$	$v_\mathrm{I}, i_\mathrm{II}$
$v_\mathrm{I}, i_\mathrm{II}$	$i_\mathrm{I}, v_\mathrm{II}$
$v_\mathrm{I}, i_\mathrm{I}$	$v_\mathrm{II}, i_\mathrm{II}$
$v_\mathrm{II}, i_\mathrm{II}$	$v_\mathrm{I}, i_\mathrm{I}$

The first four of these sets have the common property that one of the dependent variables is measured at the input port and the other at the output port, with the independent variables measured similarly. This is a useful choice for studying the signal-transmission properties of transistors and vacuum tubes, and we shall limit our considerations to these sets of variables.

11·6·2   Relationships between the two-port variables.   We are primarily interested in developing the small-signal relationships which exist between the dependent and independent variables listed in the previous table. Since the currents and voltages given there are total quantities, we again look to a Taylor's series expansion to arrive at our small-signal representation. Let us consider the first set of variables. i_I and i_II are thought of as being dependent on v_I and v_II. This means that, in principle, i_I and i_II can be expressed as

$$i_\mathrm{I} = f(v_\mathrm{I}, v_\mathrm{II}) \qquad i_\mathrm{II} = g(v_\mathrm{I}, v_\mathrm{II}) \qquad (11 \cdot 14)$$

For an intrinsic transistor or an ideal triode, the functions f and g are known. However, this is not necessary in the development to be given. Instead, we can simply visualize the functions f and g as arbitrary surfaces over the v_I-v_II plane, as shown in Fig. 11 · 15.

Now, we shall assume that the independent variables v_I and v_II can be divided into a constant term and a time-dependent term:

$$v_\mathrm{I} = V_1 + v_1(t) \qquad v_\mathrm{II} = V_2 + v_2(t) \qquad (11 \cdot 15)$$

Circuit models for transistors and vacuum tubes   *419*

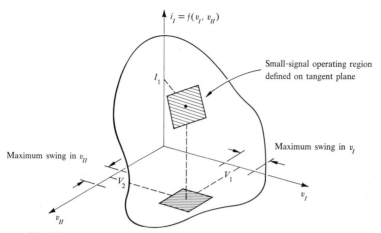

$$i_I = f(v_I, v_{II})$$

FIG. 11·15   GEOMETRICAL ILLUSTRATION OF TWO-PORT RELATIONSHIPS

and we shall assume that $v_1(t)$ and $v_2(t)$ are "small signals." The permitted sizes of $v_1(t)$ and $v_2(t)$ must be established later.

The constant terms V_1 and V_2 define an operating point

$$I_1 = f(V_1, V_2) \qquad I_2 = g(V_1, V_2) \qquad (11 \cdot 16)$$

as shown in Fig. 11·15. The small-signal swings about this operating point will produce excursions in i_I and i_{II} about their respective operating points. The magnitude of these swings can be estimated by assuming that the actual surfaces f and g can be adequately represented by their tangent planes at the points (V_1, V_2). This is the small-signal approximation (tangent plane = actual surface).

From a detailed inspection of the actual surface and its tangent plane, we could draw on the actual surface a closed curve surrounding the operating point such that, within the closed curve, the tangent plane predicts values of i_I and i_{II} which are within some fixed small percentage of the actual values (say 10 percent). This closed curve could then be used to define the maximum sizes which $v_1(t)$ and $v_2(t)$ may have for the small-signal approximation to be valid (see Fig. 11·15). Of course, the small-signal approximation may be meaningless if the bias point is located on a ridge or other discontinuity in the surface.

The mathematical expression of these geometrical ideas is contained in the Taylor's series expansion of a function of two variables. We have

$$i_I = f(V_1, V_2) + \left[\frac{\partial f}{\partial v_I} \bigg|_{V_1, V_2} \right] (v_I - V_1) + \left[\frac{\partial f}{\partial v_{II}} \bigg|_{V_2, V_1} \right] (v_{II} - V_2) +$$

$$\text{small terms} < 10 \text{ percent} \qquad (11 \cdot 17a)$$

$$i_2 = g(V_1, V_2) + \left[\frac{\partial g}{\partial v_I} \bigg|_{V_1, V_2} \right] (v_I - V_1) + \left[\frac{\partial g}{\partial v_{II}} \bigg|_{V_2, V_1} \right] (v_{II} - V_2) +$$

$$\text{small terms} < 10 \text{ percent} \qquad (11 \cdot 17b)$$

Equations (11·17) can be recast by using the equations

$$i_\mathrm{I} = I_1 + i_1(t) \qquad i_\mathrm{II} = I_2 + i_2(t) \qquad (11\cdot18)$$

together with Eqs. (11·16). The result is

$$i_1(t) = \left[\frac{\partial f}{\partial v_\mathrm{I}}\bigg|_{V_1,V_2}\right] v_1(t) + \left[\frac{\partial f}{\partial v_\mathrm{II}}\bigg|_{V_2,V_1}\right] v_2(t)$$

$$i_2(t) = \left[\frac{\partial g}{\partial v_\mathrm{I}}\bigg|_{V_1,V_2}\right] v_1(t) + \left[\frac{\partial g}{\partial v_\mathrm{II}}\bigg|_{V_2,V_1}\right] v_2(t) \qquad (11\cdot19)$$

Of course, the functions f and g have the dimensions of current in this example. Therefore, derivatives of f or g with respect to voltage have the dimensions of admittance. We then define a set of *admittance* or *y param-eters* by the relations

$$y_{11} = \frac{\partial f}{\partial v_\mathrm{I}}\bigg|_{V_1,V_2} \qquad y_{12} = \frac{\partial f}{\partial v_\mathrm{II}}\bigg|_{V_2,V_1}$$

$$y_{21} = \frac{\partial g}{\partial v_\mathrm{I}}\bigg|_{V_1,V_2} \qquad y_{22} = \frac{\partial g}{\partial v_\mathrm{II}}\bigg|_{V_2,V_1}$$

and recast Eqs. (10·19) in the form

$$i_1 = y_{11}v_1 + y_{12}v_2 \qquad i_2 = y_{21}v_1 + y_{22}v_2 \qquad (11\cdot20)$$

where we have dropped the explicit t dependence of the variables.

11·6·3 *Network representation of two-port equations: y parameters.* We can draw a circuit model which represents Eqs. (11·20) quite simply. This is shown in Fig. 11·16. Two controlled sources are used to account for the possible dependence of i_1 on v_2 and the possible dependence of i_2 on v_1.

A careful inspection of Fig. 11·16 will reveal that it is really nothing more than an application of Norton's theorem. The output port is repre-sented by a current generator $y_{21}v_1$ in parallel with an admittance y_{22}. Similarly, the input port is represented by a current generator $y_{12}v_2$ in parallel with an admittance y_{11}.

11·6·4 *Measurement of the y parameters.* In order to make the nature of the small-signal approximation clear, we have previously derived the y param-

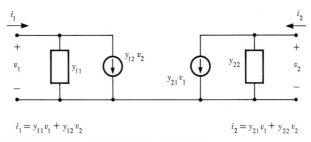

$$i_1 = y_{11}v_1 + y_{12}v_2 \qquad\qquad i_2 = y_{21}v_1 + y_{22}v_2$$

FIG. **11·16**   CIRCUIT REPRESENTATION OF y-PARAMETER EQUATIONS

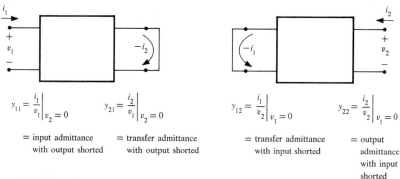

$$y_{11} = \left.\frac{i_1}{v_1}\right|_{v_2=0}$$
= input admittance with output shorted

$$y_{21} = \left.\frac{i_2}{v_1}\right|_{v_2=0}$$
= transfer admittance with output shorted

$$y_{12} = \left.\frac{i_1}{v_2}\right|_{v_1=0}$$
= transfer admittance with input shorted

$$y_{22} = \left.\frac{i_2}{v_2}\right|_{v_1=0}$$
= output admittance with input shorted

FIG. **11·17**   THE MEASUREMENT OF y PARAMETERS

eters by considering a large-signal surface and a plane which is tangent to it at the operating point. However, in practice we assume that a Taylor's series exists at the operating point, and then simply define the linear equations (11·20) for small-signal operation. The y parameters are then defined by a set of small-signal measurements:

$$y_{11} \equiv \left.\frac{i_1}{v_1}\right|_{v_2=0} = \text{input admittance with output shorted}$$

$$y_{12} \equiv \left.\frac{i_1}{v_2}\right|_{v_1=0} = \text{transfer admittance with input shorted}$$

$$y_{21} = \left.\frac{i_2}{v_1}\right|_{v_2=0} = \text{transfer admittance with output shorted}$$

$$y_{22} = \left.\frac{i_2}{v_2}\right|_{v_1=0} = \text{output admittance with input shorted}$$

These definitions suggest the simple measurement schemes given in Fig. 11·17. It is assumed that the short circuits shown there are *ac* short circuits only. The dc biases V_1 and V_2 have to be supplied by means which are not shown.[1]

The measurements suggested in these figures can be made without any specific knowledge of the large-signal characteristics of the device, or indeed any idea about the physical construction of the device. If the large-signal relationships are known in either a mathematical or graphical form, then, of course, they can be differentiated to obtain the parameters. Usually, however, the use of two-port parameters suggests that one is employing measured small-signal parameters to represent the two-port network.

11·6·5   *Two-port parameters for other sets of independent variables.* We can, of course, repeat the previous development for other selections of inde-

---

[1] Bridge techniques for measuring two-port parameters are described on pp. 219 and 220 of Linvill and Gibbons, "Transistors and Active Circuits." (See references at the end of this chapter.)

pendent and dependent variables. For example, if i_1 and i_2 are the independent variables, then v_1 and v_2 will be expressed in terms of them by the small-signal equations

$$v_1 = z_{11}i_1 + z_{12}i_2 \qquad v_2 = z_{21}i_1 + z_{22}i_2 \qquad (11 \cdot 22)$$

The two port is now characterized by its z or *impedance parameters.*

Similarly, if i_1 and v_2 are selected as independent variables, there results

$$v_1 = h_{11}i_1 + h_{12}v_2 \qquad i_2 = h_{21}i_1 + h_{22}v_2 \qquad (11 \cdot 23)$$

On the other hand, if i_2 and v_1 are selected as the independent variables,

$$i_1 = g_{11}v_1 + g_{12}i_2 \qquad v_2 = g_{21}v_1 + g_{22}i_2 \qquad (11 \cdot 24)$$

These two last sets of parameters are called *hybrid* parameters because the two independent variables do not have the same dimensions. As a result, the dimensions of the h parameters are mixed (h_{11} is in ohms, h_{12} and h_{21} are dimensionless, and h_{22} is in mhos), as are those of the g parameters.

11·6·6   *Network representation of z, h, and g parameters.* As with the y-parameter characterization, we can draw a circuit which represents the two-port equations for z, h, or g parameters by appropriate use of Thévenin's or Norton's theorem. These circuit models are shown in Figs. 11·18a, 11·18b, and 11·18c. For the z-parameter representation, Thévenin equivalent networks are used at both input and output. For the mixed sets, mixed representations are necessary. The h-parameter network has a Thévenin equivalent input loop and a Norton equivalent output, and vice

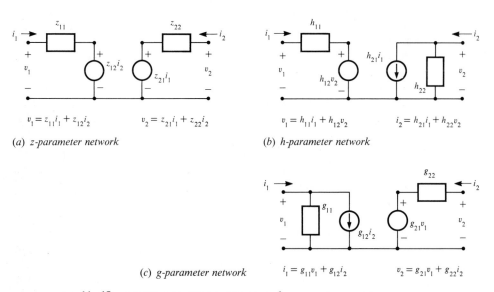

(a) *z-parameter network*

(b) *h-parameter network*

(c) *g-parameter network*

FIG. 11·18   CIRCUIT REPRESENTATION OF z-, h-, AND g-PARAMETER NETWORKS

Circuit models for transistors and vacuum tubes   423

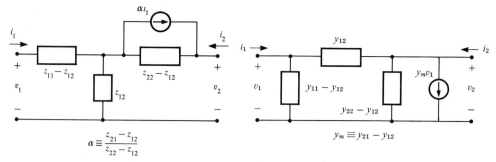

$$\alpha \equiv \frac{z_{21} - z_{12}}{z_{22} - z_{12}}$$

$$y_m \equiv y_{21} - y_{12}$$

FIG. 11·19 ALTERNATE CIRCUIT REPRESENTATIONS OF *z*- AND *y*-PARAMETER EQUATIONS

versa for the *g*-parameter circuit model. The *h*-parameter network is identical to the small-signal transistor model shown in Fig. 11·5c.

While the networks shown in Fig. 11·18 are the most direct representations of the two-port equations, it is always possible to draw other networks which exhibit the same equations. Two such networks are shown in Fig. 11·19 for *z* and *y* parameters. The *z* network is called a T-equivalent network, for obvious reasons, and can be simply related to the normal circuit model for common-base operation of a transistor. The *y* network is called a hybrid-π network.

11·6·7 *Two-port models for ideal amplifiers.* The utility of discussing four different characterizations (y, z, h, g) is simply that the parameters of a given network are usually more readily expressed in one parameter set than another. For example, the ideal common-base transistor amplifier is most easily expressed using h parameters, for then $h_{11} = 0$ (that is, $r_e + r_b$ is small, ideally zero), $h_{12} = 0$ (r_b is ideally zero), $h_{21} = \alpha$, and $h_{22} = 0$ (output admittance is ideally infinite). Thus, the h parameters for the ideal common-base transistor are

$$h_{11} = 0 \qquad h_{12} = 0 \qquad h_{21} = -\alpha \qquad h_{22} = 0$$

and the corresponding circuit model is simply that shown in Fig. 11·20a.

A vacuum triode is most simply expressed by the g parameters (Fig. 11·20b):

$$g_{11} = 0 \qquad g_{12} = 0 \qquad g_{21} = -\mu \qquad g_{22} = 0$$

And an ideal vacuum pentode (or a transistor in the common-emitter connection) is best expressed using y parameters (Fig. 11·20c):

$$y_{11} = 0 \qquad y_{12} = 0 \qquad y_{21} = g_m \qquad y_{22} = 0$$

11·6·8 *Conversions among the parameter sets.* The fact that two-port networks may sometimes be more simply expressed in one parameter set than another does not mean that the simplest parameter set must be used. Any parameter set can be used to describe any two-port, (though the results may sometimes be useless: for example, if the ideal pentode were expressed in

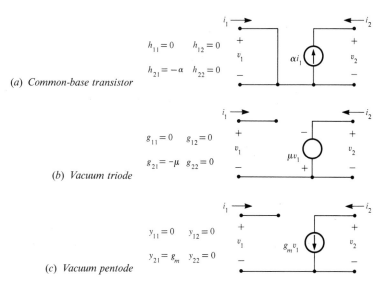

$h_{11} = 0$ $h_{12} = 0$

$h_{21} = -\alpha$ $h_{22} = 0$

(a) Common-base transistor

$g_{11} = 0$ $g_{12} = 0$

$g_{21} = -\mu$ $g_{22} = 0$

(b) Vacuum triode

$y_{11} = 0$ $y_{12} = 0$

$y_{21} = g_m$ $y_{22} = 0$

(c) Vacuum pentode

FIG. 11·20   IDEAL AMPLIFIER ELEMENTS EXPRESSED BY TWO-PORT RELATIONS

z-parameter language, we would have $z_{11} \to \infty$, $z_{22} \to \infty$). Equivalently, it is possible to transform y parameters into z, h, or g parameters, provided that certain of the parameters (or the quantity $y_{11}y_{22} - y_{12}y_{21}$) are not zero.

As an example of the algebra involved in such a conversion, let us consider the y-to-h transformation. Suppose we have measured the y parameters of a network, and we wish to find the h parameters. That is, we know the constants in

$$i_1 = y_{11}v_1 + y_{12}v_2 \tag{11·25a}$$

$$i_2 = y_{21}v_1 + y_{22}v_2 \tag{11·25b}$$

We want to rearrange this equation so that i_1 and v_2 are the independent variables. This means Eq. (11·25a) should be rewritten as

$$v_1 = \frac{1}{y_{11}}i_1 - \frac{y_{12}}{y_{11}}v_2 \tag{11·26a}$$

Using Eq. (11·26a) in (11·25b), we can now find

$$i_2 = \frac{y_{21}}{y_{11}}i_1 + \frac{y_{11}y_{22} - y_{12}y_{21}}{y_{11}}v_2 \tag{11·26b}$$

These manipulations are correct so long as y_{11} is not zero. Therefore, given that $y_{11} \neq 0$, the h parameters may be expressed in terms of the y parameters by

$$h_{11} = \frac{1}{y_{11}} \qquad h_{12} = -\frac{y_{12}}{y_{11}}$$

$$h_{21} = \frac{y_{21}}{y_{11}} \qquad h_{22} = \frac{y_{11}y_{22} - y_{12}y_{21}}{y_{11}}$$

11·7 TWO-PORT PARAMETERS FOR TYPICAL ACTIVE DEVICES

The use of two-port parameters for representing active devices is so common that manufacturers will frequently specify the two-port parameters for typical bias conditions, or as a function of bias. Of course, the two-port parameters will usually depend on frequency as well, so they may also be plotted as functions of frequency.

11·7·1 Two-port parameters for a typical transistor. As an example, the h parameters of a 2N1613 are plotted against the appropriate bias variables in Fig. 11·21. Figure 11·21a gives the parameters for the CB connection,

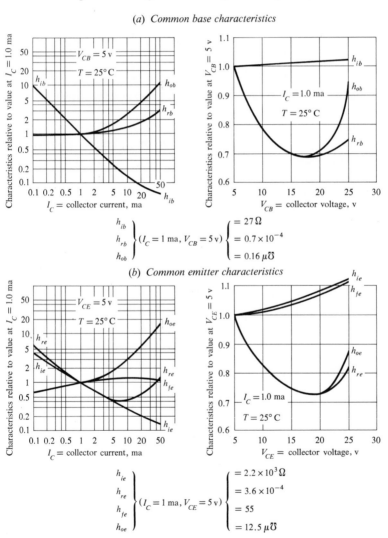

(a) Common base characteristics

$$\left.\begin{array}{c} h_{ib} \\ h_{rb} \\ h_{ob} \end{array}\right\} (I_C = 1 \text{ ma}, V_{CB} = 5 \text{ v}) \left\{\begin{array}{l} = 27\,\Omega \\ = 0.7 \times 10^{-4} \\ = 0.16\,\mu\mho \end{array}\right.$$

(b) Common emitter characteristics

$$\left.\begin{array}{c} h_{ie} \\ h_{re} \\ h_{fe} \\ h_{oe} \end{array}\right\} (I_C = 1 \text{ ma}, V_{CE} = 5 \text{ v}) \left\{\begin{array}{l} = 2.2 \times 10^{3}\,\Omega \\ = 3.6 \times 10^{-4} \\ = 55 \\ = 12.5\,\mu\mho \end{array}\right.$$

FIG. **11·21** h PARAMETERS FOR A 2N1613 VS. BIAS VARIABLES

426 Semiconductor electronics

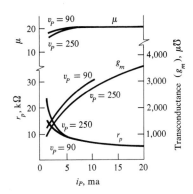

FIG. 11·22   TWO-PORT PARAMETERS FOR A TYPICAL VACUUM TRIODE (6SN7)

while Fig. 11·21*b* gives the *CE* parameters. In these parameters the sub-scripts 11, 12, 21, 22 are replaced by *ie*, etc. It is assumed that the user is familiar with the general theory; all he needs to know is which terminal is grounded during the measurements. Thus, in the common-base configuration, h_{11} is denoted by h_{ib}. The subscripts mean *input, base common*. h_{11} for the common-emitter orientation is then h_{ie}.

The other parameters are denoted by

$h_{12} \rightarrow h_r$ (reverse)

$h_{21} \rightarrow h_f$ (forward)

$h_{22} \rightarrow h_o$ (output)

Transformations exist for converting the *h* parameters of one configuration (*CE*) into those of another (*CB*), but they are messy and not usually used.

11·7·2   *Two-port parameters for typical vacuum tubes.* Since, under normal operating conditions, the grid-cathode circuit of a vacuum tube appears to be a nearly open circuit at low frequencies, vacuum-tube manufacturers need only provide curves showing how the parameters of the output circuit vary with bias. For a triode, we need to know $g_{21}(\equiv \mu)$ and $g_{22}(\equiv r_p)$. Alternately, we can express the behavior of the triode in terms of $g_m(\equiv y_{21})$ and $g_p(=y_{22} = 1/r_p)$. Typical curves are shown in Fig. 11·22.

11·8   DEVICE THEORY AND TWO-PORT MODELS

Usually, the physical theory of a device is concerned with prediction of terminal relationships from the physical properties of the device. The representation of these terminal relationships in a circuit model can be accomplished using the two-port formulation. As an example, we saw in Chap. 9 that the physical theory of the intrinsic transistor leads to the relations

$$i_E = a_{11}p_n(e^{qv_{EB}/kT} - 1) + a_{12}p_n(e^{qv_{CB}/kT} - 1)$$
$$i_C = a_{21}p_n(e^{qv_{EB}/kT} - 1) + a_{22}p_n(e^{qv_{CB}/kT} - 1)$$
$$(11·27)$$

Circuit models for transistors and vacuum tubes   427

where $a_{11} = a_{22} = \dfrac{qAD}{L_p} \coth \dfrac{W}{L_p}$ $\qquad a_{12} = a_{21} = -\dfrac{qAD_p}{L_p} \operatorname{csch} \dfrac{W}{L_p}$

We can develop a two-port small-signal circuit model directly from these equations. The equations are most suitable for y-parameter representation:

$$y_{11} = \left.\frac{\partial i_E}{\partial v_{EB}}\right|_{v_{CB}=\text{const}} = \frac{q}{kT} a_{11} p_n e^{qv_{EB}/kT} \qquad (11 \cdot 28a)$$

$$y_{12} = \left.\frac{\partial i_E}{\partial v_{CB}}\right|_{v_{EB}=\text{const}} = \frac{q}{kT} a_{12} p_n e^{qv_{CB}/kT} \qquad (11 \cdot 28b)$$

$$y_{21} = \left.\frac{\partial i_C}{\partial v_{EB}}\right|_{v_{CB}=\text{const}} = \frac{q}{kT} a_{21} p_n e^{qv_{EB}/kT} \qquad (11 \cdot 28c)$$

$$y_{22} = \left.\frac{\partial i_C}{\partial v_{CB}}\right|_{v_{EB}=\text{const}} = \frac{q}{kT} a_{22} p_n e^{qv_{CB}/kT} \qquad (11 \cdot 28d)$$

These equations may be simplified by recognizing that under normal bias conditions v_{CB} is large and negative, so y_{12} and y_{22} are essentially zero. Also, the dc emitter current is

$$I_E = a_{11} p_n (e^{qv_{EB}/kT} - 1)$$

and for the usual condition $e^{qv_{EB}/kT} \gg 1$, y_{11} can be written

$$y_{11} = \frac{qI_E}{kT}$$

Thus we find

$$y_{11} = \frac{qI_E}{kT} \qquad y_{12} = 0 \qquad y_{21} = \frac{a_{21}}{a_{11}} y_{11} \qquad y_{22} = 0$$

We can use the y-to-h transformation developed in Sec. $11 \cdot 6 \cdot 7$ to find

$$h_{11} = \frac{kT}{qI_E} \qquad h_{12} = 0 \qquad h_{21} = \frac{a_{21}}{a_{11}} = \alpha = \operatorname{sech} \frac{W}{L_p} \qquad h_{22} = 0$$

The circuit model for these h parameters is shown in Fig. $11 \cdot 23$ and is identical to the small-signal model for the common-base orientation developed in Sec. $11 \cdot 3 \cdot 4$.

We may also include base-width modulation effects in the two-port formulation. To do this we observe that the y_{12} and y_{22} are the only partial derivatives with respect to v_{CB}. Hence

$$y_{12} = \left.\frac{\partial i_E}{\partial v_{CB}}\right|_{v_{EB}} = \frac{q}{kT} a_{12} p_n e^{qv_{CB}/kT} + p_n(e^{qv_{CB}/kT} - 1)\frac{\partial a_{12}}{\partial v_{CB}} \quad (11 \cdot 29a)$$

$$y_{22} = \left.\frac{\partial i_C}{\partial v_{CB}}\right|_{v_{EB}} = \frac{q}{kT} a_{22} p_n e^{qv_{CB}/kT} + p_n(e^{qv_{CB}/kT} - 1)\frac{\partial a_{22}}{\partial v_{CB}} \quad (11 \cdot 29b)$$

The second terms in these equations account for the fact that, because the effective width of the base layer is a function of v_{CB}, so are a_{12} and a_{22}.

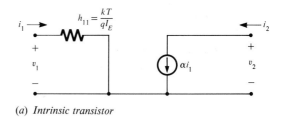

(a) Intrinsic transistor

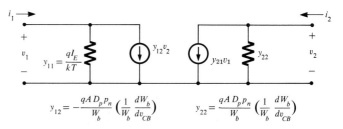

$$y_{12} = -\frac{qAD_pp_n}{W_b}\left(\frac{1}{W_b}\frac{dW_b}{dv_{CB}}\right) \qquad y_{22} = \frac{qAD_pp_n}{W_b}\left(\frac{1}{W_b}\frac{dW_b}{dv_{CB}}\right)$$

(b) Transistor with base-width modulation included

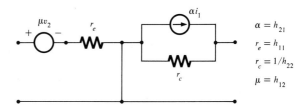

$$\alpha = h_{21}$$
$$r_e = h_{11}$$
$$r_c = 1/h_{22}$$
$$\mu = h_{12}$$

(c) Model of (b) converted to h parameters and drawn using standard transistor symbols

FIG. 11·23 SMALL-SIGNAL CIRCUIT MODEL FOR CB TRANSISTOR FROM TWO-PORT RELATIONS

As before, we may drop the exponentials, because v_{CB} is large and negative. Now, however,

$$y_{12} = -p_n\frac{\partial a_{12}}{\partial v_{CB}} \cong -\frac{qAD_pp_n}{W_b}\frac{1}{W_b}\frac{dW_b}{dv_{CB}}$$

$$y_{22} = -p_n\frac{\partial a_{22}}{\partial v_{CB}} \cong \frac{qAD_pp_n}{W_b}\frac{1}{W_b}\frac{dW_b}{dv_{CB}}$$

These values require that a current generator be added across the input terminals and an output conductance be added across the output terminals. The resulting model is shown in Fig. 11·23b. It can be converted to the model shown in Fig. 11·23c by a y-to-h parameter conversion.

Common-emitter models. Base-width modulation effects in common-emitter models can be generated by circuit transformation of the models shown in Figs. 11·23b and 11·23c, or directly from the appropriate set of equations. A simplified derivation can be based on the equations

$$i_B = -(1-\alpha)I_{ES}(e^{qv_{EB}/kT} - 1) \qquad i_C = -\alpha I_{ES}(e^{qv_{EB}/kT} - 1)$$

Circuit models for transistors and vacuum tubes 429

The effects of base-width modulation enter through derivatives of α and I_{ES} with respect to v_{CB}. The details are considered in the problems at the end of this chapter.

11·9 THE TWO-PORT INTERPRETATION OF THE RECIPROCITY THEOREM

In addition to its use in circuit characterization of active devices, the two-port formalism provides a useful insight into network theorems and network behavior. A good example is the reciprocity theorem. The usual statement of the reciprocity theorem is: "If a network is reciprocal, then the positions of an impedanceless generator and ammeter can be interchanged without affecting the reading of the ammeter." In two-port language, this is what we should call a y-parameter statement of the reciprocity theorem. We simply let the point of application of the impedanceless generator (a voltage source) be the input port, and consider the ammeter to be connected across the output port (see Fig. 11·24). The ammeter then reads

$$-i_{A1} = y_{21}v_s$$

When the positions of the source and ammeter are interchanged,

$$-i_{A2} = y_{12}v_s$$

and a reciprocal network is then one for which

$$y_{12} = y_{21}$$

If $y_{12} \neq y_{21}$ (as is true in vacuum tubes and transistors under normal operating conditions), the network is said to be nonreciprocal. Ideal amplifiers are nonreciprocal.

We may generalize the reciprocity theorem by simply recognizing that, except for highly special cases,

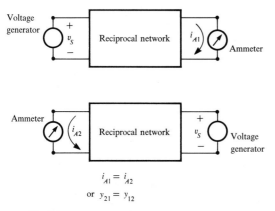

FIG. 11·24 THE RECIPROCITY THEOREM

$$y_{12} = y_{21} \Rightarrow h_{12} = -h_{21}$$
$$z_{12} = z_{21}$$
$$g_{12} = -g_{21}$$

Different ways of stating the reciprocity theorem result from the equality of transfer parameters in each set. For example, the statement $z_{12} = z_{21}$ is equivalent to the statement "The positions of an ideal current source and an ideal, or infinite impedance, voltmeter can be interchanged without affecting the voltmeter reading."

The other two statements of the reciprocity theorem substantiate the rule that in the process of interchanging sources and measuring instruments, one *must not* change the impedances in which the ports are terminated (short circuits or open circuits). If a voltage source and voltmeter are used for one measurement, the reciprocal measurement must employ a current source and ammeter.

$11 \cdot 10$ SUMMARY

A circuit model for an active device is a three-terminal network containing R's, L's, C's, controlled sources, and diodes, which are connected so that the electrical characteristics of the circuit model at its terminals approximate those of the real device with sufficient accuracy for purposes of analysis. In this chapter we have studied various procedures for making circuit models. For a convenient overview of the process, these procedures are summarized in the chart below, using the transistor as an example.

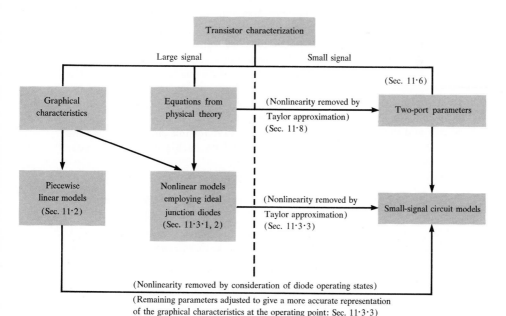

REFERENCES

Zimmerman, H. J., and S. J. Mason: "Electronic Circuit Theory," John Wiley & Sons, Inc., New York, 1962, chaps. 5 and 6.

Gray, P. E., D. DeWitt, A. R. Boothroyd, and J. F. Gibbons: "Physical Electronics and Circuit Models of Transistors," SEEC Notes, vol. 2, John Wiley & Sons, Inc., New York, 1964, chaps. 8 and 9.

Searle, C. L., A. R. Boothroyd, E. J. Angelo, Jr., P. E. Gray, and D. O. Pederson: "Elementary Circuit Properties of Transistors," John Wiley & Sons, Inc., New York, 1964, Chaps. 2–4.

Ryder, J. D.: "Electronic Engineering Principles," 3d ed., Prentice-Hall, Inc., Englewood Cliffs, N. J., Chaps. 7–9.

Linvill, J. G., and J. F. Gibbons: "Transistors and Active Circuits," McGraw-Hill Book Company, New York, 1961, chap. 9.

Cote, A. J., Jr., J. B. Oakes: "Linear Vacuum Tube and Transistor Circuits," McGraw-Hill Book Company, New York, 1961, Chaps. 2–3.

DEMONSTRATION

The construction of circuit models is a pencil-and-paper activity which leads to the representation of a device in terms of certain parameters. It is useful to show how these various parameters can be obtained from different sources, such as a transistor data sheet, a transistor curve tracer, a commercial tester (such as the Owens Laboratory Transistor Test Set Type 210), or a series of bridge measurements.

PROBLEMS

11·1 A germanium-pnp-alloy transistor has the following parameters: $\beta = \alpha/(1 - \alpha) = 50$, $r_c = 200$ kohms, $I_0 = 1$ μa.

1. Construct a piecewise-linear model for this transistor of the type shown in Fig. 11·2c. Sketch and dimension the output characteristics of this transistor. Use 5-volt increments in V_{CB} from 0 to -25 volts, and 1-ma increments in I_E from 0 to 6 ma.

2. Construct a load line on these output characteristics for a collector supply voltage of 20 volts, load resistance of 5 kohms.

3. Sketch and dimension the current-transfer curve for this circuit. Indicate the condition (ON or OFF) of the diodes in the model for each part of this transfer characteristic.

4. Repeat for an npn transistor with the same parameters.

11·2 Develop a piecewise-linear model whose terminal characteristics approximate those shown in Fig. 10·1b. Neglect the breakdown voltage, for simplicity.

11·3 Using the ideal diode model shown in Fig. 11·8a, with β unspecified, sketch and dimension the current-transfer curve for the circuit shown in Fig. P11·3.

11·4 The common-emitter current-amplification factor β of a 2N1613 varies with collector current, in the manner shown in Fig. P11·4.

1. Sketch piecewise-linear output characteristics (i_C vs. v_{CE}) for this transistor. Use i_C increments of 1 ma in the range 1 ma $<$ i_C $<$ 10 ma.

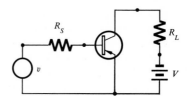

FIG. P11·3

2. Using the load line for a collector supply voltage of 20 volts and a load resistor of 2 kohms, sketch and dimension the current-transfer curve (i_C vs. i_B) for this amplifier.

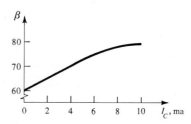

FIG. P11·4

11·5 Using the ideal diode model shown in Fig. 11·8b, sketch and dimension the current-transfer curve for the circuit shown in Fig. P11·5. Indicate the condition of each diode on each part of the transfer curve.

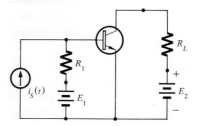

FIG. P11·5

11·6 Sketch and dimension the common-base and common-emitter characteristics for an intrinsic transistor with $I_{ES} = I_{CS} = 10^{-9}$ amp, $\alpha_F = \alpha_R = 0.98$.

11·7 A transistor has a common-emitter current gain β of 50. It is operated at a bias current of 0.5 ma. Assuming $r_b = 50$ ohms, $r_c = 1$ megohm, construct a small signal for the common-base orientation of the transistor. Calculate the input impedance of the model when the output is terminated in a 5 kohm resistor.

11·8 Construct the hybrid-π model shown in Fig. 11·9 for the transistor of Prob. 11·6. Calculate the input impedance when the output is terminated in a 5-kohm resistor.

11·9 A transistor similar to the one whose current gain β is plotted in Fig. P11·4 has $r_b \simeq 50$ ohms, $r_c \simeq 5$ megohms. Construct a hybrid-π model for this transistor, listing the maximum and minimum values of each element for the current range 1 ma $< I_c < 10$ ma. What is the corresponding range of base current?

Circuit models for transistors and vacuum tubes **433**

11 · 10 A transistor voltage regulator is shown in Fig. P11 · 10. The input to the regulator is a full-wave, capacitor-filter rectification circuit in which $C_L = 250$ μfarads. The output voltage of the power transformer is 35 volts peak (from each end to the center tap). The diode shown in the base lead of the transistor is a Zener diode with a breakdown voltage of 25 volts. The resistor R insures that this diode is always in the avalanching condition, so it may be replaced in the circuit by a 25-volt battery.

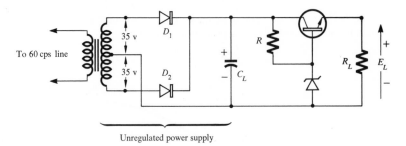

Unregulated power supply

FIG. P11 · 10

1. Using an ideal diode model of the transistor, explain the manner in which the circuit operates to regulate the output voltage R_L.

2. Assuming $R_L = 250$ ohms, sketch and dimension (a) the input voltage to the regulator, (b) the pulses of current which flow in the rectifier diodes, (c) the collector-to-base voltage across the transistor, and (d) the current flowing through the transistor.

3. Calculate the power dissipated in the transistor.

4. What is the maximum load current which this circuit can deliver before it fails to regulate properly? What is the corresponding minimum value of R_L and the power dissipated in the transistor?

5. A more accurate representation of the transistor than the ideal diode model can be obtained as follows. Assume it has a β of 50 and a base resistance of 50 ohms. Represent the emitter-base diode by an ideal diode in series with a battery (correctly polarized) of 0.6 volt (as in Sec. 7 · 1). Under these conditions what is the output voltage of the regulator?

11 · 11 In making circuit models for real devices, diodes are used to eliminate the *v-i* characteristic of a controlled source in certain quadrants. What unphysical property would the circuit models exhibit if diodes were not used? (This shows that if active devices are linear in one operating region, they must be nonlinear somewhere outside this region).

11 · 12 The plate current of a hypothetical triode can be written as

$$i_P = 4 \times 10^{-4}(v_G + 0.05\ v_P)^{3/2}\qquad \text{amp}$$

1. Sketch the plate characteristics of this triode and develop a piecewise-linear model for it, similar to the one shown in Fig. 11 · 10c.

2. Sketch a load line on these characteristics for a plate supply voltage of 400 volts and a load resistance of 100 kohms. From this load line, sketch and dimension the i_P-v_G transfer characteristic of the circuit.

3. Determine r_p and g_m at the point on this load line where $V_P = 200$ volts.

11·13 The circuit (Miller bridge) shown in Fig. P11·13 can be used to measure the μ of a triode. The measurement consists in adjusting R_2 and R_1 to get a null in the detector. Using a small-signal model of the triode, show that a null occurs when $R_2/R_1 = \mu$.

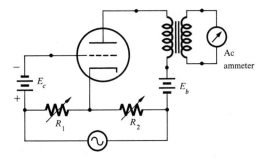

FIG. P11·13

11·14 Find the h parameters for the common-base circuit model shown in Fig. P11·14. After the h's have been expressed in terms of r_e, r_b, α, and r_c, find their values for $\alpha = 0.98$, $r_c = 500$ kohms, $r_e = 25$ ohms (that is, $I_E = 1$ ma), $r_b = 50$ ohms. Draw an h-parameter circuit representation using these values.

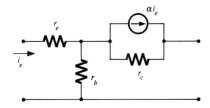

FIG. P11·14

11·15 As with the y parameters, the h parameters are basically defined by partial derivatives. Write down explicitly what derivatives the four h parameters represent, in terms of transistor-terminal variables, for both common-emitter and common-base cases.

11·16 The operating point for a transistor is $I_E = 3$ ma, $V_{EB} = 0.4$ volt, $I_c = 2.8$ ma, $V_{CB} = 10$ volts. At this point, the common-base small-signal h parameters are $h_{ib} = 50$ ohms, $h_{rb} = 3 \times 10^{-3}, h_{fb} = -0.95, h_{ob} = 10^{-5}$ volt. Construct the collector and emitter families as accurately as you can from this information. (The emitter family is a set of i_E vs. v_{EB} curves with v_{CB} const.)

11·17 Show that a π network similar to the one shown in Fig. P11·17 can be used to represent a set of y parameters in which $y_{12} = y_{21}$. Can you modify the π network to include the case where $y_{12} \neq y_{21}$?

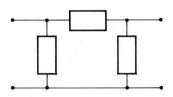

FIG. P11·17

11·18   Calculate the z parameters for a transformer which has a primary induct-ance L_1, a mutual inductance M, and a secondary inductance L_2. Draw a circuit model similar to that shown in Fig. 11·14 for the transformer.

11·19   The output characteristics of the 6SN7 triode are parametric representations of the i_P vs. (v_P, v_G) surface. Sketch the actual surface (that is, using three axes) for the curves given in Fig. 10·12. Use the output characteristics to estimate μ, r_p, and g_m at $V_p = 90$ volts and 250 volts, and compare the values with those given in Fig. 11·22.

11·20   Show directly from the ideal triode law

$$i_p = B\left(v_G + \frac{v_P}{\mu}\right)^{3/2}$$

that

$$g_m r_p = \mu$$

11·21   Compute g_m as a function of I_p for an ideal triode. Select as a match point the value of g_m at $i_p = 10$ ma, given in Fig. 11·22, and then plot the theoretical g_m and the measured g_m on the same set of axes. Comment on your results.

11·22   Express the z parameters in terms of the y parameters.

11·23   Which y parameters of a transistor are affected by the presence of base-width modulation? Calculate the theoretical values of these parameters in terms of dW_b/dV_C, the rate of change of base-width with collector-base voltage. Draw a y-parameter model for a transistor which includes these effects.

11·24   The h parameters of a network can be computed with the aid of nothing but driving-point measurements. The measurements are shown in Fig. P11·24. Determine the h parameters from these measurements.

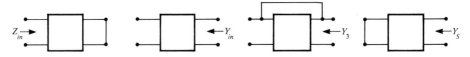

FIG. P11·24

11·25   A three-terminal network may have its terminal voltages defined with respect to an arbitrary ground, as shown in Fig. P11·25. One can then define the admittance parameters for this network by means of

$$I_a = Y_{aa}V_a + Y_{ab}V_b + Y_{ac}V_c$$
$$I_b = Y_{ba}V_a + Y_{bb}V_b + Y_{bc}V_c$$
$$I_c = Y_{ca}V_a + Y_{cb}V_b + Y_{cc}V_c$$

Y_{aa}, Y_{bb}, and Y_{cc} are driving-point parameters, and the rest are transfer parameters.

1. Show that the sum of the admittances in any row or column equals zero. (Hint: set $V_b = V_c = 0$, and use Kirchhoff's current law to show that $\sum\limits_{i=a,b,c} Y_{ia} = 0$; set $V_a = V_b = V_c$ to show $\sum\limits_{j=a,b,c} Y_{aj} = 0$).

2. If the C terminal is connected to the arbitrary ground, a two-port network results. How do the two-port parameters of this two-port arise from the generalized three-port admittances given above?

3. Draw a circuit representation for the three-port equations given above.

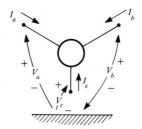

11 · 26 Construct the low-frequency hybrid-π model for a transistor from the approximate equations

$$i_B = -(1 - \alpha)I_{ES}(e^{qv_{EB}/kT} - 1) \qquad i_C = -\alpha I_{ES}(e^{qv_{EB}/kT} - 1)$$

Include the effects of base-width modulation in your calculations. Does your model agree with the one shown in Fig. 11 · 9*d*? If not, why?

12 BIASING NETWORKS

A REASONABLE PLACE to begin a study of transistor applications is with an inquiry into how transistors may be biased to achieve and maintain a desirable operating point. Unfortunately, while the biasing problem can be stated simply, its solution can become rather involved. As a result, there is a sizable literature on transistor biasing, containing many solutions to the biasing problem which are appropriate for particular cases. Under these circumstances we shall limit ourselves to defining the biasing problem and discussing those solutions to it which prove to be of widest application. We shall also discuss briefly the most common techniques for biasing vacuum tubes, where the problem and its solution are simpler than for the transistor.

12·1   THE BIASING PROBLEM FOR TRANSISTORS

The principal problem in designing a biasing network for a transistor is that some of the parameters of the transistor are extremely temperature-sensitive. Since transistor circuits usually have to operate in an environment where temperature changes may occur, the biasing network must be designed so that these temperature-sensitive parameters have a minimum effect on the operating point. In addition, if a circuit is to be produced in quantity, the desired operating point should be approximately maintained when normal manufacturing tolerances on the network components and transistor parameters are permitted. Of course, the biasing network must achieve these objectives with a minimum influence on the desired circuit behavior. That is, if we are building an amplifier, we do not want the biasing network to seriously limit the amplifier gain or frequency response.

These objectives can usually be met by designing bias circuits which stabilize the dc values of i_C (or i_E) and v_{CB} (or v_{CE}). We will see how this is done after a brief discussion of transistor parameter variations with temperature.

12·2   VARIATION OF TRANSISTOR PARAMETERS WITH TEMPERATURE

The variation of transistor parameters with temperature can be traced to four main sources:

1. The intrinsic carrier density n_i is an exponential function of temperature. The saturation currents in ideal junction diodes and transistors depend on $n_i{}^2$, and are therefore extremely temperature-sensitive. On this account alone, junction saturation currents will approximately double for every 10°C change in the temperature of a germanium transistor, and every 7°C in silicon transistors.

2. The junction laws

$$p_E(0) = p_n(e^{qv/kT} - 1) \sim n_i{}^2(e^{qv/kT} - 1)$$
$$n_E(0) = n_p(e^{qv/kT} - 1) \sim n_i{}^2(e^{qv/kT} - 1)$$

are also very sensitive functions of temperature, due both to the dependence of $n_i{}^2$ on T and the equally temperature-sensitive factors $(e^{qv/kT})$.[1]

3. The minority-carrier diffusion constant and lifetime are temperature dependent, so the diffusion length is also. This means that α and β for a given transistor will be functions of temperature.

4. Both the forward and reverse characteristics of pn junctions are sensitive functions of surface conditions, which in turn depend on temperature. In transistors whose surfaces have been carefully "passivated," these effects may be negligible, though incorrect operation of the transistor, even momentarily, may destroy the surface passivation.

---

[1] The $n_i{}^2$ factor in the junction law is responsible for the effects described in (1) above.

The effects of these parameter variations may be visualized either as changes in the graphical characteristics of the transistor or as changes in the parameters of a circuit model.

12·2·1 *Silicon transistors.* The manner in which the graphical characteristics of an *npn* silicon transistor change with temperature is shown in Fig. 12·1. Figures 12·1a, 12·1b, and 12·1c show the variation of the common-emitter output characteristics with temperature; Fig. 12·1d shows the variation of the input characteristic with temperature for $V_{CE} > 1$ volt; and Fig. 12·1e shows the typical variation of the collector-base diode saturation current I_{CBO} with temperature.

Output characteristics. To discuss the output characteristics shown in Figs. 12·1a–12·1c, we first recall that the collector current for an *npn* transistor in the *CE* connection, normal-active mode, is expressible as

$$i_C = \beta i_B + (\beta + 1)I_{CO} \qquad (12 \cdot 1)$$

According to this equation, the variations of i_C with temperature can be attributed to variations in $\beta(T)$ and $I_{CO}(T)$. The variation of β with temperature can be established directly from Figs. 12·1a–12·1c. Reading ΔI_C at $V_{CE} = 5$ volts for a ΔI_B of 0.02 ma, we find $\beta(-55°C) \sim 60$; $\beta(25°C) \sim 100$; $\beta(100°C) \sim 175$. The variation of β with temperature is therefore roughly linear for this transistor (and most other low-to-medium power silicon transistors).

The variation of I_{CO} with temperature is much more rapid, as suggested in Fig. 12·1e: I_{CO} changes by two orders of magnitude over the temperature range $25°C < T < 100°C$. However, because I_{CO} is so small, the effects of its great variation with temperature are not apparent in the output characteristics (even when multiplied by $\beta + 1$). This is reflected in Figs. 12·1a–12·1c, where the $I_B = 0$ curve is indistinguishable from the V_{CE} axis for low voltages at all temperatures. On account of this, we generally neglect the variation of I_{CO} in studying the biasing problem for silicon transistors (except for specialized applications where I_C is very low).

Input characteristic. The variation of the input characteristic with temperature can be understood as follows. For the normal-active mode in a silicon transistor, in which I_{CO} is negligible, the relation between i_B and v_{BE} is theoretically of the form

$$i_B = (1 - \alpha)I_{ES}(e^{qv_{BE}/kT} - 1) \qquad (12 \cdot 2)$$

where I_{ES} is the saturation current of the emitter-base diode with the collector-base diode shorted. For a given temperature, the I_B-V_{BE} curve will be the familiar diode curve. As temperature is varied, the diode curves are simply displaced on the V_{BE} axis, as shown in Fig. 12·1d.

The amount of the displacement can be calculated by finding how much V_{BE} changes with temperature for a fixed value of I_B. To do this, we rewrite

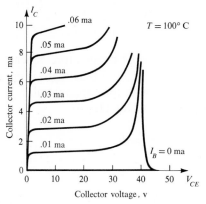

(a) Common-emitter output characteristics at 100° C

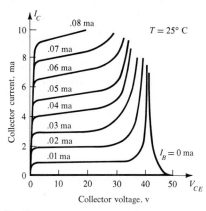

(b) Common-emitter output characteristics at 25° C

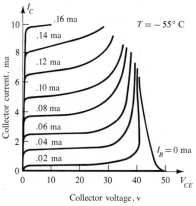

(c) Common-emitter output characteristics at −55° C

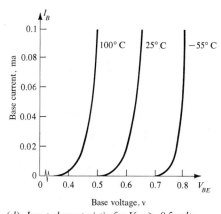

(d) Input characteristic for $V_{CE} > 0.5$ volts

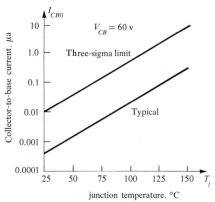

(e) Collector saturation current as a function of temperature

FIG. 12·1 **THE EFFECT OF TEMPERATURE ON THE GRAPHICAL CHARACTERISTICS OF AN** *npn* **SILICON TRANSISTOR (2N1613)**
[*Data courtesy of Fairchild Semiconductor.*]

Eq. (12·2) in the form

$$I_B = Ae^{-qE_g/kT}(e^{qV_{BE}/kT} - 1) \simeq Ae^{-q(E_g - V_{BE})/kT} \qquad (12\cdot3)$$

In Eq. (12·3) we have assumed that the principal temperature dependence of I_{ES} arises from the exponential factor in n_i^2, and that there is enough forward bias to neglect the 1 in comparison with the exponential.

To keep i_B fixed means to fix the exponent in Eq. (12·3):

$$\frac{q(E_g - V_{BE})}{kT} = \text{const} = B \qquad (12\cdot4)$$

From this equation we see that

$$\frac{dV_{BE}}{dT} = -B\frac{k}{q} \qquad (12\cdot5)$$

We evaluate dV_{BE}/dT in Eq. (12·5) by observing that the forward bias on the emitter-base junction of a silicon transistor will be roughly 0.6 volt for nearly any current level, and that E_g is 1.2 volts for silicon. Thus, at $T = 300°C$, we obtain $dV_{BE}/dT \simeq -2$ mv/°C. The curves of Fig. 12·1d show that this theoretical prediction fits experimental data rather well.

12·2·2 Germanium transistors. All the temperature-dependent effects that occur in silicon transistors also occur in germanium transistors. β changes roughly linearly with temperature, and V_{EB} decreases at the rate of about 2 mv/°C. However, the collector saturation current I_{CO} of a germanium transistor is several orders of magnitude higher than I_{CO} for a silicon transistor. This may or may not be important in determining the collector

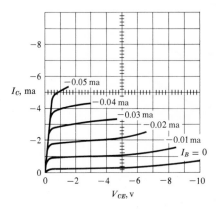

(a) *Common-emitter output characteristics of a 2N324 taken at T = 25°C*

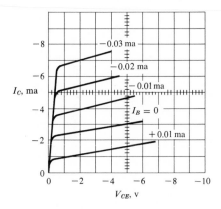

(b) *Common-emitter output characteristics of a 2N324 (pnp germanium alloy transistor) taken at T = 50°C*

FIG. **12·2**   THE TEMPERATURE DEPENDENCE OF THE COMMON-EMITTER OUTPUT CHARAC-TERISTICS OF A TYPICAL *pnp* GERMANIUM ALLOY TRANSISTOR

characteristics at room temperature; but it is quite significant at elevated temperatures, because of the rapid escalation of I_{CO} with temperature.

This point is illustrated in Fig. 12·2, where the common-emitter characteristics for a 2N324 (germanium-*pnp*-alloy transistor) are plotted at two temperatures, 25°C and 50°C. The values of I_{CO} are 3 μa at 25°C and about 20 μa at 50°C. The values of β at the two temperatures are $\beta \sim 100$ at 25°C, $\beta \sim 150$ at 50°C. The most conspicuous manifestation of the increase in I_{CO} is that at 50°C the transistor will exhibit normal amplification properties for *positive* base currents, as shown in Fig. 12·2b. While changes in β are important in predicting the exact characteristics, the change of I_{CO} with temperature dominates the behavior of the output characteristics of most germanium transistors.

12·3 TEMPERATURE-DEPENDENT CIRCUIT MODELS OF TRANSISTORS

The effects just described can be included in transistor circuit models in several ways. The most useful models are shown in Fig. 12·3. Part *a* of this figure shows the T model of a *pnp* transistor specialized to the normal-active range. The collector-base diode is replaced by a controlled source whose output current is I_{CO}. The control variable for this source is the temperature of the transistor. The emitter-base diode is an ideal junction diode D_{Ej} with a temperature-dependent saturation current, and the transistor α is also temperature dependent.

In Fig. 12·3b, the ideal junction diode is replaced by an ideal diode in series with a battery of voltage ϕ, making the model a piecewise-linear one. The battery is used to represent the fact that in a junction diode several tenths of a volt must be applied to cause appreciable current to flow. ϕ is about 0.3 volt for germanium transistors and 0.6 volt for silicon transistors. In both cases, ϕ is a function of temperature, decreasing at the rate of about 2 mv/°C. In Fig. 12·3c, the output network is simply expressed in terms of the base current i_B.

Generally, in these models the temperature dependence of α is neglected because α, which is near unity for any temperature, just gets closer to unity as T increases. However, since β depends on the deviation of α from unity, it is quite temperature-sensitive. For most transistors the temperature dependence of β can be represented roughly by

$$\beta(25°C + \Delta T) \simeq \beta(25°C)(1 + K \Delta T)$$

This approximate rule applies for both silicon and germanium transistors, with $K \sim (\frac{1}{75})°C^{-1}$ for silicon and $(\frac{1}{50})°C^{-1}$ for germanium.

With the aid of these models we can now proceed to the design of biasing networks. We shall first consider how the operating point should be selected to minimize the importance of temperature-dependent transistor parameters, and then how networks may be designed to achieve this operating point.

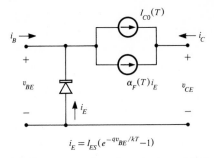

$$i_E = I_{ES}(e^{-qv_{BE}/kT} - 1)$$

(a) Temperature-dependent circuit model of a pnp transistor in normal-active mode

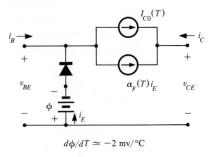

$$d\phi/dT \simeq -2 \text{ mv}/°C$$

(b) Piecewise linear model of (a) using ideal diode and battery with temperature-dependent voltage

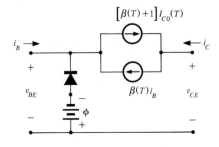

(c) Model of (b) with output network expressed in terms of i_B and β

Notes for all models:

For Ge transistors: 1. I_{CO} and $I_{ES}(T)$ double for every 10°C increase in T above 25°C

2. $\phi \simeq 0.3 \text{ v}; d\phi/dT \simeq -2 \text{ mv}/°C$

3. $\beta(25°C + \Delta T) \simeq \beta(25°C)[1 + \Delta T/50°C]$

For Si transistors: 1. $I_{CO}(T)$ and $I_{ES}(T)$ double for every 7°C increase in T above 25°C

2. $\phi \simeq 0.6 \text{ v}; d\phi/dT \simeq -2 \text{ mv}/°C$

3. $\beta(25°C + \Delta T) \simeq \beta(25°C)[1 + \Delta T/75°C]$

FIG. 12·3   TEMPERATURE-DEPENDENT CIRCUIT MODELS OF A *pnp* TRANSISTOR. FOR *npn* MODELS INTERCHANGE DIRECTION OF DIODE AND $I_{CO}(T)$.

12·4   FACTORS AFFECTING THE SELECTION OF AN OPERATING POINT

In Chap. 10 we described briefly the procedure by which one locates the permissible area of the i_C-v_{CE} plane for load lines. A vertical line is drawn at the maximum voltage rating, a horizontal line is drawn at the maximum current rating, and the maximum-rated power-dissipation parabola is sketched in. The area enclosed by these curves and the axes is the permissible area. These general rules apply at any temperature, but since the maximum ratings depend on the operating junction temperature, the permissible load-line area will generally shrink with increasing temperature, as suggested in Fig. 12·4. Thus, one needs to know or estimate the maximum temperature which the transistor will reach before the permissible load-line area can be delineated.

The maximum operating temperature of the transistor is determined by two factors: the ambient temperature and the amount of power being dissipated in the transistor. Using these two factors, transistor applications can be roughly divided into two categories. In the simpler one, the effects of power dissipation within the transistor have a negligible effect on the operating temperature of the transistor. In the more complicated case, power-dissipation effects must be included. To minimize the initial complexity of the problem, we will first consider the simpler category in which the junction operating temperature and the ambient temperature are essentially the same. Most amplifiers which drive a resistive load fall in this category. For this case, the permissible load-line area can be drawn (if necessary) and a load line can be constructed which reflects the nature of the application (*cf*. Sec. 10·1·1). A desired operating point (I_C, V_{CE}) may be located on this load line, the operating point reflecting the nature of the application (*cf*. Sec. 10·1·3). In most cases, our problem is then to design a biasing network which will maintain the desired values of I_C and V_{CE} when the ambient temperature is varied.

To emphasize the importance of stabilizing I_C and V_{CE} rather than other possible quantities (such as I_B), let us consider the simple amplifier shown in Fig. 12·5. We shall assume that a collector supply voltage of 10 volts and a load resistance of 2 kohms are dictated by the circuit requirements. We want to use the 2N324 shown in Fig. 12·2, and obtain an output-current swing of 2 ma peak-to-peak. If we construct the load line on the characteristics shown in Fig. 12·2a, we see that an operating point at $I_C = -2.5$ ma, $V_{CE} = -5$ volts, is a reasonable choice. The dc base current required is about $I_B = -0.025$ ma, and the input current swing $i_s(t)$ (see Fig. 12·2) should be about 0.02 ma peak-to-peak. The maximum collector current will then be about 3.5 ma, the minimum about 1.5 ma.

Now, we can achieve this condition by simply returning the base to the -10-volt lead through a 0.4-megohm resistor (see Fig. 12·5) which fixes I_B at -0.025 ma. This works properly at $T = 25°C$. However, when the temperature rises to 50°C, the characteristics shown in Fig. 12·2b suggest that the operating point will move into the saturation region. The collector-base junction will be forward-biased, and the output waveform will be

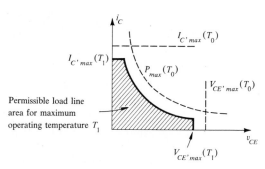

FIG. 12·4 THE CHANGE OF PERMISSIBLE LOAD-LINE AREA WITH OPERATING TEMPERATURE

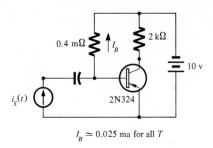

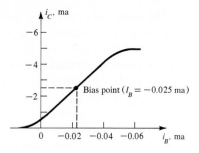

$I_B \simeq 0.025$ ma for all T

(a) *Amplifier biased with constant base current*

(b) *Current-transfer curve at 25°C*

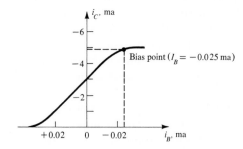

(c) *Current-transfer curve at 50°C*

FIG. 12·5 SIMPLE AMPLIFIER FOR STUDYING BIAS PROBLEM

severely distorted. Current-transfer characteristics are drawn in Fig. 12·5 to emphasize this point. On the other hand, the amplifier will still operate quite satisfactorily at 50°C if, instead of fixing I_B, we fix I_C and V_{CE}; that is, if we fix the operating point in the I_C-V_{CE} plane, not the base bias current. (Stabilizing I_C is equivalent to stabilizing V_{CE} for this simple case.) Of course, the absolute levels of base current will change with temperature, but this is not important.[1] The output signal will still be a reasonably pure sine wave, centered at $I_C = -2.5$ ma. The only change is that if the input-signal amplitude remains at 0.01 ma peak, then the increase in β at 50°C will give a somewhat greater output swing.

We shall therefore consider a sequence of possible circuits which stabilize the values of I_C and V_{CE}. Actually, the circuits tend to stabilize I_E and V_{CB}, so there is some slight motion of the operating point with temperature due to the fact that α and V_{BE} are functions of temperature. This is a minor effect, however.

12·5 PRACTICAL BIASING CIRCUITS FOR TRANSISTOR AMPLIFIERS

To stabilize I_E, we need to provide a dc current source for the emitter, as is suggested in Fig. 12·6a. However, since this would eliminate vari-

---

[1] As a matter of fact, even the direction of flow of I_B changes (see Fig. 12·2 and Demonstration 12·3).

ations in the emitter current, and therefore in the collector current, we must provide a bypass capacitor around this current source.

The current source can be approximated with a battery whose voltage V_1 is several times V_{EB} and a series resistor R_E. The base resistance R_B provides a path for the dc base current to return to ground. Since the base current will be approximately I_E/β, we want

$$\frac{R_B I_E}{\beta} \ll R_E I_E \qquad R_B \ll \beta R_E \qquad (12 \cdot 6)$$

to ensure that R_E and V_1 determine the dc emitter current.

We also want R_B to be large compared to the incremental input impedance of the transistor, so that signal current will not be lost in R_B.

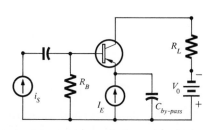

(a) Bias stabilization with a current source

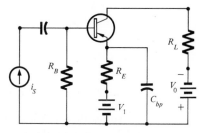

(b) Approximation to (a) using V_1 and R_E to approximate current source: $(V_1/R_E) \simeq I_E$

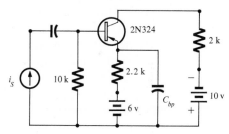

(c) Circuit of (b) with values selected to give $I_E \simeq 2.5$ ma, $|V_{CE}| \simeq 5$ volts

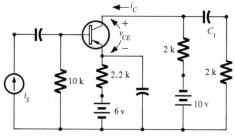

(d) Basic amplifier with additional ac-coupled load

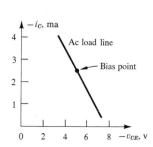

FIG. **12·6** DEVELOPMENT OF A TWO-BATTERY BIASING CIRCUIT

Assuming that the capacitance C_{bp} is a short circuit at the signal frequency, the dynamic input impedance of the transistor will be, from the π model of Fig. 11·10,

$$R_{in} \cong \frac{kT}{qI_E}(\beta + 1) + r_b \qquad (12 \cdot 7a)$$

If we reconsider our previous example with the 2N324, where an emitter current of $I_E = 2.5$ ma was required, we would have

$$R_{in} \cong \frac{25 \times 10^{-3}}{2.5 \times 10^{-3}} \times 100 = 1 \text{ kohm} \qquad (12 \cdot 7b)$$

We then need to select an R_B of about 10 kohms. Using the second inequality of Eq. (12·6), this means $R_E \gg 100$ ohms, a requirement which can be met by using $R_E > 1$ kohm. If we select R_E of 2.0 kohm, then we would need a battery voltage of

$$V_1 \cong I_E R_E + V_{EB} + \frac{I_E}{\beta} R_B$$

$$\cong 2.5 \text{ ma} \times 2.0 \text{ kohms} + 0.3 \text{ volt} + 25 \text{ }\mu\text{a} \times 10 \text{ kohm} = 5.55 \text{ volts}$$

We would then use a 6-volt battery (standard size) and an emitter resistor of 2.2 kohm (standard size) to obtain the circuit shown in Fig. 12·6c.

It should be mentioned that since an emitter battery and resistance have been added, the *dc load line* can no longer be constructed as simply as before. However, we can construct an *ac load line*, which describes how the circuit behaves at signal frequencies. To do this, we suppose that the emitter bypass capacitor C_{bp} is so large that the emitter is effectively grounded at signal frequencies. We can then draw an ac load line corresponding to $R_L = 2$ kohms through the known operating point.

It may also happen that a part of the load resistance itself is only coupled to the transistor at the signal frequencies. For example, in Fig. 12·6d we show an additional 2-kohm load coupled to the collector through a capacitor C_1. For such a case, the operating point of the transistor does not change, but the ac load line now corresponds to a collector load of 1 kohm. A 1-kohm load line has been constructed through the point ($I_C = 2.5$ ma, $V_{CE} = 5.5$ volts) on the output plane in Figure 12·6d.

12·5·1 *A one-battery biasing network.* For completeness, we should now proceed to study how the bias current I_E in this circuit would change if the ambient temperature were changed to 50°C. Since the emitter current is determined primarily by a battery voltage and a resistor, however, we shall leave this analysis as a problem and go on to a somewhat more useful circuit, which employs only one battery.

The one-battery circuit is simply a rearrangement of the circuit shown

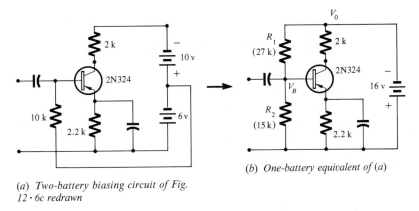

(a) Two-battery biasing circuit of Fig. 12·6c redrawn

(b) One-battery equivalent of (a)

FIG. 12·7   DEVELOPMENT OF A ONE-BATTERY BIASING NETWORK FROM AN EQUIVALENT TWO-BATTERY NETWORK

in Fig. 12·6c. The emitter and collector batteries of Fig. 12·6 are connected in a series-aiding fashion, so as far as the emitter and collector circuits are concerned, we can simply move the ground connection and use a new battery of 16 volts, as is suggested in Fig. 12·7a.

The base resistor R_B cannot readily be returned to a 6-volt tap on the battery, however, since none would ordinarily be provided. We remedy this by simply setting up the Thévenin equivalent base network shown in Fig. 12·7b. The resistors R_1 and R_2 are determined by the facts that (1) the voltage from the base terminal to ground in Fig. 12·7a should be

$$V_B = -(I_E R_E + V_{EB})$$

in the presence of I_B, and (2) the combined resistance of R_1 and R_2 in parallel should equal R_B.

On the assumption that the current flow through R_1 and R_2 will be large compared to the dc base current ($=25$ μa), these conditions can be written as

$$\frac{V_0 R_2}{R_1 + R_2} = V_B \qquad \frac{R_1 R_2}{R_1 + R_2} = R_B \qquad (12·7)$$

Using $V_0 = 16$ volts, $V_B \simeq 6$ volts, $R_B = 10$ kohms, we obtain

$$R_1 = 26.6 \text{ kohms} \qquad R_2 = 16 \text{ kohms}$$

Usually, 10 percent–tolerance resistors come in the sizes 15 kohms and 27 kohms, which would be used. This will, of course, decrease I_C somewhat.

The biasing circuit shown in Fig. 12·7b is one of the most frequently used biasing circuits for single-stage applications. We shall call it the "standard" one-battery biasing network for reference, although we shall generate other one-battery biasing networks later.

12·6 BIAS STABILITY OF THE STANDARD
ONE-BATTERY BIASING NETWORK

We now wish to study the stability of the emitter bias current under various operating conditions. For the present, we are interested primarily in ambient temperature changes and the interchangeability of transistors. To give a general analysis, we draw the bias network with a circuit model of a *pnp* transistor in Fig. 12·8. We can write the following equations for this network:

$$V_B = \phi + I_E R_E \qquad (12·8a)$$

$$\frac{V_B}{R_2} = I_B + \frac{V_0 - V_B}{R_1} \qquad (12·8b)$$

$$I_B + I_C + I_E = 0 \qquad (12·8c)$$

$$I_C = -\alpha I_E - I_{CO} \qquad (12·8d)$$

The positive directions of currents and voltages are defined in Fig. 12·7. Equation (12·8a) is Kirchhoff's voltage law around the bottom loop, with V_B simply defined as the base-to-ground voltage; Eq. (12·8b) is Kirchhoff's current law at the base terminal; Eq. (12·8c) is Kirchhoff's current law applied to the transistor; and Eq. (12·8d) is the expression for I_C provided by the transistor (that is, the transistor's "contract").

To determine I_E from this set of equations, we use Eq. (12·8c) in Eq. (12·8d) to obtain

$$I_B = -I_E(1 - \alpha) + I_{CO} \qquad (12·9)$$

Then Eq. (12·9) is used to substitute for I_B in Eq. (12·8b), and Eq. (12·8a) is used to substitute for V_B in Eq. (12·8b). When these manipulations are carried out, and some rearranging is done, we find

$$I_E = \frac{I_{CO} + (V_0/R_1) - (\phi/R_B)}{(1 - \alpha) + (R_E/R_B)} \qquad (12·10a)$$

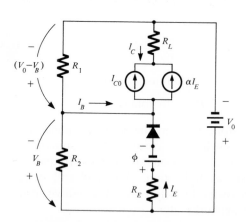

FIG. 12·8 CIRCUIT MODEL FOR ANALYSIS OF STANDARD BIAS NETWORK

where
$$R_B = \frac{R_1 R_2}{R_1 + R_2}$$

as before.

We may now use Eq. (12·10a) to check the bias condition established earlier and also to study the variation of I_E with various parameters.

Bias check. To check the bias condition obtained for the 2N324 in the previous section, we set $V_0 = 16$ volts, $\phi = 0.3$ volt, $R_B = 10$ kohms, $R_1 = 27$ kohms, $R_E = 2.2$ kohms, $I_{CO} = 3$ μa, $(1 - \alpha) \sim 1/\beta = 100$, these being the 25°C values. We then obtain

$$I_E = \frac{0.003 + 0.59 - 0.03}{0.01 + 0.22} = 2.45 \text{ ma} \qquad (12 \cdot 10b)$$

The slight difference between this value and the 2.5 ma of the previous example arises because R_1 has been set equal to 27 kohms instead of 26.6 kohms. Of course, the three-place accuracy in this calculation is of questionable validity.

It is useful to note that the bias current is approximately

$$I_E \simeq \frac{V_0/R_1}{R_E/R_B} = \left(\frac{V_0 R_2}{R_1 + R_2}\right)\frac{1}{R_E} \cong \frac{V_B}{R_E}$$

The term within the brackets is just the voltage from base to ground arising from the voltage-divider action of R_1 and R_2, assuming I_B is small compared to the current flowing in the divider. If ϕ is neglected, all this voltage appears across R_E, so $I_E = V_B/R_E$. The bias current is determined primarily by circuit components other than the transistor, which is a desirable situation for accommodating large changes in transistor parameter values.

12·6·1 *Transistor interchangeability.* To pursue this point, let us consider how much I_E would be expected to change if different 2N324 transistors are substituted in the proposed circuit. Manufacturers' specifications for this transistor, at 25°C, are

Parameter	Min.	Typical	Max.	Units
β	80	100	200	
$|I_{CO}|$		3	16	μa

From Eq. (12·10a), we see that the greatest change in I_E would occur if we happened to select a transistor with the greatest possible values of β and I_{CO}.[1] For this condition ($\beta = 200$, $I_{CO} = 16$ μa), Eq. (12·10a) gives

---

[1] In some of the literature, the change in I_E is computed from an equation similar to

$$\Delta I_E = \frac{\partial I_E}{\partial I_{CO}} \Delta I_{CO} + \frac{\partial I_E}{\partial \beta} \Delta \beta$$

though for our problem ΔI_{CO} and $\Delta \beta$ are rather large for this formula to be accurate.

$$I_E = \frac{0.016 + 0.59 - 0.03}{0.005 + 0.22} = 2.56 \text{ ma}$$

The change in I_{CO} produces about a 3 percent increase in the numerator, while the change in β produces a 2 percent decrease in the denominator. The result is a 5 percent increase in I_E in the worst case of transistor-parameter substitution at 25°C.

12·6·2   Temperature stability.   The stability of the operating point with change in ambient temperature may also be studied with the aid of Eq. (12·10a). Basically, we proceed as before, recognizing that in addition to I_{CO} and β, ϕ will also change. For a "typical" 2N324, the changes in β and I_{CO} with temperature may be read directly from Fig. 12·2:

Parameter	25°C value	50°C value
β	100	150
I_{CO}	3 μa	20 μa
ϕ	0.3 volt	~0.25 volt

ϕ is estimated from the formula $d\phi/dT \sim -2$ mvolt/°C. When the 50°C values are substituted into Eq. (12·10a), there results

$$I_E = 2.58 \text{ ma}$$

which is again considered to be acceptable stability for I_E.

12·6·3   Worst-case conditions.   One useful way of evaluating the bias stability of a proposed circuit is to calculate the extreme values which I_E could have under *worst-case conditions* for the transistor parameters. While this is frequently an overly pessimistic viewpoint, there are some cases (for example, a satellite communication system) where it is valid. As an example, we shall calculate the worst-case values of I_E for the previous circuit.

The minimum value of I_E will occur when I_{CO} and β have their minimum values and ϕ has its maximum value. Including both the 25 to 50°C temperature range and the manufacturing tolerances on β and I_{CO}, we have

$$\left.\begin{array}{l} (I_{CO})_{min} \sim 0 \\ \beta_{min} \sim 80 \\ \phi_{max} \sim 0.3 \end{array}\right\} \Rightarrow I_E = 2.41 \text{ ma}$$

The largest value of I_E will occur when I_{CO} and β are a maximum and ϕ is a minimum. These values would arise if a 2N324 with $\beta(25°\text{C}) = 200$ and $I_{CO}(25°\text{C}) = 16$ μa were used, and the circuit operating temperature were 50°C. For this temperature $\phi \sim 0.25$ volt. To estimate the values of I_{CO} and β at 50°C, we recall that β for the "typical" 2N324 increases by 50 percent over this temperature range. Also, I_{CO} for germanium transistors doubles for each 10°C rise in temperature. Hence, the appropriate values of β, I_{CO}, and ϕ are approximately

$$\left.\begin{array}{c} \beta_{max} \sim 300 \\ (I_{CO})_{max} \sim 83 \text{ } \mu a \\ \phi_{min} \sim 0.25 \text{ volt} \end{array}\right\} \Rightarrow I_E = 2.9 \text{ ma}$$

Therefore, for any 2N324 operated in the standard circuit over the temperature range $25°C < T < 50°C$, we expect I_E to fall in the range

$$2.4 < I_E < 2.9 \text{ ma}$$

This range represents quite adequate stability of the bias point, especially since the maximum values of β and I_{CO} will rarely occur together (see Prob. 12·9).

12·7 A SIMPLE FEEDBACK-BIASING CIRCUIT

A second very useful type of biasing circuit for a transistor is the feedback-biasing circuit shown in Fig. 12·9. In this circuit a base bias current is derived from the value of V_{CE} (or actually V_{CB}). Therefore, an increase in β or I_{CO}, which would increase I_C, also has the effect of decreasing V_{CE} and therefore decreasing the base bias current. There is thus a self-regulating action which tends to keep the bias constant.

We may give a simplified analysis of this circuit as follows. Let us suppose that a silicon *npn* transistor is to be used, so I_{CO} can be neglected. Furthermore, we shall assume $|V_{CE}| \gg |\phi|$, so the dc base current in Fig. 12·8 is simply

$$I_B \cong \frac{V_{CE}}{R_F} \qquad\qquad (12·11a)$$

The collector current is then

$$I_C = \beta I_B = \beta \frac{V_{CE}}{R_F} \qquad\qquad (12·11b)$$

Since $(I_C + I_B)$ flows in R_L, we have

$$V_{CE} = V_0 - (I_B + I_C)R_L \qquad\qquad (12·11c)$$

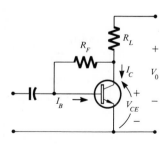

FIG. 12·9   SIMPLE FEEDBACK BIASING CIRCUIT. *npn* TRANSISTOR SHOWN.

Since $\beta \gg 1$, Eq. (12·11c) simplifies to

$$V_{CE} \simeq V_0 - I_C R_L \qquad (12·11d)$$

Using Eq. (12·11b) in Eq. (12·11c), we obtain, after some rearrangement,

$$V_{CE} = \frac{V_0}{1 + (\beta R_L / R_F)}$$

As an example, if we are using a transistor with $\beta = 100$ in a circuit, where $V_0 = 12$ volts and $R_L = 5$ kohms, and we want $V_{CE} = 6$ volts, we should select an R_F to satisfy

$$\frac{\beta R_L}{R_F} = 1 \Rightarrow R_F = 500 \text{ kohms}$$

As before, the resistor R_F will have an undesirable effect on the transmission of small signals through the circuit, since it will tend to hold v_{CE} constant all the time. To remedy this, the required value of R_F is usually split in half and a bypass capacitor is added to remove the feedback at signal frequencies. The resulting network is shown in Fig. 12·10. It is useful to note that, because R_F is large, the value of the bypass capacitor is usually small.

12·7·1 *More complicated feedback-biasing circuits.* The reader can show that the simple circuit shown in Fig. 12·9 is rather sensitive to variations in β (though it is still adequate for many purposes). To reduce this sensitivity, the more complicated circuit shown in Fig. 12·11 is sometimes used. By adding the resistors R_E and R_B, this circuit provides additional flexibility in obtaining a bias point which is insensitive to variations in the transistor parameters.

Feedback biasing is also used in multistage amplifiers with great success. One possible arrangement is shown in Fig. 12·12. Here the single transistor of Fig. 12·9 is replaced by a multistage amplifier. The bias levels of each transistor are fixed by the overall feedback. This arrangement has many advantages; for example, very few components are used to establish

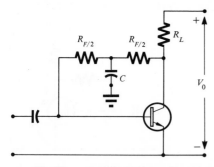

FIG. 12·10   SIMPLE FEEDBACK-BIAS CIRCUIT WITH BYPASS CAPACITOR TO ELIMINATE AC FEEDBACK

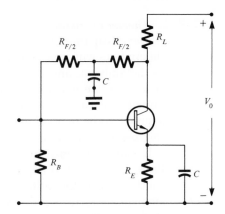

FIG. 12 · 11   FEEDBACK-BIASING CIRCUIT WITH R_E ADDED TO REDUCE OPERATING POINT
SENSITIVITY

the bias levels of several transistors. We shall consider it later, when we have
studied the properties of multistage amplifiers and the concept of feedback.

12 · 8   THE EFFECT OF POWER DISSIPATION ON BIAS STABILITY

Our study of biasing networks has thus far been limited to cases where
the effect of power dissipation at the collector junction of a transistor
could be neglected. We have not given a criterion for deciding whether
this condition is met in a given circuit, nor how one proceeds to design
biasing circuits if it is not met. In this section we shall briefly discuss the
manner in which the bias point (I_C, V_{CE}) may shift due to the effect of
power dissipation at the collector junction of the transistor. To do this we
shall first develop the concept of thermal resistance.

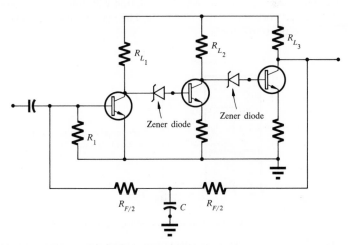

FIG. 12 · 12   FEEDBACK BIASING IN A MULTISTAGE AMPLIFIER

12·8·1 *Thermal resistance.* The biases applied to the transistor will result in power dissipation within the collector-base transition region. This power dissipation will raise the temperature of the collector junction relative to the ambient, and heat will flow from the collector junction to the heat sink. A sketch illustrating the flow of heat from the collector junction of a biased transistor to its heat sink is shown in Fig. 12 · 13. The collector is assumed to be in good thermal contact with an infinite heat sink, the heat sink being at the ambient temperature T_A.

It can be simply shown from the basic law of heat conduction (see Prob. 12 · 14) that if the power dissipated at the collector junction is held constant, then the collector-junction temperature T_j will rise above the ambient temperature T_A by an amount proportional to the power being dissipated, P:

$$T_j - T_A = \Delta T = \theta P \qquad (12 \cdot 12)$$

θ is called the *thermal resistance* of the transistor. Its dimensions are given in degrees centigrade per watt, and its value usually lies in the range

$$1 < \theta < 500$$

The lower limit is the θ for a power transistor, where the collector body is in good thermal contact with its heat sink. Transistors intended for low-level amplification of high-frequency input signals may have θ in the range of 500, primarily because the physical mounting arrangement is chosen for its electrical, rather than its thermal, properties.

12·8·2 *Calculation of operating-junction temperature.* Transistor manufacturers specify an absolute maximum junction-operating temperature which should not be exceeded. We can use the thermal resistance of the transistor to check whether this temperature limit is approached in a given application.

EXAMPLE An example of such a calculation will illustrate the procedure. Consider the bias network shown in Fig. 12 · 7b. The maximum power which the battery can supply to the transistor is

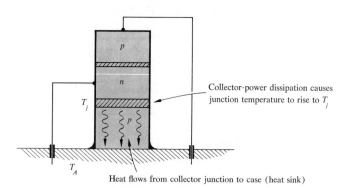

Collector-power dissipation causes junction temperature to rise to T_j

Heat flows from collector junction to case (heat sink)

FIG. **12 · 13**   THE EFFECT OF COLLECTOR POWER DISSIPATION ON COLLECTOR JUNCTION TEMPERATURE

$$P_{max} = \frac{V_0{}^2}{4(R_L + R_E)} \sim 16 \text{ mw}$$

The thermal resistance of the 2N324 is 250°C/watt, and the maximum junction temperature is 100°C.

Since 16 mw is the maximum power which can be dissipated in the transistor, the maximum rise of collector-junction temperature above the ambient temperature is

$$\Delta T_{max} = \theta P_{max} = 4°\text{C}.$$

Hence, the circuit shown in Fig. 12·7b can be operated up to an ambient temperature which is essentially equal to the maximum junction temperature of the transistor (in this case 100°C).

When the dc resistances in the collector and emitter circuits are made very small (for example, $R_E = 0$ and R_L is coupled to the transistor through a transformer), the junction temperature may rise significantly above the ambient, and in fact a *thermal runaway* situation may exist. To explain the thermal runaway phenomenon, let us suppose that a transistor is initially at room temperature when bias is applied to it. The bias power dissipated at the collector junction will produce a temperature rise, increasing I_{co} and I_C. If the circuit is such that an increase in I_C means that more power will be dissipated at the collector than before, then the collector-junction temperature will rise further, and I_C will increase more. If this sequence of events continues, it is possible to bias the transistor into the saturation region, or even destroy it by excessive heat dissipation.

When we design a biasing network, we want to know how much effect self-heating will have on the bias point, and we want to avoid the conditions for thermal runaway. This assurance can be based on a calculation of the operating temperature of the collector junction in the proposed circuit.

To make the calculation as general as possible, let us first simply express the dc collector current I_C as

$$I_C = I_{CA} + \Delta I_C \qquad (12·13a)$$

where I_{CA} is the value I_C would have if the collector junction were at the ambient temperature, and ΔI_C is the increment caused by self-heating. Note that ΔI_C need not be small in this equation.

Similarly, we shall represent V_{CB} by

$$V_{CB} = V_{CBA} + \Delta V_{CB} \qquad (12·13b)$$

Using Eqs. (12·3), the power being dissipated at the collector junction may be written as

$$P_C = I_C V_{CB} = (I_{CA} + \Delta I_C)(V_{CBA} + \Delta V_{CB}) \qquad (12·13c)$$

If we now utilize the thermal-resistance concept, we can write the actual junction temperature T_j as

$$T_j = T_A + \theta P_C$$
$$\Delta T = \theta[I_{CA} + \Delta I_C(\Delta T)][V_{CBA} + \Delta V_{CB}(\Delta T)] \qquad (12·14)$$

Equation $(12 \cdot 14)$ is an implicit equation for the unknown ΔT. For a given circuit, we can calculate (and plot if necessary) I_C, V_{CB}, etc. as functions of the collector-junction temperature T_j. We can then use these formulas (or graphs) to find the value of ΔT which satisfies Eq. $(12 \cdot 14)$ (by trial and error, if not otherwise).

EXAMPLE Let us calculate ΔT for the biasing network of Fig. $12 \cdot 7$. The emitter bias current for this circuit was calculated to be

$$I_E = \frac{I_{CO} + (V_0/R_1) - (\phi/R_B)}{(1 - \alpha) + R_E/R_B} \qquad (12 \cdot 15)$$

This formula must, of course, be true at any collector-junction temperature T_j. Also, the collector current is

$$I_C = -\alpha I_E - I_{CO} \qquad (12 \cdot 16)$$

at any temperature. Now, for a given transistor, $\Delta\beta$ and ΔI_{CO} can be calculated for any given ΔT, and from these values we can calculate ΔI_E and ΔI_C.

To be specific, the following results were obtained in Sec. $12 \cdot 6 \cdot 2$ for a typical 2N324:

	25°C	50°C
I_E	2.5 ma	2.58 ma
I_{CO}	3 μa	20 μa
$|I_C|$	2.503 ma	2.6 ma

While the change in I_C is not precisely linear with temperature, we shall assume it to be for simplicity in this example. Then ΔI_C may be expressed as

$$\Delta I_C \sim 0.1 \frac{\Delta T}{25°C} \text{ ma} \qquad 25°C < T < 50°C, \quad 0 < \Delta T < 25°C \qquad (12 \cdot 17)$$

The collector-to-base voltage for the circuit can be written

$$V_{CB} = V_0 - I_C R_L - I_E R_E - \phi \qquad (12 \cdot 18)$$

Again, because the change in each term on the right-hand side can be calculated for any given ΔT, so can ΔV_{CB}. For our example

$$\Delta V_{CB} = -R_L \Delta I_C - R_E \Delta I_E - \Delta\phi \qquad (12 \cdot 19)$$

Using the values given in Sec. $12 \cdot 6 \cdot 2$, we have

$$\Delta V_{CB} \sim -0.3 \text{ volts}$$

over the 25 to 50°C interval, or

$$\Delta V_{CB} \sim -0.3 \frac{\Delta T}{25°C} \text{ volts} \qquad 25°C < T < 50°C, \quad 0 < \Delta T < 25°C \qquad (12 \cdot 20)$$

When Eqs. $(12 \cdot 17)$ and $(12 \cdot 20)$ are substituted into Eq. $(12 \cdot 14)$, we find for an ambient temperature of 25°C,

$$\Delta T = \theta \left(2.5 \text{ ma} + 0.1 \frac{\Delta T}{25 \text{ ma}}\right)\left(6 - 0.3 \frac{\Delta T}{25}\right) \qquad (12 \cdot 21)$$

The typical thermal resistance of a 2N324 quoted by the manufacturer is 250°C/watt. Using this value in Eq. (12·21), we find

$$\Delta T = 3.8°C$$

or the actual operating temperature of the collector junction is about 31°C. The actual values of I_C and V_{CB} are then insignificantly different from the design values.

Generalization. We can generalize this example rather easily to obtain a formula for estimating ΔT for any transistor in the standard biasing circuit. We first assume

$$\Delta I_C \simeq \Delta I_E$$

and neglect $\Delta\phi$ so that

$$\Delta V_{CB} = -\Delta I_C(R_L + R_E) \tag{12·22}$$

Using Eq. (12·22) in Eq. (12·14), we have

$$\Delta T = \theta(I_{CA} + \Delta I_C)[V_{CBA} - \Delta I_C(R_L + R_E)] \tag{12·23}$$

If we again assume

$$\Delta I_C = K\Delta T \tag{12·24}$$

then Eq. (12·23) can be rewritten in the form

$$\Delta T = \frac{\theta I_{CA} V_{CBA}}{1 - \theta K[V_{CBA} - I_{CA}(R_L + R_E)]} \tag{12·25}$$

Now, V_{CBA} can be written in terms of I_{CA} as

$$V_{CBA} \cong V_0 - I_{CA}(R_L + R_E)$$

so we finally have

$$\Delta T = \frac{\theta P_A}{1 - \theta K[V_0 - 2I_{CA}(R_L + R_E)]} \tag{12·26}$$

where $P_A = I_{CA}V_{CBA}$.

As a rule, any circuit which exhibits good temperature stability of the bias point will also exhibit good stability against self-heating effects. Generally, such circuits contain some emitter resistance R_E, and sometimes also a collector-to-base feedback resistance R_F. Thermal runaway is a problem only when both these resistors are absent. An interesting demonstration of thermal runaway is suggested in Demonstration 12·2 at the end of this chapter.

12·9 BIASING CIRCUITS FOR VACUUM TUBES

Biasing a vacuum tube usually proves to be simpler than biasing a transistor. In most cases the cathode-operating temperature is 1000°C or more, so environmental variations in ambient temperature usually have an

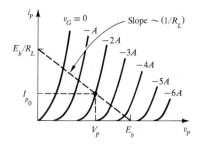

FIG. 12 · 14   TRIODE-PLATE CHARACTERISTICS WITH ASSUMED LOAD LINE

insignificant effect on the circuit properties of the tube. Under these circumstances, we shall consider only the simplest methods for obtaining a desired operating point.

Naturally, before a bias network can be chosen, one must select an operating point. This is done by first selecting a plate supply voltage and load resistor according to the ideas presented in Sec. 10 · 2 · 5. A load line can then be drawn on the plate characteristics of the tube, as suggested in Fig. 12 · 14. An operating point is selected on this load line, this selection reflecting the nature of the signals to be amplified (*cf.* Sec. 10 · 2 · 5). To be specific we have assumed on Fig. 12 · 14 that an operating point near the center of the load line is desired.

12 · 9 · 1   Cathode-bias resistor.   A common and very useful way of obtaining this operating point is to place a resistor R_K in the cathode lead, as shown in Fig. 12 · 15.[1] The grid-cathode voltage obtained in this way is simply

$$V_G = -R_K I_K \qquad (12 \cdot 27)$$

where I_K is the cathode current and R_K is the cathode resistor. For a triode, I_K and the dc plate current I_p are the same. Thus, to achieve the bias condition required in Fig. 12 · 14, we simply set

$$R_K = -\frac{V_G}{I_P} = \frac{2A}{I_{Po}}$$

For pentode tubes, it is necessary to know or estimate the screen current before R_K can be calculated.

Of course, putting a resistor in the cathode circuit will produce changes in the load line and changes in the small-signal circuit properties of the tube (see Prob. 12 · 16). However, because R_K is much smaller than R_L, its effect on the load line is usually neglected. Changes in the small-signal circuit properties may be avoided (if need be) by providing a bypass

---

[1] It is also necessary to connect the grid to ground through a resistor R_G, so that those few electrons which are intercepted by the grid can return to the cathode. The very small current which flows in R_G on this account does not disturb the bias condition as long as the grid-cathode diode is reverse-biased. However, without R_G, or even with too large a value of R_G, the bias condition is upset.

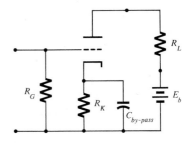

FIG. 12·15   THE USE OF A CATHODE RESISTOR TO OBTAIN BIAS

capacitor, as shown in Fig. 12·15. (This point will be developed further in the next chapter.)

EXAMPLE   Let us suppose that we want to use a plate supply voltage of 300 volts and a load resistor of 30 kohms in the simple triode amplifier shown in Fig. 12·16. The i_P-v_P characteristics for the triode to be used (12AU7) are shown in Fig. 12·16b. To locate our operating point approximately in the center of that part of the load line lying below $v_G = 0$, we want to have $I_P \cong 4$ ma and $V_G \simeq -7.5$ volts. The required R_K is then about 1.9 kohms. The nearest standard size is 1.8 kohms, which we would use.

In studying a proposed or already constructed amplifier, we may want to determine the bias produced by a given cathode resistor. We may do this graphically by sketching a *bias curve* on the plate characteristics of the tube. The bias curve is simply a plot of the relation

$$V_G = -R_K I_P \qquad (12\cdot28)$$

on the plate characteristics. The curve can be drawn for a given value of R_K by simply selecting values for I_P, computing the corresponding values for V_G from Eq. (12·28) and locating the resulting (I_P, V_G) points on the plate characteristics. The curve through these points is usually quite close to being a straight line.

EXAMPLE   Let us find the bias curve and operating point for an R_K of 2 kohms in the previous circuit. We again draw the load line corresponding to a 300-volt plate

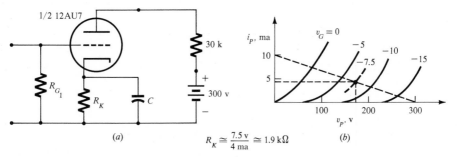

FIG. 12·16   THE CALCULATION OF R_K FROM PLATE CHARACTERISTICS AND A DESIRED OPERATING POINT

Biasing networks   461

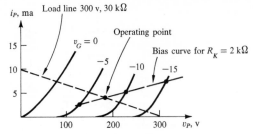

supply, 30-kohms load combination, as in Fig. 12 · 17. We then locate the points $(I_P = 5$ ma, $V_G = -10$ volts$)$, $(I_P = 2.5$ ma, $V_G = -5$ volts$)$, and $(I_P = 7.5$ ma, $V_G = -15$ volts$)$, each point representing a possible solution to Eq. (12 · 28). The bias curve drawn through these points intersects the load line at the operating point. Other graphical procedures for locating the operating point of a given circuit are given in the problems.

12·9·2 *Grid-leak bias.* In some situations, particularly where high gain is required, a different biasing technique may be used. The circuit is shown in Fig. 12 · 18, and is called a *grid-leak* biasing arrangement. The coupling capacitor C_c and grid resistor R_G, together with the grid-cathode diode of the tube, form a clamping circuit (*cf.* Sec. 7 · 5). The grid-cathode voltage is clamped at essentially zero, so the effective grid-cathode bias is a function of the form and amplitude of the input signal, as shown in Fig. 12 · 18.

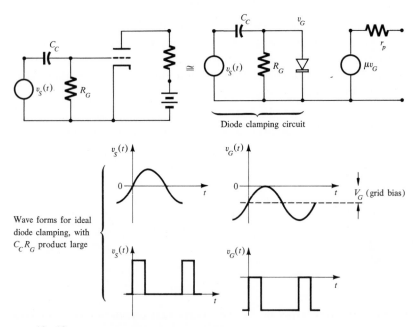

FIG. **12 · 18** THE GRID-LEAK BIASING TECHNIQUE

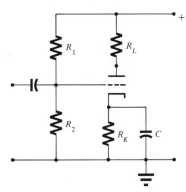

12·9·3   *Vacuum tube interchangeability.* As with transistors, the "typical" characteristics presented by vacuum tube manufacturers are not intended to give more than a reasonable approximation to the characteristics of any given tube. Generally, vacuum tube manufacturers do not provide as much information about the variation of tube parameters as transistor manufacturers do. However, this should not be interpreted to mean that there is less variability from tube to tube than there is from transistor to transistor.

When it is desired to design a bias network to maintain an operating point in the face of manufacturing variability, more complicated networks than those already described are used. One of the most common, particularly for high-g_m tubes, is shown in Fig. 12·19. Except for the sizes of the resistors and the supply voltage, this circuit is identical to the standard biasing network for an *npn* transistor.

12·9·4   *Biasing circuits for pentodes.* In general, the procedures for selecting a load line and an operating point are the same for triodes and pentodes. The new problems in designing a pentode biasing circuit are (1) to supply the required voltage to the screen and (2) to account properly for screen current in computing the value of the cathode-bias resistor, if this biasing method is used.

Designing a screen-biasing circuit usually requires that we know the screen current for the desired bias condition. This will usually not be given but can be estimated as follows.

First, in the region where the plate characteristics are essentially straight parallel lines, the ratio of the plate current to the screen current is nearly constant, independent of the electrode voltages. Thus, if the plate and screen currents are known for any one bias condition, the screen current can be estimated from the plate current for any other. For example, for a 6AU6, the manufacturer indicates under "typical operating conditions" that when $V_P = 250$ volts, $V_S = 125$ volts, $V_{G_1} = -1$ volt, $I_P = 7.5$ ma and $I_{G_2} = 3$ ma. Hence, for this tube $I_P/I_{G_2} \sim 2.5$ for essentially any linear amplifier bias condition.

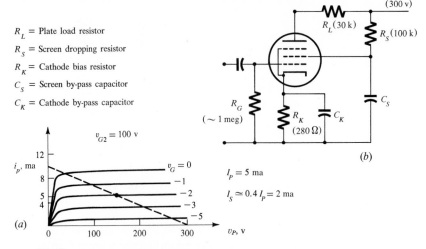

R_L = Plate load resistor

R_S = Screen dropping resistor

R_K = Cathode bias resistor

C_S = Screen by-pass capacitor

C_K = Cathode by-pass capacitor

$v_{G2} = 100$ v

$I_p = 5$ ma

$I_s \simeq 0.4\, I_p = 2$ ma

FIG. 12·20   BIASING ARRANGEMENT FOR PENTODES

We can use this idea to estimate the circuit components for a pentode biasing network as follows. We consider the 6AU6 pentode amplifier shown in Fig. 12·20. The plate characteristics are given in Fig. 12·20a for $v_{G_2} = 100$ volts. We assume the application requires us to use the plate supply voltage of 300 volts and a load resistor of 30 kohms. The corresponding load line is constructed in Fig. 12·20. The desired operating point is at $I_P = 5$ ma, $V_{G_1} = -2$ volts, $V_{G_2} = V_S = 100$ volts. Using the previous rule, we estimate the screen current to be

$$I_S = \frac{3}{7.5} \times 5 \text{ ma} = 2 \text{ ma}$$

In order to bias the tube at this point, we connect a *screen-dropping resistor* between the screen grid and the plate supply, as shown in Fig. 12·20. The screen-dropping resistor must absorb 200 volts when the screen current is 2 ma, that is, when $R_S = 100$ kohms. The cathode-bias resistor should produce a voltage drop of 2 volts when $I_K = I_P + I_S = 7$ ma is flowing through it. Hence, $R_K = \frac{2}{7}$ kohms $\simeq 280$ ohms. The final values are shown in Fig. 12·20.

Note that a screen grid bypass capacitor C_S is used in the circuit. This is to ensure that the screen voltage remains constant when signal voltages are applied to the grid. Without this capacitor, the voltage gain of the pentode amplifier is seriously reduced.

12·10   SUMMARY

The problem in designing a biasing network for a transistor is to choose a network configuration and element values which will ensure that those bias variables which affect the circuit performance of the transistor are held relatively constant when the ambient temperature changes or when transistors with normal manufacturing tolerances are substituted. In achieving these objectives the bias network should have a minimum influence on the desired properties of the circuit.

For amplifier service, we want to stabilize I_E and V_{CB}. This conclusion is reached either by direct inspection of the variation in the collector characteristics with temperature, or by observing that the parameters of the small-signal circuit models depend on these two variables. I_E and V_{CB} are usually stabilized by ensuring that they are determined primarily by the supply voltages and circuit resistances, rather than the transistor parameters.

Since vacuum tube characteristics are insensitive to typical variations in the ambient temperature, the design of biasing networks is somewhat simpler. The interchangeability problem is still present, however, and much the same measures are taken to solve it that are used with transistors.

REFERENCES

Searle, C. L., A. R. Boothroyd, E. J. Angelo, Jr., P. E. Gray, and D. O. Pederson: "Elementary Circuit Properties of Transistors," SEEC, vol. 3, John Wiley & Sons, Inc., New York, 1964, chap. 5.

Zimmerman, H. J., and S. J. Mason: "Electronic Circuit Theory," John Wiley & Sons, Inc., New York, 1962, chap. 6.

Angelo, E. J.: "Electronic Circuits," 2d ed., McGraw-Hill Book Company, New York, 1964, chaps. 6 and 9.

DEMONSTRATIONS

12·1 BIAS STABILIZATION An interesting demonstration of the effect of bias stabilization on signal amplification may be obtained using the circuits shown below. The signal is a 1-kc sine wave with the input level adjusted to produce

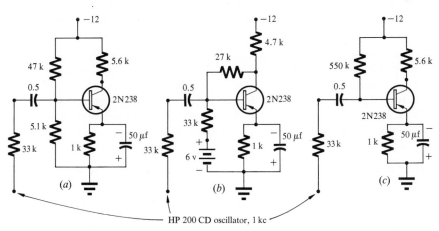

FIG. **D12·1**

a collector-voltage swing slightly below the level at which distortion due to saturation becomes obvious. A heat source, such as a soldering iron, is then applied to the transistor and the signal-output waveform is observed on a CRO. In Fig. D12 · 1, the circuit shown in (c) will very quickly begin to clip the output signal, and it will take a long time to recover when the soldering iron is removed. The relatively stable configurations represented by (a) and (b) require considerably more heat to reach the signal-clipping condition.

12 · 2 THERMAL RUNAWAY Generally, any modern transistor can be quite simply biased to avoid thermal runaway. However, the existence of the phenomenon is interesting and can be simply demonstrated. A possible circuit is shown below. The transistor is a 2N652 which should be selected to have a room temperature β of about 300 and as high an I_{CO} as possible.

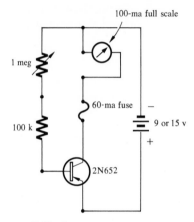

100-ma full scale

1 meg

100 k

60-ma fuse

9 or 15 v

2N652

FIG. **D12 · 2** THERMAL RUNAWAY DEMONSTRATION

The 60-ma fuse might represent the transformer primary in a transformer-coupled amplifier. With the 9-volt battery, the circuit is reasonably stable up to an I_C of about 30 ma. Beyond this, thermal runaway sets in. The runaway can be stopped by putting a small finned heat sink around the transistor case. If the 9-volt battery is replaced with a 15-volt battery, thermal runaway can only be avoided by removing an ammeter lead in the collector circuit before it is too late.

12 · 3 BASE-CURRENT CHANGES IN A STABLE BIAS ARRANGEMENT. The circuit shown in Fig. D12 · 3, p. 467, provides an instructive demonstration of the variation of base current with temperature in a stabilized bias current. A soldering iron can be used to heat the transistor. The current flowing in the base circuit will change sign and become very large, with only a relatively small shift in the emitter current (measured by the voltmeter connected across R_E).

EXPERIMENTAL PROJECT

Using one of the transistors supplied to you, design and build a standard one-battery biasing network which will achieve a bias point of $I_E = 1$ ma $\pm$ 10 percent, $V_{CE} = 6$ volts $\pm$ 10 percent. The collector load resistance should be 2 kohm. Use the power supply you constructed in the experimental project at the end of Chap. 7.

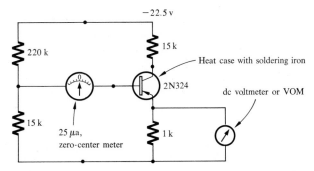

FIG. D12·3 DEMONSTRATING BASE-CURRENT FLUCTUATIONS IN A STABILIZED BIAS CIRCUIT

The bias current I_E must not change more than 10 percent over the ambient temperature limits 25 to 50°C. You will have to determine the transistor parameters (including whether the transistor is made from germanium or silicon).

PROBLEMS

12·1 1. Show that the saturation current of an ideal *pn*-junction diode or transistor increases with temperature T according to

$$I_{CO}(T) = I_{CO_A}e^{(T-T_A)/K}$$

where T_A is a reference, or ambient, temperature and I_{CO_A} is the value of I_{CO} at this temperature.

2. Evaluate K for silicon and germanium.

3. Compare the formula obtained in (1) with the data given in Fig. 12·1e for a real silicon transistor.

12·2 The transistor input characteristics given in Fig. 12·1d should ideally be of the form

$$i_B = I_0(e^{qv/kT} - 1)$$

where I_0 is a constant at a given temperature. Compare the data with this ideal result for the three temperatures given in Fig. 12·1d. Compute I_0 at each temperature and discuss its variation with the aid of the formula given in Prob. 12·1.

12·3 Do the data given in Fig. 12·2 support the theoretical prediction that I_{CO} in a germanium transistor doubles for every 10°C change in temperature?

12·4 Construct temperature-dependent circuit models for *npn* transistors similar to those given in Fig. 12·3.

12·5 A certain transistor has a $V_{CE,max}$ of 30 volts, this being reasonably independent of operating temperature. The maximum power dissipation capability is 150 mw for any ambient temperature up to 25°C. The dissipation capability is then derated by 2 mw per degree centigrade of ambient temperature above 25°C.

1. Plot the maximum-power-dissipation curve as a function of temperature.

2. Sketch the maximum-power-dissipation hyperbolas on a set of i_C-v_{CE} coordi-

nates for $T_A = 25°C$, $50°C$, and $75°C$, and estimate the minimum-collector load resistance which can be used with this transistor, if a collector-to-emitter supply voltage of 25 volts is to be used.

12·6 What is the operating point (that is, I_E) for the circuit shown in Fig. 12·6 when a 2N324 transistor is used? How much will this operating point shift under worst-case conditions over the ambient temperature range $25°C < T < 50°C$?

12·7 A 2N1613 (characteristics given in Fig. 12·1) is to be used in an amplifier with a supply voltage of 20 volts and a load resistance of 5 kohms. Select components for the standard bias network to produce $V_{CE} \approx 6$ volts, $I_C \approx 2$ ma. The design must utilize standard resistors, which come in sizes 1, 1.2, 1.5, 1.8, 2.2, 2.7, 3.3, 3.9, 4.8, 5.6, 6.8, and 8.2 $\times 10^n$ ohms, with $n = 1, 2, 3, 4, 5,$ and 6.

12·8 The *changes* of the bias point in a transistor amplifier due to changes in the ambient temperature can be calculated using a small signal (or incremental model) in which a current source is inserted in the collector to account for changes in I_{CO}. The appropriate incremental model of the transistor in the standard biasing circuit is shown in Fig. P12·8. In this figure $i_T = \Delta I_{CO}$. Compute i_e (the change in emitter bias) from this circuit. Verify your result by appropriate use of Eq. (12·10a).

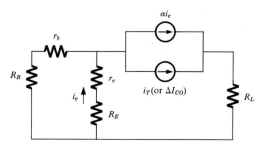

FIG. P12·8

12·9 Include changes in the emitter-base voltage ϕ in the model of Fig. P12·8 and then rework Prob. P12·7.

12·10 Suppose that transistors manufactured by a certain process have β in the range 80 to 200 and I_{CO} in the range 3 to 15 μa. The values of β and the values of I_{CO} can be assumed to be independent of each other. The approximate distribution of β and I_{CO} values at room temperature among a group of 100 transistors is given below:

β	80–100	100–120	120–140	140–160	160–180
Transistors with given value of β	15	45	30	7	3

I_{CO}, μa	3–6	6–9	9–12	12–15
Transistors with given value of I_{CO}	50	35	12	3

1. Assuming these data are representative of the production process, what is the probability of getting a transistor with $\beta > 160$, $I_{CO} > 12$ µa?

2. The manufacturer will sell transistors at the following prices:

	$80 < \beta < 120$	$80 < \beta < 140$	$80 < \beta < 160$	$80 < \beta < 200$
3 µa $< I_{CO} <$ 6 µa	$1.00	$0.90	$0.80	$0.70
3 µa $< I_{CO} <$ 9 µa	0.90	0.80	0.70	0.60
3 µa $< I_{CO} <$ 12 µa	0.80	0.70	0.60	0.55
3 µa $< I_{CO} <$ 15 µa	0.70	0.60	0.55	0.50

We want to use the biasing circuit of Fig. 12·7b and can accept a bias range for I_E of $2.4 < I_E < 2.5$ ma at room temperature. What is the cheapest transistor that will do this job with certainty?

12·11 Much of the literature on transistor biasing is concerned with calculating what is called the bias *stability factor*. The stability factor is defined by

$$S \equiv \frac{\partial I_C}{\partial I_{CO}}$$

Calculate S for the standard one-battery biasing circuit.

12·12 Calculate S for the simple feedback-biasing circuit shown in Fig. 12·9. Give a qualitative explanation of the "feedback" which exists.

12·13 A circuit which achieves bias stability by feedback is shown in Fig. P12·13. Using $\alpha = 0.98$, $V_o = -25$ volts, $R_E = 1$ kohm, $R_F = 500$ kohms, $R_B = 50$ kohms, $R_L = 10$ kohms, calculate I_E, V_{CE}, and S for this circuit. (S is defined in P12·11.)

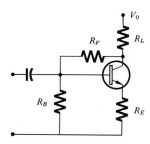

FIG. P12 · 13

12·14 Approximately what are the dc operating potentials at each element of both transistors in the circuit shown in Fig. P12·14?

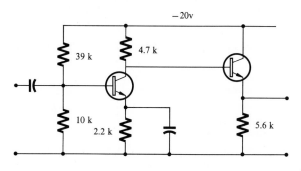

FIG. P12 · 14

12·15 An amplifier using two identical transistors is shown in Fig. P12·15. Using the model for the transistors shown, evaluate the changes in the collector currents of the two transistors resulting from changes in the saturation currents of ΔI_{CO}.

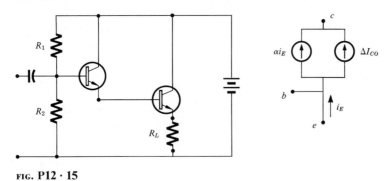

FIG. P12 · 15

12·16 The rate of heat flow H through a slab of area A is

$$H = -KA \frac{dT}{dx}$$

where $T(x)$ is the temperature at the point x and K is the thermal conductivity of the material.

1. Using this law, develop a formula for the thermal resistance of the transistor shown in Fig. 12·3.

2. Silicon has a thermal conductivity of about 0.8 watt/cm°C. What would the thermal resistance of the transistor be if the collector body has an area of 5×10^{-3} cm² and a length of 0.025 cm? One end of the collector body can be assumed to be attached directly to an infinite heat sink.

3. Suppose the collector body is not attached directly to the heat sink, but instead there is ¼ in. of wire between the heat sink and the collector body. The wire has a thermal conductivity of 50 watts/cm°-C and a cross sectional area of 10^{-4} cm². Estimate the new thermal resistance.

12·17 In this problem we will consider a graphical procedure for locating the operating point of a simple triode amplifier. In this procedure, one constructs the i_P-v_G transfer characteristic from the given load line. Then on this same set of axes, one draws a line representing the equation $V_G = -I_P R_K$. The intersection of this line with the transfer curve gives the bias point. Use this procedure to find the bias point for the amplifier shown in Fig. 12·16. Use $R_K = 2$ kohms.

12·18 If we wish to include the effects of R_K on the load line, we may proceed as follows:

1. Construct a load line with slope $(R_L + R_K)^{-1}$.
2. Use the procedure of Prob. 12·17 to locate the bias point.
3. When the bias point has been located, draw an ac load line of slope $(R_L)^{-1}$ through the operating point.
   a. Give reasons that justify these steps.
   b. Apply these three steps to the amplifier shown in Fig. 12·16, using $R_K = 2$ kohms, and compare your results with those obtained in the text.

13

AMPLIFIER DESIGN

THE AMPLIFICATION of weak signals into stronger ones is of fundamental importance in almost any electronic system. A phonograph cartridge, for example, will deliver a signal voltage of perhaps 30 mv into a 10-kohm load, representing a signal-power output from the cartridge of about 10^{-7} watt. This signal must be amplified to a power level of about one watt to give a suitable volume of sound from a loudspeaker. Similarly in communication systems such as radio, television, and radar, one must select and amplify the weak signals received at an antenna and then present the amplified signals appropriately (via a loudspeaker or CRT). Amplification also appears in electronic instrumentation and computation systems, where it is necessary to insure that the signals (usually pulses) which represent the data do not become so small that they cannot be distinguished from the circuit "noise."

Because of the general importance of amplification, the study and design of amplifiers occupies a central position in the field of electronics. In this chapter we shall begin with a consideration of the properties and design of amplifiers in which the active elements can be adequately represented by the circuit models developed in Chap. 11. For the most part we will study amplifiers whose applications are relatively clear without a discussion of communication system philosophy (for example, low-pass amplifiers for audio and video systems). However, the basic concepts and definitions will apply to the more sophisticated amplifiers described in Chap. 18.[1]

13·1 BASIC AMPLIFIER DEFINITIONS[2]

An amplifier is designed for the purpose of increasing the level of signal current, signal voltage, and/or signal power. The amount of this increase is called the *gain* of the amplifier. The gain of an amplifier will depend on the load which it drives and the source from which it is driven. Accordingly, we must specify the source and load when we specify the gain, as well as other characteristics, of the amplifier. These points are illustrated in the general network shown in Fig. 13·1. The figure shows no dc power sources, emphasizing that we are considering strictly *signal* gains.

The basic types of gain are defined on Fig. 13·1. The definitions are made with the assumption that the source voltage has a sinusoidal waveform.

The signal *power gain* is defined by

$$G = \frac{P_{out}}{P_{in}} \qquad (13 \cdot 1a)$$

where P_{out} is the signal power delivered to the load and P_{in} is the signal power absorbed at the amplifier input. Both P_{out} and P_{in} could be measured in principle by placing wattmeters in the actual circuit.

The *transducer gain* is defined by

$$G_T = \frac{P_{out}}{P_{avs}} \qquad (13 \cdot 1b)$$

where P_{avs} is the *power available from the signal source* ($P_{avs} = |V_s|^2/4R_s$). P_{avs} cannot be measured unless the signal source is correctly matched to the amplifier input. However, G_T is still a useful measure of performance, perhaps even more so than G. For instance, a telephone "repeater" amplifier in a cross-country transmission link will have a very limited amount of power available from its source. The amplifier must deliver more than

---

[1] Amplifier transient response is considered in Sec. 18·1, which may be read after Sec. 13·4 if desired.

[2] This topic will be continued in Sec. 13·5.

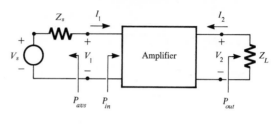

$$G = P_{out}/P_{in} \qquad\qquad G_T = P_{out}/P_{avs}$$

$$A_i = I_2(j\omega)/I_1(j\omega) \qquad\qquad A_v = V_2(j\omega)/V_s(j\omega)$$

FIG. 13·1 GAIN DEFINITIONS. ALL VOLTAGES AND CURRENTS ARE ASSUMED TO BE SINUSOIDAL. *rms* AMPLITUDES ARE SHOWN ON THE FIGURE.

a certain minimum power to a specified line impedance. For this application we want to design the amplifier to provide a transducer gain which is greater than a prescribed minimum.

The *voltage gain* of an amplifier is defined by either

$$A_v = \frac{V_2(j\omega)}{V_s(j\omega)} \qquad\qquad (13 \cdot 2a)$$

or

$$A_v = \frac{V_2(j\omega)}{V_1(j\omega)} \qquad\qquad (13 \cdot 2b)$$

where V_2 is the complex load voltage

$$V_2 = |V_2(\omega)| e^{j\phi_2(\omega)}$$

and V_s and V_1 are the complex source voltage and complex input voltage. In both cases A_v will have both a magnitude and a phase.

The *current gain* of the amplifier is defined by

$$A_i = \frac{I_2(j\omega)}{I_1(j\omega)} \qquad\qquad (13 \cdot 3)$$

where I_2 and I_1 are the complex load and input currents, respectively.

Stages of amplification. Usually the gain required in an amplifier will be more than we can obtain with a single active element. When this is the case, several *stages* of amplification will be *cascaded* to obtain the desired result, as suggested in Fig. 13·2. The manner in which an amplifier is divided into stages is arbitrary, though normally a stage will consist of an active element plus the circuit components which are associated with it to produce the desired bias condition and signal-transfer characteristics. The source and load impedances for each stage of a multistage amplifier will, of course, be determined in part by the succeeding and/or previous stages.

The progressive increase in signal level in the successive stages of a multistage amplifier implies a possible change in operating condition in each stage. Some stages may be operating under *small-signal conditions,*

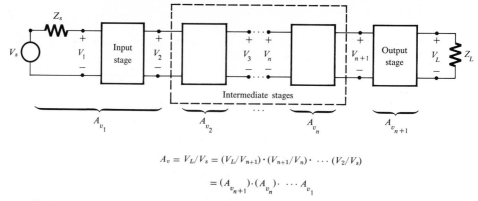

$$A_v = V_L/V_s = (V_L/V_{n+1}) \cdot (V_{n+1}/V_n) \cdot \cdots (V_2/V_s)$$

$$= (A_{v_{n+1}}) \cdot (A_{v_n}) \cdot \cdots A_{v_1}$$

FIG. 13·2 A CASCADE OF AMPLIFIER STAGES TO BUILD UP THE OVERALL AMPLIFIER GAIN.

where the excursions of current and voltage about the operating point are small enough to use the small-signal models of Chap. 11. Other stages in the amplifier may not meet this condition. For the first part of this chapter we shall consider only small-signal amplifiers.

Linearity. We shall assume that small-signal conditions imply that a stage is *linear;* that is, the functional relationship between the input signal and output signal does not depend on the amplitude of either signal. When this is not the case, the amplifier is said to be nonlinear. We shall consider nonlinearity in amplifiers at a later point.

Transfer function. When an amplifier is linear, or nearly so, its current and voltage gains can conveniently be specified as functions of the complex frequency variable s: $A_i(s)$ and $A_v(s)$.[1] In this form the gains are called *transfer functions.*

Frequency response. Speaking in terms of steady-state sine-wave signals, we cannot build (and for most purposes would not want) an amplifier in which the gain was absolutely constant with respect to frequency. However, there will usually be some band of frequencies over which we want the magnitude of the gain to be sensibly constant, with the gain falling off on either side of this frequency band. This results in the typical curve of gain vs. frequency shown in Fig. 13·3. Such a curve is generally referred to as a *frequency-response* curve. The gain in the flat region of this curve is called the *midband gain.*

If power gain is being plotted, then there will be only one "frequency-response curve." If the current or voltage gain is being plotted, then curves

---

[1] The use of the complex frequency variable for describing network behavior is discussed in most texts on network analysis. The reader who is unfamiliar with this subject is referred to the 72-page booklet by E. J. Angelo and A. Papoulis, *Pole-Zero Patterns,* McGraw-Hill Book Company, New York, 1964, for a simple introduction to the subject which will be adequate for the purposes of this chapter.

must be drawn showing the variation of both the amplitude and phase of the gain vs. frequency, as in Fig. 13·3b and 13·3c. These curves are called the *amplitude-response* and *phase-response* curves for the amplifier (or the stage). In most cases the amplitude response can be determined from the phase response and vice versa, so frequently only one of these curves is plotted.

The frequencies at which the power gain drops to one half its midband value are called the *band-edge, cutoff,* or *half-power frequencies.* The lower cutoff frequency is usually denoted by f_l, the upper cutoff frequency by f_u. The *bandwidth B* of the amplifier is the difference between these two frequencies:

$$B = f_u - f_l \tag{13·4}$$

Decibel gains. For reasons which will appear shortly, it is frequently convenient to plot gain vs. frequency on log-log scales. The magnitude of the gain is then expressed in *decibels* (db). The decibel is defined by the relation

$$db = 10 \log G \tag{13·5}$$

where log means $\log_{10}$, and G is the power gain defined in Eq. (13·1a).

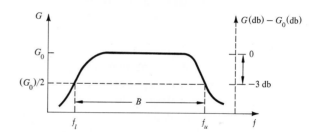

(a) *Power gain vs. frequency*

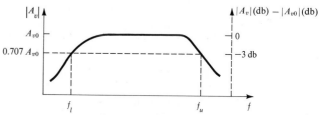

(b) *Magnitude of voltage gain vs. frequency, or amplitude response curve. If $R_L = R_{in}$, f_l and f_u are the same as in (a).*

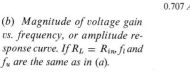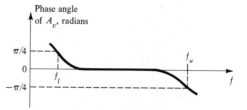

(c) *A commonly encountered phase response curve for an amplifier*

FIG. 13·3 **HYPOTHETICAL FREQUENCY RESPONSE CURVES FOR THE GENERALIZED AMPLIFIER OF FIG. 13·1.**

Since

$$P_{out} = |I_2|^2 R_L \qquad P_{in} = |I_1|^2 R_{in}$$

(see Fig. 13·1 for definitions)

$$G = \left|\frac{I_2}{I_1}\right|^2 \frac{R_L}{R_{in}}$$

and the power gain of the amplifier in decibels is

$$G \text{ (db)} = 20 \log_{10}\left|\frac{I_2}{I_1}\right| + 10 \log_{10}\frac{R_L}{R_{in}} \qquad (13\cdot6)$$

If $R_L = R_{in}$, then Eq. (13·6) reduces to

$$G \text{ (db)} = 20 \log_{10}\left|\frac{I_2}{I_1}\right| \qquad (13\cdot7a)$$

Generally, the input and load resistances are not equal. Nevertheless, it is customary to *define* the *decibel current gain* of the amplifier by

$$|A_i| \text{ (db)} = 20 \log_{10}\left|\frac{I_2}{I_1}\right| \qquad (13\cdot7b)$$

Similarly, the decibel voltage gain is defined by

$$|A_v| \text{ (db)} = 20 \log_{10}\left|\frac{V_2}{V_1}\right| \qquad (13\cdot7c)$$

In terms of the decibel notation, the power gain at the band-edge frequencies of an amplifier will be 3 db lower than the midband power gain. The band-edge frequencies are then sometimes called the *3-db frequencies*. At these frequencies the voltage gain will be 0.707 of its midband value. (Strictly speaking, we must have $R_L = R_{in}$ to make this statement, though we usually refer to the 0.707 voltage-gain condition as the 3-db point, without regard for the input and load resistances, as in Fig. 13·3b.)

13·2 MID-BAND PROPERTIES OF SINGLE-STAGE AMPLIFIERS

The basic task in designing an amplifier is to proceed from a given set of specifications (for example, power output, gain, source and load characteristics, frequency response, etc.) to an amplifier network which will realize the required performance. The quality of the design is determined by its simplicity, practicality, and the degree to which it fulfills the design specifications. If the specifications are exacting, then we shall most probably not meet them on our first attempt. It is therefore preferable to envisage the design of an amplifier in two steps: (1) an initial design in which simple but perhaps inaccurate models of the active elements are used to establish a desirable orientation of the active elements and the approximate values of various circuit components, and (2) a final design where more accurate models of the active elements are used and more nearly optimum circuit components can be obtained.

In the initial stages of the design we must decide on which configuration of the active elements (for example, common base, common emitter, or common collector for transistor amplifiers) or which combination of them will perform the given circuit function most efficiently. A qualitative knowledge of the input impedance, output impedance, current gain, voltage gain, and power gain of each orientation in the midband frequency range will aid us in making these decisions. We will develop these properties in this section.[1] We shall always assume the active element is biased in its normal amplifying region and that the small-signal models derived in Chap. 11 are applicable. To simplify some of our results we will find it convenient to make approximations based on the order of magnitude of typical parameters. The typical parameters assumed for transistors are

$$r_b = 50 - 200 \text{ ohms} \qquad \alpha \geq 0.98$$
$$r_c = 1 - 10 \text{ megohms} \qquad \beta = 50 - 200$$
$$r_e = 25 \text{ ohms at } I_E = 1 \text{ ma}$$

13·2·1 *The common-emitter orientation.* To have a specific problem before us, let us consider the single-stage *CE* amplifier shown in Fig. 13·4a. This figure shows the bias impedances and batteries necessary to establish the operating point. In Fig. 13·4b we show the amplifier with bias impedances and batteries omitted, on the assumption that these impedances do not affect signal transmission properties in the mid-frequency range. We will have to validate this assumption when we come to specific designs (or else analyze the circuit without making the assumption).

There are two ways to calculate the small-signal properties of this amplifier. One method is to substitute a circuit model for the transistor and then analyze the resulting network. Another method is to think of the transistor as being characterized by a set of two-port parameters and then

---

[1] Since this entire section deals with midband properties, we will simply use A_i and A_v to denote current and voltage gains. A_{i0} and A_{v0} will be used later when it is necessary to distinguish between midband gain and the gain at some other frequency.

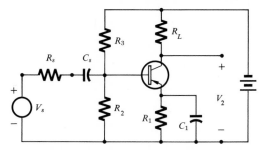

(a) *Amplifier with signal source and all bias components*

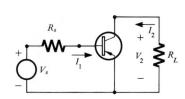

(b) *Circuit components affecting midband gain of the basic CE stage*

FIG. **13·4**   A SINGLE-STAGE, COMMON-EMITTER AMPLIFIER

calculate the small-signal properties of the amplifier in terms of these parameters (for example, h_{11}, h_{12}, h_{21}, h_{22}). Both techniques are commonly used and complement each other nicely. We shall consider both briefly.

To use the *circuit-model approach*, we redraw the small-signal circuit shown in Fig. 13·4b with a small-signal circuit model substituted for the transistor, as in Fig. 13·5. We now assume that I_1 is known and find the other currents by using Kirchhoff's current law at each node, assuming that the voltage drop across r_e is small compared to that across R_L. We may then find the input impedance, current gain, and power gain as follows:

$$R_{in} = \frac{V_1}{I_1} = r_b + \frac{r_e(r_c + R_L)}{r_c(1 - \alpha) + R_L} \simeq r_b + \frac{r_e}{1 - \alpha}$$
$$\text{if } R_L \ll r_c(1 - \alpha) \quad (13·8a)$$

$$A_i = \frac{I_2}{I_1} = \frac{\alpha r_c}{r_c(1 - \alpha) + R_L} \simeq \beta \qquad \text{if } R_L \ll r_c(1 - \alpha) \quad (13·8b)$$

$$G = \frac{I_2^2 R_L}{I_1^2 R_{in}} \simeq \frac{\beta^2 R_L}{r_b + r_e/(1 - \alpha)} \qquad \text{if } R_L \ll r_c(1 - \alpha) \quad (13·8c)$$

$$G_T = G \frac{P_{in}}{P_{avs}} = G \frac{4R_s R_{in}}{(R_s + R_{in})^2} \simeq \frac{4\beta^2 R_L R_s}{[R_s + r_b + r_e/(1 - \alpha)]^2}$$
$$\text{if } R_L \ll r_c(1 - \alpha) \quad (13·8d)$$

To obtain the voltage gain, we must first agree on the definition to be used. If we adopt

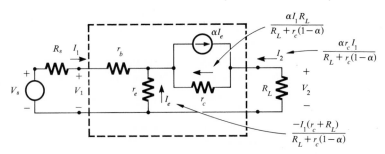

(a) *In this circuit the transistor is represented by the T model*

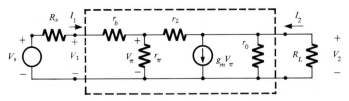

(b) *Alternate circuit with transistor represented by hybrid-π model. V_π denotes the voltage drop across r_π.*

FIG. 13·5 CIRCUIT MODELS FOR THE BASIC CE STAGE SHOWN IN FIG. 13·4b. ALL VOLTAGES AND CURRENTS ARE GIVEN AS *rms* AMPLITUDES.

$$A_v = \frac{V_2}{V_s}$$

then
$$A_v = \frac{-I_2 R_L}{I_1(R_{in} + R_s)} = -A_i \frac{R_L}{R_{in} + R_s} \tag{13 $\cdot$ 8e}$$

$$A_v = \frac{-\beta R_L}{r_b + R_s + r_e(1 - \alpha)} \tag{13 $\cdot$ 8f}$$

In Eq. (13 $\cdot$ 8f), the r_b and R_s terms have been grouped together in the denominator to emphasize that they are connected in series in the circuit. If we chose to define A_v by V_2/V_1, then this would be equivalent to setting $R_s = 0$ in Eq. (13 $\cdot$ 8e), [or conversely, we can obtain V_2/V_s from V_2/V_1 by visualizing r_b to be $(r_b + R_s)$].

Usually $R_L \ll r_c(1 - \alpha)$ for the intermediate stages of a cascade of CE amplifiers (in fact, $R_L \leq R_{in}$). However, the load resistance for the last stage may not meet this approximation, so the approximate formulas may be inaccurate for this case.

EXAMPLE   To estimate the properties of the CE stage, we use the "typical" parameters given earlier: $r_b = 100$ ohm, $r_e = 25$ ohm, $\beta \simeq (1 - \alpha)^{-1} = 100$, $r_c = 1$ megohm. We also set $R_s = 1$ kohm, $R_L = 1$ kohm $< r_c(1 - \alpha) = 10$ kohm. Then

$$R_{in} \simeq 100 + 2,500 = 2,600 \text{ ohms}$$
$$A_i \simeq \beta = 100$$
$$A_v = \frac{V_2}{V_s} = -100 \frac{1 \text{ kohm}}{3.6 \text{ kohm}} \simeq 28$$
$$G \simeq 3,800$$
$$G_T = 0.8G = 3,100$$

The power gain can be expressed in decibels as

$$G_{db} = 10 \log_{10} 3,850 = 30 + 10 \log_{10} 3.85 = 35.8 \simeq 36 \text{ db}$$

It is a good idea to avoid stating gains, impedances, etc. to greater than two-place accuracy in these calculations, since it helps to emphasize the approximate nature of the models and analysis.

13 $\cdot$ 2 $\cdot$ 2   Two-port formulation.   The circuit model approach to the problem of calculating the midband properties of the CE stage has the very important advantage of directness and simplicity. Naturally, it also has certain disadvantages. For one thing, if we want to study how the performance of the CE stage changes with changes in the bias condition, we need to know how all the parameters in the transistor circuit model vary with bias. The manufacturer does not supply information in this form, and even if he did, he would probably use a more complicated model than the one we selected for analysis. Another disadvantage of the method is that it lacks generality: a new analysis is necessary for each configuration of the transistor and each new model that is used.

These objections are not present if the amplifier properties are developed in terms of the two-port parameters of the active device. Using a two-port

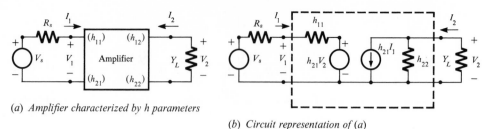

(a) Amplifier characterized by h parameters

(b) Circuit representation of (a)

FIG. 13·6 A GENERALIZED SMALL SIGNAL AMPLIFIER CHARACTERIZED BY TWO-PORT PARAMETERS

formulation, we can derive expressions for the amplifier properties without specifying the transistor configuration or the circuit model to be used. Furthermore, manufacturers do show the variation of the h parameters with bias for at least some transistors (see, for example, Fig. 11·21), so we can study practical problems with this technique, which are harder to study otherwise. Finally, the two-port approach is also of interest as a useful tool in the systematic analysis of networks.

To pursue the two-port approach, we show in Fig. 13·6 a two-port network characterized by its h parameters with a source and load attached.[1] We use a *voltage source* with a *series source impedance* and a *load admittance* to conform to the nature of the h-parameter system (that is, the input loop is a series circuit and the output loop a parallel circuit for h parameters. See Fig. 13·6). This will simplify analysis and also make it easy to generalize the results to the z-, y-, and g-parameter systems.

Input impedance. The input impedance of the amplifier can be calculated from the following equations:

Two-port: $V_1 = h_{11}I_1 + h_{12}V_2$ (13·9a)

Two-port: $I_2 = h_{21}I_1 + h_{22}V_2$ (13·9b)

Load condition: $I_2 = -V_2Y_L$ (13·9c)

Definition: $Z_{in} = \dfrac{V_1}{I_1}$ (13·9d)

The calculation proceeds as follows. From Eqs. (13·9a) and (13·9d)

$$Z_{in} = \frac{V_1}{I_1} \qquad (13·10a)$$

$$Z_{in} = h_{11} + h_{12}\frac{V_2}{I_1} \qquad (13·10b)$$

[1] Transistors are usually characterized by h or y parameters. We will develop the properties of a two-port amplifier in terms of the h parameters. A more general discussion is given in J. G. Linvill and J. F. Gibbons, "Transistors and Active Circuits," McGraw-Hill Book Company, New York, 1961, chap. 10.

From Eq. $(13 \cdot 9c)$ and $(13 \cdot 9d)$ we get

$$-V_2 Y_L = h_{21} I_1 + h_{22} V_2 \qquad (13 \cdot 10c)$$

$$\frac{V_2}{I_1} = -\frac{h_{21}}{h_{22} + Y_L} \qquad (13 \cdot 10d)$$

Hence

$$Z_{in} = h_{11} - \frac{h_{12} h_{21}}{h_{22} + Y_L} \qquad (13 \cdot 10e)$$

Output admittance. To find the output *admittance* we set the source voltage $V_s = 0$. Then, the appropriate equations are

Two-port: $\qquad\qquad\qquad V_1 = h_{11} I_1 + h_{12} V_2 \qquad (13 \cdot 11a)$

Two-port: $\qquad\qquad\qquad I_2 = h_{21} I_1 + h_{22} V_2 \qquad (13 \cdot 11b)$

Input condition: $\qquad\qquad V_1 = -I_1 Z_s \qquad\qquad (13 \cdot 11c)$

Definition: $\qquad\qquad\qquad Y_{out} = \dfrac{I_2}{V_2} \qquad\qquad (13 \cdot 11d)$

We can follow a set of steps identical to those in Eqs. $(13 \cdot 10a)$–$(13 \cdot 10e)$ to find Y_{out}:

$$Y_{out} = \frac{I_2}{V_2} \qquad (13 \cdot 12a)$$

$$Y_{out} = h_{22} + h_{21} \frac{I_1}{V_2} \qquad (13 \cdot 12b)$$

$$-I_1 Z_s = h_{11} I_1 + h_{12} V_2 \qquad (13 \cdot 12c)$$

$$\frac{I_1}{V_2} = -\frac{h_{12}}{h_{11} + Z_s} \qquad (13 \cdot 12d)$$

$$Y_{out} = h_{22} - \frac{h_{12} h_{21}}{h_{11} + Z_s} \qquad (13 \cdot 12e)$$

Note the symmetry in both the method of calculation and the final results for input impedance and output admittance. The symmetry of results would be less apparent if we were calculating Z_{out}, because Y_{out} is the natural quantity for the *h*-parameter system (it has the same dimensions as h_{22}). The output impedance is

$$Z_{out} = \frac{1}{Y_{out}} = \frac{h_{11} + Z_s}{h_{22}(h_{11} + Z_s) - h_{12} h_{21}} \qquad (13 \cdot 13)$$

Current gain. To calculate the current gain of the amplifier, we note that the output current of the network shown in Fig. $13 \cdot 6b$ is

$$I_2 = (h_{21} I_1) \frac{Y_L}{h_{22} + Y_L} \qquad (13 \cdot 14a)$$

Hence

$$A_i = \frac{I_2}{I_1} = \frac{h_{21} Y_L}{h_{22} + Y_L} \qquad (13 \cdot 14b)$$

Voltage gain. Defining the voltage gain as the ratio V_2/V_s, we have

$$V_2 Y_L = I_2 \tag{13·15a}$$

$$V_s = I_1(Z_s + Z_{in}) \tag{13·15b}$$

$$\frac{V_2}{V_s} = \frac{I_2}{I_1} \frac{1}{Y_L(Z_s + Z_{in})} \tag{13·15c}$$

$$\frac{V_2}{V_s} = \frac{h_{21}}{(h_{22} + Y_L)(Z_s + Z_{in})} \tag{13·15d}$$

By substituting Z_{in} from Eq. (13·10e), Eq. (13·15d) can also be written

$$A_v = \frac{V_2}{V_s} = \frac{h_{21}}{(h_{11} + Z_s)(h_{22} + Y_L) - h_{12}h_{21}} \tag{13·15e}$$

Power gain. The power gain is defined by

$$G = \frac{|V_2|^2 Y_{Lr}}{|I_1|^2 R_{in}} = \frac{|A_i|^2 Y_{Lr}}{|Y_L|^2 R_{in}} = |A_i|^2 \frac{R_L}{R_{in}} \tag{13·16a}$$

in which Y_{Lr} is used to denote the real part of Y_L. By making appropriate substitutions, G can be recast in the form

$$G = \frac{|h_{21}|^2 Y_{Lr}}{|h_{22} + Y_L|^2 \, \text{Re} \, [h_{11} - h_{12}h_{21}/(h_{22} + Y_L)]} \tag{13·16b}$$

Re [ ] denotes the real part of the quantity in brackets, which is the input impedance.

The transducer gain can be expressed in the form

$$G_T = \frac{4|h_{21}|^2 Y_{Lr} Z_{sr}}{|(h_{11} + Z_s)(h_{22} + Y_L) - h_{12}h_{21}|^2} \tag{13·16c}$$

EXAMPLE   As a first application of these results, let us consider the common-emitter connection of a 2N1613, driven from a 1 kohm source and driving a 5-kohm load. The h parameters for the transistor are given in Fig. 11·21b. If we suppose $I_C = 1$ ma and $V_{CE} = 5$ volts, then

$$h_{11} = 2.2 \times 10^3 \text{ ohms} \qquad h_{12} = 3.6 \times 10^{-4}$$
$$h_{21} = 55 \qquad h_{22} = 12.5 \times 10^{-6} \text{ mhos}$$

Using these data in the previous formulas, we find

$$Z_{in} = h_{11} - \frac{h_{12}h_{21}}{h_{22} + Y_L} = \left[2.2 \times 10^3 - \frac{18 \times 10^{-3}}{(12.5 + 200)10^{-6}} \right] \text{ohms}$$
$$\cong 2.19 \times 10^3 \text{ ohms} \simeq h_{11}$$

$$Y_{out} = h_{22} - \frac{h_{12}h_{21}}{h_{11} + Z_s} = \left(12.5 \times 10^{-6} - \frac{18 \times 10^{-3}}{3.2 \times 10^3} \right) \text{ mhos} = 6.1 \times 10^{-6} \text{ mhos}$$

$$A_i = \frac{h_{21}Y_L}{h_{22} + Y_L} = h_{21} \frac{1}{1 + h_{22}/Y_L} = \frac{55}{1.062} \simeq 52$$

$$A_v = \frac{h_{21}}{(h_{22} + Y_L)(Z_s + Z_{in})} = \frac{55}{(212.5 \times 10^{-6})(3.2 \times 10^3)} = 81$$

$$G = \frac{|h_{21}|^2 Y_{Lr}}{|h_{22} + Y_{Lr}|^2 \operatorname{Re}[h_{11} - h_{12}h_{21}/(h_{22} + Y_L)]} = \frac{(55)^2 2 \times 10^{-4}}{4 \times 10^{-8} \times 2 \times 10^3}$$

$$= 6.7 \times 10^3$$

$$G_T = \frac{4|h_{21}|^2 Y_{Lr} Z_{sr}}{|(h_{11} + Z_s)(h_{22} + Y_L) - h_{12}h_{21}|^2}$$

$$= \frac{4(55)^2(2 \times 10^{-4})(10^3)}{|(3.2 \times 10^3)(2 \times 10^{-4}) - 55(3.6 \times 10^{-4})|^2} = 6.3 \times 10^3$$

The effect of connecting several common-emitter stages in cascade can be studied by setting $R_L = R_{in}$; that is, finding the current gain and power gain which a stage will exhibit when its load is the input impedance of another stage. For the parameters given earlier, $R_{in} \simeq h_{11}$, so $Y_L = 1/h_{11}$. The current gain then becomes

$$A_i \simeq \frac{h_{21}/h_{11}}{h_{22} + 1/h_{11}} = \frac{h_{21}}{1 + h_{22}h_{11}} \simeq \frac{55}{1.03} = 53.5$$

The common-emitter connection is the only connection of the transistor which exhibits the property of maintaining its power gain in simple cascade connections. It is therefore the principal configuration which is used in amplifier design.

A second way in which the general two-port results are used is in studying the variation in amplifier performance with bias. Transistor manufacturers frequently give curves showing how the h or y parameters of a particular transistor vary with bias. If the h parameters are read off at several bias points, the general formulas obtained earlier can be used to study the variation in amplifier properties with bias changes.

EXAMPLE   To study the variation of the amplifier properties with bias condition, we can use Fig. $11 \cdot 21$ to obtain the h parameters at the required bias point and then the previous formulas to calculate the properties of the stage. As an example, if we decrease the bias current in the stage to 0.1 ma, and readjust the circuit constants to obtain $V_{CE} = 5$ volts, we have

$$h_{11} = 8.8 \times 10^3 \text{ ohms} \qquad h_{12} = 21.6 \times 10^{-4}$$
$$h_{21} = 33 \qquad h_{22} = 7.5 \ \mu\text{mhos}$$

Assuming a source resistance of 1 kohm and a load resistance of 5 kohms, the small-signal properties of the stage are then

$$Z_{in} = 8.5 \times 10^3 \text{ ohm}$$
$$Y_{out} = 0.25 \ \mu\text{mhos}$$
$$A_i = 32$$
$$A_v = \frac{V_2}{V_s} = 16.8$$

Thus, a 10:1 change in bias current produces significant changes in the properties of the stage. While this is not unexpected, it does underline the necessity of having a stabilized bias situation.

$13 \cdot 2 \cdot 3$ *The common-base orientation.*   The general properties of the common-base orientation can be derived from the circuit model approach, the two-port approach, or a combination of them. For interest, we shall use a

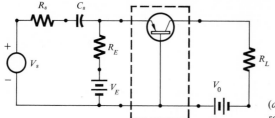

(a) CB stage with bias components and source

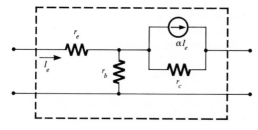

(b) Circuit model to be used for transistor

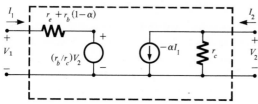

(c) Approximate h parameter equivalent of (b)

FIG. 13·7 A COMMON BASE AMPLIFIER STAGE

combination of these techniques. We shall consider the amplifier circuit shown in Fig. 13·7, and use the simple T model shown in Fig. 13·7b to represent the transistor. The h parameters for this circuit model are

$$h_{11} = r_e + r_b(1 - \alpha)\left(1 + \frac{r_b}{r_c}\right) \cong r_e + r_b(1 - \alpha) \qquad (13 \cdot 17a)$$

$$h_{12} = \frac{r_b}{r_c + r_b} \simeq \frac{r_b}{r_c} \qquad (13 \cdot 17b)$$

$$h_{21} = -\alpha - \frac{(1 - \alpha)}{[1 + (r_c/r_b)]} \simeq -\alpha \qquad (13 \cdot 17c)$$

$$h_{22} = \frac{1}{r_c + r_b} \cong \frac{1}{r_c} \qquad (13 \cdot 17d)$$

Substituting these results into the general h-parameter formulas, we find

$$Z_{in} = R_{in} = r_e + r_b(1 - \alpha) \qquad \frac{R_L}{r_c + R_L} \ll (1 - \alpha) \qquad (13 \cdot 18a)$$

$$Z_{out} = R_{out} = \frac{1}{Y_{out}} = r_c \qquad r_b < R_s < r_c \qquad (13 \cdot 18b)$$

$$A_i = -\alpha \qquad R_L \ll r_c \qquad (13 \cdot 18c)$$

$$A_v = \frac{-\alpha R_L}{R_s + r_e + r_b(1 - \alpha)} \qquad R_L \ll r_c \qquad (13 \cdot 18d)$$

$$G = \frac{\alpha^2 R_L}{r_e + r_b(1 - \alpha)} \qquad (13 \cdot 18e)$$

Thus, the common-base stage has a low input impedance, a high output impedance, and a current gain of slightly less than unity. The power gain and the voltage ratio (V_2/V_1) may be written

$$A_v \simeq G \simeq \frac{R_L}{R_{in}}$$

which shows that voltage and power gain arise only when the impedance level in the output circuit is high compared to the input impedance. In particular, this means that common-base stages cannot be cascaded and still give power gain, unless we use transformers between stages. This is usually an unwarranted complication, so common-base stages are used primarily to change impedance levels in a circuit.

13·2·4 *The common-collector orientation.* Proceeding exactly as before, we can find the h parameters for the common-collector orientation of the T model and then obtain the properties for the common-collector amplifier stage. The circuit, with biases, is shown in Fig. 13·8, the load resistor being placed in the emitter circuit. The required formulas are

$$R_{in} = \frac{r_c(r_e + R_L)}{r_c(1 - \alpha) + R_L} + r_b \simeq \frac{R_L}{1 - \alpha} \simeq \beta R_L \qquad R_L \gg r_e \quad (13 \cdot 19a)$$

$$R_{out} = (R_s + r_b)(1 - \alpha) + r_e \simeq R_s(1 - \alpha) \qquad r_c(1 - \alpha) > (R_s + r_b)$$
$$(13 \cdot 19b)$$

$$A_v \simeq \frac{R_L}{(R_s + r_b)(1 - \alpha) + r_e + R_L} \simeq 1 \qquad R_L \gg r_e + (R_s + r_b)(1 - \alpha)$$
$$(13 \cdot 19c)$$

$$G = |A_v|^2 \frac{R_{in}}{R_L} \simeq \frac{1}{1 - \alpha} \simeq \beta \qquad (13 \cdot 19d)$$

From Eqs. (13·19a) and (13·19b) we can see that the common-collector

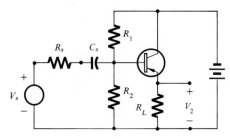

FIG. 13·8 BASIC COMMON COLLECTOR (CC) STAGE. THE STAGE IS ALSO CALLED AN "EMITTER FOLLOWER"

stage has the properties of an "impedance transformer." The input impedance is approximately βR_L, while the output impedance is approximately R_s/β. However, the impedance transformation is accomplished with a voltage gain of almost unity and considerable power gain.

Because of its unity voltage gain and impedance-transformation properties, the common-collector stage is useful for coupling an amplifier to a variable-impedance load. For example, suppose we require an amplifier to have a certain fixed voltage gain but to be capable of driving a variable load impedance (perhaps a variable number of loudspeakers are to be driven). Suppose the load impedance may be anywhere in the range 250 ohms $\leq R_L \leq 1$ kohm. We could build a common-emitter amplifier (possibly multistage) with the required voltage gain and then couple it to the load by means of a common-collector stage. If the common-collector stage uses a transistor with $\beta = 100$, the input impedance of the stage will be 25 kohms $\leq R_{in} \leq 100$ kohms. This will not seriously load the common-emitter amplifier, which might have a typical load impedance of perhaps 5 kohms. At the same time, the output impedance of the common-collector stage is about 50 ohms, so the voltage supplied to its load will be essentially constant, independent of the load resistance.

13·3   FREQUENCY CHARACTERISTICS OF THE SINGLE-STAGE AMPLIFIER

We shall now turn our attention to the frequency characteristics of a single-stage common-emitter amplifier. The amplifier to be studied is shown in Fig. 13·9a. The voltage gain vs. frequency characteristic of this amplifier will be similar to that shown in Fig. 13·9b. In the midband range the properties of the amplifier will be given by the formulas obtained earlier. At low frequencies the amplifier gain will be reduced, due to the effects of the coupling capacitor C_s and the bypass capacitor C_1. The high-frequency cutoff arises from the input capacitance of the transistor as we shall see later.

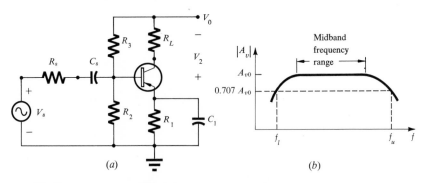

(a)   (b)

FIG. 13·9   SINGLE STAGE CE AMPLIFIER AND AMPLITUDE RESPONSE CURVE

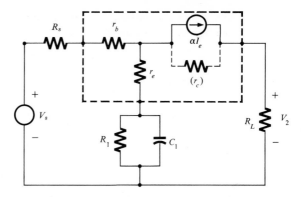

FIG. **13·10**   CIRCUIT MODEL FOR STUDYING THE EFFECT OF C_1 ON THE FREQUENCY RESPONSE OF A CE STAGE

13·3·1   *Factors affecting the low-frequency cutoff of an audio amplifier.*   We shall first study the effect of the emitter bypass capacitor C_1 on the frequency response, assuming C_s to be infinite. To begin the study, we must choose a model to represent the transistor in Fig. 13·9a. Either the hybrid-π or the T model could be used, though the T model is more appropriate for this problem, for reasons that will be apparent shortly. The amplifier circuit with a T model for the transistor is shown in Fig. 13·10.

With the circuit before us, we can now determine V_2/V_s as a function of the complex frequency variable s. There are several ways to proceed at this point:

1. Employ a standard network analysis scheme. In this case, we could define loop currents in the input and output loops and then write down Kirchhoff's voltage law for each loop. The two equations so obtained can be solved for the loop currents in terms of V_s, and then V_2 can be found from the current in the output loop.

2. Generalize the midband-gain expression. This can be very simply done by recognizing that the impedance in the emitter lead is in series with r_e in the T model. If we substitute the actual impedance in the emitter lead for r_e in the midband-gain expression, we can obtain an expression for $A_v(s)$.

3. Write down a general expression for the gain, in the form

$$A_v(s) = K \frac{s + s_0}{s + s_p}$$

The pole s_p in this expression is the natural frequency of the network and can be found by the methods of the Appendix. The values of s_0 and K are found by making sure that the general expression has the correct value as $s = j\omega \rightarrow 0$ and $s = j\omega \rightarrow \infty$. This technique is discussed in the Appendix.

We shall use the scheme indicated in step 2 for this analysis. To do so, we first write down the expression for the voltage gain at midband,

$$A_{v0} = \frac{V_2}{V_s} = \frac{-\alpha R_L}{r_e + (r_b + R_s)(1 - \alpha)} \qquad (13 \cdot 20)$$

The effect of adding an impedance Z_1 in series with r_e is studied by making the substitution

$$r_e \rightarrow r_e + Z_1$$

so that
$$A_v = \frac{-\alpha R_L}{Z_1 + r_e + (r_b + R_s)(1 - \alpha)} \qquad (13 \cdot 21)$$

At high frequencies, where C_1 has a very low impedance, Z_1 may be neglected. At low frequencies, where C_1 has essentially infinite impedance,

$$Z_1 \rightarrow R_1 \qquad A_v \rightarrow \frac{-\alpha R_L}{R_1}$$

which will typically be 4 or 5, in comparison to a midband gain of 50 to 200.

To discuss the behavior of A_v in more detail, we substitute for Z_1 its formula in terms of R_1, C_1, and the complex frequency variable s.

$$Z_1 = \frac{R_1}{1 + sR_1C_1} \qquad (13 \cdot 22)$$

$$A_v = \frac{-\alpha R_L(1 + sR_1C_1)}{R_1 + [r_e + (r_b + R_s)(1 - \alpha)](1 + sR_1C_1)} \qquad (13 \cdot 23a)$$

For purposes of interpretation, we recast Eq. $(13 \cdot 23a)$ in the form

$$A_v(s) = A_{v0} \frac{s + 1/R_1C_1}{s + 1/R_{eq}C_1} \qquad (13 \cdot 23b)$$

where A_{v0} is the midband gain [Eq. $(13 \cdot 20)$] and R_{eq} is the parallel combination of R_1 and $[r_e + (r_b + R_s)(1 - \alpha)]$.

The effects of Z_1 on A_v are contained in the second factor in Eq. $(13 \cdot 23b)$. As stated earlier,

$$\lim_{s \to 0} A_v(s) = A_{v0} \frac{R_{eq}}{R_1} \simeq \frac{-\alpha R_L}{R_1} \qquad \lim_{s \to \infty} A_v(s) \simeq A_{v0}$$

Equation $(13 \cdot 23b)$ may be represented in the complex s plane by the pole-zero diagram shown in Fig. $13 \cdot 11a$. A_v has a zero at

$$s_0 = -\frac{1}{R_1C_1} \qquad (13 \cdot 24a)$$

and a pole at

$$s_p = -\frac{1}{R_{eq}C_1} \qquad (13 \cdot 24b)$$

Hence
$$A_v(s) = A_{v0} \frac{s - s_0}{s - s_p} \qquad (13 \cdot 25)$$

Since $R_{eq} < R_1$, $|s_p| > |s_0|$.

We are interested in the variation of the magnitude and phase of A_v as

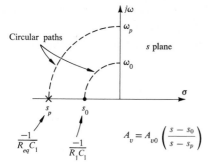

(a) Pole-zero pattern for $A_v(s)$

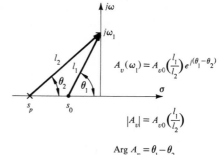

(b) Use of pole-zero pattern for determining $|A_v|$ and Arg A_v at $s = j\omega_1$

FIG. 13·11 THE POLE-ZERO REPRESENTATION OF A_v AND ITS USE AS A TOOL FOR CALCULATION

a function of the *real* frequency variable ω. A sketch of this variation may be made directly from the pole-zero pattern in the manner indicated in Fig. 13·11b. We select a test frequency $j\omega_1$ and evaluate the lengths and angles of the phasors which connect s_0 and s_p to $j\omega_1$. We have

$$j\omega_1 - s_0 = l_1 e^{j\theta_1} \qquad j\omega_1 - s_p = l_2 e^{j\theta_2}$$

$$A_v(j\omega_1) = A_{v0} \frac{l_1}{l_2} e^{j(\theta_1 - \theta_2)}$$

or

$$|A_v(j\omega_1)| = A_{v0} \frac{l_1}{l_2} \qquad \text{Arg } A_v(j\omega_1) = \theta_1 - \theta_2$$

When this is done at several test frequencies, we can sketch the behavior of $|A_v(j\omega)|$ and Arg $A_v(j\omega)$ as functions of ω.

13·3·2 Log-log plots for amplifier-gain functions. Unfortunately, direct computation of A_v in the manner just given is a tedious job if reasonable accuracy is required, and some means of simplifying and speeding-up the sketching process is desired. A very useful simplification may be obtained by plotting $|A_v(j\omega)|$ in db versus $\log_{10} \omega$. When this is done, the effects of the zero and the pole are *additive,* and the approximate behavior of $|A_v|$(db) may be rapidly sketched.

To see this, we first rewrite Eq. (13·25) in the form

$$A_v(s) = A_{v0} \frac{R_{eq}}{R_1} \frac{1 + sR_1C_1}{1 + sR_{eq}C_1} \qquad (13 \cdot 26)$$

We then form the function

$$|A_v|(\text{db}) = 20 \log_{10} |A_v(s)|$$

$$= 20 \log_{10} A_{v0} \frac{R_{eq}}{R_1} + 20 \log_{10} |1 + sR_1C_1| - 20 \log_{10} |1 + sR_{eq}C_1|$$

$$(13 \cdot 27)$$

Amplifier design   489

Eq. (13·27) shows that when $|A_v|$ is expressed in db, the only basic difference between the zero and the pole is a $+$ or $-$ sign.

To determine the real-frequency behavior of $|A_v|$(db), we set $s = j\omega$ in Eq. (13·27):

$$20 \log_{10} |A_v| = 20 \log_{10} A_{v0} \frac{R_{eq}}{R_1}$$

$$+ 20 \log_{10} \left|1 + \frac{j\omega}{\omega_0}\right| - 20 \log_{10} \left|1 + \frac{j\omega}{\omega_p}\right| \quad (13·28)$$

where

$$\omega_0 = \frac{1}{R_1 C_1} = |s_0| \qquad \omega_p = \frac{1}{R_{eq} C_1} = |s_p|$$

ω_0 and ω_p may be visualized as the "circular projections" of s_0 and s_p onto the real frequency axis, as in Fig. 13·11b.

It is apparent from Eq. (13·28) that if we can sketch the function $20 \log_{10} |1 + j(\omega/\omega_a)|$, we can sketch the entire behavior of $|A_v|$(db). Hence we consider for the moment the function $T = 1 + s/\omega_a$.

Low-frequency behavior. For $s = j\omega \ll \omega_a$, we have

$$20 \log_{10} |T| = 20 \log_{10} \left|1 + \frac{j\omega}{\omega_a}\right| \simeq 20 \log_{10} 1 = 0 \quad (13·29a)$$

Hence, when $\omega \ll \omega_a$, T is about equal to 1, or equivalently, $20 \log_{10} |T|$ tends toward a 0-db asymptote, as shown in Fig. 13·12.

High-frequency behavior. For $\omega \gg \omega_a$, we have

$$20 \log_{10} |T| = 20 \log_{10} \left|1 + j\frac{\omega}{\omega_a}\right| \simeq 20 \log_{10} \frac{\omega}{\omega_a} \quad (13·29b)$$

Thus, when $\omega \gg \omega_a$, $|T|$ in decibels is proportional to $\log_{10} \omega$. This means that if both $|T|$ and ω are plotted on logarithmic scales, or $|T|$ in decibels is plotted against ω on a logarithmic scale, the high-frequency

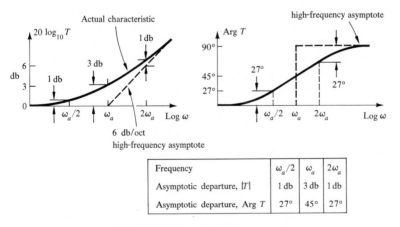

Frequency	$\omega_a/2$	ω_a	$2\omega_a$
Asymptotic departure, $|T|$	1 db	3 db	1 db
Asymptotic departure, Arg T	27°	45°	27°

FIG. **13·12** PLOTS OF **20 $\log_{10}$ $|T|$** AND **Arg T** vs. **log ω**

asymptote will be a straight line. This straight line asymptote has the property that when ω is doubled, $20 \log_{10} |T|$ changes by 6 db [a fact which may be demonstrated directly from Eq. (13·29b)].

Asymptotic departures. The transition of the amplitude characteristic between its low- and high-frequency asymptotes is very simple. At $\omega = \omega_a$, the "break frequency," the actual value of $20 \log_{10} |T|$ is

$$20 \log_{10} |T| = 20 \log_{10} |1 + j1| \simeq 3 \text{ db}$$

When $\omega = 2\omega_a$ or $\omega_a/2$, the actual characteristic is 1 db above its asymptote.

The asymptotic plot of $|T|$, together with its exact form, are given in Fig. 13·12. The asymptotic plot consists of two straight lines. The low-frequency asymptote is horizontal at 0 db. The high-frequency asymptote is a rising straight line with a slope of 6 db/octave,[1] drawn through the point $\omega = \omega_a$. At $\omega = 2\omega_a$ and $\omega = \omega_a/2$, the actual characteristic is 1 db away from its asymptote. The actual amplitude characteristic associated with the factor $1 + s/\omega_a$ can thus be sketched with negligible effort.

The phase-shift characteristic associated with a factor $1 + s/\omega_a$ has equally simple properties. The low- and high-frequency asymptotes of the phase shift are 0 and 90°, respectively. At the break frequency ω_a, the phase shift is 45°, and the phase characteristic is symmetrical about the break frequency. These facts are plotted in Fig. 13·12, together with a table of asymptotic departures.

Since we have been considering the function $T = 1 + s/\omega_a$, we have implicitly been considering a zero of A_v. If we wish to consider a pole of A_v, we would have the function

$$T = \frac{1}{1 + s/\omega_a} \qquad (13 \cdot 29c)$$

$$20 \log_{10} |T| = -20 \log_{10} \left| 1 + \frac{s}{\omega_a} \right|$$

which is the same as before, except for the minus sign. The sign of the slope

[1] An octave is a 2:1 change in frequency.

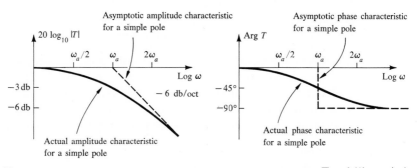

FIG. 13·13   AMPLITUDE AND PHASE CHARACTERISTICS FOR THE FUNCTION $T = 1/(1 + s/\omega_a)$

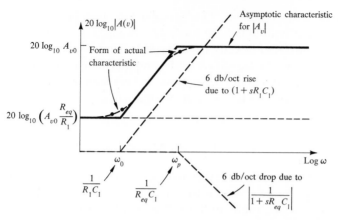

FIG. 13·14 ASYMPTOTE PLOT FOR $|A_v| = A_{v0}\dfrac{R_{eq}}{R_1}\left|\dfrac{1 + sR_1C_1}{1 + sR_{eq}C_1}\right|$

on the high-frequency amplitude asymptote is then -6 db/octave. The asymptotic departures also change sign as shown in Fig. 13·13. In particular, the high-frequency phase angle is asymptotic to $-90°$.

13·3·3 *Low-frequency cutoff due to bias impedance* Z_1. We now return to our problem of plotting

$$A_v = A_{v0}\frac{R_{eq}}{R_1}\frac{1 + sR_1C_1}{1 + sR_{eq}C_1} \tag{13·30}$$

We see that on log-log paper, the $(1 + sR_1C_1)$ factor will produce an asymptotic characteristic which is zero up to a frequency $\omega_0 = 1/R_1C_1$ and then rises at 6 db/octave (see Fig. 13·14). The factor $(1 + sR_{eq}C_1)^{-1}$ will be zero up to the frequency $\omega_p = 1/R_{eq}C_1$ and then it will fall at 6 db/octave. The overall asymptotic characteristic will be constant at the value $20 \log_{10} (A_{v0}R_{eq}/R_1)$ to a frequency $\omega_0 = 1/R_1C_1$, will rise at 6 db/octave to the frequency $\omega_p = 1/R_{eq}C_1$, and will then be constant for $\omega > 1/R_{eq}C_1$. The asymptotic departures may be inserted to produce the exact characteristic shown in Fig. 13·14. The magnitude of the rise is exactly enough to bring $|A_v|$ to A_{v0}. If ω_0 and ω_p are widely separated, as they usually will be, the lower cutoff frequency will be the same as ω_p.

Experimental verification of results. To illustrate the foregoing theory, the amplifier shown in Fig. 13·15 was built and tested. The experimental results of the test exhibit the general characteristics shown in Fig. 13·14. To see how precisely the theory predicts the experimental results, we give the measured values of the relevant circuit and transistor parameters below:

$$I_E = 1 \text{ ma} \qquad R_1 = 1 \text{ kohm} \qquad R_L = 2.7 \parallel 15 = 2.3 \text{ kohms}$$
$$\beta = 162 \qquad R_s = 1 \text{ kohm}$$

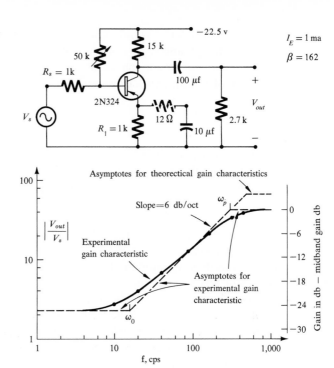

FIG. 13·15 EXPERIMENTAL AMPLIFIER AND ITS GAIN CHARACTERISTIC

On the basis of these values and an assumed r_b of 50 ohms, the theoretical midband gain is

$$|A_v| = \frac{R_L}{r_e + (r_b + R_s)(1 - \alpha)} = \frac{2.3 \text{ kohms}}{32} = 72$$

To check this value, the output voltage at $f = 2$ kcps was measured for a source voltage of 1.8 mv. The resulting measured midband gain was

$$A_{v,meas} = \frac{80 \text{ mv}}{1.8 \text{ mv}} = 44$$

representing a considerable discrepancy from the theory.

The source of this discrepancy lies in two places: (1) the ideal transistor model gives a somewhat inaccurate representation of the real transistor and (2) the bypass capacitor has some equivalent series resistance which has not been included.

To account for these discrepancies, let us consider the theoretical expression for A_v at midband, amended to include some equivalent series resistance r_{C_1} in the bypass capacitor,

$$|A_v| = \frac{R_L}{r_e + (r_b + R_s)(1 - \alpha) + r_{C_1}}$$

In this expression, R_L and $(r_b + R_s)/(\beta + 1)$ are known with reasonable accuracy. However, r_e is only known if the transistor behaves according to the ideal model, and r_{C_1} is not known at all.

Measurements made on r_e for a typical 2N324 show the following variation of r_e with emitter bias current:

I_E(ma)	0.25	0.5	1	2
r_e(ohms)	120	60	30	16

These measurements indicate that r_e is inversely proportional to I_E, as it should be, but the constant of proportionality is 30 mv. If we check the dc emitter-base diode curve, we also find

$$I_E = I_0(e^{qV_{EB}/nkT} - 1)$$

where n must be set equal to about 1.2 to fit the data.[1] Hence, to use the transistor model correctly, we must set $r_e = 30$ ohms$/|I_E$(ma)$|$. This factor causes the denominator of A_{v0} to be about 5 ohms larger than the ideal transistor theory suggests.

The other source of difficulty is in the bypass capacitor. We need a relatively large bypass capacitor (10 to 100 μfarads) if we want the lower cutoff frequency to be below 300 cps or so. Electrolytic capacitors provide the greatest capacitance per cubic inch and the lowest cost per microfarad of any of the types commonly used and are normally selected as bypass capacitors. However, these capacitors have a series resistance which is typically between 1 and 10 ohms, and may sometimes be as high as 25 ohms. The series resistance arises primarily from conductive losses in the electrolyte (see Fig. 13·16) and is moderately constant at low frequencies, though quite temperature-sensitive. The resistance can be measured by measuring the Q of the capacitor on a standard RLC bridge.

The series resistance of the 10-μfarad capacitor used in the amplifier was 12 ohms. When this resistance is included in the denominator of the theoretical gain expression, the predicted midband gain becomes 47.5, which is adequately close to the measured gain.

Let us now see how well the critical frequencies are predicted by the theory. The theoretical break frequency ω_p for a 10-μfarad capacitor is given by

$$f_p = \frac{1}{2\pi C_1[r_e + (r_b + R_s)(1 - \alpha)]} = \frac{1}{2\pi \times 10^{-5} \text{ farad} \times 31 \text{ ohms}} = 515 \text{ cps}$$

The measured break frequency for a 10-μfarad capacitor is at 300 cps instead of 515 cps. However, when we include the fact that the 10-μfarad capacitor has a 12-ohm series resistance and $r_e = 30$ ohms instead of 25, f_p becomes

$$f_p = \frac{1}{2\pi \times 10^{-5} \text{ farad} \times 48.5 \text{ ohms}} = 328 \text{ cps}$$

which is again reasonably close to the measured value.

[1] The reason for this behavior is discussed in Sec. 9·6·3.

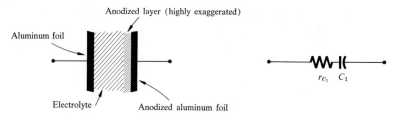

FIG. **13 · 16** ELECTROLYTIC CAPACITOR AND CIRCUIT MODEL FOR IT

The lower critical frequency, at which the gain flattens out, should be at

$$\omega_0 = \frac{1}{2\pi \times 10^{-5} \text{ farad} \times 1 \text{ kohm}} = 16 \text{ cps}$$

and the gain should become asymptotic to the value

$$A_v(f \ll f_0) = \frac{2.3 \text{ kohm}}{1 \text{ kohm}} = 2.3$$

These values are acceptably close to the measured values. Of course, the series resistance of C_1 is now unimportant.

Note that the db difference between the measured midband gain and the measured gain at very low frequencies is 25.8 db. The reader can show that this also confirms the existence of series resistance in the 10-μfarad capacitor and an increased value of r_e.

If we replace the 10-μfarad capacitor with a 37.5-μfarad capacitor having only a 2-ohm series resistance, the midband gain rises to 60, and the gain vs. frequency curve changes to the one shown in Fig. 13 · 17. The calculated cutoff frequency for this case is 110 cps, which compares favorably with the measured value of 116 cps. However, 37.5 μfarads is only the nominal value of the capacitor. When the capacitance value was measured, it was found to be 35 μfarads; the calculated lower cutoff frequency is then 118 cps.

A gain vs. frequency curve is also shown for a nominal 100-μfarad

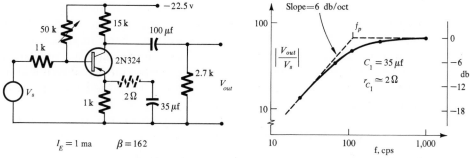

FIG. **13 · 17** EXPERIMENTAL DEMONSTRATION OF THE EFFECT OF THE BYPASS CAPACITOR C_1 ON THE GAIN CHARACTERISTIC OF AN AMPLIFIER

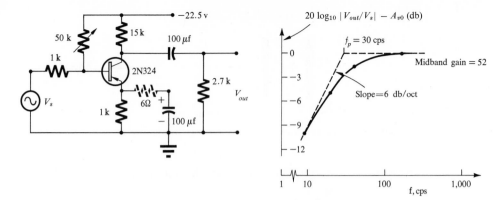

FIG. 13 · 18   LOW-FREQUENCY GAIN CHARACTERISTIC FOR A 100-μfd BYPASS CAPACITOR

capacitor in Fig. 13 · 18. The actual capacitance value is 120 μfarads, and the series resistance is 6 ohms. The calculation of midband gain and cutoff frequency for this case are left as an exercise.

The conclusions we can draw from these experimental results are

1. The transistor models which we have used give an adequate first approximation of the transistor at low frequencies, though measurements of the parameter values may be necessary if reasonably exact agreement between theory and experiment is required.

2. The low-frequency cutoff f_l is the frequency at which C_1 has a reactance equal to $r_e + (r_b + R_s)(1 - \alpha)$ *plus* any equivalent series resistance C_1 may have. f_l is usually much higher than the frequency $f_0 = 1/2\pi R_1 C_1$.

3. Electrolytic capacitors commonly have some equivalent series resistance. When these capacitors are used as bypass capacitors in the emitter lead, their series resistance will reduce the midband gain and reduce the lower cutoff frequency. Capacitors made especially for service as bypass capacitors in transistor circuits can be obtained. These capacitors have very low series resistance (~ 1 ohm).

13·3·4   *Low-frequency cutoff due to coupling capacitor C_s.*   Because the source impedance affects the low-frequency cutoff, the effects of the coupling capacitor C_s and the emitter impedance Z_1 interact to some extent. Consequently, the exact low-frequency cutoff cannot be found by considering the effects of C_1 and C_s separately, as we have been implicitly doing. However, if the cutoff frequencies due to the coupling and bias impedances are quite different, the approximate low-frequency cutoff may be found by considering C_s to be infinite, as we have done. Then the coupling-circuit cutoff frequency is found by regarding C_1 as infinite, with C_s finite.

To compute the cutoff frequency due to C_s (with the assumption that $C_1 \rightarrow \infty$) we make the substitution

$$R_s \rightarrow R_s + \frac{1}{sC_s} = \frac{1 + sC_sR_s}{sC_s} \qquad (13·31)$$

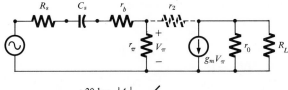

(a) Circuit for determining the low-frequency cut-off due to the coupling capacitor C_s

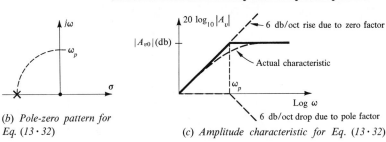

(b) Pole-zero pattern for Eq. (13·32)

(c) Amplitude characteristic for Eq. (13·32)

FIG. 13·19   LOW-FREQUENCY GAIN CHARACTERISTIC DUE TO C_s

in the formula for A_{v0}:

$$A_v(s) = A_{v0} \frac{sC_s(R_s + r_b + r_\pi)}{1 + sC_s(R_s + r_b + r_\pi)} \tag{13·32}$$

Once again the complex roots of A_v exhibit a pole and a zero (see Fig. 13 · 19). The zero is at zero frequency, corresponding to the fact that C_s will not pass dc. The pole is at a frequency

$$s_p = -\frac{1}{C_s} \frac{1}{(R_s + r_b + r_\pi)} \tag{13·33}$$

The circular projection of s_p onto the $j\omega$ axis gives the lower cutoff frequency.[1]

The log $|A_v|$–log ω plot for Eq. (13·32) is similar to the plot for Eq. (13·28), though now the zero is at $s = 0$, instead of on the negative σ axis. This means that the response rises at 6 db/octave from $\omega = 0$ (which is not visible on log-log paper). The pole at s_p gives a compensating 6 db/octave roll-off at the frequency ω_p (shown in Fig. 13 · 19), and the cutoff frequency is just

$$\omega_l = \omega_p$$

If the cutoff frequency thus obtained is considerably different from that computed for the bias impedance Z_1 (say, at least four times), then the cutoff frequency of the entire stage may be taken as the higher of the two cutoff frequencies. If the two cutoff frequencies nearly coincide, curves of the response due to the coupling and bias impedances considered separately may be sketched and added together (on a db scale) to find the approximate cutoff frequency for both circuits taken together. The cutoff frequency (at the 3-db point) thus estimated will be about 20 percent lower than the actual cutoff frequency.

[1] The frequency s_p is called the *natural frequency* of the circuit shown in Fig. 13 · 19. The concept of natural frequencies in linear systems and their relation to amplifier performance is discussed in the Appendix.

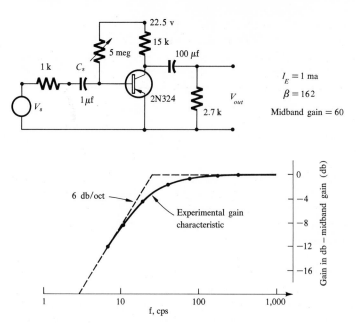

FIG. **13·20** **EXPERIMENTAL DEMONSTRATION OF** C_s **ON THE LOW-FREQUENCY GAIN CHARACTERISTIC**

Experimental verification of results. To check the theory for the low-frequency effects of C_s, the amplifier shown in Fig. 13·20 was built and tested. The measured break frequency was 27.5 cps and the measured midband gain was 60.

The theoretical gain and cutoff frequency depend on the values of r_e which are used. If we use the "ideal" value of 25 ohms at $I_E = 1$ ma, the theoretical gain is 74, and the theoretical cutoff frequency is 36 cps. If we use the measured value for r_e (30 ohms at $I_E = 1$ ma), then the theoretical expressions yield a midband gain of 63 and a cutoff frequency of 26.5 cps, which compare quite favorably with the measured values.

Again, the lesson to be learned from this experiment is that the ideal transistor model gives only an approximation to a real transistor. The approximation is generally valid, though it is useful to check the parameter values if accurate agreement between theory and experiment is required. Frequently only minor modifications, such as changing r_e from 25 mv/I_E to 30 mv/I_E, will make the model surprisingly accurate.

13·3·5 *High-frequency cutoff of a transistor amplifier.* Earlier in this chapter, we indicated that amplifier design usually proceeds in two steps. In the first step, one uses simple models of the transistor to estimate circuit behavior and establish approximate values of the circuit components. In the second step, more accurate models are used to predict the circuit performance more exactly. This design philosophy can be seen in the examples of the previous two sections. A simple transistor model used with ideal resistors and capacitors predicts a generally correct circuit response, though the theoretical values of the midband gain and critical frequencies may be significantly different from measured values. By evaluating the parameters of the transistor model and circuit components more accurately, a reasonable fit between theory and experiment can be obtained.

This same design philosophy will be even more apparent in this section. Very instructive estimates of circuit behavior can be made with simple models, but increasingly complicated models are required if we want to fit experimental results exactly (that is, to within 5 percent or so).

Transistor model. The high-frequency cutoff of the amplifier we are considering is determined primarily by the input capacitance of the transistor. The input capacitance can be estimated with various degrees of refinement, each refinement improving the accuracy of fit between theory and experiment.

As a starting point, consider the hybrid-π model shown in Fig. 13·21. The parameters which arise from intrinsic transistor behavior are drawn in heavy, and several parasitic elements are drawn in dotted. The parasitic elements are real enough, of course, but they complicate our understanding of the basic mechanism which limits high-frequency response. We therefore begin with a model containing only C_π, r_π, and the current generator.

Theoretical circuit response using intrinsic transistor model. The amplifier which we wish to study is shown in Fig. 13·22a. The circuit model is shown in Fig. 13·22b. The voltage and current gains of this amplifier can be obtained by a straightforward analysis. The current flowing in the input loop is

$$I_1 = \frac{V_s}{R_s + Z_\pi} \qquad (13\cdot34a)$$

The control voltage V_π is therefore

$$V_\pi = Z_\pi I_1 = \frac{V_s Z_\pi}{R_s + Z_\pi} \qquad (13\cdot34b)$$

The output current is

$$I_2 = g_m V_\pi \qquad (13\cdot34c)$$

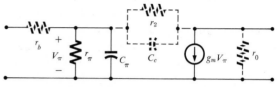

FIG. 13·21 HYBRID-π TRANSISTOR MODEL FOR COMPUTING HIGH-FREQUENCY CUTOFF OF A TRANSISTOR AMPLIFIER. PARASITIC ELEMENTS ARE DOTTED. SOLID LINES REPRESENT CIRCUIT ELEMENTS ARISING FROM INTRINSIC TRANSISTOR ACTION.

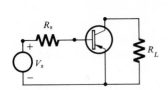 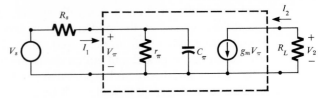

(a) Amplifier with bias components omitted

(b) Model for (a) using only intrinsic elements in the transistor model

FIG. 13·22   CALCULATION OF HIGH-FREQUENCY CUTOFF IN A SIMPLE AMPLIFIER

so the current gain is

$$A_i = \frac{I_2}{I_1} = g_m Z_\pi \tag{13·35}$$

and the voltage gain is

$$A_v = \frac{V_2}{V_s} = \frac{-g_m R_L Z_\pi}{R_s + Z_\pi} \tag{13·36}$$

If we substitute for Z_π the expression

$$Z_\pi = \frac{r_\pi}{1 + s C_\pi r_\pi}$$

and use the relation

$$g_m r_\pi = \beta$$

we find

$$A_i = \frac{\beta}{1 + s C_\pi r_\pi} \tag{13·37}$$

$$A_v = \frac{-\beta R_L}{R_s + r_\pi} \frac{1}{1 + s C_\pi R_s r_\pi / (R_s + r_\pi)} = \frac{A_{v0}}{1 + s C_\pi R_{\|}} \tag{13·38}$$

$R_{\|}$ is the equivalent resistance of R_s and r_π in parallel, and A_{v0} is the midband gain.

Experimental results.   Equation (13·37) provides a very convenient means of checking the accuracy of the model for predicting actual behavior. According to Eq. (13·37), we expect the current gain to be *independent* of load resistance. The midband current gain should be β, and the upper cutoff frequency should be

$$f_u = \frac{1}{2\pi C_\pi r_\pi}$$

The current gain should fall at 6 db/octave for $f \geq 2f_u$.

To check these predictions, the amplifier shown in Fig. 13·23 was built and tested. The 2N324 transistor used had a measured β of 125 at 1 kcps.

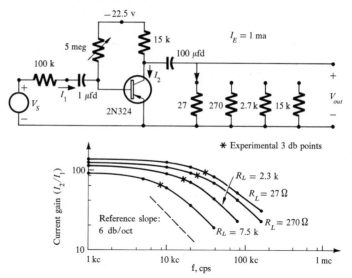

FIG. 13·23 EXPERIMENTAL AMPLIFIER TO CHECK THEORETICAL PREDICTIONS OF CURRENT GAIN AND FREQUENCY RESPONSE

The curves show that the current gain of the transistor in the circuit is indeed 125 for $R_L = 27$ ohms. Let us use this as temporary grounds for assuming that Eq. (13·37) is valid for the case where $R_L = 27$ ohms. Then, the measured 3-db frequency is 34 kcps, corresponding to a $C_\pi r_\pi$ time constant of

$$C_\pi r_\pi = \frac{1}{2\pi f_u} = \frac{10^3}{2\pi \times 34} = 4.7 \ \mu\text{sec}$$

According to the simple intrinsic transistor theory, this time constant is the lifetime of minority carriers in the base region.

To evaluate C_π and r_π individually, we recall that

$$r_\pi = (\beta + 1)r_e$$

and that for the 2N324, $r_e \simeq 30/|I_E(\text{ma})|$. At a bias current of 1 ma, then, $r_e = 30$ ohms and $r_\pi \simeq 3.8$ kohms. Hence, $C_\pi = 1,250$ pfarads.

Unfortunately, all the curves do not exhibit the same 3-db frequency nor the same midband gain. The midband gain decreases by about 25 percent as R_L is increased from 27 ohms to 7.5 kohms, while the cutoff frequency *decreases by a factor of about 4* over the same range of R_L. It is clear that the most significant departure from the simple theory is in the cutoff-frequency prediction. Hence, we need to look again at the transistor model to see whether modifications can be made to account for this discrepancy.

But first, we shall indicate how values of C_π and r_π can be estimated from the transistor data sheet.[1] The manufacturer quotes the value of β, though

---

[1] Similar calculations are given in Sec. 9·5·3, where the relation between the hybrid-π model and the physical theory of the transistor is developed.

not usually the $C_\pi r_\pi$ time constant. Instead, he will quote the *frequency at which the short circuit CE current gain is unity*. This frequency is denoted by f_T. To the extent that Eq. (13 · 37) correctly represents the current gain, f_T should (for reasonably large β) be given by

$$2\pi f_T C_\pi r_\pi = \beta$$

or

$$C_\pi r_\pi = \frac{\beta}{2\pi f_T}$$

Since $r_\pi \simeq \beta r_e$,

$$C_\pi = \frac{1}{2\pi f_T r_e}$$

For a 2N324, the typical data show $f_T \simeq 4.0$ mc. Using our knowledge that $r_e = 30$ ohms at $I_E = 1$ ma (which needs to be verified by experiment) we find

$$C_\pi \text{ (typ)} = 1,325 \text{ pfarads}$$

The transistor model again. Let us now return to our problem of understanding why the measured cutoff frequency for A_i is so drastically in error, by considering the possible effect of the parasitic elements in the transistor model of Fig. 13 · 21. The most obvious choice for explaining

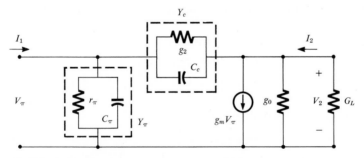

(a) *Hybrid-π model with g_c, C_c and g_0 parasites added*

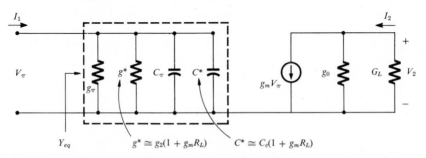

(b) *Circuit model equivalent to (a) if $g_m \gg Y_c$ and $G_L \gg Y_c$*

FIG. 13 · 24 CIRCUIT MODEL FOR CALCULATING THE EFFECT OF C_c AND r_2 ON THE CURRENT GAIN OF A CE STAGE

erroneous frequency dependence would be the capacitor C_c. This capacitor couples the input of the amplifier to the output. Of course, $C_c \ll C_\pi$, but also the output voltage of the amplifier is much greater than the control voltage V_π. Hence, *the signal voltage applied to C_c will be much greater than the signal voltage applied to C_π*. The actual signal current supplied to C_c may therefore be comparable to that flowing in C_π.

To study the effects of r_2 and C_c on the performance of the transistor, we shall determine the input admittance of the transistor model at the terminals shown in Fig. 13·24. We apply a voltage V_π to these terminals and calculate the input current. The result is

$$I_1 = (Y_\pi + Y_c)V_\pi - V_2 Y_c \qquad (13 \cdot 39a)$$

where $$Y_\pi = g_\pi + sC_\pi \qquad Y_c = g_2 + sC_c$$

The voltage V_2 can be determined by writing Kirchhoff's law at the output node:

$$0 = (g_m - Y_c)V_\pi + (Y_c + G_L + g_0)V_2 \qquad (13 \cdot 39b)$$

Now, in a typical transistor, g_m is on the order of 40×10^{-3} mhos, and a typical load will be $G_L \sim 10^{-3}$ mhos. By contrast, g_2 is about 10^{-7} mhos, C_c is between 2 and 20 pfarads, and $g_0 \sim 5 \times 10^{-5}$ mhos. Therefore, at audio frequencies (and well beyond), Eq. (13·39b) can be well approximated by

$$-g_m V_\pi = G_L V_2$$

When this result is used in Eq. (13·39a), we obtain

$$I_1 = (Y_\pi + Y_c + g_m R_L Y_c)V_\pi \qquad (13 \cdot 40a)$$

or the equivalent input admittance is

$$Y_{eq} = Y_\pi + (1 + g_m R_L)Y_c = Y_\pi + Y^* \qquad (13 \cdot 40b)$$

This equation shows that, even though the g_2 and C_c elements are themselves quite small, their effect at the input may be quite sizeable, because they appear multiplied by $(1 + g_m R_L)$. Furthermore, this effect looks promising for explaining the experimental data, because the portion of the input admittance which is due to Y_c also depends on R_L.

The result given in Eq. (13·40b) can be visualized in the manner shown in Fig. 13·24b. The transistor is again represented by a simple π model, but the element $g_2(1 + g_m R_L)$ appears in parallel with g_π and $(1 + g_m R_L)C_c$ appears in parallel with C_π.

The procedure which we have just used to account for the effects of an admittance connecting the input and output of an amplifier was first used by Miller in studying the effect of grid-plate capacitance in a triode on frequency response. The admittance $Y_c(1 + g_m R_L)$ is frequently called the *Miller admittance,* for convenience. In particular, the capacitance $C_c(1 + g_m R_L)$ is called the *Miller capacitance* and sometimes is denoted by C_m.

Let us now investigate the modifications which these new elements produce in our previous gain expressions. If we use the modified transistor model shown in Fig. $13 \cdot 24b$, the current gain should be

$$A_i = \frac{g_m V_\pi}{I_1} \frac{G_L}{g_0 + G_L} = g_m Y_{eq} \frac{G_L}{g_0 + G_L} \qquad (13 \cdot 41a)$$

Using Eq. $(13 \cdot 40b)$ in Eq. $(13 \cdot 41a)$, we obtain

$$A_i = \frac{\beta G_L}{g_0 + G_L} \frac{1}{1 + g_2 r_\pi (1 + g_m R_L^*) + S[C_\pi r_\pi + C_c(r_\pi + \beta R_L^*)]} \qquad (13 \cdot 41b)$$

where $R_L^* = 1/(g_0 + G_L)$. For the moment let us neglect the $g_2 r_\pi (1 + g_m R_L^*)$ term. The current gain then becomes

$$A_i = \frac{\beta}{1 + sC_\pi r_\pi [1 + C_c(1 + g_m R_L)/C_\pi]} \frac{G_L}{g_0 + G_L}$$

The typical data on the 2N324 indicate that $C_c \simeq 18$ pfarads. If we use this value together with $C_\pi = 1{,}250$ pfarads, $g_m = 40 \times 10^{-3}$ mhos, $g_0 = 5 \times 10^{-5}$ mho then we can construct the following table:

R_L, ohms	270	2.2 kohms	7.5 kohms
$1 + \dfrac{C_c(1 + g_m R_L)}{C_\pi}$	1.17	2.2	4.2
f_u, kcps	29	15.5	8.1
A_{i0}	123	114	98

Returning to Fig. $13 \cdot 23$, we see that addition of g_0 and the capacitor $(1 + g_m R_L)C_c$ explain the dominant features of the variation of f_u and A_{i0} with R_L. The cutoff frequencies are still not exact, but they are now off by about the same percentage as the midband gain.

A closer approximation can be obtained by including the effect of g_2. The typical data do not provide the value of g_2, so we have to use the data in Fig. $13 \cdot 23$ to select a value of g_2. We can do this by observing that the decrease in midband current gain, when $R_L^* = 5.5$ k, can be used to evaluate g_2.

$$A_{i0}(R_L^* = 5.5 \text{ kohms}) = \frac{A_{i0}(R_L = 27 \text{ ohms})}{1 + g_2 r_\pi (1 + g_m R_L^*)}$$

When this value of g_2 is used, the other values of midband A_i can be calculated to check for consistency. Furthermore, the presence of g_2 will reduce G_{eq} and will therefore raise the cutoff frequencies slightly. The resulting modifications give a satisfactory agreement between the experimental and theoretical results.

Factors affecting the upper cutoff frequency for A_v. The modifications which need to be made in the voltage-gain formula [Eq. (13 · 38)] can be inferred from the previous discussion. The cutoff frequency for A_v in Eq. (13 · 38) is related to the basic $C_\pi R_\parallel$ time constant in the input circuit. The actual input capacitance is not C_π, however, but

$$C_{eq} = C_\pi\left[1 + \frac{C_c(1 + g_m R_L)}{C_\pi}\right]$$

Also, $R_\parallel$ is not the equivalent resistance of r_π and R_s in parallel, but that of r_π, r^*, and R_s in parallel (r^* is defined on Fig. 13 · 24b). The resulting 3-db frequency is

$$f_u = \frac{1}{2\pi C_{eq} R_\parallel}$$

The principal point of interest in this expression is the presence of R_s in determining $R_\parallel$. If R_s is large compared to r_π, the voltage gain will exhibit the same 3-db point as the current gain. However, if R_s is comparable to or smaller than r_π, the 3-db frequency will be increased. Thus the bandwidth of the stage can be increased by driving it from a low-impedance source. When we want to build a wideband amplifier, we shall usually find it necessary to ensure that the source impedance driving a transistor is low.

13 · 3 · 6 Summary and generalization. Before proceeding to a discussion of multistage amplifiers, it might be well to summarize the ideas of the past several sections and indicate their generalizations. First of all, the general amplifier-gain function for the amplifier shown in Fig. 13 · 9 will be of the form

$$A_v = K\frac{s(1 + s/\omega_{0,1})}{(1 + s/\omega_{p,1})(1 + s/\omega_{p,2})(1 + s/\omega_{p,3})} \tag{13 · 42}$$

where the $\omega_{0,1}$, $\omega_{p,1}$, $\omega_{p,2}$, and $\omega_{p,3}$ may have different values from those given in the text, due to interactions which were neglected there.

The gain function given in Eq. (13 · 42) has two zeros and three poles in the finite s plane, and one zero at $s = \infty$. The two finite zeros and two of the poles are associated with the presence of C_1 and C_s. The remaining pole arises from the presence of the input capacitance C_{eq}.

A $\log |A_v|$–$\log \omega$ plot can be constructed in the manner shown in Fig. 13 · 25.[1] The contributions from each zero and pole may be added together to produce the final asymptotic characteristic. The actual characteristic may be drawn from these asymptotic characteristics by using a table of asymptotic departures, when the exact values of $\omega_{0,1}$, $\omega_{p,1}$, $\omega_{p,2}$, and $\omega_{p,3}$ are known.

The actual characteristic will exhibit a constant gain over some frequency range and will then drop at both high and low frequencies. The points at

[1] This plot is drawn with the assumption that the break frequency due to C_s is the lowest critical frequency.

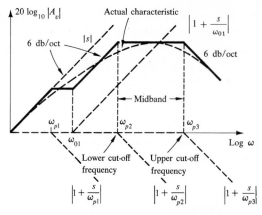

FIG. 13 · 25   AMPLITUDE CHARACTERISTIC FOR EQ. (13 · 42)

which the characteristic begins to drop are specified by quoting the cutoff frequencies for the amplifier. The cutoff frequencies are, by convention, equal to the frequencies where $|A_v|$ is 3 db down from its midband value. The phase shift at these frequencies will be about 45° in magnitude (relative to the midband phase angle).

The low-frequency cutoff will be determined primarily by circuit capacitances which are external to the transistor (C_s and C_1). By selection of appropriate *capacitance values,* the low-frequency cutoff can be placed at essentially any frequency required, with no effect on the level of midband gain.

The capacitance which enters into the determination of the high-frequency cutoff is internal to the transistor. Adjustment of the high-frequency cutoff is then usually accomplished by selection of appropriate *circuit resistances,* which *do* affect the level of the midband gain. It is important to recognize that the transistor itself places a limitation on the gain we can obtain when the upper cutoff frequency is specified, or vice versa. We shall develop the nature of this limitation more fully as the subject unfolds.

13 · 4   CASCADED (OR MULTISTAGE) AMPLIFIER CIRCUITS

In many practical situations, the gain requirements of an amplifier cannot be met with a single stage of amplification. When this occurs one usually considers the possibility of cascading amplifier stages to meet the design requirements. The design process in such cases involves a combination of common sense, experience, and analytical ability. There is no valid general prescription for designing multistage amplifiers, though there are some general rules that are of great value if they are properly understood and not applied too strictly. We shall develop some of the general ideas in this section, using simple models for the transistors.

13·4·1 *Functions of the stages in a multistage amplifier.* It is customary to consider a multistage amplifier as being composed of three parts: an input stage, an output stage, and a cascade of intermediate stages. An amplifier is separated into these parts to simplify and organize the design problems.

We shall refer to the intermediate cascade as the "main amplifier." Transistors will ordinarily be connected in the *CE* configuration in the main amplifier. The gain and bandwidth of the overall amplifier are frequently determined by the gain and bandwidth of the individual stages in the intermediate cascade. Accordingly, the number of intermediate stages and their design are usually based on the gain and bandwidth specifications required in the final amplifier.

The input stage is designed to properly couple the signal source to the main amplifier. If we are interested in obtaining a large transducer gain, we may design the input stage so that its input impedance approximately matches that of the source, if possible. Alternately, if we are interested in obtaining maximum current from the source, we design the input stage to have a low impedance compared to that of the source. Similarly, the output stage is designed to couple the main amplifier to the load. Again, we may be interested in matching the amplifier to the load or we may purposely mismatch to obtain some desirable properties in the amplifier. Of course, in addition to the impedance characteristics of the input and output stages, we must also consider their effects on the overall gain and bandwidth of the amplifier.

13·4·2 *Gain and bandwidth of cascaded stages.* The bandwidth of a cascade of stages will be less than that of the individual stages in the cascade. This bandwidth shrinkage is the price that has to be paid to increase the overall gain; in fact, for a given type of amplifying element there is a limit on the number of gain-producing stages which can be cascaded for a given overall bandwidth.

To develop the properties of a cascade amplifier chain, we shall consider the typical stage shown in Fig. 13·26. The interstage resistance R_i is the parallel combination of r_0, R_L, and the base bias resistors R_1 and R_2. The load resistance for the stage will be the parallel combination of R_i and r_π.[1]

The midband voltage gain of the stage, V_2/V_1, may be written as

$$A_{v0} = \frac{V_2}{V_1} = -g_m(R_i \parallel r_\pi) \tag{13·43}$$

The equivalent input capacitance of the typical stage is

$$C_{eq} = C_\pi + [1 + g_m(R_i \parallel r_\pi)]C_c \tag{13·44a}$$

It is useful to note that the equivalent input capacitance of the stage can

---

[1] We neglect r_b in this analysis, though we shall include it in Chap. 18 when we study more sophisticated amplifier design techniques.

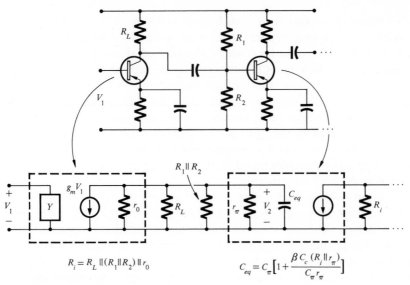

$$R_i = R_L \parallel (R_1 \parallel R_2) \parallel r_0$$

$$C_{eq} = C_\pi \left[1 + \frac{\beta C_c (R_i \parallel r_\pi)}{C_\pi r_\pi}\right]$$

(a) *Circuit model for the typical interstage*

FIG. 13·26 THE TYPICAL INTERSTAGE IN A CE CASCADE AMPLIFIER CHAIN

be written in terms of A_{v0} as

$$C_{eq} = C_\pi + (1 + |A_{v0}|)C_c$$

Since we are usually interested in situations where $|A_{v0}| \gg 1$, C_{eq} can be written in the two useful forms

$$C_{eq} = C_\pi + |A_{v0}|C_c \tag{13·44b}$$

$$C_{eq} = C_\pi + \beta C_c \frac{R_i}{R_i + r_\pi} \tag{13·44c}$$

The voltage gain for the typical stage shown in Fig. 13·25b, at frequencies $s = j\omega$ which are in or above the midband range, may be written as

$$A_v(s) = \frac{V_2(s)}{V_1(s)} = \frac{A_{v0}}{1 + s/s_p} \tag{13·45a}$$

where

$$s_p = -\frac{1}{C_{eq}} \frac{R_i + r_\pi}{R_i r_\pi}$$

The corresponding upper cutoff frequency is

$$f_u = \left| \frac{s_p}{2\pi} \right| = \frac{R_i + r_\pi}{2\pi C_{eq} R_i r_\pi} \tag{13·45b}$$

The actual bandwidth of the typical stage will be

$$B = f_u - f_l$$

But since $f_l \ll f_u$ for the amplifier we are considering, the f_u given in Eq. $(13 \cdot 45b)$ is loosely called the bandwidth of the typical stage.

From Eqs. $(13 \cdot 43)$ and $(13 \cdot 45b)$, we see that the midband gain of the stage is proportional to the factor $R_i r_\pi / (R_i + r_\pi)$, while the stage bandwidth is inversely proportional to this factor. It is convenient to express this fact in the form of a *gain-bandwidth product* for the stage. We have

$$|A_{v0}| |B| \approx \frac{g_m}{2\pi C_{eq}} \qquad (13 \cdot 46a)$$

If we multiply numerator and denominator of Eq. $(13 \cdot 46a)$ by r_π and use the relation

$$f_T = \frac{\beta}{2\pi r_\pi C_\pi}$$

we find
$$|A_{v0}| |B| = f_T \left(\frac{C_\pi}{C_{eq}} \right) < f_T \qquad (13 \cdot 46b)$$

Since $C_\pi \le C_{eq}$, we see that f_T is an upper bound on the gain-bandwidth product for the stage.

Some examples will illustrate the significance of Eqs. $(13 \cdot 46)$ in studying the gain and bandwidth of the typical stage in a multistage amplifier.

1 MAXIMUM MIDBAND GAIN The midband gain of the typical stage is, from Eq. $(13 \cdot 43)$,

$$A_{v0} = -g_m \frac{R_i r_\pi}{R_i + r_\pi}$$

To maximize A_{v0}, we want to use $R_i \gg r_\pi$. This then gives

$$|A_{v0}| = g_m r_\pi = \beta$$

The gain-bandwidth product can be no greater than f_T, so the bandwidth (or, more properly, the upper cutoff frequency) of the typical stage with $R_i \gg r_\pi$ can be no greater than

$$f_u \le \frac{f_T}{\beta}$$

For example, if a transistor has $f_T = 4$ mcps and $\beta = 100$, then in a cascade of identical stages where $R_i \gg r_\pi$, the midband gain of each stage will be 100 and the bandwidth of each stage can be no larger than 40 kcps. If a typical silicon planar transistor having $\beta = 50$ and $f_T = 750$ mcps is used in a cascade of identical stages (with $R_i \gg r_\pi$), the midband gain per stage will be 50 and the stage bandwidth will be no greater than 15 mcps.

These upper bounds on stage bandwidth for maximum gain will prove to be useful for guiding our thinking about multistage amplifiers. However, the upper bounds frequently give a very optimistic estimate of stage bandwidth, and at some point in the design process we shall want to refine our estimate of the stage bandwidth. We may do this by calculating C_π and C_{eq} and using Eq. $(13 \cdot 46b)$ to estimate the actual gain-bandwidth product. To evaluate C_π and C_{eq}, we must know β, f_T, and C_c for the transistor we are going to use, plus the load resistance on each stage ($R_i \| r_\pi$). For the moment, we shall assume that $R_i \gg r_\pi$, as before.

If we are using a transistor with $\beta = 100, f_T = 4$ mcps, and $C_c = 20$ pfarad, then

$$C_\pi = \frac{\beta}{2\pi r_\pi f_T} \simeq 1{,}300 \text{ pfarads}$$

Using Eq. $(13 \cdot 43c)$, with $R_i \gg r_\pi$, gives

$$C_{eq} = C_\pi + \beta C_c = 1{,}300 + 2{,}000 \text{ pfarads}$$

The gain bandwidth product is then

$$|A_{vo}|\,|B| = f_T\left(\frac{1{,}300}{3{,}300}\right) \cong 0.4\,f_T$$

The typical stage in a multistage amplifier with $R_i \gg r_\pi$ will then give a gain of 100 and a bandwidth of 0.4×4 mcps/100 = 16 kcps. The upper bound on the bandwidth previously estimated was 40 kcps which is a factor of 4 too high.

The difference between the actual gain-bandwidth product and its upper bound is even more pronounced in transistors intended for very high-frequency amplifier service, where C_c may dominate the equivalent input capacitance. For example, if a typical silicon planar transistor with $\beta = 50, f_T = 750$ mcps, and $C_c = 1$ pfarad is used in a multistage amplifier, with $R_i \gg r_\pi$, then the equivalent input capacitance of a stage is

$$C_{eq} = C_\pi + \beta C_c = 8 + 50 = 58 \text{ pfarads}$$

Notice that most of the input capacitance for this case is due to the Miller effect (that is, βC_c).

The gain-bandwidth product is

$$|A_{vo}|\,|B| = f_T(\%_8) \cong 107 \text{ mcps}$$

for this transistor, assuming $R_i \gg r_\pi$. This means that the typical stage will exhibit a gain of 50 and a bandwidth of about 2 mcps. The upper bound on the stage bandwidth is $(f_T/\beta) = 15$ mcps.

It appears from these examples that the upper bound on stage bandwidth may give a quantitatively misleading estimate of the actual stage bandwidth. This is true when both the gain per stage and C_c are high, so that the Miller capacitance $(1 + |A_{vo}|)C_c$ is a large part of the equivalent input capacitance. However, if we are willing to settle for less gain per stage, we can reduce the importance of the Miller capacitance and use the transistor more efficiently; that is, build an amplifier in which the gain-bandwidth product for each stage more nearly approaches the upper bound f_T. This is usually what we must do when we want to build an amplifier with a large bandwidth.

2 WIDEBAND AMPLIFIERS To study the problem of building a wideband amplifier stage, it is convenient to combine Eqs. $(13 \cdot 44b)$ and $(13 \cdot 46)$ to write the gain-bandwidth product in the form

$$|A_{vo}|\,|B| = \frac{f_T}{1 + A_{vo}C_c/C_\pi}$$

For the transistor with $C_c = 20$ pfarads, $f_T = 4$ mcps, $\beta = 100$, we find $C_c/C_\pi \simeq \frac{1}{60}$ at $I_E = 1$ ma. The gain-bandwidth product is therefore

$$|A_{vo}|\,|B| = \frac{f_T}{1 + A_{vo}/60}$$

If we use $R_i \gg r_\pi$, so $|A_{v0}| = \beta$, the gain-bandwidth product will have the value $f_T/2.4$. However, if we can accept a midband gain of only 10 per stage, the gain-bandwidth product increases to $f_T/1.16$. The actual stage bandwidth for a midband gain of 10 would then be 344 kcps.

For the transistor with $C_c = 1$ pf, $\beta = 50$, and $f_T = 750$ mcps, the gain bandwidth product goes from $f_T/7$ when $A_{v0} = 50$ (that is, $R_i \gg r_\pi$) to $f_T/2.1$ when $A_{v0} = 10$. Again, we are using the gain-bandwidth capabilities of the transistor more efficiently by accepting a lower gain per stage.

Bandwidth shrinkage in a cascade. Let us now turn to the question of the overall gain and bandwidth of a cascade of identical stages. To begin, we shall write the total voltage gain of a cascade of n stages as

$$A_{v,T}(s) = \frac{V_{n+1}(s)}{V_1(s)} = \frac{V_2(s)}{V_1(s)} \frac{V_3(s)}{V_2(s)} \frac{V_4(s)}{V_3(s)} \cdots \frac{V_{n+1}(s)}{V_n(s)}$$

$$= A_1(s) A_2(s) A_3(s) \cdots A_n(s) \qquad (13 \cdot 47)$$

where $A_{v,T}(s)$ is the gain of the cascade and $V_i(s)$ is the small-signal input voltage measured from base-to-emitter terminals of the ith transistor.

The voltage gain of the ith stage is approximately

$$A_{v,i}(s) = \frac{V_{i+1}(s)}{V_i(s)} = \frac{A'_{v0}}{1 + s/s'_p} \qquad (13 \cdot 48)$$

where A'_{v0} is the midband gain of a single stage in the cascade and $|s'_p/2\pi| = f'_u$ is the upper cutoff frequency for the single stage.

The gain of a cascade of identical stages can then be written as

$$A_{v,T}(s) = \frac{(A'_{v0})^n}{(1 + s/s'_p)^n} \qquad (13 \cdot 49)$$

from which we see that the midband gain of the cascade is $(A'_{v0})^n$.

The bandwidth of the cascade can be conveniently studied by normalizing the gain to its midband value:

$$\frac{A_{v,T}(s)}{(A'_{v0})^n} = \frac{1}{(1 + s/s'_p)^n} \qquad (13 \cdot 50)$$

Expressing the normalized gain on a decibel basis, we have

$$20 \log_{10} \left| \frac{A_v(j\omega)}{(A'_{v0})^n} \right| = -20 \log_{10} \left| 1 + \frac{j\omega}{s'_p} \right|^n = -20 \log_{10} \left| 1 + \left(\frac{f}{f'_u} \right)^2 \right|^{n/2}$$

$$= -10n \log_{10} \left| 1 + \left(\frac{f}{f'_u} \right)^2 \right| \qquad (13 \cdot 51)$$

where f'_u is the upper cutoff frequency of a single stage in the cascade of identical stages.

The asymptotes of Eq. $(13 \cdot 51)$ are shown in Fig. $13 \cdot 27$ for various values of n. The breakpoint in all cases is at $|f/f'_u| = 1$, the asymptotic departure at this point being $-3n$ db. The slope of the characteristic at high frequencies is $-6n$ db/octave.

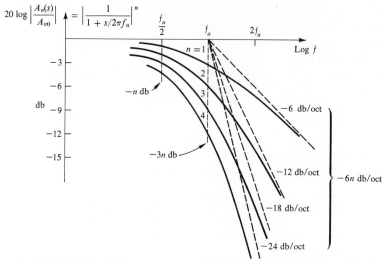

FIG. 13·27 HIGH-FREQUENCY AMPLITUDE CHARACTERISTIC FOR A CASCADE OF IDENTICAL AMPLIFIER STAGES

To find the cutoff frequency for the cascade, we have to find the frequency at which Eq. (13·51) equals -3 db. This is

$$1 + \left(\frac{f_u}{f_u'}\right)^2 = 10^{3/10n} = 2^{1/n}$$

or

$$\frac{f_u}{f_u'} = \sqrt{2^{1/n} - 1} \qquad (13·52)$$

The shrinkage in bandwidth arising from the cascading is apparent when we substitute various values for n into Eq. (13·52). The results are

n	1	2	3	4	5	6
f_u/f_u'	1.0	0.64	0.51	0.44	0.39	0.35

The use of this table should not be extended beyond the point where $f_u' \sim f_T/2$.

13·4·3 *Applications of gain-bandwidth and bandwidth-shrinkage concepts in multistage-amplifier design.* The gain-bandwidth expression given in Eq. (13·46) are the bandwidth-shrinkage table just developed provide two extremely powerful tools for thinking about multistage-amplifier problems. If the gain and bandwidth required in an amplifier are known, we can use these two concepts to determine the number of stages and the type of transistor required. If the midband gain and transistor type are specified, we can determine the amplifier bandwidth as a function of the number of stages used. We can also obtain approximate values for the interstage resistance in a given application. We shall consider several numerical examples to illustrate the application of these results.

1 **MIDBAND GAIN, BANDWIDTH, AND TRANSISTOR SPECIFIED** Suppose we are asked to design an amplifier with a midband voltage gain of 2,000 and an upper cutoff frequency of 100 kcps. We are given transistors having $\beta = 100$, $f_T = 4$ mcps, and $C_c = 20$ pfarads (2N324's). We must decide whether these transistors will do the job, and if they will, how many transistors we need and what the interstage resistance R_i should be.

First approximation. We shall proceed by a series of successive approximations. In the first approximation, we assume that each stage in the cascade will have a gain-bandwidth product of f_T. To begin, note that the midband gain of each stage in the cascade will be [Eq. (13·43)]

$$A_{v0} = -g_m(R_i \parallel r_\pi) = -\beta \left(\frac{R_i \parallel r_\pi}{r_\pi} \right)$$

The second form of this equation shows that A_{v0} can be no larger than β if we really use a cascade of identical stages. If we select $R_i \gg r_\pi$, the midband gain of each stage will be β. Therefore, we need at least two stages to get a midband gain of 2,000, if $\beta = 100$.

Let us consider a two-stage amplifier. To get a gain of 2,000, we adjust R_i to give a midband gain in each stage of $\sqrt{2,000} = 45$. The previous equation for A_{v0} shows that the required value of R_i is about $0.8r_\pi$. Then what will the bandwidth of the amplifier be? The gain-bandwidth product for each stage is no greater than f_T, so each stage will have a bandwidth of no more than

$$f'_u < \frac{f_T}{A_{v0}} = \frac{4 \times 10^6}{45} \cong 90 \text{ kcps}$$

The bandwidth-shrinkage table shows us that the two-stage amplifier would have a bandwidth of

$$f_u = 0.64 f'_u < 60 \text{ kcps}$$

Hence, we cannot get a midband gain of 2,000 and a bandwidth of 100 kcps using two identical stages.

To be sure, we need not use identical stages. Perhaps we could use a larger load resistor in the last stage and thereby obtain more midband gain. This would allow us to reduce the gain in the first stage and increase its bandwidth. However, this "solution" isn't as simple as it sounds, because increasing the load resistance of the last stage reduces the actual gain-bandwidth product of the stage. Hence, we are interested in exploring any other possibilities that are open to us.

There are two things we can do. We can return to the customer and say, "Get a transistor with $f_T = 8$ mcps or so and we can *perhaps* do the job with two transistors." If his reply is, "No, I get these transistors from a friend . . . ," then we must consider increasing the number of stages in the amplifier. Increasing the number of stages will reduce the A_{v0} required in each stage and thereby give us some additional bandwidth in each stage.

To pursue this approach, we consider a cascade of three identical stages. The midband gain of each stage needs to be

$$\sqrt[3]{2000} \simeq 13$$

Using $f_T = 4$ mcps, each stage would then have a bandwidth of

$$f_u \leq \frac{4 \times 10^6}{13} \simeq 300 \text{ kcps}$$

The bandwidth shrinkage table shows that the maximum amplifier bandwidth would then be about 150 kcps. The three-stage cascade therefore looks promising, and a further evaluation of it seems warranted.

Second approximation. In the second approximation, we shall improve our estimate of the gain-bandwidth product we should really expect. To do this, we note that a midband gain of 13 requires an interstage resistance of

$$\frac{R_i \parallel r_\pi}{r_\pi} = \frac{13}{\beta} \simeq \frac{1}{7} \Rightarrow R_i \parallel r_\pi \simeq \frac{r_\pi}{7}$$

The total input capacitance of each stage is

$$C_{eq} = C_\pi + C_c[1 + g_m(R_i \parallel r_\pi)] \simeq C_\pi + \frac{\beta C_c}{7}$$

If we arbitrarily choose a bias point[1] of $I_E = 1$ ma, then $C_\pi \simeq 1{,}300$ pfarads, and

$$C_{eq} = 1{,}300 + 288 \sim 1{,}600 \text{ pfarads}$$

The gain-bandwidth product for each stage should therefore be

$$|A_{v0}| \; |f_u| = \frac{f_T}{1 + 288/1{,}300} = 0.82 f_T$$

With a midband gain of 13, we can therefore expect a stage bandwidth of

$$f_u = \frac{0.82 f_T}{13} \simeq 250 \text{ kcps}$$

The three-stage amplifier will then have a bandwidth of about 125 kcps.

This is probably sufficiently close to the required bandwidth that further calculation is not necessary or justified. We need some margin of safety to account for variations in transistor parameters, etc., and the transistor model we have used is not much more accurate than this. In a practical situation we could therefore proceed to design biasing networks to obtain $I_E = 1$ ma for each transistor. We would then choose collector load resistors so that the total resistive load on each collector is

$$R_i \parallel r_\pi = \frac{r_\pi}{7} \simeq \frac{2{,}500 \text{ ohms}}{7} \simeq 360 \text{ ohms}$$

and use a source impedance of $R_i = 420$ ohms.[2] Bypass and coupling capacitors are then selected to give the desired low-frequency response. Experimental adjustments of the amplifier will be necessary if the theoretical design work is terminated at this point.

2 MIDBAND GAIN AND BANDWIDTH SPECIFIED Frequently, we are faced with the problem of choosing a transistor we can use to build an amplifier with given specifications. Generally, choosing a transistor means specifying β, f_T, and C_c. In such a case, we can proceed as follows. If the required midband gain is A_0, then the gain of each stage is an *n*-stage cascade will be

$$A_{v0} = A_0^{1/n}$$

We shall need enough stages to ensure that $A_{v0} \leq \beta$. Hence

---

[1] There is usually some optimum bias point at which the gain-bandwidth product is a maximum, so the selection of a bias point will not be entirely arbitrary in a real case.

[2] This is low enough to indicate that the effects of r_b should be included in further analysis.

$$A_{vo}{}^{1/n} \leq \beta \qquad n \geq \frac{A_{vo}(\mathrm{db})}{\beta(\mathrm{db})}$$

If we want to build an amplifier with a midband gain of 10,000 and are contemplating using transistors with $\beta = 100$, then we need at least two identical stages.

To choose the f_T which we need in the transistor, we need to know the required amplifier bandwidth. There will be several possibilities for f_T, depending on how many stages we are willing to use. For example, if we want an overall amplifier bandwidth of 1 mcps and we want to use the minimum number of stages (two), then the stage bandwidth must be 1 mcps/0.64 $\simeq$ 1.6 mcps. The gain-bandwidth product for the stage is then 160 mcps. However, the transistor will need to have an f_T much larger than this (perhaps nearly a factor of 10 larger), because, as we saw earlier, the gain-bandwidth product of a stage is significantly less than f_T when we attempt to get a gain of β per stage (that is, use $R_i \gg r_\pi$). The search for a suitable transistor would, therefore, begin by finding types which have $\beta \simeq 100$, $f_T \simeq 1000$ mcps. When such a transistor is found, its C_c is used to refine the gain-bandwidth estimate to see whether it will really do the job.

If we are willing to use four stages, then the gain per stage drops to 10, and the stage bandwidth rises to 1 mcps/0.4 $\simeq$ 2.5 mcps. The gain-bandwidth product for each stage is now 25 mcps instead of 160 mcps. Furthermore, if we specify $\beta = 100$, then the gain per stage is only $\beta/10$, so the gain-bandwidth product of each stage will more nearly approach f_T. We can now begin looking for available transistors with f_T in the 50 mcps range and then refine the gain-bandwidth estimate when C_c of a potentially appropriate type is known.

3 MULTISTAGE AMPLIFIERS EMPLOYING TRIODE VACUUM TUBES The most useful small-signal circuit model for a vacuum tube (either triode or pentode) can be obtained by simplifying the hybrid-π transistor model. As a result, all of the foregoing theory can be applied in the design of multistage amplifiers which employ vacuum tube triodes. To specialize the hybrid-π model, we simply make the following identifications:

$$r_\pi \to \infty \qquad r_2 \to \infty \qquad g_m \to g_m \qquad r_0 \to r_p \parallel C_{pk}$$
$$C_\pi \to C_{gk} \qquad C_c \to C_{gp} \qquad r_b \to 0$$

The gain-bandwidth product is

$$|A_{vo}|\,|B| \simeq \frac{g_m}{2\pi C_{eq}}$$

and C_{eq} can be written as

$$C_{eq} = C_{gk} + C_{pk} + C_{gp}(1 + g_m R_L)$$

The capacitance C_{gp} couples the input and output circuits exactly as before, and its effect on the input is magnified by the same factor as before.

Because the interelectrode capacitances are all roughly the same, we cannot neglect the $g_m R_L$ term in a particularly meaningful first approximation. An example will illustrate this point.

A 12AX7 has $C_{gp} = 1.7$ pfarads, $C_{gk} = 1.6$ pfarads, $C_{pk} = 0.5$ pfarad. At an operating current of 0.5 ma, $r_p \simeq 80$ kohms, and $g_m = 1.2 \times 10^{-3}$ mhos. What bandwidth can we expect in a cascade of two identical stages with this tube, if the desired midband gain is 2,000? To get a midband gain of 2,000 requires a midband gain of 45 in each stage. The interstage load resistance is then

$$R_i = \frac{A_{v0}}{g_m} = \frac{45}{1.2} \text{ kohms} = 37.5 \text{ kohms}$$

Note that, since g_m is about a factor of 40 lower for this tube than for a transistor (at $I_E = 1$ ma) R_i is about a factor of 40 higher for the same gain. In other words, vacuum tube circuits typically operate at higher impedance levels than transistor circuits.

To obtain the amplifier bandwidth, we first estimate C_{eq}.

$$C_{eq} = 1.6 + 0.5 + 1.7\,(46) \simeq 80 \text{ pfarads}$$

The gain-bandwidth product for each stage is

$$|A_{v0}|\,|B| \simeq \frac{1.2 \times 10^{-3}}{2\pi \times 80 \times 10^{-12}} = 2.4 \times 10^6$$

(compare this with the f_T of a 2N324, a transistor intended for roughly similar applications). The stage bandwidth will therefore be

$$f_u' = \frac{2.4 \times 10^6}{45} = 53.3 \text{ kcps}$$

and the bandwidth of the two-stage cascade will be

$$f_u = 0.64\,f_u' \simeq 34 \text{ kcps}$$

Here again, as in the transistor, the capacitance which couples the input to the output, C_{gp}, has a tremendous influence on C_{eq} and on the gain-bandwidth product.

4 MULTISTAGE AMPLIFIERS EMPLOYING PENTODE VACUUM TUBES The invention of the pentode arose from the problems created by grid-plate capacitance in a triode. The pentode structure provides a means of virtually eliminating C_{gp} (cf. Sec. 10·3) and therefore eliminating the effect of the gain of a stage on its gain-bandwidth product.

The gain-bandwidth product for a pentode can be well approximated by the expression

$$|A_{v0}|\,|B| = \frac{g_m}{2\pi(C_{in} + C_{out})}$$

where C_{in} is the input capacitance measured between the grid and all other electrodes except the plate (that is, the screen, suppressor, and cathode electrodes are tied together and used as one terminal for measurement of C_{in}). C_{out} is the output capacitance similarly measured between the plate and all other electrodes except the grid.

As an example, the capacitances for a 6CB6 pentode are $C_{in} = 6.3$ pfarads, $C_{out} = 2$ pfarads. At an operating plate current of 10 ma, the 6CB6 has $g_m = 6.2 \times 10^{-3}$ mhos. The resulting gain-bandwidth product is 125 mcps. If we want to use 6CB6's to build a two-stage amplifier with a 1-mcps overall bandwidth, each stage must have a bandwidth of 1.6 mcps. The stage gain is therefore about 80. The gain and bandwidth of this amplifier are approximately the same as the two-stage transistor amplifier of Example 2, in which we specified $\beta = 100, f_T \simeq 1,000$ mcps, and $C_c = 1$ pfarad.

This example provides a valuable point of comparison between pentodes and transistors. Because of the importance of C_c in determining C_{eq}, a transistor with an f_T of about 1,000 mcps is required to build a two-stage amplifier which is comparable to a two stage pentode amplifier using tubes with a gain-bandwidth product of only 125 mcps. As the gain is reduced, however, the considerably greater gain-bandwidth potential of the transistor becomes increasingly apparent.

5 DESIGNING AN AMPLIFIER FOR A GIVEN BANDWIDTH In some cases, we may decide on the number of stages in a cascade and the type of transistor and then we want to perform a more exact calculation of the gain and bandwidth which we may expect from the amplifier. As an example, suppose we are using 2N324's in a two-stage amplifier that is to have an overall bandwidth of 40 kcps. Then, using the band-width-shrinkage table, the bandwidth of each stage is

$$f_u' = \frac{f_u}{0.64} = \frac{40 \text{ kcps}}{0.64} = 62.5 \text{ kcps}$$

To obtain this bandwidth we must choose an appropriate value for R_i. Equation $(13 \cdot 45b)$ gives f_u' in terms of R_i as

$$f_u' = \frac{R_i + r_\pi}{2\pi C_{eq} R_i r_\pi}$$

Recalling that C_{eq} also depends on R_i, we have

$$f_u' = \frac{1}{2\pi} \frac{R_i + r_\pi}{R_i r_\pi} \frac{1}{C_\pi + \beta C_c R_i / (R_i + r_\pi)}$$

This equation can be solved for R_i when f_u' is known. A convenient normalized form of solution is

$$x = \pm \frac{\sqrt{a^2 + 4ab - 4a} - (2 - a)}{2(1 - ab)}$$

where
$$x = \frac{R_i}{r_\pi} \qquad a = 2\pi f_u' C_\pi r_\pi \qquad b = \frac{\beta C_c}{C_\pi} + 1$$

To be specific, we again use the parameters for a 2N324, biased at $I_E = 1$ ma: $C_\pi = 1{,}250$ pfarads, $\beta \simeq 130$, $C_c = 18$ pfarads, $r_\pi = 3.9$ kohms. These values give $a = 1.92$ and $b = 2.88$. Solving the previous equation for x yields

$$x = \frac{R_i}{r_\pi} = 0.49 \qquad R_i \simeq 2 \text{ kohms}$$

Thus, an R_i of 2 kohms will give each stage the required bandwidth. This means that

1. The generator resistance should be 2 kohms and the parallel combination of base bias resistor for the first stage should be much higher than 2 kohms.

2. The parallel combination of the interstage collector resistance R_L (see Fig. $13 \cdot 25$), the base bias resistors for the second stage, and the output resistance of the first transistor r_0 should be 2 kohms.

3. The final collector load resistor should be 1.35 kohms (the parallel combination of R_i and r_π) so that the correct input capacitance to the second stage is achieved.

Having selected R_i, we can evaluate the midband gain of the amplifier as follows. The first is driven from a source resistance R_i and drives a load of R_i in parallel with r_π. The midband gain of this stage is therefore

$$A_{v0} = \frac{\beta(R_i \parallel r_\pi)}{R_i + r_\pi} \simeq \frac{130 \times 1.35 \text{ kohms}}{6 \text{ kohms}} = 29.2$$

The second stage has a gain of $g_m(R_i \parallel r_\pi) = 42.8$. The two-stage gain from source to load is therefore 1250.

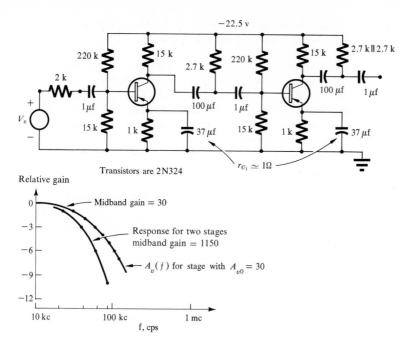

FIG. 13·28   EXPERIMENTAL TWO-STAGE AMPLIFIER AND GAIN CHARACTERISTIC

Experimental results.   As an approximate experimental check on these results, two 2N324 transistors were connected in the circuit of Fig. 13·28. Measurements showed that the two transistors had different β's (126 and 132) and somewhat different emitter resistances. However, they are as similar as two transistors may be expected to be.

The experimental amplifier shows a two-stage gain of 1150 and a cutoff frequency of 38 kcps. The experimental frequency response is shown in Fig. 13·28. The frequency-response curves for the individual stages as single-stage amplifiers are also shown in this figure. They are each normalized to a midband gain of 0 db, though the stage gains are not exactly equal (they are 26 and 30). The normalized frequency-response curves for the individual stages are sufficiently similar that they have not been separately plotted.

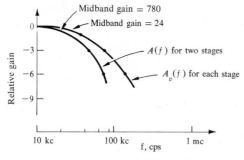

FIG. 13·29   FREQUENCY RESPONSE OF THE AMPLIFIER OF FIG. 13·28 WITH TWO
DIFFERENT TRANSISTORS AND EMITTER-BYPASS CAPACITORS

Figure $13 \cdot 29$ shows the response of the amplifier when two transistors with β's of 138 and 126 are used, and the emitter bypass capacitors are changed to 100 μfarad types, which exhibit more series resistance.

$13 \cdot 4 \cdot 4$ *Other factors affecting the number of stages which can be cascaded.* In addition to gain and bandwidth considerations, there are two other factors which affect the number of stages which can successfully be cascaded. These are (1) the progressive increase in signal level through the cascade and the attendant problems of signal distortion and (2) the noise voltages which are generated in the amplifier and in the source which supplies signals to the amplifier. We will consider these factors in this section.

Signal-level considerations. If we are interested in maintaining small-signal conditions, then the maximum voltage swing which can be permitted at the base-emitter terminals of the last transistor is about 5 mv ($v_{be} < kT/q$). The maximum signal current which can be supplied to the base of the last transistor depends on the bias current, as the following calculation shows:

$$v_{be} \simeq i_b r_\pi = i_b \frac{kT}{qI_B} = \frac{i_b}{I_B} \frac{kT}{q}$$

To keep $|v_{be}|$ less than 5 mv requires that i_b/I_B be less than $\frac{1}{5}$. Of course, placing a limit on i_b/I_B places this same limit on i_c/I_C.

This calculation suggests that, if we want to maintain strictly linear operating conditions in each stage of an amplifier, we must determine the maximum signal-current swing expected at each collector and then use a collector bias current at least five times this large. If we also are interested in minimizing power drain, then each stage will have a different bias point, reflecting the progressive increase in signal level through the amplifier. This refinement is usually not necessary, however, since 1 ma or so of bias current does not represent an excessive power drain in most circumstances and can be used in stages where the signal current is only 1 μa or so.

Of course, a 1-ma bias current will only allow an output signal of about 0.2 ma peak, which may not be enough for the last stage of the main amplifier in some applications. In such cases, we can either increase the bias current in the last stage or not insist on strictly linear operating conditions. The latter case, which involves making a large excursion over the output load line, will be treated thoroughly in Sec. $13 \cdot 5$ on power amplifiers. However, a brief comment is appropriate at this point.

We pointed out in both Chaps. 10 and 11 that, as long as a transistor is not driven out of its normal-active mode, its collector current will be almost sinusoidal if the base current supplied to the transistor is sinusoidal, even though the base-emitter signal-voltage swing might be larger than 5 mv. Hence, if we ensure that the collector signal current of the next-to-last transistor is supplied entirely to the base of the last transistor, we can relax the "strictly linear" restriction, without affecting the output current or voltage waveform significantly. This means, of course, that the load resistance for the next-to-last stage will be approximately r_π, so the bandwidth of this last stage will perhaps be lower than that of preceding stages, unless we select a transistor with a higher f_T for it.

If this is acceptable, then a bias point for the last stage which minimizes power drain and uses the transistor output characteristics efficiently can be determined from the maximum signal swing desired. As an example, suppose we want an amplifier to be capable of delivering an essentially distortion-free output signal of 2 volts (rms) into an ac collector load resistance of 1 kohm. To do this, we choose a bias point so that the minimum $|v_{CE}|$ will be 1 volt (to avoid driving into the saturation region), and the minimum $|i_C|$ will be 0.1 ma (on the assumption that the current gain will drop when the current decreases below this value). Then, by sketching a set of i_C-v_{CE} coordinates, marking off the 1 volt and 0.1 ma margins, and using some simple geometry, we conclude that a satisfactory bias point would be at $|I_C| \simeq 3$ ma, $|V_{CE}| \simeq 3.8$ volts.

Maximum input signal amplitude. Of course, regardless of the precautions we take to avoid it, distortion will always become evident at a sufficiently large input signal level. A rather dramatic illustration of this can be obtained from the experimental two-stage amplifier described earlier. To see this, let us first calculate the maximum source voltage which this amplifier can accept if strictly linear operating conditions are to be maintained (that is, v_{be} at the base-emitter terminals of the last stage is to be no greater than 5 mv). The midband-voltage gain from the base-emitter terminals of the first transistor to the base-emitter terminals of the second transistor is

$$g_m(R_i \parallel r_\pi) \simeq 54$$

so the maximum signal that can be applied to the base-emitter terminals of the first transistor is about 5 mv/54 ~ 0.1 mv. Since $r_\pi \simeq 4$ kohms and $R_s = 2$ kohms, this suggests a maximum source voltage of 130 μv, if linearity of response is required.

We can also estimate the maximum input signal in another way by considering the amplifier output. The transistor is biased at $I_E = 1$ ma, $v_{CE} = -5$v, and the ac load resistance on this stage is 1.35 kohms. By constructing a load line on the i_C-v_{CE} plane, we can conclude that the maximum output swing which we can obtain before the output transistor reaches cutoff is 1 ma $\times$ 1.35 kohms = 1.35 volts. The amplifier gain is about 800, so the maximum input voltage for a relatively distortion-free output is no *more* than 1.35 volts/800 = 170 μv.

Experimental results. Photographs of the output waveform of the two-stage amplifier as displayed on an oscilloscope are shown in Fig. 13·30. In Fig. 13·30a the input signal level is 200 μv. Some distortion can be detected by inspection of this waveform.[1] The distortion is readily apparent

---

[1] A Fourier analysis of the signal shows that the amplitude of the second harmonic voltage is about 3 percent of the fundamental voltage amplitude.

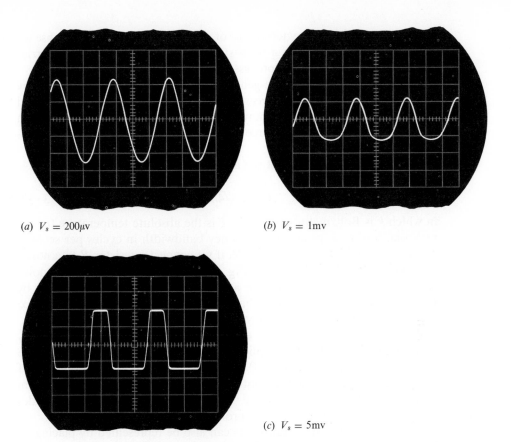

(a) $V_s = 200\mu v$

(b) $V_s = 1mv$

(c) $V_s = 5mv$

FIG. 13·30 OUTPUT VOLTAGE WAVEFORMS OF THE TWO-STAGE AMPLIFIER SHOWN IN FIG. 13·28 FOR DIFFERENT INPUT VOLTAGE. THE OSCILLOSCOPE GAIN IS DECREASED AS THE INPUT SIGNAL AMPLITUDE IS INCREASED.

in Fig. 13 · 30b, where an input signal of 1 mv is applied. An input drive of 5 mv is enough to drive the final stage from cutoff to saturation.

Noise considerations and the minimum input signal amplitude. The minimum signal that can be successfully amplified depends on the magnitude of the noise voltages which are generated in the amplifier and in the signal source. The signal must be greater than these noise voltages in each amplifier stage; or, conversely, the circuit noise will set a limit on the smallest signal which the amplifier can handle. We shall discuss this point briefly in this section.

The output of an amplifier when there is no signal input is called *noise.* Noise manifests itself in the hiss and crackle heard in an audio amplifier when the volume control is turned fully on. It is also responsible for the "snow" that appears on a television screen when the receiver is tuned to a distant station. Noise arises because the particles which carry electricity are discrete. In a resistor the random thermal motion of electrons produces a noise voltage across the terminals of the resistor even in the absence of any applied currents. In a transistor, the random nature of the carrier diffusion and recombination processes will also produce noise.

For circuit purposes, noise-generating processes are characterized by connecting suitable *noise sources* to the circuit models of resistors and active devices. These sources may be either voltage or current sources, in which the rms value of the source is determined by the process which produces the noise.

In a resistor it is found that the noise voltage has an rms value which is given by

$$V_n = \sqrt{4kTBR} \qquad (13 \cdot 53)$$

in which k is Boltzmann's constant, T is the absolute temperature (of the resistor), B is an appropriate frequency bandwidth in cycles per second, and R is the resistance. For example, if we connect a resistor R to a noiseless filter of bandwidth B, as in Fig. $13 \cdot 31a$, the rms value of the noise voltage which we would observe at the filter *output* would be given by Eq. $(13 \cdot 53)$. If the filter had a bandwidth of 40 kcps and the resistance $R = 1$ kohm, then the rms noise voltage would be 0.8 μv.

We can develop a Thevenin equivalent network for this situation by replacing the resistor with a "quiet" resistor in series with a noise-voltage source having an rms value of 0.8 μv (Fig. $13 \cdot 31b$). The "waveform" of the source is not specified, except to say that it consists of some random pattern of positive and negative spikes, the positions and amplitudes of the spikes being determined by probabilistic considerations.

The noise generated in transistors can also be represented by placing an "equivalent noise source" in series with the transistor input terminals, as in Fig. $13 \cdot 32a$. The rms value of this noise-voltage source depends on the resistance of the source driving the transistor and the transistor bias conditions.[1]

Usually there will be some optimum source resistance and bias condition (I_E) which will minimize the rms value of this equivalent noise source. Transistor manufacturers usually quote a *noise figure F* for their transistors, assuming that these optimum conditions will be used.

[1] For a more thorough treatment of noise mechanisms and their representation in transistors, see J. G. Linvill and J. F. Gibbons, "Transistors and Active Circuits," McGraw-Hill Book Company, New York, 1961, chap. 17.

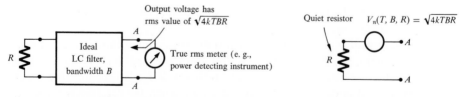

(a) *Resistor and noiseless filter for observing noise voltage*   (b) *Thévenin equivalent for (a)*

FIG. $13 \cdot 31$ CIRCUIT REPRESENTATION OF NOISE GENERATED IN A RESISTOR

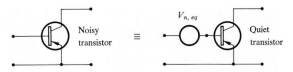

(a) *Representation of transistor noise by an equivalent noise source at the transistor input*

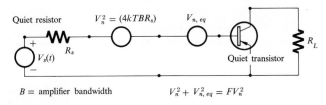

$B =$ amplifier bandwidth $V_n^2 + V_{n,eq}^2 = FV_n^2$

$F =$ noise figure of transistor when driven from source
with resistance R_s, bias current I_E

(b) *Amplifier with signal and noise voltages appearing at its input. Amplifier bandwidth is used
to calculate $V_n{}^2$ since our ultimate interest is in the noise appearing at the amplifier output*

FIG. 13·32   REPRESENTATION OF NOISE SOURCES IN A TRANSISTOR AND AN AMPLIFIER

The noise figure F is defined as

$$F = \frac{\text{mean square value of equivalent noise source}}{\text{mean square value of noise due to source resistance}} + 1 \quad (13 \cdot 54)$$

According to this definition, the total mean square noise voltage $V_{n,T}{}^2$ appearing at the input terminals of the amplifier in Fig. $13 \cdot 32b$ is

$$V_{n,T}{}^2 = FV_n{}^2(R_s)$$

where $V_n(R_s)$ is given by Eq. $(13 \cdot 53)$. F is usually quoted in decibels:

$$10 \log_{10} F = 20 \log_{10} \frac{V_{n,T}}{V_n(R_s)}$$

and will be on the order of 2 to 3 db for a low-noise transistor. The optimum source impedance is usually close to 1 kohm, so

$$V_{n,T} \simeq 1.4 V_n(R_s = 1{,}000 \text{ ohms})$$

For an amplifier with a bandwidth of 40 kcps,

$$V_n(1{,}000 \text{ ohms}) = 0.8 \ \mu v \qquad V_{n,T} \simeq 1.14 \ \mu v$$

Experimental results. The existence of noise in an amplifier implies an ultimate limitation on the size of signals which an amplifier can effectively handle. To demonstrate this, we show in Fig. $13 \cdot 33$ a series of oscilloscope traces of the output of the two-stage amplifier under varying signal-level conditions. Figure $13 \cdot 33a$ shows the noise output with no input signal. The vertical scale on this figure is 2 mv/cm. The measured rms noise output voltage is 1.67 mv. Since the midband voltage gain of the amplifier is 780, the equivalent input rms noise voltage is 2.13 μv. The rms noise voltage

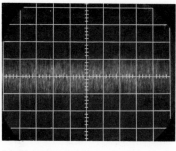

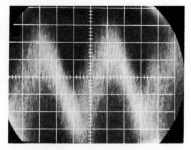

(a) $V_s = 0$

(b) $V_s = 1\mu v$

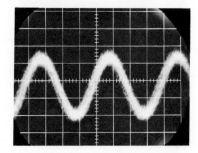

(c) $V_s = 5\mu v$

FIG. 13·33   OUTPUT VOLTAGE OF THE TWO-STAGE AMPLIFIER SHOWN IN FIG. 13·28
FOR SMALL INPUT SIGNALS. THE OSCILLOSCOPE GAIN IS DECREASED AS THE INPUT SIGNAL
AMPLITUDE IS INCREASED.

generated in the 2-kohm source resistor is 1.1 μv over the 40 kcps band-width, so the noise added by the amplifier is about 1.0 μv (rms). From these data we conclude that the measured noise figure for the amplifier is 5.6 db.

Figures 13·33b and 13·33c show oscilloscope traces of the amplifier output when a small input signal is applied. For these cases we must add the signal and noise voltages at the input to obtain the total drive applied to the amplifier.[1]

The oscilloscope traces suggest that if we want a relatively "clean" output signal in the presence of the noise, we must apply at least 5 μv of signal at the amplifier input. We would then have a *signal-to-noise voltage ratio of 5 to 1* at the signal source. If the signal-to-noise ratio were 2:1, we would still have a readily detectable output signal. By using sophisticated detection techniques we can pick out a signal with considerably less than a 1:1 signal-to-noise ratio.

Dynamic range. We can summarize the preceding discussion by saying that noise considerations determine the minimum input signal level, and that either the inherent nonlinearities in the transistor or the saturation char-acteristics of the last stage determine the maximum input signal level. For

---

[1] Noise voltages generated in the first stage of a multistage amplifier are usually the domi-nant factor in determining the output noise, because they are multiplied by the full amplifier gain.

the two-stage voltage-amplifier cascade, the minimum input signal might be about 5 μv (for a minimum signal-to-noise voltage ratio of about 5 to 1); the maximum is about 150 μv. The ratio of these signal levels is called the *dynamic range* of the amplifier and is usually quoted in db. For the two-stage voltage amplifier cascade, the dynamic range would be

$$20 \log_{10} \frac{150}{5} \cong 30 \text{ db}$$

The dynamic range will, of course, depend on what one is willing to accept as a signal-to-noise ratio, and how much signal distortion (due to nonlinear operation) is permissible.

13·4·5 *Input-stage design.* The problem in designing the input stage of a multistage amplifier is to insure that the source is properly coupled to the amplifier. Generally, we want the source to deliver the maximum possible signal to the amplifier, consistent with maintaining the required amplifier bandwidth. There are usually several alternatives for the input stage design, the choice between them being dictated by economic and other considerations (for example, minimizing the number of transistor types in the amplifier).

To organize the design problems, we will consider three different types of sources: V_s low (a few μv up to 1 mv) and R_s low (less than the interstage resistance R_i); V_s low and R_s high (greater than R_i); and V_s high (1 mv to 0.5 v) and R_s high.

V_s low, R_s low. When V_s is less than 1 mv, the small-signal approximations are valid in the input stage. Therefore, we can couple the source directly to the main amplifier (usually through an input capacitor) without loss of linearity. The fact that $R_s < R_i$ means that the bandwidth of the first stage will be greater than that of the other stages in the main amplifier.

V_s low, R_s high. Here linearity of operation is still maintained, but the effect of the input stage on the overall bandwidth needs to be considered. If the basic bandwidth of the transistor as a current amplifier is adequate for the overall amplifier, then again the source may be coupled directly to the main amplifier.

If the basic bandwidth of the transistor as a current amplifier is adequate but we wish to maximize the power transfer from the source to the amplifier, then a common-collector stage is sometimes used. The input resistance of a *CC* stage which drives a load resistance R_L is about $(\beta + 1)R_L$ at low frequencies (up to f_T/β). Its output resistance will be about R_s/β over this same frequency range. Thus, given a transistor with $\beta = 100$, we can match a 100 kohm source to a 1-kohm load, over a bandwidth of $f_T/100$.

Unfortunately, $f_T/100$ will not be a large bandwidth, unless we use a considerably better transistor in the input stage than in the main amplifier. If bandwidth is a problem in the input stage, the *CC* stage is not a particularly valid option. A more common solution is to employ a *CE* stage with an appropriately chosen feedback resistor connected between collector and base. In this scheme (to be discussed in detail in the next chapter), the input resistance is maintained at a low value ($\sim$300 ohms or so is typical) over a wide bandwidth. We sacrifice impedance matching for the bandwidth in this case. If the loss in signal-power level on this account is really significant, we may have to add a stage in the main amplifier to make it up.

V_s *high, R_s high.* When the input-signal voltage is high, we must pay attention to the problem of maintaining linearity in the overall transfer characteristic. This can be done in two ways. One is to use a *CE* stage, adjusting the bias on the transistor so that its input impedance is small compared to R_s. Then the input stage acts as a linear current amplifier, and the bandwidth burden is placed entirely on the transistor.

If bandwidth is not a problem, then again we can utilize a *CC* stage to change impedance levels. More commonly, the feedback techniques to be described in the next chapter are used.

To summarize, in practical amplifiers the input stage can frequently be the same as the first stage of the main amplifier cascade, with perhaps an adjustment of bias to account for the source impedance. When this solution is not adequate, a *CE* stage with appropriate feedback (or sometimes a *CC* stage) can be used to maintain the overall amplifier bandwidth and linearity while transforming the source impedance to a more satisfactory level for the main amplifier.

13·4·6 *Output-stage design.* The design of the output stage of an amplifier poses some very interesting problems to the designer. He is faced with the primary problem of delivering a specified output power to a specified load impedance. The required load power may be anything from milliwatts to hundreds of watts or more; while the load impedance may be a pair of earphones, a loudspeaker, a servomotor, a telephone transmission line, a line feeding power to an antenna, or deflection coils for a picture tube, to name only a few of the possibilities. The designer must, of course, consider the effect of the output stage on the overall amplifier gain and bandwidth. In addition, he may have to consider some or all of the following questions.

1. How to maintain linearity or, at least, how to minimize signal distortion in the output stage, while supplying large swings of voltage and current to the load.

2. How to ensure that the active devices are operated within their current, voltage, and power ratings, sometimes under awkward environmental conditions (such as a large, ambient temperature range and/or very small available volume for the stage).

3. How to effect an efficient conversion of the available dc power into signal power, thereby minimizing the dc power requirements (and usually the size and weight) of the power supply.

These considerations are of particular importance if the required power is high or if the amplifier must operate from a source of limited power, such as a battery or solar cell (for example, in hearing aids, transistor radios, instrumentation for satellites, etc.). The effects these conditions have on the design are so pronounced that one can frequently determine the intended uses of an amplifier by examining its output stage.

In some cases, a satisfactory output stage can be designed by making only minor modifications on the basic designs just considered. For instance, when the required output power is a few tens of milliwatts or less, when dc power dissipation is unimportant, and when the load is resistive and comparable to the interstage resistance in the main amplifier, the output stage may be similar to the previous stages in overall design. If the load is resistive but has a value unsuited to the main amplifier or if dc power consumption is a problem, then the load may be coupled to the amplifier through a transformer (see Fig. 13·34a and Sec. 13·5·4.) If the load impedance is variable and we want to maintain a constant output voltage, then a CC output stage is an interesting possibility (see Fig. 13·34b).

However, when the desired output power exceeds 100 mw or so, all the factors listed earlier will need to be considered, and simple variations of the main amplifier design will not usually be adequate. In the next section we will examine the basic properties of several amplifier circuits which have been developed for these cases. These circuits have the common objective of delivering large output power, so they are generally called *power amplifiers*.

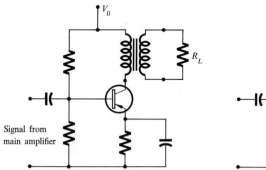

(a) A transformer-coupled load eliminates dc power loss in R_L

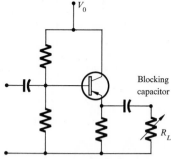

(b) A CC stage provides a convenient means of coupling the main amplifier to a variable impedance load

FIG. 13·34 OUTPUT STAGES THAT ARE SIMPLE MODIFICATIONS OF THE MAIN AMPLIFIER DESIGN

We shall begin the study of power amplifiers by increasing our vocabulary of basic amplifier definitions. The definitions to be given apply to all amplifiers, though their importance is usually restricted to stages which deliver appreciable output power.

To make the definitions as general as possible, we shall refer to the *generalized* active device shown in Fig. 13 · 35. The instantaneous value of the total output current is denoted by i_{II} and the instantaneous total voltage at the output port is v_{II}.

13·5·1 *Basic amplifier definitions (continued): class A, B and C amplifiers.*

When a sinusoidal input signal is applied to the active device (either i_{I} or v_{I}), i_{II} and v_{II} will vary with time. If the bias condition and input-signal amplitude are such that i_{II} is never zero (Fig. 13 · 35b), then the device is said to be operating under *class A* conditions. The amplifier stages which we have considered so far in this chapter are called class A stages.

If the bias condition and input signal are such that i_{II} flows for only one-half cycle of the input sinusoid, the operating condition is termed *class B* (Fig. 13 · 35c). If i_{II} is nonzero for more than one-half cycle but less than a full cycle, the operating condition is called *class AB*. We will see shortly that transistor power amplifiers are usually operated in the class AB condition, when the efficient conversion of dc power into signal power is important.

When the output current i_{II} is zero for less than one-half cycle then we have a *class C* operating condition (Fig. 13 · 35d). This is the typical

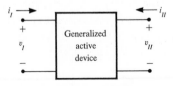

(a) *Generalized active device with total currents and voltages defined*

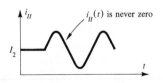

(b) *Output current for class A amplifier*

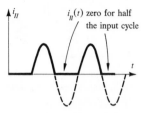

(c) *Output current for class B amplifier*

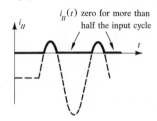

(d) *Output current for class C amplifier*

FIG. 13 · 35 OUTPUT CURRENT WAVEFORMS IN CLASS *A*, *B*, AND *C* AMPLIFIERS operating mode for a device used to supply a large output power to a tuned circuit (for example, in the output of a radio transmitter).

Efficiency. The *efficiency* of an amplifier is defined to be the ratio of the signal power delivered to the load to the dc power supplied to the circuit:

$$\eta = \frac{\text{signal power to load}}{\text{dc power input}} \times 100 \qquad (13 \cdot 55)$$

Usually we are concerned primarily with the *output circuit efficiency,* η_{II}, which we define to be the signal power delivered to the load divided by the dc power chargeable to the flow of current in the output circuit of the device (or devices):

$$\eta_{II} = \frac{\text{signal power to load}}{\text{dc power supplied to output circuit of device}} \times 100$$

The definitions of efficiency are *not* based on the assumption of a sinusoidal signal, though we shall be primarily concerned with this case.

Nonlinear distortion. An amplifier stage is said to be *nonlinear* when its output waveform contains frequency components which are not present in the input waveform. The distortion of the output waveform which results from amplifier nonlinearity is called *nonlinear distortion.*

The basis for nonlinear distortion is suggested in Fig. 13 · 36. Here, the output characteristics of the generalized device are shown, together with a resistive load line. The output characteristics are exaggerated, to emphasize that they are not perfectly horizontal, uniformly spaced curves.

In Fig. 13 · 36b a transfer characteristic has been sketched for the given load line. If the amplifier were perfectly linear, the transfer characteristic would be a straight line. The nonlinearities in the active device cause the actual transfer characteristic to deviate from the ideal.

The actual transfer characteristic can be expressed quantitatively in a Taylor series. Using the bias point as the point of reference, we have

$$i_{II} = I_{II} + a_1 x_1 + a_2 x_1^2 + a_3 x_1^3 + \cdots \qquad (13 \cdot 56)$$

where $x_1 = i_1$ or v_1, whichever is used as the input variable.

The nonlinearity of the transfer characteristic will cause two types of distortion:

1. If a sinusoidal input signal is applied, then the output signal will contain harmonics of the input signal frequency. Mathematically, the time variable portion of the output current can be expressed in a Fourier Series

$$i_2 = A_1 \sin (\omega t + \phi_1) + A_2 \sin (2\omega t + \phi_2) + \cdots \qquad (13 \cdot 57)$$

in which the amplitudes of the harmonics are related to the coefficients in the Taylor's series expansion of the transfer characteristic *and* to the input signal amplitude.

A convenient measure of the distortion arising in this case is the *percent total harmonic distortion D. D* is defined by the equation

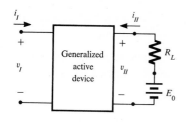
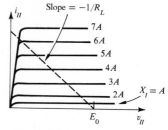

(a) *Generalized active device with DC source and load.* X_I *represents either* i_I *or* v_I, *depending on the device.*

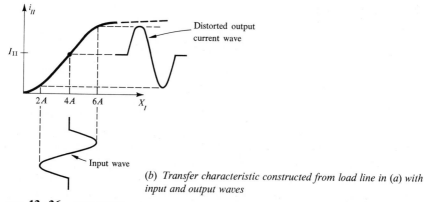

(b) *Transfer characteristic constructed from load line in (a) with input and output waves*

FIG. 13·36   NONLINEAR DISTORTION AND ITS RELATION TO THE TRANSFER CHARACTER-
ISTIC OF A STAGE

$$D = \frac{\sqrt{A_2^2 + A_3^2 + \cdots}}{A_1} \times 100 \qquad (13 \cdot 58)$$

From this definition, the ratio of the total harmonic power developed in the load to the fundamental power developed in the load is $D/100$. A total harmonic distortion of about 1 percent is perceptible in an audio system; 5 percent harmonic distortion is classed as "objectionable."

2. If the input signal contains two or more sinusoidal components, then the nonlinear transfer characteristic leads to a second type of distortion, called *intermodulation distortion.* For example, if the transfer characteristic can be adequately represented by the Taylor series

$$i_{II} = I_{II} + a_1 x_1 + a_2 x_1^2 \qquad (13 \cdot 59)$$

and the input signal is

$$x_1 = x_1 \sin \omega_1 t + x_2 \sin \omega_2 t \qquad (13 \cdot 60)$$

then the Fourier series for the output waveform will contain frequency components at ω_1, $2\omega_1$, ω_2, $2\omega_2$, $(\omega_1 + \omega_2)$ and $(\omega_1 - \omega_2)$. The appearance of the sum and difference frequencies is characteristic of intermodulation distortion.

In an audio system, the sum and difference frequencies produced by intermodulation distortion are the principal source of distortion noticed by the listener. The effect is to produce a harsh, unpleasant sound.

There are many ways of defining and measuring intermodulation distortion, making it impractical to consider the quantitative aspects of the subject here.[1]

13·5·2   Class A transistor power amplifiers.   In this section we shall study the main properties of class A transistor power amplifiers. The theoretical relations to be obtained are of interest both for design purposes and as standards of comparison for other types of amplifiers. In practice, class A power amplifiers are frequently employed to drive class AB or class B final output stages. The design principles which we shall discuss will be limited to the use of transistors as the active elements, though they apply equally well to vacuum tubes.

13·5·2·1   Direct-coupled resistive load.   It is of interest to begin by calculating the maximum efficiency which can be obtained when the load is coupled directly to the output terminals of the active device, as in Fig. 13·37. For this purpose we show an idealized version of the CE output characteristics of the transistor in Fig. 13·37b, together with a resistive load line. The operating point is chosen so that the output signal can swing equally far in the positive and negative direction.

The output current in this circuit will be expressed as

$$i_C(t) = I_C + i_c(t) \tag{13·61a}$$

and the output voltage as

$$v_{CE}(t) = V_{CE} + v_{ce}(t) \tag{13·61b}$$

I_C and V_{CE} represent the collector bias current and the collector-to-emitter bias voltage, while $i_c(t)$ and $v_{ce}(t)$ represent the time-varying components.

For the present analysis we assume that $i_c(t)$ and $v_{ce}(t)$ are sinusoidal:

$$i_c(t) = \sqrt{2}I_c \sin \omega t \qquad v_{ce}(t) = \sqrt{2}V_{ce} \sin \omega t \tag{13·62}$$

where I_c and V_{ce} are the rms values of the output quantities. With these definitions, the average signal power developed in R_L is

$$P_L = I_c V_{ce} \tag{13·63}$$

The average power delivered by the battery to the output circuit is

$$P_B = V_0 I_C \tag{13·64}$$

To maintain class A operating conditions, we require

$$\sqrt{2}I_c \leq I_C \qquad \sqrt{2}V_{ce} \leq \frac{V_0}{2} \tag{13·65}$$

---

[1] See "Radiotron Designer's Handbook," Wireless Press, Sydney, Australia, 1953, pp. 612ff.

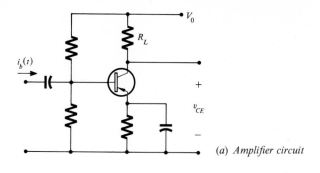

(a) Amplifier circuit

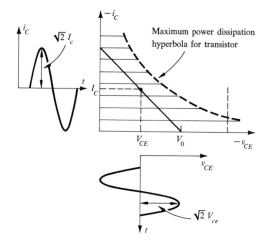

(b) Output waveforms is an ideal class A amplifier, resistive load

FIG. 13·37   A SIMPLE CLASS A TRANSISTOR POWER AMPLIFIER

Using these relations in Eq. (13 · 63), we find

$$P_L \leq \frac{I_C V_0}{4} \tag{13·66}$$

The collector-circuit efficiency of this circuit *for a sinusoidal input* is therefore

$$\eta_C = \frac{P_L}{P_B} \leq 25 \text{ percent} \tag{13·67}$$

To obtain maximum efficiency we must use the maximum permissible drive. Note that in this operating mode the power leaving the battery is a constant, independent of the input-signal amplitude. When no signal is present, half this power is dissipated in the load resistor and half in the device. When an input signal is applied, the power dissipated in the load *increases* and the power dissipated in the device *drops*. Under maximum signal conditions, the active device converts half the dc power supplied to it into signal power, the signal power being dissipated in the load.

13·5·2·2 *Transformer-coupled resistive load.* The efficiency of the previous circuit can be increased by a factor of nearly 2 by using a transformer to couple the load to the active output device (Fig. 13·38a). This simply eliminates the dissipation of dc power in the load. In other respects the operation is similar to the previous case. Since practical class-A-driver stages usually employ transformer coupling, we will develop the principal design relations for the circuit in some detail.

In the transformer-coupled amplifier, essentially the full battery voltage is applied to the device output terminals under no-signal conditions. The power supplied by the battery is still constant and equal to

$$P_B = V_0 I_C \tag{13·68}$$

However, under maximum input-signal conditions the transformer action causes the maximum collector voltage to be $2V_0$. Therefore,

$$\sqrt{2} V_{ce} \leq V_0 \tag{13·69}$$

from which we obtain

$$P_L = I_c V_{ce} \leq \frac{I_c V_0}{2} \tag{13·70}$$

or

$$\eta_c \leq 50 \text{ percent} \tag{13·71}$$

Effect of device ratings on load impedance. Even with an efficiency of 50 percent, a considerable amount of power will be dissipated in the device if a reasonably large output power is required. Accordingly, we must ensure that the circuit operation does not carry the device beyond its ratings and that the biasing network will maintain the bias point in the face of considerable internal heating. The bias-stability considerations are not different from those given in Chap. 12, except that we must note that the dc collector load resistance is very small (winding resistance of transformer primary) so the burden of operating-point stability is placed entirely on the emitter and base bias resistors.

The effect of the current, voltage, and power ratings on the circuit

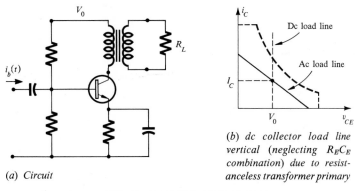

(a) Circuit

(b) dc collector load line vertical (neglecting $R_E C_E$ combination) due to resistanceless transformer primary

FIG. 13·38 A CLASS A AMPLIFIER WITH TRANSFORMER-COUPLED LOAD

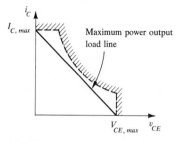
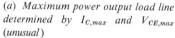

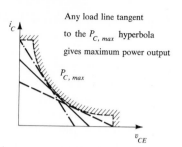

(a) *Maximum power output load line determined by $I_{C,max}$ and $V_{CE,max}$ (unusual)*

(b) *Maximum power output determined by $P_{C,max}$*

FIG. **13·39**   EFFECT OF DEVICE RATINGS ON THE POWER OF A CLASS A TRANSISTOR POWER AMPLIFIER

behavior is indicated in Fig. 13·39. In Fig. 13·39a we show the somewhat unusual condition (for a transistor intended as a class A driver) where $I_{C,max}$ and $V_{CE,max}$ determine the load line which gives maximum power output. Figure 13·39b shows the condition where the maximum power-dissipation hyperbola determines the maximum power-output load line. In either case, the load line which we select must lie in the permitted area of the V_{CE}-I_C plane.

Design theory.   Several factors need to be considered to choose a load line which is appropriate for a given application. It is simplest to begin a design with a consideration of the required power output, which we denote by P_L.

For a start, we assume the efficiency of the stage is 50 percent. Then, under no signal conditions, the transistor must dissipate $2P_L$ watts. Figure 13·40 shows a hyperbola constructed from this condition:

$$I_C V_{CE} = 2P_L \qquad (13\cdot72)$$

We may choose a load line by drawing a tangent to the constant power-dissipation hyperbola. The point of tangency is the bias point. There are,

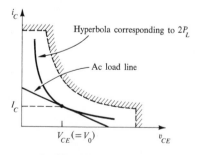

FIG. **13·40**   DESIGN OF A TRANSFORMER-COUPLED CLASS A OUTPUT STAGE FOR A GIVEN POWER OUTPUT, P_L

of course, an infinite number of possible load lines, *all* of which have the following properties:

1. The power dissipated in the transistor is $2P_L$ watts under no-signal conditions.

2. The output power is 50 percent $\times 2P_L = P_L$ watts when the input drive is sufficient to produce a full excursion over the load line.

In practice, load lines in the high-voltage low-current portion of the plane are preferred because (1) waveform distortion is worse at high currents than at high voltages and (2) a smaller input-signal-current swing will produce the required output power.

If we specify both the required output power P_L and the battery voltage V_0, then the load line is completely determined. The load line is tangent to the hyperbola at the point

$$V_{CE} = V_0 \tag{13·73}$$

$$I_C = \frac{2P_L}{V_0} \tag{13·74}$$

The slope of the load line gives the load resistance. This is simply

$$R_L' = \frac{V_{CE}}{I_C} = \frac{V_0^2}{2P_L} \tag{13·75}$$

This, of course, is the "reflected" load resistance. The turns ratio of the transformer is selected so that the actual load resistor R_L is transformed into R_L' in the transformer primary circuit.

With the transformer thus selected, all that remains to finish the initial design is to select components for the biasing network which will give the required value of I_C.

EXAMPLE   An inexpensive portable transistor radio is operated from a 9-volt battery and will deliver about 500 mw of audio power. The final output stage of the radio will have a power gain of about 500, so the driver for the output stage needs to supply about 1 mw of signal power. We shall design a class A transformer-coupled driver for this application.

First, we allow for signal-power losses in the transformer by assuming that it is only 75 percent efficient (this is a typical efficiency for an inexpensive transformer). Then the class A driver must feed a signal power of 1.33 mw to the transformer primary. Therefore, $P_L = 1.33$ mw. The battery voltage $V_0 = 9$ volts. Hence

$$I_C = \frac{2 \times 1.33 \times 10^{-3}}{9} = 0.296 \text{ ma} = 0.3 \text{ ma}$$

The load resistance as viewed from the transformer primary should be

$$R_L' = \frac{V_{CE}}{I_C} = 30 \text{ kohms}$$

(Note that this is a rather large R_L', so amplifier bandwidth may be seriously affected by this stage. In an inexpensive receiver we would not be concerned about this. In other cases, feedback techniques are used to obtain bandwidth in the stage.)

We shall see in the next section that the final output stage will have an input impedance on the order of 2 kohms. The transformer is selected to give this required impedance transformation. (Usually the transformer turns ratio is not specified, but the impedance transformation which it produces is. The present transformer would be specified as 30/2 kohms.)

We now select the bias components to obtain $I_C = 0.3$ ma. We want to use as small an R_E as possible so that we do not waste battery voltage. A typical value is 470 ohms. For 0.3-ma bias current this means a voltage drop of about 0.5 volts. If the driver transistor is germanium, we need to add another 0.3 volts for junction drop. Then the base-to-ground voltage is 0.8 volts. This gives, in Fig. 13·41,

$$0.8 = 9\,\frac{R_2}{R_1 + R_2}$$

Furthermore, for bias stability, we want

$$\frac{R_1 R_2}{R_1 + R_2} \simeq 10\,R_E = 4.7 \text{ kohms}$$

Using these two conditions gives

$$R_1 = 53 \text{ kohms} \qquad R_2 = 5.2 \text{ kohms}$$

A 100-μfarad bypass capacitor is selected for convenience. The completed design is shown in Fig. 13·41.

The input-signal swing required to produce the required output can be estimated when a transistor has been selected. If we use our old friend the 2N324, $\beta \simeq 120$. The required peak collector-current swing is 0.3 ma, so the input current has a peak value of 0.3 ma/120 = 2.5 μa.

13·5·3 Class B (and class AB) transistor power amplifier.

When we attempt to extend the application of class A power amplifiers to situations requiring more than about 100 mw of output power we encounter two major difficulties. First, the class A amplifier cannot deliver an output power of more than $P_{C,max}/2$, where $P_{C,max}$ is the maximum permissible collector dissipation. Devices with a $P_{C,max}$ of 1 watt or more are expensive, so their use must usually be carefully justified. Second, class A operation is also

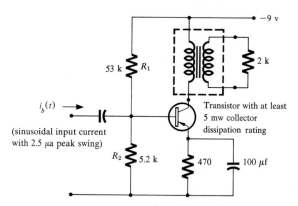

FIG. 13·41 DESIGN EXAMPLE: CLASS A DRIVER DELIVERING ABOUT 1.0 mw TO A 2k LOAD

wasteful, because the dc power supply must *continuously* supply twice as much power to the circuit as will be needed for *maximum* input-signal conditions.

The power supply requirements are graphically illustrated by the following example. The power supply for a typical inexpensive transistor radio is a 9-volt battery rated at about 500 ma-hr. It will deliver, at most, 4.5 watt-hr of energy. A typical listening level is 250 mw of signal power, so if the output stage is class A, the battery power dissipation would be 500 mw at all times. The radio could therefore give a maximum of nine continuous hours of service (actually much less, due to changes in the battery voltage with "age"—that is, after the first three hours).

Again, if very high output power is required (say, hundreds of watts), the power supply requirements pose cost, weight, and size problems. These problems are circumvented by operating transistors in a class B mode in the power-output stage. As we shall see shortly, the theoretical efficiency of a class B stage is 78.5 percent. Additionally, and in many cases more importantly, the power drain is very small (ideally zero) when there is no signal applied.

The fact that the transistor is biased at cutoff in the class B mode introduces a new problem, however: it only delivers an output current for one half-cycle of the input waveform, so severe distortion results. This problem is circumvented by using two transistors and taking advantage of the symmetry of the input-signal waveform. We can use one transistor to supply the load current during half of the input cycle—say, the positive half-cycle—and the other transistor to supply the load current during the negative half-cycle.

One way of doing this is to use a *push-pull* circuit configuration. The development of the push-pull circuit is illustrated in Fig. 13·42. In Fig. 13·42a, we show a load coupled to two transistors via a three-winding transformer. The transistors are driven by sources which are 180° out of phase. Each transistor supplies current to the load for half of an input cycle, the transformer windings being of such polarities that a sine-wave signal is reconstructed in the load. In Fig. 13·42b we show a more practical arrangement, in which the input drives for each transistor are obtained from a common source, the appropriate phase relationship being obtained by the input transformer. The circuit is drawn to emphasize its natural symmetry. The final arrangement is shown in Fig. 13·42c, where some of the transformer windings have been internally connected and only one bias battery is used.

Another possible arrangement, which employs *complementary* symmetry (that is, the existence of *pnp* and *npn* transistors), is shown in Fig. 13·43. Here, no transformer is required, which is an advantage.

Of course, even when two transistors are used, there is still some distortion in a practical class B amplifier, related primarily to the change of conduction from one transistor to the other. This distortion is aptly called *crossover* distortion, and is minimized in practice by using a class AB

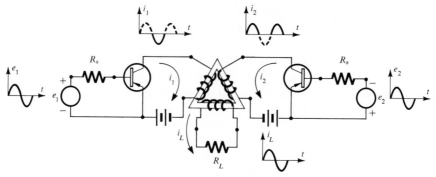

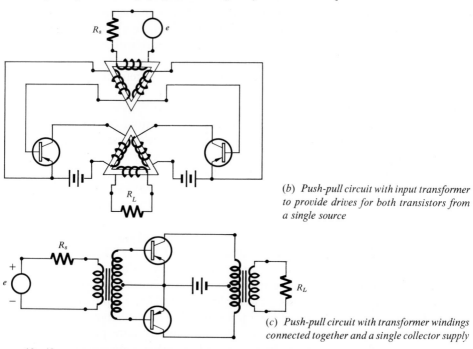

(a) Push-pull output circuit employing three winding transformers and two inputs

(b) Push-pull circuit with input transformer to provide drives for both transistors from a single source

(c) Push-pull circuit with transformer windings connected together and a single collector supply

FIG. 13·42   DEVELOPMENT OF THE PUSH-PULL AMPLIFIER CIRCUIT

bias condition. We shall discuss this briefly after we have considered the theoretical basis of ideal class B operation.

Operating characteristics of an ideal class B push-pull stage. The natural symmetry of a push-pull amplifier leads to some interesting ways of visualizing its operation and to simplicity in calculating its performance. A very instructive graphical presentation based on the symmetry of the circuit is shown in Fig. 13·44. Here, the characteristics of the two transistors are aligned so that the bias points for each transistor ($V_{CE} = V_0$, $I_C = 0$) coincide. A load line is drawn through this bias point. The load line extends into the normal-active region of one transistor and the cutoff

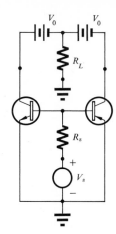

FIG. 13·43   PUSH-PULL AMPLIFIER EMPLOYING *npn* AND *pnp* TRANSISTORS IN COMPLEMENTARY SYMMETRY

region of the other. Hence, each transistor conducts for only one-half of the input cycle.

To choose a load line for a given application, we again begin the reasoning with a consideration of the required output power, P_L. At the outset we shall neglect the saturation characteristics of the transistors and assume that each transistor can be driven to its i_C axis. Then for a full swing over the arbitrary load line shown in Fig. 13·44, the rms value of the output signal voltage is

$$V_{ce} = \frac{V_0}{\sqrt{2}} \qquad (13·76)$$

and the rms output current is

$$I_c = \frac{V_{ce}}{R_L} = \frac{V_0}{\sqrt{2}R_L} \qquad (13·77)$$

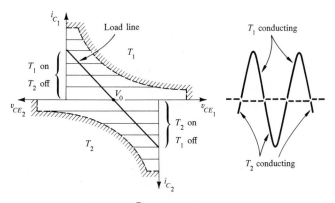

FIG. 13·44   GRAPHICAL DESCRIPTION OF CLASS *B* PUSH-PULL OPERATION

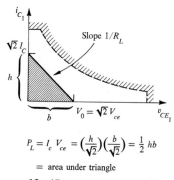

$$P_L = I_c V_{ce} = \left(\frac{h}{\sqrt{2}}\right)\left(\frac{b}{\sqrt{2}}\right) = \tfrac{1}{2} hb$$

$$= \text{area under triangle}$$

FIG. 13·45   RELATION BETWEEN POWER OUTPUT AND LOAD LINE FOR A CLASS B PUSH-PULL AMPLIFIER. THE FORMULA FOR P_L GIVEN IN THIS FIGURE IS THE TOTAL POWER OUTPUT, NOT THE POWER OUTPUT FROM ONE TRANSISTOR.

where R_L is the load resistance. The output power is then

$$P_L = I_c V_{ce} = \frac{V_0^2}{2R_L} \qquad (13 \cdot 78)$$

Of course, if the input signal does not cause a full swing over the load line, then the output power will be less than that given in Eq. (13·78).

Using the definition

$$P_L = I_c V_{ce} \qquad (13 \cdot 79)$$

and some simple geometry, we can arrive at the very useful result that the output signal power obtained in a class B circuit is equal to the area under the triangle shown in Fig. 13·45.

In many applications the desired P_L and a prescribed V_0 are given. This specifies the area under the triangle and the point of intersection of the load line with the voltage axis. There will, of course, be many load lines for which the circuit will deliver the required output power, two of which are shown in Fig. 13·46. However, if we want to minimize the input drive and maximize the circuit efficiency, we should choose the load line of least slope (that is, most nearly horizontal) which has the required area.[1] This will clearly minimize the required input base current, since it minimizes the peak output current. To show that it will also maximize the efficiency, we consider the current waveforms shown in Fig. 13·47. Here, we see that the battery current is a full-wave-rectified sine wave. The peak value of the battery current is $\sqrt{2}I_c$, where I_c is the rms output current. The dc battery current is therefore

$$I_0 = \frac{2}{\pi}(\sqrt{2}I_c) \qquad (13 \cdot 80)$$

---

[1] In the ideal case being considered, the load triangle of least slope will have one of its vertices on the i_C axis. In a more realistic case, the load triangle of least slope will have one of its vertices on the vertical line representing $v_{CE} \simeq 1$ volt to avoid the distortion which occurs when we drive the transistor into saturation.

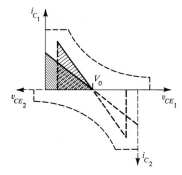

FIG. 13·46 TWO LOAD LINES WITH DRIVE ADJUSTED TO GIVE EQUAL AREAS IN THE TRIANGLES, AND THEREFORE EQUAL OUTPUT POWER

and the power supplied by the battery to the circuit is

$$P_B = V_0 I_0 = \frac{2}{\pi} (\sqrt{2} I_c) V_0 \qquad (13 \cdot 81)$$

From Eq. (13·81), we see that the power supplied to the circuit is minimized by minimizing I_c; that is, minimizing the height of the load triangle for given area, or using the load line of least slope for the given P_L.

For this condition, the efficiency of the circuit is

$$\eta = \frac{P_L}{P_B} = \frac{I_c V_{ce}}{\frac{2}{\pi} (\sqrt{2} I_c) V_0} = \frac{I_c (V_0 / \sqrt{2})}{\frac{2}{\pi} (\sqrt{2} I_c) V_0} = \frac{\pi}{4} \qquad (13 \cdot 82)$$

or 78.5 percent.

To summarize, the most efficient class B circuit for given P_L and V_0 is designed by choosing

$$R_L = \frac{V_0^2}{2 P_L} \qquad (13 \cdot 83)$$

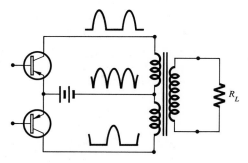

FIG. 13·47 CURRENT FLOW IN THE CLASS B AMPLIFIER. EACH TRANSISTOR CONDUCTS FOR ½ THE INPUT CYCLE. THE BATTERY SUPPLIES CURRENT DURING BOTH HALVES OF THE CYCLE.

The input base-current drive required to obtain the desired P_L then has an rms value of

$$I_b = \frac{I_c}{\beta} = \frac{\sqrt{2}}{\beta} \frac{P_L}{V_0} \tag{13$\cdot$84}$$

and the circuit efficiency is ideally 78.5 percent.

Maximum power output from a class B push-pull amplifier. The maximum power which a class B push-pull amplifier can deliver is determined by the maximum power the transistor can dissipate, $P_{C,max}$. Figure 13$\cdot$48 illustrates the reasoning behind this statement. The load triangle of maximum area must have its hypotenuse tangent to the $P_{C,max}$ hyperbola if we want to ensure that the instantaneous dissipation never exceeds the rated dissipation. This is a conservative design condition, since the average power dissipation will be less than the maximum instantaneous power dissipation. However, transistors are physically small devices, so they do not have a very large capacity for storing heat, and the conservative philosophy is usually adopted.

Now, all load lines which are tangent to the hyperbola enclose the same area within the load triangle, so they all produce the same output power. Furthermore, this output power is

$$P_{L,max} = 2P_{C,max} \tag{13$\cdot$85}$$

so two transistors operating in an ideal class B push-pull amplifier will deliver four times as much power as a single transistor operating in a class A mode.

To avoid distortion which occurs at high peak currents, the maximum power-output load line is usually constructed from a bias point of about $V_{CE,max}/2$. This, of course, minimizes the input drive for the required output.

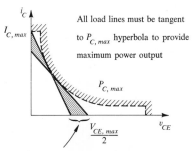

Load lines must originate within this
interval to satisfy device ratings

FIG. 13$\cdot$48 **LOAD LINE FOR MAXIMUM POWER OUTPUT FROM A CLASS B AMPLIFIER IN WHICH THE TRANSISTOR DISSIPATION NEVER EXCEEDS $P_{C,max}$**

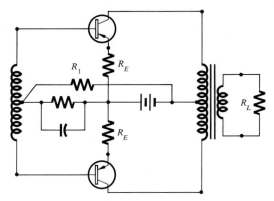

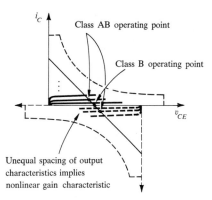

(a) *Addition of resistors to class B circuit to obtain class AB operating condition*

(b) *Location of class B and class AB operating points*

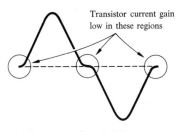

(c) *Output waveform exhibiting crossover distortion*

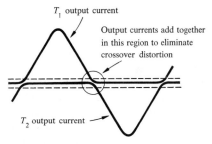

(d) *Output waveforms for the desired class AB bias condition*

FIG. 13 · 49   ILLUSTRATING CROSSOVER DISTORTION AND ITS ELIMINATION BY PROPER BIASING

Distortion.   The major sources of distortion in a class B amplifier arise from driving the transistors too near saturation and from imperfect transfer of conduction from one transistor to the other, near the zero crossing of the input waveform (called crossover distortion).

The first of these problems is avoided by using a combination of drive and load resistance which does not cause the transistor to swing too near saturation. The second problem is associated with a drop in β at low collector currents. To circumvent this, the two transistors may be biased slightly into their normal-active region, as shown in Fig. 13 · 49. This means the operating mode is now class AB, instead of class B, so the circuit will absorb dc power when no signal is applied. In addition, the emitter resistors used to obtain the bias cannot be bypassed, because the rectified currents which flow in the emitters of each transistor (see Fig. 13 · 47) would charge the bypass capacitor. The bias level would then fluctuate with signal level. Of course, an unbypassed R_E means a loss of signal-power gain, so the efficiency of the circuit is decreased. Hence, the cost of removing crossover distortion is a decrease in efficiency and the presence of some standby power drain. However, these factors are partially offset by the bias stability which is provided by the emitter resistors.

Crossover distortion is exhibited by Fig. 13·49 and experimentally demonstrated very effectively by the simple circuit shown in Demonstration 13·2. One can show very nicely with this demonstrator that the ear is very sensitive to the presence of crossover distortion.

EXAMPLE   Let us return to our design of a power amplifier for a transistor radio. We shall design a class AB final-output stage to deliver a power output of 250 mw to a loudspeaker. The battery voltage is 9 volts.

To begin the design we account for signal-power losses in the output transformer (the main source of power loss) by assuming that the transistor must deliver a power of 250 mw/transformer efficiency = 250 mw/0.75 = 333 mw to the collector circuit. The load resistance for this case is

$$R_L = \frac{V_0{}^2}{2P_L} = 121 \text{ ohms}$$

Note that this is the load which is reflected from the secondary into *half* the primary. The collector-to-collector resistance is *four* times R_L because the transformed resistance depends on the turns ratio squared. If the loadspeaker voice coil is an 8-ohm type, then the output transformer would be specified as a 484-ohm CT/8; the nearest standard transformer is a 500-ohm CT/8.

The selection of the biasing resistors depends on the characteristics of the transistors near cutoff and is usually best done experimentally. Usually a forward bias of a few percent of the peak current (on each transistor) is adequate. For our problem the peak output current is $V_0/R_L \cong 70$ ma, so we need about 5 ma for each stage. This bias is obtained with the smallest practical R_E which will maintain stability of the operating point. If we set $R_E = 10$ ohms, then at peak current there will be a voltage drop of 0.7 volts across it. This will help limit distortion, but it also reduces the power output by about 0.7/9, or 8 percent; we accept this as being satisfactory. The other elements of the bias network are selected to produce the desired I_E and to maintain bias stability.

Since the current flowing in the ON transistor will be large compared to 1 ma, except when the transistor is being driven toward cutoff, the input impedance of the conducting transistor will be essentially $\beta R_E = 10\beta$ over most of the conducting cycle. Since β lies in the range of 100 to 200 for typical medium-power transistors, the input impedance of each stage will be about 1 to 2 kohms. For a peak output current of 70 ma, we need an input voltage of

$$V_s = \frac{70 \times 10^{-3}}{\beta} \beta R_E = 700 \text{ mv}$$

from the driver stage. This corresponds to a power from the driver of

$$P_{driver} - \frac{[(700/\sqrt{2}) \times 10^{-3}]^2}{10^3} = 0.25 \text{ mw}$$

for an output power of about 200 mw (allowing for a 10 percent loss in the bias resistors). The actual power gain of the output stage is then nearly 10^3, or 30 db.

The design of an amplifier to meet given specifications is usually accomplished by a series of successive approximations. In the first approximation, the amplifier is mentally divided into an output stage (or possibly a driver stage and an output stage), a main amplifier cascade, and an input stage. An approximate design of each of these pieces is then undertaken.

When a large power output is required (more than 100 mw or so), the output stage is frequently designed first. The first order of business in designing an output stage is to choose the class of operation, by considering the amount of signal-power output required, the capabilities of the available active elements, the efficiency desired, the amount of distortion that is allowable, and so on. When the class of operation and active elements have been chosen, one proceeds to design the stage to obtain an appropriate bias point and an appropriate network for coupling the active element to the load. Finally, one evaluates the input-drive requirements and the input impedance.

The design techniques for a power amplifier are largely graphical, so the principles employed in this chapter apply equally well to transistors, vacuum tubes, or any other active device. Since the design centers on power output, one is usually forced to take what he gets in frequency response from the power stage. This power stage frequently provides the main limitation in overall frequency response of an amplifier.

The output requirements for the main amplifier are known when the power-output stage has been designed. By comparing the output requirements of the main amplifier to those of the signal source, we can determine the gain and bandwidth needed in the main amplifier. From the gain and bandwidth requirements the designer proceeds to estimate the number of stages required in the main amplifier and to select a suitable active element for it. These estimates are made by employing the gain-bandwidth concept for the individual stages and the bandwidth-shrinkage table for cascaded identical stages. An approximate value of R_i can also be obtained in this way.

If necessary, more exact calculations of the characteristics of the main amplifier are made using circuit models for the active devices in which the element values may be determined by experiment.

The source impedance is then compared to the input impedance of the main amplifier to determine whether a special input stage is required. Signal-to-noise ratio and other factors may also have a bearing on the input-stage design.

Circuit models for active devices employ a rather interesting role in amplifier design. For the most part, very simple models are adequate. For example, in the main amplifier the gain-bandwidth and bandwidth-shrinkage ideas frequently provide enough information to design an amplifier to greater accuracy than is warranted because of variations in the active devices from unit to unit. If more precise calculations are necessary, then the hybrid-π model can usually be made to fit the active device accurately enough, but measured parameters will most probably be necessary (particularly r_π). If even greater accuracy is necessary, it is usually preferable to resort to measured two-port parameters and general two-port analysis techniques, rather than attempt to fit measured parameters to a circuit model.

REFERENCES

Linvill, J. G., and J. F. Gibbons: "Transistors and Active Circuits," McGraw-Hill Book Company, New York, 1961, part II, and chaps. 17 and 19.

Woll, H. J., and J. B. Angell: "Handbook of Semiconductor Electronics," L. P. Hunter, Editor, McGraw-Hill Book Company, New York, 1956, chaps. 12 and 13.

Pettit, J. M., and M. McWhorter: "Electronic Amplifier Circuits," McGraw-Hill Book Company, New York, 1961, chap. 3.

Angelo, E. J.: "Electronic Circuits," McGraw-Hill Book Company, New York, 1964, chaps. 16 and 17.

Thornton, R. D., et al.: "Multistage Transistor Circuits," SEEC Notes, vol. 1, John Wiley & Sons, Inc., New York, 1963, chap. 2.

DEMONSTRATIONS

13·1 EXPERIMENTAL CHARACTERISTICS OF SINGLE-STAGE AND MULTISTAGE AMPLIFIERS
The amplifier shown below can be used to demonstrate all the basic ideas pertaining to small-signal amplifiers which were developed in this chapter. Provision is made in the input stage to short out C_s or C_1 to verify the low-frequency characteristics of a single-stage amplifier. 2N324's are used to obtain cutoff frequencies in the range of 100 kcps or lower, for convenience in measurement and to minimize the importance of parasitic wiring capacitances, etc.

13·2 CLASS B AND AB AMPLIFIERS   The waveforms, efficiency, and crossover distortion in a class B amplifier can be conveniently demonstrated with the circuit shown. It is particularly interesting to simultaneously listen to the amplifier output and observe it on an oscilloscope while the crossover distortion is varied by the potentiometer.

EXPERIMENTAL PROJECT

Using two of the transistors supplied to you, build a two-stage amplifier which uses the bias resistors you chose for the experimental project in Chap. 12. Select emitter bypass and coupling capacitors to obtain an overall low-frequency 3-db point at 20 cps, using a source resistance of 10 kohms. The cutoff frequencies of the amplifier can be measured with a signal generator (e.g., HP 610a) and a voltmeter (e.g., HP 400D); use an oscilloscope to monitor the output voltage. Alternately, a suitable signal generator will be developed in the experimental project at the end of the next chapter.

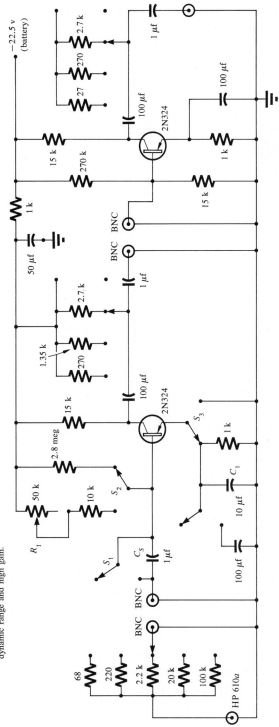

Notes: 1. Mount the amplifier above a sheet metal plate on 1 in. stand-offs to eliminate hum pickup in the output. The sheet metal plate should be connected to the ac line ground terminal, if one is available.

2. BNC connections at the output and input of each stage permit separate determination of response characteristics of each stage.

3. A third stage identical to the second can be added to illustrate the relationship between dynamic range and high gain.

A. To check frequency response with $C_1 \rightarrow \infty$, operate S_3

B. To check frequency response with $C_S \rightarrow \infty$, operate S_1 and S_2 and set R_1 to give desired bias condition

FIG. D13·1   TWO-STAGE DEMONSTRATION AMPLIFIER

Another stage identical to this one can be connected to the cascade to give greater gain

Note: Mount the amplifier above a sheet metal plate on 1 in. stand-offs to eliminate hum pickup in the output. The sheet metal plate should be connected to the ac line ground terminal, if one is available.

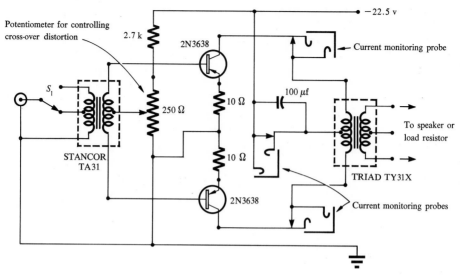

FIG. **D13 · 2** CLASS AB POWER AMPLIFIER DEMONSTRATOR, WHICH CAN BE DRIVEN FROM A SINE WAVE GENERATOR OR FROM THE AMPLIFIER SHOWN IN FIG. **D13 · 1**

PROBLEMS

13 · 1 1. A certain phonograph cartridge delivers an open-circuit signal voltage of 10 mv and has an internal impedance of 1 kohm. What is the available power from this cartridge?

2. If we want to drive an amplifier with this cartridge and have the amplifier supply 10 watts of output-signal power to a loudspeaker with an impedance of 16 ohms (resistive), what will the transducer gain of the system be? Express your answer in both numerical and decibel form.

3. If the input impedance of the amplifier is 5 kohms (resistive), what power gain would the amplifier have to produce? Express your answer in both numerical and decibel form. How does this compare with the transducer gain computed in (2)?

4. Find the magnitude of the current and voltage gains required for the system described in (3). Express your answer in both numerical and decibel forms. Comment on any differences which exist between the db gain obtained in (3) and those just obtained.

5. Suppose a transistor stage can deliver roughly a 30-db power gain. What is the least number of stages you might expect to have in the preceding amplifier?

13 · 2 Compute the midband properties of the CE stage using the hybrid-π model (*cf.* Fig. 13 · 5*b*) and compare your answers with Eqs. (13 · 8*a*)–(13 · 8*f*).

13 · 3 Compute the h parameters for the hybrid-π model used in Fig. 13 · 5*b* and then use the general two-port results given in Sec. 13 · 2 · 2 to obtain the properties of the stage.

13·4 Compute Y_{in}, Y_{out}, and the current gain for a two-port network that is characterized by y parameters. The load is G_L; the input is driven from a current source in parallel with a source conductance G_s.

13·5 A given two-port network has the following h parameters: $h_{11} = 25$ ohms, $h_{12} = 10^{-4}$, $h_{21} = -1$, $h_{22} = 10^{-6}$ volt.
1. Compute the input impedance, output admittance, current gain, voltage gain, power gain, and transducer gain for this two-port network, assuming that $R_s = R_L = 1$ kohm.
2. What value of R_L will maximize the power gain, and how large is this maximum value?
3. Evaluate the input impedance of the amplifier when R_L has been selected as in (2). What source impedance is required to make $G_T = G$? If a given source did not already have this impedance, what would you do?

13·6 A two-port network is represented by h parameters: h_{11}, h_{12}, h_{21}, h_{22}. The parameters are all real. Find the source impedance and load admittance which must be used if the network is to be matched at both ends: that is, $R_s = R_{in}$, $G_L = G_{out}$. What happens when $h_{11}h_{22} = h_{21}h_{12}$? (If you want to follow this up, see J. G. Linvill and J. F. Gibbons, "Transistors and Active Circuits," McGraw-Hill Book Company, New York, 1961, pp. 241–246).

13·7 Compute the h parameters for the T model of the transistor oriented in the CC configuration, and then verify the formulas given in Sec. 13·2·4.

13·8 Construct a table of asymptotic departures for the function
$$T = \frac{1}{1 + s/\omega_a}$$
Compute departures for both $|T|$ and Arg T for $\omega = \omega_a/4$, $\omega_a/2$, ω_a, $2\omega_a$, and $4\omega_a$. Prove that the amplitude and phase departures are equal to two frequencies ω_1 and ω_2 whose geometric mean is ω_a (that is, $\omega_1\omega_2 = \omega_a^2$).

13·9 The single-stage amplifier shown in Fig. P13·9 is to have a midband voltage gain $|V_2/V_s|$ of 50. The known circuit values are on the diagram. The transistor is to operate with $I_E = -0.5$ ma, $V_{CE} = 8.5$ volts. The transistor parameters are $r_b = 100$, $r_c = \infty$, $\beta = 100$, $f_T = 4$ mcps.
1. Give the values for R_L, R_1 and R_2.
2. How low can the low-frequency cutoff be made by increasing C_s?
3. What is the upper cutoff frequency?

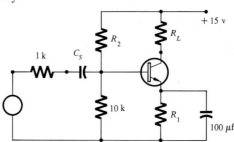

FIG. P13·9

13·10 From the data given in Fig. 13·15 [$A_v(\omega \to 0)$ and $A_v(\omega \to \infty)$] compute the apparent resistance in the emitter lead at high frequencies and compare this value given in the text.

13·11 Derive a formula for the upper cutoff frequency of the amplifier shown in Fig. P13·11, neglecting C_c and r_c. What is the upper cutoff frequency when $|I_E| = 1$ ma, $r_b = 50$ ohms, $\beta = 100$, $R_s = 1$ kohm, $R_1 = 100$ ohms, $R_L = 3$ kohms, and $f_T = 4$ mcps?

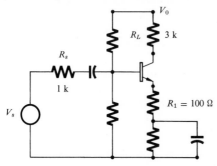

FIG. P13 · 11

13·12 An amplifier has a gain characteristic given by

$$A_v(s) = 4 \times 10^6 \frac{s(s + 10)}{s(s + 30)(s + 300)(s + 10^5)}$$

1. Estimate the upper and lower cutoff frequencies and the midband gain for this amplifier directly from the expression for A_v.
2. Plot the pole-zero diagram for $A_v(s)$.
3. Sketch and dimension the asymptotic curves for $|A_v|$ and Arg A_v vs. log ω. In both cases, show each factor separately, together with the total curve.

13·13 You are asked to design a transistor amplifier with a midband gain of 50 and an upper cutoff frequency of 20 kcps. The amplifier is fed from a 1-kohm source and drives a 2-kohm load. Assume the transistor has $\beta = 100$. What bias current would you select? How big should f_T be to meet the requirements, assuming C_c is negligible? How big should f_T be if $C_c = 15$ pfarads?

13·14 A 12AX7 vacuum triode is to be used in the circuit shown in Fig. P13 · 14. The values given achieve a bias point at which $g_m = 1250$ μ℧, $r_p = 80,000$ ohms. The interelectrode capacitances are $C_{gp} = 2$ pf, $C_{gk} = 2$ pf. C_{pk} can be neglected.
1. Compute the lower cutoff frequency for this amplifier.
2. Compute the upper cutoff frequency neglecting C_{gp}. Repeat including C_{gp}.

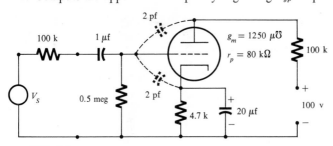

FIG. P13 · 14

13·15 A typical transistor intended for audio applications has $f_T = 5$ mcps, $C_c = 20$ pf, $\beta = 100$, $r_0 = 20$ kohms. Using a 15-volt supply voltage, design a two-stage amplifier which has its upper cutoff frequency at 50 kcps and its lower cutoff frequency at 100 cps. The source has an internal resistance of 2.5 kohms and must be coupled to the amplifier through a capacitor. Select bias components to give $|I_E| = 1$ ma, $v_{CE} = 6$ volts. You may assume the transistor is "theoretical" (that is, $r_e = kT/q|I_E|$), and that the capacitors you select are perfect.

13·16 *Theorem.* The midband voltage gain of a *CE* amplifier cannot exceed 40 per volt of collector supply voltage (that is, $A_{vo} < 40 \ V_0$).
     Proof....

13·17 Assume the small-signal properties of a transistor in the *CE* connection can be adequately represented by a hybrid-π model which includes r_π, C_π, g_m, and C_c.
    1. Determine the *h* parameters for this model as functions of the complex frequency variable *s*.
    2. Use the *h* parameters to determine $A_v(s)$ if the transistor is driven from a resistive source and drives a resistive load.

13·18 A common-collector stage (less biasing components) is shown in Fig. P13·18. The transistor can be represented by a hybrid-π model with $f_T = 10$ mcps, $\beta = 50$, $C_c = 10$ pfarads. The transistor is biased at $|I_E| = 1$ ma.

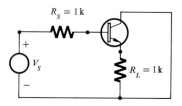

FIG. P13·18

    1. Compute V_2/V_s as a function of the complex frequency variable *s*. Plot the poles and zeros of V_2/V_s for the element values given.
    2. Sketch the asymptotic forms of $|A_v(s)|$ and Arg $|A_v(s)|$ vs. log ω.
    3. What is the approximate bandwidth of this circuit?
    4. What is the bandwidth if $R_s = 10$ kohms?

13·19 Repeat Prob. 13·18 for a common-base amplifier, using $R_s = R_L = 1$ kohm only.

13·20 What is the low-frequency current (I_2/I_s) gain of the amplifier shown in Fig. P13·20? What is the approximate bandwidth of the amplifier assuming that the admittance of C_c is small compared to G_L at the frequencies of interest.

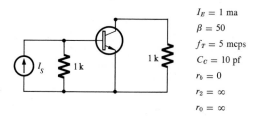

$I_E = 1$ ma
$\beta = 50$
$f_T = 5$ mcps
$C_c = 10$ pf
$r_b = 0$
$r_2 = \infty$
$r_0 = \infty$

FIG. P13·20

13·21 The circuit shown in Fig. P13·21 is sometimes used when a transistor amplifier must drive a relatively high-impedance load (5-10 kohms or more) and the amplifier bandwidth must be maintained.

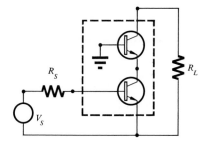

1. How would you bias this combination of transistors to keep both of them in their normal-active range?
2. Compute the gain and upper cutoff frequency for the circuit assuming $|I_E| = 1$ ma, $f_T = 10$ mcps, $C_c = 10$ pfarads, $R_s = 1$ kohm, $R_L = 10$ kohms.
3. Compare the results of (2) with the results obtained if only one transistor were used in the *CE* connection.

13·22 A two-stage amplifier has the following properties: Stage 1: $f_l = 100$ cps, $f_u = 100$ kcps, $A_{vo} = 50$. Stage 2: $f_l = 60$ cps, $f_u = 75$ kcps, $A_{vo} = 60$. Sketch and dimension the db vs. log f curves for each stage and for the total amplifier. Determine f_l, f_u, and the midband gain for the two-stage amplifier.

13·23 An application requires an amplifier with a midband gain of 5000 and an upper cutoff frequency of 1 mcps. Can this application be met using a transistor with $\beta = 100, f_T = 5$ mcps? If so, how many stages are required? If silicon transistors having $\beta_{min} = 30, f_{T,min} = 500$ mcps are used, how many transistors are required? State your assumptions clearly, including any values of C_c or other parameters that seem reasonable.

13·24 A cascade of two identical *CE* stages is to have an overall bandwidth of 4 mcps. What bandwidth is required in the individual stages? What value of interstage resistance R_i is needed to achieve this bandwidth if the transistors have $\beta = 50, f_T = 100$ mcps, $C_c = 2$ pfarads and $|I_E| = 1$ ma.

13·25 Estimate the f_T required to build an amplifier with a midband gain of 10,000 and a bandwidth of 500 kcps. The transistors can be assumed to have $\beta = 50$, and the minimum number of transistors are to be used. Assume the cascade will be made up of identical stages for simplicity.

13·26 A cascade of two identical *CE* stages is to have a bandwidth of 100 kcps. The transistors have $\beta = 100, f_T = 10$ mcps, $C_c = 10$ pfarads. Select a generator resistance and load resistance to insure that the stages are identical. What is the midband gain of the resulting amplifier? Sketch the asymptotic gain and phase characteristics for the frequency range

$$25 \text{ kcps} < f < 400 \text{ kcps}$$

Be careful to include all − signs in computing the phase characteristic.

13·27 The upper cutoff frequency f_u of a cascade of identical stages is related to the cutoff frequency of the individual stages f_u' by

$$f_u = f_u' \sqrt{2^{1/n} - 1}$$

Show that

$$f_u \simeq \frac{0.83}{\sqrt{n}} f_u' \qquad n \geq 2$$

by direct comparison for $n = 2$, 4, and 8.

13·28 *Practically useless theorem.* If the gain-bandwidth product of a single stage is constant ($A_{vo} f_u = K$), and it is desired to build a cascade amplifier having a fixed bandwidth B, then to obtain maximum midband gain one should use n stages, where

$$\sqrt{n} = \frac{0.83 K}{\sqrt{eB}}$$

e is the base of natural logarithms. The gain per stage will be $e^{1/2}$. Give the proof.

13·29 A proposed two-stage amplifier is shown in Fig. P13·29. The transistors have $\beta = 100$, $f_T = 4$ mcps, $C_c = 20$ pfarads.
1. Find the bias points for the two transistors.
2. Compute the output-current swing of the amplifier for the given source voltage. Comment on your result (that is, why is the bias point for the second transistor different from that for the first?).

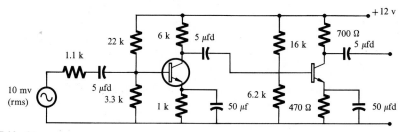

FIG. P13·29

13·30 An application requires an amplifier with a midband voltage gain of 60 db, an upper cutoff frequency of 50 kcps, and a lower cutoff frequency of 100 cps. Can the amplifier be constructed from transistors which have $f_T = 5$ mcps, $C_c = 20$ pfarads, $\beta = 100$? If so, design an amplifier to fulfill the specifications. If not, specify a larger f_T and proceed to design the amplifier.

13·31 A two-stage amplifier has a voltage gain V_2/V_s of 2,500 and a bandwidth of 100 kcps. It is driven by a source with 10-kohm internal impedance. The amplifier has a noise figure of 4 db when this source is used.
1. What is the minimum signal which can be applied if we require a signal-to-noise ratio of 2:1 at the output?
2. If the maximum input signal for distortion-free amplification is 200 μv, what is the dynamic range of this amplifier (in db)?
3. If the output transistor in this amplifier is biased at $|I_E| = 2$ ma and drives an ac load resistance of 2 kohms, approximately what is the maximum input signal? The dynamic range?

13·32 An amplifier has a voltage gain of 5,000 and a bandwidth of 1 mcps. It is driven from a source with 1-kohm internal resistance, and the amplifier input is matched to the source. Compute the rms noise voltage appearing at the output, assuming that the first transistor has a noise figure of 4 db at its operating point.

13·33 Two 12AX7 triodes are to be used in the amplifier shown in Fig. P13·33. The triodes may be represented by a hybrid-π model in which $C_\pi \to C_{gk} = 2$ pf, $C_c \to C_{gp} = 2$ pf, $r_\pi \to \infty$, $r_b \to 0$, $r_c \to \infty$, $r_0 = 100$ kohms, $g_m = 10^{-3}$ mhos. In addition, there is a 5-pf wiring capacitance to be associated with the input to each stage. Determine the midband gain and high-frequency cutoff of this amplifier. What effect would a comparable wiring capacitance have on a transistor amplifier designed for this same frequency range?

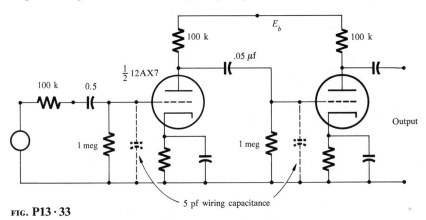

FIG. P13·33

13·34 A 12AX7 is a vacuum tube which is roughly equivalent to a 2N324 in overall characteristics, as the following problem will show:

The small-signal circuit model for a vacuum tube is generated from the hybrid-π model by making the identifications

$$r_\pi \to \infty \qquad C_c \to C_{gp} \qquad g_m \to g_m \qquad r_0 \to r_p \parallel C_{pk}$$
$$C_\pi \to C_{gk} \qquad r_c \to \infty \qquad r_b \to 0$$

The 12AX7 has $C_{gp} \simeq 2$ pfarads, $C_{gk} \simeq 2$ pf, $C_{pk} \simeq 0.5$ pfarad. At an operating plate current of 0.5 ma, $g_m = 1.2 \times 10^{-3}$ mho, $r_p \simeq 80,000$ ohms.

1. Estimate the plate load resistors required to build a two-stage amplifier with a midband gain of 2,000.

2. Compute the gain-bandwidth product for each stage, and estimate the bandwidth of the two-stage amplifier.

13·35 The tremendous influence of C_{gp} on the gain-bandwidth product is what led to the development of the pentode.

1. A 6CB6 is a common pentode used for television video-amplifier service. It has $C_{gp} = 0.02$ pfarad, $C_{gk} = 6.3$ pfarads, $C_{pk} = 3$ pfarads. At a plate current of 10 ma it has $g_m = 6.2 \times 10^{-3}$ mho, $r_p = 0.62$ meg. How much gain can be obtained in a three-stage amplifier using the 6CB6 if the overall amplifier bandwidth is to be 3 mcps?

2. A 2N708 has $\beta = 50$, $f_T = 1,000$ mcps and $C_c = 1$ pfarad. Compare the 3-mcps-wide three-stage amplifier that can be built with this transistor with the one using 6CB6's.

13·36 A 2N1613 silicon planar transistor has $V_{CE,max} = 40$ volts, $P_{c,max} = 800$ mw

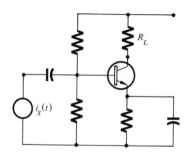

FIG. P13·36 CLASS *A* POWER AMPLIFIER

(with no heat sink), and $I_{c,max} \simeq 0.5$ amp. The transistor also has $\beta \simeq 100$. β will be assumed to be constant in this problem.

1. On an i_C-v_{CE} plane sketch the permissible load-line area for this transistor.

2. What is the maximum output-signal power that this transistor can deliver in the circuit of Fig. P13·36?

3. Construct the maximum and minimum resistance collector load lines that can be used to obtain the maximum power output, and state the values of $R_{L,max}$ and $R_{L,min}$ (as an approximation, neglect the voltage drop across R_E in constructing the relevant load lines).

4. Locate the bias point (I_C, V_{CE}) which must be obtained to allow maximum power output in each case.

5. Compute the base-current swing necessary in each case to produce full output.

6. Estimate the power input and power gain on the assumption that the input resistance is r_π.

13·37 Derive an approximate formula for the power gain of a transistor in the circuit of Fig. 13·34 when it is delivering maximum power output. Express your result in terms of $P_{C,max}$ and the bias variables.

13·38 Design a class A transformer-coupled power amplifier to supply a power output of 50 mw to a 10-ohm load. The transistor may be assumed to have $\beta = 50$. Specify the transformer and the transistor ratings you need. The supply voltage is a 9-volt battery. Take bias-stability considerations into account in your design.

13·39 A class A driver for a class B final stage is shown in Fig. P13·39. The input to the class B stage is represented by the ideal diodes and 100-ohm resistors. Only one diode conducts during each half cycle.

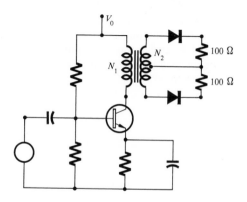

FIG. P13·39

The transistor has $V_{CE,max} = 40$ volts, $I_{C,max} = 0.5$ amp, $P_{C,max} = 0.5$ watt. Determine the values of V_0, I_C, and N_1/N_2 which will insure that the driver delivers maximum power (and therefore maximum current) to the class B input with minimum input signal. If the transistor has $\beta = 50$, what is the minimum input base drive (approximately)?

13·40 Plot the efficiency of a transformer-coupled class A amplifier as a function of $(P_0/\hat{P}_0)$ where P_0 is the actual output power and $\hat{P}_0$ is the maximum output power that can be obtained for the given load resistor.

13·41 An inexpensive transistor radio uses a class A driver to supply power to a class AB final-output stage. The circuit is similar to the one shown in Fig. P13·39, with the 100-ohm resistors replaced by 500-ohm resistors. Assuming $V_0 = 9$ volts, design a driver stage to supply 5 mw of power to the class AB circuit. The transformer efficiency is 75 percent, and the transistor β is 100. In your design be sure to specify the $V_{CE,max}$ and $P_{C,max}$ ratings you need for the transistor and design the biasing circuit to obtain the proper bias point. The emitter bypass capacitor should be selected to ensure a low-frequency cutoff of 200 cps. Do not waste more than 1 volt of the battery voltage in the emitter resistor and emitter-base junction.

13·42 What is the relationship between area under the triangle formed by the load line and the i_C-v_{CE} axes and power output in a class A stage?

13·43 Two transistors having $v_{CE,max} = 50$ volts and $P_{C,max} = 1$ watt are to be used in an ideal class B amplifier. The amplifier is to operate from a 22.5-volt supply.

1. What is the maximum output power that can be obtained and the load resistance needed to obtain it?
2. What input-current swing is required if $\beta = 50$ for the transistors.
3. Plot the efficiency of this circuit as a function of the power output.

13·44 A class B amplifier stage is to be designed to give an output power of 500 mw to an 8-ohm loudspeaker. The available voltage supply is a 9 volt battery. The transformer can be assumed to be 75 percent efficient. To minimize distortion it is necessary to limit the drive so that the minimum v_{CE} is 1 volt. Select a transformer turns ratio for this application.

13·45 A class B transistor power amplifier is to deliver 5 watts to a 100-ohm load with a sinusoidal input signal. The circuit is to operate at the highest possible efficiency and use the minimum possible drive. Specify the transformer which should be used if $v_{CE,max} = 25$ volts? 50 volts? What is the maximum collector current in each case? If the sinusoidal signal can have any amplitude from zero to the value required to give 5 watts of output power, what is the greatest *average* power dissipated in each transistor (for $v_{CE} = 50$ volts)? What is the output power under this latter condition?

14

INTRODUCTION TO FEEDBACK

IN THE PREVIOUS chapter we studied the circuit properties of some basic amplifier configurations. We saw that the gain and bandwidth of a stage could be partially controlled by selecting appropriate source, load, and interstage impedances. However, there are limits on the amount of control which we can exercise over a simple *CE* stage by the methods described in the last chapter. For one thing, a biasing network which maintains adequate bias stability may absorb more signal power than we would like it to. The biasing network may also make it difficult to achieve the desired low-frequency response, because of the bypass and coupling capacitors. More importantly, we have no way of accounting for the variability of gain and bandwidth which will arise from variations in the parameters of the active devices from unit to unit or with age.

When these problems (and others to be mentioned later) are important, the *intentional* use of feedback in a stage or perhaps around an entire amplifier sometimes gives the designer additional freedom in controlling its properties. The word "intentional" is emphasized because feedback through parasitic elements (such as C_c and r_2 in a CE stage) always exists.

We usually apply feedback to an amplifier by returning a portion of its output signal to the input circuit. The actual input to the amplifier is then determined by both the signal from the source and the fed-back signal, as in Fig. 14·1. At any given frequency, the fed-back signal and the source signal can both be represented by complex numbers, such as $M_s(\omega)e^{j\phi_s(\omega)}$ and $M_f(\omega)e^{j\phi_f(\omega)}$. M is used for generality to represent either a current or a voltage. The combination of the fed-back signal and the source signal can be visualized as an equivalent source which delivers to the amplifier input a signal of

$$M_s(\omega)e^{j\phi_s(\omega)} + M_f(\omega)e^{j\phi_f(\omega)}$$

If $\phi_f(\omega)$ and $\phi_s(\omega)$ are related by

$$\phi_f(\omega) = \phi_s(\omega) \pm 180°(\pm n \times 360°)$$

in some frequency range, then in that frequency range the equivalent source amplitude is

$$M_s(\omega) - M_f(\omega)$$

Since $M_f(\omega)$ tends to cancel $M_s(\omega)$, the feedback is termed *negative* in this frequency range. The desirable amplifier characteristics mentioned earlier are obtained by proper use of negative feedback.

If $\phi_f(\omega)$ and $\phi_s(\omega)$ are related by

$$\phi_f(\omega) = \phi_s(\omega) \pm n \times 360°$$

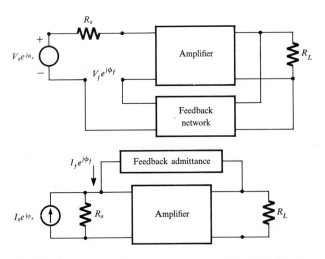

FIG. 14·1 TWO OF THE MANY POSSIBLE TYPES OF FEEDBACK

in some frequency range, then the equivalent source amplitude is

$$M_s(\omega) + M_f(\omega)$$

$M_f(\omega)$ now tends to reinforce $M_s(\omega)$ and the feedback is termed *positive* in this frequency range. Positive feedback can also be used to control the characteristics of an amplifier, though great care is necessary here because too much positive feedback will cause an amplifier to break into oscillation.

The principal problem associated with the use of feedback is that the phase of the fed-back signal relative to the signal from the source changes with frequency. A feedback path which gives negative feedback in the midband-frequency range of an amplifier may give positive feedback and the potential for oscillation in some "out-of-band" frequency range. To avoid this possibility, it is usually necessary to know the gain and phase characteristics of the basic amplifier over a wide frequency range and possibly to adjust these characteristics so that when the contemplated feedback is applied the conditions for oscillations are not met.

Perhaps the second most important problem associated with feedback is how to study it. Some of the most powerful techniques for analyzing feedback amplifiers (or feedback systems in general) treat the amplifier as simply a black box with a given transfer function. These techniques give a good deal of insight into feedback methods and problems, but they do not always lead a circuit designer to an appreciation of how to apply feedback in some simple cases and what to expect if he does. For this reason we shall divide the study of feedback amplifiers into two parts. In the first part of this chapter we shall concentrate on the properties of simple feedback amplifiers. In particular, we shall first examine the characteristics that can be obtained by the use of negative feedback around a single stage of amplification, the emphasis being on how feedback can be used to trade gain and bandwidth and related matters of practical importance. Similarly we shall introduce positive feedback by first studying several practical types of sine-wave oscillators that are made by using frequency-selective positive feedback. We shall then briefly consider how a generalized treatment of feedback amplifiers can be formulated to embrace feedback of arbitrary magnitude and phase.

14·1 PARALLEL NEGATIVE FEEDBACK AND THE STABILIZATION OF A_i

One of the desirable effects of negative feedback is that the midband gain of an amplifier stage can be *stabilized;* that is, can be made nearly independent of variations in transistor parameters from unit to unit (or with age). The type of feedback network determines whether the current gain or the voltage gain (or perhaps a transfer immittance) is stabilized.

As a first example, consider the circuit shown in Fig. 14·2. A conductance G_f is added between the collector and base terminals of a simple CE

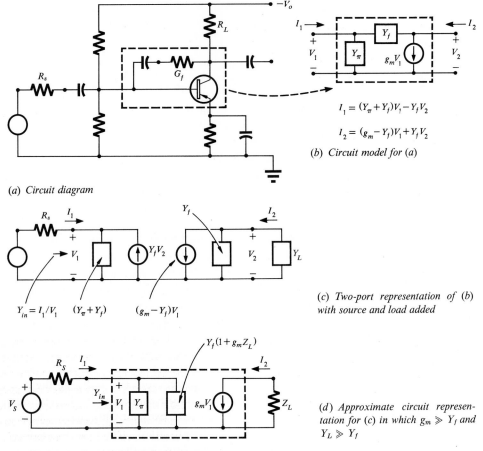

$$I_1 = (Y_\pi + Y_f)V_1 - Y_f V_2$$

$$I_2 = (g_m - Y_f)V_1 + Y_f V_2$$

(b) Circuit model for (a)

(a) Circuit diagram

$Y_{in} = I_1/V_1 \quad (Y_\pi + Y_f) \quad (g_m - Y_f)V_1$

(c) Two-port representation of (b) with source and load added

(d) Approximate circuit representation for (c) in which $g_m \gg Y_f$ and $Y_L \gg Y_f$

FIG. 14·2   COLLECTOR-TO-BASE FEEDBACK CONDUCTANCE AROUND A SINGLE CE STAGE

stage. A large capacitor is also connected in series with G_f so that the bias condition is not affected.[1]

Now, from the point of view of the signal, there is a 180° phase reversal between the input and output voltages of the stage; the feedback is therefore *negative,* at least at frequencies for which the transistor models we shall use are valid. To study the properties of the stage at signal frequencies, we redraw the network in the form shown in Fig. 14·2b. For simplicity, we neglect r_b and lump any finite output conductance r_0 of the transistor in with Y_L. Y_f is considered to include both G_f, g_c, and sC_c.

If we consider the transistor with the feedback conductance G_f as a new two-port device, we can write the following equations relating the rms signal currents and voltages at the terminals shown in Fig. 14·2b:

[1] By redesigning the bias network to account for the presence of G_f, this capacitor can be removed.

$$I_1 = (Y_\pi + Y_f)V_1 - Y_f V_2 \qquad I_2 = (g_m - Y_f)V_1 + Y_f V_2 \quad (14 \cdot 1)$$

These equations can be visualized in the manner shown in Fig. $14 \cdot 2c$. Here, two controlled sources are used to represent the dependence of I_1 on V_2 and the dependence of I_2 on V_1.

In most practical situations, Y_f will be very small compared to g_m and Y_L. When this is the case, the output-signal voltage of the stage is

$$V_2 = -g_m Z_L V_1 \qquad (14 \cdot 2)$$

where $Z_L = 1/Y_L$. The input current is therefore

$$I_1 = [Y_\pi + Y_f(1 + g_m Z_L)]V_1 \qquad (14 \cdot 3)$$

and the input admittance is

$$Y_{in} = \frac{I_1}{V_1} = Y_\pi + Y_f(1 + g_m Z_L) \qquad (14 \cdot 4)$$

Equation $(14 \cdot 4)$ is identical to the equations which were developed in Chap. 13 to study the effects of C_c on the upper cutoff frequency of a stage. To show the connection, we redraw the two-port network in Fig. $14 \cdot 2d$. Here the input admittance contains a factor $Y_f(1 + g_m Z_L)$, this being due to the coupling of output to input through Y_f. Remember that this representation of the effect of Y_f is based on the assumption that $g_m \gg Y_f$ and $Y_L \gg Y_f$. This is called the *Miller approximation*.

To study the effects of feedback on the properties of the stage, we first write out the current gain:

$$A_i = \frac{I_2}{I_1} = \frac{g_m V_1}{Y_{in} V_1} = \frac{g_m}{Y_\pi + Y_f(1 + g_m Z_L)} \qquad (14 \cdot 5)$$

We shall assume that the load is purely resistive for the moment, and that $g_m R_L \gg 1$, as it will be in a practical amplifier. We also assume that $G_f \gg g_c$, so that $Y_f = G_f + sC_c$. Then Eq. $(14 \cdot 5)$ becomes

$$A_i \cong \frac{g_m}{g_\pi + g_m R_L G_f + sC_\pi + sC_c g_m R_L}$$

$$= \frac{g_m}{g_\pi + g_m R_L G_f} \frac{1}{1 + s(C_\pi + C_c g_m R_L)/(g_\pi + g_m R_L G_f)} \qquad (14 \cdot 6)$$

The midband current gain is therefore

$$A_{i0} = \frac{g_m}{g_\pi + g_m R_L G_f} \qquad (14 \cdot 7a)$$

and the upper cutoff frequency is

$$f_u = \frac{g_\pi + g_m R_L G_f}{2\pi(C_\pi + C_c g_m R_L)} \qquad (14 \cdot 7b)$$

The product of the midband gain and upper cutoff frequency is

$$|A_{i0}|\,|f_u| = \frac{g_m}{2\pi(C_\pi + C_c g_m R_L)} \tag{14·7c}$$

which is the same as the gain-bandwidth product without feedback [cf. Eq. (13·46)].

14·1·1 *Stabilization of the midband-current gain.* Let us now use these results to study the performance of the circuit. We begin with Eq. (14·7a). Using the equality

$$g_m r_\pi = \beta$$

we can rewrite Eq. (14·7a) in the form

$$A_{i0} = \frac{\beta}{1 + \beta R_L G_f}$$

This equation shows that $A_{i0} = \beta$ when $G_f = 0$, and

$$A_{i0} = \frac{1}{R_L G_f}$$

when

$$\beta R_L G_f \gg 1$$

When this latter condition is met, the midband-current gain of the stage is *independent of the β of the transistor*. The midband gain is therefore said to be *stabilized*.

EXAMPLE   As an illustration of the effect of this stabilization, let us consider what happens when two transistors, which are nominally the same but actually have β's of 100 and 200, are used in the stage. We shall assume $R_L = 1$ kohm. To achieve a stabilized midband gain for the transistor with $\beta = 100$, we need $\beta R_L G_f \gtrsim 10$, or

$$G_f \simeq \frac{10}{\beta R_L} = \frac{10}{10^2 \times 10^3} = 10^{-4} \text{ mhos} \qquad R_f \simeq 10 \text{ kohms}$$

With this value of G_f, the midband-current gain is 9.1. When the transistor with $\beta = 200$ is substituted, the midband gain changes to

$$A_{i0} = \frac{200}{1 + 200 \times 10^{-1}} = 9.5$$

Thus, when β is doubled, the midband gain changes by 6 db without feedback and by only 0.36 db when feedback is applied. Of course, we sacrifice a great deal of midband gain (about 20 db) to obtain this insensitivity of gain to β variations. However, this is frequently not too great a price to pay. The gain can be recovered by cascading stages.

System theory interpretation of Eq. (14·7a). An interpretation of Eq. (14·7a), or its alternate form

$$A_{i0} = \frac{\beta}{1 + \beta R_L G_f}$$

can be given which highlights the "self-adjusting" nature of A_{i0} in the presence of the variable β. We shall develop this interpretation briefly for the insight it gives into a different point of view about feedback.

To do this we define

$$R_L G_f = \mu$$

and rewrite A_{i0} in the form

$$A_{i0} = \frac{\beta}{1 + \mu\beta} = \frac{1}{\mu} \frac{1}{1 + (1/\mu\beta)}$$

The interpretation now proceeds as follows. Suppose we want to build an amplifier to give a midband gain of $1/\mu$. The amplifiers available to us can perhaps be adjusted to give this gain, but they have parameters which vary from unit to unit and with age. This will cause undesirable variations in A_{i0}. To avoid this, we build an amplifier with a midband-current gain β, such that $\mu\beta \gg 1$. Then, by setting $G_f = R_L/\mu$, we obtain the required gain, and it is independent of the variations of parameters of the active devices.

14 · 1 · 2   *Effect of feedback on the stage bandwidth.* The addition of the feedback conductance G_f also has an interesting effect on the bandwidth of the stage. This can be most simply studied by considering the gain-bandwidth product given in Eq. (14 · 7c):[1]

$$|A_{i0}| \, |B| = \frac{g_m}{2\pi(C_\pi + C_c g_m R_L)}$$

An inspection of this equation reveals the following important facts:

1. To the extent that approximations made in the analysis are valid, the gain-bandwidth product is independent of G_f. This means that we can *trade midband gain for bandwidth by adjusting G_f.*

2. The gain-bandwidth product is the same as it is for an amplifier without feedback.

Thus the gain-bandwidth product of an amplifier or, more particularly, of an active device begins to emerge as a rather fundamental parameter. We found this parameter to be of considerable value in Chap. 13 when we were estimating the properties of multistage amplifiers. Here again the gain-bandwidth product turns up as the governing factor in trading gain for bandwidth. Notice that to the extent that our approximations are valid, we have not sacrificed anything in terms of the gain-bandwidth product of a stage by using feedback, and we have added the feature of gain stabilization.

EXAMPLE   To study the effect of the feedback conductance in a practical example, let us use the parameters which apply for a 2N324 in the gain-bandwidth expression.

---

[1] We are again assuming that $f_u \gg f_l$, so the upper cutoff frequency is essentially the same as the bandwidth.

Since we found earlier that $r_e = 30$ ohms$/|I_E(\text{ma})|$, we have $g_m = 33 \times 10^{-3}$ mho at $I_E = 1$ ma; $C_\pi \cong 1,300$ pfarads, $C_c \simeq 18$ pfarads. We shall choose a load resistance of 270 ohms, for reasons which will appear shortly. Using these numbers, the gain-bandwidth product is

$$|A_{i0}| \, |f_u| = \frac{33 \times 10^{-3}}{2\pi(1300 + 150) \times 10^{-12}} = 3.6 \times 10^6$$

If $G_f = 0$, $A_{i0} = \beta \simeq 130$, so $f_u = 28$ kcps. If A_{i0} is reduced to 10 by feedback, f_u increases to 360 kcps.

Experimental results. The circuit shown in Fig. 14·3 was constructed to check these predictions experimentally. The results are in satisfactory agreement with the theory. Each data point gives the midband-current gain and upper cutoff frequency for the given value of G_f. Since the product of midband gain and cutoff frequency is theoretically independent of G_f, the plot should be a straight line with a slope of -1:

$$\log A_{i0} = \log K - \log f_u \qquad \text{or} \qquad y = C - x$$

The frequency at which the current gain is unity should be approximately the same as f_T for the transistor. For the particular 2N324 tested, f_T was about 4 mcps.

It should be noted that the results shown in Fig. 14·3 validate the previous theory more accurately than one should expect in general. Usually, when R_f

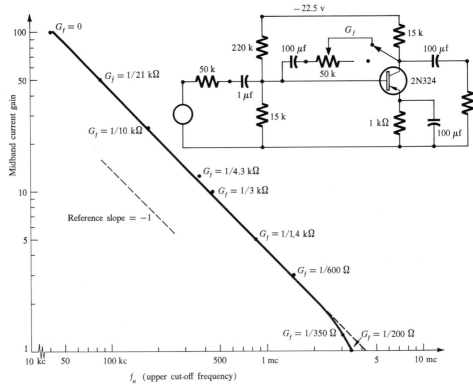

FIG. **14·3**   EXPERIMENTAL DATA TO SUPPORT THE THEORY OF PARALLEL FEEDBACK THROUGH G_f

564   Semiconductor electronics

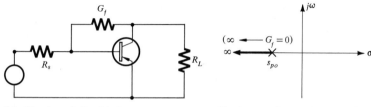

(a) Circuit excluding biasing components

(b) Locus of s_p as G_f varies from 0 to ∞

FIG. 14·4 ROOT-LOCUS DIAGRAM FOR A SIMPLE CE STAGE WITH A FEEDBACK CON-DUCTANCE ADDED FROM COLLECTOR TO BASE

becomes comparable to R_L, there will be a more noticeable decrease in the gain-bandwidth product. Thus, we cannot trade gain for bandwidth as efficiently at high bandwidths as at lower bandwidths. However, there are other things we can do to attempt to keep the gain-bandwidth product at high frequencies (a common technique is to use an inductance in series with R_f to decrease the feedback at high frequencies). These measures introduce extra frequency dependence in the transfer function of the amplifier, however, and they must be carefully studied to ensure that they do not set up oscillation conditions at some unsuspected frequency.

Interpretation of bandwidth changes on a pole-zero diagram: root-locus. It is also interesting to visualize the change in bandwidth with changes in G_f in terms of the pole-zero diagram for the stage. If we again limit our considerations to the upper cutoff frequency, the expression for $A_i(s)$ given in Eq. (14·6) exhibits a single pole on the real s axis at the position

$$s_p = -\frac{g_\pi + g_m R_L G_f}{C_\pi + C_c g_m R_L} = \frac{-(1 + \beta R_L G_f)}{2\pi r_\pi C_\pi (1 + C_c g_m R_L / C_\pi)}$$

When $G_f = 0$, the pole is located at

$$s_{p0} = \frac{-1}{2\pi r_\pi C_\pi (1 + C_c g_m R_L / C_\pi)}$$

As G_f is increased, the pole moves along the negative real axis, with $s_p \to \infty$ as $G_f \to \infty$. This behavior is shown in Fig. 14·4.

Because s_p is the root of the denominator of $A_i(s)$, the trajectory of s_p as G_f is varied is termed the *root-locus*. Figure 14·4 is a root-locus diagram. Root-locus diagrams provide an interesting and powerful tool for studying feedback systems, as we shall see later.

14·1·3 *Effect of feedback on input admittance.* In addition to affecting the gain and bandwidth of the stage, the addition of the feedback conductance G_f also affects the admittances of the stage at each port. The relation between feedback and input admittance is most readily expressed by Eq. (14·5):

$$A_i = \frac{g_m}{Y_{in}} \tag{14·8a}$$

Introduction to feedback 565

This equation shows that the product of current gain and input admittance is constant:

$$A_i Y_{in} = g_m \qquad (14 \cdot 8b)$$

From this condition, we can see that a reduction in A_i by a factor of 10 will cause an increase in Y_{in} by the same factor. Alternately, we can "trade" A_i and Y_{in}, in the same sense that midband gain and bandwidth can be traded.

EXAMPLE If we again use $g_m = 33 \times 10^{-3}$ mho, corresponding to $I_E = 1$ ma, then

$$A_i Y_{in} = 33 \times 10^{-3}$$

If the addition of G_f produces a midband gain of 10, then $Y_{in} = 3.3 \times 10^{-3}$, or $R_{in} = 300$ ohms. If we were to build a cascade of identical stages from this basic CE stage with G_f feedback, all stages except the last one would have an ac load resistance of no more than 300 ohms in the midband range. This accounts for our use of $R_L = 270$ ohms in a previous example.

It is also important to note that the control on input admittance which is afforded by the use of feedback is useful in designing input stages in a multistage amplifier. If we want to ensure that the source delivers a current to the input stage, then we will be interested in ensuring that the input stage has a low impedance compared to the source. The circuit under study is also of interest if we want to match the input to a 300-ohm transmission line. For such a case, it maintains as large a bandwidth as the transistor will allow.

Effect of the feedback on low-frequency response. In addition to its effect on midband gain, bandwidth, and input admittance, the feedback network may have some effect on the low-frequency response of the amplifier. If the biasing network is designed to give the correct bias point with G_f included, then the gain stabilization feature will be present right down to zero frequency, though R_L will perhaps be different at dc than in the midband range if the stages are capacitively coupled. If a dc-blocking capacitor C_f is used in series with G_f, however, then the low-frequency gain will be considerably different from the midband gain.

This effect can be studied by substituting

$$Y_f = \frac{sC_f}{1 + sC_f R_f}$$

into Eq. (14 · 5), and using $Y_\pi = g_\pi$, $Z_L = R_L$, in recognition of the fact that typical values of C_f will affect only the low-frequency characteristics. With these approximations the current gain can be written as

$$A_i(s) \simeq \frac{\beta(1 + sC_f R_f)}{1 + sC_f R_f(1 + \beta R_L G_f)}$$

for frequencies in the midband range and below. The current gain is thus β at $s = 0$, as it should be. The feedback introduces a pole which causes the gain to begin to drop at 6 db/octave at a frequency

$$f_p = \frac{1}{2\pi C_f R_f} \frac{A_{i0}}{\beta}$$

There is also a zero at

$$f_0 = \frac{1}{2\pi C_f R_f}$$

so that the asymptotic gain characteristic again flattens out at its midband value of

$$A_{i0} = \frac{\beta}{1 + \beta R_L G_f}$$

Comparison of cascade-amplifier chains with and without feedback. With the influence of feedback on gain, bandwidth, and input admittance known, we can evaluate in a preliminary way the effect of cascading several CE stages, each of which has a feedback conductance G_f, and compare the result with the no-feedback case. An example will serve to illustrate the main points.

EXAMPLE Suppose we want to use a 2N324 to build an amplifier with a midband-current gain of 1,000. We will consider a cascade of identical stages. Since the 2N324 has a typical β of 100, we need at least two stages. If we want to use feedback effectively, then we want each stage to have a gain which is appreciably less than β. We shall therefore consider three identical stages, each with a gain of 10.

The formula for midband gain with feedback is

$$A_{i0} = \frac{\beta}{1 + \beta G_f R_L}$$

so for a typical β of 100 we need

$$G_f R_L = 0.09$$

We shall assume that the biasing network is designed so that the ac load resistance on each collector is simply the input resistance of the next stage $(R_i \gg R_{in})$. The required input resistance can be found from Eq. $(14 \cdot 8a)$ as

$$A_{i0} = \frac{g_m}{Y_{in}} \Rightarrow R_L = \frac{A_{i0}}{g_m}$$

or $R_L = 300$ ohms at $I_E = 1$ ma. The required value of R_f is then

$$R_f = \frac{R_L}{0.09} \simeq 3.3 \text{ kohms}$$

This R_f will be connected from collector to base of each stage (with a suitable blocking capacitor), and a final load resistance of 300 ohms will be used.

The bandwidth of each stage can be estimated as

$$|A_{i0}||B| \simeq \frac{f_T}{1 + (g_m R_L C_c / C_\pi)} = \frac{f_T}{1 + (A_{i0} C_c / C_\pi)} \sim \frac{f_T}{1.14}$$

Using $f_T = 4$ mcps gives a gain-bandwidth product of 3.5 mcps, and therefore a stage bandwidth of 350 kcps. The three-stage amplifier will have a bandwidth of about one-half this, or 175 kcps.

If three transistors with $\beta = 200$ are used in the circuit, the midband gain will change to 11 per stage, or the overall gain will change to 1,331. The gain-bandwidth product will change (assuming f_T is the same in all transistors) to 3.15 mcps, so the stage bandwidth becomes 286 kcps, with the overall bandwidth about 140 kcps.

Of course, we can also design a three-stage amplifier without feedback to give an overall gain of 1,000 and a bandwidth of 175 kcps. However, if three transistors with $\beta = 200$ are substituted into this circuit, the midband gain will change to $2^3 \times 1,000 = 8,000$, and the bandwidth will shrink drastically. Herein lies the utility of feedback.

14·1·4 *Stabilization of bias by feedback.* If we eliminate the dc blocking capacitor in series with the feedback conductance G_f, then the feedback will be effective at dc. If sufficient feedback is used to make the current gain of the stage independent of the transistor parameters, we can obtain bias stability with the same network that stabilizes the signal gain.

For example, if we select a dc collector load resistance of 1 kohm and a feedback resistance of $R_f = 10$ kohms, the midband-stage gain will be 10. If we then supply a base bias current of 0.1 ma to the stage, the collector current will be 1 ma. A simple technique for doing this is shown in Fig. 14·5. If the signal is direct-coupled to the transistor in this circuit (via an appropriate R_s to maintain the correct bias condition), the circuit will amplify signals down to zero frequency.

If we want the feedback to be effective at dc but *not* at signal frequencies (so as to stabilize only the bias point), then we can split the feedback resistor into two parts and add a shunt capacitance, as shown in Fig. 14·6. R_f is usually a rather large resistor, so the shunt capacitance used to eliminate feedback at signal frequencies can be much smaller than the bypass capacitance needed in the emitter circuit.

Because large capacitors with very low series resistance are hard to obtain, collector-to-base feedback is frequently preferred for obtaining bias stability. This type of bias stabilizing may also be applied around an entire amplifier if there are no dc blocking capacitors in it. Bias stability in such circuits is obtained with a minimum of components.

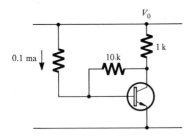

FIG. 14·5 USE OF COLLECTOR-BASE FEEDBACK TO STABILIZE BIAS

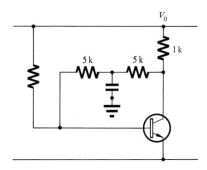

FIG. **14·6** ADDITION OF A CAPACITOR IN THE FEEDBACK NETWORK TO ELIMINATE FEEDBACK AT SIGNAL FREQUENCIES

14·1·5 *Frequency-dependent feedback networks.* If the feedback conductance is replaced by a more general feedback admittance, further control over the gain and bandwidth properties of the stage can be obtained. One interesting example is given in Fig. 14·7, where the feedback network is designed to compensate for the response characteristics of a ceramic phonograph cartridge. The feedback network also determines the bias condition in this amplifier.

14·1·6 *Two-port formulation of G_f feedback.* We conclude our present discussion of feedback through G_f by expressing the properties of the amplifier stage with feedback in two-port language. To study the effect of G_f on the two-port network, we first split the network into two parts, as shown in Fig. 14·8. We characterize the active two-port network by its y parameters:

$$I_1 = y_{11a}V_1 + y_{12a}V_2 \qquad I_2 = y_{21a}V_1 + y_{22a}V_2 \qquad (14·9)$$

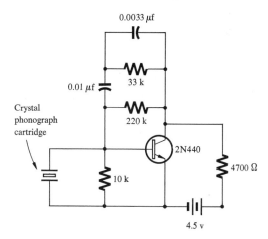

FIG. **14·7** CIRCUIT FOR EQUALIZING PICK-UP AND RECORD CHARACTERISTICS FOR CERAMIC OR CRYSTAL PHONOGRAPH CARTRIDGES
[*Used with permission of H. C. Lin, CBS Hytron*]

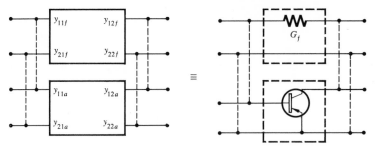

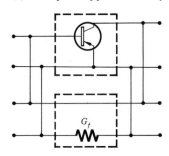

(a) Development of feedback theory in two-port language

(b) Combination two-port in which the parameters are not the sum of the parameters of the individual two-ports

FIG. 14·8 COMBINATION TWO-PORT NETWORKS AND FEEDBACK THEORY

The feedback network is characterized by a similar set of two-port equations:

$$I_1 = y_{11f}V_1 + y_{12f}V_2 \qquad I_2 = y_{21f}V_1 + y_{22f}V_2 \qquad (14 \cdot 10)$$

where $y_{11f} = y_{22f} = G_f$ and $y_{12f} = y_{21f} = -G_f$.

Now, when the two networks are connected together, we can show by simple algebraic manipulations that the two-port parameters for the combination two-port will be the sum of the two-port parameters of the component two-ports:

$$\begin{aligned} y_{11c} = y_{11a} + y_{11f} \qquad y_{12c} = y_{12a} + y_{12f} \\ y_{21c} = y_{21a} + y_{21f} \qquad y_{22c} = y_{22a} + y_{22f} \end{aligned} \qquad (14 \cdot 11)$$

Physically, we are allowed to simply add the parameters, because the independent variables used to describe the separate networks are the same, *both before and after the interconnection.*[1] This condition would not be satisfied if we attempted to characterize the transistor with h parameters and the feedback network with y parameters. Nor would it be satisfied if we connected the feedback two-port network in the manner shown in Fig. 14·8b. Here both two-ports are shorted out by the interconnection, so $V_1 = V_2$ after the interconnection but not before.

Because the effects of G_f (or more generally, Y_f) are most simply included

[1] A more detailed discussion of the conditions which must apply before one can add the two-port parameters of two interconnected two-port structures is given in J. G. Linvill and J. F. Gibbons, "Transistors and Active Circuits," McGraw-Hill Book Company, New York, 1961, pp. 222ff.

in a y-parameter formulation, the addition of a feedback admittance from the output port to the input port is frequently described as y-parameter feedback.

If we use a simple intrinsic transistor model containing only Y_π and g_m, the y parameters of the active device are

$$y_{11a} = Y_\pi \qquad y_{12a} = 0 \qquad y_{21a} = g_m \qquad y_{22a} = 0$$

When a feedback admittance Y_f (including both G_f and sC_c) is added, we have

$$y_{11c} = Y_\pi + Y_f \qquad y_{12c} = -Y_f \qquad y_{21c} = g_m - Y_f \qquad y_{22c} = Y_f$$

which are identical to Eqs. (14·1).

The most significant effect of G_f is in the y_{12c} term. The feedback conductance is added to control this reverse-transmission property of the combination two-port system.

The circuit properties of the combination two-port can be written down directly from its y parameters. For example, the current gain is

$$A_i = \frac{I_2}{I_1} = \frac{g_m(Y_L/Y_f + Y_L)}{Y_{in}} \simeq \frac{g_m}{Y_{in}}$$

as before. The input admittance of the combination two-port is

$$Y_{in} = y_{11c} - \frac{y_{12c}y_{21c}}{y_{22c} + Y_L} \cong y_{11c} + \frac{g_m Y_f}{Y_L} = Y_\pi + Y_f(1 + g_m Z_L)$$

where we have employed $g_m \gg Y_f$ and $Y_L \gg Y_f$ in the approximate form of this equation. These conditions are fulfilled in many practical situations. Note that the effect of adding Y_f is to *increase* the input admittance. The output admittance will also be increased by this type of feedback. If we want to increase the input impedance, we must select another form of feedback.

Visualizing the effects of feedback in terms of combination two-port parameters is of interest because of its generality. We can study the effects of feedback around a multistage amplifier by simply defining the y parameters of the active network appropriately and then using the appropriate formulas to find gain, impedances, etc.[1]

14·2 SERIES NEGATIVE FEEDBACK IN THE EMITTER CIRCUIT AND THE STABILIZATION OF A_v

If an unbypassed resistor is placed in the emitter lead of the simple common-emitter stage, the following effects will occur: (1) the voltage gain

---

[1] J. G. Linvill and J. F. Gibbons, "Transistors and Active Circuits," McGraw-Hill Book Company, New York, 1961, chap. 20.

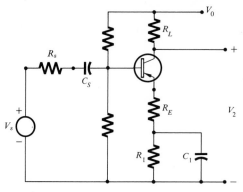

(a) General circuit

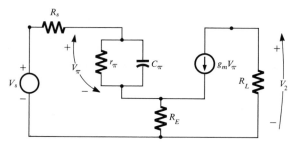

(b) Circuit model for (a) in and above the midband frequency range, neglecting C_c and r_b

FIG. 14·9 CE STAGE WITH SERIES EMITTER FEEDBACK

of the stage will be reduced and perhaps stabilized, (2) the stage bandwidth will increase, and (3) the input and output impedances will increase. These effects give us additional freedom in adjusting the circuit properties of the stage, though our control is more limited in this case than it is for y-parameter feedback. In this section we shall briefly develop the foregoing points.

The general circuit to be studied is shown in Fig. 14·9. In the circuit model shown in Fig. 14·9b, we have neglected parasitic elements (most importantly, C_c) to simplify the analysis. As a result, only the midband properties of the stage will be accurately predicted. The frequency-response predictions will give only a rough estimate of the actual circuit characteristics.

14·2·1   *Effect of feedback on input impedance.* It is convenient to begin the analysis by developing the approximate input impedance of the stage. To do this, we assume I_1 in Fig. 14·10 is known and calculate the input voltage of the circuit. The result is

$$V_{in} = I_1(Z_\pi + R_E + g_m Z_\pi R_E) \qquad (14·12a)$$

from which

$$Z_{in} = \frac{V_{in}}{I_1} = Z_\pi + R_E(1 + g_m Z_\pi) \tag{14 \cdot 12b}$$

Using the relations

$$Z_\pi = \frac{r_\pi}{1 + sr_\pi C_\pi} \qquad g_m r_\pi = \beta \qquad r_\pi = (\beta + 1)r_e$$

we may recast Eq. (14 · 12b) in the form

$$Z_{in} = \frac{(\beta + 1)(R_E + r_e) + R_E s(r_\pi C_\pi)}{1 + sr_\pi C_\pi} \tag{14 \cdot 13}$$

Eq. (14 · 13) shows that the low-frequency ($s \to 0$) input impedance of the stage is

$$Z_{in}(s \to 0) = (\beta + 1)(R_E + r_e)$$

which can be increased above r_π by choosing a high R_E. In this respect the circuit is similar to the CC stage. The bandwidth over which this high impedance is maintained is not particularly large, however, being only $f = (2\pi r_\pi C_\pi)^{-1}$ (or the β cutoff frequency). This means that if we want the circuit to maintain a high input impedance over a wide bandwidth, we simply have to look for a good transistor. We cannot effectively trade voltage gain for impedance in the same way that current gain and input admittance can be traded in the y-parameter-feedback arrangement.

14 · 2 · 2   Stabilization of A_v.   To compute the voltage gain for the stage, we first determine the input current:

$$I_1 = \frac{V_s}{R_s + Z_{in}} \tag{14 \cdot 14}$$

The output voltage is

$$V_2 = -g_m(Z_\pi I_1)R_L \tag{14 \cdot 15}$$

so the voltage gain can be written in the form

$$A_v = \frac{V_2}{V_s} = \frac{-g_m Z_\pi R_L}{R_s + Z_{in}} \tag{14 \cdot 16}$$

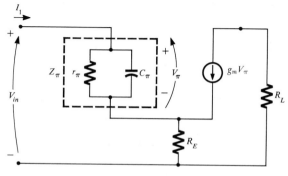

FIG. 14 · 10   CIRCUIT MODEL FOR COMPUTING APPROXIMATE INPUT IMPEDANCE

Using Eq. (14·13) for Z_{in}, A_v can be manipulated to yield

$$A_{v0} = \frac{\beta R_L}{R_s + (\beta + 1)(R_E + r_e)}$$

$$f_u = \frac{R_s + (\beta + 1)(R_E + r_e)}{2\pi r_\pi C_\pi (R_s + R_E)}$$

$$|A_{v0}|\,|f_u| = \frac{1}{2\pi r_\pi C_\pi}\frac{\beta R_L}{R_s + R_E} = f_T\frac{R_L}{R_s + R_E}$$

To stabilize the midband gain, we want to make A_{v0} independent of β. This is done by choosing R_E so that

$$(\beta + 1)(R_E + r_e) \gg R_s$$

Then the midband voltage gain is

$$A_{v0} \simeq \frac{R_L}{R_E + r_e}$$

The upper cutoff frequency then becomes

$$f_u \sim \frac{\beta + 1}{2\pi r_\pi C_\pi}\frac{R_E + r_e}{R_s + R_E}$$

The gain bandwidth product,

$$|A_{v0}|\,|f_u| = f_T\frac{R_L}{R_s + R_E}$$

now depends on the emitter resistance R_E. Therefore, as we increase R_E, the voltage gain will drop and tend to become stabilized. At the same time, the increase in R_E will reduce the gain-bandwidth product.

Experimental results. To check the predictions of the theory, the circuit shown in Fig. 14·11 was constructed. In this stage the ac collector load resistance is composed of 4.7 kohms in parallel with the 15-kohm bias resistor and the output

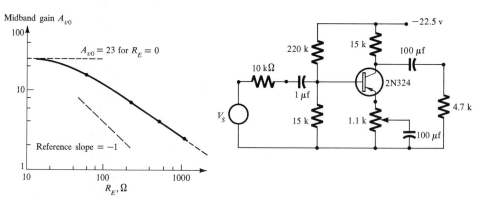

FIG. 14·11 THE DEPENDENCE OF A_{v0} ON R_E, THE SERIES FEEDBACK RESISTOR IN THE EMITTER CIRCUIT. THE SLOPE OF THE CURVE SHOULD APPROACH -1 WHEN $\beta R_E \gg R_S$.

resistance of the transistor, which is about 20 kohms. The net collector load is about 3 kohms.

The measured midband-voltage gain of the circuit is plotted in Fig. 14·11 as a function of the unbypassed emitter resistance. In the stabilized region, the gain should become inversely porportional to R_E; that is, the slope of log A_{v0} vs. log R_E should approach -1. Also, the gain should be unity when $R_E \simeq R_L$, which the curve verifies.

The bandwidth of the circuit, which is not shown in Fig. 14·11, is expected to be erroneous, since the effects of C_c have been neglected. The circuit actually exhibits an approximately constant $|A_{v0}||f_u|$ product of about 400 kcps, as long as $A_{v0} \geq 4$. The theory suggests the gain-bandwidth product should be about 1.2 mcps, as long as $R_E \leq 1$ kohm.

14·3 POSITIVE FEEDBACK AND OSCILLATION

Positive feedback occurs in an amplifier when a signal is returned to the input which reinforces the signal already there at that moment. One possible form of such a circuit is shown in Fig. 14·12. Here, we show a two-stage amplifier in which we have assumed that $R_i \gg r_\pi$, so the current gain per stage is β. We will assume that the signal frequency lies in the midband range, for simplicity.

The amplifier output current is $\beta^2 I_1$. A fraction $Y_f/(Y_L + Y_f)$ of this flows through the admittance Y_f when the switch S_1 is in the position shown. Let us assume for simplicity that Y_f is much smaller than the input admittance of the amplifier, so that when S_1 is operated, I_f does not change. Then, when S_1 is operated, the input current to the amplifier is

$$I_1 = I_s + I_f = I_s + \frac{\beta^2 I_1 Y_f}{Y_L + Y_f}$$

Solving for I_1, we obtain

$$I_1 = \frac{I_s}{1 - \beta^2 Y_f/(Y_L + Y_f)}$$

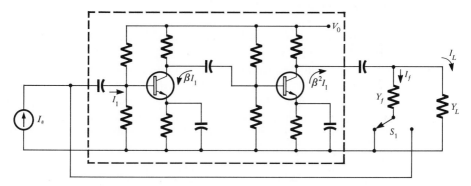

FIG. 14·12   TWO-STAGE AMPLIFIER FOR ILLUSTRATING POSITIVE FEEDBACK

The load current I_L becomes

$$I_L = \frac{\beta^2 I_1 Y_L}{Y_f + Y_L} = \frac{\beta^2 I_s}{1 - \beta^2 Y_f/(Y_L + Y_f)} \frac{Y_L}{Y_f + Y_L}$$

If we define the current gain to be I_L/I_s, then we see that if Y_f and Y_L are real, and $\beta^2 Y_f/(Y_f + Y_L) < 1$, the gain with feedback is greater than the gain without feedback. In particular, as

$$\frac{\beta^2 Y_f}{Y_L + Y_f} \to 1$$

the gain of the amplifier approaches infinity. And when $\beta^2 Y_f/(Y_f + Y_L)$ reaches unity, the system will produce an output with no externally applied input signal.[1]

If the condition

$$\frac{\beta^2 Y_f}{Y_f + Y_L} = 1$$

is met at one frequency only, (by making Y_f a series RLC circuit, for example), then the system will tend to produce sinusoidal oscillations at that frequency. The system is then called a sine-wave oscillator. If the oscillation condition is met over a band of frequencies, then the system may produce a distinctly nonsinusoidal waveform.

Of course, values of Y_f and Y_L can be chosen so that the positive feedback will not produce oscillation. We could use negative feedback around each stage to stabilize its current gain, and then positive feedback around the two-stage cascade to increase the over-all gain. Great care is necessary, however, to ensure that oscillations do not develop at a frequency where we do not expect them.

While we have described a two-stage circuit for simplicity, positive feedback can occur in both single-stage and multistage circuits. We shall discuss several examples of both types in this section. Our discussion will be limited to the use of positive feedback in building sine-wave oscillators, though the reader should bear in mind that positive feedback is also used together with negative feedback to build stable amplifiers.

14·3·1 *The Colpitts oscillator.* A particularly useful type of sinusoidal oscillator circuit is the Colpitts oscillator, shown in Fig. 14·13. With proper precautions, the amplitude and frequency of oscillation of a Colpitts oscillator can be made quite stable. As a result, Colpitts circuits find frequent application as local oscillators in superheterodyne receivers, telemetering equipment, etc.

To emphasize the nature of the feedback condition, the Colpitts circuit is redrawn in Fig. 14·14 as a normal CE amplifier driving a complex

---

[1] The condition $\beta^2 Y_f/(Y_f + Y_L) = 1$ does not mean $I_L \to \infty$, since previous equations involve dividing by zero when $\beta^2 Y_f/(Y_f + Y_L) = 1$. Instead, it means a finite I_1 and I_L are possible for zero I_s.

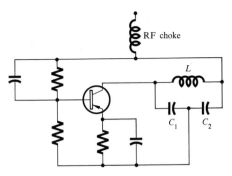

FIG. 14·13   COLPITTS OSCILLATOR CIRCUIT

load. The load resistance is r_π, and the final capacitor C_2 has been divided into two parts: C_2' and C_π.

The circuit is further redrawn in Fig. 14·14b, using a circuit model for the intrinsic transistor, to aid in the analysis. Parasitic elements will be neglected to simplify the initial presentation of ideas.

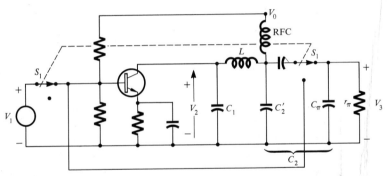

(a) "Colpitts" amplifier driving a load of $r_\pi \| C_\pi$ to simulate transistor input

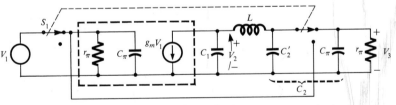

(b) Small signal circuit model for (a) neglecting biasing components and using intrinsic tranistor model only

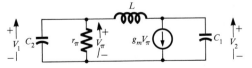

(c) Circuit model for (a) when S_1 has been operated to convert amplifier into oscillator

FIG. 14·14   CIRCUIT MODELS FOR THE COLPITTS OSCILLATOR

Introduction to feedback   577

Let us consider applying a sine-wave voltage source of amplitude V_1 at the input. The collector current is then sinusoidal and has the value $g_m V_1$. The output voltage V_3 of the amplifier can be found by straightforward network-analysis procedures. Its general form will be

$$V_3 = \frac{-r_\pi}{1 + a_1 s + a_2 s^2 + a_3 s^3} g_m V_1$$

where s is the complex frequency variable and the a_i are functions of L, C_1, and C_2. The a_i are positive numbers.

The voltage gain from the input terminals of the amplifier to the output is then

$$\frac{V_3}{V_1} = \frac{-\beta}{1 + a_1 s + a_2 s^2 + a_3 s^3}$$

Now, from this expression, or by inspection of Fig. 14·14, it can be seen that at low frequencies the output voltage V_3 of this amplifier will be 180° out of phase with the base-emitter input voltage V_1. However, as the frequency of the input signal V_1 is increased, phase shift will occur as the signal is transmitted through the LC network, and both the magnitude and phase of the voltage gain V_3/V_1 will change.

We can study the changing phase of the output voltage with respect to the input voltage by writing the voltage gain of the circuit of Fig. 14·14b in the form

$$\frac{V_3}{V_1} = \frac{-\beta}{(1 - a_2 \omega^2) + j\omega(a_1 - a_3 \omega^2)}$$

Since we are interested in real frequencies, we have made the substitution $s = j\omega$. Remember that the a_i are positive numbers which can be evaluated by network analysis in terms of L, C_1, and C_2.

Now, in the previous equation, at the frequency

$$\omega_0{}^2 = \frac{a_1}{a_3}$$

the imaginary part of the denominator vanishes, and the gain becomes

$$\frac{V_3}{V_1}\left(\omega_0 = \sqrt{\frac{a_1}{a_3}}\right) = \frac{-\beta}{1 - a_2 a_1 / a_3}$$

This means that V_3 is either in phase or 180° out of phase with V_1 at $\omega = \omega_0$. Furthermore, if a_2 can be adjusted so that

$$\frac{a_2 a_1}{a_3} = \beta + 1$$

then $$\frac{V_3}{V_1} = 1$$

and the output voltage of the amplifier will be identical to its input voltage.

Under this condition, we can operate the switch shown in Fig. 14·14b, and the new system will, in principle, continue to produce the same output voltage as before. However, now its input is supplied from its own output. Such a system is called an *oscillator,* and the condition $V_3/V_1 = 1$ is called the *oscillation condition.*[1]

There are many different ways of expressing the oscillation condition. For example, a person accustomed to feedback terminology might say, "The open-loop gain (meaning the *magnitude* of V_3/V_1 with the source connected in Fig. 14·14) should be unity and the open-loop net phase shift should be zero." There are other equivalent forms of the oscillation condition which prove to be useful in studying particular oscillator circuits. We shall consider some of these in the subsequent development and also remove some of the idealizations and assumptions which were used in the development just given. Before we do this, however, it is worthwhile to make a few remarks about practical oscillator circuits, to help put the oscillation condition in a proper perspective.

If an amplifier-oscillator circuit combination is actually set up with switches, as in Fig. 14·14, it is sometimes possible to find conditions of bias and signal amplitude where the circuit will produce almost pure sinusoidal oscillations, if it is first driven from a source before the feedback loop is closed; but this same circuit may not start oscillating on its own if bias power is applied to the circuit with the feedback loop connected. Or if it does, the amplitude of the oscillations may vary with time, sometimes settling down to a stable value and sometimes not.

Such a circuit can be made "self-starting" by adjusting the magnitude of V_3/V_1 to be "slightly greater than unity" at the desired oscillation frequency, which is the recommended practice. Let us see how this self-starting condition will lead to the build-up of oscillations in the circuit.

Suppose we adjust a_1, a_2, and a_3 in a Colpitts circuit to make $V_3/V_1 = 1.01$ when the circuit is operating as an amplifier. Now we remove the bias power and operate the switch to set up the oscillator circuit. When the bias power is reapplied, the step excitation will produce damped oscillations in the tuned circuit. As a result, we can suppose that at some time t a small voltage V_1, of the proper frequency for oscillation $(a_1/a_3)^{1/2}$, appears at the input terminals of the transistor. An amplified replica of this voltage, of magnitude $1.01V_1$, will appear at the output terminals (that is, across C_2') in the time it takes the signal to propagate through the circuit. Since there is a 360° phase shift in this circuit, the propagation time is simply the period of the oscillation cycle, T.

Now, the output of the circuit is connected directly to its input, so the above reasoning leads us to conclude that if we begin with an input voltage of V_1, then after T sec the input voltage will be $1.01V_1$. After $2T$ sec,

---

[1] Note that the $C_\pi r_\pi$ combination appears as a load on the $C_1 L C_2'$ network in either position of the switch. The oscillation condition assumes that impedance conditions are properly taken into account.

the input voltage will be $(1.01)^2 V_1$, and after nT sec, the input voltage will be $(1.01)^n V_1$. The voltages at all points in the circuit thus build up exponentially with time if $V_3/V_1 > 1$ at the oscillation frequency.

An interesting measure of the build-up time is the time required for the oscillation amplitude to increase by an order of magnitude. To find this time we set

$$(1.01)^n V_1 = 10 V_1$$

This gives $\qquad n \log_{10} 1.01 = \log_{10} 10 \qquad n = 230$

Thus, it takes $230T$ sec for the oscillation amplitude to increase by one decade, where T is the period of the oscillation.[1] This assumes, of course, that $V_3/V_1 = 1.01$. If, instead, $V_3/V_1 = 1 + \delta$, then we would have

$$n = \frac{2.3}{\delta}$$

and the time required for the oscillations to increase their amplitude by a decade would be

$$n = \frac{2.3T}{\delta}$$

In the Colpitts circuit, if the oscillation frequency is 1 mcps and we have set $V_3/V_1 = 1.01$, then the build-up time is 230 μsec/decade; or a 1-mv input signal would increase to 1 volt in 690 μsec.

This behavior cannot continue indefinitely, of course. It is based on the assumption that the circuit parameters do not change as the signal amplitude increases, an approximation which we know to be invalid in any amplifier. For example, the β of a transistor is not perfectly constant, a fact which will directly affect the previous analysis.

Now, changes in the circuit parameters require us to reevaluate the oscillation condition to see if the build-up condition still applies. This requires nonlinear analysis, since we are dealing with a system where the parameters are functions of the signal level. However, for the present discussion, we need not resort to such sophistication. We shall assume (and in practice the assumption is usually justified) that variations in β, and other effects of this type, are of minor importance and can be neglected in a first approximation. Under these conditions, the oscillations will build up until the transistor is either driven into saturation or cutoff (or perhaps both) at the peak of the signal. The amplifier gain is then abruptly reduced and the oscillation conditions are no longer met. Of course, the energy stored in the frequency-determining elements will cause the oscillations to continue, but the amplitude of the oscillations will be damped out until the transistor recovers from its saturated or cutoff condition to reestablish the conditions for build up.

The effect of this behavior on the oscillator waveform can take on many

[1] A low-frequency oscillator for demonstrating this build up is described in Demonstration 14 · 2 at the end of this chapter.

facets. If V_3/V_1 is small (usually 1.005 or less), the oscillations will build up to a value where the peak oscillation amplitude is just large enough to drive the transistor into saturation or cutoff. Such an oscillation is called a *near-class-A oscillator,* since the transistor is inactive for only a very small portion of each cycle.

If larger values of V_3/V_1 are used (1.02 or more), there will usually be some very evident distortion of the oscillator waveforms in at least some parts of the oscillator, though the frequency-determining elements may be of sufficiently high quality to filter out most of this distortion, so that a relatively pure sine wave can be observed in the tuned network (shifted somewhat in frequency from the ideal value, though the shift may be imperceptible).

If extremely large values of V_3/V_1 are used, then the transistor can remain saturated and therefore inactive for many cycles of the oscillation. Usually this is not desirable. However, it *is* frequently desirable, particularly when highly efficient operation is required, to adjust V_3/V_1 so that the transistor supplies a short pulse of current to the tuned circuit (at the collector-voltage minimum) during each cycle. This type of oscillator is called a *class C oscillator* and is used when we want to obtain the highest possible power output from a transistor with a given collector-power-dissipation rating. We rely on the quality of the elements in the resonant circuit to provide a sufficiently pure sine wave under these "impulse excitation" conditions.

Resume. Let us now review the preceding discussion and restate it more compactly and in different terms:

1. If an amplifier circuit can be operated at a frequency where its output voltage and its input voltage are equal (apart from multiples of 360° phase shift), then the output can be connected to the input to form an ideal oscillator. Of course, this assumes that impedance conditions are properly accounted for, so that the connection of output to input does not invalidate the oscillation condition.

2. In a practical oscillator, the condition just described does not ensure the build up of oscillations. This objection is met by making the overall amplifier gain (including the as-yet-unconnected feedback loop) slightly larger than unity. The result of this step is that when the feedback loop is connected the amplitude of the oscillations should *theoretically* build up exponentially with time *indefinitely.* Of course, such behavior cannot occur in a physical system. The indefinitely long exponential build-up is predicted only for an *idealized linear system.* The real network exhibits nonlinearities which will ultimately set in to limit the amplitudes of the signals that can circulate in the system. *In all cases, system nonlinearities are responsible for the amplitude limitation in a self-starting oscillator.*

3. Depending on the precise value of V_3/V_1 at the zero-net-phase condition, the oscillator network can operate as a near-class-A oscillator or a class C oscillator.

4. The system nonlinearity which produces the amplitude limitation will also result in "frequency pulling," though this can be minimized by proper precautions.

We shall attempt to keep all these facts in mind by referring to the oscillation condition as being a *linear* oscillation condition, if it is derived with the aid of a linear model for the circuit. In this chapter we shall study some of the most widely used oscillator circuits, employing simple models for the active devices and the ideal linear oscillation condition to obtain an *approximate* condition for oscillation. We shall also study our results to determine ways in which the circuit can be modified to make it self-starting. In practice, this type of analysis usually enables the designer to build an oscillator which can then be refined by a combination of experimental and analytical techniques to give satisfactory performance.

Linear oscillation conditions in a Colpitts oscillator. To design a practical Colpitts (or any other) oscillator circuit, we first need to know how the ideal conditions for oscillation are related to the circuit parameters. The design relations can be determined by several different but equivalent analysis techniques. We shall consider some of the most important techniques using the Colpitts oscillator as a running example.

We shall first analyze the Colpitts oscillator by determining the conditions which must exist for the circuit to deliver an output when no external input signal is present. We shall use the circuit model shown in Fig. 14·14c. Since our purpose is to answer some basic questions, we can neglect the biasing-network parameters and use the simplest possible model of the transistor which has a chance of explaining the circuit behavior. Thus, we neglect r_b, r_0, and g_2 in the transistor model and assume that C_c is incorporated into the feedback inductance L. C_π is incorporated into C_2.

We may write two nodal equations for the circuit shown in Fig. 14·14b:

$$V_1\left(g_\pi + sC_2 + \frac{1}{sL}\right) - V_2\left(\frac{1}{sL}\right) = 0$$

$$V_1\left(g_m - \frac{1}{sL}\right) + V_2\left(sC_1 + \frac{1}{sL}\right) = 0$$

(14·17)

These equations can be true if (1) both V_1 and V_2 are identically zero, which means no oscillations, or (2) if the determinant of the coefficients of V_1 and V_2 is zero.[1] Hence, if it is possible for the *system* shown in Fig. 14·14 to oscillate, we must have

---

[1] The frequencies s at which V_1 and V_2 can exist without an external drive are called the natural frequencies of the system. For an ideal oscillator, we want the natural frequencies to lie on the $(j\omega)$ axis of the complex frequency plane. The determinant given in Eq. (14·18) provides one procedure for finding the natural frequencies. Others are described in Sec. 14·4 and in the Appendix.

$$\begin{vmatrix} g_\pi + sC_2 + \dfrac{1}{sL} & -\dfrac{1}{sL} \\[3mm] g_m - \dfrac{1}{sL} & sC_1 + \dfrac{1}{sL} \end{vmatrix} = 0 \qquad (14 \cdot 18)$$

The word "system" in the preceding paragraph is most important. It implies that all loads (in this case, none) which are to be attached are being considered. If we analyze a system, find that it will oscillate, and then put on additional loading, we cannot be sure that the oscillation conditions will then be met.

Proceeding with the analysis, we compute the determinant:

$$\Delta = \left(g_\pi + sC_2 + \frac{1}{sL}\right)\left(sC_1 + \frac{1}{sL}\right) + \frac{1}{sL}\left(g_m - \frac{1}{sL}\right) = 0 \quad (14 \cdot 19)$$

We now set $s = j\omega$ and multiply out Δ. We find

$$\omega^4 L^2 C_1 C_2 - \omega^2 L(C_1 + C_2) + j[\omega L(g_\pi + g_m) - \omega^3 L^2 C_1 g_\pi] = 0 \quad (14 \cdot 20)$$

Equation $(14 \cdot 20)$ has both real and imaginary parts. If it is to be zero, both the real and imaginary parts must individually be zero. Thus

$$\omega^4 L^2 C_1 C_2 = \omega^2 L(C_1 + C_2) \qquad (14 \cdot 21a)$$

$$\omega L(g_\pi + g_m) = \omega^3 L C_1 g_\pi \qquad (14 \cdot 21b)$$

From Eq. $(14 \cdot 21a)$ we find

$$\omega^2 = \frac{C_1 + C_2}{L C_1 C_2} = \frac{1}{LC} \qquad (14 \cdot 22a)$$

where C is the equivalent capacitance of C_1 and C_2 in series. Equation $(14 \cdot 22a)$ gives the *frequency of oscillation*.

From Eq. $(14 \cdot 21b)$, we find

$$\omega^2 L C_1 = 1 + \frac{g_m}{g_\pi}$$

which may be used with Eq. $(14 \cdot 22a)$ to obtain

$$\frac{C_1}{C} = 1 + \frac{g_m}{g_\pi} \qquad \text{or} \qquad 1 + \frac{C_1}{C_2} = 1 + \beta \qquad (14 \cdot 22b)$$

Hence, C_1/C_2 should be equal to β. This is sometimes loosely called the oscillation condition. It relates the network parameters to the current gain of the amplifier.

If C_1/C_2 is equal to β, the idealized circuit will produce a pure sine-wave oscillation. In a real circuit, C_1/C_2 will have to be somewhat less than β to account for losses in r_b, r_0, L, etc. and to provide enough margin for a self-starting condition.

If C_1/C_2 is gradually reduced from β until a stable self-starting condition is reached, the oscillator will be operating in a class A or near-class-A condition. If the transistor were perfectly linear over its entire normal-active region, then in the near-class-A condition oscillations would build up until (1) the excursions of the emitter current had an amplitude approximating the emitter bias current I_E and/or (2) the excursions of the collector-base voltage approximated the collector-base bias V_{CB}. The excursion into the cutoff or saturation region required to provide the necessary non-linearity for limiting is slight, so that the circuit waveforms would be very nearly sinusoidal.

If C_1/C_2 is reduced well below β, the emitter-base junction is driven into cutoff during an appreciable portion of the cycle, and the emitter current flows in pulses. In the limit, the transistor provides an impulse excitation to the LC tank circuit during each cycle, and the oscillator is operating under class C conditions. This type of operation is required when the ac power to be obtained from the oscillator is comparable to the maximum permissible collector dissipation of the transistor.

It is worthwhile to summarize the procedure followed in analyzing the circuit, since it is useful for many oscillator problems:

1. Draw the *complete system* envisaged and make a circuit model for it. The circuit models used for the devices should be sufficiently precise to include the principal details of device behavior without unnecessary complications.

2. Write a complete set of equations for the system. This set of equations will always be of the form given in Eqs. (14 · 17), in that there will be no *external* driving forces, so the right-hand sides will always be zero.

3. Write down the circuit determinant and set it equal to zero. This will yield a polynomial (generally unsolvable) with both real and imaginary parts. The real and imaginary parts must separately equal zero. From this fact one can, in principle, ascertain a *frequency of oscillation* and a minimum *"condition"* for oscillation. In simple cases, conditions for "self-starting" can be deduced from these results.

Naturally, the accuracy of the predicted oscillation frequency and oscillation condition will depend on the accuracy of the model used. For our

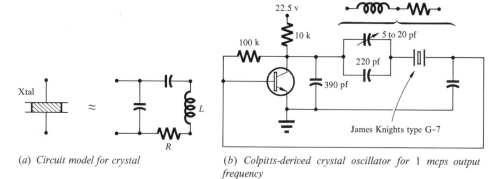

(a) *Circuit model for crystal*

(b) *Colpitts-derived crystal oscillator for 1 mcps output frequency*

FIG. 14 · 15   COLPITTS OSCILLATOR WITH FREQUENCY DETERMINED BY PIEZOELECTRIC CRYSTAL

case, C_1/C_2 will have to be less than β to provide for losses in r_b and r_0 which were not included in the analysis and to provide for self-starting. We also expect r_b and r_0 to enter into the determination of the oscillation frequency. In fact the problem quickly becomes quite complicated algebraically, though the principles are still simple.

14·3·2 *Colpitts-derived crystal oscillator circuits.* When an oscillator is required to produce a very exact output frequency, a quartz crystal which has accurately known vibration frequencies may be used in the frequency-determining network. In one of its vibration modes, a quartz crystal can be represented by the circuit model shown in Fig. 14·15a. For this mode of vibration, the crystal can be used to replace the inductance in a Colpitts oscillator. A practical example of such an oscillator intended to operate at 1 mcps is shown in Fig. 14·15b. Oscillators of this type can have an output-frequency deviation of a few parts in 10^8 or less, for short (15-min) time periods.

14·4 ALTERNATE CRITERIA FOR OSCILLATION

In addition to the positive-feedback point of view, the conditions for oscillation in a circuit can be viewed in other illuminating ways. One particularly useful technique will be developed in this section.

If we reorient our thinking about oscillators, we can see that there are three essential parts in any oscillator circuit:
1. A power source
2. A means of storing energy
3. An amplifier (or a dynamic negative resistance)
If the means of storing energy is an LC circuit, then the interplay of the three parts is relatively straightforward. The LC circuit has some natural oscillation frequency at which energy that is stored in the circuit is transferred back and forth between the L and the C. Resistive losses in the circuit absorb some of this circulating energy, however, so the oscillations of current and voltage in the LC circuit are damped out. The role of the amplifier is to release energy from the power source to the LC network, to make up for these energy losses.

The mechanisms by which energy is released to the LC circuit can take several forms. In some oscillators a sharp pulse of current is supplied to the LC tank circuit every cycle, to make up for energy losses during the cycle. In other circuits, an amplifier with an appropriate feedback network provides a *dynamic* negative resistance or conductance at a pair of terminals. An LC circuit can be attached to this terminal pair and the system will produce oscillations under proper conditions. These two possibilities are shown in Figs. 14·16a and 14·16b.

Circuits that operate in this latter category can be analyzed using small-signal models for the active devices. We shall show how this is done, using

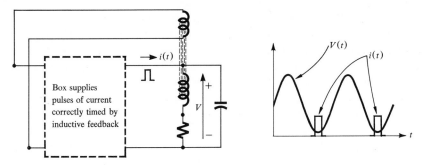

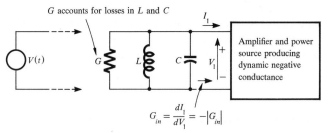

(a) *Use of current pulses to supply energy to a lossy LC tank circuit*

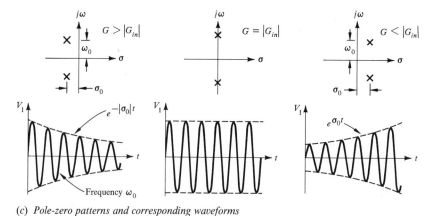

(b) *Use of negative dynamic conductance to account for circuit losses*

(c) *Pole-zero patterns and corresponding waveforms*

FIG. 14·16 *LC* OSCILLATORS WITH NETWORKS ADDED TO SUPPLY ENERGY LOSSES INHERENT IN THE *L* AND *C* ELEMENTS

a Colpitts oscillator as an example. To do so, we need to understand the concept of the natural frequencies of a system (*cf.* the Appendix). The natural frequencies of a system are defined to be the frequencies at which signals can exist in a system without applied sources.[1] If we can produce a box such as the one shown in Fig. 14·16b, with G and G_{in} equal in magnitude, then a sinusoidal voltage V_1 could exist without an external

[1] Or, alternately, the frequencies which appear in the voltages and currents in a network for $t > 0$, when it is excited by an impulse at $t = 0$.

drive. The natural frequency for this case would simply be $\sqrt{LC}/2\pi$. If $|G_{in}|$ is less than G, natural frequencies still exist, but now they correspond to sinusoidal waveforms whose amplitudes decay exponentially with time. Similarly, if $|G_{in}| > G$, natural frequencies exist which correspond to sinusoidal waveforms with amplitudes that increase with time. These possibilities are shown in Fig. 14·16c, along with possible positions of the natural frequencies in the complex s plane for each case.

To build an ideal oscillator, we are interested in ensuring that the natural frequencies be on the $j\omega$ axis of the complex s plane. To do this, we need a simple technique for determining the natural frequencies of a proposed system. A convenient procedure for our future use is to write down the total nodal admittance Y_N seen across *any* node pair in the system and set this admittance equal to zero.[1] The values of the complex frequency variable s which cause Y_N to be zero are the natural frequencies.

We can prove this statement rather simply, as follows. Suppose the system shown in Fig. 14·16b is responding at its natural frequency with energy stored at an earlier time. We now bring a voltage source $V(t)$ up to the node pair, as shown. Suppose this voltage source has been adjusted to have exactly the same waveform as V_1 (sinusoid with exponentially decaying amplitude, or whatever). Then, when the source is attached, it will deliver no current to the network. Therefore, the nodal admittance at that (complex) frequency must be zero.

We can use this result in oscillator analysis as follows. By computing the admittance seen across a convenient node pair in an oscillator network, we can determine natural frequencies of the network. We can then obtain the conditions that must hold to ensure that the natural frequencies are in s-plane positions that correspond to sustained oscillations. We shall perform such a calculation for a Colpitts oscillator.

The network model is redrawn in Fig. 14·17. The nodal admittance of this network can be split into two parts. The two components sC_2 and g_π are shown in Fig. 14·17b, along with an equivalent admittance Y^* reflected to the V_1 node through L. To compute Y^*, we use Kirchhoff's law at the V_2 node to give

$$V_1\left(g_m - \frac{1}{sL}\right) + V_2\left(sC_1 + \frac{1}{sL}\right) = 0 \qquad (14\cdot23)$$

Using this equation in the definition of admittance

$$Y^* = \frac{I^*}{V_1} = \frac{V_1 - V_2}{sL}\frac{1}{V_1}$$

gives

$$Y^* = \frac{sC_1 + g_m}{1 + s^2LC_1} \qquad (14\cdot24)$$

This admittance can be "realized" in the form shown in Fig. 14·17c.

---

[1] This assumes that the system does not have isolated parts.

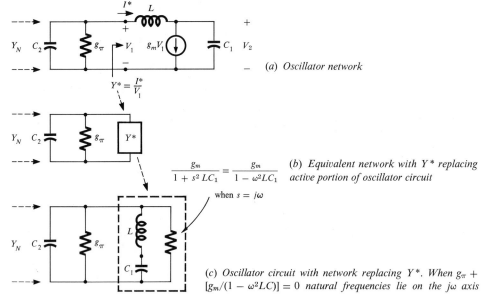

(a) Oscillator network

$Y^* = \dfrac{I^*}{V_1}$

(b) Equivalent network with Y^* replacing active portion of oscillator circuit

$$\frac{g_m}{1 + s^2\, LC_1} = \frac{g_m}{1 - \omega^2 LC_1}$$

when $s = j\omega$

(c) Oscillator circuit with network replacing Y^*. When $g_\pi + [g_m/(1 - \omega^2 LC)] = 0$ natural frequencies lie on the $j\omega$ axis

FIG. 14·17   COLPITTS-OSCILLATOR ANALYSIS USING NODAL-ADMITTANCE CONCEPTS

From this figure it can be seen that the nodal admittance will have its zeros at real frequencies ($s = j\omega$) when (1) the LC_1 branch has an inductive susceptance which just cancels the capacitive susceptance due to sC_2, and (2) the ($g_m/1 + s^2LC$) branch has a negative conductance which will just cancel g_π. This can occur at a frequency $s = j\omega$ if

$$-\frac{g_m}{1 - \omega^2 LC_1} = g_\pi$$

which is equivalent to the oscillation condition obtained earlier.

The nodal-admittance criterion for visualizing oscillator behavior thus pictures the role of the active element and some of the associated components as providing a negative conductance to make up energy losses in the circuit. In circuit-theory parlance, the dynamic negative resistance (or conductance) "produces" the energy which is dissipated in the positive resistances of the circuit. An element which is capable of converting dc power into signal power can, with appropriate feedback, provide the apparent negative resistance. We shall find the negative-resistance concept valuable in several other oscillator circuits to be explored at a later point.

14·5   OTHER TYPES OF OSCILLATOR CIRCUITS

There are several other types of oscillator circuits that find interesting applications and will be mentioned briefly. The analysis of some of these oscillators is considered in the problems at the end of this chapter.

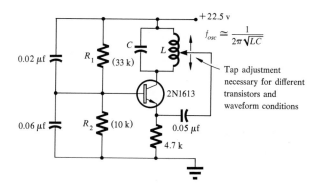

$$f_{osc} \cong \frac{1}{2\pi\sqrt{LC}}$$

Tap adjustment necessary for different transistors and waveform conditions

(a) Typical circuit. Tap adjusted on coil to allow for different transistors and obtain a sinusoidal waveform if desired.

L: 5 close turns, No. 22 copper wire, ¼ in. diameter with center tap

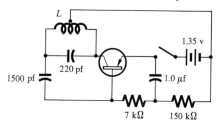

(b) Hartley oscillator delivering a 27 mcps pulsed signal. Oscillator weighs about 3 grams. (K. S. Rawson and P. H. Hartline, Science, 18 December 1964, pp. 1596–1597)

FIG. 14·18   THE HARTLEY OSCILLATOR

14·5·1   *The Hartley oscillator.* One important type of oscillator circuit is the Hartley oscillator, shown in Fig. 14·18a. This circuit is generally similar to the Colpitts circuit, except that the proper phase relation for the fed-back signal is obtained by transformer action in the tapped coil. The circuit is especially popular in vacuum tube oscillators.

A practical transistor version of the Hartley circuit used for a somewhat unusual telemetry application is shown in Fig. 14·18b.[1] This circuit was devised to study the movements of deermice as they return to their nests after being removed to distance places. The oscillator simply serves as a radio transmitter operating at 27 mcps, giving pulsed output signals (the feedback is so great that the transistor operates in a class C mode). It is sufficiently small and lightweight that it can be implanted in the mouse for the study.

14·5·2   *RC oscillators.* Usually, oscillators employ *LC* resonant circuits when the desired oscillation frequency is high enough that high-quality (low-loss) *L* and *C* elements can be used. For frequencies below a few hundred kcps, however, *RC* circuits can be used to obtain the necessary phase shift. One basic type of *RC* oscillator circuit is shown in Fig. 14·19a. The circuit is

[1] K. S. Rawson and P. H. Hartline, Telemetry of Homing Behavior of the Deermouse, *Science*, Dec. 18, 1964, pp. 1596–1597.

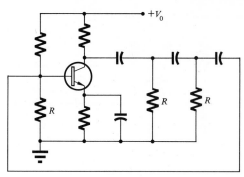

(a) Simple lag-line oscillator.

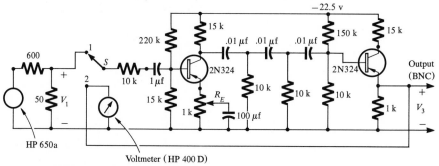

(b) 500 cps lag-line oscillator. Oscillations can be suppressed or excited by variation of R_E.

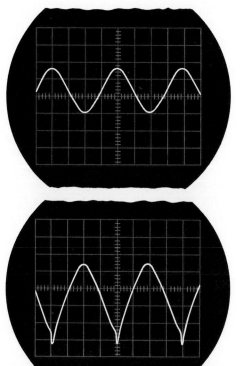

(c) Output waveform at the emitter follower when "linear" oscillation conditions are set up before S is changed to position 2. Horizontal scale: 0.5 ms/cm.

(d) Output waveform when $V_3/V_1 = 1.025$ in the circuit of (b). Horizontal scale: 0.5 ms/cm.

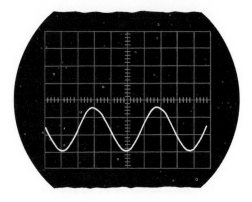

(*e*) *Output voltage at the collector of the amplifier transistor for $V_3/V_1 = 1.025$. Vertical scale 5 v/cm; horizontal scale 0.5 ms/cm. Horizontal center line is 0 volts dc. The dc voltage across R_E is 1.4 volts. An accurate measurement of the minimum voltage of the waveform above shows it to be 1.416 volts, indicating that the transistor is driven into saturation ($V_{CE} = 16$ mv) during a portion of the oscillation cycle.*

FIG. **14 · 19**   LAG-LINE OSCILLATORS AND WAVEFORMS FOR VARIOUS VALUES OF THE AMPLIFIER GAIN

sometimes called a lag-line oscillator. Roughly speaking, the amplifier produces 180° of phase shift and the three RC sections produce about 60° of phase shift apiece.

The components for a demonstration version of this oscillator, working at 500 cps, are given in Fig. 14 · 19*b* (and in Demonstration 14 · 2 at the end of this chapter). Here an emitter follower is used to isolate the RC line from the input of the amplifier transistor. The demonstration provides a convenient means of experimentally studying the validity of the linear-oscillation condition and the effect of self-starting allowances.

If the switch is in position 1, the circuit operates as a normal amplifier. If the source voltage is adjusted to 20 mv (500 cps) and the emitter resistor R_E is adjusted so that the output voltage V_3 is 19 mv, then when the switch is put in position 2, the circuit will not oscillate. If R_E is adjusted to give an emitter-follower output of 19.5 mv when the switch is in position 1, then the circuit may oscillate when the switch is operated, though its amplitude will vary erratically with time. When the emitter-follower output is readjusted to 19.8 mv, the circuit will usually oscillate when the switch is operated, though it may not restart if the bias power is removed and then reapplied.

When the emitter-follower output is increased to 20 mv, the circuit promptly oscillates when the switch is closed or when bias is applied directly to the oscillator with the feedback loop closed. The output voltage at the emitter follower is 60 mv. A reproduction of an oscilloscope trace of the output voltage for this case is shown in Fig. 14 · 19*c*. The collector voltage of the amplifier transistor is also essentially sinusoidal with a peak amplitude of about 5 volts, though some distortion is apparent upon close inspection. The distortion does not appear in Fig. 14 · 19*c* because there is considerable filtering between the output of the amplifier and that of the emitter follower.

When R_E is further reduced to the point that the emitter-follower output becomes 20.5 mv when the switch is in position 1, then the output voltage

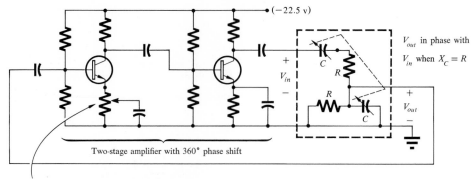

FIG. **14·20** WIEN BRIDGE OSCILLATOR

of the emitter follower changes to that shown in Fig. 14·19*d*. The voltage appearing at the collector of the amplifier transistor is shown in Fig. 14·19*e*. The flat portion occurs at low voltage where the transistor is being driven into saturation, this being the mechanism which limits the oscillation amplitude in this case (further details are provided in the caption for Fig. 14·19*e*).

The circuit can also be used to conveniently demonstrate the build up of oscillations. To do this, 0.25-μfarad capacitors are added in parallel with the 0.01-μfarad capacitors in the lag line. This reduces the oscillation frequency to about 20 cps, which means that the time required for the oscillation amplitude to build up by a decade is about 10 seconds for $V_3/V_1 = 1.01$. By careful adjustment of R_E and the supply voltage, this condition can be obtained. The build-up of oscillations can then be observed when power is applied to the circuit connected as an oscillator. Three experimental trials on this circuit give times of 8, 12, and 17 sec., respectively, for the oscillation amplitude (at the emitter follower) to build up from 5 μv to 50 μv.

Wien bridge oscillator. A second type of *RC* oscillator is shown in Fig. 14·20. Here a two-stage amplifier is used to produce a 360° phase shift, and then an *RC* Wien bridge circuit is used to select the frequency which will be returned to the amplifier input in the proper phase. An interesting variation on this circuit can be obtained by using silicon diodes to provide the resistances in the Wien bridge. The result is an oscillator which can be tuned by applying a control current to the diodes.

14·6 AMPLITUDE AND FREQUENCY STABILITY IN OSCILLATOR CIRCUITS

The amplitude and frequency stability of an oscillator are of great practical importance, though there is little that can be said without reference to a particular circuit and particular operating conditions. The following general remarks seem appropriate, however.

In the near-class-A oscillators, the oscillation amplitude is proportional to the emitter bias current I_E, if limiting is due to cutoff, while it is proportional to the dc collector voltage (V_{CB} or V_{CE}), if limiting is due to saturation. The usual stability of bias points and supply-voltage levels provides a sufficient degree of amplitude stability for most applications.

When greater amplitude stability is required, negative feedback can be used to stabilize the gain of the amplifier at the required value. One popular form of negative feedback (called automatic volume control, after its early use in radio for this purpose) is obtained by deriving a bias voltage for the amplifier from the oscillator output signal. A circuit illustrating the idea is shown in Fig. 14·21. In this circuit, a self-starting reference bias point for the oscillator transistor is established by a Zener diode. As oscillations build up, the rectifier-filter combination produces an additional bias which decreases the magnitude of bias applied to the oscillator transistor, therefore decreasing the emitter current and the g_m of the transistor. When an equilibrium is reached, variations in the oscillation amplitude will produce bias variations which tend to restore the oscillation amplitude.

The frequency stability of a near-class-A oscillator is determined primarily by the quality of the elements in the resonant network and the care taken by the designer to isolate these frequency-determining elements from those parts of the circuit where the circuit parameters may vary significantly over an operating cycle. The use of a crystal, or the common-collector buffer used in Fig. 14·19b, Fig. 14·21 and the experimental project at the end of this chapter are examples of these types of measures. Very precise frequencies can be obtained from lightly loaded near-class-A oscillators, with sufficient precautions.

When one turns to class C oscillators, however, matters change significantly. The nonlinearity which finally produces amplitude stability may

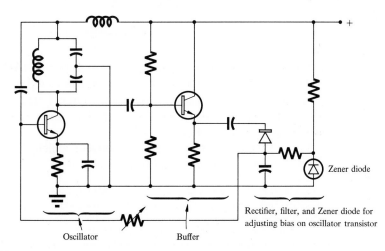

FIG. 14·21   OSCILLATOR WITH AUTOMATIC AMPLITUDE CONTROL

also introduce important reactances into the circuit, which result in serious "frequency pulling." A simple demonstration of this phenomenon is afforded by simply shorting out the 10-kohm source resistance at the input of the amplifier described in Demonstration $14 \cdot 2$. In most cases where gross nonlinearities are important, nonlinear-analysis techniques are required to predict the waveform of the circuit, the waveform predictions usually being graphical estimates of the oscillator waveshape for assumed types of nonlinearities.[1]

$14 \cdot 7$ FEEDBACK AMPLIFIER THEORY AND STABILITY CONSIDERATIONS

We saw in the early sections of this chapter that negative feedback applied around a single amplifier stage could be used to trade gain and bandwidth, and at the same time reduce the sensitivity of the stage to variations in transistor parameters. Negative feedback can also reduce distortion and noise that are generated within the basic amplifier stage. These advantages lead us to inquire whether negative feedback can be applied around several stages to obtain the same benefits in performance. This can be done, but great care must be used to ensure that the conditions for oscillation are not set up when the feedback is applied, and that the amplifier gain characteristic with feedback is suitable for the application.

In this section we shall give a more general treatment of feedback amplifier theory, valid for both positive and negative feedback, and briefly indicate how stability considerations enter into feedback amplifier design. We shall concentrate on resistive shunt feedback so that the principles can be discussed with a minimum of mathematical complexity.[2]

$14 \cdot 7 \cdot 1$ *Current gain of a transistor amplifier with resistive shunt feedback.* To begin the development, we show in Fig. $14 \cdot 22$ an amplifier whose current gain can be described by

$$A_i(s) = \frac{I_2(s)}{I_1(s)} \qquad (14 \cdot 25)$$

Reference directions for $I_2(s)$ and $I_1(s)$ are shown on the figure.

The amplifier has a feedback resistor R_f connected from input to output, with a switch S_1 which can be used to disconnect the feedback. We assume for simplicity that $R_f \gg R_L$ and $R_f \gg R_{in}$, these conditions being met in most practical situations.

For a specified current $I_1(s)$, the current which flows in the feedback resistor (with the reference direction shown) is

[1] See H. J. Zimmermann and S. J. Mason, "Electronic Circuit Theory," John Wiley & Sons, Inc., New York, 1962, chap. 10, and A. A. Andronow and C. E. Chaiken, "Theory of Oscillations," Princeton University Press, Princeton, N.J., 1949.

[2] A straightforward extension of the ideas to include a generalized feedback admittance is the subject of some of the problems at the end of this chapter. A more comprehensive treatment of feedback amplifier theory is given in references cited there.

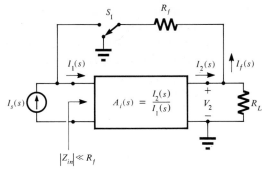

FIG. 14·22   CURRENT AMPLIFIER WITH SHUNT FEEDBACK

$$I_f(s) = A_i(s)\frac{R_L}{R_f}I_1(s) \qquad (14\cdot 26)$$

This is independent of the position of S_1, since we have assumed $R_f \gg R_{in}, R_L$.

The ratio $-[I_f(s)/I_1(s)]$ is called the *return ratio* and is denoted by $T(s)$. For the amplifier shown in Fig. 14·22, the return ratio is simply

$$T(s) = -A_i(s)\frac{R_L}{R_f} \qquad (14\cdot 27)$$

As we shall see presently, the return ratio plays a central role in describing the performance of the amplifier with feedback.

When the switch S_1 is closed, the input current to the amplifier is

$$I_1(s) = I_s(s) + I_f(s) \qquad (14\cdot 28a)$$

Using Eq. (14·26) for $I_f(s)$, we can rewrite Eq. (14·28) in the form

$$I_1(s)\left[1 - A_i(s)\frac{R_L}{R_f}\right] = I_s(s) \qquad (14\cdot 28b)$$

The amplifier output current is then

$$I_2(s) = A_i(s)I_1(s)$$

$$= \frac{A_i(s)}{1 - A_i(s)\dfrac{R_L}{R_f}}I_s(s) \qquad (14\cdot 29)$$

so the amplifier gain *with feedback*, $A_f(s)$, becomes

$$A_f(s) = \frac{I_2(s)}{I_s(s)} = \frac{A_i(s)}{1 - A_i(s)R_L/R_f}$$

$$= \frac{A_i(s)}{1 + T(s)}$$

$$= -\frac{R_f}{R_L}\frac{1}{1 + [1/T(s)]} \qquad (14\cdot 30)$$

14·7·2 *Feedback amplifier performance and the return ratio.* A very comprehensive view of feedback amplifier performance can be obtained by studying the relationship between the return ratio $T(s)$ and the amplifier gain with feedback $A_f(s)$ [Eq. (14·30)]. We shall introduce the most basic ideas in this section.

To begin, we consider the return ratio

$$T(s) = -A_i(s) \frac{R_L}{R_f}$$

From this equation, we can see that $T(s)$ will have the same frequency dependence as $A_i(s)$ for the case of resistive feedback. In particular, $T(s)$ will have the midband value

$$T_0 = -A_{i0} \frac{R_L}{R_f} \tag{14·31a}$$

and a normalized frequency response which is the same as that of $A_i(s)$:

$$\frac{T(s)}{T_0} = \frac{A_i(s)}{A_{i0}} \tag{14·31b}$$

When these properties of $T(s)$ are inserted into the final step of Eq. (14·30), we find that

1. The midband gain of the feedback amplifier A_{f0} is given by

$$A_{f0} = -\frac{R_f}{R_L} \frac{1}{1 + (1/T_0)} \tag{14·32a}$$

2. The frequency response of the feedback amplifier is determined by the factor

$$G_f(s) = \frac{1}{1 + [1/T(s)]} \tag{14·32b}$$

Equation (14·32a) reveals that the midband gain of a feedback amplifier is stabilized against fluctuations in A_{i0} by making $T_0 \gg 1$. This is one of the principal advantages of negative feedback. Equation (14·32b) reveals another aspect of feedback, however. The frequency response of the feedback amplifier is not the same as that of the amplifier without feedback. In particular, the natural frequencies of the amplifier without feedback satisfy the characteristic equation

$$A_i(s) = \infty \tag{14·33}$$

while the natural frequencies of the feedback amplifier satisfy the characteristic equation

$$1 + \frac{1}{T(s)} = 0$$

or

$$T(s) + 1 = 0 \tag{14·34}$$

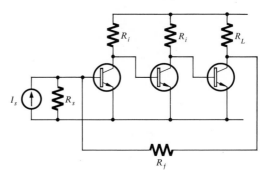

FIG. 14·23   THREE-STAGE AMPLIFIER WITH SHUNT FEEDBACK; BIASING ELEMENTS NOT SHOWN

An example will illustrate that the natural frequencies predicted by Eq. (14·34) can be considerably different from those given by Eq. (14·33), and that in fact the designer may need to exercise great care to ensure that the amplifier will not burst into oscillations when the feedback is applied.

EXAMPLE   A convenient amplifier for illustrating the main problems associated with the application of feedback around several stages is the three stage cascade shown in Fig. 14·23.[1] In this amplifier there is a sufficient number of phase reversals to produce negative feedback in the midband range, but for frequencies above the midband range, the phase of $I_f(j\omega)$ with respect to $I_s(j\omega)$ will change and oscillations can occur.

For simplicity, we assume that $R_i \gg r_\pi$, $R_L = r_\pi$, so the stages are identical. Each stage then has a current gain of the form

$$A(s) = \frac{-\beta}{1 + s/\omega_0}$$

where $\omega_0 = 1/C_{eq}r_\pi$, $C_{eq} = C_\pi + C_c(1 + \beta)$. The three-stage cascade has a current gain of

$$A_i(s) = \frac{-\beta^3}{(1 + s/\omega_0)^3}$$

and the return ratio is

$$T(s) = \frac{R_L}{R_f}\frac{\beta^3}{(1 + s/\omega_0)^3} = \frac{T_0}{(1 + s/\omega_0)^3}$$

Since β^3 will be $\sim 10^5$ to 10^6, it is a simple matter to make $T_0 \gg 1$ in the midband range. We should like to do this to achieve the desirable advantages of negative feedback. However, the natural frequencies of the amplifier with feedback are also effected by the value of T_0 we choose, and in fact the "amplifier" will oscillate if we make $T_0 \geq 8$.

To see that this is so, let us consider how the positions of the natural frequencies in the complex frequency plane change as T_0 is increased from zero (by decreasing R_f from infinity). The natural frequencies of the system are the poles of $A_f(s)$

---

[1] Application of the general theory to one- and two-stage cascades is considered in Probs. 14·16 to 14·18.

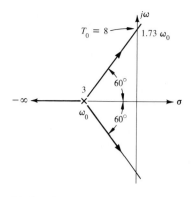

(a) Root locus

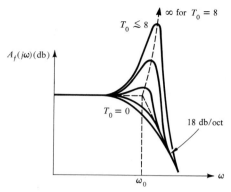

(b) Normalized amplifier frequency response as a function of T_0

FIG. 14·24 ROOT LOCUS PLOTS FOR $(s/\omega_0)^3 + 3(s/\omega_0)^2 + 3(s/\omega_0) + (1 + T_0) = 0$ AND RESULTING AMPLIFIER RESPONSES

or the roots of the equation

$$T(s) + 1 = 0$$

which for the present case is

$$(1 + s/\omega_0)^3 + T_0 = 0$$

The root-locus plot (that is, the locus of the positions which the roots of the previous equation take as T_0 is increased continuously from zero) is shown in Fig. 14·24a. For $T = 0$ (or $R_f = \infty$, so the amplifier has no feedback), the three roots are coincident on the negative real axis. As T_0 is increased, one of the roots moves out along the negative real axis, while the other two roots become a complex pair and move toward the real-frequency axis along lines which make angles of 60° with respect to the real axis in the complex-frequency plane.

The corresponding frequency responses are plotted with T_0 as a parameter in Fig. 14·24b. For $T_0 = 8$, the complex pair of natural frequencies lies on the $j\omega$ axis and the amplifier response $\to \infty$ as $\omega \to 1.73 \omega_0$. The system will ideally oscillate at the frequency $\omega = 1.73 \omega_0$ if $T_0 = 8$. If we use $T_0 < 8$, we can avoid oscillation, though if we want to avoid serious peaking in the frequency response of the amplifier with feedback, we have to use $T_0 \lesssim 4$, which is not large enough to provide many of the advantages which we hoped to obtain by using negative feedback.

14·7·3 *Stability of feedback amplifiers.* The preceding example highlights the fundamental problem associated with using feedback around several stages of amplification: we have to make T_0 large in the passband (to achieve the desirable effects of negative feedback) and at the same time ensure that the natural frequencies of the amplifier lie in desirable positions (that is, ensure that the frequency response of the amplifier with feedback is acceptable for the intended application). From a design standpoint, the variables that we can adjust to achieve these conditions are (1) the gain and frequency response of the main amplifier (we will usually not want a cascade of identical stages) and (2) the form and value of the feedback admittance (in the case of a feedback resistor only, the value of R_f).

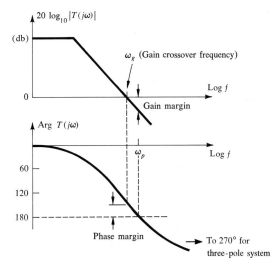

It is frequently convenient to select a suitable $A_i(s)$ and R_f in a series of two steps. In the first step, we make sure that $A_i(s)$ and R_f allow a range of values for T_0 which is sufficiently large to achieve the desirable effects of negative feedback without causing oscillation or producing an amplifier which might oscillate if some parameters were to change slightly. This aspect of feedback amplifier design is called the *stability problem*.

In the second step, we study how the natural frequencies of the feedback amplifier are related to the permissible values of T_0, and choose a desirable value of T_0 based on the positions which we want the natural frequencies to have in the complex-frequency plane.

The stability problem is usually studied by specifying that $T(s)$ satisfy certain *gain and phase margins*. We shall develop these concepts in this section. The second part of the feedback-amplifier design problem is usually studied with the aid of root-locus techniques or some equivalent method, and is beyond the scope of the present discussion.[1]

Gain and phase margins. In order to prevent oscillation, we want to ensure that the feedback amplifier cannot have a natural frequency on the $j\omega$ axis or in the right half of the complex-frequency plane. This condition is usually satisfied by ensuring that $T(j\omega)$ cannot equal -1 for any value of ω [or $T(j\omega) + 1$ cannot equal zero].

We can accomplish this by ensuring that $|T(j\omega)|$ reaches unity (or 0 db) before arg $T(j\omega)$ reaches 180°. Toward this end we define what are called *gain* and *phase margins* for $T(j\omega)$. The phase margin is the amount by which arg $T(j\omega)$ is less than 180° at the frequency ω_g, where $|T(j\omega_g)| = 1$. ω_g is called the *gain-crossover frequency*. The gain margin is the amount of

---

[1] Interested readers can consult Chap. 18 and the references at the end of this chapter.

additional gain needed to cause instability at the frequency ω_p, where arg $T(j\omega_p) = 180°$. Gain and phase margins and the critical frequencies ω_p and ω_g are illustrated in Fig. 14·25.

The purpose of establishing gain and phase margins is to make allowances for deviations in system performance due to oversimplified models for analysis, variations in parameters, and so on. Usually an amplifier designed to give a phase margin of 30° at the gain-crossover frequency willl have adequate stability. The use of gain and phase margins in guiding the design of a feedback amplifier is considered in Prob. 14·20.

14·8 SUMMARY

In this chapter we have considered some simple schemes for applying both positive and negative feedback in amplifiers. The most practical of the two negative-feedback schemes which we discussed is the one which employs a feedback conductance G_f, connected between the collector and base terminals of a stage. When this type of feedback is employed, the current gain of the stage can be stabilized without sacrifice of gain-bandwidth product in the stage. The input and output impedances of the stage are also lowered (and stabilized). A cascade of such stages can be connected together to build an amplifier with stabilized gain and as high an upper cutoff frequency as the transistors will permit.

Positive feedback provides a means of controlling the characteristics of an amplifier and a means of constructing circuits which oscillate. The ideal conditions for sinusoidal oscillation in a circuit can be determined by (1) requiring that the circuit equations yield non-zero node voltages at the desired frequency when no external drive is applied, or (2) requiring that the nodal admittance of the network at any node pair be zero at the desired frequency $s = j\omega_0$, or (3) requiring that the magnitude and phase of the voltage (or current) fed back to the amplifier input equal that required to produce the given amplifier output.

The effects of positive and negative feedback on any circuit can be visualized by considering the motion of the natural frequencies of the *entire system* with changes in the circuit parameters. This concept is useful for relating the behavior of active circuits to more general network theory.

REFERENCES

Angelo, E. J.: "Electronic Circuits," 2d ed., McGraw-Hill Book Company, New York, chap. 19.

Mason, S. J., and H. J. Zimmermann: "Electronic Circuits, Signals and Systems," John Wiley & Sons, Inc., New York, 1960, chap. 9.

Truxal, J. G.: "Control System Synthesis," McGraw-Hill Book Company, New York, 1955, chap. 4.

Bode, H. W.: "Network Analysis and Feedback Amplifier Design," D. Van Nostrand Company, Inc., Princeton, N.J., 1945.

Joyce, M. V., and K. K. Clarke: "Transistor Circuit Analysis," Addison-Wesley Publishing Company, Inc., Reading, Mass., 1961, chaps. 8–16.

Linvill, J. G., and J. F. Gibbons: "Transistors and Active Circuits," McGraw-Hill Book Company, New York, 1961, chap. 20.

DEMONSTRATIONS

14·1 SINGLE-STAGE NEGATIVE FEEDBACK AMPLIFIER The effects of feedback on gain and bandwidth of a single stage can be readily demonstrated with the circuit shown. The construction details are similar to those given for the demonstration amplifiers in Chap. 13.

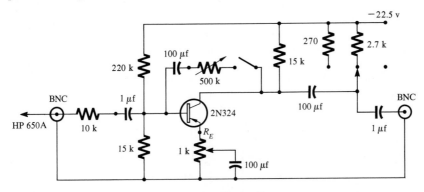

FIG. **D14·1** AMPLIFIER FOR DEMONSTRATING G_f AND R_E FEEDBACK

14·2 POSITIVE FEEDBACK AND THE CONDITIONS FOR OSCILLATION The effect of positive feedback on the gain of an amplifier and the conditions for oscillation of a circuit can be simply demonstrated with the circuit shown. The circuit employs the amplifier of D14·1 with the G_f feedback switched out and the R_E feedback

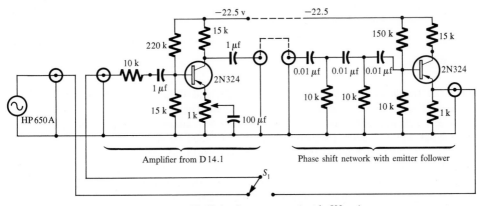

(Oscillation frequency approximately 500 cps)

FIG. **D14·2** AMPLIFIER FOR DEMONSTRATING THE EFFECT OF POSITIVE FEEDBACK ON GAIN AND THE CONDITIONS FOR OSCILLATION

used to control the stage gain. The amplifier is connected to a phase-shift network which drives an emitter follower. The output of the emitter follower can then be coupled back to the amplifier input, thus preventing the amplifier from loading the phase-shift network.

By adjusting the gain of the amplifier (R_E potentiometer) the emitter-follower output voltage can be made equal to the input-signal voltage in magnitude and phase (a dual-beam oscilloscope is useful for this purpose). The switch S_1 can then be operated, and the circuit will produce sustained oscillations. If the gain is slightly less than this, the amplifier output will be very large at the appropriate signal frequency.

EXPERIMENTAL PROJECT

Using two of the transistors supplied to you and the power supply constructed previously, construct a phase-shift oscillator which will oscillate at the frequencies 30 cps, 300 cps, 3 kcps, 30 kcps, and 300 kcps. The variation in frequency can be accomplished by switching appropriate capacitors into the phase-shift line (or changing R in the phase-shift line). To avoid loading the phase-shift line you may wish to use an emitter follower as shown in the figure. Using this oscillator, check the frequency response of the amplifier you constructed in the experimental project for Chap. 13. You may wish to experiment with feedback added to your amplifier.

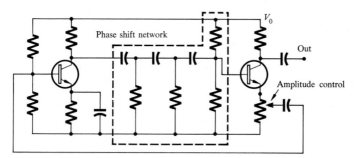

ғıɢ. E14·1

PROBLEMS

14·1 A transistor with $\beta = 100$, $f_T = 10$ mcps, and $C_c = 10$ pfarads is driven from a 300-ohm source and drives a 500-ohm load. The transistor is to be biased at $I_E = 1$ ma, $V_{CE} = 6$ volts. A 22.5-volt battery is available as a supply voltage.

1. What will the midband-current gain of the stage be if a G_f is used which matches the amplifier input impedance (in the midband range) to the source impedance? Voltage gain?
2. What value of G_f is required to obtain the impedance-matching condition?
3. What is the upper cutoff frequency for the current gain? Voltage gain?
4. What changes in midband-voltage gain and input impedance will occur if a transistor with $\beta = 200$ is substituted, using the value of G_f obtained in (2)?

14·2 1. Show that

$$\frac{dA_{i0}}{d\beta} = \left(\frac{A_{i0}}{\beta}\right)^2$$

for a simple *CE* stage with G_f feedback.

2. A single-stage *CE* amplifier with G_f feedback drives a 300-ohm resistive load. The transistor has a typical β of 100, though manufacturing tolerances are such that β_{min} of 80 and β_{max} of 200 may be expected. Approximately how much midband-current gain can be obtained if we require that $A_{i0,max} - A_{i0,min} = 0.2 \, A_{i0,typ}$? What value of G_f achieves this result? Repeat for $A_{i0,max} - A_{i0,min} = 0.05 \, A_{i0}$.

14·3 You are asked to design an amplifier using transistors whose β may fall anywhere in the range of 100 to 200. The midband-current gain of the amplifier is to be 1,000, and this gain should not vary by more than 10 percent for any combination of transistors. Assuming the stages are identical, how many stages are required, and what is the gain of each stage?

14·4 A *CE* stage is designed to give a midband-current gain of 20 when driving an R_L of 500 ohms. A transistor with $\beta = 100$ and G_f feedback is used to achieve the desired gain. Compute the variation in input impedance which you expect if β is allowed to vary from 50 to 200.

14·5 An amplifier is required which will deliver a signal amplitude of 1 volt to a 300-ohm load. The signal source has an internal impedance of 300 ohms (resistive) and an open-circuit voltage of 1 mv. Design an amplifier to meet these specifications. Transistors with $\beta_{typ} = 100$ are available, though β may vary from 50 to 150 from unit to unit. A 10 percent variation in the output amplitude is acceptable under worst-case conditions.

14·6 1. Obtain a general formula for the midband output admittance of a *CE* stage with G_f feedback. The source impedance is R_s. State assumptions and approximations clearly.

2. Using $I_E = 1$ ma, $\beta = 100$, $R_s = 1$ kohm, plot Y_{out} vs. G_f over the range of values of G_f which meet the assumptions you have used in deriving your formula.

3. What applications can you think of where this control over Y_{out} might be useful?

14·7 A transistor with $\beta = 100$, and $f_T = 100$ mcps is to be used to build an input stage for an amplifier, where an input impedance of 50 kohms is required. The load resistance on the stage is 2 kohms, and the source impedance is 50 kohms. Design a circuit with an unbypassed emitter resistor to perform the required function. What is the voltage gain of the resulting amplifier stage? Its bandwidth?

14·8 The circuit shown in Fig. P14·8 has both G_f and R_E feedback. Compute the midband input impedance for this circuit.

14·9 1. Approximately what values of C_1, C_2 are required to build a Colpitts oscillator which produces nearly sinusoidal oscillations at a frequency of 1 mcps with an inductance of 2.5 μh, using a transistor with $\beta = 100$?

2. Suppose the operating bias current for the circuit is 1 ma. A 500-ohm load resistor is now placed between the base and emitter terminals. Compute the natural frequencies of the system with the load in place. Will this system oscillate? If not, what can be done to make it oscillate?

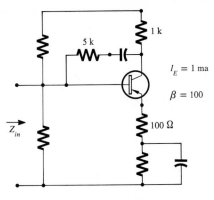

FIG. P14·8

14·10 Find the frequency of oscillation and the oscillation conditions for a Hartley oscillator.

14·11 How much gain is required in the amplifier section of a Wien bridge oscillator to place the natural frequencies of the system on the $j\omega$ axis of the complex s plane? Plot the natural frequencies of this system as a function of the gain of amplifier. You may assume for simplicity that the impedance of the frequency-determining network is sufficiently low that the amplifier input does not load the network (a CC buffer stage might be necessary in a practical situation).

14·12 An inductively loaded common-emitter stage is shown in Fig. P14·12 (the bias network is omitted for simplicity). What is the input admittance Y_{in}? What condition must be satisfied for Y_{in} to have a negative real part? [If Y_{in} has a negative real part, a tuned circuit with shunt G equal to the input conductance and susceptance equal to the negative of the input susceptance may be connected at the input to make an oscillator.]

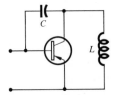

FIG. P14·12

14·13 A two-port network has the following h parameters:

$$h_{11} = 1 \text{ ohm} \qquad h_{12} = 0.1 \qquad h_{21} = 100 \qquad h_{22} = 1 \text{ mho}$$

What is the input resistance of this two-port network when $Y_L = 1$ mho? Draw a network for Z_s that will make the system composed of the two-port and its source and load oscillate at $\omega = 1$ rad/sec.

14·14 A circuit called a negative-impedance converter is shown in Fig. P14·14a. When the biasing sources are neglected and a simple T model is used for the transistor, the circuit reduces to that shown in Fig. P14·14b. Show that the dynamic input impedance (that is, small-signal input impedance) of this circuit is propor-

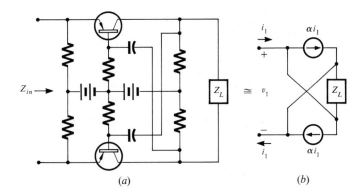

(a)   (b)

tional to $-Z_L$. (Negative-impedance converters are useful for making filters and oscillators.)

14·15 A transfer network composed of only R and C elements is shown in Fig. P14·15a (the two R's and the two C's have the same values).

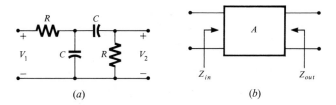

FIG. P14·15   (a)   (b)

1. Evaluate $V_2/V_1(s)$ and show that this ratio can be real (that is, V_2 in phase with V_1). For what frequency is this zero-phase condition obtained if $R = 1$ kohm, and $C = 0.001$ µfarad?

2. If you were given complete freedom to specify A, Z_{in}, and Z_{out} of the amplifier shown in Fig. P14·15b, what properties would you choose, and how would you connect the amplifier to the network in Fig. P14·15a to make an oscillator?

3. If transistor frequency response can be neglected (by using transistors with high f_T and low C_c), sketch a transistor amplifier which would meet the amplifier requirements in (2) sufficiently closely that it could reasonably be expected to work in the oscillator. Component values are not required in the entire amplifier, though you should include bias components in your sketch and give a brief discussion of how an adequate approximation for the required values for Z_{in}, Z_{out}, and A are to be obtained in your amplifier. You should also provide for the fact that A will need to be somewhat higher than the ideal value, to make the oscillator work. You may specify reasonable values of β in your arguments if it is necessary.

14·16 Use the general feedback theory given in Sec. 14·7 to derive expressions for the gain and bandwidth of a CE single-stage amplifier with resistive shunt feedback. Express both the gain and the bandwidth in terms of T_0 and specify the conditions under which gain and bandwidth can be traded by adjusting T_0. Sketch the root-locus plot for the feedback amplifier and compare it with Fig. 14·4.

14·17 Analyze the network shown in Fig. 14·7 using the general feedback theory given in Sec. 14·7, amended to include the fact that the feedback conductance has been replaced by a feedback admittance. The cutoff frequency of the common emitter amplifier can be neglected for this problem.

14·18 Using a cascade of two identical CE stages with $R_i \gg r_\pi$, $R_L = r_\pi$, write down the return ratio for a feedback amplifier with resistive shunt feedback. Sketch the root-locus plot for this amplifier and comment on your results.

14·19 Analyze an unloaded Colpitts oscillator using the general feedback theory presented in Sec. 14·7.

14·20 We saw in Sec. 14·7 that resistive shunt feedback could not be applied around a cascade of three identical stages and still obtain a usefully large value of T_0. In this problem we will consider some common procedures for changing $A_i(s)$ and R_f to effect a more suitable design.

1. *Broadbanding a stage.* It is frequently useful to "broadband" one of the stages in the basic amplifier (that is, increase the cutoff frequency of one of the stages). To study the effect of this step, let us assume that we have three CE stages, each with current gain

$$A(s) = \frac{-50}{1 + s/\omega_0}$$

We can use collector-to-base feedback around the last stage to reduce the midband gain to 5 and increase the cutoff frequency of the stage to about $10\omega_0$. The overall current gain will now be approximately

$$A_i(s) = \frac{-12,500}{(1 + s/\omega_0)^2(1 + s/10\omega_0)}$$

Write out $T(s)$ for this amplifier assuming we want to use resistive shunt feedback. What is the maximum value of T_0 that we can have to ensure a phase margin of 30° at the gain crossover frequency.

2. *Pole cancellation technique.* The improvement in useable values of T_0 obtained above is significant, though we still might wish to have larger values of T_0 in the amplifier passband. We can achieve this by connecting a capacitor C_f in parallel with the feedback resistor R_f. (*a*) Write out $T(s)$ for the broadbanded three-stage amplifier above with this new addition. (*b*) Assuming C_f is chosen so that $R_f C_f = 1/\omega_0$, compute the largest value of T_0 that can be used to give a phase margin of 30° at the gain crossover frequency. What are the natural frequencies of this amplifier for $T_0 = 10$? $T_0 = 20$? (Procedures for sketching the frequency response of amplifiers which have complex natural frequencies are described in Chap. 18, in the event that you want to know what the frequency response of the amplifier is for these values of T_0.)

15

SWITCHING CIRCUITS
AND SWITCHING DEVICES

THERE ARE MANY important electronic systems in which active devices are
used to build *switching circuits*. In these circuits, voltages or currents are
switched abruptly between predetermined levels, where they may remain
either indefinitely or for controlled periods of time. Systems which utilize
such circuits include digital computers, television and radar receivers,
pulse communication systems, and instrumentation for systems where data
comes in pulse form (for example, nuclear-particle counting). In this
chapter we shall study some basic types of switching circuits and briefly
develop some of their applications in these systems.

Basic states of a switch. When an active device is used as a switch, it is usually considered to have two states:[1]

1. An ON state, usually corresponding to a heavily conducting condition, perhaps carrying the device into its saturation region.

2. An OFF state, corresponding to a more lightly conducting condition, perhaps carrying the device into its cutoff region.

Each active device used in a switching circuit can be assumed to be in one of these two states, apart from short periods of time during which transitions between the states are occurring. These transition times are important, of course, since they set limits on how fast switching circuits can work. For a complete understanding of the characteristics of a particular switching circuit, it is necessary to understand both the conditions which produce the transition of the active device, or devices, between its two states and the factors which affect the transition time. However, there are many circuits in which the actual transition time between states is so short that it can be neglected. In this chapter we shall concentrate most of our attention on circuits in which the "zero-transition-time" approximation does not seriously affect the basic operating features of the circuit. This approximation will simplify the presentation of ideas, and still enable us to obtain useful theoretical and design relations for the principal types of switching circuits.

Before we begin our study, it should be emphasized that switching circuits are *nonlinear* circuits. They are usually driven with or generate waveforms of sufficient amplitude to drive the active devices from cutoff to saturation. These swings of voltage and current are so large that the small-signal linear models of devices we have used in the preceding two chapters are *no longer applicable*. We must instead resort to the use of *large-signal models* and *piecewise-linear-analysis* techniques.

15 · 1 REGENERATIVE AND NONREGENERATIVE CIRCUITS

It is useful to divide switching circuits into two categories: *regenerative* and *nonregenerative*. In a nonregenerative circuit, an active device is held in a desired state by an *externally applied drive*. A useful analogy here is that of a simple relay, where a control current must always be applied to the relay coil to keep the contacts closed. In a regenerative circuit, the switching action is *initiated* by an externally applied pulse (or by the satisfaction of an internal circuit condition), but then the circuit action itself carries the change of state to completion. An analogy for this case is a seesaw with a frictionless groove cut in its center in which a freely rolling ball rests as a weight (Fig. 15 · 1). This system has two stable states in which it can remain indefinitely. To change the state of the system, a

---

[1] There are some devices with more than two possible stable states. We shall not consider these devices in this book.

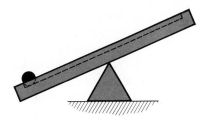

FIG. **15·1**   A MECHANICAL SYSTEM WITH TWO STABLE STATES

"switching pulse" is applied by lowering the free end of the seesaw. If the free end is lowered to a position which is slightly beyond the horizontal, then the switching force can be removed and the system will complete the change of state by an internal regenerative action: the ball moves beyond the pivot point and tilts the seesaw; this causes the ball to move down the groove; the change in position of the ball applies more torque to the seesaw, tilting it further, etc. If the switching force is not adequate to bring the seesaw to the horizontal position, then the system will return to its initial state when the switching force is removed.

15·2   THE TRANSISTOR AS A NONREGENERATIVE SWITCH

Transistors are used as nonregenerative switches in power relays, as conducting paths for communication signals, and in digital computation systems, to name just a few of the applications. The basic circuit for these applications is shown in Fig. 15·2. In this circuit the transistor is supposed to act like a knife switch, controlled by the control source.

A partial list of questions we might ask to evaluate the performance of the transistor as a controlled knife switch are as follows:

1. How much load current will the switch pass when it is ON?
2. What is the voltage drop across the switch when it is ON?

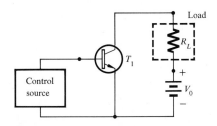

FIG. **15·2**   THE BASIC NONREGENERATIVE TRANSISTOR SWITCHING CIRCUIT. IN THE ON CONDITION, THE CONTROL SOURCE DRIVES A LARGE CURRENT INTO THE BASE OF T_1, FORCING IT INTO ITS SATURATION REGION. IN THIS CONDITION, NEARLY THE FULL POWER-SUPPLY VOLTAGE APPEARS ACROSS THE LOAD. IN THE OFF CONDITION, THE CONTROL SOURCE APPLIES A REVERSE BIAS TO THE EMITTER-BASE JUNCTION, TURNING T_1 OFF, SO THAT NO POWER IS DELIVERED FROM THE BATTERY TO THE LOAD.

3. What dynamic impedance does the switch present to the load when ON?

4. How much control power is required to turn the switch ON?

5. What are the leakage impedance and leakage current of the switch when it is OFF?

6. In the OFF state how much voltage will the switch stand off without breaking down?

7. How long does it take the transistor to switch from OFF to ON, and vice versa?

There are, of course, many other interesting questions which might be asked about the switch. However, for our present purposes the list above is adequate. We shall consider the answers to these questions by studying a sequence of different applications.

15·2·1   *The transistor switch as a power relay.* One of the interesting features of the transistor as a power relay is that the power which can be switched is many times the collector dissipation rating of the transistor. This is because in both the open and closed states, the power dissipated in the transistor is low. An example will illustrate this and other features of the operation of a transistor as a power relay.

Our example for this section will be the circuit shown in Fig. 15·3. At the command of a control signal we wish to supply power to the load resistor (perhaps a remote controlled alarm light; or a headlight dimmer in an automobile, where the load is a relay coil). For simplicity, we shall assume that the control signal is a current i_B. Let us begin by studying how the control signal turns the switch ON.

ON *condition of the power switch.*   To do this, we refer to the circuit shown in Fig. 15·4, which is a basic common-emitter amplifier. The transistor is assumed to have a common-emitter current gain (β) of 50. We suppose at the outset that the base resistor R_1 has a value of 1 megohm. The base current is therefore approximately 10 μa. The collector current is 50 times this, or 0.5 ma. The voltage across the load resistance is therefore 2.5 volts, leaving approximately 22.5 volts reverse bias across the collector-base junction. In this condition the circuit is merely acting as a dc amplifier with the operating point shown in Fig. 15·4.

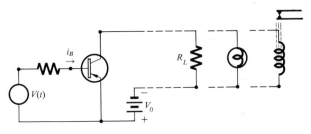

FIG. 15·3   NONREGENERATIVE SWITCHING CIRCUIT EXAMPLE

610   *Semiconductor electronics*

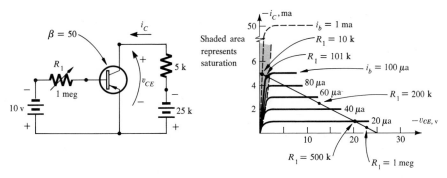

FIG. 15·4   CIRCUIT FOR STUDYING ON CONDITION OF A TRANSISTOR POWER SWITCH

Now, suppose we begin to reduce R_1. Then the operating point changes in the manner shown in Fig. 15·4. When R_1 reaches 500 kohms, the base current will be 20 μa. The collector current will then be 1 ma, and there will be a 5-volt drop across the load resistance. When R_1 reaches 101 kohms, the base current will be 0.099 ma, the collector current will be 4.95 ma, and the voltage drop across the load resistance will be 24.75 volts. Since a small forward voltage drop (say 0.25 volts) is required across the emitter junction to cause these currents to flow, the actual voltage drop from collector to base will now be very nearly zero.

Suppose now that the base resistance is further reduced, to 10 kohms, for example. This causes a base current of approximately 1 ma to flow. If the transistor were acting as a normal CE amplifier, 1 ma of base current would result in 50 ma of collector current. However, this condition cannot exist; the battery and resistance in the collector circuit limit the maximum collector current to 5 ma. As a result, 1 ma of base current drives the transistor deep into its saturation region (see Fig. 15·4).

When circuit conditions force a situation in which $|I_C| < \beta|I_B|$, the transistor will be operating in its saturation region. Both the emitter-base *and* the collector-base junctions will be forward-biased, and the voltage drop from collector to emitter will usually be very small compared to the forward-bias voltage on either junction. Thus the switch is simply an over-driven amplifier. The base current drives the transistor into saturation to establish the ON state.

The state of affairs inside the transistor is shown in Fig. 15·5. Figure 15·5a shows the hole distribution across the base when $i_B = 10$ μa. Figures 15·5b and 15·5c show the hole distribution for $i_B = 20$ μa and 99 μa, respectively. For the first two cases the collector-base voltage is negative, so the *actual* hole density (p_C') at the right-hand edge of the base is essentially zero for each of these figures. For Fig. 15·5c, the collector-base voltage is zero, so $p_C' = p_n$. Notice that as more base current is required, the area under the hole-density curve must increase. This is because the base current is recombination current, and the recombination current depends on the average excess density in the base (or the total number of minority carriers stored in the base region).

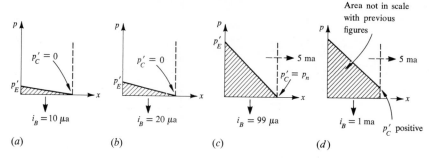

FIG. 15·5 HOLE-DENSITY PROFILES IN THE BASE OF A *pnp* TRANSISTOR FOR VARIOUS BASE DRIVES; (*d*) REPRESENTS SATURATION

To increase the base current beyond 0.099 ma, we must again increase the area under the carrier-density curve. However, since essentially 5 ma of current is flowing in the collector circuit for Fig. 15·5*c*, we cannot increase the slope of the density profile in order to get more charge stored in the base and therefore more base current. Instead, we must allow the hole density at the collector-base junction to become positive. From the law of the junction, this is, of course, equivalent to allowing the collector-base voltage to become positive. That is, the collector-base junction is now *forward-biased.* This is shown in Fig. 15·5*d*.

Calculation of junction voltage drops in the ON *state.* We can calculate the junction voltage drops by calculating the values of p_E and p_C which permit a base current of -1 ma and a collector current of -5 ma. To do this, recall that the relations between terminal currents and excess-minority-carrier densities in the transistor base region are entirely *linear:*[1]

$$i_E = I_{ES}\frac{p_E}{p_n} - \alpha_R I_{CS}\frac{p_C}{p_n}$$

$$i_C = -\alpha_F I_{ES}\frac{p_E}{p_n} + I_{CS}\frac{p_C}{p_n}$$

(15·1)

When i_E and i_C are known, these two equations will yield p_E/p_n and p_C/p_n, and then the law of the junction can be used to obtain the junction voltages.

When Eqs. (15·1) are solved for p_E/p_n and p_C/p_n, there results

$$\frac{p_E}{p_n} = \frac{i_E + \alpha_R i_C}{I_{ES}(1 - \alpha_R\alpha_F)}$$

$$\frac{p_C}{p_n} = \frac{i_C + \alpha_F i_E}{I_{CS}(1 - \alpha_F\alpha_R)}$$

(15·2)

[1] I_{ES}, α_F, α_R, and I_{CS} are, by definition, *positive* numbers. For an *npn* transistor, Eqs. (15·1) become

$$i_E = -I_{ES}(e^{-qV_{EB}/kT} - 1) + \alpha_R I_{CS}(e^{-qV_{CB}/kT} - 1)$$
$$i_C = \alpha_F I_{ES}(e^{-qV_{EB}/kT} - 1) - I_{CS}(e^{-qV_{CB}/kT} - 1)$$

The junction laws then give

$$v_{EB} = \frac{kT}{q} \ln\left(\frac{p_E}{p_n} + 1\right) = \frac{kT}{q} \ln\left[\frac{i_E + \alpha_R i_C}{I_{ES}(1 - \alpha_R \alpha_F)} + 1\right] \quad (15 \cdot 3a)$$

$$v_{CB} = \frac{kT}{q} \ln\left[\frac{i_C + \alpha_F i_E}{I_{CS}(1 - \alpha_R \alpha_F)} + 1\right] \quad (15 \cdot 3b)$$

Because p_E/p_n and $p_C/p_n \gg 1$ for the case under discussion, we can also write

$$v_{EB} - v_{CB} \cong \frac{kT}{q} \ln\left(\frac{i_E + \alpha_R i_C}{i_C + \alpha_F i_E} \frac{I_{CS}}{I_{ES}}\right) \quad (15 \cdot 3c)$$

EXAMPLE Let us now compute the junction voltage drops which we expect in a practical example. We shall suppose the transistor used in the circuit of Fig. 15·4 is a 2N1303 (a *pnp* germanium-alloy transistor intended for computer and switching applications). This transistor has $\alpha_F \simeq \alpha_R \simeq 0.98$, $I_{CO} \simeq I_{EO} = 2$ μa. From Eq. (9·49), we find

$$I_{CS} = \frac{I_{CO}}{1 - \alpha_F \alpha_R} = 50 \text{ μa}$$

The terminal currents in the ON state are $i_E = 6$ ma, $i_C = -5$ ma.

Hence,

$$v_{EB} = \frac{kT}{q} \ln\left[\frac{6 - (0.98)5}{50(0.04)} \times 10^3\right] = \frac{kT}{q} \ln 550 \simeq 158 \text{ mv}$$

$$v_{CB} = \frac{kT}{q} \ln\left[\frac{-5 + (0.98)6}{50(0.04)} \times 10^3\right] = \frac{kT}{q} \ln 440 \simeq 153 \text{ mv}$$

The collector-to-emitter voltage v_{CE} is therefore

$$v_{CE} = v_{CB} - v_{EB} = -5 \text{ mv}$$

As suggested earlier, the two junction voltages are nearly equal, so v_{CE} is much smaller than either junction voltage.

A simple circuit for experimentally verifying these results is given in the Demonstration 15·1 at the end of this chapter. The experimentally determined values of v_{CE} are always in qualitative agreement with the previous theory; and the quantitative agreement is also good so long as $|I_C| \gtrsim |\beta I_B|/5$. However, beyond this point the injection levels are so high in a transistor that is turned ON that the theory of operation of the transistor must be amended to include high-level injection effects (that is, degradation of emitter efficiency, change of carrier lifetimes and recombination processes, etc.). In such cases, the foregoing theory predicts junction voltage drops that are only roughly correct; and the v_{CE} obtained from the preceding argument will also be in error (more nearly 25 mv than 5 mv for the preceding example).[1]

Saturation conditions in the large-signal circuit model. It is also useful to interpret the internal behavior of the transistor in its saturated condition

---

[1] Readers interested in a treatment of the problem which includes high-level injection effects are referred to C. A. Mead, The operation of junction transistors at high currents and in saturation, *Solid State Electronics,* vol. 1, pp. 211–225, July, 1960, and J. F. Gibbons, Super-saturated transistor switches, *IEEE* Prof. Group on Electron Devices, November, 1961.

in terms of the large-signal circuit model shown in Fig. 15·6. If the transistor is operating under conditions which put it in the *normal-active* operating region, the currents flowing in D_{Cj} and its associated controlled source may be neglected. The currents flowing in D_{Ej} and the associated controlled source are nearly equal, both being about β times the base current.

In the saturated state, the currents in D_{Ej} and its associated controlled source will still be on the order of β times the actual base current; but D_{Cj} will be ON and very little of the current $\alpha_F i_{Ej}$ will flow out into the collector circuit. To be more specific, we may use Kirchhoff's current law at the emitter and collector terminals to obtain

$$i_C = i_{Cj} - \alpha_F i_{Ej} \tag{15·4a}$$

$$i_E = -\alpha_R i_{Cj} + i_{Ej} \tag{15·4b}$$

Equations (15·4) can be thought of as two equations for the unknowns i_{Ej} and i_{Cj} in terms of i_C and i_E. If we again use $i_C = -5$ ma, $i_E = 6$ ma, $\alpha_F = \alpha_R = 0.98$, we find $i_{Ej} = 27.5$ ma, $i_{Cj} = 22$ ma. The voltage drops in the diodes D_{Cj} and D_{Ej} are found from

$$v_{EB} = \frac{kT}{q} \ln\left(\frac{i_{Ej}}{I_{ES}} + 1\right) \qquad v_{CB} = \frac{kT}{q} \ln\left(\frac{i_{Cj}}{I_{CS}} + 1\right)$$

the results being identical to those obtained previously. In terms of the circuit model, the saturation state is entered when D_{Cj} turns ON.

Maximum current in the ON state. We have just seen that in its ON state, a transistor switch has a v_{CE} drop which is on the order of a few millivolts or perhaps tens of millivolts, if i_C is large and $<|\beta i_B|$. Since the power supply voltages will be large compared to this v_{CE}, we can calculate i_C with quite reasonable accuracy by assuming $v_{CE} = 0$.

As a result of the small v_{CE}, very little power will be dissipated in the transistor when it is ON. Hence, the maximum load current which the transistor can handle will usually not be determined by the collector

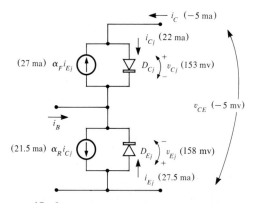

FIG. 15·6   EBERS-MOLL MODEL AND THE INTERPRETATION OF SATURATION

power-dissipation rating but rather by the degradation of performance under high-level injection conditions at the junctions (for example, emission crowding may produce destructively high current densities; *cf.* Fig. 9·30).

EXAMPLE The fact that the maximum load current for a transistor switch is determined by factors other than collector power dissipation is illustrated by the following example. A 2N1303 has an $I_{C,max}$ rating of 300 ma. At this current $V_{CE} = 0.35$ volts, and the power dissipated in the structure is about 100 mw. For comparison, the maximum collector dissipation rating is 150 mw.

OFF *condition of the power switch.* The transistor switch is turned OFF by applying a reverse bias to the emitter-base junction. In a *pnp* transistor, this will correspond to reducing the *actual* stored charge of holes in the base region to essentially zero. The collector circuit battery will also reverse bias the collector-base junction, so as to establish an operating condition in the *cutoff* region.

The "leakage" currents which the transistor will permit for this condition will be $i_C \simeq I_{CO}$, $i_E \simeq I_{EO}$. For a 2N1303, this means that the switch will allow about 2 μa of current to flow through it when it is OFF.

Since the load resistance will be small enough so that $R_L I_{CO}$ represents a small voltage drop, essentially the full collector supply voltage will be applied between the collector-to-emitter terminals of the switch when it is OFF. The maximum collector supply voltage which can be used is then determined by the collector voltage rating of the transistor, $V_{CE,max}$. The average collector-power-dissipation rating will be of secondary importance.

Let us now summarize the performance features of the transistor power switch, using some of the questions formulated in the introduction to Sec. 15·2.

1. The switch will pass $I_{C,max}$ when ON.
2. The voltage drop across the switch when ON will be a few millivolts or perhaps tens of millivolts (more for silicon transistors).
3. The switch will pass a collector current of about I_{CO} when OFF.
4. The maximum collector supply voltage which can be applied to the transistor in the OFF state is $V_{CE,max}$.
5. The control power required to turn the switch ON is simply V_{BE} times the base current. For our running example, this amounts to 1 ma $\times$ 158 mv, or 0.158 mw. The power that the load receives when the switch is ON is simply the (supply voltage)^2/load resistance, or 125 mw.

Using the facts established in (1) and (4) above, we can see that the maximum power which can be delivered to the load in a simple power-switching circuit is

$$P_{max,switchable} = V_{CE,max} I_{C,max}$$

For a 2N1303 this product is 25v $\times$ 300 ma = 7.5 watts, which is 50 times larger than the collector-power-dissipation rating.[1]

---

[1] Of course, the rate at which this 7.5 w can be switched ON and OFF depends on the collector-power-dissipation rating.

15·2·2 *The transistor as a switch for communication signals.* A transistor which is used in a communication-signal switching circuit is shown in Fig. 15·7. In addition to the dc voltages and the control source required to close the switch, there is an ac-coupled path for a communication signal in this circuit. When the switch is ON, the communication signal is fed to the ac load, represented by the terminated transmission line. As is suggested in Figs. 15·7b and 15·7c, the significant properties of the switch are now its small-signal ON and OFF impedances. We can calculate these with the aid of the formulas obtained earlier.

ON *impedance.* The impedance between the collector and emitter terminals of a saturated transistor switch will be essentially a pure resistance up to a frequency of about f_T. Since this represents most of the useful frequency range of the transistor, we shall assume for simplicity that the highest frequency present in the communication signals is less than f_T. Then the ON resistance is simply defined by

$$r_{ON} = \left| \frac{dv_{CE}}{di_C} \right| \qquad (15 \cdot 5)$$

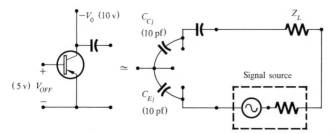

(a) *Basic circuit*

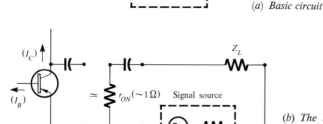

(b) *The saturated switch presents a small-signal resistance r_{ON} to the communication signal*

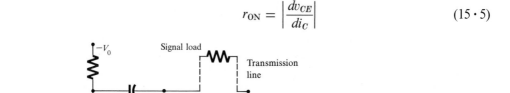

(c) *At communication signal frequencies the cutoff transistor switch provides a capacitance to ground, which permits some signal current to flow*

FIG. **15·7**   THE TRANSISTOR AS A SWITCH FOR COMMUNICATION SIGNALS

The magnitude sign is used in Eq. (15·5) to cover both *pnp* and *npn* cases. To calculate r_{ON}, we first recast Eq. (15·3c) for v_{CE} in a more convenient form. We can use

$$i_E = -i_C - I_B$$

in Eq. (15·3c) to obtain

$$v_{CE} = \frac{kT}{q} \ln \left[\frac{(1 - \alpha_R)i_C + I_B}{\alpha_F I_B - i_C(1 - \alpha_F)} \frac{I_{CS}}{I_{ES}} \right] \qquad (15·6)$$

We use the symbol I_B to denote that the control source is assumed to deliver a constant current to the transistor to turn it ON. Hence there are no variations in i_B.

It is convenient to carry out the differentiation of Eq. (15·6) in symbolic terms as far as possible:

$$\frac{dv_{CE}}{di_C} = \frac{kT}{q} \frac{d}{di_C} \left[\ln \frac{x(i_C)}{y(i_C)} + \ln \frac{I_{CS}}{I_{ES}} \right]$$

$$= \frac{kT}{q} \frac{d}{di_C} (\ln x - \ln y + \text{const})$$

$$= \frac{kT}{q} \left(\frac{1}{x} \frac{dx}{di_C} - \frac{1}{y} \frac{dy}{di_C} \right)$$

$$= \frac{kT}{q} \left[\frac{1 - \alpha_R}{(1 - \alpha_R)i_C + I_B} + \frac{1 - \alpha_F}{\alpha_F I_B - (1 - \alpha_F)i_C} \right] \qquad (15·7)$$

r_{ON} will be the magnitude of the number calculated from Eq. (15·7).

EXAMPLE Using the numerical data from the power-switching example, we find

$$r_{ON} = 25 \times 10^{-3} \left| \frac{0.02}{(-0.1 - 1)10^{-3}} + \frac{0.02}{(-.98 + 0.1)10^{-3}} \right|$$

$$= 25 \times 0.02 \times \left(\frac{1}{1.1} + \frac{1}{.88} \right) = 0.97 \text{ ohm}$$

Because of inadequacies in the model under high-level injection conditions, this result is not exactly correct, though it does give the correct order of magnitude. The dynamic ON resistance of a transistor switch is usually on the order of a few ohms (*cf.* Demonstration 15·2).

OFF *impedance.* In the OFF condition, both junctions of a transistor switch are reverse-biased. The dynamic resistance of the junctions will then be extremely large, and the OFF impedance of the switch to communication signals will be determined primarily by the junction capacitances. The situation is shown in Fig. 15·7c.

EXAMPLE For a 2N1303 transistor the junction capacitances with 5 volts reverse bias applied to the junctions are about 10 pfarads each. The switch then presents a capacitance of about 5 pfarads to ground for the communication signal. This is not so small as to be insignificant, as it provides a reactance of only 30 kohms at a frequency of 1 mcps. If the transmission line has a characteristic impedance of

30 ohms, then there will be a signal voltage appearing across it in the OFF condition which is only about 1/1,000th that of the ON condition. The ratio of these signal voltages is usually quoted in db with the ON condition voltage used as a reference. For our case there is about a 60-db reduction of the signal fed to the transmission line when the switch is turned OFF. There are applications in which this is not sufficient reduction, and of course the problem is magnified if the signal contains higher frequencies than 1 mcps. When more signal reduction is needed, we must usually seek a transistor with a lower junction capacitance (when the signal bandwidths are sufficiently narrow, the capacitance can sometimes be incorporated into a tuned network to alleviate the problem).

15·2·3 *Closing and opening times in transistor switches.* Of the questions which were formulated in Sec. 15·2 to serve as a basis for studying the transistor as a nonregenerative switch, only number 7 remains to be considered. How long does it take to switch the transistor from OFF to ON, and vice versa? This question is of extreme importance in digital instrumentation and computer applications, since it will set an ultimate limit on the speed of response of the entire system. In this section we will give a physical discussion of the factors which affect the closing and opening times, together with an approximate mathematical analysis.[1] A more complete treatment is given in the references at the end of this chapter.

Turn-ON time. Let us suppose that the switch is initially in the OFF position. At $t = 0$, the control circuit injects a current through the emitter-base path, tending to increase the flow of collector current and (ultimately) bring the transistor into the saturation condition. How long does it take for the transistor to reach the steady-state ON condition?

The answer to this question is most conveniently phrased in charge-control terms. The base current is represented physically by the flow of majority carriers into the transistor base layer. This will provoke a neutralizing injection of minority carriers from the emitter and a build-up of the minority-carrier charge stored in the base region. Equilibrium is reached when a sufficient excess charge of both minority and majority carriers has been stored in the base to account for base current flow by recombination only.

A pictorial representation of the process is given in Fig. 15·8. The diagrams shown there are to be visualized as a sequence of snapshots of the excess-hole-density profile in the base region of the *pnp* switching transistor during the turn-ON transient. This figure can be divided into two major parts. Figures 15·8a–15·8d illustrate the build-up of the hole charge stored in the base from $t = 0$ to $t = t_1$, at which time the collector-base junction reaches zero bias. The hole-density profile for this period of time appears to be a straight line pivoted at a hole density of essentially zero at the collector edge of the base. The motion of the hole-density line is like the motion of a dump truck bed as it is unloading.

---

[1] Simple equipment for demonstrating the closing and opening times of transistor switches is described in Demonstration 15·3.

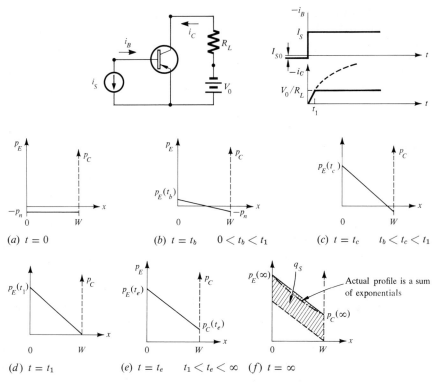

FIG. **15·8**   ILLUSTRATING THE TURN-ON PROCESS FOR A TRANSISTOR SWITCH

At $t = t_1$, the collector-base junction reaches the zero-bias condition, and the transistor enters its saturation region. In this condition the voltage drop from collector to emitter is small, and the collector current is therefore *fixed by the external circuit*. From the point of view of the load, the current is essentially equal to its final value, and the switch appears to be closed. However, the base region of the transistor has not yet reached a steady-state condition. Instead, during the time interval $t_1 < t < \infty$, the hole charge stored in the base continues to build up (Figs. 15·8e and 15·8f) until there is enough charge in the base region to account for the entire base current by recombination only. The difference between the stored charge at $t = \infty$ and that at $t = t_1$ is a measure of the depth of saturation achieved by the given base drive. We shall denote this difference by q_s; q_s is shown graphically in Fig. 15·8f (is the straight-line approximation for the hole-density profile a good one in Fig. 15·8f?)

Mathematics of the turn-ON transient.   A simple analytical treatment of the build-up of stored charge and its relation to the terminal currents during the turn-ON transient can be obtained from the stored-charge formulation of transistor behavior. We shall use this method to study some of the simpler aspects of the turn-ON transient for the intrinsic transistor.[1]

---

[1] More general treatments will be found in the references at the end of this chapter.

In the stored-charge formulation, the terminal currents in the transistor are expressed as functions of the minority-carrier charge stored in the base region. The fundamental relation is

$$-i_B = \frac{q_B(t)}{\tau_p} + \frac{dq_B(t)}{dt} \tag{15·8}$$

where i_B is the base current and q_B is the total excess charge of minority carriers stored in the base (holes, in this case).

Since the waveform of i_B is known during the entire turn-ON transient, we can solve Eq. (15·8) for $q_B(t)$ during the entire transient. At the beginning of the transient, the excess density in the base is negative (see Fig. 15·8), and there is a small reverse bias applied to the base-emitter junction and a small positive base current I_{BO} flowing in the *pnp* transistor, as in Fig. 15·8. At $t = 0$, the base current is abruptly changed to $-I_B$, and stored charge begins to build up in the base. The solution to Eq. (15·8) which fits these conditions is

$$q_B(t) = [(I_B + I_{BO})\tau_p](1 - e^{-t/\tau_p}) - I_{BO}\tau_p \tag{15·9}$$

where $I_{BO}\tau_p = Q_0$ accounts for the area under the excess-density curve in Fig. 15·8a.[1] $q_B(t)$ exhibits the simple exponential behavior shown in Fig. 15·9c during the entire transient. Equation (15·9) can be used to determine the time at which the stored charge has the value implied in *any* one of the sketches shown in Fig. 15·8.

To compute the collector and emitter currents during the transient, we must divide q_B into a sum of forward- and reverse-charge components, $q_F(t)$ and $q_R(t)$ (*cf.* Sec. 9·5):

$$q_B(t) = q_F(t) + q_R(t) \tag{15·10}$$

The collector and emitter currents are expressed in terms of $q_F(t)$ and $q_R(t)$ by

$$i_E = q_F\left(\frac{1}{\tau_t} + \frac{1}{\tau_p}\right) + \frac{dq_F}{dt} - q_R\left(\frac{1}{\tau_t}\right) \tag{15·11a}$$

$$i_C = -\frac{q_F}{\tau_t} + q_R\left(\frac{1}{\tau_t} + \frac{1}{\tau_p}\right) + \frac{dq_R}{dt} \tag{15·11b}$$

These equations are similar to the equations developed in Sec. 9·5, except that dq_F/dt and dq_R/dt terms have been added to account for changes in q_F and q_R with time.[2]

Now, during the first phase of the transient $0 < t < t_1$, q_R is a small negative number, $-qAWp_n/2$, which does not vary with time. We may neglect q_R (and Q_0) during this phase without loss of understanding. Then, from Eqs. (15·9) and (15·10),

$$q_F(t) = q_B(t) = I_B\tau_p(1 - e^{-t/\tau_p}) \tag{15·12}$$

---

[1] We are neglecting emitter junction-capacitance effects temporarily.
[2] This simple addition of the dq_F/dt and dq_R/dt will be justified in Chap. 17.

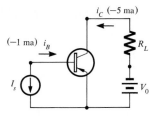

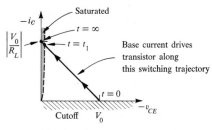

(a) Simple switching circuit

(b) Switching trajectory in $i_C - v_{CE}$ plane

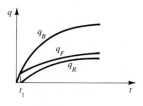

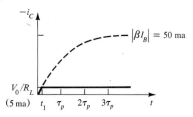

(c) Build-up of base charge with time for a step change in i_B

(d) Idealized collector current waveform for a step change in i_B with $|\beta I_B| > V_0/R_L$

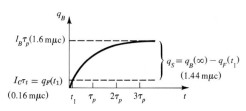

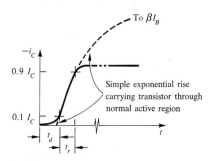

(e) q_F follows q_B up to $t = t_1$. After $t = t_1$, q_R begins to increase. q_B is a simple exponential at all times, but q_F and q_R are not. However, $q_F + q_R = q_B$ at all times.

(f) Actual waveform of i_C during the turn-ON transient.

FIG. 15·9 CURRENT AND CHARGE WAVEFORMS DURING THE TURN-ON TRANSIENT IN THE INTRINSIC TRANSISTOR

for $0 \leq t \leq t_1$. Also, from Eq. (15·11b), the instantaneous collector current during this phase is given approximately by

$$i_C = -\frac{q_F(t)}{\tau_t} \qquad (15 \cdot 13)$$

To compute the time t_1 at which the transistor enters the saturation region, we set

$$i_C(t_1) = -\frac{V_0}{R_L} = -I_C$$

Then, using (15·13)

$$q_F(t_1) = I_C\tau_t \qquad (15 \cdot 14)$$

Substituting this result into Eq. (15 · 12) and solving for t_1 then yields

$$t_1 = -\tau_p \ln\left(1 - \frac{I_C}{\beta I_B}\right) \qquad (15 \cdot 15a)$$

The idealized collector-current waveform during this phase of the transient is shown in Fig. 15 · 9d. The collector current simply increases from zero toward βI_B along the exponential curve shown in Fig. 15 · 9d. In a switching application, however, βI_B is a larger current than the collector circuit can supply, so the transistor enters its saturation region after t_1 seconds.

If $I_C/\beta I_B \ll 1$, only a small portion of the exponential shown in Fig. 15 · 9d is used in traversing the active region of the device characteristic. For this latter case we can expand Eq. (15 · 15a) to give

$$t_1 = \tau_p \frac{I_C}{\beta I_B} = \frac{q_F(t_1)}{I_B} \qquad (15 \cdot 15b)$$

Equation (15 · 15b) is interpreted by saying that when $I_C \ll \beta I_B$, the time required for the transistor to enter the saturation region is simply the time it takes for the current I_B to deliver the required charge to the base. For a given I_C, t_1 is minimized by using the largest I_B possible.

Actual waveform of i_C during turn-ON. In a real transistor there will of course be departures from the idealized waveform just predicted. For one thing, the base current must charge the emitter-base junction capacitance before significant minority-carrier injection can occur. The collector-base junction capacitance must also be charged, so the current actually delivered to the circuit will be less than the previous calculation suggests. In addition, the β of a real transistor will be low at low collector currents.

On account of these factors, some time will elapse between the time the base current is applied and the time that the transistor can be considered to have entered its normal-active region. The actual collector current will have a waveform like that shown in Fig. 15 · 9f. The time required for the collector current to increase from zero to 10 percent of its final value is called the *delay time,* and is denoted by t_d. The time required for the collector current to increase from 10 percent of its final value to 90 percent of its final value is called the *rise time* and is denoted by t_r. Equations (15 · 15) give expressions which can be used to estimate the rise time of the collector current. The delay time can be estimated from junction capacitances and a knowledge of the low current β, as in the following example.

As a practical matter, manufacturers usually quote both t_d and t_r on transistors which are intended for switching applications.

EXAMPLE 1   To illustrate the previous theoretical results, let us consider the circuit shown in Fig. 15 · 10. Here, the emitter-base junction is initially reverse-biased by a 5-volt source. To turn the switch ON, the base supply voltage is switched to -20 volts, which with the 20-kohm base resistor provides approximately -1 ma of base current for the *pnp* transistor.

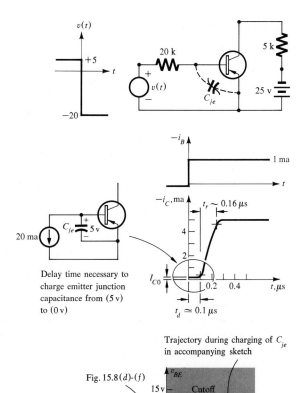

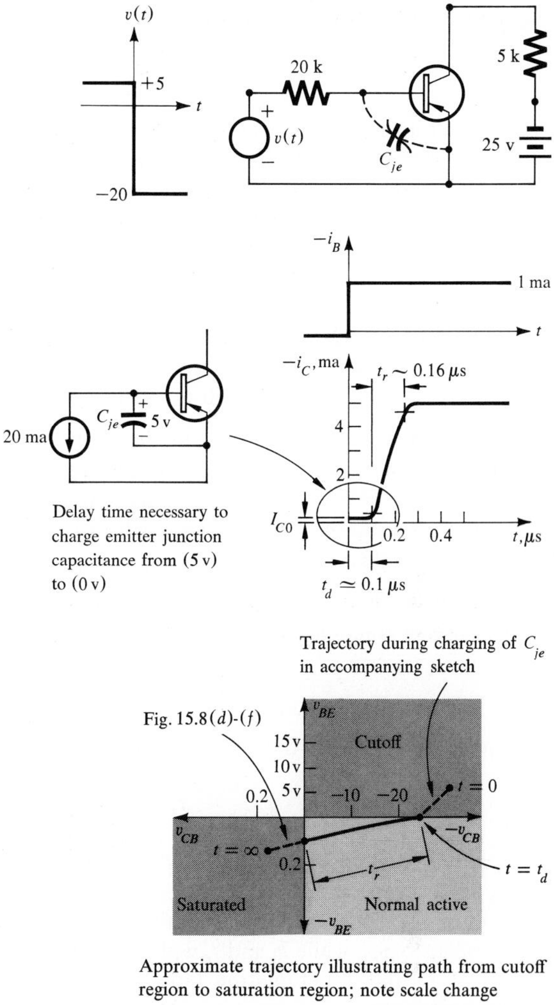

Fig. 15.8(d)-(f)

Approximate trajectory illustrating path from cutoff region to saturation region; note scale change

FIG. 15·10 THEORETICAL EXAMPLE OF A TURN-ON TRANSIENT IN A NONREGENERATIVE SWITCH

The 2N1303 transistor has an emitter-base junction capacitance of approximately 10 pfarad at reverse bias of 5 volts. If this capacitance were constant, it would take the 1-ma base-charging current a time

$$t = \frac{C \, \Delta V}{I_B} = \frac{10 \times 10^{-12} \times (5 \text{ volts})}{10^{-3}} = 50 \text{ nsec}$$

to bring the emitter-base junction to a zero-bias condition, (that is, bring the transistor from its cutoff state to the edge of its normal-active state). Since the junction capacitance increases as the junction approaches zero bias (that is, C_{je} is a nonlinear function of v_{BE}), more time than this will be necessary to bring the transistor into its normal-active region. The precise time can be calculated if the emitter doping profile and geometry are known, though it is customary to apply a factor of 1.5 to 2 to this estimate and arrive at a delay time of

$$t_d \sim 2 \frac{C \, \Delta V}{I_B} = 0.1 \; \mu\text{sec}$$

Thus, the transistor enters its normal-active region about 0.1 μsec after the base voltage is switched (see the i_C waveform in Fig. 15·10).

To estimate the rise time, we calculate t_1. To find t_1, we need to know the carrier lifetime τ_p. This can be found from the 2N1303 data sheet as

$$C_\pi r_\pi \simeq \tau_p = \frac{\beta}{2\pi f_T} \cong \frac{50}{2\pi \times 5 \times 10^6} = 1.6 \; \mu\text{sec}$$

Using this value, we find

$$t_1 = \frac{1.6 \times 10^{-6} \times -5 \; \text{ma}}{50 \times -1 \; \text{ma}} = 0.16 \; \mu\text{sec}$$

For comparison, the 2N1303 data sheet gives a *maximum* rise time of 0.4 μsec for this transistor, when $I_C = 10$ ma and $I_B = 2$ ma. The maximum rise time allows for $\beta_{min} = 20$ and $f_{T,min} = 3$ mcps. If we use these values in the previous calculation, we obtain $t_1 = 0.53$ μsec, which is in satisfactory agreement with the published rise time.

EXAMPLE 2   To find t_1 in the previous example, we obtained f_T from the 2N1303 data sheet and computed τ_p from it. Since we used

$$\tau_p = \frac{\beta}{2\pi f_T}$$

we could as well rewrite the idealized turn-ON time t_1 in terms of f_T:

$$t_1 = \frac{I_C}{I_B} \frac{1}{2\pi f_T} \tag{15·15c}$$

Usually a transistor is designed with some particular set of applications in mind, so the data sheets will describe some transistors as "intended for amplifier service" and others as "intended for switching applications." f_T (or f_α) will be quoted on amplifier types, while t_d and t_r will be quoted on switching types (with specified values of I_C and I_B). Equation (15·15c) is useful for estimating t_r when f_T is known, or vice versa.

As an example of this latter calculation, the data sheet for a 2N3639 silicon switching transistor gives $t_r = 10$ nanoseconds $= 10^{-8}$ sec for $I_C = 10$ ma, $I_B = 0.5$ ma. According to Eq. (15·15c), this transistor should have an f_T of about 320 mcps. (The manufacturer happens to quote $f_{T,min}$ of 300 mcps for this transistor, though both f_T and t_r are frequently not quoted.) As a matter of interest, t_d for these same values of I_C and I_B is 15 nsec.

> *The final approach to equilibrium in a turn-ON transient.*   As indicated before, the transistor base region is not in equilibrium at the time t_1, even though i_C has reached its final value. Instead, the stored charge continues to build up until a steady-state charge of $q_s + q_F(t_1)$ exists. During this time ($t_1 < t < \infty$), both q_F and q_R increase with time.
>
> The charge q_s which must be stored during the interval $t_1 < t < \infty$ can be obtained quite simply from our previous results. Referring to Fig. 15·9c or Eq. (15·9), we see that

$$q_B(\infty) = I_B \tau_p \qquad (15 \cdot 16a)$$

Also, from Eq. (15 · 14), we have

$$q_F(t_1) = q_B(t_1) = I_C \tau_t \qquad (15 \cdot 16b)$$

From these equations we determine q_s as

$$q_s = q_B(\infty) - q_B(t_1) = I_B \tau_p - I_C \tau_t = \left(I_B - \frac{I_C}{\beta}\right)\tau_p \qquad (15 \cdot 17)$$

q_s is shown on Fig. 15 · 9c. Figure 15 · 9e gives $q_F(t)$ and $q_R(t)$, each of which has an equation of the form

$$A + Be^{-t/\tau_p} + Ce^{-(2\beta+1)t/\tau_p}$$

for the time interval $t_1 < t < \infty$. This form of solution can be established directly from Eqs. (15 · 11) with the use of the identity

$$q_B(t) = q_F(t) + q_R(t)$$

While the actual time dependence of q_R and q_F contains two exponentials, the $e^{-(2\beta+1)t/\tau_p}$ term is usually not important because of its rapid decay with time. Except for a short period of time near $t = t_1$, both q_F and q_R can be well approximated by simple exponentials having the time constant τ_p.

From the simple exponential behavior of $q_B(t)$ we can estimate the time required for the transistor to come into equilibrium with the base-current drive. In a total elapsed time of $3\tau_p$ sec after application of the base drive, the base-stored charge will have reached 95 percent of its final value, and the switching transient will have been essentially completed. This is about 5 μsec for our 2N1303 running example.

The turn-OFF transient. Of course, the relatively long time interval required for the base current to establish the charge q_s in the base region is not apparent in the current waveforms of the circuit during the turn-ON transient. The existence of the saturating charge q_s has a great influence on the circuit currents in the turn-OFF transient, however.

In order to turn the switch OFF, we must remove all of the stored charge in the base region. Again, there will be two phases in the turn-OFF transient. The transistor must first be brought out of its saturation region, which involves reducing q_s to zero. The time required for this is called the *storage-delay time* t_s. During the time interval $0 < t < t_s$, the collector current remains constant at the value $|V_0/R_L|$.

When q_s reaches zero, the transistor enters its normal-active region (Fig. 15 · 11b) and then proceeds to the cutoff state, the collector current decreasing to zero. The time required for the collector current to decrease from 90 percent of its ON value to 10 percent of its ON value is called the *fall time* t_f.

A turn-OFF transient can be initiated by simply setting i_B to zero. The appropriate charge and current waveforms are shown in Fig. 15 · 11. In this case we simply wait for recombination to reduce q_s to zero. The

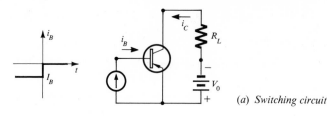

(a) Switching circuit

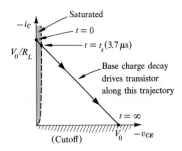

Saturated
$t = 0$
$t = t_s (3.7\,\mu s)$

Base charge decay
drives transistor
along this trajectory

$t = \infty$

(Cutoff)

(b) Switching trajectory in $i_C - v_{CE}$ plane

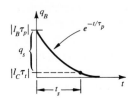

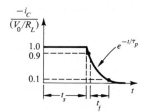

(c) Waveform of total base-
stored charge

(d) Collector-current waveform
during turn-OFF transient

FIG. 15·11   CHARGE AND CURRENT WAVEFORMS IN A TURN-OFF TRANSIENT

pertinent mathematical relationships are

$$\frac{q_B}{\tau_p} + \frac{dq_B}{dt} = 0 \tag{15·18}$$

$$\Rightarrow q_B(t) = q_F(t) + q_R(t) = |I_B\tau_p| e^{-t/\tau_p} \tag{15·19}$$

where $|I_B\tau_p|$ is the total excess-charge stored in the base region when the turn-OFF transient begins.

The collector current is held constant at the value $|V_0/R_L|$ until $q_B(t)$ reaches the value

$$q_B(t_s) = |I_C|\tau_t = \frac{V_0}{R_L}\tau_t \tag{15·20}$$

After the storage delay time, the transistor is once again in a condition where $q_R \cong 0$ and $q_B(t) = q_F(t)$. During this phase of the transient the collector current can be expressed once again as

$$i_C = -\frac{q_F(t)}{\tau_t}$$

626   *Semiconductor electronics*

The collector current decays exponentially toward zero during the time interval $t_s \leq t \leq \infty$. This behavior is sketched in Fig. 15·11d.

The storage-delay time for the case where the base drive is reduced to zero to initiate the turn-OFF process is found by combining Eqs. (15·19) and (15·20) and then solving for t_s. The result is

$$t_s = \tau_p \ln \left| \frac{\beta I_B}{I_C} \right| \qquad (15·21)$$

For our 2N1303 running example ($\tau_p = 1.6$ μsec, $\beta = 50$, $I_B = 1$ ma, $I_C = 5$ ma), the storage-delay time is about 3.7 μsec.

It is clear from Eq. (15·21) that increasing I_B to obtain a short turn-ON time has an adverse affect on the storage-delay time, and vice versa. To eliminate this problem we can take a more positive approach to the turn-OFF problem. We initiate the turn-OFF by using a reverse base current I_{B2}. This extracts majority carriers from the base region rather than waiting for recombination to reduce the excess carrier density. If a large reverse base drive is applied, the storage-delay time can be simply estimated by

$$t_s \simeq \frac{q_s}{|I_{B_2}|} \qquad (15·22)$$

which is just the time required to remove q_s, if recombination is neglected. We can express q_s as

$$q_s = |I_{B_1}\tau_p| - \frac{V_0}{R_L}\tau_t \qquad (15·23)$$

so that Eq. (15·22) becomes

$$t_s = \tau_p\left(\left|\frac{I_{B_1}}{I_{B_2}}\right| - \frac{|I_C|}{|\beta I_{B_2}|}\right)$$

If we use $|I_{B_2}| = |I_{B_1}|$, t_s will be very nearly τ_p. If $|I_{B_2}| = 10\,|I_{B_1}|$, t_s is essentially $0.1\,\tau_p$.

The use of a reverse base drive will also reduce the fall time of i_C after the storage-delay phase is completed. In this instance the collector current appears to be headed toward βI_{B_2}, as shown in Fig. 15·12, and a definite cutoff of i_C is obtained (rather than an exponential approach to cutoff).

EXAMPLE We will again use data on a 2N3639, published by the manufacturer, to check Eq. (15·23). The quoted storage time for this transistor is 17 nsec when a base current of 0.5 ma is reversed to turn off a collector current of 10 ma. According to our previous calculations, $f_T \sim 300$ mcps, and the data sheet gives $\beta = 60$. Hence,

$$\tau_p = \frac{\beta}{2\pi f_T} = \frac{60}{2\pi \times 300} \times 10^{-6}$$
$$\sim 30 \text{ nsec}$$

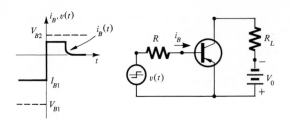

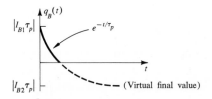

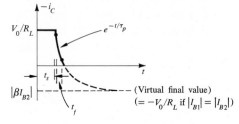

FIG. 15·12   CHARGE AND CURRENT WAVEFORMS DURING A TURN-OFF TRANSIENT IN WHICH THE DIRECTION OF i_B IS REVERSED TO DECREASE THE STORAGE TIME

The formula for t_s then gives

$$t_s = \tau_p\left(1 - \frac{20}{60}\right) \sim 30 \text{ nsec} \times \frac{2}{3} = 20 \text{ nsec}$$

which is adequately close to the published t_s.

The *fall time* for the collector current, denoted by t_f on Fig. 15·12 can be seen to be approximately the time required for a curve of the form e^{-t/τ_p} to reach one-half its final value (because $|I_{B_1}| = |I_{B_2}|$). Thus

$$t_f \sim 0.7 \ \tau_p = 21 \text{ nsec}$$

The data sheet gives $\tau_f = 20$ nsec for this transistor.

15·2·4   *Experimental switching waveforms in a nonregenerative switching circuit.* To help clarify some of the previous theory and to give some experimental indication of its range of validity, the circuit shown in Fig. 15·13 was built. The circuit is a simple, nonregenerative switching circuit using a *pnp* transistor (2N655).[1] It is driven from a voltage source which produces a pulse going negative from an absolute value of 0 volts.

---

[1] This transistor was selected to simplify the demonstration (give long storage times, etc.); it is not typical of high-speed switching transistors in exact details, though the principles to be established apply equally well in both cases.

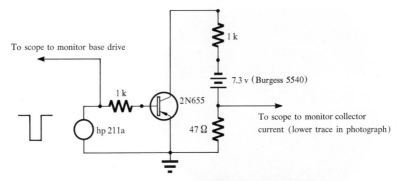

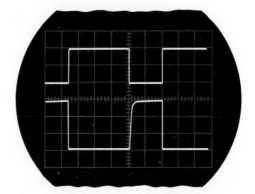

(a) Circuit parameters for demonstrating switching circuit waveforms

(b) Upper trace: generator output pulse: 0.1 v/vertical division. The pulse goes negative from a base line of 0 v.

Lower trace: collector current in 47Ω resistor: 0.05 v/vertical division. The pulse goes positive from a base line of 0 v.

Time scale: 50 μs/cm

FIG. 15·13   EXPERIMENTAL SWITCHING TRANSIENTS IN A NONREGENERATIVE SWITCHING CIRCUIT

A 47-ohm resistor is inserted in the collector lead to monitor the collector current. The collector battery provides a voltage of 7.3 volts when a 1-kohm resistor is attached across it.

The series of photographs given in Fig. 15·13 show the effects of various levels of base drive. In each figure the upper trace is the output voltage of the voltage generator, and the lower trace is the corresponding collector current. The relevant voltage and time scales are given with each photograph. The numbers given in the following discussion are obtained from the data given in the figure captions.

In Fig. 15·13b, the output pulse of the generator has an amplitude of 0.2 volt and a duration of 100 μsec. The collector current during the pulse in this photograph is about 3 ma, which is not enough to drive the transistor out of its normal-active range. Note that the time scale on this figure is 50 μsec/cm, and that, except for the leading edge, the transistor amplifies the input pulse rather well.[1]

Figure 15·13c shows what happens when the pulse width is changed to 10 μsec, everything else remaining the same. The transistor is still in its normal-active region, though it cannot respond rapidly enough to

---

[1] This collector-current photograph simply gives the step response of the amplifier, a topic which we shall take up further in Chap. 18.

Switching circuits and switching devices   629

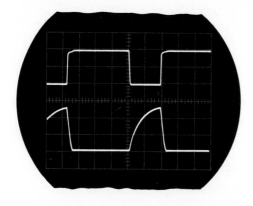

FIG. 15·13 (CONT'D.)

(c) Upper trace: generator output pulse: 0.1 v/cm vertical, 5 μs/cm horizontal. The pulse goes negative from a base line of 0 v.

Lower trace: collector current in 47Ω resistor: 0.05 v/cm vertical, 5 μs/cm horizontal. The pulse goes positive from a base line of 0 v.

follow the input pulse. Instead, the collector current is an exponential curve, heading toward a *final* current of 3 ma. This can be verified from the collector-current waveform shown in part (*b*) by estimating the height of the collector-current pulse 10 μsec after the base pulse is applied [that is, after the first, small horizontal division in part (*b*)]. The collector current in part (*c*) is simply an expanded view of part (*b*) for the first 10 μsec, except that in part (*c*) the drive is removed after 10 μsec, so the collector current never attains its final value.

In Fig. 15·13*d*, the pulse generator output is 0.3 volt, and the collector current reaches a height of 6.4 ma, at which point the current saturates. However, the level of saturation is not very great, as is evidenced by these facts:

1. The leading edge of the collector-current waveform still has the characteristic $1 - e^{-\alpha t}$ rise for an appreciable part of the pulse, and in fact the collector current would only extrapolate to a value of about 7 to 8 ma if the base pulse were indefinitely long.

2. The measured v_{CE} is 0.18 volt, which indicates that the collector-base junction is only forward-biased to a voltage of about 0.12 volt. This is not considered to be "heavily saturated."

3. There is very little storage time evident at the end of the base pulse, suggesting that q_s is not very large for this level of base drive.

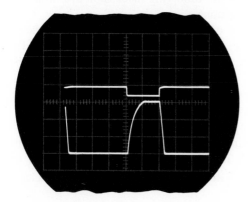

FIG. 15·13 (CONT'D.)

(d) Upper trace: generator output pulse: 0.5 v/cm vertical, 5 μs/cm horizontal. The pulse goes negative from a base line of 0 v.

Lower trace: collector current in 47Ω resistor: 0.1 v/cm vertical, 5 μs/cm horizontal. The pulse goes positive from a base line of 0 v.

FIG. 15·13   (CONT'D.)

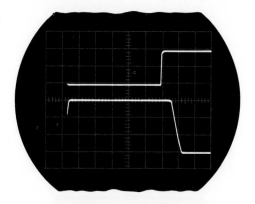

(e) Upper trace: generator output pulse: 0.5 v/cm vertical 2 μs/cm horizontal. The pulse goes negative from a base line of 0 v.

Lower trace: collector current in 47Ω resistor: 0.1 v/cm vertical, 2 μs/cm horizontal. The pulse goes positive from a base line of 0 v.

In Fig. 15·13e, the pulse voltage is increased to 1 volt. The base-emitter voltage drop during the pulse is about 0.3 volt, so the base current is approximately 0.7 ma. The transistor has a measured β of 100 at $I_E = 6$ ma, so the transistor is now heavily saturated by this base drive. Note that:

1. The turn-ON time for the collector current has now decreased to about 0.3 μsec.

2. The storage time is very apparent, being nearly 1 μsec. The fall time has also increased.

The minimum value of the collector-emitter voltage (reached at the end of the base pulse) is 60 mv, so the collector-base and emitter-base junction voltages are more nearly cancelling each other.

The collector-current waveform in Fig. 15·13e is the waveform which is usually observed when transistors are operated as saturated switches in the simple circuit of Fig. 15·13a, though by proper design of the transistor the rise, storage, and fall times can be decreased to 10 to 100 nsec.

Speed-up capacitors in switching circuits. The foregoing theory and experimental results show that when a transistor is turned ON with a constant-current base drive, and turned OFF by reducing the base drive abruptly to zero, turn-ON time and storage-delay time have to be "traded." A large I_B will produce a very short turn-ON time; but it will also require a large q_s and therefore a long storage-delay time, and vice versa.

In practice, this may be avoided by using very high-speed transistors, some of which are purposely treated to give short minority-carrier lifetimes and therefore low storage times. However, for a given transistor, "speed-up" capacitors can also be inserted in the circuit to minimize turn-ON time and storage-delay problems. The effects of these speed-up capacitors are quite pronounced, as the following experiment shows.

The circuit to be studied is shown in Fig. 15·14. It is the same circuit as that in Fig. 15·13, except that a capacitor has been added in parallel with R_s. The function of this capacitor is most simply studied by assuming that the forward voltage drop across the emitter-base junction is zero

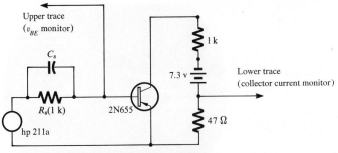

(a) *Nonregenerative switching circuit with speed-up capacitor*

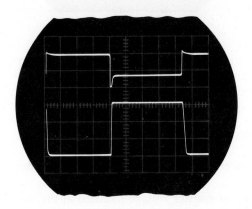

(b) *Waveforms using a speed-up capacitor of 0.004 μfd*

Upper trace: v_{BE} vs. t with 0.5 v/cm vertical, 2 μs/cm horizontal. The pulse goes negative from a base line of 0 v.

Lower trace: collector current in 47Ω resistor: 0.1 v/cm vertical, 2 μs/cm horizontal. The pulse goes positive from a base line of 0 v.

FIG. 15·14   **THE USE OF SPEED-UP CAPACITORS IN SWITCHING CIRCUITS**

in the conducting state, (that is, assuming the emitter-base diode is an ideal diode). Then, when the base generator produces an output pulse of amplitude V_B, two things happen:

1. There is (ideally) an impulse of charge supplied to the base to charge the capacitor to the pulse-voltage amplitude. The impulse of charge has the value $C_s V_B$ coul.

2. A steady base current of V_B/R_s is supplied to the base for the duration of the pulse.

If we set the value of C_s so that the impulse of charge supplied to the base equals the total charge demanded by the base for a steady base current of V_B/R_s, then the turn-ON time should be *zero*.[1] The value of C_s is simply found from the foregoing argument. We want

$$C_s V_B = I_B \tau_p = \frac{V_B}{R_s} \tau_p \Rightarrow C_s R_s = \tau_p$$

(for the 2N655, $\tau_p \simeq 4$ μsec).

[1] To the extent that the stored-charge approximation is valid; that is, to the extent that the hole-density profile in the base region is always a straight line. Actually, the impulse of charge is initially localized near the emitter, and the collector-current rise time will depend on the transit time for holes across the base region. After about one-fifth of the transit time, the hole-density profile will be sufficiently straight to justify the stored-charge approximation.

Before proceeding to the experimental results, we should inquire about the effect of this "speed-up" capacitor on the storage and fall times, because a deleterious effect here would nullify any potential decrease in turn-ON time for most applications. However, we can show that the speed-up capacitor also aids in the recovery phase of the transient. The argument goes as follows. When the base-drive voltage is reduced to zero, the voltage on C_s tends to reverse bias the emitter-base junction. But in order to reverse bias the emitter-base junction, the charge stored in the base region must be removed. Ideally, the charge will be removed in an impulse once again, and both the storage-delay and fall time will be reduced to zero. In practice, of course, this does not occur, though very significant decreases in turn-ON time, storage time, and fall time are obtained.

We show in Fig. 15·14b the response of the circuit shown in Fig. 15·14a for $R_s = 1$ kohm and $C_s = 0.004$ μfarad. The upper trace is the voltage seen at the emitter-base terminals of the transistor when the generator-output pulse has an amplitude of 1 volt. The voltage is negative-going from a base line of zero volts. The lower trace is the collector current monitored in the 47-ohm resistor. Note that the collector current turn-ON is as rapid as that of the pulse source, and the storage and fall times are very small. The time scale on the figure is 2 μsec/cm (horizontal); the circuit is, in fact, exactly the same as the one shown in Fig. 15·13a, except for C_s. The addition of C_s changes the collector-current response from the one shown in Fig. 15·13e to that shown in Fig. 15·14b. In addition, the collector-to-emitter voltage drop promptly assumes its final value of 60 mv when a C_s of 0.004 μfarad is used, rather than gradually approaching the final value.

15·2·5 *The application of transistor switches in logic circuits; and microelectronics.* The availability of fast, highly accurate digital computers for business, science, and engineering needs has already had a very important influence on our society. As computers become less expensive, smaller, lighter, etc., their applications become more widespread; and their influence on the future of our society will be even greater than it is now, perhaps little short of fantastic.

Digital computers and other digital systems are representative of a class of systems in which relatively simple building blocks are organized in a certain way to perform a required function. A digital computer is composed almost entirely of three basic building blocks, these being the *and gate, or gate,* and *not gate.*[1] Transistors are used as nonregenerative switches in these basic circuits.

Most of the effort involved in designing a digital computer is to conceive of an efficient way of interconnecting these blocks to achieve the desired results. Accordingly, to understand the design of a computer, it is for the most part unnecessary to consider what particular active element is used to build the basic gates. We could

[1] The reader interested in a treatment of how the basic blocks are interconnected to do computation is referred to the demonstrations at the end of this chapter and to R. S. Ledley, "Digital Computer and Control Engineering," McGraw-Hill Book Company, New York, 1960, chap. 2.

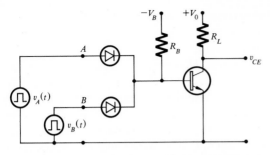

FIG. 15·15   OR CIRCUIT MADE FROM TWO DIODES AND ONE TRANSISTOR. THE TRANSISTOR IS CUT OFF BY THE BASE BIAS UNLESS A POSITIVE SIGNAL VOLTAGE IS APPLIED TO EITHER A OR B. A POSITIVE SIGNAL (~ 1 VOLT) AT A OR B WILL TURN THE TRANSISTOR ON, AND ITS COLLECTOR VOLTAGE WILL THEN DROP TO A LOW VALUE.

visualize the gates as being synthesized from relays, vacuum tubes, transistors, or other possible devices. As a practical matter, however, transistors are used (and in fact, the growth of the digital-computation field can be intimately related to the development of transistors, and conversely). Transistors provide active elements which are small and light weight, have long life and high reliability, and require only a small amount of power for operation. In addition, the processes for manufacturing transistors and other semiconductor components are such that many transistors, diodes, and resistors can be made on a single piece of silicon and connected together to obtain some desired circuit function. An entire circuit may be on a wafer of silicon which is no larger than $0.03 \times 0.05 \times 0.1$ in. Semiconductor fabrication techniques are presently at the point where a major problem in increasing the number of circuit functions performed on a wafer is that of having enough space to make the required number of contacts to the external "world."

Microelectronics is the name given to the fabrication and application of these semiconductor circuits.[1] They find applications in many systems besides digital computers, though we shall concentrate on the digital computation circuits, because this promises to be the area of their greatest importance. We will describe the processes by which a relatively simple microcircuit OR gate is made. The processes will be suggestive of the versatility of the fabrication techniques.

The OR gate which we wish to consider is shown in Fig. 15·15. In this circuit, the base is returned to a voltage $-V_B$, reverse biasing the emitter-base and collector-base junctions and the two diodes. Accordingly, the collector-to-ground voltage v_{CE} is essentially equal to the collector supply voltage V_0. When a positive voltage of about 1 volt magnitude is supplied to either diode A or B (or a current which is greater than $|V_B/R_B|$), the transistor will turn ON and v_{CE} will drop to zero. The value of v_{CE} therefore indicates whether a pulse exists at A or B at the time in question.

Let us now see how the circuit of Fig. 15·15 can be built on a wafer of silicon.

---

[1] The terms *integrated* electronics, *molecular* electronics, etc., are used to describe somewhat more general possibilities than the use of semiconductors only. As a matter of interest, the availability and future prospects of microcircuits has also caused and will continue to cause a change in the education of electronics scientists. A perusal of the literature will show the developing trend. In particular, *cf.* J. G. Linvill, et al., Integrated electronics vs. electrical engineering education, *Proc. IEEE,* December, 1964, pp. 1425–1429.

The steps are diagrammed in Fig. 15 · 16. We begin with a wafer of p-type silicon which has had an n-type region grown on top of it. The n-type region is lightly doped. We now grow an oxide layer several thousand angstroms thick on the surface of this wafer and paint the top of it with a photosensitive emulsion.

To expose the emulsion selectively, we first paint a masking pattern on a large piece of paper and then photograph it. A replica of the photographed mask, shown in Fig. 15 · 16d, is then placed over the wafer and the photosensitive emulsion is exposed to light. This hardens the emulsion beneath the clear areas in the mask. We can now rinse off the unexposed emulsion, leaving the silicon wafer in the condition shown in part e of Fig. 15 · 16. The oxide is now removed everywhere except in those areas beneath the hardened emulsion (by a chemical etch).

The wafer is now inserted into a diffusion furnace, and a p-type dopant is diffused into the semiconductor. The diffusion time is long enough to allow the p-type dopant to penetrate right through the n-type layer on the surface of the wafer. After the diffusion, only n-type islands remain, these being beneath the areas where the oxide was not removed prior to diffusion. A top view of the wafer is shown in Fig. 15 · 16h, with the p-type surface areas shaded. The large n-type island will contain the resistors, the smaller island on the left will be the collector of the transistor, and the smaller island on the right will contain the two diodes.

The sample is now reoxidized, recoated with the photosensitive emulsion, and placed beneath the mask shown in Fig. 15 · 16j. This mask is registered so that its dark areas fall properly over the islands. Now, when the wafer is exposed to light, only the emulsion in selected parts of the n-type islands will be unexposed. Again the unhardened emulsion is rinsed off and the oxide removed in the unexposed areas. The wafer is reinserted into a diffusion furnace, and a shallow, p-type diffusion is performed. This time the p-type dopant is not allowed to penetrate all the way through the n-type layer. When this diffusion is completed, we have made four p-type resistors on the large island, a base region on the small island furthest left, and the p regions for the two diodes on the small island furthest to the right. The state of the wafer is illustrated in Fig. 15 · 16j.

After another series of oxidation, emulsion coating, etc., steps, we expose the wafer under the mask shown in part (k) of Fig. 15 · 16, and proceed to do an n-type diffusion. This time we form the emitter of the transistor, and an n contact for the diodes (common to both diodes.) The state of the wafer is illustrated in Fig. 15 · 14l.

We then reoxidize the wafer and expose it under a contact mask. When the oxide is removed after this exposure, we are left with small holes in the oxide above the resistor terminals, the emitter, base, and collector terminals, and the diode terminals. The state of the wafer after this operation is shown in Fig. 15 · 16m.

Aluminum is now evaporated over the entire slice, the aluminum making an ohmic contact to all the terminals previously bared.[1] The slice is once again coated with emulsion and exposed beneath the mask shown in Fig. 15 · 16n. Now the aluminum can be removed everywhere except beneath the hardened emulsion. Thus, we finally arrive at the structure shown in Fig. 15 · 16o, in which there are six aluminum contact pads and aluminum wiring connections to a set of resistors, diodes, and a transistor. The surface area of the entire circuit is 0.03×0.05 in.

---

[1] Aluminum is a p-type doping agent, but where it contacts n-type material, the n-type doping is so heavy that a diode with very high saturation current results. This is equivalent to an ohmic contact.

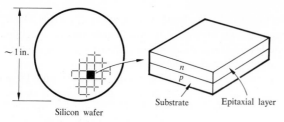

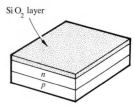

(a) *p-type silicon wafer with n-type epitaxial layer*

(b) *Oxide layer grown on top of* (a)

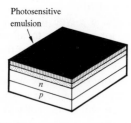

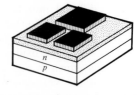

(c) *Photosensitive emulsion applied*

(d) *Mask for exposing* (c) *to light*

(e) *Wafer after exposure to light and subsequent rinse for removing unexposed emulsion*

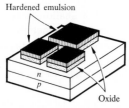

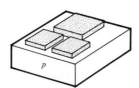

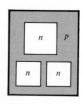

(f) *Oxide-removal step leaves oxide islands with hardened emulsion on top*

(g) *Appearance of wafer after removal of hardened emulsion and p-type diffusion (n-type material remains only under oxide islands)*

(h) *Top view of* (g) *with oxide removed (p-type surface shaded for contrast)*

p-type areas within n-type islands do *not* penetrate through n-type material

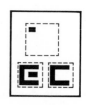

(i) *Mask for base and resistor diffusion*

(j) *Wafer after a shallow p-type diffusion using mask in* (i); *all darkened areas are p-type*

(k) *Emitter-diffusion mask*

FIG. 15·16   THE STEPS IN CONSTRUCTING A MICROCIRCUIT OR GATE

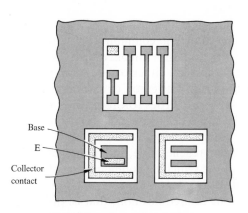

Base

E

Collector contact

(l) Wafer after emitter diffusion; shaded areas are p-type; dotted areas are heavily doped n-type; clear areas are lightly doped n-type

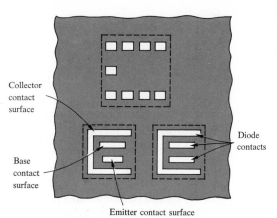

Collector contact surface

Base contact surface

Emitter contact surface

Diode contacts

(m) After the contact mask has been applied and subsequent rinsing and etching completed, the entire surface of the wafer is coated with oxide, except beneath the unshaded areas shown above

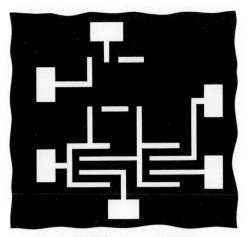

(n) Mask for use in removing aluminum from undesired areas

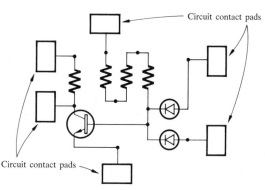

Circuit contact pads

Circuit contact pads

(o) The final circuit, drawn with normal circuit symbols

(p) Circuit mounted on header

Switching circuits and switching devices   637

Regenerative switching circuits have much in common with nonregenerative circuits, at least in so far as the transistor is concerned. Except for the closing and opening times, transistors are either OFF or ON in regenerative circuits (in the simplest cases, this means cutoff or saturated) and in these two states they display the same properties (ON resistance, ON voltage drop, leakage current when OFF, etc.) that we have just studied for nonregenerative circuits.

The significant difference between regenerative and nonregenerative switching circuits lies in the form of the switching commands to which the circuit responds. In a nonregenerative circuit, command signals are applied continuously to keep the switch in the desired position; that is, the switch is simply an overdriven amplifier. In regenerative circuits the switching commands are given to the circuit in the form of *triggering pulses*. Upon receiving a triggering pulse, the circuit itself will establish the base drives which are necessary to hold the transistor or transistors in the desired condition. Ideally, the triggering pulses have no influence on the length of time that a transistor is maintained in the ON or OFF state.

Circuits which operate in this manner have a number of very important functions. They are useful in performing logic in a digital computer and in counting and information storage. They play an important role in the deflection circuitry which causes the electron beam to scan across the picture tube in a television or radar receiver. They are also useful in instrumentation and test equipment for generating special waveforms (such as "square waves" with variable symmetry, triangular waves, parabolic waves, and so on).

Regenerative switching circuits have much in common with the sinusoidal oscillators which were described in Chap. 14. We require for a regenerative switching circuit (1) a power supply, (2) an amplifier or a dynamic negative-resistance characteristic, (3) energy-storage elements, and (4) nonlinearity. As was true for sinusoidal oscillators, amplifiers with positive feedback provide the required negative resistance.

The role of nonlinearity in a regenerative switching circuit is as follows.[1] To switch a regenerative circuit from one of its stable states (or perhaps temporarily stable states) to another, we apply a trigger pulse or otherwise satisfy a circuit condition which brings the transistors in the circuit into their normal-active range. In this condition, a regenerative circuit can be thought of as a linear amplifier with enough positive feedback to cause the natural frequencies of the system to be in the right half of the complex-frequency plane. The unstable transients which result will drive the transistors into a cutoff or saturation condition, at which point the amplifier gain becomes zero and the transient ceases; that is, the circuit parameters

---

[1] See the discussion of sinusoidal oscillator amplitude stability for the role of nonlinearity in these circuits.

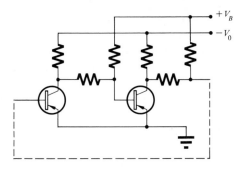

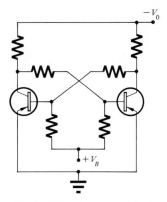

(a) *Two-stage dc-coupled amplifier using two batteries for bias adjustment*

(b) *Part (a) redrawn to emphasize symmetry*

FIG. 15·17 SOME EQUIVALENT FORMS OF THE BASIC FLIP-FLOP

change abruptly and the natural frequencies move into the left half of the complex frequency plane.[2]

15·3·1 *Classification of regenerative switching circuits.* It is convenient to divide regenerative switching circuits into three principal categories, according to whether the circuit can remain indefinitely in one or both of its stable states when free of external triggering pulses. If the circuit can remain in *either* state permanently, it is called *bistable;* if it can remain in only *one* state permanently, it is called a *monostable* circuit; and if it cannot remain permanently in either state, it is called an *astable* circuit.

Bistable circuits are frequently referred to as *flip-flops,* while astable circuits are called *multivibrators.* However, this terminology is not universally adopted, since some people refer to flip-flops as bistable multivibrators, while others refer to multivibrators as astable flip-flops.

15·3·2 *The basic flip-flop.* In some respects the flip-flop is like a nonregenerative switch, in that it has two stable states. However, the flip-flop circuit does not require a continuous, external control signal to hold it in either of its stable states. Rather, the circuit configuration itself ensures this condition. We shall see how this property can be put to use later. For the moment, let us consider the basic flip-flop circuit shown in Fig. 15·17, to understand how the two stable states arise.

The flip-flop has been drawn in two ways to illustrate different aspects of the circuit. In Fig. 15·17a we view the flip-flop as a two-stage, dc-coupled amplifier with its output fed back to its input. Without this feedback (and with appropriate resistance values) the circuit would operate as an amplifier. In Fig. 15·17b, the flip-flop is redrawn to emphasize the

---

[2] In some circuits the abrupt change in circuit constants is provided by diodes which are included for this purpose, so that the transistors need not be driven from cutoff to saturation.

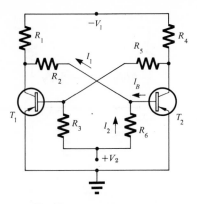

FIG. 15·18   A POSSIBLE SET OF STEADY-STATE CONDITIONS IN A FLIP-FLOP

symmetry of the circuit. As is always true, symmetry in a physical problem can be used to advantage in understanding the behavior of the system and in simplifying the mathematics needed to quantitatively describe it. We shall attempt to exploit this symmetry in what follows.

The stable states of a flip-flop. Let us first verify that the circuit shown in Fig. 15·18 has two stable states. To do this, we will use a self-consistent argument; that is, we shall assume that a stable state exists with one transistor ON (saturated) and the other OFF (cutoff), and then show that circuit conditions justify this assumption.

Assume that T_1 is OFF and T_2 ON. Then all leads which make direct contact to T_2 are effectively connected to ground. Therefore, the current flowing through R_1 and R_2 is

$$I_1 = \frac{V_1}{R_1 + R_2} \tag{15·24}$$

and the current flowing through R_6 is

$$I_2 = \frac{V_2}{R_6} \tag{15·25}$$

the positive directions of I_1 and I_2 being indicated on the figure.

The base current for T_2 is then

$$I_B = I_1 - I_2 = \frac{V_1}{R_1 + R_2} - \frac{V_2}{R_6} \tag{15·26}$$

Now, if we select R_4 so that

$$\frac{V_1}{R_4} < \beta I_B$$

the base current will be sufficient to maintain T_2 in the saturated state, as assumed.[1] Usually the circuit is symmetrical ($R_1 = R_4$, $R_2 = R_5$, $R_3 = R_6$),

---

[1] We have neglected the small current component due to flow from V_2 through R_3 and R_5, though this only improves our argument.

so the condition for saturation is

$$\frac{V_1}{R_1 + R_2} - \frac{V_2}{R_3} > \frac{V_1}{\beta R_1} \qquad (15 \cdot 27)$$

This condition applies to either transistor, of course.

To show that T_1 is OFF when T_2 is ON, we observe that the voltage source V_2, resistively divided through R_3 and R_5, provides a positive bias voltage at the base of T_1, thus ensuring that it is cutoff. The base-emitter voltage is simply

$$V_{BE} = \frac{V_2 R_5}{R_5 + R_3} \qquad (15 \cdot 28)$$

A moderately large V_{BE} is useful to reduce the probability of noise triggering in the circuit and to maintain the tolerance of the operating conditions to changes in I_{CO}.

It will be clear from symmetry that a stable condition also exists when T_1 is ON and T_2 OFF. It can also be argued from symmetry that a possible state exists in which T_1 and T_2 conduct equally. Such a state does exist, but it is unstable. Any slight disturbance driving one transistor into slightly higher conduction will, by virtue of the positive feedback, drive the circuit toward one of its truly stable circuits.

In the seesaw analogy mentioned in the first part of this chapter, placing the freely rolling ball in the exact center of the seesaw will produce an unstable equilibrium which is equivalent to that just described.

EXAMPLE   To illustrate the foregoing ideas, we shall calculate the voltages and currents at significant points in the circuit of Fig. $15 \cdot 19$. The transistor is a 2N1303, with a typical β of 50.

The base current for the ON transistor will be computed first. Using Eq. $(15 \cdot 26)$ we find

$$I_B = \frac{12 \text{ volts}}{3 \text{ kohms} + 6.8 \text{ kohms}} - \frac{6 \text{ volts}}{15 \text{ kohms}} = 1.22 \text{ ma} - 0.4 \text{ ma} \sim 0.8 \text{ ma}$$

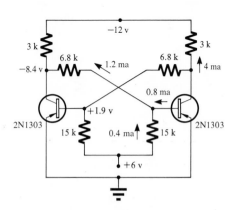

FIG. $15 \cdot 19$ THE STEADY-STATE CONDITIONS IN A TYPICAL FLIP-FLOP

The collector current is approximately

$$|I_C| = \frac{12 \text{ volts}}{3 \text{ kohms}} = 4 \text{ ma}$$

Since $\beta I_B = 40$ ma, and the collector current is only 4 ma, the ON transistor is certainly in saturation. The guaranteed minimum β for the transistor is 20, so any 2N1303 transistor will function satisfactorily in the circuit.

The OFF transistor has a reverse bias of

$$V_{BE} = \frac{6 \times 6.8 \text{ kohms}}{(6.8 + 15) \text{ kohms}} = 1.87 \text{ volts} \simeq 1.9 \text{ volts}$$

applied to the base-emitter junction. The collector-to-ground voltage is

$$V_{CE} = \frac{V_1 R_2}{R_1 + R_2} = \frac{-12 \times 6.8 \text{ kohms}}{9.8 \text{ kohms}} = -8.4 \text{ volts}$$

Hence, the collector-base junction has a reverse bias of $8.4 + 1.9 = 10.3$ volts applied to it, also.

Nonsaturated flip-flops. It is possible to design flip-flops which do not allow enough current to flow in the flip-flop to maintain one of the transistors cutoff and the other saturated. In one type of nonsaturating circuit (shown in Fig. 15 · 20), a resistor is added in the common-emitter lead to limit the circuit current. The capacitor C_E is added to smooth out voltage fluctuations during switching transients. The circuit now operates from a single supply voltage, which is sometimes an advantage. A stable state now exists in which one transistor simply conducts more current than the other one.

One of the principal reasons for using such a circuit is that switching speeds can be kept high if the transistors are always operated in their normal-active region. This avoids the storage-delay problem, which is one

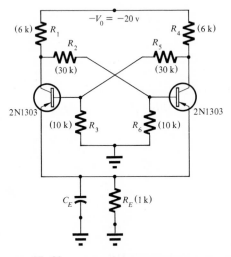

FIG. 15 · 20   A FLIP-FLOP IN WHICH THE ON TRANSISTOR IS NOT SATURATED

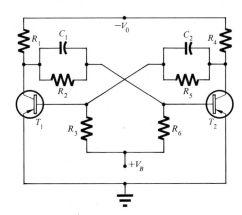

FIG. 15·21   A FLIP-FLOP WITH SPEED-UP ON COMMUTATING CAPACITORS

of the factors that limits the operating speed in a saturated flip-flop. Nonsaturated flip-flops are usually more sensitive to noise triggering, however.[1]

It is possible to choose sufficiently low values of R_E to obtain the T_1-saturated–T_2-cutoff condition for the stable states. For such cases the emitter resistor is added primarily for the purpose of eliminating the base supply voltage needed in the basic circuit.

Regeneration analysis of the flip-flop.   The speed with which a flip-flop switches from one state to another is an important consideration, both from the standpoint of obtaining the best possible circuit performance and for the purpose of assuring that the circuit will really switch. In this section we shall describe the *regenerative mechanism* which can, when properly set in motion, produce the desired change of state. Later we will consider the *triggering requirements;* that is, how to set the regenerative action in motion.

To facilitate rapid switching action in a flip-flop, two capacitors are added to the circuit, as shown in Fig. 15·21. These capacitors, called "speed-up" or "commutating" capacitors, serve to couple changes in v_{CE} from one transistor to the base of the other without the resistive dividing action which would otherwise exist.

To give a qualitative demonstration of the regenerative action in this circuit, we assume that both transistors are momentarily in their normal-active regions. Now, suppose there is an increase in v_{BE} for T_2. This increases the collector current in T_2, thereby causing v_{CE} of T_2 to drop. This drop in v_{CE} is coupled to the base-emitter terminals of T_1 through the commutating capacitor C_2, and results in a drop of v_{BE} for T_1. The collector current of T_1 therefore decreases, its v_{CE} increases, and this increase is coupled back to the base of T_1 via the commutating capacitor C_1. In progressing around this loop, the initial disturbance has been amplified by a two-stage amplifier, so the voltage which is returned to the

---

[1] The proof that the circuit of Fig. 15 · 20 is a nonsaturating type is left for the problems.

base of T_2 after transversing the amplifier loop is significantly larger than the initial disturbance and in phase with it (that is, we have a positive feedback loop). The disturbance will therefore grow until T_2 saturates, at which point the amplifier gain drops to zero, with the system remaining in the T_2-ON–T_1-OFF condition.

From a circuit theory standpoint, the regenerative action is associated with the presence of a right-half plane natural frequency in the flip-flop, when both transistors are in their active regions. We shall give a brief analysis, which is inaccurate in detail but accurate in concept, to demonstrate this point.

Our problem is to find the natural frequencies of the circuit shown in Fig. 15·22. In drawing the circuit model in Fig. 15·22b, we have assumed that the collector load resistors are large compared to r_π and that the commutating capacitors are short circuits at the natural frequencies. These gross approximations are made to remove nonlinearities from the analysis. (The fact that r_π and C_π are functions of the emitter current causes problems unless they always appear multiplied together.)

The equilibrium equations for the network shown in Fig. 15·22b are

$$Y_\pi V_1 + g_m V_2 = 0 \qquad g_m V_1 + Y_\pi V_2 = 0 \qquad (15 \cdot 29)$$

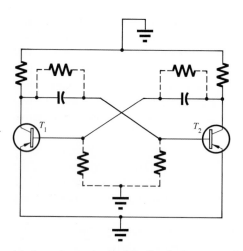

(a) Approximate circuit of flip-flop during regeneration cycle

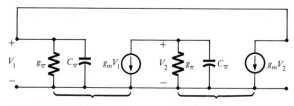

(b) Circuit model for (a)

FIG. 15·22   REGENERATION ANALYSIS FOR THE FLIP-FLOP

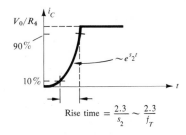

Rise time $= \dfrac{2.3}{s_2} \sim \dfrac{2.3}{f_T}$

FIG. 15·23 COLLECTOR CURRENT OF A TRANSISTOR BEING SWITCHED ON IN A FLIP-FLOP

The natural frequencies are the frequencies at which V_1 and V_2 can have nonzero values without external drive. Formally, the natural frequencies are the values of s which satisfy the determinantal equation

$$\begin{vmatrix} Y_\pi & g_m \\ g_m & Y_\pi \end{vmatrix} = 0 \qquad (15 \cdot 30a)$$

Equation $(15 \cdot 30a)$ leads to

$$(g_\pi + sC_\pi)^2 - g_m^2 = 0 \qquad (15 \cdot 30b)$$

which can be recast in the form

$$s^2 \tau_p^2 + 2s\tau_p + (1 - \beta^2) = 0 \qquad (15 \cdot 30c)$$

where $\tau_p = r_\pi C_\pi$. The roots of Eq. $(15 \cdot 30c)$ are

$$s_1 = \frac{-\beta - 1}{\tau_p} \qquad s_2 = \frac{\beta - 1}{\tau_p} \qquad (15 \cdot 31)$$

Therefore, the currents and voltages in the unexcited network are of the form

$$I = Ae^{s_1 t} + Be^{s_2 t} \qquad (15 \cdot 32)$$

The presence of the right-half-plane natural frequency s_2 ensures growing transients in the network (in this case, a simple exponential increase with time).

Rise time. When the natural frequencies are known (or can be estimated), then the rise time of the output waveform can be calculated (or estimated). In the case just described, where there is only one right-half-plane natural frequency, the collector current of the transistor which is being switched ON will be similar to that sketched in Fig. 15·23. If we define the rise time (RT) as the time required for the current to increase from 10 percent of its final value to 90 percent of its final value, then

$$RT \simeq \frac{2.3}{s_2} \qquad (15 \cdot 33)$$

for the simple exponential rise (independent of the final current). Using the value of s_2 given in Eq. (15·31) yields for the rise time

$$RT = \frac{2.3\tau_p}{\beta} \qquad (15 \cdot 34a)$$

which may also be written as

$$RT = \frac{2.3}{f_T} \qquad (15 \cdot 34b)$$

It is clear from this analysis that if we want a flip-flop to have a short rise time, we have to use transistors with high values of f_T.

Selection of commutating capacitors. In an actual circuit, it is necessary to choose the commutation capacitors carefully to maintain as large a bandwidth in the two-stage amplifier as is practical. The simplest technique for doing this is to choose C_1 so that the C_1R_2 combination has about the same time constant as the $r_\pi C_\pi$ combination; namely, τ_p. The C_1R_2 and $C_\pi r_\pi$ networks then form a compensated attenuator which can pass a square wave without distortion. In practice, C_1 is usually increased by a factor of 2 or 3 above this ideal value. This provides some overshoot in the square-wave response of the compensated attenuator and a corresponding overdrive of the transistor in the flip-flop.[1] The result is a slight reduction in the rise time.

The circuit shown in Fig. 15·24 is the flip-flop described earlier with commutating capacitors selected according to the discussion given above. The $r_\pi C_\pi$ time constant for the 2N1302 has already been calculated to be 1.6 μsec, so a C_1R_2 time constant of 3.4 μsec has been selected. With 6.8 kohms for R_2, the required value of C_1 is 500 pfarads.

Triggering requirements and methods. To change the state of a flip-flop, we must, at a minimum, drive the circuit just beyond the point where the collector currents of the two transistors are equal. At the end of the trigger pulse, the transistor which we wish to have ON should have a larger collector current than its companion. If the trigger pulse fails to achieve this condition (because it is too short, for example), then the flip-flop will revert to its previous state (*cf.* the ball-and-seesaw analogy given earlier).

Generally, it is advisable to arrange the trigger circuit to effect a turn-OFF rather than a turn-ON condition. A commonly used circuit for this purpose is shown in Fig. 15·25. Here, a positive-going trigger pulse will pass through the diode connected to the OFF transistor. The trigger pulse will be coupled to the ON transistor base via C_1. The trigger pulse must remove enough charge from the ON base to bring the ON transistor to the appropriate point in the normal-active region (that is, where its i_C and the i_C of its companion are equal). When this condition has been achieved, the

[1] The larger value of C_1 also qualitatively accounts for the effect of C_c in increasing the actual input capacitance.

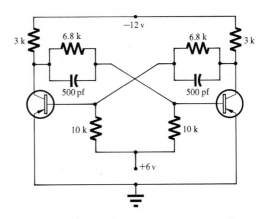

FIG. 15·24 FLIP-FLOP WITH COMMUTATING CAPACITORS CHOSEN TO MAKE $C_1 R_2 \simeq 2 C_\pi r_\pi$

trigger can be removed and the circuit will regeneratively tend to the new state.

A useful feature of Fig. 15·25 is that, so long as the trigger voltage amplitude is less than the voltage swing at the collector of the ON transistor, no trigger will be applied to the ON collector. The diodes therefore serve two purposes: they disconnect the trigger circuit except when it is actually firing the circuit, and they steer the trigger pulse to the proper point of the circuit.

Applications of flip-flops. If a flip-flop is driven by a regular train of pulses, then the output voltage at the collector of either transistor will be a square wave with a repetition rate which is half that of the input-pulse train. If this output voltage is used to trigger another flip-flop, then the switching

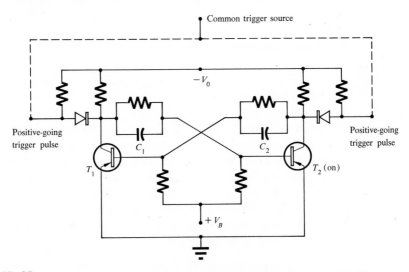

FIG. 15·25 A METHOD FOR TRIGGERING A FLIP-FLOP BETWEEN ITS STABLE STATES

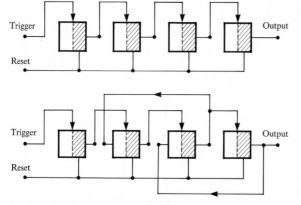

(a) Counter giving one output pulse for every 16 input pulses. All flip-flops are set to have the right-hand transistors ON at the outset, and to require a positive-going pulse for changing state.

(b) Circuit of (a) with feedback connections for advancing the count after the eighth input pulse to make a scale-of-ten counter.

FIG. 15·26   THE CONSTRUCTION OF A SCALE-OF-TEN COUNTER FROM FOUR BINARY COUNTERS.

rate of the second circuit is reduced by a factor of 2 compared to the first. If we look at the output voltage of one of the transistors in the second flip-flop, it will be a square wave with one-fourth the repetition frequency of the input pulses. Accordingly, we may view the pair of flip-flops as an electronic pulse counter which gives an output pulse for every four input pulses. By connecting n flip-flops in tandem, we can obtain an output-pulse rate which is $\frac{1}{2}^n$ times the input-pulse rate.

For many purposes, it is desirable to have a *decimal* counter (that is, a circuit which gives an output pulse for every 10 input pulses) rather than the *binary* counter which we have described above. For this purpose the circuit shown in Fig. 15·26 can be used. In Fig. 15·26a we show four flip-flop circuits connected in tandem. The output-pulse rate in this circuit will be $\frac{1}{16}$th of the input-pulse rate. In order to make the circuit a scale-of-ten counter, feedback connections are added as shown in Fig. 15·26b. The effect of the feedback connections can be briefly explained as follows.[1] The circuit operates as a normal binary counter for the first eight input pulses. This means that at the end of eight input pulses, the output voltage from the last flip-flop has risen. Without the feedback connections, it will take another eight input pulses to cause the output voltage of the last flip-flop to fall. However, the feedback connections serve to "advance the count" of the circuit internally; that is, set the second and third flip-flops in a condition which would correspond to fourteen input pulses in the circuit without feedback. The internal count is advanced after the eighth input pulse and before the ninth input pulse. Accordingly, the ninth pulse puts the counter into a state corresponding to 15 pulses in the binary counter without feedback. When the tenth input pulse arrives it will bring all the flip-flops back to their initial states. A complete output pulse is therefore formed for every ten input pulses.

[1] A thorough exposition is given in J. Millman and H. Taub, "Pulse and Digital Circuits," McGraw-Hill Book Company, New York, chap. 11.

By suitable use of AND circuits, it is possible to obtain ten outputs from the circuit to indicate the intermediate counts. These outputs can be displayed by numbered lamps or other devices which continuously "read" the state of the counter. Of course, n scale-of-ten circuits can be connected in tandem to count by 10^n, and the output of the last decimal counter can be used to drive a mechanical counter, if necessary.

Pulse counters have many important industrial and scientific applications. They are used with some type of electronic transducer which converts the number of objects to be counted to an equal number of pulses. A photoelectric cell and light source mounted on opposite sides of a conveyor belt is one possibility; the output pulses from a particle detector (such as a Geiger counter) is another. In these applications, the counters are gated ON for a precise period of time to measure the number of pulses received in a fixed time. The counter can then be cleared and re-gated ON. The counter thus provides a measure of the number of pulses received per second.

Counters can also be used to measure quantities which do not come in "pulses." For example, if we want to measure a dc voltage very accurately, then we could consider building an oscillator whose output frequency is controlled by the unknown voltage. The output frequency can be measured by gating a counter ON for a precisely known period of time (say, 1 sec). We can then determine the input voltage from an experimentally determined calibration curve of input voltage vs. output frequency.

15·3·3   The basic multivibrator.   From the point of view of its stable states, the multivibrator is the opposite extreme of the flip-flop. In the multivibrator, the two states are only temporarily stable. Circuit action causes the multivibrator to switch back and forth between its two states periodically, so that the multivibrator output resembles a square wave. A multivibrator is, in fact, a nonlinear oscillator, and its applications can be traced to this property.

The basic multivibrator circuit which we will study is shown in Fig. 15·27, together with some representative circuit values. As with the flip-flop, there are two well-defined stable states, each with one transistor heavily conducting and the other cut off. However, there are two important differences between the multivibrator shown in Fig. 15·27 and the flip-flops discussed earlier: (1) the base bias battery now has a polarity which tends to turn the transistors ON and (2) there is *no dc coupling* between the two transistors. As a result the fact that one transistor is ON cannot have a long-term effect on its companion.[1]

The fact that the base and collector bias supplies are now of the same polarity means that separate supplies need not be used. The resistors R_1 and R_4 can simply be adjusted to the proper values.

[1] This also allows for the possibility that both transistors will be ON, which represents a circuit failure. Usually, component tolerances will cause one transistor to go ON first when power is applied, or element values can be purposely selected to achieve this result.

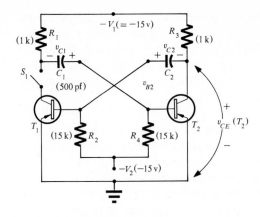

(a) Multivibrator circuit

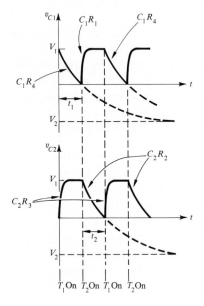

(b) Voltage waveform across C_1

(c) Voltage waveform across C_2

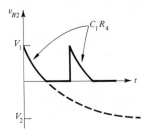

(d) Base emitter voltage for T_2

(e) Collector-to-emitter voltage for T_2

FIG. 15·27   THE BASIC MULTIVIBRATOR

650   *Semiconductor electronics*

The operation of the circuit can be explained as follows. Let us imagine that a knife switch has been placed in the collector lead of T_1, and that the switch is initially open. Then the equilibrium conditions for the circuit are

1. T_2 is ON. This is because its base current is V_2/R_4 ($=1$ ma), while the maximum collector current it can have is V_1/R_3 ($=15$ ma). The β of the transistor is assumed to be large enough to insure that T_2 is in its saturated state.

2. A current of V_2/R_2 ($=1$ ma) is flowing through the base-emitter junction of T_1. There is enough charge in the base region of T_1 to saturate it, but the collector of T_1 is floating at the moment.

3. C_1 is charged to essentially V_1 volts; C_2 has essentially zero charge.

Now we close the knife switch. This turns T_1 ON, and connects the left-hand end of C_1 to ground. The base voltage of T_2, which is now equal to the voltage across C_1, is driven to $+|V_1|$ volts, and T_2 rapidly cuts off. However, C_1 now begins to charge toward V_2 through R_4 and T_1; and when v_{C_1} reaches approximately zero volts, T_2 will turn ON again.[1] During this time interval, C_2 charges to V_1 volts through T_1 and R_3; so that when T_2 turns ON, C_2 will be connected across the base-emitter terminals of T_1, cutting it OFF. There follows a rapid charging of C_1 to V_1 through R_1 and the base lead of T_2, plus a slower charging of C_2 toward V_2 through R_2 and the collector lead of T_2. When v_{C_2} reaches zero, T_1 turns back ON, C_1 turns T_2 OFF, etc. Waveforms illustrating these details are shown in Fig. 15·27.

The important new function in the multivibrator as contrasted to the flip-flop is that the time duration between switching actions is controlled by element values in the circuit itself. To develop general expressions for the timing intervals, we note that, when T_1 switches ON, the base of T_2 is driven to $+|V_1|$ by C_1. C_1 then charges toward V_2 along the exponential curve shown in Fig. 15·27b. The voltage at the base of T_2 is therefore

$$v_{B_2}(t) = -|V_2| + |V_1 + V_2|e^{-t/R_4C_1} \qquad (15\cdot35)$$

(This equation is written by inspection of the waveform shown in Fig. 15·27b).

T_1 turns OFF when the base voltage of T_2 reaches zero. Setting Eq. (15·35) equal to zero, we can solve for the conduction time for T_1:

$$t_1 = R_4C_1 \ln\frac{V_1 + V_2}{V_2} \qquad (15\cdot36a)$$

Similarly, the conduction time for T_2 is

$$t_2 = R_2C_2 \ln\frac{V_1 + V_2}{V_2} \qquad (15\cdot36b)$$

---

[1] We are assuming that the emitter-base voltage necessary to cause currents to flow in the transistor is small enough in comparison with V_1 and V_2 to be neglected in this analysis.

These results are based on the assumptions

$$R_1C_1 \ll R_2C_2 \qquad R_3C_2 \ll R_4C_1$$

this being necessary to ensure that the capacitor effecting the turn OFF is fully charged to $|V_1|$ when the base of the ON transistor reaches zero volts.

If the resistors R_2 and R_4 are returned to V_1, then Eqs. (15·36) become

$$t_1 \simeq 0.7\ R_4C_1 \qquad t_2 \simeq 0.7\ R_2C_2$$

The total cycle time is $t_1 + t_2$ in any case.

If $R_4 = R_2$ and $C_1 = C_2$, then $t_1 = t_2$, and we obtain an overall cycle time of 1.4 R_2C_2, with the output voltage approximating a symmetrical square wave. However, we need not use this condition. We could instead use $C_1 = C_2$, but allow $R_4 \neq R_2$. If we allow unequal values of R_4 and R_2, but arrange matters so that $(R_4 + R_2)$ is a constant (by using two potentiometers connected to the same shaft, for example), then the cycle time will be 0.7 $(R_4 + R_2)C_1$, and the circuit generates variable-width pulses at a fixed repetition rate.

Applications of multivibrators. Applications of multivibrators fall into two main classes: they are useful in generating many types of nonsinusoidal waveforms and for various switching functions. As we have seen, multivibrators by themselves can produce a "square wave" with controllable symmetry. The spikes and rounded edges can be eliminated by appropriate clipping and clamping circuits. In addition, the output of the multivibrator can be passed through appropriately designed *RLC* networks to produce triangular waveforms, parabolic waveforms, and a variety of pulse trains.

The output voltage of a multivibrator can also be used to gate other waveform-generating circuits. For example, the circuit shown in Fig. 15·28 can be used to generate a linearly decreasing output voltage (or sweep voltage) if it is driven from a pulse of the correct amplitude and duration.

The gating function provided by the multivibrator is also useful in many switching applications. For example, if we want to display two

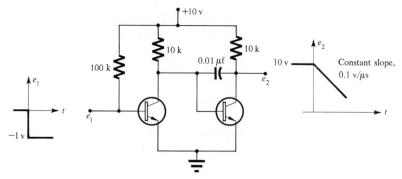

FIG. 15·28   CIRCUIT FOR GENERATING OUTPUT VOLTAGE WHICH DECREASES LINEARLY WITH TIME

waveforms on an oscilloscope screen, we can use a multivibrator together with a pair of nonregenerative switches to switch the oscilloscope input back and forth between the two waveforms that we wish to display. We can also use this same scheme to switch a voltage supply alternately to two different circuits, etc.

15·3·4   *The monostable multivibrator.* There are certain applications for gating circuits which cannot be met with a multivibrator, however. In these applications we want to generate a pulse with controllable dimensions, but which is synchronized with some external pulse source. For such cases, a monostable form of regenerative circuit is required.

The basic two-transistor monostable circuit, shown in Fig. 15·29, is made from a simple combination of the preceding ideas. One of the cross connections between transistors is as in the flip-flop, while the other is characteristic of the multivibrator. In the circuit shown in Fig. 15·29, there is one stable state with T_2 ON and T_1 OFF. A positive-going trigger pulse of sufficient amplitude succeeds in turning T_2 OFF, and the regenerative circuit action turns T_1 ON. When T_1 goes ON, the left-hand terminal of C_1 is effectively connected to ground, so the base of T_2 is driven positive by an amount $|V_0|$. However, C_1 charges toward $-V_{B_2}$ through R_6, and when the base of T_2 reaches approximately zero volts, T_2 turns to ON again.

The circuit output is normally taken to be the collector-to-emitter voltage of T_2. It consists of a pulse of height $-V_0$ and width

$$t = R_6 C_1 \ln \frac{V_0 + V_{B_2}}{V_{B_2}}$$

The width may be adjusted by varying R_6, C_1, or V_{B_2}. Further details on the circuit action will be left for the problems.

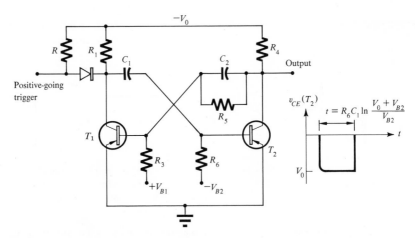

FIG. 15·29   A MONOSTABLE MULTIVIBRATOR

One interesting application of the monostable multivibrator is in pulse regeneration in a digital transmission system. A pulse which starts out as a rectangular pulse of accurately known height and width may, in the course of transmission through the system, degenerate into a bell-shaped pulse. If the bell-shaped pulse is used to trigger a monostable multivibrator, then the output of the monostable circuit gives a new pulse which can be sent on through the system. "Amplification" in the digital transmission system is thus provided by monostable circuits.

Another useful application of the monostable circuit is in the generation of a delayed pulse or a delayed gating waveform. In this application, the output pulse of a monostable circuit is differentiated in a suitable RC circuit, such as the one shown in Fig. 15·30. The pulse derived from the trailing edge of the waveform is delayed from the input trigger by a circuit-determined time T. If a delayed-gating waveform (that is, a rectangular pulse) is required, this delayed pulse is used to trigger a second monostable circuit which operates from a negative trigger. If a positive-going trigger is required, the delayed pulse can be amplified in a single-stage amplifier, which will invert it.

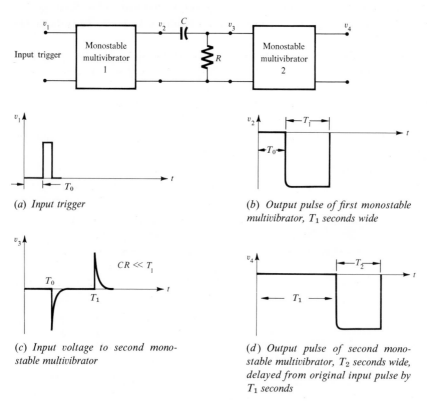

(a) Input trigger

(b) Output pulse of first monostable multivibrator, T_1 seconds wide

(c) Input voltage to second monostable multivibrator

(d) Output pulse of second monostable multivibrator, T_2 seconds wide, delayed from original input pulse by T_1 seconds

FIG. 15·30 THE USE OF TWO MONOSTABLE MULTIVIBRATORS TO OBTAIN A DELAYED OUTPUT PULSE

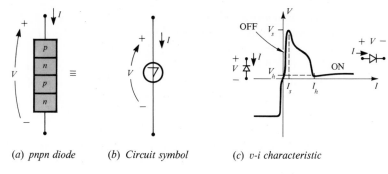

(a) pnpn diode   (b) Circuit symbol   (c) v-i characteristic

FIG. 15·31   A *pnpn* DIODE AND ITS *v-i* CHARACTERISTIC

15·4   SPECIAL DEVICES FOR SWITCHING APPLICATIONS

In addition to junction transistors there are several other semiconductor devices which have terminal characteristics that are nearly ideal for many switching applications. From a commercial standpoint, the most important of these special devices is the *pnpn* structure. In this section we will outline the principal characteristics and indicate some of the major applications for *pnpn* devices.

A *pnpn* diode, together with its circuit symbol and reference directions for voltage and current, is shown in Fig. 15·31. The device is sometimes called a four-layer diode, and the circuit symbol for the device has been derived from this name—it is the numeral 4 with the point of the 4 indicating the direction of positive current flow.

The *v-i* characteristic for the device, shown in Fig. 15·31c, exhibits two stable states. In the OFF state, the structure behaves very much like a single *reverse-biased* diode. In this state, the center junction of the structure controls the terminal properties of the device. If the avalanche breakdown voltage of the center junction is high, the device will be able to stand off high voltages in the OFF state, while passing very little current.

In the ON state, the terminal properties of the device are similar to those of a single *forward-biased* diode. As a result, the structure can pass a large current in the ON state with only a small voltage drop across it. Between the ON and OFF states there is a region of negative dynamic resistance which can be used to switch the device between its ON and OFF states.

Compared to fast-switching transistors, the switching speeds of *pnpn* devices are rather long (~ 0.1 μsec turn-ON time, ~ 1–10 μsec turn-OFF time), so *pnpn* devices are primarily useful in very high-power, moderately low-frequency applications. Such applications include timing and time-delay circuits, pulse generators and ring counters for high-power operation, dc-to-ac converters, motor-control and light-dimming circuits, and many other power-switching applications. Within this range of applications, *pnpn* devices are nearly ideal switching elements and enjoy wide useage.

15·4·1 *The v-i characteristic of a pnpn diode.* To explain the *v-i* characteristic of a *pnpn* diode, we will first employ an artifice to fix some basic ideas. For this purpose, we show in Fig. 15·32 a *pnpn* diode with contacts made to all four layers, and shorting switches connected around the two outer junctions. The shorting switches are the contacts of an ideal, normally closed relay which is placed in series with the device, so that the shorting switches will open when the current flowing through the device reaches a certain level.

To study the *v-i* characteristic of this device, we consider the circuit shown in Fig. 15·32b. Here, a battery of variable voltage and a large resistance are connected to the device. As the battery voltage is increased from zero, most of the voltage is absorbed across the reverse-biased center junction of the *pnpn* device. Only a small constant current I_0 (the saturation current of the center junction) flows in the circuit. This state of affairs persists until the applied voltage approaches the avalanche voltage of the center junction, V_B. Further increases in the applied voltage are absorbed in the series resistor and increase the current flowing in the circuit, in accordance with

$$\frac{V_A - V_B}{R} = I$$

When V_A has been increased to the point where the circuit current is equal to the relay-tripping current I_R, the shorting switches open, and the *pnpn* device assumes a new state (its ON state) in which the voltage drop across the device, v, assumes a very small value.

We can study the characteristics of this new state with the aid of a two-transistor analog, shown in Fig. 15·33. The transistors shown in

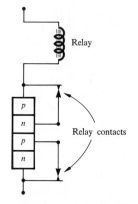

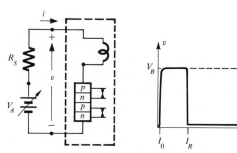

(a) *pnpn device with shorting switches around outer junctions*

(b) *v-i characteristic of modified pnpn device with test circuit*

FIG. **15·32** A *pnpn* STRUCTURE WITH RELAY-CONTROLLED SWITCHES AROUND THE OUTER JUNCTIONS

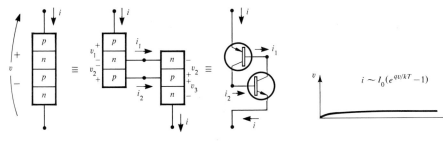

(a) Two-transistor analog for pnpn device; all junctions are forward-biased because the transistors are saturated

(b) v-i characteristic for (a) when transistors have $\beta \gg 1$

FIG. 15·33   pnpn DIODE IN ITS ON STATE

Fig. 15·33 are connected in a way that ensures that they will both be in saturation as long as their β's are at least 1.0.[1] To prove this we note that, from the symmetry of the pnpn structure, the base current of the pnp unit is the same as the collector current of the npn unit and vice versa. If we assume the pnp and npn transistors are identical, this condition implies that each transistor is driven with a base current equal to its collector current, which is a saturating level of base drive if $\beta \geq 1$. The resulting v-i characteristic for the structure will be similar to the one shown in Fig. 15·33. The voltage drop across the saturated collector junction essentially cancels the voltage drop across one of the outer junctions, and the entire structure behaves approximately like a single forward-biased diode. The voltage drop of the device is therefore reduced to a few tenths of a volt, and the v-i characteristic can be approximated by an

$$i = I_0(e^{qv/kT} - 1)$$

law.

The pnpn device plus series relay exhibits the basic action that occurs in the pnpn diode. Minority carrier injection at the two outer junctions is inhibited at low currents but not at high currents. In order to produce a current large enough to turn the outer junctions ON, a voltage approaching the avalanche voltage of the center junction must be applied. Once the device has switched ON, however, it will remain ON as long as a current greater than its holding current is passed through it.

Practical pnpn devices are, of course, not made with relays and shorting switches. They are made by either diffusing the required number of p and n layers into a silicon wafer or connecting a pair of transistors in the analog configuration suggested in Fig. 15·33a.

If the analog configuration is used, resistors are connected between the emitter-base terminals of each transistor to approximate the effect of the shorting switches (see Fig. 15·34). These resistors will short out the emitter-base junctions at current levels where the dynamic resistance of the

[1] We assume the transistors have identical characteristics, for simplicity. Otherwise, the saturation condition is $\beta_{pnp} + \beta_{npn} \geq 2$.

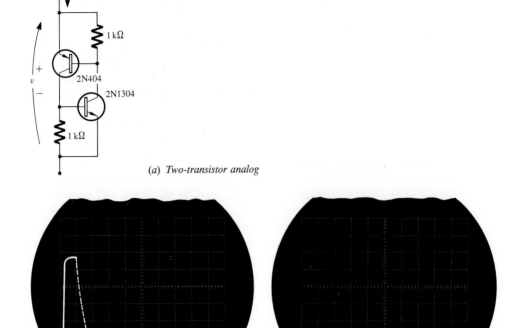

(a) *Two-transistor analog*

(b) *v-i characteristic of (a)*
Vertical scale: 5 *v/cm*
Horizontal scale: 0.1 *ma/cm*

(c) *v-i characteristic with* 1KΩ *resistors removed*
Vertical scale: 5 *v/cm*
Horizontal scale 0.1 *ma/cm*

FIG. 15·34   A TWO-TRANSISTOR ANALOG OF A *pnpn* DEVICE AND ITS *v-i* CHARACTERISTIC

junction $(\sim kT/qI)$ is large compared to the shunting resistance, but they will have no effect at current levels where the dynamic resistance of the junctions is small compared to the shunt resistance. For example, 1-kohm resistors will give effective shorting at currents of 1 μa or less, but not at currents of 1 ma or more.

A two-transistor analog using a 2N404 (*pnp*) and a 2N1304 (*npn* complement to the 2N404) with 1-kohm shorting resistors has the *v-i* characteristic shown in Fig. 15·34. The *v-i* characteristics, reproduced from a transistor curve tracer, are plotted both with the resistors and without them to verify the features just described.

Commercial *pnpn* devices are made of silicon, where the nonunity injection efficiency plays the role of the shorting resistance (*cf.* Sec. 9·6·3). The injection of minority carriers across the outer junctions is low at low current levels because of recombination in the junction-transition region. At higher currents the injection of carriers across the junction becomes the dominant mechanism for current flow through the junction. In actual

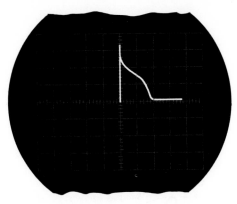

Vertical scale: 10 volts/cm
Horizontal scale: 1 ma/cm

FIG. 15·35   REPRODUCTION OF THE v-i CHARACTERISTIC OF A COMMERCIAL $pnpn$ DIODE

silicon $pnpn$ structures, the injection efficiency increases gradually with current, so the negative resistance region usually has a more gradual slope than is indicated in the analog device of Fig. 15·34. The v-i characteristic of a commercial $pnpn$ diode, shown in Fig. 15·35, illustrates this point.

15·4·2   Switching action in pnpn diodes.   An interpretation of the v-i characteristic which focuses attention on the switching characteristics of the device may be made with reference to Fig. 15·36. In part a of this figure, the first quadrant of the v-i plane has been divided into two regions by the dc device characteristic. Specifying the voltage across the device and current through it locates a point in the v-i plane. Any point which lies on the dc characteristic represents a condition in which the device can remain indefinitely under proper circuit conditions. Thus, if a $pnpn$ diode is driven from a current source of strength I_1 (that is, a vertical load line in Fig. 15·36a), the voltage across the device is uniquely determined (V_1) and the conditions described by the point (V_1,I_1) will persist indefinitely. A characteristic such as this is frequently called a "current-stable" or "open-circuit-stable" characteristic.

When circuit conditions define a voltage and current which is not on the dc characteristic, the circuit is in a transient condition which will carry the device back to the dc characteristic. The exact trajectory which is mapped out as this transient proceeds depends on both the circuit and the device.

As an example, let us consider the switching trajectory shown in Fig. 15·36b, which is typical of the switching-ON portion of a sawtooth oscillator circuit to be described later. The switching trajectory carries the device from a point in region I of its characteristic to a point in region III. An intermediate point on the switching trajectory with coordinates (V_0,I_0) is noted on Fig. 15·36b.

Switching circuits and switching devices   659

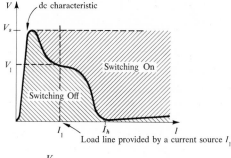

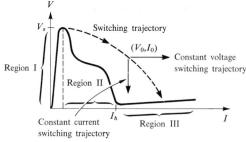

(a) *dc characteristic with switching* ON *and* OFF *regions defined*

(b) *Switching trajectory for a turn-*ON *transient*

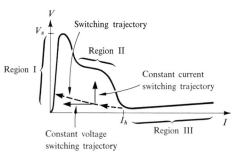

(c) *Switching trajectories during a turn-*OFF *transient*

FIG. 15·36   SWITCHING TRAJECTORIES FOR THE *pnpn* DIODE

If we suppose for a moment that the voltage were being held constant at V_0, then the current would build up exponentially with time, along the horizontal constant-voltage trajectory shown in Fig. 15·36b, approximately tripling its value every 50 nsec.[1] If this condition persisted, of course, the current would become so large that the device would quickly be destroyed.

If, instead of holding the voltage constant at V_0, the current were held constant at I_0, then the switching transient would follow the vertical constant-current trajectory indicated in Fig. 15·36b. In this case the voltage across the device would steadily decrease to the point indicated on the dc characteristic.

In a sawtooth oscillator, a capacitor (which is neither a voltage source nor a current source) is discharging through the device. It produces a

[1] This is the approximate transit time for minority carriers across the base layers of a commercial *pnpn* diode.

switching trajectory similar to that shown in Fig. 15·36b. The curvature of this characteristic indicates the tendency of a capacitor to act momentarily as a voltage source.

The time required for the switching trajectory to proceed from its starting point in region I to the dc characteristic in region III is called the *switching time* or turn-ON time and is a function of both the circuit and the device.

Similar but oppositely directed switching trajectories are obtained in the switching-OFF region. If the instantaneous voltage and current (V_0,I_0) on a switching trajectory define a point such as that shown in Fig. 15·36c, then the constant voltage trajectory is a horizontal line which carries the device to region I of its dc characteristic. Along this trajectory, the current decreases at a rate which depends on minority-carrier lifetimes in the four-layer diode base layers. Typical lifetimes are on the order of 0.1 to 1 μsec, so the current will decrease by a factor of e in this time interval.

The constant-current trajectory now proceeds vertically upward to region II of the dc characteristic, and a typical combination of circuit and device will have a curved trajectory heading toward lower current and higher voltage. Here again the switching time or turn-OFF time is determined by the time required for the trajectory to proceed from its starting point in region III to the dc characteristic in region I and is a function of both the device and the circuit.

From a "device-terminals" standpoint, the ideas contained in the preceding two paragraphs may be compactly summarized as follows:

1. When the device is OFF (in region I), it may be turned ON by (a) causing the voltage across it to exceed V_s, which initiates a turn-ON switching transient, and (b) ensuring that after the switching transient is completed, a steady current of at least I_h is furnished to the diode.

2. When the device is ON, it may be turned OFF by (a) reducing the current through it below I_h, which initiates a turn-OFF transient, and (b) ensuring that after the switching transient is completed, the external circuit provides a voltage and current appropriate to region I.

15·4·3 *Sawtooth oscillator and pulse generator circuits utilizing the pnpn diode.* Two of the fundamental applications for *pnpn* diodes are in relaxation oscillator and pulse generator circuits. These circuits will be described as design examples. Most of the applications of *pnpn* diodes which have appeared in the literature are variations on the basic sawtooth oscillator to be described.

Sawtooth oscillator circuit description. The fundamental relaxation circuit is the sawtooth oscillator shown in Fig. 15·37a. The operating cycle for this circuit is as follows: We assume the capacitor C_1 to be discharged initially and the four-layer diode to be OFF. Under this condition, current flowing through R_1 from V_0 charges C_1 toward V_0 along an exponential curve, as in Fig. 15·37b. When the voltage across C_1 reaches the diode

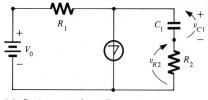

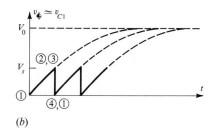

(a) Basic sawtooth oscillator circuit

(b)

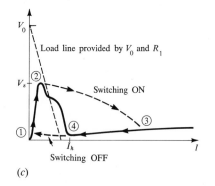

(c)

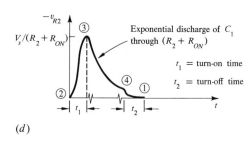

(d)

FIG. 15·37 THE BASIC *pnpn* DIODE SAWTOOTH OSCILLATOR CIRCUIT

switching voltage,[1] a switching transient is initiated which will start to switch the four-layer diode ON. The capacitor then begins to discharge through the four-layer diode. If C_1 is big enough, the four-layer diode will turn fully ON before the capacitor discharges appreciably. The capacitor is therefore essentially "switched" across the current-limiting resistor R_2 and begins to discharge toward V_h, the holding voltage of the *pnpn* voltage.[2] During this time the current flowing in the *pnpn* diode has two components:

1. A current of $V_0 - V_h/R_1$ flowing in the battery–*pnpn*-diode loop
2. A current of $[(V_s - V_h)/R_2]e^{-t/R_2C}$ flowing in the capacitor–*pnpn*-diode loop (assuming $R_2 \gg R_{on}$)

To make the circuit work as a relaxation oscillator, we must select R_1 and/or V_0 so that V_0/R_1 is less than I_h. When this condition obtains, the sum of the currents (1) and (2) given above will ultimately drop below I_h, since (2) would ultimately drop to zero, and (1) is less than I_h.

When the current flowing in the four-layer diode is reduced to I_h, the *pnpn* diode will start to turn off. The capacitor will keep the voltage across the *pnpn* diode at a relatively low value during this turn-OFF time. When the diode has been turned off, the capacitor will begin to charge toward V_0 once more, and a new cycle is initiated. The resulting waveform of voltage across the capacitor is shown in Fig. 15·37b.

[1] This will usually be at essentially the same time that V_c reaches V_s, since R_2 is only a small resistor provided to limit the capacitor discharge current and it has only an insignificant effect on the voltage which appears across the four-layer diode during the charging cycle.

[2] V_h is the device voltage drop at the point labelled ④ on Fig. 15·37c.

It is also instructive to interpret this operating cycle on the v-i characteristic and thereby focus attention on the switching properties of the circuit. This is done in Fig. 15·37c. The charging of the capacitor in region I corresponds to movement from ① to ②. When the voltage across the four-layer diode reaches V_s, a switching-ON transient is initiated. For the first part of this switching transient, the capacitor appears to be a voltage source, so the switching trajectory initially proceeds from ② horizontally out into the v-i plane. After a short time, the current will have removed enough charge from C_1 to cause its voltage to drop appreciably, so the switching trajectory bends downward. If C_1 is small enough, the curvature of the trajectory will be sufficient to terminate the switching-ON transient at a peak current (represented by point ③) which is within the rating of the four-layer diode. In this case, the turn-ON time is a function of only C_1 and the characteristics of the *pnpn* diode. For larger capacitors than this, a current-limiting resistor R_2 must be provided to limit the peak current; in this case the shape of the switching trajectory at high currents and the switching time are determined by C_1, R_2, and the characteristics of the *pnpn* diode.

The ③ to ④ portion of the V-I portrait given in Fig. 15·37c represents the exponential discharge of C_1 through R_2 and the series resistance of the *pnpn* diode. When the circuit reaches the condition represented by ④ there is insufficient current to keep the *pnpn* diode in its saturated state, and it begins to turn off. The capacitor once again looks momentarily like a constant-voltage source, and the horizontal switching-OFF transient is initiated. Some of the current flowing from V_0 through R_1 charges C_1 during the turn-OFF period, however, so the actual trajectory curves upward to point ①, as shown in Fig. 15·37c.

The reference points ①, ②, ③, and ④ are shown on Fig. 15·37b to indicate when they occur on the sawtooth waveform. The typical appearance of the sawtooth is such that ② and ③ appear to be together and ④ and ① appear to be together, because the turn-ON and turn-OFF times are so short. By viewing the voltage across R_2, some fine detail in the switching trajectories may be observed. A typical plot of voltage across R_2 as a function of time is shown in Fig. 15·37d with the reference points ② and ③, and ④ and ① located on it.

EXAMPLE As a numerical example, we will design a sawtooth oscillator to provide a 50-volt sawtooth at a repetition frequency of 7 kcps. The 50-volt amplitude requirement specifies a four-layer diode with a switching voltage V_s of 50 volts. In addition, we choose a holding current of $I_h = 10$ ma for this example.

We must select a battery voltage greater than 50 volts if we want the circuit to oscillate. We choose $V_0 = 100$ volts. To choose R_1, we note that $V_0/R_1 < I_h$, which means R_1 must be greater than 10 kohms. We select $R_1 = 20$ kohms. To choose C_1, we note that it must charge through 20 kohms to 50 volts in $1/(7 \times 10^3)$ sec. This requires $C_1 = .01$ μfarad. To choose R_2, we observe that the peak current in the capacitor–*pnpn*-diode loop will occur when the diode has just been switched ON, the capacitor's voltage at that point being 50 volts. An R_2

of 50 ohms will limit this peak current to 1 amp, a satisfactory level for normal operation. With these circuit constants, the sawtooth-repetition frequency will be 7 kcps, and a 50-volt pulse will be generated across R_2 each cycle.

Pulse generator circuit. While our considerations to this point have been limited to sawtooth oscillator circuits, the circuit of Fig. 15·37*a* may be very simply modified to make a monostable circuit, or pulse generator, as shown in Fig. 15·38. A conventional diode has been added in series with the *pnpn* diode to provide a high-impedance triggering point. The supply voltage V_0 must now be selected to be less than V_s, for example, 35 volts. In this situation, the capacitor C_1 will charge up to V_0 and the circuit will come to rest. When a negative-going trigger pulse of magnitude greater than $V_s - V_0$ is applied to the trigger point, the *pnpn* diode will be driven into region II, and C_1 will then discharge. Once again, since R_1 will not pass holding current, the *pnpn* diode ultimately shuts off, after which C_1 recharges to V_0 and awaits the next trigger pulse.

If the conventional diode and the *pnpn* diode are interchanged, the circuit will operate on positive pulses, and if the conventional diode is replaced by a second *pnpn* diode, the circuit will fire on either polarity of pulse, although the pulse amplitude and V_0 will now have to be readjusted.

The high peak current which the *pnpn* diode will pass makes this circuit, or a modification of it, of considerable interest in kilowatt pulse modulator circuits as well as in many lower-power pulse generator and pulse amplifier situations.

15·4·4 *pnpn* **triode, or controlled rectifier.** A *pnpn* diode with a base lead attached to one of the central layers (usually the *p* layer) is called a *pnpn* triode, or controlled rectifier. The circuit symbol for the controlled rectifier, together with reference directions for voltage and current, is shown in Fig. 15·39.

The addition of this gating lead causes the *v-i* characteristics to assume the form shown in Fig. 15·39. Each characteristic is generally similar to a

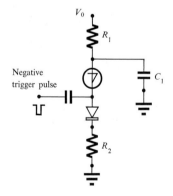

FIG. 15·38   A MONOSTABLE CIRCUIT (PULSE GENERATOR) EMPLOYING THE *pnpn* DIODE

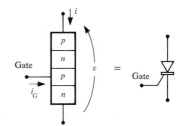

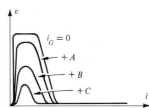

(a) *Schematic representation of device, and circuit symbol*

(b) *v-i characteristics of the pnpn triode with gate current as a parameter*

FIG. **15·39**   THE *pnpn* TRIODE, OR CONTROLLED RECTIFIER

pnpn diode, except that the switching voltage and holding current are now functions of the gate current i_G.

The reason for this behavior can be understood by referring to the two-transistor analog shown in Fig. 15·40a. Here, a gate current is supplied to the base lead of the *npn* transistor. This gate current will produce a forward bias across the shunt resistor and thereby lower the terminal current *i* which is necessary to turn the outer junctions ON and therefore to switch the device into its low-impedance state. The behavior of the analog on a transistor curve tracer is shown in Fig. 15·40b and is very similar to the behavior of a commercial *pnpn* triode.

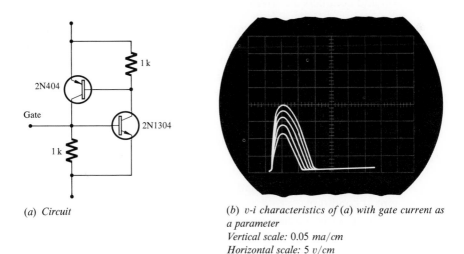

(a) *Circuit*

(b) *v-i characteristics of (a) with gate current as a parameter*
Vertical scale: 0.05 ma/cm
Horizontal scale: 5 v/cm

FIG. **15·40**   A TWO-TRANSISTOR ANALOG FOR A *pnpn* TRIODE

Switching circuits and switching devices *665*

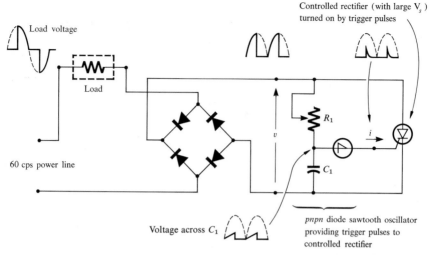

FIG. 15·41   FULL WAVE POWER CONTROL CIRCUIT USING *pnpn* DIODES AND TRIODES

Like the *pnpn* diode, commercial *pnpn* triodes are usually made by diffusing the *p* and *n* layers into silicon. Again, the variation in injection efficiency with current level which is typical of silicon devices is responsible for the device behavior. The gate current is effective, because it provides external control over the injection efficiency of one of the outer junctions.

Normally, *pnpn* triodes are made so that their switching voltage with zero gate current may be several hundred volts. The triode is used in circuits where the applied voltages are much less than this. To switch the device ON, a pulse of gate current is applied to lower the switching voltage below the voltage actually applied to the device. The triode then switches ON and conducts a current determined by the external circuit.

Motor control and light-dimming applications.   Because the entire area of a *pnpn* device is conducting current when the device is ON,[1] these devices can pass very large currents in the ON state. (*pnpn* triodes with 100-amp current ratings are readily available.) As a result, these devices find interesting applications in high-power motor control, light-dimming, industrial heating, and similar situations.

An interesting circuit which demonstrates a full-wave power control is shown in Fig. 15·41. Here a load (such as a motor) is connected to the power lines through a diode bridge. The alternate terminal pair of the diode bridge is connected to a *pnpn* diode-triode control circuit, and ac power is fed to the load only when the control circuit is in a short-circuit condition.

The control circuit operates as follows. The diode bridge passes a full-wave rectified sine wave to the control circuit. The peak voltage

---

[1] That is, there is no lateral base-current flow which crowds the emission of minority carriers into a small area of the emitter, as there is in a transistor.

in this wave is not sufficient to fire the controlled rectifier, but it is sufficient to fire the *pnpn* diode. The resistor R_1 and capacitor C_1 (two components of a *pnpn*-diode sawtooth-oscillator circuit) determine the time at which the *pnpn* diode fires. By making R_1 variable, this firing time can be set to be nearly anywhere in the half cycle of the applied waveform.

When the *pnpn* diode turns ON, the discharge of C_1 through it and the *pnpn* triode turns the triode ON. When the triode turns ON, the control circuit assumes a short-circuit condition and ac power is passed to the load. The diode bridge arrangement ensures that the load waveform contains no dc component, which is necessary in many types of ac loads (dc currents cause unwanted heating in ac motors, for example). A complete catalog of circuit applications of *pnpn* triodes is available from manufacturers of these devices.[1]

15·5 SUMMARY

In this chapter we have seen that all four regions of operation of the transistor are important in switching applications. The cutoff and saturation regions do not serve as simply boundaries which limit the output swing of an amplifier. On the contrary, in switching applications transistors frequently have their stable states in the saturated and/or cutoff regions, and the normal-active or inverted-active regions simply provide a means of switching between the two states. The ON and OFF properties of the switch are determined by the behavior of the transistor in its saturated and cutoff regions.

In nonregenerative switches, the switching commands are provided by an external control signal, with the transistor simply responding as an overdriven amplifier. The properties of a transistor switch are in many respects very nearly ideal, so transistor switches find wide application in all types of electronic equipment. Usually the circuit currents and voltages in these switching circuits can be analyzed, at least to a first approximation, by considering the transistors as three-terminal piecewise-linear elements: either completely open or a three-terminal short circuit.

Regenerative switching applications are similar to nonregenerative circuits in that the stable (or temporarily stable) states can still be analyzed with simple OFF or ON models for the transistors. The principal difference between the regenerative circuit and the nonregenerative circuit is that the switching command comes in the form of a pulse, and the circuit itself establishes the necessary base drives to hold the transistors in their stable states.

Switching circuits, both regenerative and nonregenerative, find their principal applications in digital transmission, computation and control, and instrumentation. The effectiveness of transistor circuits in these applications,

[1] See, for example, the General Electric SCR (Silicon Controlled Rectifier) Manual.

coupled with the rapid technological advances in microelectronics, gives great promise for the future of long-life, trouble-free solid-state control in many important sections of our economy.

In addition to the field of microelectronics, however, which we have described as a "miniaturization" of circuits developed for existing applications, there is the possibility of inventing new types of devices, which would be important in their own right and which would not consist of a simple scaling-down of existing circuits. The *pnpn* family of devices is perhaps a forerunner of this type. While the device can be made from conventional components, commercial *pnpn* devices are made from silicon and rely on internal effects to produce the nonunity injection efficiency that causes the two outer junctions to assume different behavior at different current levels. The *pnpn* diode thus performs the same circuit function as a pair of complementary transistors and two resistors. The *pnpn* diode is not simply a "scaled-down version" of an existing circuit to miniature proportions; it is a device which uses physical phenomena within semiconductor materials (silicon) to obtain directly a desired circuit function.

Because of their low leakage current and relatively high switching voltage in the OFF state, and their high current-handling capability and low voltage drop in the ON state, *pnpn* devices make very good high-power switching devices for low- and medium-frequency applications ($\lesssim$ 100 kcps).

REFERENCES

Linvill, J. G., and J. F. Gibbons: "Transistors and Active Circuits," McGraw-Hill Book Company, New York, 1961, chaps. 21 and 22.
Gray, P. E., D. DeWitt, A. R. Boothroyd, and J. F. Gibbons: "Physical Electronics and Circuit Models of Transistors," SEEC vol. 2, John Wiley & Sons, Inc., New York, 1964, Chap. 10.
Pettit, J. M.: "Electronic Switching and Timing Circuits," McGraw-Hill Book Company, New York, 1959.
Millman, J., and H. Taub: "Pulse and Digital Circuits," McGraw-Hill Book Company, New York, 1965.
Ledley, R.: "Digital Computer and Control Engineering," McGraw-Hill Book Company, New York, 1960.
Littauer, R.: "Pulse Electronics," McGraw-Hill Book Company, New York, 1965.
Gentry, F. E., et al.: "Semiconductor Controlled Rectifiers," Prentice-Hall, Inc., Englewood Cliffs, N.J., 1965.

DEMONSTRATIONS

15·1 ON VOLTAGE OF A TRANSISTOR SWITCH   The concept of saturation is sometimes a difficult one to grasp. The demonstration shown here is included for this reason. In this demonstration a high-impedance dc voltmeter (for example, hp 425A) is used

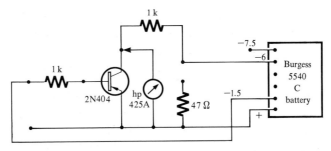

Note: If two sockets are used the transistor can be wired in the normal connection in one and the inverted connection in the other and saturation voltages for the two connections can then be compared.

FIG. D15·1

to measure the junction voltages for saturating levels of base drive. (The circuit board illustrated has several connections intended for subsequent demonstrations.) A 2N404 in this circuit will exhibit a v_{CE} of about 23 mv, while v_{CB} and v_{EB} are about ten times larger than this.

15·2   ON RESISTANCE OF A TRANSISTOR SWITCH   The ON resistance of the saturated switch can be measured with the set-up shown in demonstration 15·1. If the collector lead is changed from the 6-volt tap to the 7.5-volt tap, a 1.5-ma change in I_C will result. The corresponding change in v_{CE} can be measured ($\sim$4 mv) and the ON resistance calculated (about 2.5 ohms). It is also interesting to change the base drive tap, to see how v_{CE} and r_{ON} change.

15·3   TURN-ON AND TURN-OFF TRANSIENTS IN A NONREGENERATIVE TRANSISTOR SWITCH
The transient phenomena associated with the turn-ON and turn-OFF of a transistor switch can also be very conveniently demonstrated with the circuit given in Fig. D15·3. The circuit is the same as that of demonstration 15·1, with the

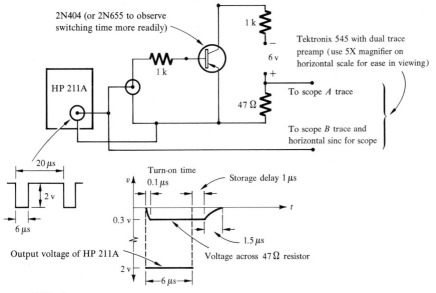

FIG. D15·3

inclusion of a square-wave generator and a dual-trace oscilloscope. The square-wave generator has a symmetry control so that its (negative-going) output pulse can be made to be about 10 μsec wide at a repetition rate of 50 kcps. The amplitude of the output pulse can be adjusted to affect the turn-ON and storage times.

It is interesting to substitute a transistor intended for amplifier service in this circuit to see the changes that result. A 2N655 will exhibit about 5 μsec storage time.

15·4 Another interesting demonstration of the charge stored in the base of a transistor and various ways of removing it can be seen by connecting the transistor as a diode and measuring its recovery time. By using the two different connections shown in Fig. D15·4, the effect of forward bias on the collector can be observed. The recovery in A will have an $e^{-2\pi f_T t}$ dependence, while the recovery of B will have an e^{t/τ_p} dependence. The recovery time set-up is similar to the one given in Chap. 8.

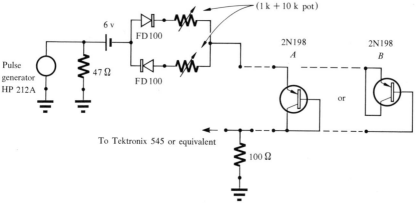

FIG. D15·4 THE RATIO OF RECOVERY TIMES FOR A AND B, WHICH IS β, WILL DEPEND ON THE CURRENT BEING SHUT OFF

15·5 The flip-flops, multivibrators, etc., in the text have workable values given in the associated figures and can be made into demonstrations quite readily. However, the applications of these circuits are sometimes less obvious than other demonstration circuits. As a result, it has proven very instructive and successful to tell students in the class how to act as "flip-flops", gates, etc. and then to "program them" into a simple computer. The technique for doing this is described, together with some simple examples, in "Games that Teach the Fundamentals of Computer Operation," an article published by D. J. Engelbart in the *IRE Transactions on Electronic Computers,* March, 1961, pp. 31–41. Engelbart has used his method with laymen programmed to act as flip-flops, shift registers, etc., arranged to perform some simple computations. This is a delightful as well as instructive demonstration, which also works well at some parties.

15·6 The analog *pnpn* diode and triode described in the text are interesting for visual demonstration on a transistor curve tracer. The real *pnpn* devices are of sufficient commercial significance that a demonstration of these analog devices is very worthwhile. Varying the emitter bypass resistors will give some idea of the various shapes of device characteristics that can be obtained. Both diode and triode characteristics can be viewed directly on a transistor curve tracer by attaching a gate lead to the 2N1304 base and returning this gate to the base terminal of the transistor curve tracer.

15·1   A 2N1306 is a nearly symmetrical *npn* switching transistor with $\beta_{typ} = 100$ and $I_{CO} = 3$ μa.

1. Develop a circuit like that of Fig. 15·4 to obtain a base current of 2 ma and a collector current of 10 ma. Calculate the junction voltage drops in this condition.

2. In the final steady-state condition for this switch, is the straight-line approximation for the minority-carrier density variation across the base region a good one? Why?

15·2   A certain silicon switching transistor has $\alpha_F = 0.98$, $\alpha_R = 0.9$, $I_{CO} = 10$ na. If this transistor is used in the circuit of Fig. 15·4 what will the collector-to-emitter voltage drop (v_{CE}) be? Draw a large-signal circuit model for the transistor (specifying I_{ES} and I_{CS} for the diodes) and calculate the currents flowing in each element of the circuit model.

15·3   A transistor switch is to be driven with a base current of 5 ma. It is desired to keep v_{CE} less than 10 mv when 20 ma of collector current is flowing through the switch. What should α_F be in a symmetrical transistor to ensure this condition?

15·4   A transistor can be used in the simple, nonregenerative switching circuit (Fig. 15·4) in either of two ways: (1) the terminal marked "emitter" can be connected as shown in Fig. 15·4 or (2) the terminal marked "emitter" can be connected to the load (that is, the transistor can be operated in an inverted connection). Which connection will give the small v_{CE} under the conditions $\dfrac{i_C}{i_B} \leq 1$, $\alpha_F > \alpha_R$? Is this true for all $\dfrac{i_C}{i_B} > 1$?

15·5   Calculate the ON resistance of the transistor given in Prob. 15·2 in the switching circuit of Fig. 15·7. The dc currents are as specified in Prob. 15·2.

15·6   Calculate the ON resistance of the transistor given in Prob. 15·2 if it is operated in the inverted connection (in the circuit given in Prob. 15·5).

15·7   The silicon switching transistor of Prob. 15·2 has $C_c = 2.5$ pf, $C_{Ej} = 1.8$ pf. If the transmission line shown in Fig. 15·7 has a characteristic impedance of 50 ohms (that is, it behaves like a 50-ohm resistor), at what frequency will there be a 60-db reduction of the signal fed to the transmission line when the switch is turned OFF? Repeat for a characteristic impedance of 300 ohms in the transmission line.

15·8   A transistor with $\beta = 50$, $f_T = 300$ mcps is to be switched ON by a base current of 5 ma. The collector circuit will supply a current of 50 ma.

1. Estimate the rise time required for the collector current of this transistor to reach saturation after the base drive is applied. Be sure to state your assumptions clearly.

2. Estimate q_S for this situation.

3. Approximately how much time elapses from the time that I_B is applied until the base charge has reached 98 percent of its final value.

15·9   A transistor manufacturer gives the following data on a switching transistor:

Rise time, nsec	2	5	10	20
Turn-ON base current, ma	1.8	0.9	0.46	0.22

In all cases the collector current in the ON condition is 5 ma. How well do these data fit the simple theory described in the text?

15·10 A transistor with $\beta = 50$, $f_T = 300$ mcps is turned OFF by reducing the base current to zero. In the ON condition, the terminal currents were $I_B = 0.5$ ma, $I_C = 10$ ma. Estimate the storage time and the 90 to 10 percent fall time for the collector current under these conditions.

15·11 Estimate t_s and t_f for the transistor in Prob. 15·10 if it is turned OFF by reversing the base current (until i_C reaches zero).

15·12 A transistor with $\beta = 50$, $f_T = 300$ mcps has $I_C(\text{ON}) = 10$ ma. To turn this transistor OFF, a reverse base current of 2 ma is applied. Plot (approximately) τ_R and τ_S as a function of I_{B_1}, the base drive in the ON condition. What value of I_{B_1} minimizes the sum of τ_R and τ_S.

15·13 A transistor switch is to be closed by a voltage source feeding an RC circuit in the base lead (Fig. P15·13).

1. Select a value of C such that the steady-state charge on it is equal to the excess density of holes stored in the base in the ON condition. Neglect junction voltage drops.

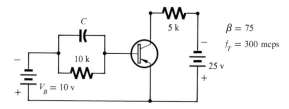

FIG. P15·13

2. The drive condition in (1) is equivalent to supply an impulse of charge (CV_B) and then a steady-state current to maintain this charge. What response would you expect to see in the collector current under these circumstances?

15·14 1. An *npn* transistor has an emitter junction capacitance of 2 pf at a reverse bias of 5 volts on the emitter-base junction. The transistor has $\beta = 50$, $f_T = 100$ mcps. The transistor is to be turned ON with a base current of 2 ma, the ON collector current being 10 ma. Estimate the delay time and the rise time for this application.

2. If $\alpha_F = 0.98$ ($\beta = 50$), $\alpha_R = 0.85$, $I_{CO} = 10^{-10}$ amp, what will the values of v_{CB}, v_{EB}, and v_{CE} be in the steady ON condition?

15·15 If the transistor of Prob. 15·14 is turned OFF with a reverse base drive of 5 ma, estimate the storage and fall times.

15·16 The 2N1303 data sheet provides the following information on stored charge vs. ON base current for the transistor: $q_S = 1,200$ $\mu\mu c$ for $I_B(\text{ON}) = 1$ ma. The typical β for this transistor is 50, and the typical f_α is 4.5 mcps. The minimum β is 20 and the minimum f_α is 3 mcps. How do these β and f_T figures agree with the quoted q_S and I_B?

15·17 It is desired to design a flip-flop which will deliver essentially a positive 10-volt output to a 1-kohm load. Thus, in Fig. 15·17 the collector load resistors should be 1 kohm and the collector supply voltage 10 volts, and *npn* transistors should be used. The base battery voltage available is −5 volts. Choose the remaining circuit values so that the OFF transistor has an emitter-base reverse bias of at least 2.5 volts and that the ON transistor will be saturated when a transistor with $\beta_{min} = 20$ is used in the circuit.

15·18   A single-battery version of the circuit suggested in Prob. 15·17 is shown in Fig. P15·18. Here, a common-emitter resistor is used to develop the required base voltage of 5 volts when 10 ma collector current is flowing. Compute the value of R_E required in this circuit. Note that the collector supply voltage is increased to 15 volts to accommodate this change.

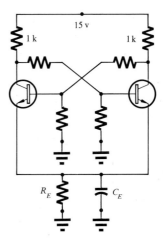

FIG. P15·18

15·19   1. Consider the circuit shown in Fig. P15·19. Calculate the dc currents in all branches of this circuit, and show that the ON transistor ($\beta = 50$) in saturation.
   2. The transistors have $f_T = 50$ mcps. Approximately what values of speed-up capacitors would you expect to use in this circuit?
   3. How much R_E is needed to replace the 1.5-volt base battery (as in Fig. P15·18)? What other change is necessary if the operating conditions are to remain unaffected?

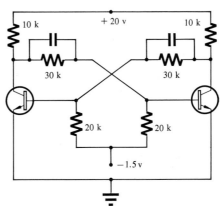

FIG. P15·19

15·20 Show that the circuit of Fig. 15·20 in the text is a nonsaturating flip-flop (that is, the ON transistor is not in saturation).

15·21 A flip-flop is shown in Fig. P15·21. Assume that each transistor has $\beta = 50$, and that T_1 is ON, T_2 OFF.

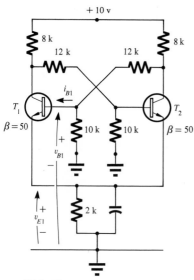

FIG. P15·21

1. Calculate the voltage drop from the collector of T_1 to ground, assuming the voltage drop v_{CE} for T_1 is small.
2. Calculate i_{B_1}.
3. Is T_1 saturated?
4. Calculate v_{B_1} and v_{E_1}.

15·22 A 2N3638 has $f_{T,min} = 300$ mcps. If it is used in the circuit of Fig. 15·24 (in the text), about what values should the speed-up capacitors have to be placed in parallel with the 6.8-kohm resistors to facilitate switching? Approximately how long will it take the flip-flop (with these capacitors) to change state when a switching pulse brings the OFF transistor up to 10 percent of its final ON current?

15·23 A basic saturated multivibrator is shown in Fig. P15·23. Sketch and

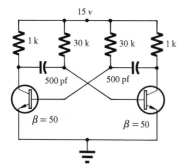

FIG. P15·23

674 Semiconductor electronics

dimension the collector-to-ground and base-to-ground waveforms for one of the transistors for a complete cycle of operation. Neglect the switching times of the transistors in these calculations.

15·24 The circuit shown in Fig. P15·24 employs two diodes to prevent the ON transistor from being driven into saturation. Sketch and dimension the collector and base waveforms for one complete cycle of operation. Verify whether the diodes do indeed prevent saturation.

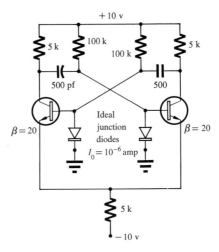

15·25 Design a symmetrical multivibrator of the type shown in Fig. P15·23, which can have a period which is continuously variable from 20 μsec to 20 msec. Capacitors can be switched into the circuit and potentiometers mounted on the same shaft are available. Switching times in the transistors may be neglected.

15·26 The monostable multivibrator shown in Fig. 15·29 in the text has $R_1 = 500$ ohms, $R_3 = 33$ kohms, $R_4 = 500$ ohms, $R_5 = 6.8$ kohms, $C_2 = 500$ pf, $C_1 = 1000$ pf, $R_6 = 6.8$ kohms, $V_{B_1} = +6$ volts, $V_{B_2} = -6$ volts, $V_0 = -10$ volts.
   1. Determine the currents that flow in the various branches of the circuit in the stable state.
   2. Compute and sketch the output pulse of this circuit when a trigger pulse which is adequate for switching is applied.
   3. How large should the triggering pulse be?

15·27 A nonsaturating monostable multivibrator is shown in Fig. P15·27. Compute and sketch the collector and base waveforms for T_2.

15·28 The period of the circuit shown in Fig. P15·27 can be adjusted by returning the upper end of the 30-kohm resistor to an adjustable voltage E (instead of $+10$ volts). Compute the period as a function of E over its useful range.

15·29 The circuit shown in Fig. P15·29 is a direct-coupled monostable multivibrator.
   1. Determine the conditions which apply in the stable state of this circuit.
   2. Show that the circuit delivers a 1-μsec-wide output pulse. (C_1 may be changed to provide various pulse widths.)

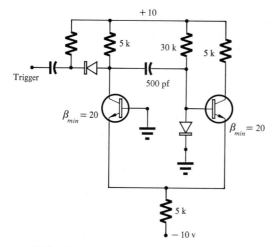

FIG. P15·27

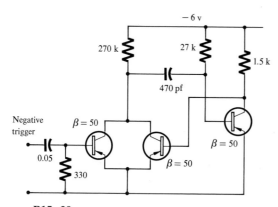

FIG. P15·29

15·30 A *pnpn* sawtooth oscillator circuit is shown in Fig. P15·30. The *pnpn* diode has ($V_s = 40$ volts, $I_s = 10$ μa) and ($V_h = 1$ volt, $I_h = 10$ ma). Design a sawtooth-oscillator circuit with a 1 kcps repetition frequency. Compute and

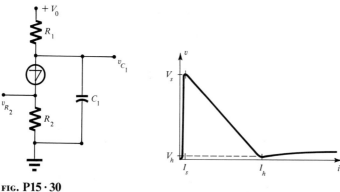

FIG. P15·30

676 *Semiconductor electronics*

sketch and voltage across the sawtooth capacitor and the series resistor R_2, neglecting switching times in the *pnpn* device. If the pulse obtained across R_2 were rectangular, how much power would be delivered to R_2 per pulse?

15·31 What changes are necessary in the circuit of Fig. 15·30 to make a triggered pulse generator (rather than an oscillator) out of it?

15·32 To achieve linearity in a sawtooth oscillator, the circuit shown in Fig. P15·32 is proposed. Will this circuit produce superior sawtooth linearity to the simpler one shown in Fig. P15·30? What tolerances must be placed on the circuit values to ensure proper operation as an oscillator?

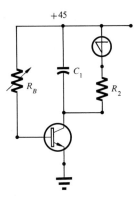

FIG. P15·32

15·33 A *pnpn* multivibrator is shown in Fig. P15·33. Neglecting switching times in the *pnpn* devices, compute and sketch the voltage waveform across each *pnpn* diode (the conventional series diodes are added to prevent reverse breakdown of the *pnpn* diodes).

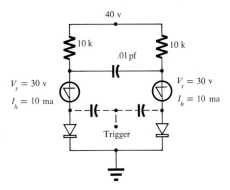

FIG. P15·33

15·34 Explain the action of the circuit of Fig. P15·33 if the supply voltage were lowered to 25 volts and negative trigger pulses of 10 volts magnitude were applied to the trigger point. Assume one side is ON in beginning your explanation.

15·35 It is sometimes necessary to build a circuit with a very long time delay (of minutes). The circuit shown in Fig. P15·35 can be used for this purpose quite

Switching circuits and switching devices 677

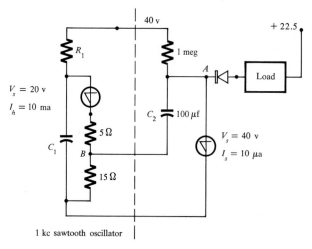

FIG. P15·35

nicely (to obtain up to 30 minutes time delay). R_1 and C_1 are adjusted to produce a 1 kc oscillator, giving pulses at B at a 1 kcps repetition rate. The main diode has a 40-volt V_s, but it cannot be turned ON from the 40-volt supply because a large charging resistor (1 megohm) is required to produce a long charging time for the 10 μfarad capacitor. The leakage current of the *pnpn* diode is such that, without the 1-kc sawtooth oscillator, the voltage at A can only rise to about 30 volts. However, the 1-kc sawtooth circuit produces pulses which can effect a turn-ON of the diode at A, thus switching the 22.5-volt supply across its load.

1. Select R_1 and C_1 to produce the 1-kc oscillator.

2. Compute and dimension the voltage at A in the presence of the 1-kc sawtooth oscillator. Assume the main diode acts as a constant 10-μa drain during the charging of C_2.

3. At what time do you expect the main diode to turn ON?

4. If it fails to turn ON at that time, when will the next turn-ON opportunity be?

15·36 Consider the circuit of Fig. 15·41 in the text. Assume that a capacitor of 1,000 pfarad is adequate with a 20-volt *pnpn* diode to turn-ON the controlled rectifier.

1. What maximum value should the potentiometer have to ensure that the circuit fires on every half-cycle?

2. What value of resistance is required to supply the voltage to the load shown in Fig. 15·39?

3. If the potentiometer is completely out of the circuit (zero resistance), the circuit will fire when the input voltage reaches 20 volts. If the input voltage is a 115-volt rms sine wave, how much power is fed to a resistive load in this condition, compared to a true full wave of power?

16

COMMUNICATION SYSTEM PHILOSOPHY

A COMMUNICATION SYSTEM is designed to provide a means of sending information from one point to another. As we pointed out in the introduction,[1] a typical electronics communication system will have a transducer at the input to convert the information-bearing signal into an equivalent electrical signal. The electrical signals are then sent to the desired destination and reconstructed by means of an output transducer so that an observer can understand the message.

If we wish to communicate over long distances, such as in radio, television, long-distance telephony, and so on, then it proves to be advan-

---

[1] A rereading of the introduction and Secs. 7·3 and 7·6 might be helpful before beginning this chapter.

tageous for several reasons to transmit the information-bearing signals on a carrier. The resulting communication system is called a *carrier transmission system* and is one of the basic tools of the communication systems engineer.

In this chapter we will examine the philosophy and develop the basic properties of some of the more common carrier transmission systems. From this discussion we will deduce performance requirements for various subsystems that are needed in carrier transmission systems. These subsystem requirements will be of considerable use to us in the analysis and design work which form the subject matter of the final chapters of this book.

In addition, in the present chapter we shall define more precisely what we mean by an information-bearing signal and expand on the relation between information-bearing signals and their Fourier or sine-wave representations. We shall also show that signal processing in a carrier transmission system can very profitably be studied by considering the changes in the frequency spectrum of an information-bearing signal as it is transmitted through the system.

16 · 1   INFORMATION-BEARING SIGNALS

In the introduction to this book we described very briefly the way in which information is transmitted and received in an AM radio broadcast. It will be convenient to review this process here, using the signals which a particular radio station broadcasts to fix some basic ideas about information-bearing signals and how we interpret them.

The station we will consider is WWV. This station is operated by the National Bureau of Standards for the purpose of maintaining a set of frequency and time standards which are generally available at any point in the United States. The station transmits standard time intervals, time announcements, and standard audio frequencies on a series of standard radio-frequency carriers.

One of these carrier frequencies is 10 mcps, which is maintained to within at least two parts in 10^{11} at all times. This carrier is amplitude-modulated by either a 440-cps or a 600-cps tone. The audio frequencies are given alternately, starting precisely on the hour with 600 cps. After 3 min, the 600-cps tone is interrupted for 2 min. The 440-cps tone is then transmitted for 3 min and interrupted for 2 min. Each 10-min period is the same.

In addition, a pulse or tick, which sounds very much like a pendulum clock, is transmitted at intervals of precisely 1 sec and a voice announcement of U.S. Eastern Standard Time is given each 5 min.

A block diagram illustrating how the radiated signal may in principle be constructed at the transmitter is given in Fig. 16 · 1. Here a modulator[1] is

[1] *cf.* Sec. 7 · 6.

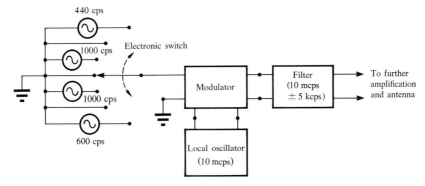

(a) *Partial block diagram of transmitter*

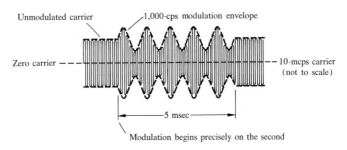

(b) *Radiated signal giving one-second ticks during periods of silence*

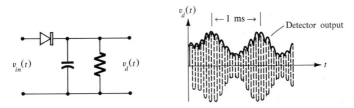

(c) *Output of envelope detector during the period when the input voltage has a sine-wave modulation envelope*

FIG. 16·1 SCHEMATIC ILLUSTRATION OF *WWV* SIGNAL CONSTRUCTION AND DETECTION

supplied with a 10-mcps carrier-frequency signal and a series of signals which provide the audio tones and the one second ticks.

The manner in which the audio tones and 1-sec ticks are fed to the modulator deserves careful attention, since it gives considerable insight into the meaning of the term "information-bearing signal" and the requirements of a system intended to transmit sounds.

Let us assume that we tune to the 10-mcps WWV carrier during one of the 2-min periods in which no audio tone is being transmitted. During most of this time, the audio input to the modulator is zero, the modulator output being a fixed-amplitude 10-mcps sine wave. The modulator output is amplified and can readily be transmitted by driving a simple vertical antenna whose height is a quarter of the free-space wavelength of a 10-mcps sine wave (that is, 7.5 m).

The 1-sec ticks which are transmitted during the 2-min period of silence are obtained by electronically switching one pair of the modulator-input terminals to a 1,000-cps tone generator. The switch is operated precisely on the second and terminated 5 msec after the precise second, as shown in Fig. 16 · 1b. The radiated signal now consists of a 10-mcps carrier whose amplitude or envelope now varies in accordance with the 1-kcps tone for a period of 5 msec. The 10-mcps carrier is said to be *amplitude modulated* by the 1-kcps signal. Notice carefully that this represents a change in the transmitted signal.

At the receiver the radiated signal is picked up, amplified and then fed to the envelope detector shown in Fig. 16 · 1c,[1] where the amplitude variations of the carrier are recovered. The ac component of the detector output is an approximate reproduction of the 5-msec burst of the 1,000-cps tone which may be amplified and fed to a loudspeaker.

An observer listening to the radio receiver hears a soft tick, sounding very much like the tick of a pendulum clock, when the 5-msec tone burst is applied to the speaker. There are two points that are important about this observation:

1. The observer does *not* identify the tone as a 1,000-cps tone even with the best acoustical reproduction system.

2. The observer does identify a change in the received signal, however, and *it is this change in received signal which conveys information to him.* Specifically it conveys the fact that 1 sec has elapsed since the last tick.

Let us now assume that at the end of the 2-min period, which is silent except for the 1-sec ticks, a 440-cps tone comes on. At the modulator, this is accomplished by electronically switching the modulator audio input to a 440-cps-tone generator. The connection is made to the 440-cps generator exactly on the second. The radiated signal now consists of a 10-mcps carrier, amplitude-modulated by a 440-cps sine wave. This modulation continues for 1 sec − 10 msec at which time the modulator audio input is switched off for 10 msec. Precisely on the second, the modulator audio input is switched to the 1,000-cps-tone generator for 5 msec to produce the tick, then back off for 25 msec, and then back to the 440-cps-tone generator until 10 msec before the next second. This sequence of switching operations is then repeated for 3 min, after which another 2-min period of silence with 1-sec ticks begins.

During the time that this sequence of signals is being transmitted, the following signal is heard from the loudspeaker at the receiver output:

1. An apparently *continuous* tone which we can readily identify as a 440-cps tone (by comparing the received tone with that generated by a tuning fork, if necessary). This tone appears to last for a full period of 3 min, though it is actually interrupted to produce the 1-sec ticks.

[1] *cf.* Sec. 7 · 6.

2. One-sec ticks which are apparently superimposed on the 440-cps tone.

Note that the beginning of the 440-cps modulation again marks a change in the transmitted and received signal, and again this change transmits information to us. In particular, it signifies that the time is now an odd multiple of 5 min after the last hour.[1]

There are two main features of these signals we receive from WWV that need to be emphasized:

1. Information is transmitted by an *unexpected change* in the signal. This idea is basic to the definition of an information-bearing signal. If we could predict the exact behavior of the signal for the entire future, it would transmit no new information to us. It could only verify something we already know.[2]

2. The ear responds to sine-wave bursts of different durations in different ways. We can identify a tone burst as an "apparently pure sine wave" if it lasts long enough. However, there is some minimum time which is necessary to identify the pitch of a tone. This is a fact of great importance in communication systems intended to transmit speech and music, as we shall see in the succeeding paragraphs.

16 · 2 FOURIER ANALYSIS OF INFORMATION-BEARING SIGNALS

In the preceding section we have attempted to describe some relatively simple transmitted and received signals entirely in terms of their variation with time. This provides a correct view of the overall signal processing which occurs in an AM transmission and reception system—one on which many ingenious and important patents have been conceived. However, the terms which we used to describe the signal carry implications which do not stand up very well under close inspection. In particular, we talked about an amplitude-modulated carrier. For the most part we avoided calling this an amplitude-modulated *sine wave,* though this is the picture which the discussion (and Fig. 16 · 1) conjures up. We also talked about identifiable tones as being sine-wave signals of sufficiently long duration.

Now, in the interests of precision in the description of information-bearing signals and signal-processing, we have to recognize that there is no such thing as an amplitude-modulated sine wave. If a waveform is to be a sine wave, then it must have a constant amplitude and have a sinusoidal variation with time which extends from the infinite past to the infinite future.

[1] Actually, there is also a voice announcement of the time, which gives the hour and the minute at which the 440-cps tone will begin.

[2] In the language of information theory, the WWV signal does not communicate very much information to us, because the changes which occur in the signal are highly predictable. For example, we can synchronize a reasonably accurate wrist watch with the WWV signal and then predict the future variations in the received signal (audio tones and 1-sec ticks) for several hours in advance to within an accuracy of a quarter of a second. Or, we can make considerably better predictions with a more accurate frequency standard.

While this means that no one has ever seen, heard, or generated a true sine wave, we can nevertheless justifiably use sine waves for the analysis and design of systems. We can also be quite sure that a well-made sine-wave generator produces an essentially pure sine wave and that sine-wave testing means what it appears to mean.

We can understand the basis for these important facts by a repeated and judicious use of the idea that a *sine-wave burst can be visualized as a true sine wave multiplied by a pulse which is zero for all times t except those at which we wish the sine wave to appear.* The pulse is said to "modulate the amplitude" of the sine wave.

In the WWV transmission we saw that modulation of the carrier amplitude was necessary to obtain transmission of information. We will see later that the frequency of a carrier can also be modulated, so we do not want to think of the modulation process too narrowly. However, for the next several paragraphs we will concentrate on amplitude modulation.

We will begin by analyzing the effects of modulating a sine wave with the modulating function shown in Fig. 16·2a. We shall call this modulating function the *basic modulating pulse*. We will concentrate primarily on a mathematical analysis at the outset, giving only enough physical interpretation to suggest reasons for some of the steps we shall take. An elaboration on physical details will be given when some basic mathematical facts are established.

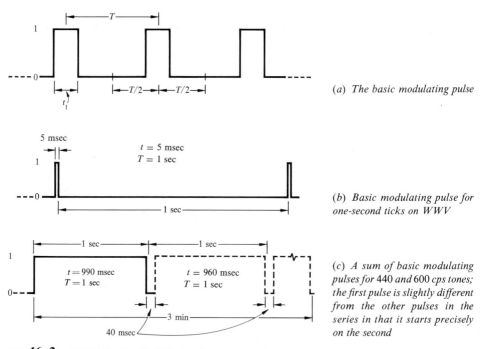

(a) The basic modulating pulse

(b) Basic modulating pulse for one-second ticks on WWV

(c) A sum of basic modulating pulses for 440 and 600 cps tones; the first pulse is slightly different from the other pulses in the series in that it starts precisely on the second

FIG. 16·2   THE BASIC MODULATING PULSE AND SOME EXAMPLES OF ITS APPLICATION

16·2·1 *Fourier analysis of the basic modulating pulse.* The basic modulating pulse has a width t_1 and a period T. Both t_1 and T are left unspecified for generality. The height of the modulating pulse will be assumed to be unity. The utility of the basic modulating pulse in considering actual waveforms in communication systems is suggested in Figs. 16·2b and 16·2c. Figure 16·2b shows the basic pulse with the dimensions required to modulate a 1,000-cps sine wave to produce the 1-sec ticks heard over WWV. Figure 16·2c shows a series of basic pulses with dimensions required for modulating a 440-cps sine wave to produce the 440-cps tone heard over WWV for 3 min.

To Fourier analyze the basic modulating pulse we set the time origin in the center of the pulse and use as the basic period the time interval $-T/2$ to $T/2$, as suggested in the central pulse shown in Fig. 16·2a. The function is then an even function of time and can be represented by a series of cosines:

$$f(t) = \Sigma a_n \cos (2n\pi f_p t) \tag{16·1}$$

where $f_p = 1/T$ is the pulse-repetition frequency of the basic modulating pulse. The Fourier coefficients for $f(t)$ can be found by the standard techniques. They are

$$a_0 = t_1/T \tag{16·2a}$$

$$a_n = \frac{2t_1}{T} \frac{\sin(\pi n t_1/T)}{\pi n t_1/T} \qquad n > 0 \tag{16·2b}$$

The appearance of the $(\sin x)/x$ function in Eq. (16·2b) will be of considerable use to us in future discussions. In particular, the following properties will be important:

1. The zeros of the function occur when

$$\frac{n t_1}{T} = \frac{m}{2} \qquad m = 1, 3, 5, 7, \ldots \tag{16·3}$$

where n is the order of the harmonic.

2. The spacing between harmonics is $1/T$ on the frequency scale. Note that *this is independent of the value of t_1.*

3. The zeros of the function occur on the frequency scale at the points

$$f = \frac{m}{2t_1} \qquad m = 1, 3, 5, 7 \tag{16·4}$$

Note that *the positions of the zeros on the frequency scale are independent of the value of T, and depend on the reciprocal of t_1.*

These properties are illustrated in Fig. 16·3 for the two specific cases of a pulse which has a duration of 1 sec and occurs once every 10 min (roughly the basic modulating function for the 440-cps tone on WWV) and a pulse with duration of 5 msec occuring once every second (the basic 1-sec ticks on WWV).

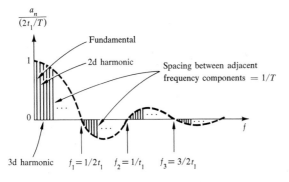

(a) *General form of Fourier components* a_n *for n > 0. Each vertical line represents the amplitude of the appropriate harmonic in the Fourier expansion of the basic modulating pulse.*

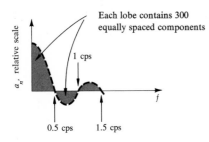

(b) *Fourier spectrum for a basic modulating pulse with* $t_1 = 1$ *sec,* $T = 10$ *min* (a_n *for n > 0*).

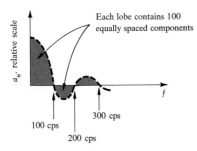

(c) *Fourier components for a basic modulating pulse with* $t_1 = 5ms$, $T = 1$ *sec* (a_n *for n > 0*).

FIG. 16·3   FOURIER SPECTRUM FOR THE BASIC MODULATION PULSE

Energy content in the Fourier components. We can relate the energy content of a waveform to the coefficients in its Fourier expansion by the usual rule of adding the squares of the ac components together and dividing by 2:

$$\text{Energy content} \sim a_0{}^2 + \frac{1}{2} \sum_{n=1}^{\infty} a_n{}^2$$

This step is justified by visualizing the waveform to be the voltage drop across a 1-ohm resistor.

For the basic modulating pulse, we can simply square the $(\sin x)/x$ envelope of the Fourier components to highlight the frequency range in which the majority of the signal energy is located. The sketch given in Fig. 16·4 shows that the great majority of the energy is contained in frequency components which lie below the first zero of the $(\sin x)/x$ function. This fact will be of considerable use to us in studying the properties of signals.

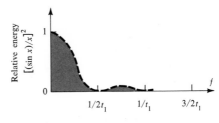

16·2·2 *Modulation of a sine wave with the basic modulating pulse.* Let us now analyze the effect of modulating a sine wave with frequency ω_0 with the basic modulating pulse. Mathematically, the modulated sine wave can be written in the form

$$g(t) = f(t) \sin \omega_0 t \qquad (16·5)$$

where $f(t)$ is the basic modulating function. Using Eq. (16·1), $g(t)$ may be rewritten as

$$g(t) = \sum_{n=0}^{\infty} a_n \cos n\omega_p t \sin \omega_0 t \qquad (16·6)$$

We can find the Fourier components of the modulated sine wave by considering the first two terms in the expansion of Eq. (16·6):

$$g(t) = a_0 \sin \omega_0 t + a_1 \cos \omega_p t \sin \omega_0 t + \cdots \qquad (16·7)$$

The first term in Eq. (16·7) is simply a sine wave at the frequency ω_0, having an amplitude of a_0. The second term can be converted into two sine waves by using the trigonometric identity

$$\sin (a + b) = \sin a \cos b + \cos a \sin b$$

The result is

$$a_1 \cos \omega_p t \sin \omega_0 t = \frac{a_1}{2} [\sin (\omega_0 + \omega_p)t + \sin (\omega_0 - \omega_p)t] \qquad (16·8)$$

Thus, the dc component in the Fourier expansion of the basic pulse gives rise to a sine wave at frequency ω_0 in the expansion of $g(t)$, while the first harmonic in the Fourier expansion of the basic pulse gives rise to *two* sine waves in the Fourier expansion of $g(t)$. These sine waves lie at the frequencies $\omega_0 \pm \omega_p$ and have an amplitude of $a_1/2$. In general Eq. (16·8) can be used to show that the nth harmonic in the Fourier expansion of the basic modulating pulse will be transformed into two sine waves at the frequencies $\omega_0 \pm n\omega_p$, having amplitudes $a_n/2$.

The Fourier spectrum of $g(t)$ which we obtain from these considerations is shown in Fig. 16·5. This spectrum is developed from that of the basic modulating pulse $f(t)$ by simply dividing the amplitude coefficients of $f(t)$

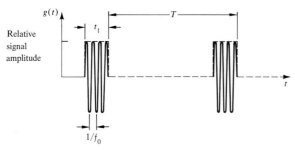

(a) Time waveform obtained by multiplying $\sin 2\pi f_0 t$ *by the basic modulating pulse*

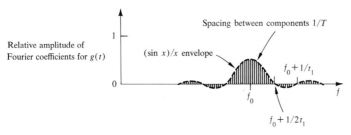

(b) Fourier spectrum of the time signal g(t)

FIG. 16·5   RESULTS OF MODULATING A SINE WAVE WITH THE BASIC MODULATION PULSE

by two (for $n > 0$) and then reproducing the spectrum of $f(t)$ on either side of the frequency f_0.[1]

Note carefully that *the modulation of a sine wave at frequency f_0 with the basic modulating pulse $f(t)$ can be viewed as a frequency-shifting operation* in which the basic spectrum of $f(t)$ is shifted to a new position on the frequency scale.

16·2·3   *Physical interpretation and applications.*

The theory just presented can readily be generalized in a way which will enable us to understand the signal processing that occurs in an AM transmitting and receiving system. However, we will defer this discussion to the next section in order to discuss some important physical implications of the results just obtained.

---

[1] Alternately, we could visualize the construction of Fig. 16·5b by first dividing the coefficients given in Fig. 16·3a by 2, reflecting this spectrum about the origin, and then shifting the entire spectrum to a center frequency f_0. Or, by including positive and negative frequencies in the Fourier representation (that is, using the $-j\omega$ axis in the complex frequency plane), the Fourier series can be written as

$$f(t) = \sum_{n=-\infty}^{\infty} C_n e^{j(2\pi nt_1/T)}$$

where

$$C_n = \frac{t_1}{T} \frac{\sin(\pi nt_1/T)}{\pi nt_1/T}$$

In this case, $f(t)$ is already seen to be a $(\sin x)/x$ function centered at $n = 0$ (dc), with positive and negative frequency components.

Sine-wave generators. We mentioned earlier that it is impossible to generate a real sine wave because it must exist from the infinite past to the infinite future. However, we can buy test instruments that are called sine-wave generators, and use them for sine-wave testing. The preceding results show why this is possible.

To understand this, let us suppose that we turn on a sine-wave generator at $t = 0$. The frequency dial on the generator is set at 1,000 cps. Then we can view this process as using a basic modulation pulse whose width t_1 increases with time to modulate a true 1,000-cps sine wave.[1]

Now the positions of the zeros of the $(\sin x)/x$ curve on the frequency scale depend on how long the generator has been on. For example, after 1 msec, the first zeros will occur at ± 500 cps. This means that the energy output of the generator can be concentrated in a frequency range of *no less than* about $1,000 \pm 500$ cps after 1 msec of operation. When the generator has been on for 10 msec, the zeros of the $(\sin x)/x$ have moved in to ± 50 cps, and when the generator has been on for 1 sec, the zeros of the $(\sin x)/x$ will have moved in to ± 0.5 cps.

After the generator has been allowed to run for 10 min, its energy output can be concentrated in a frequency range of about $1,000 \pm 0.0008$ cps. In other words, it takes 10 min for the oscillator frequency to stabilize to 1 part in 10^6. This is a *theoretical* limit, established by the fact that transients associated with turning the generator on initially require many frequency components to represent the discontinuity in the generator output waveform.

As a practical matter, the inherent frequency stability of the test oscillator may not be as good as 1 part in 10^6; and even if it were, we would probably be content to assume that the generator output was a very good approximation to a sine wave after 10 min of warm-up time, because in our observations of circuit response (or whatever else we are testing with the generator) we would probably not be able to measure the difference between the actual circuit output and the theoretical output of the circuit if it were driven from a mathematically pure sine wave.

Physical justification of Fourier analysis. A critical reader may reasonably object to the preceding example by saying we are essentially using Fourier analysis techniques to justify the use of Fourier methods. This objection is well taken and can only be justified in the following way.

A Fourier analysis of any real signal provides us with nothing more than a *mathematical model* of it. The model uses a series of waveforms which have only a mathematical reality to represent a real, physically observable waveform. The validity of the mathematical model rests on the fact that *the observer is an important part of the overall system.*[2] If there is no measur-

---

[1] We do not care what the repetition period T of the basic modulating pulse is, since it will not enter the argument. We can assume that it is one year, or the time since the generator was last used, etc.

[2] There is a close relationship between this particular problem and the basic quantum-mechanical question of the effect of the observer on the system.

able difference between the response he calculates for a system, using a series of pure sine waves, and that which he obtains experimentally, using real signal generators (perhaps a large number of them), then a "Fourier model" for the real signal is valid. The argument which we use here is conceptually the same as the one which we use when we substitute a model of a transistor for the real device. It is of course the basis on which any mathematical model for any real process rests.

An analysis of pitch determination.[1] Naturally, the point at which differences between the measured response of a system and the calculated response of its model become insignificant depend on the accuracy required by the analyst. Of more importance, however, is the fact that they also depend ultimately on the measuring instruments. The response of the ear to the tone bursts heard on WWV provides an interesting example of this.

The 1-sec ticks which we receive can be visualized as a 1,000-cps sine wave modulated by a 5-msec pulse occuring once every second. The first zero of the $(\sin x)/x$ function for this basic modulating pulse occurs at a frequency of 100 cps. The sound energy received by the ear is therefore distributed over a frequency range of at least $1,000 \pm 100$ cps, a range which is too broad for the ear to clearly perceive a pitch.

On the other hand, the 440-cps tone is on for one second. After the first 20 msec of this tone, the sound energy received by the ear is distributed over a frequency range of about 440 ± 25 cps, which is sufficient for the ear to perceive a reasonably accurate pitch.

Of course, the tone continues for nearly 1 sec, so the energy is ultimately distributed over a frequency range of 440 ± 0.5 cps. However, the ear is not sensitive to such a fine difference in pitch. The best resolution is about ± 3 cps, which is reached after about the first 200 msec of the pulse.

Stated somewhat differently, the ear cannot distinguish the difference in pitch between a sine-wave tone burst which is on for 200 msec and one which is on for a minute, an hour, a day, a week, ... ($\rightarrow$ a pure sine wave). Therefore, tone bursts of 200 msec or longer can be analyzed as if they were mathematically pure sine waves, as far as the ear is concerned.

Fourier analysis of speech and music. The idea that a minimum time is necessary for an observing instrument to "perceive" a sine wave, or a generating instrument to generate one (or several simultaneously) also provides the physical justification for the Fourier analysis of speech and music, and an interesting insight into musical composition and performance.

On an organ, for example, each note of a trill in the upper register of the instrument lasts on the order of 30 to 40 milliseconds, due to the limitations of the organist in moving his fingers. The sound-pressure waveforms created by these trills will not be sine waves, though each note in the trill

---

[1] The data presented in this section were obtained using the simple tone-burst generator described in Demonstration 16 · 1. The demonstration was inspired by experiments described in D. Gabor, Theory of communication, Inst. of Electrical Engineers Journal, vol. 93, pp. 429ff, 1946, and the results essentially duplicate those reported in his paper.

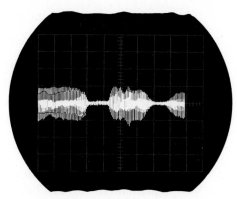

will appear to be a sine wave, because there is adequate time for the ear to clearly distinguish the pitch of the note. Trills or runs in the lower register of the organ must be played more slowly to be clearly discernible.

Speech signals also have a minimum time duration which is more than long enough for the ear to perceive sine-wave components in the speech signals. An example is shown in Fig. 16·6, where an oscilloscope trace of a 1-sec segment of a news broadcast (obtained from the audio output of an AM receiver) is reproduced. The central wave pattern in this figure is the word "why." Notice that the word lasts for several hundred milliseconds, which is adequate time for each sine-wave component in the Fourier series needed to build up this signal to be resolved to less than ±2 cps.

16·3   SIGNAL PROCESSING IN SYSTEMS EMPLOYING

AN AMPLITUDE-MODULATED CARRIER

Let us now turn to a more detailed examination of some practical carrier transmission systems which employ an amplitude-modulated carrier to transmit information from one point to another. We will accept as a postulate that these systems must be capable of faithfully reproducing any sine wave in the frequency range from 0–5 kcps. This frequency range is typical of that used for AM broadcasting and telephony. The ear can, of course, distinguish sounds up to frequencies of 15 kcps or more, so the arbitrary 5-kcps frequency limit will not result in a particularly "high-fidelity" sound transmission. However, it is adequate for reproduction of speech and most music.

16·3·1   *AM broadcasting.*   Figure 16·7 shows a general modulating signal $f_s(t)$ and the carrier envelope corresponding to it. The amplitude-modulated carrier can be written in the form

$$f_c(t) = A[1 + mf_s(t)] \sin \omega_c t \qquad (16 \cdot 9)$$

We will consider two examples of signals $f_s(t)$:

1. $f_s(t) = \sin \omega_s t$ ($\omega_s \ll \omega_c$; otherwise, an envelope cannot be simply described). Then

$$f_c(t) = A[1 + m \sin \omega_s t] \sin \omega_c t$$
$$= A \sin \omega_c t + \frac{A}{2} m \cos (\omega_c - \omega_s)t - \frac{A}{2} m \cos (\omega_c + \omega_s)t \qquad (16 \cdot 10)$$

using trigonometric identities given earlier. We assume $m \leq 1$.

Equation $(16 \cdot 10)$ can be represented by the frequency spectrum given in Fig. $16 \cdot 8$, where the original frequency f_s has been shifted up in frequency and split into two side frequencies, one on each side of the carrier. m is called the modulation index and determines the fractional change in carrier amplitude that will be produced. For undistorted transmission $0 < m < 1$, and as a practical matter m is usually limited to 0.8 in an AM broadcast.

In the frequency spectrum of the modulated carrier, the amplitude of the side frequencies is $m/2$ times the amplitude of the carrier. Notice that amplitude-modulating the carrier has not changed the amplitude of the carrier frequency component in the frequency spectrum. This means that the energy which appears in the side bands comes from the modulating source.

2. $f_s(t)$ is a function of time that can be developed in a Fourier series using sine waves lying in the 0–5 kcps frequency band. In this case we first analyze the message into its appropriate Fourier components and then express $f_s(t)$ as a Fourier series (within the limits described earlier). A hypothetical example of the result is shown in Fig. $16 \cdot 8b$, where the original frequency spectrum has been shifted in frequency as before.

A *lower sideband* and an *upper sideband* have been defined on Fig. $16 \cdot 8b$, these being the frequency ranges to which the original spectrum has been shifted. Notice that the information transmitted is carried in the sidebands and that there is enough information in *either* sideband (that is, enough

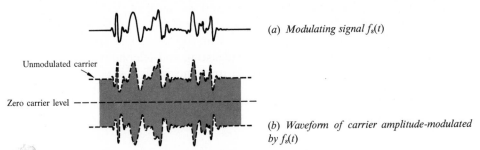

(a) *Modulating signal* $f_s(t)$

Unmodulated carrier

Zero carrier level

(b) *Waveform of carrier amplitude-modulated by* $f_s(t)$

FIG. 16 · 7 A GENERAL MODULATING SIGNAL $f_s(t)$ AND THE CORRESPONDING AMPLITUDE-MODULATED CARRIER

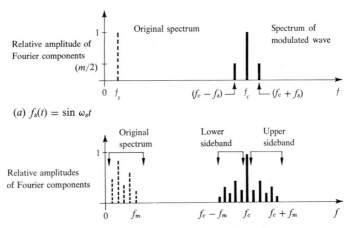

(a) $f_s(t) = \sin \omega_s t$

(b) $f_s(t)$ *represented by a series of sine waves lying below some maximum frequency* f_m *(0–5 kcps for AM broadcast)*

FIG. 16·8 **FREQUENCY SHIFT EFFECT OF AM (ONLY POSITIVE FREQUENCIES ARE SHOWN HERE)**

Fourier components) to reconstruct the original message. We will describe systems which take advantage of this fact later.

By requiring AM broadcast stations to limit their signal inputs to the frequency range 0–5 kcps, the Federal Communications Commission can establish several radio stations in any given area by assigning different carrier frequencies to each station, spacing the carrier frequencies so that the frequency spectra of the transmitted signals do not overlap.

Some basic types of modulators. Practical modulators take on several physical forms depending on the desired output waveform. In the cases to be described here, the modulator exhibits nonlinearity to effect the desired frequency shifting.[1]

The importance of nonlinearity in modulators was discussed in Chap. 7, and a practical diode modulator which finds application in carrier telephony systems was described there.

Another common method of accomplishing modulation is to use the modulating signal to vary the collector (or plate) voltage of a class B or class C amplifier, as shown in Fig. 16·9. Since the peak amplitude of the signal voltage is very nearly equal to the supply voltage in these cases, the amplitude of the carrier output voltage follows the variations in voltage supplied by the modulator quite closely (*cf.* Prob. 16·12).

A third useful type of modulator is the *balanced modulator* shown in Fig. 16·10.[2] Here, the carrier is applied in phase to the two transistor

---

[1] Time-varying *linear* modulators can also produce frequency *shifting;* for example, the gain of an amplifier can be varied as $A_0 + A_1 \sin \omega_c t$. If the input to such an amplifier is a sine wave at frequency ω_s, its output will contain components at $\omega_c \pm \omega_s$.

[2] Balanced modulators can also be made with diodes. *cf.* H. J. Zimmermann and S. J. Mason, "Electronic Circuit Theory," John Wiley & Sons, Inc., New York, 1962, pp. 142ff.

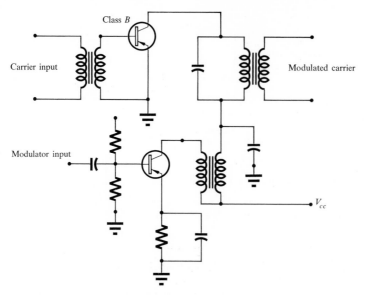

FIG. **16·9**   CIRCUIT FOR COLLECTOR MODULATION

inputs, while the modulating signal is applied in opposite phase to the two inputs. The transistors are operated in a class B mode so that modulation can take place in each transistor.

If the circuit is perfectly symmetrical, the output will contain only the sidebands (*cf.* Prob. 16 · 14). The carrier can be reinserted in the output if desired, though there are some communication facilities to be described later where this is not done.

Carrier frequency amplifiers. Usually, the output of a modulator will contain many signals in addition to the desired one. Some type of filtering is then necessary to insure that only the desired signal is transmitted.

The transmission characteristics for a suitable filtering network are very simply visualized in the frequency plane, as shown in Fig. 16 · 11. Ideally, for a standard AM broadcast we want to transmit all frequency components in the range $f_c \pm 5$ kcps and reject all others.

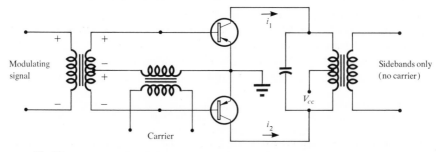

FIG. **16·10**   A BALANCED MODULATOR CIRCUIT

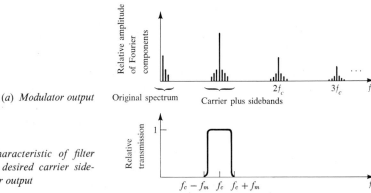

(a) Modulator output — Original spectrum — Carrier plus sidebands

(b) Transmission characteristic of filter designed to extract desired carrier sidebands from modulator output

FIG. 16·11   TYPICAL OUTPUT OF A MODULATOR AND THE NEED FOR FILTERING

In simple cases, an *LC* network designed to give the desired transmission characteristic can be placed directly between the modulator and the antenna, though usually some amplification of the modulator output is necessary, so a *band-pass amplifier* will be built to select and amplify the desired signal. The band-pass amplifier must provide an amplification characteristic which is uniform over the frequency band $f_c \pm 5$ kcps and then drops off rapidly for frequencies outside this range.

Band-pass amplifiers are also called *filter amplifiers* or *carrier-frequency amplifiers* and find wide application in carrier communication systems. We shall discuss the design of these amplifiers in Chap. 18.

16·3·2   Receivers for amplitude-modulated waves.   In the Introduction we showed two block diagrams illustrating the manner in which an AM broadcast signal could be detected. We now wish to expand on this description, showing in particular how changes in the frequency spectrum of the signal occur in the receiver.

Detection.   Perhaps the basic function of an AM receiver is that of *detection.* We have already described the envelope-detection process and discussed the basic envelope detector (a half-wave rectifier with an appropriate *CR* load) in Sec. 7·6·1. We found that the design of such a detector could be based on a knowledge of the modulation index and the audio frequencies which we wish to reproduce.

It is apparent from this discussion that a satisfactory detector can be designed by visualizing its function as that of recovering the time-varying envelope of the received waveform. However, we can also give an interesting interpretation of the detection process in terms of the changes which must occur in the frequency spectrum of the signal.

Basically, the job the detector must perform is to shift the Fourier components of the information-bearing signal from the frequency range $f_c \pm 5$ kcps back to the audio range (0 to 5 kcps). As we pointed out in Chap. 7, frequency shifting can be accomplished by the appropriate use of a nonlinear element.[1] When such an element is driven with two sine waves, its output contains not only the two sine waves but, in principle, all harmonics of each sine wave and sine waves at all the possible difference frequencies.

This suggests that passing the received signal through a diode will accomplish the required frequency shifting, and we can then use a low-pass filter to pass only the audio-frequency components of the diode output. This scheme can be simply put into practice in the elementary receiver shown in Fig. 16·12. Here, a receiving antenna (a few feet of wire in simple cases) is simply transformer-coupled to a diode. A tuning capacitor at the input is used to select the desired station, and a parallel CR network provides the filtering function. An AM detector which uses the nonlinearity inherent in a class B amplifier is the subject of Prob. 16·17.

Gain and selectivity. While detection can be regarded as the basic signal-processing step in an AM receiver, the elementary receiver shown in Fig. 16·12 needs to be surrounded by other signal-processing blocks if the receiver is to be more than a toy. In particular, the wide usage of radio communications equipment is based on the fact that considerable gain and selectivity can be obtained in a radio receiver.

Gain is required because the direct detection of a received signal provides far too little power to drive a loudspeaker or even an efficient pair of ear-phones, if the receiver is far from the transmitter. Generally we want to provide a voltage of at least 0.5 to 1 volt to the voice coil of an 8-ohm speaker, whereas the received signal may be only a microvolt or so. This means that we need a voltage gain of roughly 10^5 to 10^6 in the receiver.

---

[1] A linear time-varying element (or amplifier) can also perform the frequency-shifting function.

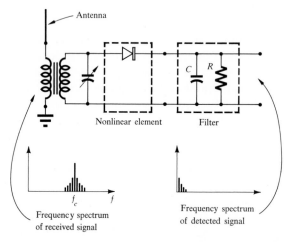

FIG. **16·12** AN ELEMENTARY AM RECEIVER ILLUSTRATING THE BASIC DETECTION PROCESS

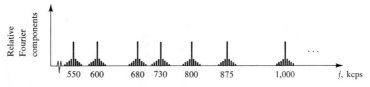

(a) *Signals are received by the antenna from all broadcasting stations*

(b) *Transmission characteristic of filter adjusted to select 680 kcps carrier and sidebands*

FIG. 16 · 13   **AM BROADCAST SIGNALS AND CORRESPONDING FILTERING CHARACTERISTICS REQUIRED IN AN AM RECEIVER**

We would of course like more, though noise voltages will set a limit on sensitivity at some point.

Selectivity is required to insure that signals from adjacent stations are not received simultaneously. In order to obtain selectivity, we have to incorporate some type of filtering into the receiver which will allow the desired signal to pass and will reject all others. In Fig. 16 · 13b the transmission characteristic of a desirable filter is plotted as a function of frequency. The filter is positioned to pass the Fourier components of signals arriving at the 680-kcps carrier frequency and reject all other signals.

We can build a reasonable approximation to such a filter by a repeated use of the idea that the output voltage of an *LC* tank circuit will have a band-pass characteristic when it is driven from a current source. However, as suggested in Fig. 16 · 14a, the filter characteristic will not have the required shape.

To improve the shape, we can cascade a number of these networks, using transistors between stages in the manner shown in Fig. 16 · 14b. By proper adjustment of the resonance peaks in the individual circuits, we can obtain the required selectivity and in the process obtain some of the gain required in the receiver.

Unfortunately, since we wish to receive any one of several stations, we have the problem of tuning the filter so that its center frequency can be placed in any desired position of the broadcast band. To do this, we have to tune each resonant circuit, making sure that their relative positions on the frequency scale are always correctly maintained. It is possible to do this, and some receivers employ only this type of filtering prior to detection. These receivers are called *tuned radio frequency* receivers.

However, a more widely used solution to the problem of making a tunable filter (with gain) is to use the superheterodyne principle invented by E. H. Armstrong, shown in the block diagram of Fig. 16 · 15a. Here we

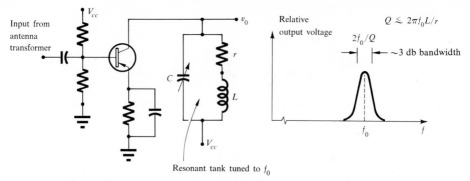

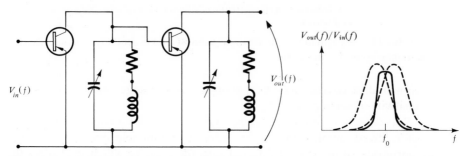

(a) Use of a single tuned circuit to approximate desired filter characteristic

(b) Use of a cascade of appropriate single tuned circuits to approximate the desired filter characteristic. Capacitors are tuned to select desired station.

FIG. 16·14 DEVELOPMENT OF APPROXIMATIONS TO THE FILTER CHARACTERISTIC REQUIRED IN AN **AM** RADIO

build a local oscillator which is always tuned to a frequency 455 kcps higher than the carrier frequency of the station we wish to receive. The local oscillator is tuned at the same time that the station is "tuned in."

The received signal and that from the local oscillator are then fed to a nonlinear element (either a diode or a transistor operated over a large portion of its active region), which, as always, produces harmonics and sum and difference frequencies of all signals present. In particular, the modulator output for a carrier f_c has *the original side bands reproduced* about 455 kcps, $f_c - 455$ kcps, f_c, $f_c + 455$ kcps, etc.

Now, a cascade of stages with filters which pass only 455 ± 5 kcps can be built to select and amplify the signal at 455 ± 5 kcps. Of course, because local oscillator tuning and station tuning are accomplished at the same time, the filter amplifier operates at 455 ± 5 kcps for all stations and need not be tuned.

In effect, the superheterodyne principle simply shifts the information-bearing signal to a new carrier frequency, this new carrier frequency being in the center of the pass band of a fixed-tuned filter. The "internal carrier frequency" in the radio is called an *intermediate frequency;* the cascade of stages which provides the selectivity function (and most of the required

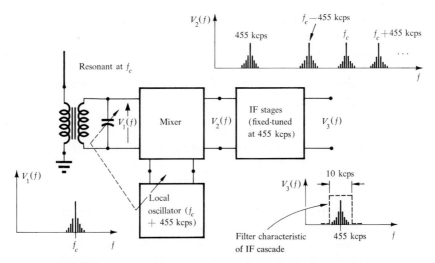

Resonant at f_c

Mixer

IF stages (fixed-tuned at 455 kcps)

$V_2(f)$

$V_1(f)$

$V_3(f)$

Local oscillator (f_c + 455 kcps)

Filter characteristic of IF cascade

Approximate frequency spectrum at input. Antenna tuning provides a rough filter to reduce unwanted broadcast signals, though this is not necessary.

(*a*) *Filtering is obtained by using a fixed-tuned filter (IF) and frequency-shifting the desired input signal to the passband of the fixed-tuned filter*

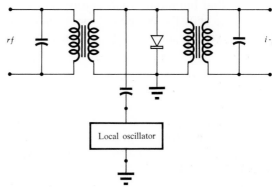

rf

i-f

Local oscillator

(*b*) *A simple mixer utilizing diode non-linearity for frequency conversion*

FIG. 16·15   THE SUPERHETERODYNE PRINCIPLE

gain) is called an *intermediate frequency amplifier.*[1]

In comparison to the first scheme mentioned for achieving selectivity, in which a filter is tuned or scanned across the broadcast band, the super-heterodyne receiver, in effect, scans the broadcast past a fixed-tuned filter, achieving the same overall result with fewer variable tuning elements.

The nonlinear subsystem which performs the frequency-shifting function in a superheterodyne receiver is called a *mixer,* as indicated in Fig. 16·15*a,* or sometimes a *first detector.*[2] The signal emerging from the IF amplifier

[1] A two-stage, 455-kcps IF amplifier is designed in Sec. 18·2, which may be read at this point.

[2] Both terms are perhaps less appropriate than "local modulator," since the frequency-shifting process does not produce signal detection, and linear addition could also be called mixing.

will have a sufficient amplitude that the envelope detector which follows it will usually produce enough audio signal to drive a class A driver for the final class B push-pull output stages in a typical entertainment-type receiver.

Simple mixers can be made which utilize the nonlinearity inherent in a diode. Such a mixer is shown in Fig. 16·15b.

In entertainment-type receivers, the local-oscillator function and the mixing function are frequently combined in one transistor. Such a circuit is sometimes called an autodyne frequency converter. A practical circuit of this type is shown in Fig. 16·16. The dashed lines between the capacitors indicate that they are ganged; that is, their rotors are connected to the same shaft.

Unfortunately, the autodyne frequency converter introduces noise into the signals at all the output frequencies, so a weak incoming signal may be heard against a background of considerable noise. When this is a problem, the mixer is preceded by a tuned radio frequency stage to improve the signal-to-noise ratio to the point where the noise introduced by the frequency converter is less important (*cf.* Prob. 16·19).

Résumé. We can briefly summarize the preceding two sections by saying that the radio broadcasting and receiving process is based on our ability to shift the frequency spectrum of the information-bearing signal at will without affecting the information content of the signal. The frequency-

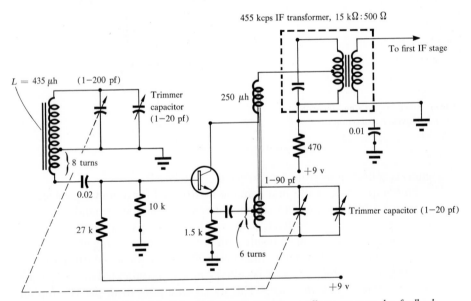

Input tuning accomplished with 435 μh 11(1–200) pf. The oscillator operates by feedback from 250 μh collector coil to tuned circuit in the transistor emitter lead. The IF output is selected by a tuned transformer mounted in a shielded can.

FIG. 16·16   A PRACTICAL OSCILLATOR-MIXER CIRCUIT

shifting operations usually employ some type of nonlinear element, which must be followed by some type of filtering, in order to obtain the desired signal and reject other signals which are produced by the nonlinear element. Radio receivers and transmitters are built by properly combining the ideas of frequency shifting, selective filtering, and gain.

16·3·3 *Single side band philosophy and its application to carrier telephony.*
As we have stated several times previously, the side bands carry the information in a modulated wave. Since each side-band contains enough information (Fourier components) to reconstruct the original signal, it is not really necessary to transmit both side bands, and it is not necessary to transmit the carrier. By using a balanced modulator and appropriate filtering, we can transmit only one sideband and thus reduce the power and bandwidth requirements for a specified transmission effectiveness. Such a scheme is known as a *single-side-band* (SSB) *system.*

One of the most interesting applications of this principle is in carrier telephony. The frequency bandwidth allotted to a given telephone conversation is 200 to 3,500 cps. This frequency band is referred to as a *channel.*

In order to make long-distance telephony economically feasible, many telephone conversations must be sent simultaneously over one transmission system (either a radio link, cross-country transmission lines, or a coaxial cable). To do this, frequency shifting and single side-band techniques are employed so that a large number of telephone channels can be stacked one above another in frequency. In one particular system, twelve channels (which are called a *group*) are stacked into a band lying from 60 to 108 kcps. Then, five of these groups are shifted to form a *super group* occupying a frequency band from 312 to 552 kcps. Finally, ten super groups, representing 600 separate conversations, are placed in the frequency range of 64 to 2,788 kcps to form a combined signal for transmission.

At the receiving end, of course, the various groups and super groups will have to be separated by selective filtering and ultimately each conversation will have to be frequency-shifted back to the 200-to-3,500-cps range and sent to its final destination.

16·4 **FREQUENCY MODULATION**

We pointed out at the beginning of this chapter that to transmit information a waveform must change with time in some manner which cannot be predicted in advance. In these terms, true sine waves transmit no information, but "sine-wave bursts" can. Stated differently, true sine waves must be modulated if they are to transmit information.

We described in the last section a system for transmitting information by amplitude-modulating a carrier. In that system, the desired information is transmitted in the amplitude fluctuations of the carrier; or, from the standpoint of frequency analysis, the desired information can be thought of as being concentrated in side bands about the carrier frequency.

We will now describe a system in which information is transmitted by modulating the frequency of the carrier (FM); we will compare briefly the AM and FM systems as we go.

To fix an initial picture in mind, we show in Fig. 16·17 a simple Colpitts oscillator, in which a portion of one of the oscillator capacitances is a capacitance microphone. The oscillator *frequency* is therefore signal-dependent, though to a reasonable approximation the oscillation amplitude is not. For mathematical convenience we will assume that the oscillator output (for example, v_{CE}) has a constant amplitude and a variable frequency. The rate of variation in frequency must be small compared to the no-signal oscillator frequency to justify this approximation.

To give a mathematical representation of such a signal, we begin by observing that the sine function is mathematically defined in terms of an angle θ. A generalized sinusoidal function can be therefore defined by

$$f_c(t) = A_c(t) \sin \theta_c(t) \tag{16·11}$$

A true sinusoidal *waveform* is obtained when we make A_c constant and allow $\theta_c(t)$ to vary linearly with time:

$$\theta_c(t) = \omega_c t + \phi \tag{16·12}$$

We can then define the angular frequency of the sine wave to be

$$\omega_c = \frac{d\theta_c}{dt} \tag{16·13}$$

Equation (16·11) is a sine wave only if ω_c is constant.

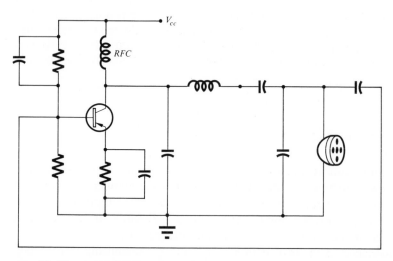

FIG. 16·17 A COLPITTS OSCILLATOR WITH A CAPACITANCE MICROPHONE SERVING AS PART OF THE TUNED CIRCUIT

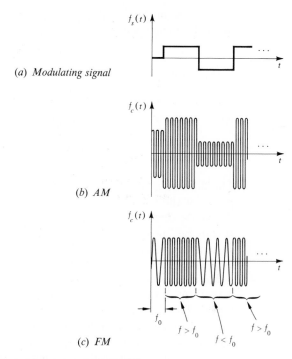

(a) *Modulating signal*

(b) *AM*

(c) *FM*

FIG. 16·18   AM AND FM MODULATION OF A CARRIER $f_c(t)$

When we wish to transmit information using a "sine-wave carrier" we must cause either A_c or θ_c to vary in a different way than they would in a true sine wave. That is, we must cause A_c to vary with time, or cause $d\theta_c/dt$ to vary with time. The two possibilities are shown in Fig. 16·18 for a square-wave modulating waveform. In both cases the resulting waveforms are not sine waves, but they can be expressed in a series of sine waves.

Modulating A_c is, of course, the AM system philosophy. Modulating $d\theta_c/dt$ is called an FM system philosophy, though there are two slightly different possibilities:

1. We can modulate $\phi(t)$ in accordance with an information-bearing signal $f_s(t)$ to obtain

$$\phi(t) = \phi_0 + m_1 f_s(t) \qquad \theta(t) = \omega_c t + \phi_0 + m_1 f_s(t) \qquad (16\cdot14)$$

This is called *phase modulation*.

2. We can define an *instantaneous frequency* by Eq. (16·13), and modulate this instantaneous frequency in accordance with the desired signal:

$$\frac{d\theta_c}{dt} = \omega_c + m_2 f_s(t) \qquad \theta_c = \omega_c t + \theta_0 + m_2 \int f_s(t)\, dt \qquad (16\cdot15)$$

This is called *frequency modulation*.

Both phase modulation and frequency modulation are special cases of *angle modulation* $\theta_c(t)$ and are not essentially different from each other.

To see this more clearly, we can assume that the modulating signal is an audio-frequency sine wave

$$f_s(t) = B \sin \omega_s t \qquad (16 \cdot 16)$$

Then, in the phase-modulated system

$$\theta_c(t) = \omega_c t + \phi_0 + m_1 B \sin \omega_s t \qquad (16 \cdot 17)$$

while in the frequency-modulated system

$$\theta_c(t) = \omega_c t + \theta_0 - \frac{m_2 B}{\omega_s} \cos \omega_s t + \theta_1 \qquad (16 \cdot 18)$$

The difference between these two equations is simply that in one case the phase of the carrier varies with the modulating signal while in the other the phase of the carrier varies with the integral of the modulating signal. It can be shown[1] that there is no basic difference between these two systems, so we shall refer to both of them as "FM" in what follows.

16·4·1 *Fourier spectrum of an FM signal.* As in the AM case, the transmitted waveform is not a sine wave, but it can be expanded in a series of sine waves. The Fourier expansion is important in both cases because it defines the frequency bands in which the information-bearing signal is carried and helps to clarify and specify system requirements for transmitting, receiving, and detecting the information-bearing signals.

To analyze a frequency-modulated wave by the Fourier method, we substitute Eq. (16·17) for $\theta_c(t)$ into Eq. (16·9):

$$f_c(t) = A_c \sin (\omega_c t + \phi + m \sin \omega_s t) \qquad (16 \cdot 19)$$

where we have incorporated $m_1 B$ into a single factor m which we call the *modulation index*. (We shall relate the modulation index to the properties of actual modulators later.) For convenience, we choose $\phi = \pi/2$ and $A_c = 1$ to obtain

$$f_c(t) = \cos (\omega_c t + m \sin \omega_s t) \qquad (16 \cdot 20)$$

We now wish to find the Fourier expansion of $f_c(t)$ given in Eq. (16·20). We can make some progress in this direction by expanding Eq. (16·20) using trigonometric identities:

$$f_c(t) = \cos \omega_c t \cos (m \sin \omega_s t) - \sin \omega_c t \sin (m \sin \omega_s t) \quad (16 \cdot 21)$$

From this point onward the analysis becomes somewhat more difficult, though the basic idea is simple. Let us consider for the moment the function $\cos (m \sin \omega_s t)$. The argument of this function, $m \sin \omega_s t$, is clearly a periodic function of time. In fact, it is simply a sine wave with a period $2\pi/\omega_s$, and an amplitude of m radians. The function is plotted in Fig. 16·19a for one period.

We have also plotted $\cos (m \sin \omega_s t)$ in Fig. 16·19 for various values of m. These plots point up several general properties of the function:

[1] H. S. Black, "Modulation Theory," D. Van Nostrand, Princeton, N.J., 1953, Chaps. 3, 12.

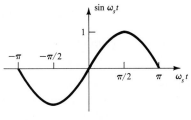

(a) $\sin \omega_s t$ vs. $\omega_s t$

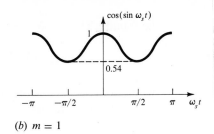

(b) $m = 1$

(c) $m = 2.5$

(d) $m = 5$

FIG. 16·19 APPLICATION OF FOURIER ANALYSIS IN THE MODULATION OF AN **FM** WAVE

1. $\cos (m \sin \omega_s t)$ is a periodic function of the variable $(\omega_s t)$. It can therefore be expanded in a Fourier series in the variable $(\omega_s t)$.

2. $\cos (m \sin \omega_s t)$ is an *even* function of time, because $\cos \theta = \cos (-\theta)$. Therefore the Fourier expansion of $\cos (m \sin \omega_s t)$ will contain only cosines:

$$\cos (m \sin \omega_s t) = \sum_{n=0}^{\infty} a_n(m) \cos n\omega_s t \qquad (16 \cdot 22)$$

3. On the $(\omega_s t)$ scale, $\cos (m \sin \omega_s t)$ has a period of π, not 2π. Therefore, *only even harmonics* will appear in the expansion; that is, $n = 0, 2, 4, \ldots$ in Eq. $(16 \cdot 22)$.

4. The form of the function, and therefore the Fourier coefficients $a_n(m)$, are sensitive functions of m, as shown in Figs. $16 \cdot 19b$, $16 \cdot 19c$, and $16 \cdot 19d$. (The reader can sketch other interesting and instructive examples by setting $m = \pi, 2\pi, 3\pi$.)

To find the Fourier coefficients, we have to evaluate the integrals

$$a_0(m) = \frac{1}{\pi} \int_{-\pi}^{\pi} \cos (m \sin \theta) \, d\theta$$

$$a_n(m) = \frac{2}{\pi} \int_{-\pi}^{\pi} \cos n\theta \cos (m \sin \theta) \, d\theta \qquad (16 \cdot 23)$$

for given n and m. The values of these integrals cannot be expressed in closed form, though a series expansion for the coefficients $a_n(m)$ can be found.[1] The expansions are

$$a_n(m) = 2\left[\frac{(\tfrac{1}{2}m)^n}{0!\,n!} - \frac{(\tfrac{1}{2}m)^{n+2}}{1!(n+1)!} + \frac{(\tfrac{1}{2}m)^{n+4}}{2!(n+2)!} - \cdots\right] \quad (16 \cdot 24)$$

with the factor of 2 before the brackets being absent for $n = 0$.

These functions also arise when one attempts to solve Laplace's equation in cylindrical coordinates. They are called *Bessel functions of the first kind* and are represented in the Bessel function literature by the symbol J. For our problem

$$J_n(m) = \frac{a_n(m)}{2} \quad n > 0$$

$$J_0(m) = a_0(m) \tag{16 \cdot 25}$$

These functions are tabulated for various values of n and m, though the series expansion given in Eq. $16 \cdot 24$ converges rapidly and is not difficult to use directly.

Using either tables or Eq. $(16 \cdot 24)$, the following Fourier expansions can be obtained for the functions drawn in Fig. $16 \cdot 19b$, $16 \cdot 19c$, and $16 \cdot 19d$:

$$\cos(\sin \omega_s t) = 0.77 + .23 \cos 2\omega_s t + .005 \cos 4\omega_s t \tag{16 \cdot 26a}$$

$$\cos(2.5 \sin \omega_s t) = -0.05 + .87 \cos 2\omega_s t + 0.15 \cos 4\omega_s t$$
$$+ .005 \cos 6\omega_s t \tag{16 \cdot 26b}$$

$$\cos(5 \sin \omega_s t) = -0.18 + 0.09 \cos 2\omega_s t + 0.78 \cos 4\omega_s t$$
$$+ 0.26 \cos 6\omega_s t + 0.036 \cos 8\omega_s t + 0.01 \cos 10\omega_s t \tag{16 \cdot 26c}$$

For later purposes, it will be useful to point out that the coefficients with maximum amplitude appear approximately at the points where $n = m$. It has been empirically established that $a_n(m) \lesssim 0.05$ for $n = m + 1$.

Let us now attempt to pick up the main thread of the analysis once again. In order to find the Fourier expansion of

$$f_c(t) = \cos(\omega_c t + m \sin \omega_s t)$$

we were led to consider the function $\cos \omega_c t \cos(m \sin \omega_s t)$. We have now found that this latter function can be expressed as

$$\cos \omega_c t \cos(m \sin \omega_s t) = \cos \omega_c t \sum_{n=0}^{\infty} a_n(m) \cos n\omega_s t) \tag{16 \cdot 27}$$

where we know the values of the $a_n(m)$ when m is specified.

Now, we can use trigonometric identities to rewrite Eq. $(16 \cdot 27)$ in the form

---

[1] Hint for the reader interested in the mathematical derivation of Eq. $(16 \cdot 24)$: Expand $\cos x$ in a power series in Eqs. $(16 \cdot 23)$ and integrate the series term by term. The equality

$$\int_0^{2\pi} \sin^k x \, dx = 2\pi \frac{1 \cdot 3 \cdot 5 \cdots (k-1)}{2 \cdot 4 \cdot 6 \cdots k} \quad \text{with } k \text{ even}$$

will be helpful in the integrations.

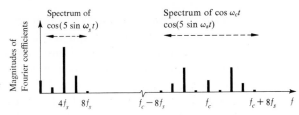

FIG. 16·20   FOURIER SPECTRUM OF $\cos (5 \sin \omega_s t)$ AND $\cos \omega_c t \cos (5 \sin \omega_s t)$

$$\cos \omega_c t \cos (m \sin \omega_s t) = a_0(m) \cos \omega_c t + \frac{a_2(m)}{2} [\cos (\omega_c + 2\omega_s)t$$

$$+ \cos (\omega_c - 2\omega_s)t] + \frac{a_4(m)}{2} [\cos (\omega_c + 4\omega_s)t$$

$$+ \cos (\omega_c - 4\omega_s)t] + \cdots \quad (16·28)$$

From Eq. (16·28) we can see that the term $\cos \omega_c t \cos (m \sin \omega_s t)$ requires for its Fourier expansion a term at the carrier frequency plus *side bands at all even harmonics of the sine-wave-modulating signal*, these side bands being symmetrically situated about the carrier. However, the most important terms in the Fourier expansion of $\cos (m \sin \omega_s t)$ lie in the frequency range of 0–$(m + 1)f_s$ cps, so multiplying $\cos (m \sin \omega_s t)$ by $\cos \omega_c t$ simply *shifts the principal components of the Fourier expansion of $\cos (m \sin \omega_s t)$ from a frequency band 0 to $\sim(m + 1)f_s$ cps to a frequency band of $f_c \pm \sim(m +1)f_s$ cps.* This is shown in Fig. 16·20 for the term $\cos \omega_c t \cos (5 \sin \omega_s t)$.

Unfortunately, the analysis is only half done, since we must also perform a Fourier analysis of the second term in Eq. (16·21). However, this analysis is not different in principle from that just finished, so we will be content to observe that the function $\sin (m \sin \omega_s t)$ can be expanded in a Fourier series in *odd* powers of ω_s:

$$\sin (m \sin \omega_s t) = \sum_{n=0}^{\infty} a_n(m) \sin n\omega_s t \qquad \text{with } n \text{ odd} \qquad (16·31)$$

and that the effect of multiplying $\sin (m \sin \omega_s t)$ by $\sin \omega_c t$ is to shift the frequency spectrum implied in Eq. (16·31) to the point where it is centered about f_c:

$$\sin \omega_c t \sin (m \sin \omega_s t) = \frac{a_1(m)}{2} [\sin (\omega_c + \omega_s)t + \sin (\omega_c - \omega_s)t]$$

$$+ \frac{a_3(m)}{2} [\sin (\omega_c + 3\omega_s)t + \sin (\omega_c - 3\omega_s)t] + \cdots \quad (16·32)$$

Again the coefficients $a_1(m)$, $a_3(m)$, ... can be calculated from the series given in Eq. (16·24) or looked up in mathematical tables.[1]

This second term now adds to the Fourier spectrum of the signal *side bands at all odd harmonics of the sine-wave-modulating signal*, these side bands again being symmetrically situated about the carrier.

[1] The standard reference is Jahnke and Emde, "Tables of Functions," Dover Publications, Inc., New York, 1945.

The overall picture which this analysis gives us is thus as follows: When a "sine wave" carrier at frequency f_c is frequency-modulated by a *single* sine wave at frequency f_s, the Fourier representation of the resulting signal shows (in general) a carrier component at f_c plus side bands at *all* the frequencies $f_c \pm nf_s$, where $n = 1, 2, 3, \ldots$. The frequency bandwidth required to represent the signal is therefore in principle infinite.

As a practical matter, however, the side band components become small after $n \simeq m + 1$, so the frequency bandwidth required to transmit the signal is limited. This is shown in Fig. 16·21 where the full Fourier-amplitude spectrum of a frequency-modulated carrier is plotted for the values $m = 1$, 2.5, and 5. There are several interesting features in this figure which deserve comment:

1. The amplitude of the *carrier* is a function of m. This can be related to the cos $(m \sin \omega_s t)$ functions shown in Fig. 16·19 quite simply. The dc component in the Fourier expansion of these waveforms gives the amplitude of the carrier in the Fourier spectrum of the transmitted signal. The dc component is clearly higher for $m = 1$ than it is for $m = 2.5$ or $m = 5$. In fact, the dc component is less than 0.4 for any $m > 1.7$.

The fact that the amplitude of the carrier component in Fig. 16·17 is a function of m is of considerable physical significance. In an FM system,

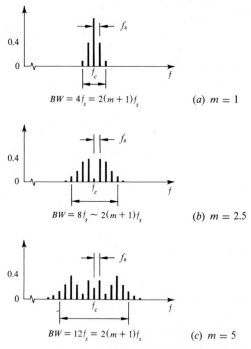

BW $= 4f_s = 2(m + 1)f_s$ (a) $m = 1$

BW $= 8f_s \sim 2(m + 1)f_s$ (b) $m = 2.5$

BW $= 12f_s = 2(m + 1)f_s$ (c) $m = 5$

FIG. 16·21   FOURIER SPECTRUM FOR AN **FM** WAVE INCLUDING BOTH EVEN AND ODD HARMONICS OF THE MODULATING FREQUENCY f_s. IN EACH CASE THE BANDWIDTH (BW) BEYOND WHICH $a_n(m) \lesssim 0.02$ CAN BE ESTIMATED FROM $BW = 2(m + 1)f_s$

the amplitude A of the transmitted signal is constant:

$$f_c(t) = A \sin \theta_c(t)$$

This means the total rms power of the FM wave is constant. When there is no modulation, all the transmitted power is radiated at the carrier frequency. When modulation is present, however, some of the carrier power is diverted into the side bands, the average radiated power still remaining *constant*. This is in sharp contrast to the AM system, where the power transmitted at the carrier frequency is not affected by the modulation (providing the modulating signal has no dc component).

2. The importance of various side-band frequencies is a function of m. Again, this is easily comprehended from Fig. 16·19. Here, the number of cycles of the function cos ($m \sin \omega_s t$) is a sensitive function of m. For $m = 5$, there are five peaks and four troughs in the interval $-\pi < \omega_s t < \pi$. Since the waveform is generally sinusoidal in nature (no steps, cusps, spikes, etc). we would expect to find the most important Fourier components at the 4th and 6th harmonics, components of less importance at the 2nd and 8th harmonics, and very little elsewhere. This is borne out by the frequency spectrum shown in Figs. 16·20 and 16·21.

3. Since the side-band frequencies in the full spectrum are spaced at intervals of f_s, and the number of important harmonics is roughly $m + 1$, the total bandwidth required is approximately $2(m + 1)f_s$.

Now, for large m and f_s, the bandwidth required to transmit the signal may become extremely large. This is not entirely undesirable, because the noise- and interference-reduction advantages which the FM system has over the AM system are directly related to the fact that each sine wave in the modulating signal is represented by several harmonics in the transmitted signal. However, the Federal Communications Commission must establish some limit on the allowable bandwidth per station to make efficient use of the frequency spectrum available for broadcasting. They do this by specifying the *maximum frequency deviation* (measured with the carrier frequency as a reference) allowed at the transmitter.

Maximum frequency deviation and the modulation index. To see why the FCC specifies the maximum frequency deviation, we need to stand back a bit so we can compare the physical idea of a frequency-modulated carrier with its mathematical representation.

We introduced the idea of an FM carrier by thinking of a Colpitts oscillator circuit with a portion of its tuning capacitance being a capacitance microphone. In such a circuit the frequency deviations of the oscillator can be simply related to the fluctuations in the microphone capacitance, and the maximum deviation of frequency from the no-signal or carrier frequency can be easily controlled.

In the mathematical representation of the signal, it is more convenient to use a modulation index m. Furthermore, in this representation the side-band amplitudes and the number of important side bands are simply related to m.

The relationship between the maximum frequency deviation and the modulation index may be established as follows. The instantaneous frequency of the frequency-modulated carrier is *defined* to be

$$\omega_i = \frac{d\theta_c}{dt} \qquad (16 \cdot 33)$$

For the carrier signal we have been considering,

$$f_c(t) = \cos{(\omega_c t + m \sin{\omega_s t})}$$

which gives

$$\theta_c(t) = \omega_c t + m \sin{\omega_s t}$$

The instantaneous carrier frequency is therefore

$$\omega_i = \omega_c + m\omega_s \cos{\omega_s t} \qquad (16 \cdot 34)$$

The maximum frequency deviation is therefore

$$\Delta f = |f_c - f_i| = mf_s \qquad (16 \cdot 35)$$

Relationship between maximum allowable frequency deviation and signal bandwidth. Since Δf is the quantity that is specified by the FCC, it is useful to express the bandwidth required to transmit an FM signal in terms of Δf.
From Eq. (16 · 35), we have

$$m = \frac{\Delta f}{f_s} \qquad (16 \cdot 36)$$

Using this expression for m and the previous idea that the important sidebands lie in the frequency range $f_c \pm (m + 1)f_s$, we see that the total bandwidth required to transmit the FM signal is

$$BW \simeq 2(m + 1)f_s = 2\left(\frac{\Delta f}{f_s} + 1\right)f_s = 2\Delta f + 2f_s \qquad (16 \cdot 37)$$

For a maximum signal frequency $f_s = 15$ kcps and a maximum allowable Δf of 75 kcps, this shows that a maximum bandwidth of about 150 kcps + 30 kcps = 180 kcps is required for transmitting an FM signal.

16 · 4 · 2   *FM detectors.* The only basic difference between an FM receiver and an AM receiver is that the signal-detection process is different. Typically, both receivers shift the received signal in frequency to an IF carrier (via a local oscillator and mixer) and then proceed to obtain the necessary frequency-selective amplification. Naturally, after detection, audio amplification is necessary in both receivers.

The FM detection process can, like the AM detection process, be viewed either in terms of converting the time-varying signal into an audio signal or by paying more attention to the frequency shifting which must occur.

Perhaps the simplest FM detector (and one of the best) can be made in the manner suggested in Fig. 16 · 22. The output of the IF amplifier is passed through a limiter (for example, back-to-back avalanche diodes) to

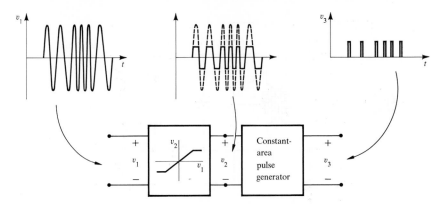

FIG. **16·22**   AN **FM** DETECTOR UTILIZING A CONSTANT-AREA PULSE GENERATOR; THE OUTPUT OF THE CONSTANT-AREA PULSE GENERATOR IS AVERAGED TO RECOVER THE AUDIO INFORMATION

give a waveform which is essentially a variable-frequency square wave. This waveform is used to trigger a circuit (monostable flip-flop) which puts out a constant-area pulse each time the square wave goes positive (that is, crosses zero). When these constant-area pulses are averaged (in a suitable RC network, for example), the audio signal will be recovered.

A second common type of detector utilizes the shift in phase which occurs when an off-resonance signal is applied to a tuned transformer. This type of detector is called a *balanced discriminator*. The circuit is shown in Fig. 16·23. In its idealized form, the diode-RC combinations serve as peak detectors. The voltage across the upper capacitor is approximately $|E_1 + E_2|_{peak}$, while that across the lower capacitor is $|E_1 - E_2|_{peak}$. With the primary and secondary both tuned to the unmodulated IF frequency, E_2 will lead E_1 by 90°, as shown in Fig. 16·23. The detector output

$$E_d = |E_1 + E_2| - |E_1 - E_2|$$

is zero for this case. As the frequency changes, however, the phase of E_2 shifts with respect to E_1, so that $|E_1 + E_2|$ and $|E_1 - E_2|$ are no longer equal and an output voltage E_d is obtained from the detector. If the frequency deviations are small, the output voltage will be proportional to the frequency deviation. This gives rise to the discriminator characteristic shown in Fig. 16·23. The discriminator output voltage is thus proportional to the deviation of the frequency from the no-signal IF frequency which is what we require to convert the frequency variations to an audio voltage.

16·5   TELEVISION SYSTEMS

As our final topic in the discussion of practical communication systems, we shall briefly discuss some basic aspects of television systems. Television

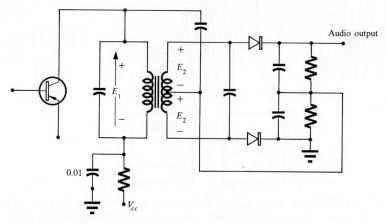

Audio output

E_1

$+$
E_2
$-$
$+$
E_2
$-$

0.01

V_{cc}

(a) Balanced discriminator

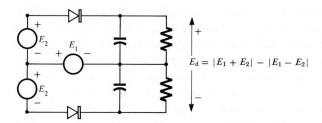

E_2

E_1

E_2

$$E_d = |E_1 + E_2| - |E_1 - E_2|$$

(b) Simplified form of (a)

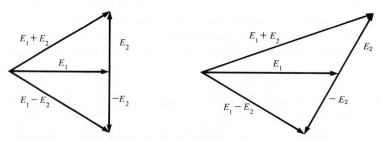

$E_1 + E_2$

E_2

E_1

$E_1 - E_2$

$-E_2$

$E_1 + E_2$

E_2

E_1

$E_1 - E_2$

$-E_2$

(c) Diode input voltages at resonance and off resonance

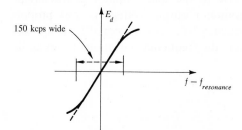

E_d

150 kcps wide

$f - f_{resonance}$

(d) Output voltage of discriminator vs. input frequency

FIG. 16·23   A BALANCED DISCRIMINATOR FOR DETECTING AN FM SIGNAL

712   Semiconductor electronics

transmission utilizes both AM and FM modulation to send the picture and sound information. The signal processing which occurs in the receiver is therefore quite interesting in the variety of subsystem functions that must be performed.

16·5·1 *Picture reproduction.* It is perhaps simplest to consider first how the picture which appears on the screen of a television receiver is reproduced. We shall consider this problem in two parts: scanning and picture formation.

Scanning. The picture on the receiver screen is "painted" there by scanning an electron beam across the inside face of the picture tube. When the electrons strike the screen it glows with a brightness which is proportional to the beam current arriving at that instant. For the moment we will assume the beam current is constant and describe only the beam-scanning pattern.

The beam is scanned across the picture tube face by a magnetic deflection system in the zigzag pattern shown in Fig. 16·24. The scan proceeds from the left, slanting downward toward the right. When the beam reaches the right edge of the screen, it is cut off and deflection currents are applied to the deflection coils to ensure that when the beam is turned back on, it will appear at the left edge of the screen. The leftward motion is called the *retrace* or *flyback*. When the retrace operation is completed, the beam is again turned on and the deflection coils are driven with currents which again produce a linearly sweeping scan, slanting across the face of the picture tube.

In the commercial TV system, the time required for the beam to scan

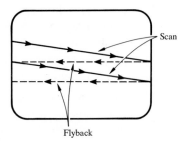

(a) *Horizontal scan and flyback*

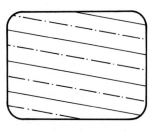

(b) *The interlaced scanning pattern. There are 262.5 solid scan lines and 262.5 broken scan lines. Each set of 262.5 lines is called a "field." The two fields are called a "frame."*

FIG. 16·24   PICTURE TUBE SCANNING IN A TELEVISION RECEIVER

across one line of the tube face once is 52.5 μsec. The actual flyback time is less than 5 μsec, so the flyback is nearly horizontal, as indicated in Fig. 16·24a.

The actual pattern of lines that appears on the screen during a picture reproduction is shown in Fig. 16·24b. The beam begins in the top center of the tube and scans half a line. The slope of the scan lines is adjusted so that the beam arrives at the lower right edge of the tube after 262.5 lines have been traced out. These 262.5 lines are called the first *field*.

The beam is then returned to the upper left corner of the tube face and a second field of 262.5 lines is traced out as shown in Fig. 16·24b. Each set of 262.5 lines is traced out in $\frac{1}{60}$ sec, so the entire pattern is traced out thirty times per second. The interlaced scan is employed to avoid the flicker which would occur if an appreciable area of the screen were illuminated only thirty times per second.

Picture formation. The picture which appears on the screen is formed by modulating the beam current during the scan of each line. The rate at which the beam needs to be modulated is determined by the scanning rate.[1] A 525-line pattern constructed in both horizontal and vertical directions divides the tube face into a checkerboard which has $(525)^2 \simeq 280{,}000$ squares. Each of these squares is called a *picture element*. The beam crosses approximately 525 picture elements during each scan line. Hence each picture element is painted approximately independently of all the rest, once every thirtieth of a second.

To do this we apply a voltage between the grid and cathode of the electron gun in the picture tube, the amplitude of the voltage being proportional to the brightness which a given picture element is supposed to have. To a first approximation, the required grid-cathode voltage can be considered to be a series of pulses, the height of each pulse representing the brightness of a picture element. Since there are 525 picture elements per line, and the scanning time per line is about 52.5 μsec, the time width of each pulse is 0.1 μsec. These details are shown in Fig. 16·25.

Signal bandwidth estimates. We can estimate some important system requirements with the aid of this number and the basic properties of our old friend, the basic modulating pulse. If each picture element is painted out once every thirtieth of a second, then the grid-cathode voltage associated with a given picture element is a basic modulating pulse with $t = 0.1$ μsec, $T = \frac{1}{30}$ sec, as shown in Fig. 16·25b. The Fourier components needed to represent this pulse will have appreciable energy in the frequency range of 0 to $\frac{1}{2\tau} \sim 5$ mcps. As a result, the amplifiers which drive the grid of the electron gun must have a bandwidth of about 5 mcps; and, of course, this bandwidth requirement will reflect itself in IF stage requirements and at other points in the reception and transmission apparatus.

---

[1] The scanning rate is, in turn, selected by considering the fineness of the fluctuations in light and shade which would be apparent to the eye; that is, the characteristics of the final observer enter again.

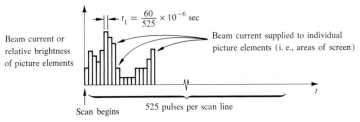

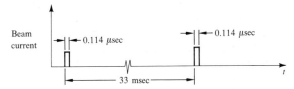

(a) Beam current during one horizontal scan line

(b) Beam current supplied to a given picture element as a function of time

FIG. **16·25** APPROXIMATE BEAM CURRENT FOR PICTURE FORMATION

16·5·2 *The television signal.* From the preceding discussion, we can see that two types of information must be transmitted to the receiver, in order for it to reproduce the scene being televised:

1. The receiver needs information to ensure that the electron beam scans the picture tube face in synchronism with a similar beam which is generating the picture information as it sweeps across the camera tube face at the studio.

2. The receiver needs information from which it can reproduce the brightness that each picture element should have.

This information is obtained by appropriately combining the output signal of the camera tube with blanking and synchronizing pulses which are generated at the transmitting station. The combination is used to amplitude-modulate a carrier wave. The modulation envelope has the general form shown in Fig. 16·26. In part *a*, we show a hypothetical signal for two lines of a field. The brightness of each picture element along a given scan line is represented by the variations of signal amplitude appearing in that portion of Fig. 16·26 labeled *picture*. The scan line occupies a time interval of 52.5 μsec.

At the end of the line, there is a *horizontal-blanking pulse* which lasts for 10 μsec. On top of the blanking pulse there is another pulse called a *horizontal-synchronizing pulse* which is used to trigger the horizontal-deflection circuitry in the receiver.

In the standard signal white corresponds to negative modulation and is represented by an amplitude of the carrier envelope that does not exceed 15 percent of the maximum carrier amplitude. Black is represented by an envelope amplitude of 75 percent of the maximum amplitude, and provides a reference level against which the background brightness of the picture is set.

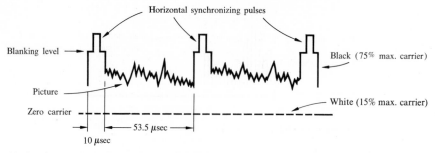

(a) Synchronization pulses and picture brightness for two adjacent scan lines in a field

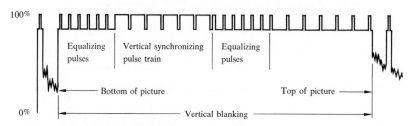

(b) Vertical synchronization information (U.S. standards)

FIG. 16·26   MODULATION SIGNAL CONTAINING PICTURE AND SCANNING INFORMATION

Part *b* of Fig. 16·26 shows the manner in which the vertical synchronization is achieved. Vertical synchronization is complicated by (1) the fact that the receiver must have some means of distinguishing between vertical and horizontal synchronizing pulses, (2) the need of simultaneously maintaining horizontal synchronization, and (3) 60-cycle irregularities introduced by the interlacing.

To solve these problems, the vertical-synchronizing pulses are given a time length corresponding to three horizontal lines. Horizontal synchronization is maintained by the serrations that break up the vertical pulse into 6 blocks, as illustrated in Fig. 16·26*b*. These serrations have twice the horizontal-line frequency and are timed so that the rise of every other serration occurs at the instant a horizontal-synchronizing pulse is required. Finally, groups of six equalizing pulses are introduced just before and after the vertical pulses, to insure that the interlacing is properly accomplished.[1]

Frequency spectrum of the transmitted signal.   As we have seen in a previous paragraph, a bandwidth of about 4 mcps is required to transmit the picture information. If double side-band AM transmission were used, the picture information would occupy a frequency band of ±4 mcps about the carrier, or a total bandwidth of 8 mcps. In order to conserve bandwidth, a system which is basically single side band is used (actually a portion of

[1] The reader interested in more details is referred to A. R. Applegarth, Synchronizing generators for electronic television, *Proc. IRE,* vol. 34, p. 128, March, 1946.

the lower side band is transmitted, so the system is called a *vestigial side-band* system). The FCC provides a total channel width of 6 mcps for each television station, as indicated in Fig. 16·27. The picture carrier is 1.25 mcps above the lower edge of this channel. If we take Channel 2 as an example, the allowed channel extends from 54–60 mcps. The picture carrier is at 55.25 mcps.

The picture information is contained in an upper side band extending from 55.25 to 59.25 mcps, and a portion of the frequency spectrum is repeated in a lower sideband extending from 55.25 to 54.50 mcps.

The sound information is transmitted on a frequency-modulated carrier which lies 4.5 mcps above the picture carrier, as shown in Fig. 16·27. The maximum allowable frequency deviation Δf for the FM sound transmission is 25 kcps, so the required bandwidth is about $2 \times 25 = 50$ kcps. The picture and sound carriers are placed sufficiently far apart that when the receiver is properly tuned no interference results.

Block diagram for a television receiver. A simplified block diagram of a television receiver is shown in Fig. 16·28. The superheterodyne principle is employed with separation of the sound and picture channels taking place after the mixer. The synchronization pulses are picked off after the picture or *video* signal has been amplified and has a dc level inserted which gives the required contrast between black and white.

The synchronizing pulses are distinguishable from each other by employing a differentiating circuit in the horizontal-deflection circuitry and an integrator in the vertical-deflection circuitry. In this way the vertical-deflection system is only triggered when the large block of synchronizing pulses comes along.

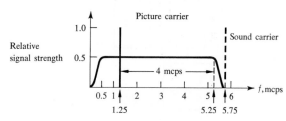

(a) *Frequency spectrum of transmitted signal relative to lower edge of allowed channel*

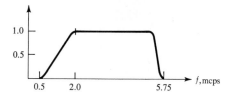

(b) *Recommended response characteristic of receiver*

FIG. 16·27 STANDARD CHANNEL ALLOWANCES FOR COMMERCIAL TELEVISION TOGETHER WITH RECOMMENDED RECEIVER CHARACTERISTICS

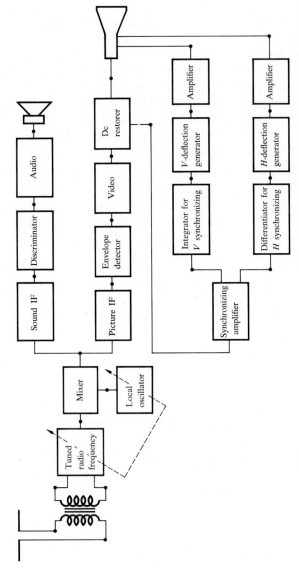

FIG. 16 · 28   A SIMPLIFIED BLOCK DIAGRAM OF A TELEVISION RECEIVER. SEVERAL VARIATIONS IN SIGNAL PROCESS-
ING ARE POSSIBLE. IN ONE CASE, THE MIXER IS FOLLOWED BY IF AMPLIFICATION AND DETECTION BEFORE THE SOUND
INFORMATION IS PICKED OFF. THE SYNCHRONIZING CIRCUITS MAY ALSO BE DRIVEN DIRECTLY FROM THE ENVELOPE
DETECTOR. (*For further details, see G. M. Glasford, "Fundamentals of Television Engineering," McGraw-Hill
Book Co., New York, 1955, Ch. 17.*)

At the beginning of this chapter we pointed out that it is an *unexpected change* in a waveform which communicates information to us. So far we have not made too much use of this fact, choosing instead to represent the information-bearing signals by their frequency spectra. In this section we return to a further consideration of information-bearing signals as time-dependent waveforms.

The unexpected changes in a waveform, which convey its information to us, are not the only unexpected changes in an actual signal. There will always be *noise* superimposed on the received signal and noise added to the signal by the receiver. Of course, noise fluctuations in a waveform also represent unexpected changes in the signal, though they do not convey information to us. To receive a signal clearly, we must make sure that the unexpected changes in the total signal which carry the desired information can be clearly distinguished from the random changes in signal due to noise, which simply frustrate our attempts to receive the desired information.

To see how these ideas relate to the design of a communication system, consider the voltage-time diagram shown in Fig. 16 · 29. This diagram can be thought of as representing the picture information for one horizontal scan line in a television transmission; T is the scan duration ($\sim$50 μsec) and t_1 is the time the electron beam spends on one picture element ($\sim$0.1 μsec).

The vertical scale of the diagram has been divided into intervals which represent the *minimum detectable change in signal amplitude.* The number of these levels n is determined by the magnitude of the noise voltages present in the system. If there were no noise present, a change in signal between adjacent time intervals which is as small as we please could in principle be detected.[1] However, in the presence of noise there are only n possible detectable signal levels.

Now, in a period of T seconds, there are T/t_1 basic time intervals, in each of which the signal may occupy any one of n possible levels. Therefore, there is only a *finite* (but perhaps very large) *number of possible detectable signal patterns which can be sent in the T second interval.* Two such patterns are shown in Fig. 16 · 29, where boxes in the voltage-time diagram are simply shaded to represent the chosen signal.

The total number of possible signal patterns is

$$n^{T/t_1}$$

since we have n choices in each of T/t_1 intervals. If each possible signal pattern is called a *message,* then there are n^{T/t_1} messages which can be sent in the time T.

In an intuitive sense, the information which we receive when we have a

---

[1] Actually, there is some minimum change which is detectable by an observer, and this should also enter the discussion. *cf.* Prob. 16 · 33.

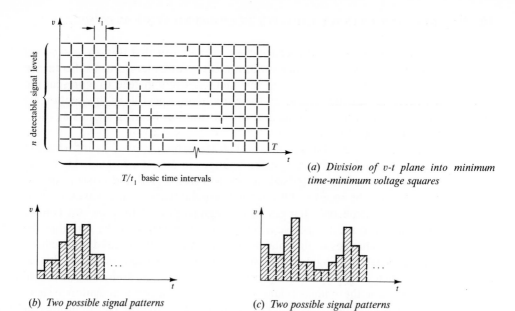

(a) Division of v-t plane into minimum time-minimum voltage squares

T/t_1 basic time intervals

(b) Two possible signal patterns

(c) Two possible signal patterns

FIG. 16·29 VOLTAGE-TIME DIAGRAM FOR DETERMINING THE NUMBER OF DETECTABLE SIGNAL PATTERNS IN T SECONDS

given signal pattern before us is related to the probability that we could have predicted the pattern beforehand. Stated in a slightly different way, the information we receive is related to the number of measurements we need to make to reconstruct the given signal.

Now this number is *not* n^{T/t_1}, as the following reasoning shows. Suppose for the sake of argument that there are 1,000 distinguishable signal levels. Then we can determine the actual signal level in any one of the basic time intervals t by making *no more than 10 measurements*. To take a specific example, suppose the signal lies in the voltage interval corresponding to $n = 31$. Then a possible sequence of measurements is as follows:

Measurement	Answer to measurement
1. Is $v > 500$	No
2. Is $v > 250$	No
3. Is $v > 125$	No
4. Is $v > 63$	No
5. Is $v > 32$	No
6. Is $v > 16$	Yes
7. Is $v > 24$	Yes
8. Is $v > 28$	Yes
9. Is $v > 30$	Yes
10. Is $v > 31$	No

Thus, looking at questions (5) and (10) we can conclude that the voltage is in the 31st distinguishable level by making 10 measurements.

We can do this because $2^{10} > 1,000$, or $10 > \log_2 1,000$. Of course, we do not actually make these measurements in the simple communication systems which we have previously described, though we could devise a system which did so (the system would utilize a series of logical gates).

Because it only requires 10 measurements to establish the voltage level, we can argue that 10 "pieces of information" are all that we need to predict the signal level. Each piece of information is called a *bit* (for *binary digit*, since we use the base 2 in the measurement process). We then say that each voltage level contains 10 bits of information, if $n = 1000$ or, in general, each voltage level contains $\log_2 n$ bits for n levels.

Thus, in a sequence of T/t intervals, the *information content can be defined to be*

$$\text{Information content} = \frac{T}{t} \log_2 n \text{ bits} \qquad (16 \cdot 38)$$

The average rate at which we receive information from the signal is called the *channel capacity* C and is defined to be

$$C = \frac{\text{Information content}}{T} = \frac{1}{t} \log_2 n \text{ bits/sec} \qquad (16 \cdot 39)$$

This very important formula, which is a part of the brilliant work of C. Shannon in communication system theory,[1] can also be rewritten in the form

$$C = B \log_2 \left(1 + \frac{S}{N}\right) \qquad (16 \cdot 40)$$

To show that the channel capacity formula can be written in this form, we first assume that the desired signal has a maximum signal power of S. The received signal contains the desired signal plus noise, the noise power being denoted by N. Now, $\sqrt{N}$ is a measure of the flucutation in signal voltage which we will observe in a given time interval, and we may therefore define $\sqrt{N}$ to be the minimum detectable change in signal level (see Fig. $16 \cdot 28a$). Since the maximum received signal power is $S + N$, the maximum number of detectable levels is

$$n = \sqrt{\frac{S + N}{N}}$$

Furthermore, the Fourier spectrum of a basic modulating pulse has its energy concentrated in a range

$$B = \frac{1}{2t_1}$$

[1] C. E. Shannon, A mathematical theory of communication, *Bell System Tech. J.,* vol. 27, pp. 379–423, July, 1948; pp. 623–656, October, 1948.

so we can rewrite Eq. 16·39 in the form

$$C = 2B \log_2 \sqrt{\frac{S+N}{N}} = B \log_2 \left(1 + \frac{S}{N}\right)$$

16·6·1   *Channel capacity and system evaluation.* The channel capacity formula is a very important tool for evaluating communication systems, because it tells us how efficient the system is. A simple example will illustrate this point.

A telephone channel has a bandwidth of about 4,000 cps. Suppose the signal-to-noise ratio for a given signal source is 100 (which is actually rather low for a typical call). Then the telephone channel is capable of transmitting

$$C = 4,000 \log_2 101 \simeq 26,600 \text{ bits/sec}$$

Now, in his early work in this field, Shannon experimentally determined that each letter or space in an English text of reasonable length gives the reader one bit of information. At a comfortable talking speed, a person generates about 10 letters (including spaces) per second, so a telephone channel which is capable of transmitting 26,600 bits per second is used at an average rate of 10 bits per second.

Why is the channel so inefficient? Basically, the answer is that it has been designed to transmit with nearly perfect fidelity *any* signal which can be represented by Fourier components lying in the frequency range of 0 to 4,000 cps. This includes music, English text, any foreign language, unintelligible babble, and so on. It was *not* designed with the idea that only English text would be transmitted.

Unfortunately, the problem of designing a system to transmit only English text is a very difficult one. The reason that each letter or space conveys only one bit of information is that there are a large number of *language constraints* that must be followed in constructing English text. These language constraints are such that a reader can guess, at any point in a given text, what the next letter will be with an accuracy of 50 percent (hence the one bit per letter estimate). To build a system for transmitting only English text, one must devise a way of encoding the text to take maximum advantage of all the language constraints. We do not know precisely how to do this, though a lot of work is being done in this direction.

Of course, what we say about English text applies to any other message or signal source. Generally, the signal source will have constraints which limit the number of messages it can generate, and the form these messages can take. In some cases, signals can be encoded to take advantage of these constraints, though even then the system required for encoding and decoding is usually more expensive than bandwidth (that is, than a system which can transmit information at a greater rate than the source can generate it). Thus, while Shannon's results give us a very useful means of measuring system capability, present systems rarely measure up to this standard. Shannon's basic work therefore still provides a challenge for future workers.

16·7 SUMMARY

In this chapter we have described the most common procedure for representing information-bearing signals (Fourier analysis) and have discussed the manner in which some of the more common carrier communication systems transmit the information-bearing signals from one point to another.

The basic philosophy from which all of these systems are designed is:

First, study the basic frequency bandwidth required for obtaining a reasonably accurate Fourier representation of the signal, using the final observer as the judge of the term "reasonably accurate."

Then choose a scheme for modulating a carrier so that the signal can be shifted to a convenient frequency range for transmission.

Next, study the Fourier spectrum of the modulated signal to determine the frequency range in which the important transmitted frequency components are located.

Then, choose a signal-processing scheme (usually from among several alternatives) from which the performance requirements of various subsystems in the transmitter and receiver can be deduced and, finally, build the system.

There are certain subsystems or functional blocks that find wide application in carrier communication systems. These are *base-band* or *signal-frequency amplifiers* (for example, 0 to 20 kcps for audio signals, 0 to 4 mcps for video signals), *oscillators, modulators, filters,* and *detectors.* A thorough grasp of the importance of frequency-shifting and its creative application in system design provides a useful insight into system operation and a useful design tool.

There still remain many communication systems that we have not discussed, most of which use digital codes to represent the signals and transmission schemes which take account of the properties of digital signals (*cf.* the Introduction to this book). There also remains the careful study of Shannon's theorem and its relation to communication systems. This chapter provides only a bare introduction to communication system philosophy and engineering.

REFERENCES

Harman, W. W.: "Principles of the Statistical Theory of Communication," McGraw-Hill Book Company, New York, 1963.

Schwarz, M.: "Information Transmission, Modulation, and Noise," McGraw-Hill Book Company, New York, 1959.

Terman, F. E.: "Radio Engineering," 3rd ed., McGraw-Hill Book Company, New York, chaps. 9, 10, 15, 17, and 18.

DEMONSTRATIONS

16·1 *Tone-burst generator.*   A convenient circuit for demonstrating the response of the ear to sine-wave bursts of differing durations is shown in Fig. D16·1. The demonstrator provides a useful focal point for a discussion of the physical basis of Fourier analysis.

The circuit is simply a multivibrator in which the ON time of the left transistor (a Darlington pair, for high gain) can be varied from about 1 to 300 msec. The voltage appearing at the collector of the right transistor (also a Darlington pair, for high gain) is used to saturate a nonregenerative switch when the left transistors in the multivibrator are ON. The nonregenerative switch simply connects an audio generator through a matching transformer to a loudspeaker.

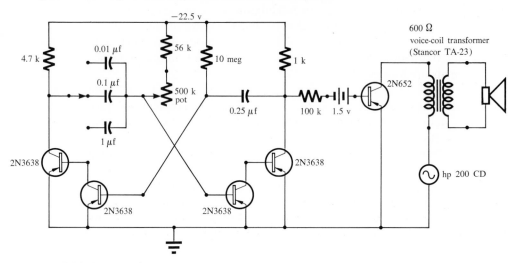

FIG. D16·1   **TONE-BURST GENERATOR**

PROBLEMS

16·1 WWVH is a station maintained by the National Bureau of Standards in Hawaii, operating on the same carrier frequencies as WWV. The audio tones transmitted by WWVH are on continuously for the relevant 3-min periods, with the 5-msec, 1,000-cps bursts being added to the 440-cps or 600-cps tone.

1. Using equal voltages for the 440-cps and 1,000-cps tones, sketch the modulating signal for a period of 10 msec, starting precisely on the second. Assume that the 440- and 1,000-cps tones are in phase.

2. Sketch the waveform of the modulated carrier for the same period of time, assuming that the minimum value of the modulated carrier is set at 50 percent of the unmodulated carrier amplitude by the modulator.

16·2 Sketch and dimension the frequency spectrum for a square wave, using Eqs. (16·2a) and (16·2b) directly. Repeat for a basic modulating pulse with $6t_1 = T$.

16·3 1. Show that the effect of shifting a basic modulating pulse so that its point of symmetry is at $t = t_0$ (instead of $t = 0$) is to multiply the Fourier coefficients of the basic modulating pulse (for $n > 0$) by a phase factor $e^{j2\pi t_0/T}$ (that is, the time origin and basic interval $-T/2$ to $T/2$ do not change, so the pulse is now *not* symmetrically situated between $-T/2$ and $T/2$).

2. Sketch a basic modulation pulse with $4t_1 = T$. On the same axes sketch another basic modulation pulse with $4t_1 = T_1$ but with the center of the second pulse shifted to the right of the center of the first pulse by $2t_1$ sec. Using the result in (1), obtain the Fourier spectrum for this combination and compare it with that of a square wave.

16·4 All WWV carrier frequencies are stated to be accurate to within one part in 10^{11}. Suppose we can build a 10-mcps oscillator which has this inherent accuracy (this is not the way WWV does it). How much time must elapse after we start the generator until it is possible that the required accuracy in generator frequency be obtained?

16·5 WWVL is a low-frequency station located in Denver operating on a carrier frequency of 20 kcps. If the carrier generator is interrupted for repairs and then restarted, how much time will elapse before the generator output will be accurate to within one part in 10^{11}.

16·6 In a certain radar system, the transmitted signal is a 1,000-mcps carrier, modulated by a basic modulating pulse with a duration of 1 μsec and a pulse-repetition frequency of 1,000 pulses per second.

1. Sketch the frequency spectrum of the transmitted wave.

2. How many frequency components are there up to the first zero magnitude component of the transmitted wave? The second zero?

16·7 A sinusoidal carrier of frequency f_c is 50 percent modulated by a sinusoidal signal of frequency f_s. (That is, $m = 0.5$). Sketch the resulting waveform and frequency spectrum ($f_m \ll f_c$). Determine the power contained in the sidebands relative to that in the carrier.

16·8 When a modulating signal has an amplitude greater than that of the carrier, (that is, $m > 1$), the carrier amplitude remains zero for all time that the modulating signal amplitude is negative and greater than the carrier. Assuming a modulating signal with $m = 2, f_s = 1$ kcps, $f_c = 10$ mcps:

1. Sketch and dimension the carrier waveform.

2. How many pairs of sidebands are produced? How many have an amplitude of more than 1 percent of the carrier?

16·9 A signal frequency band extending from 100 to 4 kcps is required to represent signals to be broadcast by amplitude-modulating a 5 mcps carrier.

1. Sketch a frequency spectrum for the modulated wave. Express the bandwidth of the modulated wave as a percentage of the carrier frequency.

2. How would the percentage change if the carrier were 50 mcps?

16·10 The equation of a modulated wave is

$$f = 10(1 + 0.5 \cos 5{,}000t - 0.3 \cos 10{,}000t) \sin 5 \times 10^6 t$$

1. What are the frequency components in the modulated wave and what is the amplitude of each of them?

2. Sketch the modulation envelope and evaluate the degree of modulation for the peaks and troughs.

16·11 A tank circuit consisting of a capacitor C in parallel with an inductance L has a resonant frequency of 1 mcps. The inductance has a series resistance r such that $\omega L/r = Q = 100$ at the resonant frequency. A current source with a waveform $i = I_0 (1 + 0.4 \cos 2\pi \times 4 \times 10^3 t) \sin 2\pi \times 10^6 t$ is applied across the tank circuit. What will be the degree of modulation of the voltage appearing across the tuned circuit with this excitation?

16·12 Assume that the transistor shown in Fig. P16·12 is ideal (that is, β constant; $i_C - v_{CE}$ characteristics perfectly horizontal from $v_{CE} = 0$ to $v_{CE} = \infty$).

1. Sketch the output voltage v_{CE} in this circuit for a carrier input voltage v_C of 0.1 volts peak. Repeat for $v_C = 10$ volts peak. Compare the results with those obtained in the diode bridge modulator of Sec. 7·6. You may assume $f_c \gg f_m$.

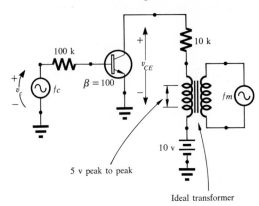

FIG. P16·12

16·13 Repeat Prob. 16·12 for $f_c = f_m$.

16·14 Assume that the output currents i_1 and i_2 of the two transistors shown in the balanced modulator circuit of Fig. 16·10 can be expressed as

$$i_1 = A_0 + A_1 v_{b_1} + A_2 v_{b_1}^2$$
$$i_2 = A_0 + A_1 v_{b_2} + A_2 v_{b_2}^2$$

1. What frequency components does the current $i_1 - i_2$ contain, assuming the transistors to be identical? The carrier input voltage is represented by $V_c \cos \omega_c t$ and the modulating signal is represented by $V_s \cos \omega_s t$.

2. If the output circuit is tuned to eliminate the modulating frequency component, what are the resulting frequency components?

3. If by further filtering the lower side band is removed and a carrier is reinserted, what would the signal waveform look like. For simplicity, let $f_c = 1$ mcps, $f_m = 1$ kcps for this sketch.

16·15 Assume that the output voltage of the amplifier shown in Fig. P16·15 can be represented by

$$v_{ce} = 20 v_{be} + 0.1 v_{be}^2$$

as long as $|v_{be}(t)| < 20$ mv

1. Determine the ratio of the second harmonic amplitude to the fundamental amplitude, if $v_{be} = 0.015 \cos \omega_s t$ volts.

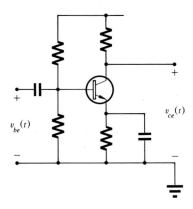

16 · 16 A voltage $v_{be}(t) = 0.005 \cos 2{,}000t + 0.01 \cos 3{,}000t$ is applied to the amplifier shown in Fig. P16 · 15. The transfer characteristic is still that given in P16 · 15. The amplifier output is applied to a filter which passes all frequencies between 100 cps and 500 cps with unity given and rejects all others. What is the output voltage of the filter?

16 · 17 Detection of an AM wave can be accomplished with the aid of a class B transistor stage such as the one shown in Fig. P16 · 17.

   1. Assuming an ideal transistor (β constant, zero saturation voltage) sketch the $i_C - i_B$ transfer characteristic for the circuit; use this characteristic to sketch v_{CE}, assuming the input is a sine-wave-modulated carrier with $m = 0.8$ and the capacitances in the output circuit can be neglected.

   2. How would you select C_1 and C_2 in a practical detector?

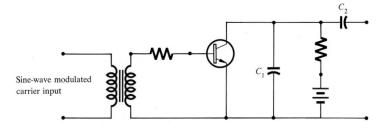

16 · 18 The signal picked up by the antenna of a given receiver is 1 μv peak for a distant station. The internal impedance of the antenna is 100 ohms. If the carrier is 50 percent modulated, how much power gain is required for the receiver to deliver an output power of 0.5 watt to an 8-ohm speaker? Assume the receiver input is matched to the antenna.

16 · 19 An AM radio broadcast station operates on a carrier frequency of 1 mcps. A receiver having the block diagram shown in Fig. P16 · 19 is used to receive the signal. The IF frequency is 455 kcps, and the station is allowed a signal frequency bandwidth of 0 to 5 kcps. Other stations in the area operate at carrier frequencies of 850 kcps, 900 kcps, and 1,100 kcps.

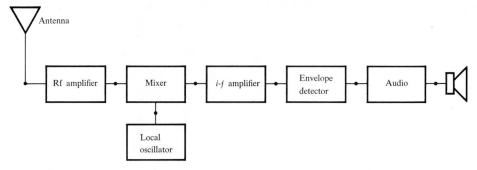

FIG. P16·19

1. Sketch the frequency spectrum of signals received by the antenna.
2. Sketch the frequency spectrum of the signals at the output of the RF amplifier, the mixer, and the IF amplifier.

16·20 A frequency converter is to be used to translate a 200-mcps carrier to 70 mcps. Give at least two local oscillator frequencies that could be used and the frequency components in the output for each local oscillator frequency.

16·21 Explain why it is more important that the frequency stability of the local oscillator be higher in a receiver intended for carrier frequencies at 30 to 100 mcps than in a receiver intended for the broadcast band.

16·22 Show that Eq. (16·24) is obtained from Eqs. (16·23a) and (16·23b), using the hint given in the text.

16·23 Find the Fourier coefficients for the function sin (m sin $\omega_s t$). Compare the results with Eq. (16·24).

16·24 A 1-kcps modulating signal has sufficient amplitude to produce a modulation index of $m_f = 4$.
1. What bandwidth is required to pass the frequency spectrum of the modulated wave, if all components with amplitude greater than 0.01 are to be passed?
2. What would be the effect of decreasing the modulation frequency to 100 cps and holding m_f fixed?
3. If the modulation *amplitude* of the 100-cps signal (and therefore the frequency deviation Δf) were the same as the 1-kcps signal amplitude), what bandwidth would be required, and approximately how many harmonic components of amplitude greater than 0.01 would there be? What conclusion do you draw from this result?

16·25 A 100-mcps carrier is to be frequency-modulated by a 10-kcps sine wave. A systems designer desiring to minimize the bandwidth required to transmit the signal arranges for a maximum frequency deviation $\Delta f = \pm 50$ cps. He then builds a receiver with an IF bandwidth of ± 1 kcps, feeling that this will be adequate to detect the signal. Is this bandwidth adequate? If not, specify the actual bandwidth.

16·26 An 80-mcps carrier having a peak amplitude of 10 volts is modulated at 10 kcps with a frequency deviation of 30 kcps. Determine the amplitude of the carrier, first-, second-, third-, fourth-, and fifth-order side-band components.

16·27 1. A modulating signal $f_s(t) = 0.1 \sin 2\pi \times 10^3 t$ is used to modulate a

10-mcps carrier in both an AM and an FM system. The 0.1-volt signal produces a 100-cps frequency deviation in the FM case. Compare the receiver IF amplifier bandwidths for the two cases.

2. Suppose the modulating signal amplitude is increased to 10 volts and the frequency deviation increases accordingly. Recompute the receiver IF amplifier bandwidths for the two cases.

16·28 The commercial FM broadcast band extends from 88 mcps to 108 mcps. The FCC allows each station a maximum frequency deviation of ± 75 kcps. If a constant-area pulse generator of the type described in the text is used to detect the signal, what is the maximum pulse width the constant-area pulse can have?

16·29 A single-tuned circuit has a phase response which is linear with frequency shift over a certain frequency range about its resonant frequency. To investigate this effect,

1. Write out the transfer function $E(s)/I(s)$ for the network shown in Fig. P16·29.

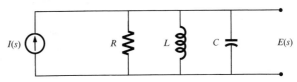

FIG. P16·29

2. Sketch and dimension a pole-zero plot for the transfer function.

3. Let $Q = \omega_c CR$, $\omega_c{}^2 = 1/LC$. Show that for $Q \gg 1$, the poles are located approximately $-\omega_c/2Q \pm j\omega_c$.

4. Show that when the current source shifts in frequency from ω_c to $\omega_c + \Delta\omega(\Delta\omega \ll \omega_c)$, the output voltage shifts in phase by

$$\phi = \tan^{-1}\left(\frac{\Delta\omega}{\omega_c/2Q}\right) \sim \frac{2Q\,\Delta\omega}{\omega_c}$$

5. If $\omega_c = 2\pi \times 100$ mcps, $\Delta f = 75$ kcps, what Q will be necessary to limit ϕ to $10°$?

6. How would you use these ideas to make a simple FM detector?

16·30 Explain why synchronizing pulses are not sent with a polarity corresponding to white.

16·31 Explain why in interlaced scanning it is necessary that the total number of lines in the picture be an odd number.

16·32 Assuming the eye can just distinguish two objects that are separated by $2'$ of arc, calculate the minimum distance from which a 6×9-in. picture can be viewed without the line structure being apparent.

16·33 Find the channel capacity required to transmit TV pictures, assuming 500,000 picture elements are required for good resolution and 10 different brightness levels are required for proper contrast. Thirty pictures per second are to be transmitted.

16·34 Assuming that a person's occupation is a word which can be found in a pocket dictionary ($\sim$300,000 entries), show that the game of 20 questions (in which you attempt to determine a person's occupation by receiving yes or no answers to questions asked) can always be won (using the dictionary).

17

LUMPED MODELS AND THE HIGH-FREQUENCY BEHAVIOR OF TRANSISTORS

IN CHAP. 9, WE DISCUSSED the basic physical theory of operation for a transistor and obtained a set of equations which can be used to describe the low-frequency performance of the intrinsic transistor in all four of its operating regions. We also saw that by making a first-order correction to the dc theory we could develop the basic T and hybrid-π models for the common-base and common-emitter connections of the intrinsic transistor. The hybrid-π model, modified by the addition of the collector junction capacitance C_c, served us well in calculating the circuit behavior of the transistor in the amplifier and oscillator circuits considered in Chaps. 13 and 14.

In modern high-frequency transistors ($f_T > 100$ mcps), however, the simple T and hybrid-π models will only give a first-order design which may require considerable refinement if measured and calculated behavior are to agree. This is particularly true in the design of band-pass amplifiers, very-high-frequency oscillators, and wideband low-pass amplifiers which, in addition to resistive loads, employ reactive elements in the interstage and feedback networks to achieve a desired response.

In this chapter we will develop an approach to transistor modelling and a general transistor model which includes all of the results previously obtained (both large and small signal) and can be simply extended into a frequency range where previous models are seriously in error. The results of this chapter will be very valuable in understanding the high-frequency circuits to be studied in the next chapter.

17·1 THE INTRINSIC TRANSISTOR

The intrinsic transistor again plays an important role in the over-all modeling process, so we will begin by developing a model for the intrinsic transistor from first principles, without restrictions as to the frequency of operation. To do this, we return to a consideration of the basic processes by which minority carriers are transported across the base region to form the collector current. We will consider a common-base structure for convenience in analysis, relating it to the common-emitter configuration at a later point.

We show in Fig. 17·1 a p^+np^+ intrinsic transistor with the base region divided into a number of elementary volumes. We assume that space-

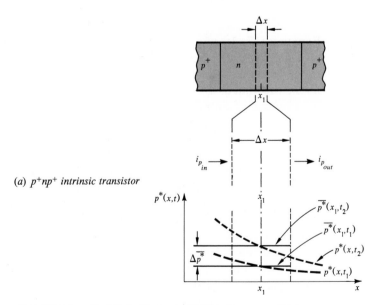

(a) p^+np^+ intrinsic transistor

FIG. 17·1 APPROXIMATE REPRESENTATION OF EXCESS HOLE DENSITIES IN A SLICE OF FINITE WIDTH Δx.

charge neutrality applies in the base, and that minority carriers flow only by diffusion. We also assume that the transistor is being driven by signals that change with time, so that the carrier densities at every point in the transistor base region are changing with time.

We now station ourselves at a plane located at x_1 and apply the ideas of carrier continuity to the volume bounded by planes located at $x_1 - \Delta x/2$ and $x_1 + \Delta x/2$, as shown in Fig. $17 \cdot 1$.[1] Curves of the excess-hole distribution $p^*(x) = p(x) - p_n$ are shown in Fig. $17 \cdot 1b$ for two times t_1 and t_2, the difference being denoted by Δt.

Now, holes which enter this volume in a time interval Δt either (1) flow out, (2) recombine, or (3) are stored. Of course, the same is true for electrons, but because we have assumed space-charge neutrality, we can deduce majority-carrier phenomena when the minority-carrier behavior has been established.

In terms of the elementary volume shown in Fig. $17 \cdot 1a$, the simple accounting for holes suggested above leads to the equation

$$i_{p,in} \, \Delta t = i_{p,out} \, \Delta t + \frac{qA \, \Delta x}{\tau_p} \bar{p}^*(x) \, \Delta t + qA \, \Delta x \, \Delta[\overline{p^*(x)}] \qquad (17 \cdot 1)$$

where $\overline{p^*(x_1)}$ is the *average* excess-hole density in the volume, and $\Delta[\overline{p^*(x_1)}]$ is the *change* in the average excess-hole density in the time interval Δt (see Fig. $17 \cdot 1b$).

Now, if we rearrange Eq. $(17 \cdot 1)$, divide by Δt and take the limit as $\Delta t \to 0$, we find

$$i_{p,in} - i_{p,out} = \frac{qA \, \Delta x}{\tau_p} \overline{p}^*(x_1) + qA \, \Delta x \, \frac{d\overline{p}^*}{dt} \qquad (17 \cdot 2)$$

Equation $(17 \cdot 2)$ is an ordinary differential equation expressing continuity of holes, but it involves the size of the slice through Δx. We can eliminate the dependence of the continuity equation on Δx by dividing by Δx and taking the limit as $\Delta x \to 0$. When we do this we obtain the partial differential equation of continuity

$$-\frac{1}{q} \frac{\partial j_p}{\partial x} = \frac{p^*(x,t)}{\tau_p} + \frac{\partial p^*(x,t)}{\partial t} \qquad (17 \cdot 3)$$

We can use this equation as a basis for developing the high-frequency terminal properties of the transistor in much the same way that the dc continuity equation [that is, Eq. $(17 \cdot 3)$ with the $\partial p^*/\partial t$ term equal to zero] was used to develop the low-frequency theory in Chap. 9.

While this is an important way to study the high-frequency properties of the structure,[2] it leads to undue mathematical complexities and tends to obscure the physical basis for the final results. This raises the question

---

[1] The analysis given here is similar to that given in Secs. $8 \cdot 4$ and $8 \cdot 5$ where diode transient performance was studied.

[2] See for example, J. G. Linvill and J. F. Gibbons, "Transistors and Active Circuits," McGraw-Hill Book Company, New York, 1961, pp. 20–24.

whether approximations can be made which will keep the principal physical features of operation before us continuously and yield quantitatively useful models for the transistor, but eliminate the necessity for solving the partial differential equation of continuity.

Such approximations can indeed be made by satisfying ourselves with an approximate calculation of $p^*(x_1 t)$ at a few well-chosen points in the base region, rather than attempting to calculate the precise behavior of $p^*(x,t)$ everywhere. To do this, we simply *leave* Δx *in Eq. (17·2) (and Fig. 17·1) finite and regard the average excess density* $\overline{p^*}(x_1)$ *as an adequate description of the excess density throughout the slice.* We then refer to the slice as a *lump.* The models which we develop using this approximation are called *lumped models.* For the present, we leave the exact size of the lump (Δx) unspecified, though when we come to actual problems we will have to choose the number and size of the lumps in some manner.

17·1·1 *A lumped model for the intrinsic transistor base region.* To develop the lumped model, we observe that since Δx is fixed, the only variable on the right-hand side of Eq. (17·2) is $\overline{p^*}(x_1)$. To emphasize this, we rewrite Eq. (17·2) in the form

$$i_{p,in} - i_{p,out} = H_c\overline{p^*}(x_1) + S_p\frac{d\overline{p^*}(x_1)}{dt} \qquad (17\cdot4)$$

where
$$H_c \equiv \frac{qA\,\Delta x}{\tau_p} \qquad S_p \equiv qA\,\Delta x \qquad (17\cdot5)$$

Note that H_c and S_p only involve geometrical factors and material parameters. Note also that, since $qA\,\Delta x$ is the volume of the slice, the average excess-hole charge stored in the volume element is simply

$$\overline{q^*} = qA\,\Delta x\,\overline{p^*}(x_1)$$

Eq. (17·4) can therefore be recast in the form

$$i_{p,in} - i_{p,out} = \frac{\overline{q^*}}{\tau_p} + \frac{d\overline{q^*}}{dt} \qquad (17\cdot6)$$

In either case [Eq. (17·4) or (17·6)], the first term on the right-hand side represents the average recombination current in the volume element at time t_1, while the second term represents the average rate at which holes are being stored in the volume element at the time t_1. Equation (17·4) stresses the linear dependence of the recombination current on $\overline{p^*}(x)$ and the linear dependence of the storage current on $d\overline{p^*}(x)/dt$.

A schematic representation of the minority-carrier behavior in the lump can be obtained by properly defining symbols and flow laws to represent the minority-carrier continuity equation in a single lump. The necessary ideas are developed in the following subsections.

The combinance. The current which results from the average recombination of excess holes and electrons in the lump is related to the average excess-hole density by

$$i_{Hc} = H_c \overline{p^*}(x_1) \tag{17.7}$$

In Fig. 17·2 we represent this current by means of a two-terminal element which we call a *combinance*. The *value* of the combinance is, in the present discussion,

$$H_c \equiv \frac{qA\,\Delta x}{\tau_p} \qquad \text{amp-cm}^3 \tag{17.8}$$

Because the recombination process involves both holes and electrons in equal numbers, its representation by a two-terminal element is quite appropriate. We think of the current in the upper lead as hole flow, whereas the equal current in the lower lead is electron flow. A box is placed at the top of the element to remind us that the current flowing through the combinance depends only on the minority-carrier density $p^*(x_1)$.

The storance. The current which results from the storage of excess carriers is related to the rate of change of the excess density by the equation

$$i_{S_p} = S_p \frac{d\overline{p^*}(x_1)}{dt} \tag{17.9}$$

In Fig. 17·3 we represent this current by means of a two-terminal element which we call a *storance*. The *value* of the storance is, in the present discussion

$$S_p \equiv qA\,\Delta x \qquad \text{coul-cm}^3 \tag{17.10}$$

Again, a two-terminal element is used to emphasize that carrier storage involves the simultaneous build-up of both hole and electron densities, as required by space-charge neutrality.

The total excess *charge* q_p stored in S at any given time is the time integral of Eq. (17·9) and is simply

$$q_P = S_p \overline{p^*}(x_1) \tag{17.11}$$

As was true for the combinance, the symbol for a storance is two-terminal but not symmetric to remind us that the current flowing in the element is dependent only on the excess-hole density.

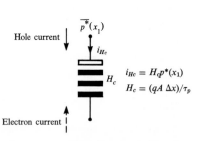

Hole current

$\overline{p^*}(x_1)$

i_{Hc}

H_c

$i_{Hc} = H_c p^*(x_1)$
$H_c = (qA\,\Delta x)/\tau_p$

Electron current

FIG. 17·2   DEFINITION AND PROPERTIES OF A COMBINANCE

$$i_{s_p} = S_p \frac{d\overline{p^*}(x_1)}{dt}$$

Hole current ↓

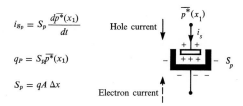

$$q_P = S_p\overline{p^*}(x_1)$$

$$S_p = qA\,\Delta x$$

Electron current

FIG. 17·3   DEFINITION AND PROPERTIES OF A STORANCE

Schematic representation of the continuity equation. We may now combine the recombination and storage elements to give a schematic representation of Eq. (17·4). This is done in Fig. 17·4. The flow laws for H_c and S_p are such that the minority-carrier continuity equation (17·4) is simply Kirchhoff's current law at the $\overline{p^*}(x_1)$ node.

Electron currents flowing into and out of the lump have also been included in Fig. 17·4. They are shown dotted, to represent the actual direction of motion of electrons. Kirchhoff's current law at the lower node in the lump is simply the *majority* carrier continuity equation for the lump.

The diffusance. Carrier-diffusion phenomena enter the lumped model as the means by which carriers are transported from one lump to the next. The diffusion process therefore provides a means of relating $i_{p,in}$ and $i_{p,out}$ to $\overline{p^*}(x_1)$ and the excess-hole densities in the two adjacent lumps.[1]

Figure 17·5 has been prepared to facilitate discussion. The actual hole-concentration profile at a given time t is dotted on this figure. The lumped approximation to it is drawn in solid.

The actual current flowing in across the left face of lump 2 is

$$i_p\bigg|_{x_0} = -qAD_p \frac{\partial p^*}{\partial x}\bigg|_{x_0} \tag{17·12}$$

Since we are representing phenomena within each lump by average values of $\overline{p^*}(x)$, we must approximate Eq. (17·12) by

$$i_p\bigg|_{x_0} = -qAD_p \frac{\overline{p^*}(x_2) - \overline{p^*}(x_1)}{\Delta x_{21}} \tag{17·13}$$

where Δx_{21} is the distance between the centers of the two lumps. We regard Δx_{21} as fixed, and rewrite Eq. (17·13) as

$$i_p\bigg|_{x_0} = H_d[\overline{p^*}(x_1) - \overline{p^*}(x_2)] \tag{17·14}$$

[1] Of course, if an electric field is present, holes can move by drift as well as diffusion. In a transistor with a uniformly doped base region at low injection levels, we can neglect this effect. However, drift motion is very important if the base is not uniformly doped, as we shall show later.

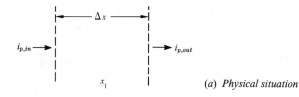

(a) Physical situation

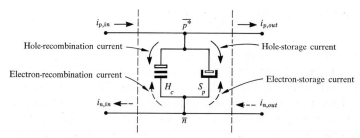

Hole-recombination current

Hole-storage current

Electron-recombination current

Electron-storage current

(b) Lumped model for (a) with elements defined to account for recombination and storage currents

$$i_{p,in} - i_{p,out} = H_c\overline{p^*} + S_p \frac{d\overline{p^*}}{dt}$$

FIG. 17·4   SCHEMATIC REPRESENTATION OF THE CONTINUITY EQUATION IN A LUMP

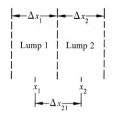

Lump 1 | Lump 2

(a) Physical arrangement

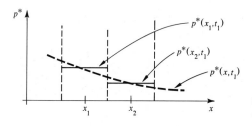

(b) Excess hole density in two adjacent lumps and lumped approximations

(c) Definition of a diffusance and its associated flow laws

$$H_d \equiv \frac{qAD}{\Delta x_{21}}$$

$$i_d \equiv H_d[\overline{p^*}(x_1) - \overline{p^*}(x_2)]$$

FIG. 17·5   LUMPED REPRESENTATION OF FLOW BY DIFFUSION

In Eq. (17·14) the only variables are $\overline{p}^*(x_1)$ and $\overline{p}^*(x_2)$. H_d is a coefficient which depends only on material parameters and geometrical factors.

We are thus led to define a third element called a *diffusance*. Diffusance is given the symbol H_d. Its *value* is

$$H_d = \frac{qAD_p}{\Delta x_{21}} \qquad (17 \cdot 15)$$

and its dimensions are in amperes-cubic centimeter. The current flowing from *left to right* in H_d is defined by

$$i_d = H_d[\overline{p}^*(x_1) - \overline{p}^*(x_2)] \qquad (17 \cdot 16)$$

Note that, in contrast to recombination and storage, diffusion involves only one carrier type. Furthermore, diffusion is a bilateral phenomenon. That is, if $\overline{p}^*(x_2)$ is greater than $\overline{p}^*(x_1)$, holes will diffuse from right to left. Hence, H_d must be represented by a symmetrical symbol, whereas H_c and S_p are represented by nonsymmetrical symbols. No distinctive marks are made on the terminals of the diffusance element.

Complete lumped model for hole flow in the transistor base region. With the aid of the diffusance element, we can draw a lumped model to represent all of the hole-flow processes in the base. This is shown in Fig. 17·6. Diffusance elements are simply used to tie adjacent nodes together.

Of course, electrons also diffuse *and* drift between lumps, so there should be elements in the electron line to account for these transport processes. However, for any problem where the terminal behavior can be obtained by studying minority-carrier processes only, this is not necessary. Majority-carrier phenomena are determined in the transistor base region from

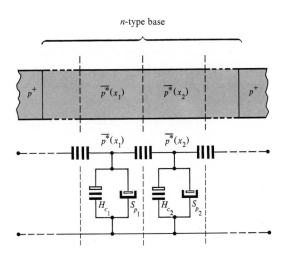

FIG. 17·6 COMPLETE LUMPED MODEL FOR HOLE FLOW IN THE INTRINSIC TRANSISTOR BASE

current-continuity and space-charge neutrality requirements. Furthermore, the electron densities are nearly the same in adjacent lumps, so we may for simplicity neglect the actual electron-transport processes and connect adjacent electron nodes together by a wire. This will then limit the lumped model to low-level injection conditions in an extrinsic semiconductor.[1]

17·1·2 *A lumped representation of the transition region.* We have now formulated a lumped representation for minority-carrier transport in a neutral semiconductor bar. We shall now develop a lumped model for the *pn* junction and show how it may be connected to the lumped model just formulated to build up a model for the entire transistor.

There are two ways in which this may be done. The most general procedure is to further develop the lumped model so that it may be extended right through the emitter and collector transition regions. To do this, we have to devise a lumped representation for the drift flow of both types of carriers and provide a lumped representation of Gauss' law (via appropriately connected interlump electric capacitances). When this is done, a complete lumped model can be constructed for the entire transistor. Separate parts of the transistor such as the neutral *p* region, the neutral *n* region and the transition regions are then distinguished only by the fact that assumptions which are made in the analysis allow one to neglect some of the elements of the lumped model in these separate regions.

This procedure is a satisfying one in the sense that one starts with a diagrammatic model which includes all the physical processes and then specializes it for individual problems. However, such a development is also very lengthy, and the principal end results can be easily understood without reference to the entire argument. For our purposes, it will suffice to summarize the properties of the lumped model of the transition region.

The necessary ideas are illustrated in Fig. 17·7. We recall that, by assumption, the voltage applied to a junction appears entirely across the transition region. The effects of this bias voltage are

1. To change the width of the space-charge layer to accommodate the applied voltage v_a

2. To establish minority-carrier densities on each side of the transition region:

$$p^*(x = 0) = p_n(e^{qv_a/kT} - 1) \qquad n^*(y = 0) = n_p(e^{qv/kT} - 1) \quad (17 \cdot 17)$$

The first effect may be accounted for as charge supplied to the junction capacitance C_j. The charging current is a majority-carrier current on each side of the junction.

The second effect serves to establish the minority-carrier densities which are required to produce a minority-carrier current. Since we are studying a

---

[1] Less restricted lumped models have been developed which characterize semiconductors without limitations on doping level or injection level. See J. G. Linvill and J. F. Gibbons, *op. cit.*, part 1.

(a) *pn junction with transition region*

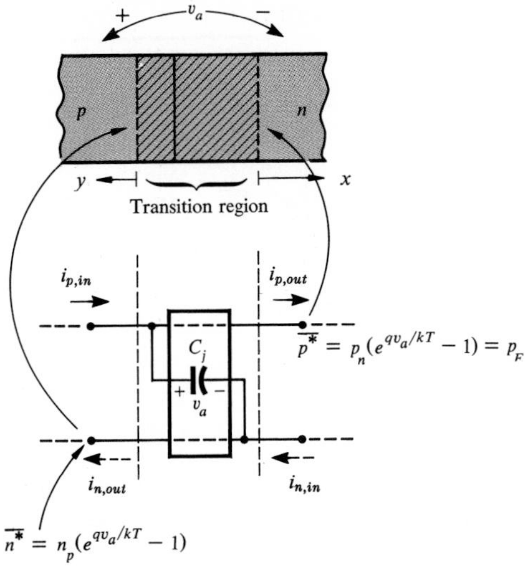

Transition region

$$\overline{p^*} = p_n(e^{qv_a/kT} - 1) = p_F$$

(b) *Lumped representation of (a) utilizing C_j to represent junction capacitance and junction laws to define excess minority carrier densities in the neutral regions on each side of the junction*

$$\overline{n^*} = n_p(e^{qv_a/kT} - 1)$$

FIG. **17·7**   LUMPED REPRESENTATION OF A *pn* JUNCTION

p^+np^+ structure, we can neglect electron injection into the emitter and collector bodies.

We may represent these effects in the lumped model suggested in Fig. 17·7. A box is drawn around the transition region. A capacitance C_j is connected from the hole-current line on the p side to the electron-current line on the n side. The current which flows in this element is

$$i_{C_j} = \frac{d}{dt}(C_j v_a) \tag{17·18}$$

where v_a is the applied voltage. As suggested in Fig. 17·7, this current is a majority current, and the *junction voltage drop appears across C_j*.

The injection effect is represented by simply defining the excess minority-carrier densities at the appropriate nodes by means of the law of the junction.

This model of the junction can now be tied directly to the lumped model for the transistor base region, as illustrated in Fig. 17·8. To do this we must, of course, set the average excess density in the lump nearest to each junction equal to the excess density at the junction. The width of the lumps at the edges must then be small enough so that the error between the actual average density and the edge density is not too large.

17·2   A TWO-LUMP MODEL FOR THE INTRINSIC TRANSISTOR

We now wish to show how the lumped model is used to calculate the performance of the transistor. The first step in such a calculation is to decide how many lumps should be used and what their sizes should be. A small number of lumps will simplify calculations but lead to only approxi-

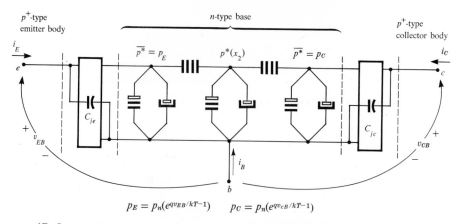

$$p^* = p_E \qquad p^*(x_2) \qquad \overline{p^*} = p_C$$

$$p_E = p_n(e^{qv_{EB}/kT-1}) \qquad p_C = p_n(e^{qv_{CB}/kT-1})$$

FIG. 17·8 COMPLETE LUMPED MODEL FOR THE INTRINSIC TRANSISTOR. THE BASE REGION IS REPRESENTED BY THREE LUMPS FOR CONVENIENCE.

mate results. Increasing the number of lumps will increase the accuracy of the results but will also require increased computational effort.

The simplest possible model for a transistor base region is a two-lump model. We require two lumps because the emitter and collector junctions must have separate control over the densities in the lumps adjacent to the junctions. Of course, a two-lump model gives only a very crude approximation to the actual distributed behavior of $p^*(x_1t)$, but it gives a very *good* approximation to the terminal characteristics of the device. In fact, the two-lump model contains all of the results of the Ebers-Moll model, the stored charge model, and the basic T and hybrid-π small-signal models. This simply means that the low- and medium-frequency terminal performance of the transistor is very insensitive to the detailed behavior of $p^*(x_1t)$ within the base region.

We will briefly show the equivalence between these various models and then show how to improve our estimates of terminal performance at medium and high frequencies by adding additional lumps to the structure.

17·2·1 Selection of parameters for a two-lump model. To select the parameters for a two-lump model, we divide the base in half, as in Fig. 17·9. This division of the base gives

$$\Delta x_1 = \Delta x_2 = \frac{W}{2}$$

for the size of each lump and

$$\Delta x_{12} = \frac{W}{2}$$

for the spacing between centers of the lumps. Using the previous formulas for the lumped-model parameters, we then have

$$H_{c_1} = \frac{qAW}{2\tau_p} \qquad H_{c_2} = \frac{qAW}{2\tau_p} \qquad H_d = \frac{2qAD}{W}$$

$$S_{p_1} = \frac{qAW}{2} \qquad S_{p_2} = \frac{qAW}{2} \tag{17·19}$$

In addition, our model of the junction requires that we assume the average density in the lump nearest the emitter to be equal to p_E, and similarly for the collector.

To estimate the accuracy which we can expect this model to give, we appeal to our condition that the actual density in each lump not be significantly different from the average value we assign to it. The effects which we can expect to occur if we use the H and S parameters just given, with p_E and p_C as the average densities in the lumps, are indicated in Fig. 17·9. First of all, it is clear that using p_E and p_C as the average densities in the two-lump model assigns too much recombination and charge storage to the emitter half of the base and too little to the collector half, though the total is correct. It is also clear that the average slope of the hole-density profile at $x = W/2$ will be estimated to be *twice* as large as it actually is, if we use $\Delta x_{12} = W/2$. In other words, the diffusion current between the lumps will be twice as large as it should be, if we set $\Delta x_{12} = W/2$.

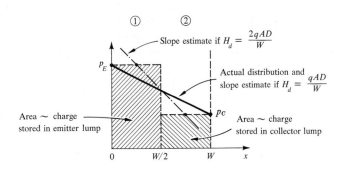

(a) *Excess hole density distribution in the intrinsic transistor base and lumped approximation to it*

(b) *Two-lump model for the intrinsic transistor with H_d selected on the basis of edge densities p_E and p_c rather than average densities*

$$H_{c_1} = H_{c_2} = \frac{qAW}{2\tau_p}$$

$$H_d = \frac{qAD}{W}$$

$$S_{p_1} = S_{p_2} = \frac{qAW}{2}$$

$i_{H_{c_1}} = H_{c_1}p_E$

$i = H_d(p_E - p_c)$

$p_E = p_n(e^{qv_{EB}/kT} - 1) \qquad p_C = p_n(e^{qv_{CB}/kT} - 1)$

$i_{S_{p_1}} = S_{p_1}\dfrac{dp_E}{dt} \qquad i_{H_{c_2}} = H_{c_2}p_C \qquad i_{S_{p_2}} = S_{p_2}\dfrac{dp_c}{dt}$

FIG. 17·9   A TWO-LUMP MODEL FOR THE INTRINSIC TRANSISTOR

Now, this diffusion current is very important. It is the mechanism which couples the two lumps together and is the basis for the transistor action. It is reasonable to expect that a more accurate estimate of the properties of the structure would be obtained by estimating the diffusion current between the lumps more accurately.

One way to do this is to add more lumps. We can also achieve the same effect in a two-lump model by noting that since we are using *edge* densities in the lumps to be representative of the average density, the average slope is more correctly calculated by setting $\Delta x_{12} = W$, thus making

$$H_d = \frac{qAD}{W} \qquad (17 \cdot 20)$$

We therefore arrive at the model shown in Fig. $17 \cdot 9b$. The currents which flow in this model are determined by the densities which exist at the *edges* of the lumps. To keep this in mind, we will refer to the model as an *edge-density model*, or a π-section model. Such a model correctly estimates the interlump diffusion current and the total recombination current and charge stored in the base, though the division of charge between the lumps is incorrect.

Determination of the lumped-model parameters. The actual values of the lumped-model parameters to be used in this model can be determined if the device geometry and material parameters are given. Thus, one can proceed from device design data directly to a model, or vice versa.

A more typical engineering situation is that one is given a device whose properties must be determined by measurements made on the device terminals. These measurements will be in the form of saturation currents, α's, β's, and so on, and can be converted to H and S parameters, if necessary. Usually the conversion step is inessential or at least requires a minimum of effort, as the following examples show.

17·2·2 *Applications of the two-lump model.* In this section we will show by means of several examples that the two-lump model is equivalent to the large-signal models we have derived previously, and we will see how the parameters of the two-lump model are related to externally measurable quantities. The reader should note carefully that each time we use the lumped model we are, in effect, solving the minority-carrier continuity equation (approximately) and applying the appropriate boundary conditions. The basic technique is therefore quite fundamental.

EXAMPLE 1. *Ebers-Moll model.* To show the connection between the Ebers-Moll model and the two-lump model, we first write the dc equations which govern the two-lump model. From Fig. $17 \cdot 10$ we have

$$i_E = (H_d + H_{c_1})p_E - H_d p_C \qquad i_C = -H_d p_E + (H_d + H_{c_2})p_C \qquad (17 \cdot 21)$$

where p_E and p_C are defined by the junction laws:

$$\frac{p_E}{p_n} = (e^{qv_{EB}/kT} - 1) \qquad \frac{p_C}{p_n} = (e^{qv_{CB}/kT} - 1) \qquad (17 \cdot 22)$$

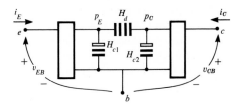

$$i_E = p_E(H_{c_1} + H_d) - p_C H_d \qquad\qquad i_C = -p_E H_d + p_C(H_d + H_{c_2})$$

(a) *General dc equations for the two-lump model*

$$p_E = p_n(e^{qv_{EB}/kT} - 1) \qquad\qquad p_E = p_n(e^{qv_{CB}/kT} - 1)$$

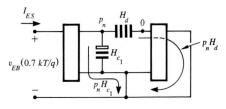

$$I_{ES} \equiv i_E \bigg|_{\substack{v_{CB} = 0 \\ p_E = p_n}} = p_n H_{c_1} + p_n H_d = p_n(H_{c_1} + H_d)$$

(b) *Lumped-model interpretation of the short-circuit emitter saturation current* I_{ES}

$$-i_C = \frac{H_d}{H_d + H_{c_1}} i_E \equiv \alpha i_E$$

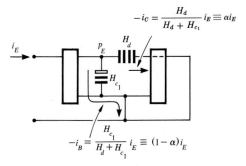

(c) *Lumped-model interpretation of* α *as the division of emitter current between* H_d *and* H_{c_1}

$$-i_B = \frac{H_{c_1}}{H_d + H_{c_1}} i_E \equiv (1 - \alpha) i_E$$

FIG. 17·10   THE EQUIVALENCE OF THE TWO-LUMP MODEL TO THE EBERS-MOLL MODEL

Now in the Ebers-Moll representation, we had [cf. Eqs. (9·45a) and (9·45b)]

$$i_E = I_{ES}\left(\frac{p_E}{p_n}\right) - \alpha_R I_{CS}\left(\frac{p_C}{p_n}\right) \qquad i_C = -\alpha_F I_{ES}\left(\frac{p_E}{p_n}\right) + I_{CS}\left(\frac{p_C}{p_n}\right) \quad (17·23)$$

which must also be used in conjunction with the junction laws, Eqs. (17·22).

The equivalence between Eqs. (17·21) and (17·23) is apparent. Formally, we obtain

$$I_{ES} = p_n(H_d + H_{c_1}) \qquad I_{CS} = p_n(H_d + H_{c_2})$$
$$\alpha_F = \frac{H_d}{H_d + H_{c_1}} \qquad \alpha_R = \frac{H_d}{H_d + H_{c_2}} \qquad\qquad (17·24)$$

The formal equivalence means, of course, that the dc properties of the two

Lumped models and the high-frequency behavior of transistors 743

models will be the same and that we can use Eqs. (17·24) to measure the H parameters if we need to know them (actually we measure p_n times each parameter, though this is not important, since the junction law ultimately takes care of the factor p_n when we relate currents to voltages, or vice versa).

Apart from its formal equivalence to the Ebers-Moll model, note the conceptual value of the lumped formulation. For example, to measure I_{ES}, we set $p_C = 0$ (this means we connect the collector to the base by a wire, so $v_{CB} = 0$) and $p_E = p_n$ (by applying $v_{EB} = 0.7\ kT/q$). We can see by inspection of Fig. 17·10b that the internal result of applying these biases is (1) a current $p_n H_d$ flows into the collector circuit and (2) a current $p_n H_{c_1}$ flows into the base.

Again, if we drive the emitter with a given current and measure the short-circuit collector current, the formal result [from Eqs. (17·21)] will be

$$i_C = -\frac{H_d}{H_d + H_{c_1}} i_E \quad \text{or} \quad \alpha_F = \frac{H_d}{H_d + H_{c_1}}$$

However, this result is more simply obtained by inspection of the lumped model shown in Fig. 17·10c. The current that flows into the emitter is simply divided between the two conductance-like elements H_d and H_{c_1}. A fraction $H_d/(H_d + H_{c_1})$ flows through H_d and a fraction $H_{c_1}/(H_d + H_{c_1})$ flows through H_{c_1}.

Note that by substituting the values of the H parameters in terms of device parameters, we find

$$\alpha_F = \frac{1}{1 + (H_{c_1}/H_d)} = \frac{1}{1 + (W^2/2D_p\tau_p)} = \frac{1}{1 + (W^2/2L_p^2)}$$

in correspondence with the dc theory presented in Chap. 9. Thus, the lumped model provides us with a convenient way of visualizing current flow in the transistor base, separating it into recombination and diffusion components and relating these current components to external circuit parameters.

EXAMPLE 2. *The stored-charge model.* The formal content of the stored-charge model is given in the set of three equations

$$i_E = q_F \left(\frac{1}{\tau_t} + \frac{1}{\tau_p}\right) + \frac{dq_F}{dt} - q_R \frac{1}{\tau_t}$$

$$i_C = -q_F \frac{1}{\tau_t} + q_R \left(\frac{1}{\tau_t} + \frac{1}{\tau_p}\right) + \frac{dq_R}{dt} \quad (17·25)$$

$$-i_B = \frac{q_R + q_F}{\tau_p} + \frac{d(q_R + q_F)}{dt}$$

where q_F and q_R are two component charges which make up the total charge stored in the base (see Fig. 17·11). τ_p is the minority-carrier lifetime in the base, and τ_t is the transit time for holes across the base region. τ_p and τ_t are called *stored-charge* parameters or *charge-control* parameters, since they are the constants that need to be known in Eqs. (17·26) for actual numerical work.

The two-lump model equivalent of these results is obtained by writing out the terminal currents in the model shown in Fig. 17·11c:

$$i_E = (H_d + H_{c_1})p_E + S_{p_1}\frac{dp_E}{dt} - H_d p_C$$

$$i_C = -H_d p_E + (H_d + H_{c_2})p_C + S_{p_2}\frac{dp_C}{dt} \quad (17·26)$$

$$-i_B = H_{c_1}p_E + S_{p_1}\frac{dp_E}{dt} + H_{c_2}p_C + S_{p_2}\frac{dp_C}{dt}$$

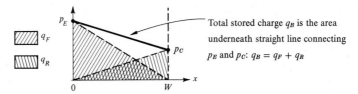

Total stored charge q_B is the area underneath straight line connecting p_E and p_C: $q_B = q_F + q_R$

(a) Division of base charge into q_F and q_R components

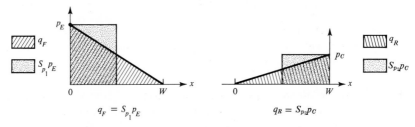

$q_F = S_{p_1} p_E$

$q_R = S_{p_2} p_C$

(b) Equivalence between q_F and $S_{p_1} p_E$ and q_R and $S_{p_2} p_C$

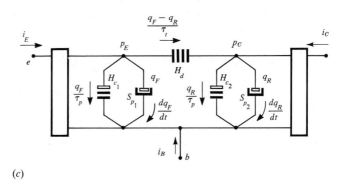

(c)

FIG. 17·11   THE EQUIVALENCE BETWEEN THE TWO-LUMP MODEL AND THE STORED-CHARGE MODEL

The charge stored in the emitter lump is defined by

$$q_E = S_{p_1} p_E \qquad (17 \cdot 27a)$$

and the charge stored in the collector lump is defined by

$$q_C = S_{p_2} p_C \qquad (17 \cdot 27b)$$

these being equivalent to q_F and q_R (see Fig. 17·11b). Furthermore, from the definitions of the lumped-model parameters for the intrinsic transistor

$$\frac{H_d}{S_{p_1}} = \frac{H_d}{S_{p_2}} = \frac{2D}{W^2} \equiv \frac{1}{\tau_t} \qquad (17 \cdot 28a)$$

$$\frac{H_{c_1}}{S_{p_1}} = \frac{H_{c_2}}{S_{p_2}} = \frac{1}{\tau_p} \qquad (17 \cdot 28b)$$

Using Eqs. (17·27) and (17·28) we can readily rewrite Eqs. (17·26) in the stored-charge form.

Notice that if one of the charge-control parameters (τ_p or τ_t) is given, then S_{p_1}

can be evaluated. Since we are dealing with an intrinsic transistor, of course, $S_{p_1} = S_{p_2}$. Note also that the ratio of Eq. (17·28a) to Eq. (17·28b) is

$$\frac{H_d}{H_{c_1}} = \beta = \frac{\tau_p}{\tau_t}$$

which was one of the first relations discussed for the stored-charge model in Sec. 9·5.

Basically the stored-charge model and the two-lump model are identical to each other, involving only a transformation of variables ($p_E \leftrightarrow q_F$, $p_C \leftrightarrow q_R$). The mathematics involved in making any calculation with either model is identical in form, though the parameters of the model look different. The similarity is stressed in Fig. 17·11c, where the currents flowing in the lumped elements are written in terms of the charge components q_F and q_R.[1]

17·2·3 Small-signal models for the intrinsic transistor.

When the lumped model is used for large-signal calculations, the relation between currents and junction voltages usually remains implicit. When the terminal currents are given, the carrier densities p_E and p_C can be calculated, and then the junction law can be used to calculate the junction voltages as functions of time, or vice versa. In a small-signal model, however, we can eliminate the intermediate step of calculating the densities. In the process, we obtain a model which has R's, C's and generators in it, and circuit calculations become considerably simpler.

We will consider two basic models in this section, the T and the hybrid-π, showing how they are both obtained from the two-lump model by making suitable small-signal approximations.

The small-signal approximations to be made involve the following ideas. First, we express the total value of any variable in the form

$$y_A(t) = Y_A + y_a(t) \qquad (17·29)$$

where Y_A is the dc part and $y_a(t)$ the small-signal part. In particular,

$$i_E(t) = I_E + i_e(t) \qquad (17·30a)$$

$$v_{EB}(t) = V_{EB} + v_{eb}(t) \qquad (17·30b)$$

$$p_E(t) = P_E + p_e(t) \qquad (17·30c)$$

The junction law relates $p_E(t)$ and $v_{EB}(t)$ by

$$p_E(t) = p_n(e^{[qv_{EB}(t)/kT]} - 1) \qquad (17·31)$$

Using Eqs. (17·30b) and (17·30c) in Eq. (17·31) we can show that

$$P_E = p_n(e^{qV_{EB}/kT} - 1) \qquad (17·32a)$$

$$p_e(t) \simeq (P_E + p_n)\frac{qv_{eb}(t)}{kT} \simeq \frac{qv_{eb}(t)}{kT}P_E \qquad (17·32b)$$

---

[1] If the lumped model or the stored-charge model is used to represent a nonsymmetrical transistor, then $H_{c_1} \neq H_{c_2}$, $S_{p_1} \neq S_{p_2}$, and it may not be correct to use τ_p for both recombination currents (see Prob. 17·5).

The approximation being valid when $|v_{eb}| \ll kT/q$.

The small-signal models are all developed for the normal-active region, where the collector-base junction is reverse-biased. As a result, Eqs. (17·32) give

$$P_C = -p_n \qquad p_c(t) = 0 \qquad (17·33)$$

The T model. Let us now consider the common-base connection of the transistor shown in Fig. 17·12. The small-signal equations for the lumped model are

$$i_e = (H_{c_1} + H_d)p_e + S_{p_1}\frac{dp_e}{dt} \qquad i_c = -H_d p_e \qquad (17·34)$$

If we use Eq. (17·32b), we can rewrite these equations to show the dependence of i_e and i_c on v_{eb}. The result is

$$i_e = (H_{c_1} + H_d)P_E\frac{q}{kT}v_{eb} + S_{p_1}P_E\frac{q}{kT}\frac{dv_{eb}}{dt} \qquad (17·35a)$$

$$i_c = -H_d P_E\frac{q}{kT}v_{eb} \qquad (17·35b)$$

Now, these equations are in the form

$$i_e = g_e v_{eb} + C_e\frac{dv_{eb}}{dt} \qquad (17·36a)$$

$$i_c = -g_m v_{eb} \qquad (17·36b)$$

where[1]

$$\frac{1}{r_e} = g_e = [(H_{c_1} + H_d)P_E]\frac{q}{kT} = \frac{qI_E}{kT} \qquad (17·37a)$$

$$C_e = [S_{p_1}P_E]\frac{q}{kT} = \frac{Q_F q}{kT} \qquad (17·37b)$$

$$g_m = [H_d P_E]\frac{q}{kT} = \alpha g_e = \frac{qI_C}{kT} \qquad (17·37c)$$

They can therefore be visualized in the manner shown in Fig. 17·12b. Here, the parameter r_e is our old friend kT/qI_E. Equation (17·37b) shows that the capacitance C_e in parallel with it, which is also bias-dependent, has a value such that it stores a charge equal to Q_F when it has a voltage of kT/q across it.

Note that the current generator in the collector circuit has a value determined by v_{eb}. While this is perfectly legitimate, this is not the usual form for the model. Since the emitter impedance is low, the emitter current will more likely be known than v_{eb}. As a result, it is customary to express the output current as a function of i_e.

On account of the frequency-dependent impedance in the emitter circuit,

---

[1] The identifications between the lumped-model parameters and the dc currents and charges are obtained from Eqs. (17·21a), (17·21b), and (17·27a).

the output-current generator will have a frequency dependence. This is shown in the following steps. First we express $i_e(t)$ and $i_c(t)$ in terms of the complex frequency variable s:

$$i_e(t) = I_e(s)e^{st} \tag{17·38a}$$

$$i_c(t) = I_c(s)e^{st} \tag{17·38b}$$

Then, from Eqs. (17·36) or Fig. 17·13b we obtain

$$I_e(s) = (g_e + sC_e)V_{eb}(s) \tag{17·39a}$$

$$I_c(s) = -g_m V_{eb}(s) \tag{17·39b}$$

The ratio of $I_c(s)$ to $I_e(s)$ is then

$$\alpha(s) = -\frac{I_c(s)}{I_e(s)} = \frac{g_m}{g_e + sC_e} = \frac{\alpha_0}{1 + s/\omega_\alpha} \tag{17·40}$$

where α_0 is the low-frequency α and ω_α is defined by

$$\omega_\alpha = \frac{1}{C_e r_e}$$

Using Eqs. (17·37a) and (17·37b), we find

$$\omega_\alpha = \frac{H_{c_1} + H_d}{S_{p_1}} \simeq \frac{2D}{W^2}$$

In the complex-frequency plane, the α given in Eq. (17·40) exhibits a single pole at $s = -\omega_\alpha + j0$. Since this is only an approximation to the true $\alpha(s)$, Eq. (17·40) is referred to as a single-pole approximation to α, or briefly a *single pole* α.

The new small-signal model based on $I_e(s)$ is shown in Fig. 17·12c. This model has added to it the junction capacitances from the lumped model in Fig. 17·12a and a parasitic base resistance r_b. Usually $C_{je} \ll C_e$, so it is neglected, and the emitter current in the external circuit is considered to be equal to the current which flows in the $r_e C_e$ combination.

Note that specifying f_α provides another means of determining S_{p_1}. Hence, a small-signal measurement can be used to determine a parameter in the lumped model, which is *not* limited to small-signal conditions. This is the basis of the method developed by Moll for computing large-signal transient performance of transistors.[1]

Note also that the lumped model gives a very simple explanation of the frequency-dependent α. By inspection of Fig. 17·12a, we can see that changes in the emitter current i_e have to charge S_{p_1} before they can change p_e and produce collector current changes.

The hybrid-π model. We can obtain the basic hybrid-π model for the common-emitter connection by a simple reorientation of the model shown

---

[1] J. L. Moll, Large signal transient behavior of junction transistors, *Proc. IRE* vol. 42, pp. 1773–1784, Dec. 1954.

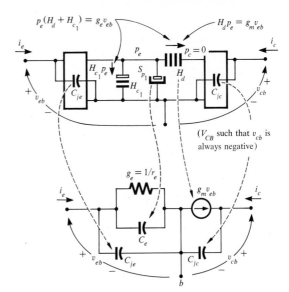

(a) Lumped model for common base small-signal operation

(V_{CB} such that v_{cb} is always negative)

(b) Small-signal model developed from equations for lumped model

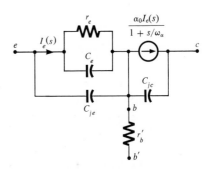

(c) Small-signal model with current generator in collector circuit expressed in terms of $I_e(s)$

FIG. 17·12   DEVELOPMENT OF SMALL-SIGNAL T-MODEL FROM TWO-LUMP MODEL

in Fig. 17·12b or 17·12c, followed by appropriate circuit transformations. However, it is simpler to return to the two-lump model and develop the hybrid-π model directly. The procedure is outlined in Fig. 17·13.

The small-signal equations for the lumped model shown in Fig. 17·13a are

$$-i_b = H_{c_1} p_e + S_{p_1} \frac{dp_e}{dt} \qquad (17·41a)$$

$$i_c = -H_d p_e \qquad (17·41b)$$

Using the p_e-to-v_{eb} transformation gives us

$$-i_b = H_{c_1} P_E \frac{q}{kT} v_{eb} + S_{p_1} P_E \frac{q}{kT} \frac{dv_{eb}}{dt} \qquad (17·42a)$$

$$i_c = -H_d P_E \frac{q}{kT} v_{eb} \qquad (17·42b)$$

Lumped models and the high-frequency behavior of transistors   **749**

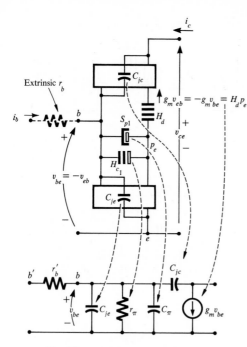

(a) Two-lump model for common emitter small-signal operation

(b) Small-signal model obtained from (a)

FIG. 17·13   DEVELOPMENT OF THE HYBRID-π MODEL FROM THE TWO-LUMP MODEL

Since the emitter is used as the common lead, it is convenient to use the base-emitter voltage v_{be} instead of v_{eb} in Eqs. (17·42). When we substitute

$$v_{be} = -v_{eb}$$

into these equations, we obtain

$$i_b = g_\pi v_{be} + C_\pi \frac{dv_{be}}{dt} \tag{17·43a}$$

$$i_c = g_m v_{be} \tag{17·43b}$$

where

$$g_\pi = \frac{1}{r_\pi} = \frac{qI_B}{kT} = \frac{qI_E}{kT}\frac{1}{\beta+1} \tag{17·44a}$$

$$C_\pi r_\pi = \frac{S_{p_1}}{H_{c_1}} = \tau_p \tag{17·44b}$$

$$g_m = \frac{\alpha_0 q I_E}{kT} \tag{17·44c}$$

The small-signal model resulting from Eqs. (17·43a) and (17·43b) is shown in Fig. 17·13b, along with parasitic junction capacitances which are brought down directly from the transition regions in the lumped model.

Notice that in both the T and hybrid-π models, the small-signal voltage-to-density transformation directly transforms H's and S's into g's, C's and generators. The values of each element can be determined from the dc conditions which exist in the lumped model.

Note also that C_π in the hybrid-π model and C_e in the T model both arise from S_{p_1}, and are equal to each other. In the hybrid-π model, because r_π is relatively large, we choose to leave the current generator in the collector circuit expressed in terms of the base-emitter voltage of the intrinsic transistor. As a result, the generator has no frequency dependence. All of the frequency dependence in the hybrid-π model is contained in the $r_\pi C_\pi$ combination, where a base current will produce a frequency-dependent base-emitter voltage.

If we wish to know the frequency-dependence of i_c on i_b, we may use the hybrid-π model directly to obtain

$$\beta(s) = \frac{I_c(s)}{I_b(s)} = \frac{g_m r_\pi}{1 + sC_\pi r_\pi} = \frac{\beta_0}{1 + sC_\pi r_\pi}$$

The frequency at which $\beta(s)$ is unity is called f_T. For reasonably large β_0, f_T is given by

$$f_T = \frac{\beta_0}{C_\pi r_\pi} = \frac{1}{C_e r_e} \left(\frac{\beta_0}{\beta_0 + 1} \right) \simeq f_\alpha$$

In other words, f_T and f_α are essentially the same in the single-pole approximation and may be used interchangeably. However, in high-frequency transistors a single-pole approximation to α is not adequate, and f_T and f_α are different, as we shall see. In such cases f_T is usually quoted instead of f_α.[1]

17·3   HIGH-FREQUENCY PERFORMANCE OF THE INTRINSIC TRANSISTOR

If we wish to know how accurately the simple T and hybrid-π models approximate the behavior of a real transistor, we can make a series of terminal measurements on the device and compare the measured performance with that computed from the models, or we can compare the simple theory with the more accurate (but still idealized) distributed theory based on the partial differential equation of continuity.

Such comparisons show that the T model requires major surgery to represent the transistor adequately at high frequencies ($f_T/5 \lesssim f \lesssim f_T$), while only minor changes are necessary in the hybrid-π model. In this section we will explore the causes for this difference and develop suitable high-frequency models for transistors which have a uniform doping in the base (for example, alloy transistors).

We shall concentrate first on the frequency dependence of the α of an intrinsic transistor. We found in Chap. 9 that the dc α of the intrinsic

[1] Instead of quoting f_T, manufacturers sometimes quote $|\beta|$ at a sufficiently high frequency f_0 that $\beta \cong \beta_0/2\pi f_0 C_\pi r_\pi$. In such cases $\beta(f_0)f_0 = f_T$.

transistor could be expressed as

$$\alpha = \operatorname{sech} \frac{W}{L_p}$$

When frequency dependence of the form e^{st} is assumed for all quantities, the partial differential equation of continuity (17·3) can be written in the form

$$\frac{\partial^2 p^*}{\partial x^2} = \frac{p^*(1 + s\tau_p)}{L_p{}^2} = \frac{p^*}{(L_p')^2}$$

and then its solution can be found as before. We can in fact simply replace the dc diffusion length by a *complex diffusion length*

$$L_p' = L_p(1 + s\tau_p)^{-1/2}$$

in all of the dc results and thereby obtain the equivalent small-signal results as a function of the complex frequency variable s. In particular, this piece of mathematical footwork gives us

$$\alpha(s) = \operatorname{sech} \left[\frac{W}{L_p}(1 + s\tau_p)^{1/2} \right]$$

which, for real frequencies, becomes

$$\alpha(j\omega) = \operatorname{sech} \left[\frac{W}{L_p}(1 + j\omega\tau_p)^{1/2} \right] \qquad (17 \cdot 45)$$

By comparison, the simple single pole α is

$$\alpha(j\omega) = \frac{\alpha_0}{1 + j\dfrac{\omega}{\omega_\alpha}} \qquad (17 \cdot 46)$$

where

$$\alpha_0 = \frac{1}{1 + \dfrac{W^2}{2L_p{}^2}} \simeq \operatorname{sech} \frac{W}{L_p}$$

An instructive comparison of these two expressions is obtained by plotting the real and imaginary parts of α (or its magnitude and phase) as a function of frequency. The results are shown in Fig. 17·14, for the idealized case where $\alpha_0 = 1$. The hyperbolic secant spirals around the origin as $\omega \to \infty$, while the simple one-pole expression gives a semicircular plot, exhibiting a maximum 90° of phase shift as $\omega \to \infty$.

The α cutoff frequency f_α is the frequency where $|\alpha| = 0.707 \, |\alpha_0|$. From Fig. 17·14 it can be seen that for the hyperbolic secant, α cutoff occurs at a phase angle of nearly 60° and a frequency somewhat greater than $2D/W^2$. The one pole α gives a reasonable approximation to the sech out to a phase angle of about 30° $[f \lesssim 0.6(2D/W^2)]$, though beyond this point the approximation becomes rather poor.

If we want to know how the transistor will behave at frequencies beyond $f_\alpha/5$ or so, we are interested in having a better approximation to the sech

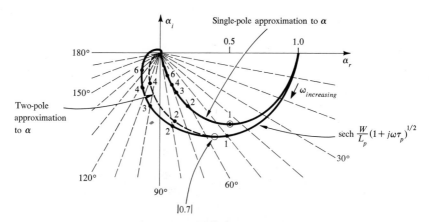

FIG. 17·14 **POLAR PLOTS OF** $1/(1 + j\omega/\omega_\alpha)$, $1/(1 + j\omega/1.16\omega_\alpha)(1 + j\omega/6.83\omega_\alpha)$, **AND SECH** $(W/L_p)(1 + j\omega\tau_p)^{1/2}$ **VERSUS FREQUENCY; THE FREQUENCY IS GIVEN IN THE DIMENSIONLESS UNITS** $\omega W^2/2D = \omega/\omega_\alpha$

function than the one pole α affords. We can obtain such an approximation very simply by using more than two lumps to approximate carrier-transport phenomena in the base region.

The simplest way to do this is to divide the base in half and use a π-section of the type previously obtained to represent *each half* of the base. We do this to be sure that we will not make a significant error by using edge densities for average densities.

The model which results from this division of the base is shown in Fig. 17·15. We refer to it as a two-π-section model to emphasize the difference between it and the three-lump model which we would obtain by dividing the base into three equal lumps.

Kirchhoff's laws at the three density nodes of this model are

$$I_e(s) = (H_d + H_{c_1} + sS_{p_1})P_e(s) - H_dP_1(s) \qquad (17\cdot47a)$$

$$0 = -H_dP_e(s) + (2H_d + 2H_{c_1} + 2sS_{p_1})P_1(s) - H_dP_c(s) \qquad (17\cdot47b)$$

$$I_c(s) = 0 - H_dP_1(s) + (H_d + H_{c_1} + sS_{p_1})P_c(s) \qquad (17\cdot47c)$$

Using these equations we can show that, for the normal-active region where $P_c(s) = 0$,

$$P_1(s) = \frac{P_e(s)}{2}\frac{1}{1 + \dfrac{1}{\beta} + s/4\omega_\alpha{}^2} \qquad (17\cdot48)$$

and

$$\alpha(s) = -\frac{I_c(s)}{I_e(s)} = \frac{\alpha_0}{(1 + s/1.16\omega_\alpha)(1 + s/6.83\omega_\alpha)} \qquad (17\cdot49)$$

where $\omega_\alpha = 2D/W^2$, as before.

Using the voltage-to-density transformation

$$P_e(s) = P_E\left(\frac{q}{kT}\right)V_{eb}(s) \qquad (17\cdot50)$$

Lumped models and the high-frequency behavior of transistors   753

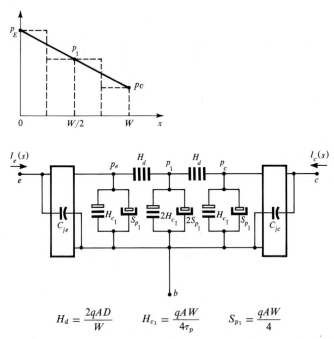

$$H_d = \frac{2qAD}{W} \qquad H_{c_1} = \frac{qAW}{4\tau_p} \qquad S_{p_1} = \frac{qAW}{4}$$

FIG. 17·15 A TWO-π-SECTION MODEL OF THE INTRINSIC TRANSISTOR BASE REGION, USING EDGE DENSITIES TO DESCRIBE THE PERFORMANCE OF EACH LUMP

and Eq. (17·48) we can develop the approximate circuit model shown in Fig. 17·16a from the two-π-section model. The emitter impedance in this model is somewhat more complicated than that of the simple T model, and the current generator in the collector circuit has two poles instead of one. However, because the emitter impedance is low, the simpler model shown in Fig. 17·16b is used, where only α is different from the simple T model described earlier. Because the factor $(1 + s/6.83\omega_\alpha)$ has a magnitude of nearly unity for $0 < \omega < \omega_\alpha$, its principal effect can also be approximated by including an *excess-phase factor* in the one-pole expression for α:

$$\frac{\alpha_0}{(1 + j\omega/1.1\omega_\alpha)(1 + j\omega/6.83\omega_\alpha)} \simeq \frac{\alpha_0 e^{-jm(\omega/\omega\alpha)}}{1 + j(\omega/\omega_\alpha)}$$

where $m \sim 0.1$.

Junction capacitances and parasitic resistances are added to the model in Fig. 17·16c, and allowance is made for the possibility that recombination within the emitter-base junction gives $r_e = nkT/qI_E$, instead of the ideal kT/qI_E. The resulting model provides a very good fit for the experimental high-frequency performance of an alloy-junction transistor (with a uniformly doped base), though as we shall see it is not a satisfactory model for a diffused-base transistor.

The hybrid-π model. Considering the discrepancy between the single pole α and the two-pole approximation, it is perhaps surprising to find that

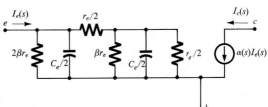

(a) Two-π-section small-signal model

$$\frac{-\alpha_0}{(1 + s/1.16\omega_\alpha)(1 + s/6.83\omega_\alpha)} = \alpha(s)$$

(b) T-model approximation to (a) typically employed; only α(s) is different from the usual T-model

$$I_c(s) = \frac{-\alpha_0 I_e(s)}{(1 + s/1.16\omega_\alpha)(1 + s/6.83\omega_\alpha)}$$

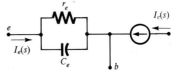

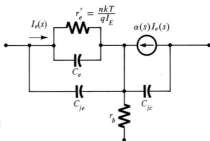

(c) High frequency T-model for a transistor with uniform base doping

FIG. 17 · 16   SMALL-SIGNAL MODEL FOR THE COMMON BASE CONNECTION OF AN INTRINSIC TRANSISTOR BASED ON A TWO-π-SECTION LUMPED MODEL

only minor changes in the hybrid π model are required for it to adequately represent the transistor. Mathematically, this fact is obtained by combining Eqs. (17 · 48) and (17 · 47c) to obtain

$$I_c = -\frac{P_e(s)}{2}\frac{H_d}{H_d + H_{c1} + sS_{p1}}$$

When the voltage-to-density transformation is employed, we obtain, after some manipulation,

$$I_c = \frac{-g_m V_{eb}}{1 + s/4\omega_\alpha} = \frac{g_m V_{be}}{1 + s/4\omega_\alpha}$$

The current generator in the collector circuit thus becomes frequency-dependent. For $f \le f_\alpha$, this change is reflected primarily in the phase angle of I_c with respect to V_{be}. We can again add an excess phase factor as follows:

$$\frac{g_m}{1 + j\omega/4\omega_\alpha} \simeq g_m e^{-jm(\omega/\omega\alpha)} \qquad \omega \le \omega_\alpha$$

where $m \simeq 0.25$. The excess phase factor can usually be ignored, unless the phase angle of I_c with respect to V_{bc} at high frequencies is important (as it is in a feedback amplifier, for example).

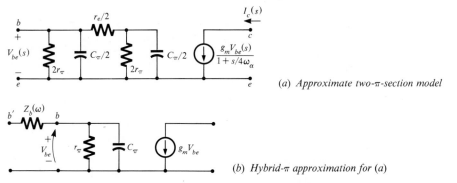

(a) Approximate two-π-section model

(b) Hybrid-π approximation for (a)

FIG. 17·17 SMALL-SIGNAL MODEL FOR THE COMMON EMITTER CONNECTION OF AN INTRINSIC TRANSISTOR BASED ON A TWO-π-SECTION LUMPED TRANSISTOR MODEL

The changes in the input impedance can be obtained from Eq. 17·46, and are also minor. An approximate model is shown in Fig. 17·17*a*, and the most commonly used model is shown in Fig. 17·17*b*. Here, a fictitious base impedance is added to the basic hybrid-π model to give an approximate accounting for that part of the input impedance that is not determined by the $C_\pi r_\pi$ combination.[1] The fictitious base impedance is usually smaller than the extrinsic base resistance which must be included in the hybrid-π model for a real transistor, so the two-π-section common-emitter model is not essentially different from the basic hybrid-π model. Again it may be necessary to compute r_π from $(\beta + 1)\, nkT/qI_E$ in some cases.

Comparison of high-frequency T and hybrid-π models. The basic difference between the hybrid-π and T models is that the collector-current response to a change in emitter-base voltage is very rapid. Charge storage near the emitter responds instantly to a change in emitter-base voltage, and the time lag for charge storage in the central portions of the base is also short ($\sim 1/4\omega_\alpha$ sec). The forward-transfer characteristics can therefore be adequately described by a frequency-independent generator in the hybrid-π representation, and the charge storage can be approximated reasonably well by a single capacitor C_π in the input circuit.

On the other hand, charge storage occurs more slowly when the emitter-base terminals are driven from a current source. Consequently, the base-charging time is more evident and its effects on circuit models more pronounced in the T models.

These basic conclusions are not changed significantly by the addition of more lumps. The input impedance of either the T or hybrid-π model approaches that of an appropriately defined RGC transmission line, and the forward-transfer characteristics also change slightly. In a real transistor, most of these effects are not evident for $\omega \lesssim \omega_\alpha$, and they are masked at

---

[1] For a more detailed discussion, including the lateral effects of the base current, see P. E. Gray, et al., "Physical Electronics and Circuit Models," SEEC vol. II, John Wiley & Sons, Inc., New York, 1964, pp. 162–168.

higher frequencies by the effects of parasitic capacitances and resistances. The most notable exception to this is the phase angle of the current generator in the collector circuit of either model, which is important at high frequencies.

17·4 THE DRIFT TRANSISTOR

Nearly all modern transistors intended for high-frequency applications are made by diffusing the base and emitter into a doped semiconductor crystal which serves as the collector. The result of the diffusion process is that the doping density in the base region is not constant, and consequently the analysis of carrier flow becomes more complicated. In this section we shall see how the new features which arise from a nonuniformly doped base can be incorporated into our previous models. We will again see that the hybrid-π model needs very little modification to represent the transistor adequately.

17·4·1 *Development of a model for analysis.* To begin, it is useful to sketch the fabrication steps that are used in making high-frequency transistors. We will concentrate on planar transistors, though the basic ideas to be described are applicable to other high-frequency transistors.

We show in Fig. 17·18a a p-type silicon crystal on which a layer of

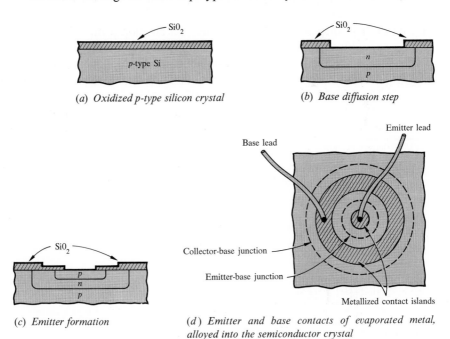

(a) Oxidized p-type silicon crystal

(b) Base diffusion step

(c) Emitter formation

(d) Emitter and base contacts of evaporated metal, alloyed into the semiconductor crystal

FIG. 17·18   STEPS IN THE FORMATION OF A PLANAR TRANSISTOR AND THE RESULTING GEOMETRY AND DOPING PROFILE

silicon dioxide has been grown (by allowing oxygen to pass over the surface of the crystal for a short time). This oxide (roughly a thin, quartz layer) forms a protective layer for the silicon through which doping impurities cannot readily diffuse.

In order to make a transistor, one therefore cuts a window in the oxide layer and then allows phosphorus to diffuse into the crystal through this window (*cf.* Sec. 5·2). The result is shown in Fig. 17·18, the diffused phosphorus layer forming the base region of the transistor. Note that the diffusion proceeds both perpendicularly to the face of the crystal and laterally underneath the oxide layer, so that where the collector-base junction comes to the surface of the crystal it is protected by the oxide layer.

To form the emitter, the surface is reoxidized, a smaller window is cut out, and boron is diffused into the crystal. The result is shown in Fig. 17·18c. Finally, by using suitable masks, the oxide is removed over a portion of the base region and ohmic contacts are made to the emitter and base by alloying evaporated metals onto the surfaces of the two regions. The transistor is mounted on a metal tab to provide a collector contact on the opposite side of the crystal.

A top view of the transistor with leads attached to the metallized layers is shown in Fig. 17·18d. Figure 17·19 shows the doping profile which is obtained along a section through the center line of the structure. Base widths on the order of 1–2μ (10^{-4} cm) can readily be fabricated by this process, the result being a very-high-frequency transistor ($f_T \sim 1,000$ Mc).

17·4·2   *The intrinsic drift transistor.* To develop a model which is generally suitable for the type of transistor just described, we again attempt to visualize the real transistor as being an intrinsic transistor embedded in a network of parasitic elements. A suitable circuit model is shown in Fig. 17·20. Here, the actual collector capacitance, which is distributed over the entire area of the collector junction, has been divided into two parts. C_{c_1} is a portion of the total capacitance which is charged directly by

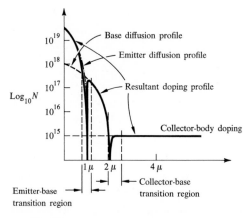

FIG. 17·19   POSSIBLE DOPING PROFILE ALONG THE CENTER LINE OF A PLANAR TRANSISTOR

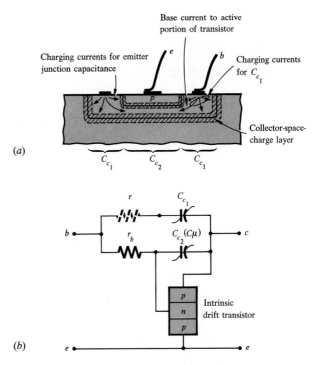

FIG. 17·20 DEVELOPMENT OF AN APPROXIMATE CIRCUIT MODEL FOR THE DRIFT
TRANSISTOR

the flow of base current. There is some small resistance associated with this current flow (shown dotted in Fig. 17·20b), though its impedance is small compared to that of C_{c_1}, so that C_{c_1} is essentially directly accessible between the base and collector terminals.

The portion of the collector capacitance which lies underneath the emitter, C_{c_2}, must be charged by current flowing in the thin part of the base region. This robs current from the active portion of the structure (that is, base current which charges C_{c_2} cannot contribute to charge storage in the neutral base region), as suggested in Fig. 17·20b. In this case, r_b is not negligible. Wavy lines are drawn through C_{c_1} and C_{c_2} to indicate that these capacitances are functions of the collector-base voltage. The remainder of Fig. 17·20b consists of an intrinsic transistor representing the active region of the device. It is assumed that low-level operating conditions apply, so that the emitter current is uniformly distributed over the emitter surface.

Drift field in the base region. The effect of the nonuniform base doping is to create an electric field in the base which *increases* the speed with which holes injected from the emitter travel across the base region. The base transit time for minority carriers is lower than it would be in a transistor of the same base width having a uniform doping in the base. The diffusion

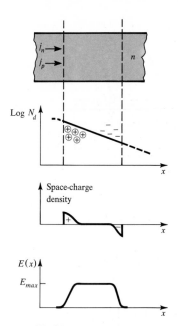

FIG. 17·21   DEVELOPMENT OF AN INTERNAL DRIFT FIELD IN A NONUNIFORMLY DOPED
n-TYPE SEMICONDUCTOR

technology thus enables us to make transistors which have thin bases and at the same time a built-in electric field to aid minority-carrier transport.

The source of this built-in field is explained more fully in Fig. 17·21. Here, we examine a slice of the material in a part of the base region which is well removed from the junctions. We assume that no biases are applied, so equilibrium conditions apply. Specifically this means that

$$n(x)p(x) = n_i^2 \qquad (17·51)$$

for thermal equilibrium. Also, since there is no current flowing, we must have

$$j_n = qD_n \frac{\partial n}{\partial x} + q\mu_n nE = 0 \qquad (17·52a)$$

$$j_p = -qD_p \frac{\partial p}{\partial x} + q\mu_p p_E = 0 \qquad (17·52b)$$

Now, in the uniform base doping case, we also have local space-charge neutrality

$$n = N_d + p \qquad (17·53)$$

and since N_d is not a function of x, then (17·51) and (17·53) give n_n and p_n, which are not functions of x. But then the diffusion current terms in Eqs. (17·52) are zero, and E must be zero.

However, when the base doping density is nonuniform, N_d varies with x, so n and p also vary with x. Thus, to maintain j_n and $j_p = 0$, a field E must be set up to counteract the diffusion tendency.

How does this field develop? The doping-density gradient provides an electron-density gradient. This gradient is in a direction to cause electrons to diffuse from left to right. As they do this, they leave behind unneutralized donor atoms and create a *slight* space charge—enough to create the field, but not enough to produce a significant change in the magnitude of $n(x)$ at any point.

The space-charge density is positive near the emitter and negative near the collector, and results in an electric field with a magnitude given by Eq. (17·55) and directed from left to right. The electric field builds up to a point where drift and diffusion tendencies for each type of carrier balance.

This balance between drift and diffusion is exactly the same type of argument which was used in Chap. 5, where we studied the properties of a *pn* junction. However, there is one significant difference between this situation and the one which applies at a junction. Here the impurity gradient is gradual (that is, N_d does not change very rapidly with x), whereas in the junction there is a much more abrupt change of doping, including a change in the charge of an ionized doping atom.

Because of this difference, we assume that space-charge neutrality is *approximately* maintained throughout the base region, and use Eq. (17·53) to give

$$n(x) \cong N_d(x) \qquad (17·54)$$

Since we are in *n*-type material, the equilibrium hole density may safely be dropped in this equation. Equation (17·52a) then requires that an electric field $E(x)$ be developed at every point whose magnitude is

$$E(x) \cong \frac{kT}{q} \frac{1}{n(x)} \frac{\partial n(x)}{\partial x} = \frac{kT}{q} \frac{1}{N_d(x)} \frac{\partial N_d(x)}{\partial x} \qquad (17·55)$$

EXAMPLE   To evaluate the drift field, we need to know $N_d(x)$, which depends on the precise doping profile. As an example, we will estimate $E(x)$ for the diffused-base transistor described in the previous section. For this purpose, $N_d(x)$ is redrawn in Fig. 17·22, and a straight line is constructed to approximate the actual doping profile. Notice that the vertical scale is logarithmic.

Now, a straight-line approximation for log $N_d(x)$ implies a constant E field:

$$\frac{d[\ln N_d(x)]}{dx} = \frac{1}{N_d(x)} \frac{dN_d(x)}{dx}$$

For the doping density shown, N_d changes by two orders of magnitude in a distance of about 1μ, so

$$\frac{d[\log_{10} N_d(x)]}{dx} \sim \frac{2}{10^{-4}} = 2 \times 10^4/\text{cm}$$

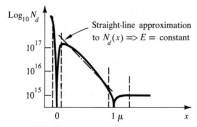

FIG. 17·22 APPROXIMATE BASE DOPING PROFILE FOR USE IN EVALUATING BUILT-IN DRIFT FIELD

The corresponding electric field is

$$E = 2.3 \frac{kT}{q} \frac{d[\log_{10} N_d(x)]}{dx}$$
$$\sim 1150 \text{ volts/cm}$$

which is a sizable electric field.

Motion of injected carriers in a drift transistor. To evaluate the detailed effect of such an electric field on injected minority carriers, we have to return to the continuity equation and include in it the fact that minority carriers can both drift and diffuse. If we want precise answers, we have to know $N_d(x)$ precisely, instead of using an approximate field such as the one just calculated. However, we can avoid these mathematical complications in a first approximation by adding an element to our lumped model to account for drift flow between lumps. This is very simply done in the following way.

Lumped representation of drift flow. We show in Fig. 17·23 a section of semiconductor which has been divided into two lumps. We will use edge densities in defining the flow law, since we know these will be most useful for the transistor base region.

The exact flow law for the drift of holes in a constant electric field E is

$$i_p(x_0) = qA\mu_p p(x_0)E \tag{17·56}$$

In lumped terms, this equation becomes

$$i_p(x_0) \cong qA\mu_p E \frac{p^*(x_1) + p^*(x_2)}{2} \tag{17·57}$$

where $p^*(x_1)$ and $p^*(x_2)$ are the excess-hole densities at the edges of the two lumps, as shown in Fig. 17·23.[1] In other words, the average flow between the lumps depends on the average density at the point of interest.

The symbol which is used to represent this flow mechanism is called a *driftance* and has the symbol and properties shown in Fig. 17·23. The

---

[1] A more accurate procedure for accounting for drift flow is described in J. G. Linvill and J. F. Gibbons, "Transistors and Active Circuits," McGraw-Hill Book Company, New York, 1961, chaps. 2 and 5. In that treatment different weight is given to $p^*(x_1)$ than $p^*(x_2)$ so that the final result conforms to the exact solution more precisely.

(a) Section of semiconductor

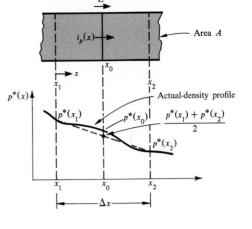

(b) Density distribution illustrating the determination of drift flow between lumps from edge densities

$$i_p(x_0) = qA\mu_p E p^*(x_0)$$

$$\simeq \left(\frac{qA\mu_p E}{2}\right)[p^*(x_1) + p^*(x_2)]$$

$$= F[p^*(x_1) + p^*(x_2)]$$

(c) Symbol and properties of a driftance

$$i_F \equiv F[p^*(x_1) + p^*(x_2)]$$

$$F \equiv \frac{qA\mu_p E}{2}$$

FIG. 17·23   THE DEVELOPMENT OF A LUMPED ELEMENT FOR CALCULATING THE BEHAVIOR OF EXCESS MINORITY CARRIERS FLOWING BY DRIFT IN A CONSTANT E FIELD

driftance is defined by

$$F = \frac{qA\mu_p E}{2} \tag{17·58}$$

A nonsymmetric symbol is used to indicate that drift flow, unlike diffusion, is direction-dependent. If we raise $p^*(x_2)$ above $p^*(x_1)$, the result is still a current flow from left to right (or in the direction of the field). The arrowhead on the driftance symbol points in the direction of the E field.

Two-lump model for the intrinsic drift transistor. We can now add the drift element to our previous two-lump model and thereby generate a two-lump model for a drift transistor. The result is shown in Fig. 17·24. Together with the junction laws,[1] this model can be used for either large- or small-signal calculations in exactly the same way that the simpler model with no drift element is used. For comparison with the uniform base transistor, we will calculate the small signal α. The appropriate equations are

$$I_e(s) = (H_d + F + H_{c1} + sS_{p1})P_e(s) \tag{17·59a}$$

$$I_c(s) = -(H_d + F)P_e(s) \tag{17·59b}$$

[1] Because p_n is a function of x in a drift transistor base region, the value of p_n to be used in calculating p_E is different from the value which should be used for calculating p_C. To be specific,

$$p_E = p_n(x = 0)(e^{qv_{EB}/kT} - 1)$$
$$p_C = p_n(x = W)(e^{qv_{CB}/kT} - 1)$$

This distinction does not manifest itself in measurable differences in terminal properties of the device, however, as it is absorbed in the values of the lumped elements during calculations.

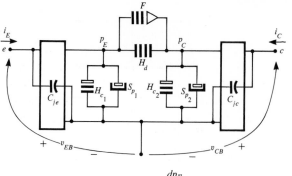

$$i_E = p_E(H_{c_1} + H_d + F) + S_{p_1}\frac{dp_E}{dt} - (H_d - F)p_C$$

$$i_C = -(H_d + F)p_E \qquad\qquad + (H_d - F + H_{c_2})p_C + S_{p_2}\frac{dp_C}{dt}$$

FIG. 17·24   TWO-LUMP MODEL FOR A TRANSISTOR WITH MINORITY-CARRIER DRIFT FLOW IN THE BASE REGION

where we have used the complex amplitudes of i_e, i_c, and p_e directly, and assumed a normal-active operating condition.

These equations yield an $\alpha(s)$ of

$$\alpha(s) = \frac{H_d + F}{H_d + F + H_{c1} + sS_{p1}} \cong \frac{1}{1 + s/\omega_{\alpha0}\left(1 + \dfrac{F}{H_d}\right)} \qquad (17·60)$$

where $\omega_{\alpha0}$ is the radian alpha cutoff frequency the transistor would have in the absence of a drift field (that is, $\omega_{\alpha0} = H_d/S_{p1} = 2D/W^2$).

Because the two-lump model still gives us a single pole α, the T and hybrid-π models do not change their form. However, the alpha cutoff frequency (or the value of C_π) does change.

The magnitude of the change can be estimated by using the value of E and other parameters for the diffused-base transistor described earlier. The increase in ω_α is given by

$$1 + \frac{F}{H} = 1 + \frac{\mu_p}{D_p}\frac{W}{2}E$$

Using $W = 1\mu = 10^{-4}$ cm and the previously calculated value of E gives

$$1 + \frac{F}{H} = 1.5$$

for this specific transistor. The value of $\omega_{\alpha0}$ which we would estimate for a 1μ base width is 25×10^8, so the radian cutoff frequency in the presence of the drift field would be 37×10^8; or $f_\alpha \sim 6 \times 10^8$ cps = 600 mcps for the two-lump model.

Excess phase shift and the basic T-model for a drift transistor. As might be expected, the addition of a second mechanism for transporting carriers between lumps makes the two-lump model less adequate as an approximation to the real structure. This is shown most graphically in Fig. 17·25 where a polar plot of α for a very-high-frequency transistor is shown. Note that this transistor shows an $|\alpha|$ of 0.8 at a frequency where the phase angle of α is 90°. The one-pole approximation for α is indeed very poor for this transistor over nearly all of its high-frequency range. The two-pole approximation [obtained from Eq. (17·48) with ω_α increased to account for the drift field] is better, though a more accurate approximation is still necessary if we expect the calculated performance to agree with experimental results at very high frequencies.

Of course adding too many lumps to the model complicates the ease with which calculations can be made, so one looks for other ways to modify the basic two-lump model without affecting its simplicity. It has been found experimentally that a reasonable approximation can be achieved by attaching a so-called *excess phase factor* to the one pole alpha:

$$\alpha(f) = \frac{\alpha_0 e^{-jmf/f_\alpha}}{1 + jf/f_\alpha} \qquad (17·61)$$

The factor m can be evaluated when the base-doping profile is known from the expression

$$m = 0.22 + 0.1 \ln \frac{N_d(x = 0)}{N_d(x = W)}$$

where $N_d(x = 0)$ is the base doping at the emitter edge of the base and $N_d(x = W)$ is the base doping at the collector edge of the base.

As is suggested in Fig. 17·25, the fit which Eq. (17·61) provides for experimental data is very good. As a result, the intrinsic transistor shown in Fig. 17·20 can be replaced by a simple T model in which the collector current generator has the excess phase factor attached. The emitter impedance is much more complicated than the simple $r_e C_e$ combination suggests, though for common-base applications this is relatively unimportant.

The hybrid-π model for a drift transistor. On the basis of our previous

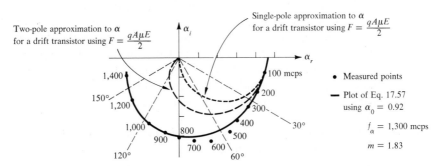

FIG. 17·25 COMPARISON OF SINGLE POLE α, MEASURED α AND EXCESS PHASE APPROXIMATION FOR AN EXPERIMENTAL COAXIAL MICROALLOY DIFFUSED BASE TRANSISTOR

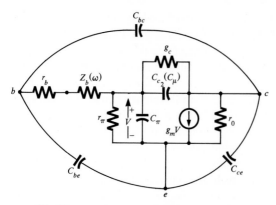

FIG. 17·26 GENERALIZED HYBRID-π MODEL FOR A TRANSISTOR. THE FORM OF THE MODEL IS THE SAME FOR ALL TRANSISTORS

Notes: 1. C_{be}, C_{ce} are header capacitances.

2. C_{bc} is the sum of the collector-base header capacitance and C_{c1} given in Fig. 17·20.

3. r_π is ideally $(\beta + 1)(kT/qI_E)$, though in some cases it may be higher because of recombination in the emitter-base junction.

4. g_c and r_0 arise from base-width modulation and have the typical values $r_0 = 20\ k$, $g_c = 2\ meg$.

5. Base impedance is a function of frequency, varying typically from 50–200Ω.

6. Emitter junction capacitance neglected, intrinsic collector junction capacitance included in C_{c_2} (also called Cu).

results, we may expect that the hybrid-π model will still give us a valid representation of the transistor, except for the phase angle of the current generator at high frequencies. This discrepancy is usually neglected, and the hybrid-π model shown in Fig. 17·26, where parasitic resistances, junction capacitances, and header capacitances have been included, is usually employed to represent a real transistor. By judiciously selecting parameters,[1] this model provides a reasonably good fit for the measured terminal performance of the transistor, in spite of the fact that r_b should really be represented by a complex impedance, and a phase factor should be included with g_m.

Unfortunately, the entire model is sufficiently complicated that it is not used in practical design work. However, it is usually possible to make simplifying approximations which eliminate many of the elements and thus produce a model which is simple enough for circuit design. When this is not possible or when we want to refine our estimates of circuit parameters made with an oversimplified model, we must resort to measured two-port parameters for the transistor, as described in Sec. 17·5.

17·4·3 *Dependence of hybrid-π parameters on bias in a drift transistor.* One of the features of the process for making diffused-base transistors which

[1] C. L. Searle, et al., "Elementary Circuit Properties of Transistors," SEEC, vol. 3, John Wiley & Sons, Inc., New York, 1964, pp. 107–118.

we have not emphasized is that the base-doping density near the emitter is large, and of course the emitter body is doped even more heavily to preserve a high-injection efficiency. The result of these high-doping densities is that the junction capacitance C_{je} associated with the emitter-base junction of a diffused-base transistor is very high.

In a circuit model such as the one shown in Fig. 17·26, this junction capacitance is included in C_π, so its effects are not immediately obvious. However, from a consideration of the lumped model shown in Fig. 17·27 we are reminded that

1. The C_π given in Fig. 17·26 consists of an intrinsic component—say, C_π'—and the emitter-base junction capacitance C_{je}.

2. C_π is related to the hole-storage process in the transistor base, and *its magnitude is directly proportional to the bias current* I_E.

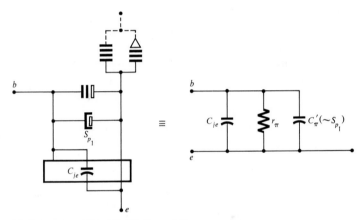

(a) *Lumped model to identify* C_{je} *and* C_π

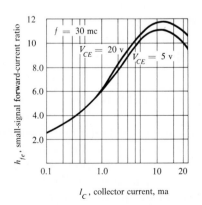

(b) *Effect of* C_{je} *on high-frequency current gain*

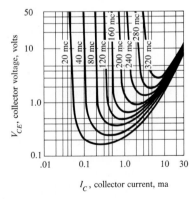

(c) *Effect of* C_{je} *on* f_T

FIG. 17·27 THE INFLUENCE OF C_{je} ON THE CHARACTERISTICS OF A DRIFT TRANSISTOR

3. The portion of $I_e(s)$ which flows in C_{je} does not result in minority-carrier injection nor contribute to transistor action.

4. C_{je} *is a slowly varying function of* I_E [it depends on $(V_{EB})^{1/3} \sim (\ln I_E)^{1/3}$].

Now, in order to obtain a reasonable value of the collector-base break-down voltage, the doping density at the collector-base junction must be small. In a transistor with a uniformly doped base, this implies that C_{je} will be small compared to C_π' for typical bias currents, so it may be neglected. In a diffused-base transistor, however, this is not so. C_{je} is large, because the base-doping density near the emitter is large even though the base-doping density at the collector-base junction is small. As a result, we must operate drift transistors at high values of I_E, in order to make $C_\pi' \gg C_{je}$ and thereby to obtain maximum performance from the transistor.

This is reflected in manufacturer's data in the manner shown in Fig. 17·27b and 17·27c. The high-frequency current gain (β_0) and the value of f_T are both functions of the bias current. A bias current of about 5 ma is necessary to reach a point where $C_{je} < C_\pi'$.

This does not affect the use of previous formulas for calculating C_π from f_T, because measurements of high-frequency β and f_T depend only on the total value of C_π. To emphasize this, we list below the procedure for obtaining the parameters for the hybrid-π model, shown in Fig. 17·26, from the manufacturer's data:

1. Choose the bias point and read off the corresponding values of f_T and β_0. If β is given at some frequency f_0 (30 Mc in Fig. 17·27b), then $\beta_0 = \beta(f_0) \times (f_T/f_0)$.

2. Calculate g_m from qI_C/kT.

3. Calculate r_π from $(\beta_0 + 1)(kT/qI_E)$.

4. Calculate C_π from $g_m/2\pi f_T$. This includes both C_π' and C_{je}.

5. Obtain C_c (C_{ob}) from the data sheet. If C_{ob} is small, corrections for header capacitances will have to be made to obtain C_c.

6. Obtain r_b from the data sheet (usually in the form of an $r_b C_c$ product) or from high-frequency input impedance data.

7. Estimate r_0 from I_C-V_{CE} characteristics, if required.

8. Estimate g_c from $1/(\beta_0 r_0)$, if required.

17·5 y PARAMETERS FOR HIGH-FREQUENCY TRANSISTORS

It is apparent from the preceding discussion that a manufacturer must give a rather large number of parameters to specify the performance of a transistor intended for very-high-frequency applications. Rather than do this, it is more common to specify the parameters needed in the basic hybrid-π model and then supplement this with measured two-port parameters for the transistor under a given bias condition and over a wide frequency range. The basic hybrid-π model is used for an initial design, and

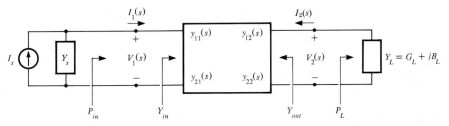

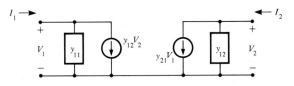

(a) General two-port defined by y-parameters

(b) Norton-equivalent representation of (a)

$$I_1 = y_{11}V_1 + y_{12}V_2 \qquad I_2 = y_{21}V_1 + y_{22}V_2$$

FIG. 17·28   DEFINITION OF y-PARAMETERS AND A SUITABLE SOURCE AND LOAD

then the two-port parameters are used to improve the estimates of the circuit performance to be expected.[1] In some situations (for example, in the design of band-pass amplifiers) measured two-port parameters are used directly in the design work, since they give a more accurate description of device behavior than circuit models can provide.

The admittance or y parameters are the most readily measured parameters over a wide frequency range, and they are accordingly the ones that are usually specified. The y parameters are defined by the equations

$$I_1(s) = y_{11}(s)V_1(s) + y_{12}(s)V_2(s)$$
$$I_2(s) = y_{21}(s)V_1(s) + y_{22}(s)V_2(s)$$

where s is the complex frequency variable, and each quantity is a complex number in general. In place of the generalized 11, 12, etc., subscripts, others are usually used:

$y_{11} = y_{ie}$ (input admittance, emitter common)

$y_{12} = y_{re}$ (reverse transfer, emitter common)

$y_{21} = y_{fe}$ (forward transfer, emitter common)

$y_{22} = y_{oe}$ (output admittance, emitter common)

---

[1] The design of amplifiers using two-port parameters is described in any of several texts on linear-active network theory; for example, J. G. Linvill and J. F. Gibbons "Transistors and Active Circuits," McGraw-Hill Book Company, New York, 1961, chaps. 9–11. Other references are listed at the end of the chapter.

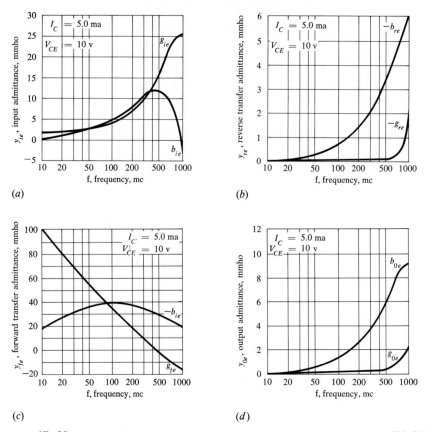

(a)

(a)

(b)

(c)

(d)

FIG. **17·29** *y*-PARAMETERS FOR THE COMMON EMITTER CONNECTION OF A 2N 918, MEASURED OVER A FREQUENCY RANGE FROM 10 mcps TO 1,000 mcps

The *y* parameters are simply interpreted as being the parameters of a Norton equivalent circuit for a two-port box as shown in Fig. 17·28. A set of typical *y* parameters for the common-emitter connection of a 2N918, measured over a frequency range of 10–1,000 mcps is shown in Fig. 17·29.

17·6 SUMMARY

In this chapter we have developed the lumped-model approach to transistor modeling and applied it in several different situations. Basically, the development proceeds directly from the continuity equation for minority carriers; as a result, every time one uses the lumped model for calculations, he solves the continuity equation (approximately) and applies the appropriate boundary conditions. The approach is an approximate one, but nevertheless very fundamental.

We saw that the lumped model provides a convenient conceptual framework for understanding the relationships between various models which were derived earlier in the book, and at the same time a convenient means for improving the accuracy of these models when this is required.

Because the modeling technique rests on making a model directly from the continuity equation, one can extend the technique to separately account for electrons and holes, include trapping phenomena, and so on. The general technique is, of course, also applicable to other physical systems (mechanical, hydraulic, thermal, etc.), where once again the modeling process rests directly on a set of appropriate continuity equations.

REFERENCES

Linvill, J. G., and J. F. Gibbons: "Transistors and Active Circuits," McGraw-Hill Book Company, New York, 1961, part 1.
Lindmayer, J., and C. Wrigley: "Fundamentals of Semiconductor Devices," D. Van Nostrand Company, Inc., Princeton, N.J., 1965, Chap. 4 and 5.

PROBLEMS

17·1 All of the current that flows in a p^+n junction diode can be accounted for by recombination in the n-type body. This suggests that the simplest lumped model for a p^+n diode has the form shown in Fig. P17·1.

1. Show that if $H_c = (qAL_p)/\tau_p$ then the v-i law for the lumped model is the same as that for the ideal diode.

2. Give an interpretation of the formula $H_c = (qAL_p)/\tau_p$.

3. Show that the reverse saturation current of the diode is $|I_0| = p_n H_c$. What direction is the current flowing in the H_c in this case and why?

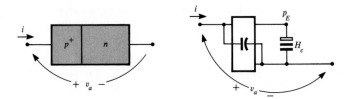

FIG. P17·1

17·2 The effect of shining a light on a p^+n diode can be described in the manner shown in Fig. P17·2. Photogenerated carriers created within about a diffusion length of the junction can diffuse to the junction and create a short-circuit current i_r. Using the lumped-model representation of this effect shown in Fig.

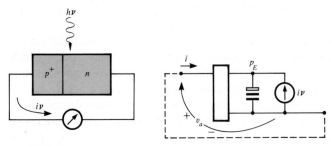

FIG. P17·2

P17·2, what open-circuit voltage would appear at the diode terminals in the presence of the light if $i_\nu = 1$ ma, $I_0 = 1$ μa?

17·3 A symmetrical p^+np^+ germanium alloy transistor has an emitter area of 10^{-3} cm², a base width of 10^{-3} cm, and a uniform base-doping density of $10^{16}/$cm³. Compute the parameters for the two-lump model of this transistor which uses edge densities (that is, Fig. 17·9b). You may neglect base-width modulation in these calculations. What effect would base-width modulation have on the lumped model parameters?

17·4 Compute I_{ES} and α_F for the transistor described in Prob. 17·3.

17·5 A p^+np^+ germanium alloy transistor has $A_E = 10^{-3}$ cm², $A_C = 2 \times 10^{-3}$ cm², a uniform base-doping density of $10^{16}/$cm³, and a base width of 5 μ. The emitter and collector are coaxial, as shown in Fig. P17·5. Give the parameters of

FIG. P17·5

a two-lump model for this transistor, using A_C to compute $H_{c,2}$ and $S_{p,2}$ and A_E to compute H_d, $H_{c,1}$ and $S_{p,1}$. Give an argument to justify the use of A_E in calculating H_d. Compute I_{ES}, I_{CS}, α_F, and α_R for this transistor and show that $\alpha_F I_{ES} = \alpha_R I_{CS}$.

17·6 What voltage appears between the emitter-base terminals of the transistor given in Prob. 17·3 when the transistor is connected as shown in Fig. P17·6? You may neglect base-width modulation.

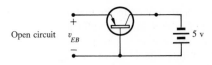

FIG. P17·6

17·7 A phototransistor (Fig. P17·7) is to be considered as a photoelectric device in the two ways shown in Fig. P17·7c and P17·7d. Evaluate the load currents in each case and explain the difference. The collector-base junction can be assumed to be heavily reverse-biased in each case.

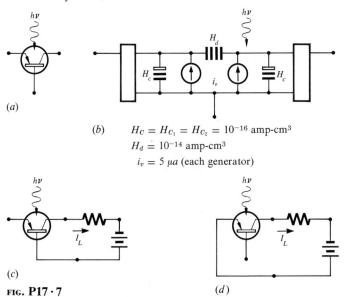

(a)

(b) $H_C = H_{C_1} = H_{C_2} = 10^{-16}$ amp-cm³
        $H_d = 10^{-14}$ amp-cm³
        $i_v = 5$ μa (each generator)

(c)

FIG. **P17·7**   (d)

17·8 Develop the T and hybrid-π models for the transistor given in Prob. 17·3. Assume a bias condition of $I_E = 1$ ma, $V_{CB} = -5$ volts. Include junction capacitances in your model, and the effect of base-width modulation on the model parameters.

17·9 A *pnp* germanium transistor has a base width which would give it a 50-mcps alpha cutoff frequency if it were a diffusion transistor. It has a resistivity of 10 ohms-cm at the collector edge of the base.

1. Using a single-pole approximation for α, determine approximately the base-doping density at the emitter which would give the transistor a 100-mcps alpha cutoff frequency.

2. Using $f_\alpha = 100$ mcps and the doping ratio just obtained, determine $\alpha(f)$ from the excess-phase approximation, Eq. (17·58). Sketch the α determined in (1) and that just found on a polar plot for comparison.

17·10 The approximate high-frequency characteristics of a 2N918 are given by $\beta(100$ mcps$) = 9$, $C_0 \simeq 1.0$ pf at $V_{CE} = 10$ volts. Construct a simple hybrid-π model for this transistor for operation at a bias point $I_C = 5$ ma, $V_{CE} = 10$ volts. You may neglect r_b. Now compute the input admittance of the model with the output short-circuited at $f = 50$ mcps, 200 mcps, 500 mcps, and 1,000 mcps, and compare the results with the typical small-signal parameters given in Fig. P17·26. Assume $I_E = I_C = 5.0$ ma in your calculations. If you are interested, see C. L. Searle, et al., "Elementary Circuit Properties of Transistors," SEEC Notes, vol. 3, John Wiley & Sons, Inc., New York, 1964, pp. 107–117, for a procedure which can be used to obtain element values for the generalized hybrid-π model (Fig. 17·27) which will bring its terminal properties into agreement with the *y*-parameter data.

18

HIGH-FREQUENCY CIRCUITS
AND TRANSISTOR LIMITATIONS

WE SAW IN THE LAST CHAPTER that an accurate characterization of transistors at high frequencies requires a circuit model with many elements or a black-box approach in which measured two-port parameters are used to characterize the device. Because of this, the accurate design of amplifiers with large bandwidths (such as a video amplifier) is best approached in two steps. An initial design using the hybrid-π model serves to suggest approximate circuit components to achieve a desired response. The design may then be refined, using measured two-port parameters, or it can simply be experimentally adjusted to give the required performance.

A similar situation exists in the design of band-pass amplifiers (also called tuned amplifiers, filter amplifiers, or IF amplifiers), where in many cases even the initial design must be based directly on measured two-port parameters.

In this chapter we will develop the principles which are used in designing practical video and band-pass amplifiers. We will see that the ultimate performance which can be obtained in these circuits is determined by limitations which arise within the transistor. We shall conclude the chapter by developing the concept of the maximum frequency of oscillation of a transistor. This maximum oscillation frequency is determined entirely by the transistor parameters, and represents the maximum frequency at which the device can be considered to be active.

18·1 VIDEO AMPLIFIERS

Generally, a satisfactory video amplifier can be designed from gain and bandwidth specifications for the over-all amplifiers. However, these specifications are not directly related to the signals which a video amplifier actually handles. Real video signals are more nearly characterized as a sequence of pulses or a series of step changes in signal amplitude, and as a result it is more logical to design a video amplifier to give a certain pulse or step response than a prescribed steady-state frequency response. Design specifications for video amplifiers are therefore frequently given in terms of the step function response which the amplifier should have, and a suitable amplifier can in some cases be designed directly from these specifications. In complicated cases, however, it is usually easier to design an amplifier for a prescribed steady-state frequency response than a specified step response, so we need to know how to convert time-response specifications to frequency-response specifications. We will take up these topics by means of several examples in the following sections.

18·1·1 *Step-function response: definitions.* The general nature of the output waveform of a video (or other low-pass) amplifier, when a step function is applied to its input, is shown in Fig. 18·1. The following quantities are defined with reference to this figure.

Rise time. Because of shunt capacitance at the input of each stage of the amplifier, the input voltage for each stage will not rise abruptly when the step input is applied. As a result, the output voltage (or current) of each stage and that of the over-all amplifier will exhibit a finite *rise time* T_R.

Since it is usually difficult to tell just when the rise begins and ends, the rise time is defined as the time required for the output to increase from 10 to 90 percent of its final value:

$$\text{Rise time} = T_R = t_{90} - t_{10} \qquad (18 \cdot 1)$$

Overshoot. In attempting to minimize the rise time of the amplifier, one frequently employs interstage or feedback networks which cause the out-

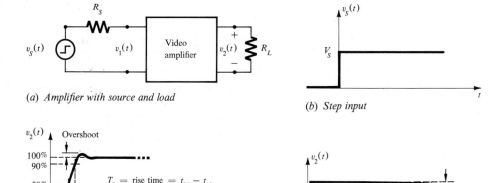

(a) Amplifier with source and load

(b) Step input

(c) Output voltage for a short period of time after step input

(d) Output voltage for a long period of time after step input

FIG. 18·1   GENERAL STEP-FUNCTION RESPONSE OF A VIDEO AMPLIFIER

put voltage to rise above the 100 percent level and then exhibit damped oscillations around it. This phenomenon is called *overshoot* and is expressed as a percent of the final output.

Sag. Unless the amplifier is designed to give constant gain all the way down to dc, the amplifier output will deviate from the 100 percent level. This deviation is called *sag* and is expressed as a percent in a given time. Of course, the presence of sag makes the exact 100 percent value somewhat uncertain, though the time scales on Figs. 18·1c and 18·1d are such that this is usually a minor problem.

18·1·2   *Step response of a single-stage transistor amplifier.*   To apply the foregoing definitions to a practical case of interest, we will calculate the main features of the step response of the single-stage amplifier shown in Fig. 18·2. The transistor in this amplifier is assumed to be a 2N916 operating at a bias point of $I_C = 5$ ma, $V_{CE} = 5$ volts. At this operating point it has

$$g_m = \frac{1}{5}\,\mho \qquad \beta_0 = 100 \qquad C_c = 3 \text{ pfarads} \qquad r_\pi = 500 \text{ ohms}$$

$$r_b = 60 \text{ ohms} \qquad f_T = 320 \text{ mcps} \qquad C_\pi = 100 \text{ pf}$$

The other values shown on the figure are chosen for convenience; the factors which affect their choice will be described later.

Rise-time calculation.   To calculate the rise time, we neglect the coupling and emitter bypass capacitors, since they are large and the voltages across them cannot change significantly in a short period of time. We use a

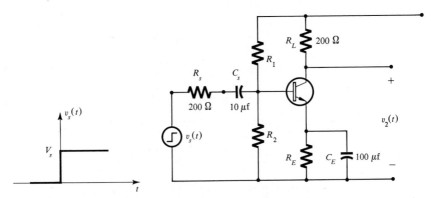

FIG. 18·2 SINGLE-STAGE AMPLIFIER WITH SMALL SIGNAL STEP-FUNCTION INPUT

hybrid-π model for the transistor, employing the Miller approximation (as in Chap. 13) to incorporate the effects of C_c into C_{eq}. Parasitic header capacitances in the transistor have been neglected for simplicity. The resulting circuit model is shown in Fig. 18·3.

The signal-input voltage to the transistor can be written by inspection of Fig. 18·3:

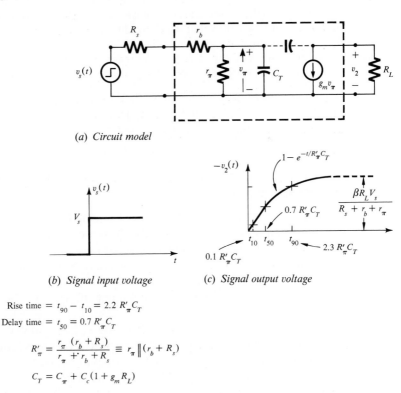

(a) Circuit model

(b) Signal input voltage   (c) Signal output voltage

Rise time $= t_{90} - t_{10} = 2.2\,R'_\pi C_T$

Delay time $= t_{50} = 0.7\,R'_\pi C_T$

where

$$R'_\pi = \frac{r_\pi(r_b + R_s)}{r_\pi + r_b + R_s} \equiv r_\pi \| (r_b + R_s)$$

$$C_T = C_\pi + C_c(1 + g_m R_L)$$

FIG. 18·3 CIRCUIT MODEL FOR CALCULATING RISE TIME AND DELAY TIME OF THE AMPLIFIER OF FIG. 18·2

$$v_\pi(t) = \frac{V_s r_\pi}{R_s + r_b + r_\pi}(1 - e^{-t/R'_\pi C_{eq}}) \qquad (18 \cdot 2a)$$

where
$$R'_\pi = \frac{r_\pi(r_b + R_s)}{r_\pi + r_b + R_s} \equiv r_\pi \| (r_b + R_s) \qquad (18 \cdot 2b)$$

$$C_{eq} = C_\pi + C_c(1 + g_m R_L) \qquad (18 \cdot 2c)$$

The amplifier output signal voltage is then

$$v_2(t) = -g_m R_L v_\pi = \frac{-\beta R_L V_s}{R_s + r_b + r_\pi}(1 - e^{-t/R'_\pi C_{eq}}) \qquad (18 \cdot 3)$$

$v_2(t)$ is shown on Fig. $18 \cdot 3c$. The 100 percent level is

$$(\beta R_L V_s)/(R_s + r_b + r_\pi)$$

The 10 and 90 percent points on the exponential curve occur at $0.1\ R'_\pi C_{eq}$ and $2.3\ R_\pi C_{eq}$, respectively, so the rise time is

$$T_R = t_{90} - t_{10} = 2.2 R_\pi C_{eq} \qquad (18 \cdot 4)$$

EXAMPLE   Using the transistor parameters given earlier and the values $R_s = R_L = 200$ ohms, we obtain $R'_\pi = 147$ ohms, $C_{eq} = 100 + 123 = 223$ pf. From Eq. $(18 \cdot 4)$ we then obtain $T_R = 72$ nanosec. Recalling that a single picture element on a TV picture tube is illuminated for 100 nanosec, we can see that the rise time of this stage is only somewhat larger than that which would be needed to paint alternate picture elements essentially black and white. This is, of course, a much more fine-grained change than the eye can detect, so an over-all amplifier rise time of 72 nanosec is less than we really need for a television video amplifier.

Delay time calculation.   The delay time of the amplifier stage (defined on Fig. $18 \cdot 1$) can be calculated directly from Eq. $(18 \cdot 3)$ by finding the time at which the exponential has the value 0.5. The result is

$$T_D = 0.7 R'_\pi C_{eq}$$

which is equal to 23 nanosec for our previous example. We will return to the significance of T_D later.

Sag due to the coupling capacitor C_s.[1]   For simplicity we will calculate the sag due to C_s and C_E separately, much as we calculated their effects on low-frequency response of the stage separately in Chap. 13. We will consider how the sags from these two causes may be combined after we have calculated the individual sags.

Since the sag is a long-time phenomenon, we can assume that the voltages across all small capacitors (C_π and C_c) have reached their final values and that their subsequent discharging as the input voltage drops (due to charge accumulating on C_s) can be neglected on the relevant time scale.

With these assumptions, the circuit model to be used for calculating the sag due to C_s is shown in Fig. $18 \cdot 4$. Here we simply have a low-frequency

---

[1] A procedure for calculating sag using Laplace transformation techniques is the subject of some of the problems at the end of this chapter.

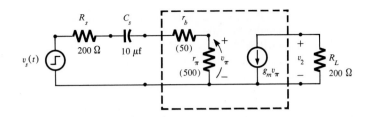

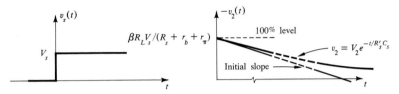

FIG. 18·4   CIRCUIT MODEL FOR COMPUTING SAG DUE TO C_s

model for the transistor with the R_sC_s input combination. Assuming C_s is initially uncharged (as far as signal changes are concerned), the input voltage to the transistor for a step change in signal voltage is, by inspection of Fig. 18·4,

$$v_\pi(t) = \frac{V_s r_\pi}{R_s + r_b + r_\pi} e^{-t/R'_s C_s} \qquad (18 \cdot 5)$$

where

$$R'_s = R_s + r_b + r_\pi$$

The amplifier output voltage is

$$v_2(t) = -\frac{\beta V_s R_L}{R_s + r_b + r_\pi} e^{-t/R'_s C_s} \qquad (18 \cdot 6)$$

where the first factor is the same as that appearing in Eq. (18·3) and is the 100 percent level.

The initial slope of $v_2(t)$ is given by

$$\frac{dv_2(t)}{dt}\bigg|_{t=0} = \frac{\beta V_s R_L}{R_s + r_b + r_\pi} \frac{1}{R'_s C_s} \qquad (18 \cdot 7)$$

or, when stated as a percentage of the 100 percent level,

$$100\% \frac{1}{v_2(0)} \left(\frac{dv_2}{dt}\bigg|_{t=0}\right) = -\frac{100\%}{R'_s C_s} \qquad (18 \cdot 8)$$

The actual amount of sag for a small time interval Δt after the application of the input signal can be computed from Eq. (18·8) by multiplying by the given time interval Δt, though this is not usually done. Instead, Eq. (18·8) serves to define the sag, as the following example shows.

EXAMPLE   If we consider a television picture in which there is a background which has uniform intensity across the screen, (for example, the wall of a room or a cloudless sky), we want the video amplifier to maintain an essentially constant output voltage for the entire period of a horizontal scan line; that is, 52.5 μsec. If we agree that the brightness of the picture element furthest right on the scan line should be within 10 percent of the brightness of the picture element furthest left, then we can tolerate a total sag of 10 percent in 52.5 μsec. This is usually expressed as

$$\text{Sag} = -\frac{10\%}{52.5\ \mu\text{sec}} = -0.19\%/\mu\text{sec}$$

If we use the circuit values given in Fig. 18·4, we find the sag due to the coupling circuit to be

$$\text{Sag}\ (C_s) = -\frac{100\%}{R'_s C_s} = \frac{1}{7.5}\%/\mu\text{sec} = 0.133\%/\mu\text{sec}$$

Sag due to emitter by-pass capacitor C_E. To compute the sag due to C_E, we proceed in exactly the same fashion as before. We neglect the effect of C_s, use a low-frequency T model to represent the transistor, and then convert the emitter impedance into an equivalent impedance seen in the base lead. The steps are outlined in Fig. 18·5.

From inspection of Fig. 18·5, it can be seen that the current flowing in the input loop for a step change in signal voltage will initially be

$$i_s(t = 0) = \frac{V_s}{R_s + r_b + r_\pi} \tag{18·9}$$

whereas, when $t \to \infty$, $i_s(t)$ becomes

$$i_s(t \to \infty) = \frac{V_s}{R_s + r_b + r_\pi + (\beta + 1)R_E} \tag{18·10}$$

The form of $i_s(t)$ between these two extremes is

$$i_s(t) = i_s(\infty) + [i_s(0) - i_s(\infty)]\,e^{-t/R_{Eq}C'_E} \tag{18·11}$$

where $\qquad R_{Eq} = (\beta + 1)R_E \| (R_s + r_b + r_\pi) \qquad C'_E = \dfrac{C_E}{(\beta + 1)}$

The amplifier output voltage is

$$v_2(t) = -\beta i_s(t)R_L \tag{18·12}$$

which starts at the same 100 percent level as before but does not decay to zero as $t \to \infty$, since some dc response $[A_v(s = 0) \cong R_L/R_E]$ remains. Again the behavior as $t \to \infty$ is not of interest to us. We want to know the initial slope, which we obtain directly from Eqs. (18·11) and (18·12):

$$100\%\frac{1}{v_2(0)}\left[\frac{dv_2(t)}{dt}\Bigg|_{t=0}\right] = -\frac{i_s(0) - i_s(\infty)}{i_s(0)}\frac{100\%}{R_{Eq}C'_E}$$

$$= -\left[1 - \frac{R_s + r_b + r_\pi}{R_s + r_b + r_\pi + (\beta + 1)R_E}\right]\frac{100\%}{R_{Eq}C'_E} \tag{18·13}$$

For the usual case, $(\beta + 1)R_E \gg (R_s + r_b + r_\pi)$, so we obtain after some manipulation

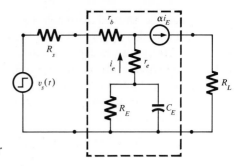

(a) Circuit employing T-model for transistor

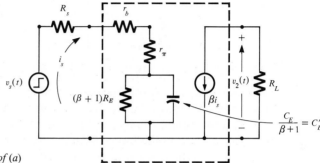

(b) Circuit-equivalent of (a)

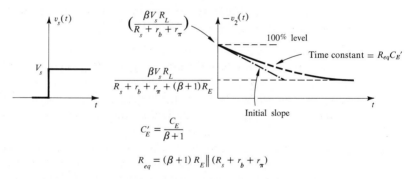

$$C_E' = \frac{C_E}{\beta + 1}$$

$$R_{eq} = (\beta + 1) R_E \| (R_s + r_b + r_\pi)$$

FIG. 18·5   CIRCUIT MODEL FOR COMPUTING SAG DUE TO C_E IN THE CIRCUIT OF FIG. 18·2

$$100\% \frac{1}{v_2(0)} \left[\frac{dv_2(t)}{dt} \bigg|_{t=0} \right] = - \frac{100\%}{C_E [r_e + (R_s + r_b)(1 - \alpha_0)]} \quad (18·14)$$

For the circuit values given in Fig. 18·5 this gives

$$\text{Sag} = 0.03\%/\mu\text{sec}$$

Sag from multiple sources. The fact that C_s and C_E both contribute to sag poses no problem in computing the total sag. We simply add the sags produced by each cause together to find the overall sag produced by the circuit. This is a valid procedure, even though the low-frequency amplitude responses due to these capacitors *cannot* be added together. This result can be proved formally by Laplace transformation techniques, though it can also be proved from the following physical argument.

To compute sag, C_s and C_E are initially assumed to be short circuits. When the step voltage is applied, the currents flowing in each capacitor can be calculated using a low-frequency small-signal model for the transistor. The resulting currents give the initial time rates of change of voltage across each capacitor:

$$\frac{dv_{C_s}}{dt}(t = 0^+) = \frac{i_{C_s}(0^+)}{C_s} \qquad \frac{dv_{CE}}{dt}(t = 0^+) = \frac{i_{CE}(0^+)}{C_E}$$

But inspection of Fig. 18·2 shows that

$$\left.\frac{dv_{be}}{dt}\right|_{t=0^+} = -\left[\left.\frac{dv_{C_s}}{dt}\right|_{t=0^+} + \left.\frac{dv_{CE}}{dt}\right|_{t=0^+}\right]$$

which shows that the sag from the two sources add directly.

Sag compensation. Sag can be eliminated by making the amplifier respond uniformly right down to dc. This involves eliminating C_s and C_E and making appropriate changes in the biasing circuitry, and is in fact a very practical solution to the sag problem.

Sag can also be compensated for short time intervals (≤ 1 msec or so) by introducing the R_d and C_d elements shown in Fig. 18·6. These elements provide other useful features as well in a multistage amplifier and will be considered further at the end of Sec. 18·1·4.

Step response of a cascade of identical RC-coupled stages. A single stage of amplification will not be adequate to obtain the gain required in a practical video amplifier, so several stages will have to be connected in cascade. The question then arises as to what the step response of the over-all system will be when the step response of individual stages is known.

This is not an easy question to answer, because of interaction among the stages and the somewhat arbitrary definitions used for rise time, delay time, and so on. However, there are a few approximate but valuable rules which a designer can use to estimate the over-all characteristics in terms of the single-stage characteristics. In particular, for a cascade of n stages of the type shown in Fig. 18·2, it can be shown that[1]

$$T_R^2 \, (n \text{ stages}) \cong \sum_{i=1}^{n} T_{Ri}^2 \qquad (18 \cdot 15a)$$

$$T_D \, (n \text{ stages}) \cong \sum_{i=1}^{n} T_{Di} \qquad (18 \cdot 15b)$$

$$\text{Sag} \, (n \text{ stages}) = \sum_{i=1}^{n} (\text{sag})_i \qquad (18 \cdot 15c)$$

In simple cases, these rules are enough to enable a designer to obtain approximate circuit values from which the individual stages in a multistage

[1] See Pettit and McWhorter, "Electronic Amplifier Design," McGraw-Hill Book Company, New York, 1961, chaps. 4 and 5. If the stages are identical, the rise time of n stages is better approximated by $T_R \, (n \text{ stages}) = 1.1 \sqrt{n} \, T_R \, (1 \text{ stage})$.

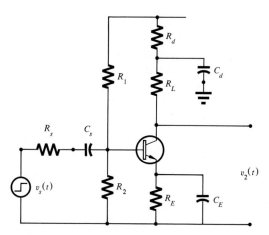

FIG. 18 · 6 SAG COMPENSATION CAN BE OBTAINED BY PROPERLY SELECTING THE COMPONENTS R_d AND C_d IN THIS CIRCUIT; THE THEORY IS DEVELOPED IN SEC. 18 · 1 · 4

amplifier can be designed. In more complicated cases, we usually fall back on the relationship between step response and frequency response provided by the Fourier (or Laplace) transformation. We will describe this relationship in the next section.

18 · 1 · 3 *Relation between step response and frequency response of a video amplifier.* In terms of its Fourier representation, a step function has components at all frequencies. An ideal amplifier for such a signal must provide constant gain to infinite frequency and must maintain the correct phase relationship among all the components (usually this means a phase shift proportional to frequency).

A practical video amplifier, of course, will not satisfy these requirements, so its output can at best be a distorted reproduction of the input. The distortion produced by the amplifier is described in the time domain by quoting the delay, rise time, overshoot, and sag of the output voltage when a step function input is applied. Amplifier distortion in the frequency domain is described by quoting the magnitude and phase of the amplifier gain as a function of frequency; that is, quoting the frequency response of the amplifier.

The step response of an amplifier can be precisely related to its frequency response by Fourier (or Laplace) transformation techniques. If one is given the step response of an amplifier, he can find the magnitude and phase of the steady-state gain function at all frequencies. Thus, step-response specifications can be converted into frequency-response specifications. Likewise, if one is given the frequency response over the frequency interval $0 \leq \omega \leq \infty$, the step response can be calculated; and so, frequency-response specifications can be converted into step-response specifications.

If extreme precision is required, the entire behavior of the amplifier gain and phase shift with frequency is necessary to compute the rise time,

time delay, etc., which the amplifier will exhibit when a step function input is applied. However, there are certain specific features of the frequency-response characteristics which largely determine specific features of the step response. For example, the rise time of a video amplifier is related to its upper cutoff frequency. Similarly, time delay is related to the slope of the phase response in the amplifier passband (or to the slope of the amplitude response outside the passband). We will present the rules pertaining to rise time and time delay in the next two paragraphs. Overshoot and sag will be described later.

Rise time and bandwidth. Since finite bandwidth and finite rise time are both caused by the existence of shunt capacitance at the input to each stage, it seems reasonable to expect that a relation should exist between these two quantities. Various empirical studies have shown that a general relationship does exist and is of the form

$$T_R B = 0.35 \text{ to } 0.45 \tag{18.16}$$

where T_R is the 10–90 percent rise time of the amplifier and B is the bandwidth from 0 to the upper cutoff frequency f_u.

This is perhaps the most useful rule relating frequency and transient response and is frequently adequate to obtain a very good first approximation to a multistage video amplifier design. From given over-all gain and rise-time specifications, the designer can use Eq. (18.16) to find over-all gain and bandwidth specifications. He can then proceed to an initial design using the methods presented in Chap. 13, plus certain refinements to be presented later in this chapter.

EXAMPLE   We can demonstrate the validity of Eq. (18.16) for the single-stage amplifier of Fig. 18.2 quite simply. The rise time is

$$T_R = 2.2 \, R'_\pi C_{eq}$$

while the amplifier bandwidth is

$$B = \frac{1}{2\pi R'_\pi C_{eq}}$$

Hence

$$T_R B = \frac{2.2}{2\pi} = 0.35$$

If we consider a cascade of two identical stages (with $R_s = R_i =$ interstage resistance), Eq. (18.15a) gives

$$T_R(2 \text{ stages}) \simeq (1.1\sqrt{2}) \, T_R(1 \text{ stage})$$

and the bandwidth-shrinkage table given in Chap. 13 gives

$$B(2 \text{ stages}) \simeq 0.64 \, B(1 \text{ stage})$$

Combining these results, we obtain[1]

[1] Similar calculations on stages which have overshoot gives the $(T_R B = 0.45)$ limit (*cf.* Pettit and McWhorter, "Electronic Amplifier Circuits," McGraw-Hill Book Company, New York, 1961, chap. 4).

$$T_R B(2 \text{ stages}) \simeq (1.6)(0.64) \; T_R B(1 \text{ stage}) = T_R B(1 \text{ stage}) = 0.35$$

Time delay and phase response. If the output of a video amplifier were a perfectly amplified replica of the signal except for that of a fixed delay T_D, then we could write the output voltage in the form

$$v_2(t) = Av_s(t - T_D) \tag{18 \cdot 17}$$

In physical terms this simply amounts to a shifting of the time origin.

In Fourier analysis, a shifting of the time origin corresponds to the addition of a phase factor which is proportional to frequency. This can be easily seen as follows. Suppose a function $f(t)$ has the Fourier expansion

$$f(t) = \Sigma a_m \cos m\omega t \tag{18 \cdot 18}$$

(For simplicity, we neglect sine terms in the expansion.) Then, shifting the time scale gives

$$
\begin{aligned}
f(t - t_0) &= \Sigma \; a_m \cos m\omega(t - t_0) \\
&= \Sigma \; a_m \cos (m\omega t - m\omega t_0) \\
&= \Sigma \; a_m \cos [m\omega t - \phi_m(\omega)] \tag{18 \cdot 19}
\end{aligned}
$$

Shifting the time scale is thus equivalent to adding a phase angle

$$\phi_m(\omega) = -m\omega t_0$$

to each cosine. Furthermore, the time delay t_0 is related to the frequency of any harmonic by the simple formula

$$t_0 = - \frac{\phi_m(\omega)}{m\omega} \tag{18 \cdot 20}$$

Now, an amplifier will add phase shift to any sine-wave signal that is passed through it. If an amplifier had the gain and phase characteristics shown in Fig. $18 \cdot 7a$, then its output would be a perfect replica of the input, but delayed by a time

$$T_D = - \frac{d\phi(\omega)}{d\omega} \tag{18 \cdot 21}$$

A real video amplifier will have a characteristic more like that shown in Fig. $18 \cdot 7c$, however. Here the normalized gain expressed as a function of the complex frequency variable s is (roughly)

$$A(s) = \frac{1}{(1 + s/\omega_0)^n}$$

for n identical stages. The corresponding phase angle is, for $s = j\omega$,

$$\phi(\omega) = -n \tan^{-1} \frac{\omega}{\omega_0}$$

which, for small ω, can be written

$$\phi(\omega) \simeq - \frac{n\omega}{\omega_0}$$

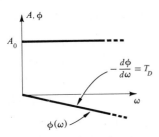

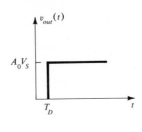

(a) *Ideal gain and phase charac-*
teristics for a video amplifier

(b) *Step response resulting*
from (a)

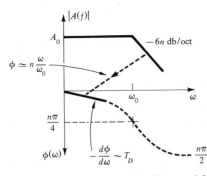

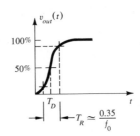

(c) *Approximation to actual video amplifier*
gain and phase characteristics

(d) *Step response resulting*
from (c)

FIG. 18·7 THE RELATIONSHIPS BETWEEN RISE TIME DELAY, PHASE RESPONSE, AMPLI-
TUDE RESPONSE IN A LOWPASS AMPLIFIER

The phase shift is linear with frequency for low frequencies, and we can
therefore define an approximate delay time by

$$T_D \simeq \frac{d\phi(\omega)}{d\omega} = -\frac{n}{\omega_0}$$

This is an interesting relation, since it gives us an estimate of the delay in
terms of the *slope of the gain characteristic outside the passband.* Remember
that this is only an estimate, though the general idea is very basic.[1]

18·1·4 *Practical problems in video amplifier design: resistive loading.* The
most basic problem to be solved in designing a video amplifier is to make
the rise time sufficiently short. Time delay is usually not crucial, and sag
can be eliminated or minimized by appropriate design techniques.

Because of the inverse relation between rise time and bandwidth in an
amplifier, a short-rise-time or "fast" amplifier is also referred to as a

---

[1] A thorough treatment of the relation between the gain and phase characteristics of
an amplifier and their relation to signal delay is given in H. W. Bode, "Network Analysis and
Feedback Amplifier Design," D. Van Nostrand Company, Inc., Princeton, N.J., 1945.

"wideband" or "broadband" amplifier, and techniques for obtaining large bandwidths are sometimes called broadbanding techniques.

A full-scale recounting of all the procedures used to design a broadband amplifier is beyond our present interests, though we do wish to point out some of the major problems and indicate the most common solutions to them.

Transistor limitations. We saw in Chap. 17 that high-frequency transistors have to be operated at larger emitter bias currents than alloy units ($\sim$5–10 ma) in order to realize the higher gain-bandwidth product which these transistors inherently have. This means that r_π will not be large enough for us to neglect r_b in the analysis, as we did in Chap. 13. As a result, there are limitations on the effectiveness with which we can trade gain and bandwidth in simple, resistively loaded stages, as we shall see presently.

Furthermore, the collector capacitance C_c (or C_μ) provides sufficient coupling between input and output at high frequencies that a Miller approximation to account for its effect is less accurate than before. We use the Miller approximation in spite of this difficulty, though we expect our analysis to be less valid than before.

Gain and bandwidth of resistively loaded stages and cascades. The nature of the transistor limitations, particularly those arising from r_b, will be developed by first obtaining formulas for the gain and bandwidth of a stage including the effects of r_b, and then giving several examples to compare the performance suggested by these formulas with that obtained from the simpler ideas developed in Chap. 13.

For this purpose, we show in Fig. 18·8 a high-frequency circuit model for an intermediate stage of a resistively loaded amplifier. We employ a Miller approximation to account for the effects of C_c. The approximate

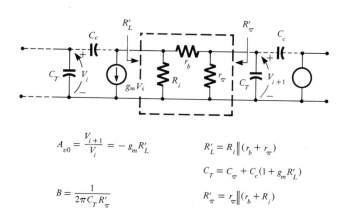

$$A_{v0} = \frac{V_{i+1}}{V_i} = -g_m R'_L \qquad\qquad R'_L = R_i \| (r_b + r_\pi)$$

$$C_T = C_\pi + C_c(1 + g_m R'_L)$$

$$B = \frac{1}{2\pi C_T R'_\pi} \qquad\qquad R'_\pi = r_\pi \| (r_b + R_i)$$

FIG. 18·8 CIRCUIT MODEL FOR AN INTERMEDIATE STAGE OF A CASCADE AMPLIFIER, INCLUDING THE EFFECTS OF r_b

load resistance for the stage is

$$R'_L = \frac{R_i(r_b + r_\pi)}{R_i + r_b + r_\pi} \tag{18·22a}$$

which can be conveniently normalized to r_π to give

$$\frac{R'_L}{r_\pi} = \frac{R_i}{r_\pi} \frac{1 + (r_b/r_\pi)}{1 + (r_b/r_\pi) + (R_i/r_\pi)} \tag{18·22b}$$

The midband voltage gain of the stage is[1]

$$\frac{V_{i+1}}{V_i} = A_{v0} = -g_m R'_L \frac{1}{1 + (r_b/r_\pi)} \tag{18·23}$$

Since we will usually be selecting an R'_L to obtain a prescribed gain, it is useful to combine Eqs. (18·23) with (18·22) to obtain

$$\frac{R_i}{r_\pi} = \frac{A_{v0}[1 + (r_b/r_\pi)]}{\beta_0 - A_{v0}} \tag{18·24}$$

The total shunt capacitance in each stage can be written in the two forms

$$C_{eq} = C_\pi + C_c(1 + g_m R'_L) = C_\pi + C_c\left[1 + A_{v0}\left(1 + \frac{r_b}{r_\pi}\right)\right] \tag{18·25}$$

and the resistance facing C_{eq} can be written in the two forms

$$R'_\pi = \frac{r_\pi(R_i + r_b)}{r_\pi + R_i + r_b} \tag{18·26a}$$

$$\frac{R'_\pi}{r_\pi} = \frac{(R_i/r_\pi) + (r_b/r_\pi)}{1 + (R_i/r_\pi) + (r_b/r_\pi)} \tag{18·26b}$$

The stage bandwidth is therefore

$$B = \frac{1}{2\pi R'_\pi C_{eq}} = \frac{1}{2\pi r_\pi C_\pi} \frac{1}{1 + (C_c/C\pi)\{1 + A_{v0}[1 + (r_b/r_\pi)]\}} \frac{r_\pi}{R'_\pi} \tag{18·27b}$$

By combining Eqs. (18·27) and (18·23), we can write the gain-bandwidth product for the stage as

$$|A_{v0}|B = \frac{f_T}{1 + (C_c/C_\pi)\{1 + A_{v0}[1 + (r_b/r_\pi)]\}} \frac{R'_L}{R'_\pi} \frac{1}{1 + (r_b/r_\pi)} \tag{18·28a}$$

$$= \frac{f_T}{1 + (C_c/C_\pi)\{1 + A_{v0}[1 + (r_b/r_\pi)]\}} \frac{R_i}{r_b + R_i} \tag{18·28b}$$

All of these expressions reduce to similar expressions developed in Chap. 13 if $r_b \to 0$. However, when r_b becomes comparable to r_π, the gain-bandwidth product is reduced.

Let us now consider several examples which point out the effect of r_b on the performance of an amplifier.

[1] The factor $1/(1 + r_b/r_\pi)$ does not apply to the last stage of a cascade.

EXAMPLE   1. Suppose we want to design a video amplifier to have a gain of 500 and an overall rise time of 0.1 μsec. Using $T_R B = 0.35$, this means we need an overall bandwidth of 3.5 mcps. If we consider a cascade of two identical stages, we need a gain of about 23 per stage and a stage bandwidth of (3.5 mcps/0.64) ≃ 5.5 mcps. The operating gain-bandwidth product for the stage thus needs to be about 126 mcps. Considering the probable influence of C_c on the gain-band product, we look for transistors with an f_T of about 250–300 mcps. A suitable choice is a 2N916 which, when biased at $I_C = 5$ ma, $V_{CE} = 5$ volts, has the typical parameters

$$f_T = 320 \text{ mcps} \qquad r_b = 60 \text{ ohms} \qquad C_\pi = 100 \text{ pfarads}$$
$$\beta = 120 \qquad r_\pi = 600 \text{ ohms} \qquad C_c = 3 \text{ pfarads}$$
$$g_m = 0.2 \text{ ℧}$$

Let us now see whether this transistor will do the job. To begin it is instructive to consider the amplifier circuit values which would be computed if we neglected r_b.

Calculations neglecting r_b.   Using Eq. (18 · 28) with a gain $A_{vo} = 23$, we find

$$A_{vo}B = \frac{f_T}{1 + C_c/C_\pi(1 + A_{vo})} = \frac{320}{1 + [3(24)/100]} = 186 \text{ mcps}$$

This transistor seems to provide a comfortable margin of safety, since we require a gain-band product of only 126 mcps per stage. Hence, we proceed to calculate the appropriate circuit values.

To do this we first use Eq. (18 · 24) to estimate R_i. We have

$$\frac{R_i}{r_\pi} \simeq \frac{23}{120 - 23} = 0.237$$

or $R_i = 142$ ohms. Now, if we use $R_s = R_i$ in a cascade of identical stages, then the voltage appearing across r_π of the first transistor will only be $r_\pi/(R_s + r_b + r_\pi)$ of the source voltage. Neglecting r_b, this means that only 600/742 of the source voltage will appear across r_π of the first stage, so the midband gain of the amplifier, measured from the base-emitter terminals of the input stage to the output, should be $A_{vo}(742/600) = 620$ to provide a source-to-load voltage gain of 500. Thus, as a reasonable approximation, we will assume that the gain given by Eq. (18 · 23) should be $\sqrt{620} = 25$ per stage. Then from Eq. (18 · 24) we find a new estimate of $R_i/r_\pi = 0.262$ or $R_i = 157$ ohms. Using Eq. (18 · 22b) we find $R_L'/r_\pi = 0.208$, or $R_L' = 125$ ohms. The two-stage amplifier should therefore have $R_s = R_i = 157$ ohms and a final load resistor of 125 ohms. We expect this amplifier to provide a mid-band source-to-load voltage gain of 500. Each stage should exhibit a gain-band-width product of 180 mcps (for $A_{vo} = 25$) and a stage bandwidth of 7.2 mcps. The two-stage bandwidth should be (7.2 × 0.64) mcps = 4.6 mcps, which is more than the required bandwidth.

Calculations including r_b.   Let us now see what performance we would expect if these circuit values were used in the real transistor, where $r_b = 60$ ohms.

First of all, Eq. (18 · 28b) shows that for a midband gain $A_{vo} = 25$, the gain-bandwidth product will only be

$$|A_{vo}|B = \frac{320}{1 + 0.03(28.5)} \frac{157}{217} = 125 \text{ mcps}$$

or just barely enough to meet the specifications. Furthermore, since R_i, R_s, and R_L

were determined without regard for r_b, we might suspect that the selected values will probably not provide the required response characteristics. To check this conclusion, we calculate the performance expected for $R_s = R_i = 157$ ohms, $R_L = 125$ ohms.

The gain of the final stage is defined in such a way that it is not affected by r_b {that is, the factor $1/(1 + r_b/r_\pi)$ in Eq. (18·23) does not apply for the final stage}. Hence, it remains at 25 for $R_L = 125$ ohms. The input capacitance to the final stage $C_{eq,2}$ is therefore 178 pf, as before. However, according to Eq. (18·26b), $C_{eq,2}$ faces R'_π/r_π of $0.36/1.36 = 0.265$, or $R'_\pi = 160$ ohms instead of the 125 ohms which Eq. (18·26b) would give with $r_b = 0$. Hence, the second stage has a bandwidth of

$$ B = \frac{10^{12}}{2\pi \times 160 \times 178} = 5.6 \text{ mcps} $$

Continuing back along the cascade, the first stage faces a normalized load resistance $R'_{L,1}/r_\pi = 0.212$ [from Eq. (18·22b)], or $R'_{L,1} = 127$ ohms. From Eq. (18·23), the stage gain is then 23.2. In addition, the fraction of the source voltage that appears across r_π of the input stage is $600/817 = 0.73$, so the midband gain of the amplifier, from source to load, is

$$ A_{v0} = 0.73 \times 23.2 \times 25 = 425 $$

instead of the required 500.

The bandwidth of the second stage is essentially the same as that of the first stage (5.6 mcps), so the two-stage cascade has a bandwidth of 3.6 mcps. In other words, the two-stage amplifier will not meet the required specifications, whereas we expected from our calculations, neglecting r_b, that there would be bandwidth to spare. It is clear from this example that wideband-amplifier calculations must include the effects of r_b if they are to give realistic estimates of circuit values.

EXAMPLE 2. To illustrate this point, we recalculate the performance we could obtain from the two-stage cascade including r_b in the determination of circuit values. We again seek an over-all voltage gain of 500. If we assume R_i will be on the order of 160 ohms, the voltage appearing across r_π of the first stage will be $600/820 \, V_s$, so we need an internal gain of about 680 to produce an over-all gain of 500. If we distribute this gain evenly between stages, we need a gain of 26 per stage. Proceeding as before, we obtain $R_s = R_i = 182$ ohms, $R_L = 130$ ohms. Including the resistive division of V_s at the input, this amplifier then provides a midband source-to-load voltage gain of 485 and an overall amplifier bandwidth of 3.2 mcps. The results of the two sets of calculations are summarized in the table below:

Table 18.1

R_s, Ω	R_i, Ω	R_L, Ω	A_{v0} (source-to-load)	B, mcps (two-stage)	$A_{v0}B$ mcps
182	182	130	485	3.2	1580
157	157	125	425	3.6	1530

In neither case does the amplifier meet the required specifications. A slight (3 percent) increase in gain-bandwidth product has been obtained by using r_b in the calculations of R_i, etc. However, reducing R_s to 150 ohms in the first row will increase the amplifier gain to 505 and the bandwidth to 3.6 mcps. Experimental adjustments of this type can usually be made to meet specifications, and other design techniques can be employed if one must obtain the ultimate in performance. However, this is an aside; the main point of these examples is that the effects of r_b must be included to obtain a realistic estimate of the performance of a wideband amplifier operating with a high I_E, where $r_b/R_i \geq 0.25$.[1]

Sag compensation. When the source, interstage, and load resistances have been selected to give an appropriate rise time (or bandwidth), there remains the task of choosing biasing elements to achieve the proper operating point for each stage. The choice of biasing elements and biasing arrangements is very important, because it determines the low-frequency response of the amplifier or, what is equivalent, the sag in the step response of the amplifier.

If the biasing is accomplished using feedback techniques and the amplifier is made to amplify uniformly right down to dc, then there will be no sag. However, if coupling and emitter bypass capacitors are used in each stage, then the sag in a multistage amplifier can become rather large, and some means of compensating for it is necessary. In this section we will consider a very practical means for obtaining sag compensation. In all cases, sag compensation involves decreasing the low-frequency cutoff of the amplifier and is therefore also called "low-frequency compensation."

The basic idea in providing sag compensation is to incorporate circuit elements in the amplifier which will provide a *positive* initial slope in the step response and then to adjust the value of this slope to compensate for the negative initial slope caused by C_s and C_E. A circuit which has the required property is the $R_d C_d$ circuit shown in Fig. 18·9. In addition to providing the sag compensation function, this circuit also provides two other useful features:

1. It allows us to split the collector load resistor into two parts. The high-frequency load resistor is R_L (plus any additional loading provided by succeeding stages in a cascade amplifier). The dc collector load is $R_d + R_L$. For short rise time, we will generally want $R_L \simeq 200$ ohms or less, but for bias stability we may want $R_d + R_L \sim 5$–10 kohms.

2. It eliminates any dependence of the stage gain on power supply impedance. This is particularly important in multistage amplifiers, where the stages can be coupled together strongly enough through a few ohms of power supply impedance to produce oscillations. The $R_d C_d$ circuit avoids this possibility and is aptly called a "decoupling circuit." Fortunately, R_d and C_d can be varied considerably without affecting the decoupling function.

To show that the circuit provides the required positive slope, we neglect C_s and C_E (a step that is justified by the argument on sag from multiple sources given earlier) and consider the nature of the time response of the

---

[1] A procedure for using emitter degeneration to decrease r_b/r_π is described in the problems. In some cases a 10 to 15 percent increase in gain-band product can be obtained using this technique.

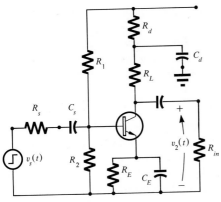

(a) Actual circuit

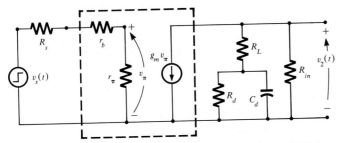

(b) Circuit model for studying sag compensation provided by R_dC_d in a single stage

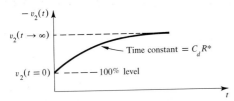

(c) Circuit response of (b) to step change in v_s

FIG. 18·9 SAG COMPENSATION OBTAINED BY PROPERLY SELECTED DECOUPLING COMPO-NENTS; ONLY THE INPUT STAGE IS CONSIDERED IN THIS EXAMPLE

amplifier output when a step current $g_m V_\pi$ is applied. By inspection of Fig. 18·9c, we can write

$$v_2(t) = v_2(0) + [v_2(\infty) - v_2(0)][1 - e^{-t/C_dR^*}] \qquad (18·29a)$$

where
$$R^* = R_d \| (R_L + R_{in}) \qquad (18·29b)$$

$$v_2(t = 0) = -g_m V_\pi (R_L \| R_{in}) \qquad (18·29c)$$

$$v_2(t \to \infty) = -g_m V_\pi (R_{in} \| R_L + R_d) \qquad (18·29d)$$

The initial slope of Eq. (18·29a) is

$$\left. \frac{dv_2}{dt} \right|_{t=0} = \frac{v_2(\infty) - v_2(0)}{C_dR^*}$$

and the sag is

$$100\% \frac{1}{v_2(0)} \left(\frac{dv_2}{dt}\Bigg|_{t=0}\right) = \frac{100\%}{C_d R^*} \left(\frac{v_2(\infty)}{v_2(0)} - 1\right)$$

which is positive as required.

EXAMPLE If we assume for purposes of illustration that $R_2 = 200$ ohms, $R_{in} = 500$ ohms (input impedance to the succeeding stage in a cascade amplifier), then $[v_2(\infty)/v_2(0)] = 500/140 = 7$, and $R^* \simeq 700$ ohms. If we assume that C_s and C_E produce a sag of 0.16 percent per μsec per stage, we want

$$6 \times \frac{100\%}{(C_d)(700 \text{ ohms})} = 0.16\%/\mu\text{sec}$$

or $C_d \simeq 5$ μfarad to provide sag compensation. In a multistage amplifier, we can compensate for several sources of sag with just one decoupling circuit, though it is better practice to distribute decoupling circuits throughout the amplifier. The input stages in particular should be provided with decoupling, since they are most sensitive to signal pick up from the power supply impedance (cf. 3-stage demonstration amplifier, Chap. 13).

18·1·5 *Broadbanding a video amplifier.* We have seen in the last several sections that video amplifiers must have a large bandwidth if they are to provide a short rise time and that, if this large bandwidth is obtained by resistive loading, the base resistance becomes an important factor in reducing the gain-bandwidth product we can obtain from a stage. One is thus tempted to search for ways to reduce the importance of r_b and making better use of the gain-bandwidth product.

Many schemes have appeared in the literature for accomplishing these ends.[1] Frequently, a particular circuit performs best with a particular type of transistor, though there are some general principles which apply in most cases. In this section we will discuss some of the basic ideas which are involved in attempting to improve the response of a video amplifier, by using more sophisticated circuits than the simple resistively loaded cascade.

Before embarking on these matters, however, a point of perspective is in order. By careful circuit design and experimental alignment, we can usually obtain a gain-bandwidth improvement of between 2 and 4 over that of a simple resistive cascade (in a transistor amplifier). In some cases this may not be a sufficiently great increase to warrant the additional effort (mostly in alignment of the final amplifier). We usually resort to these techniques only when the very best performance obtainable must be achieved.

Inductive peaking in video amplifiers. Generally, improved performance in a video amplifier can be obtained by the judicious addition of inductance in the amplifier. One popular scheme employs an inductance in series with the interstage resistance R_i, as shown in Fig. 18·10a. The idea here is to

---

[1] cf. References at the end of the chapter.

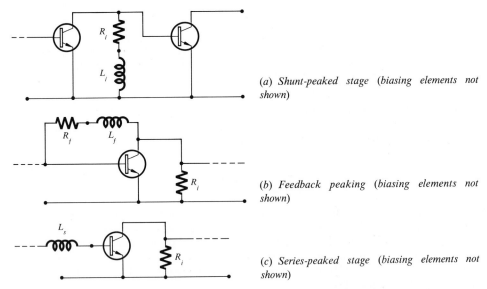

(a) Shunt-peaked stage (biasing elements not shown)

(b) Feedback peaking (biasing elements not shown)

(c) Series-peaked stage (biasing elements not shown)

FIG. 18·10   SEVERAL CIRCUITS FOR INDUCTIVE PEAKING IN VIDEO AMPLIFIERS

select an inductance which will increase the magnitude of the load impedance near the cutoff frequency of the amplifier and therefore extend the
bandwidth. This circuit is called a *shunt-peaked* circuit, since the inductance is in parallel with the transistor input capacitance.[1]

A second scheme, which is essentially equivalent to the shunt-peaked
stage, uses an inductance in series with a feedback resistor, as shown in
Fig. 18·10b. We saw in Chap. 14 (*cf.* Sec. 14·1 and Fig. 14·2) that simple
resistive feedback from collector-to-base provides a convenient means of
trading gain and bandwidth and, at the same time, making the circuit
relatively immune to variations in transistor parameters. The addition of
inductance in the feedback loop capitalizes on these advantages in the
midband-frequency range and then cuts out the feedback near the band
edge (and beyond). This once again gives an increased gain-bandwidth
product.

A third scheme, which is less popular but simpler for purposes of
exposition, is the *series-peaked* stage shown in Fig. 18·10c. The effect
of L_s in this circuit can be described as follows. The $r_\pi C_{eq}$ combination
can be expressed as an equivalent series r and C, though of course
the series-equivalent r and C will vary with frequency. However, if the
inductance is chosen to be roughly in resonance with the equivalent series
input capacitance of the transistor at the amplifier cutoff frequency, then
the current in the input loop, and therefore V_π, will be larger than it
would otherwise be. This may then extend the -3db frequency of the
amplifier.

[1] By correctly choosing the peaking inductance, the deleterious effects of r_b can also
be minimized.

The series-peaked stage. The series-peaked stage has a somewhat simpler gain function than the other two arrangements, and the responses which can be obtained from it are accordingly less flexible than those available in the other stages. However, the essential ideas involved in broadbanding an amplifier can be very simply presented using the series-peaked stage as an example, so we shall analyze this configuration in some detail.

To begin, we solve for the high-frequency voltage gain of the circuit shown in Fig. $18 \cdot 10c$ as a function of the complex frequency s. We have, from the circuit model of Fig. $18 \cdot 11$,

$$V_\pi = \frac{V_s Z_\pi}{R_s + r_b + sL_s + Z_\pi} \qquad (18 \cdot 30a)$$

where

$$Z_\pi = \frac{r_\pi}{1 + sC_T r_\pi} \qquad (18 \cdot 30b)$$

The voltage gain is then

$$A_v(s) = \frac{V_L(s)}{V_s(s)} = \frac{-g_m V_\pi R_L}{V_s}$$

$$= \frac{-\beta_0 R_L}{(R_s + r_b + r_\pi) + s[L_s + C_T r_\pi (R_s + r_b)] + s^2 L_s C_T r_\pi} \qquad (18 \cdot 31)$$

For normalization purposes, we can also rewrite Eq. $(18 \cdot 31)$ in the form

$$A_v(s) = \frac{-\beta R_L}{R_s + r_b + r_\pi} \cdot \frac{1}{1 + s\left[\dfrac{L_s + C_T r_\pi (R_s + r_b)}{R_s + r_b + r_\pi}\right] + s^2 \dfrac{L_s C_T r_\pi}{R_s + r_b + r_\pi}}$$

$$= \frac{A_{v0}}{1 + as + bs^2} \qquad (18 \cdot 32)$$

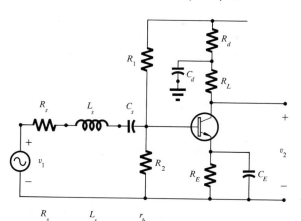

(a) *Circuit diagram of the series-peaked stage*

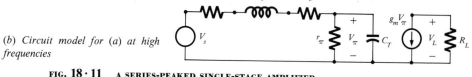

(b) *Circuit model for (a) at high frequencies*

FIG. $18 \cdot 11$ A SERIES-PEAKED SINGLE-STAGE AMPLIFIER

where

$$a = \frac{L_s + C_T r_\pi (R_s + r_b)}{R_s + r_b + r_\pi} \qquad (18 \cdot 33a)$$

$$b = \frac{L_s C_T r_\pi}{R_s + r_b + r_\pi} \qquad (18 \cdot 33b)$$

$$A_{v0} = \text{midband gain} = \frac{-\beta R_L}{R_b + r_s + r_\pi} \qquad (18 \cdot 33c)$$

It is apparent from Eq. (18·32) that when $s = j\omega$ is small, the voltage gain is the same as before. However, $A_v(s)$ now has two poles instead of one and a correspondingly greater diversity of possible responses.

Log A–log f plots for a two-pole function. One useful way of describing the variety of possible characteristics is to construct normalized plots of the gain in db vs. the logarithm of the frequency. The asymptotic features of these plots are similar to the $A(\text{db})$ vs. log ω plots given in Chap. 13 for a single pole (or zero), though the asymptotic departures are not so readily remembered, so a chart of the type given in Fig. 18·12 is useful for evaluating the performance.

The function which is plotted in Fig. 18·12 is $20 \log_{10} T$, where

$$T = \frac{1}{1 + (\sigma/\omega_0{}^2)s + (s^2/\omega_0{}^2)} \qquad (18 \cdot 34)$$

In terms of Eqs. (18·33), $a = \sigma/\omega_0{}^2$, $b = 1/\omega_0{}^2$.

The function given in Eq. (18·34) has a low-frequency asymptote of 0 db, and a high-frequency asymptote which falls at the rate of 12 db/octave (or 20 db/decade). The low-frequency and high-frequency asymptotes intersect at the frequency

$$\omega = \omega_0 = \frac{1}{\sqrt{b}}$$

The behavior of the function around its break point is determined by the ratio σ/ω_0 as shown in Fig. 18·12. Like the simpler one-pole functions described in Chap. 13, the function given in Eq. (18·34) has the property that its asymptotic departures are the same for frequencies whose geometric mean is the break frequency. That is, for $\sigma/\omega_0 = 0.1$, Fig. 18·12 shows that $20 \log_{10} T$ is 2.5 db above its low-frequency asymptote at $\omega_1/\omega_0 = 0.5$. It is therefore 2.5 db above its high-frequency asymptote at $\omega_2/\omega_0 = 2$. On account of this property, only the asymptotic departures for $\omega/\omega_0 \leq 1$ need to be shown.

EXAMPLE As an example of the changes in performance which we can obtain using a peaking inductance, we use the transistor parameters given earlier in a stage of the type shown in Fig. 18·10c where $R_L = 200$ ohms, $R_s + r_b = 200$ ohms, $r_\pi = 500$ ohms, and $C_T = 100$ pf $+ 3$ $(1 + 40)$ pf $= 223$ pf. With no inductive peaking this circuit gives a midband gain of $200/7 = 28.6$ and a (-3)-db bandwidth of 5 mcps.

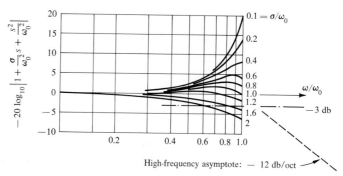

$$-20 \log_{10} \left| 1 + \frac{\sigma}{\omega_0^2} s + \frac{s^2}{\omega_0^2} \right|$$

FIG. 18·12 PLOT OF $\left| 1 + \dfrac{\sigma s}{\omega_0^2} + \dfrac{s^2}{\omega_0^2} \right|$ IN DB VERSUS LOG ω, ILLUSTRATING THE VARIETY OF RESPONSES AVAILABLE WITH A TWO-POLE RESONANT FUNCTION

If a peaking inductance of 5 μhenries is added, then from Eq. (18·33b) we have

$$b = \frac{1}{\omega_0^2} = \frac{(5 \times 10^{-6})(223 \times 10^{-12})(500)}{700} \text{ sec}^2$$

$$\simeq 800 \times 10^{-18} \text{ sec}^2$$

This gives $\omega_0 = 35.4 \times 10^6$ rad/sec, or a break frequency of $f_0 = 5.63$ mcps. The value of a is, from Eq. (18·33a),

$$a = \frac{5 \times 10^{-6} + (223 \times 10^{-12})(500)(200)}{700} = \frac{2.73 \times 10^{-5}}{700} = 3.9 \times 10^{-8} \text{ sec}$$

Since $a = \sigma/\omega_0^2$, we have

$$\frac{\sigma}{\omega_0} = a\omega_0 = 1.38$$

From the curves shown in Fig. 18·12, this value of σ/ω_0 suggests that the -3-db frequency will also be very nearly $f_0 = 5.63$ mcps. Hence, the -3-db bandwidth has been increased by about 12 percent, though the high-frequency asymptote is now -12 db/octave instead of -6 db/octave.

If larger inductances are used, the resonant effects will occur at a lower frequency, and significant peaking in the steady state response will appear. Calculations have been carried out for $L_s = 10$ and 15 μhenries to illustrate these points. The results are shown in Fig. 18·13. In all cases, the gain characteristic with L_s ultimately drops at -12 db/octave. Note that the -3-db frequencies are lower for $L_s = 10$ or 15 μhenries than for $L_s = 0$. By comparison to the resistive stage, the step response for these values of L_s will exhibit significant overshoot and ringing, in addition to an *increased* rise time—in other words, poorer performance in all respects. By contrast, the step response for $L_s = 5$ μhenries will exhibit very little overshoot, and in fact the rise time will be about 10 percent less than the rise time with no peaking inductance. However, the time delay, which is roughly proportional to the slope of the gain characteristic outside the passband, will be increased over the delay with no L_s.

Pole-zero patterns and stagger tuning in video amplifiers. The variety of responses that are available from the series-peaked stage can also be conveniently described by considering the motion of the poles of the gain

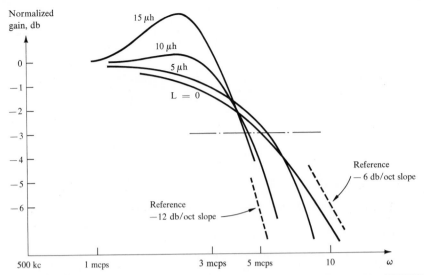

FIG. 18·13 CALCULATED HIGH FREQUENCY GAIN CHARACTERISTIC FOR AMPLIFIER SHOWN IN FIG. 18·11 FOR VARIOUS VALUES OF L_s

function as L_s is varied. To do this, we simply factor the quadratic

$$1 + as + bs^2 = 0$$

to obtain

$$s_1, s_2 = \frac{-a}{2b} \pm \frac{1}{2b} \sqrt{a^2 - 4b} \tag{18·35}$$

a and b are given by Eqs. (18·33a) and (18·33b). For $L_s = 0$, the roots are

$$s_1 = \infty$$

$$s_2 = -\frac{R_s + r_b + r_\pi}{C_T r_\pi (R_s + r_b)} = -\frac{1}{C_T r_\pi \| (R_s + r_b)}$$

This second root is simply the pole position for the simple resistive stage, and can be seen directly from Eq. (18·32).

As L_s is increased from zero, the poles s_1 and s_2 move toward each other as shown in Fig. 18·14, both remaining on the real axis. Ultimately the two poles coalesce, this condition being obtained for

$$\frac{a^2}{4b} = 1$$

$$\frac{[L_s + C_T r_\pi (R_s + r_b)]^2}{4(R_s + r_b + r_\pi)(L_s C_T r_\pi)} = 1 \tag{18·36}$$

The smaller value of L_s which satisfies Eq. (18·36) is called the critical inductance L_{crit}, and is given by

$$L_{crit} = \frac{r_\pi C_T (R_s + r_b)^2}{4(R_s + r_b + r_\pi)}$$

For the previous example $L_{crit} = 2.5$ μhenries.

As L is increased beyond L_{crit}, the solutions to Eq. (18·35) become complex, so the poles move out into the complex plane, with s_1 and s_2 being complex conjugates of each other. This behavior is shown in Fig. 18·14. Ultimately the poles come together again on the real axis, and finally, as $L_s \to \infty$, the poles approach the positions

$$s_1\Big|_{Ls\to\infty} \to 0$$

$$s_2\Big|_{Ls\to\infty} \to -\frac{1}{C_T r_\pi}$$

The positions we would like the poles to take depends on the amplifier we are considering. If we want a single stage amplifier to have minimum rise time with no overshoot, then L_{crit} is the value we should use. In a cascade of stages, however, values of $L_s > L_{crit}$ may be more appropriate. We will see why this is so in the next paragraph.

The maximally flat gain function. When we design a cascade of simple resistively loaded stages, we obtain a gain function of the form

$$\frac{A_0}{\sqrt{1 + (f/f_0)^2}}$$

for each stage. To the extent that the stages are unilateral (that is, C_c incorporated into C_T), a cascade of n such stages has the gain function

$$\left(\frac{A_0}{\sqrt{1 + (f/f_0)^2}}\right)^n$$

Each stage provides a pole in the complex frequency plane at $s = \sigma = -2\pi f_0$, the n stages providing n poles at this same position. The coincidence of all the poles leads to the bandwidth-shrinkage problem described in Chap. 13 and requires us to increase the cutoff frequency of each stage as we increase the number of stages to keep the overall -3-db bandwidth the same.

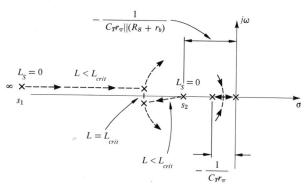

FIG. **18·14** MOTION OF THE POLES OF A SERIES-PEAKED STAGE AS L_s IS VARIED FROM 0 TO ∞

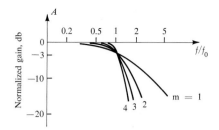

FIG. **18·15**   THE MAXIMALLY FLAT GAIN FUNCTION

There are other gain functions which do not suffer so seriously from this bandwidth-shrinkage problem, however. One interesting function of this type can be written in the normalized form

$$A(f) = \frac{1}{\sqrt{1 + (f/f_0)^{2m}}} \qquad (18\cdot37)$$

This function is called the *maximally flat* or Butterworth[1] gain function; it has the property that the first m derivatives of $A(f)$ are zero at $f = 0$. All the functions described by Eq. (18·37) have the same -3-db bandwidth, though the gain characteristic becomes more nearly rectangular as m increases (see Fig. 18·15).

Now, the question arises as to how the poles of an amplifier should be arranged to obtain such a gain characteristic.[2] The answer turns out to be extremely simple. All of the poles are distributed on a semicircle in the complex frequency plane. The semicircle is centered at the origin and has a radius of $2\pi f_0$ (see Fig. 18·16). The poles are placed on the semicircle as follows:

If we have m poles at our disposal, the first pole is placed on the semicircle at an angle $\theta = 180°/2m$ from the ω axis. The remaining poles are then placed at angular separations of $180°/m$ from each other. The construction, together with two simple examples, is given in Fig. 18·16.[3]

If we wish to obtain a very rectangular gain characteristic, then we will need to have many poles at our disposal. However, a single stage need not produce all the poles. Instead, we can build a multistage amplifier in which we assign to each stage a simple pole pattern out of the total array of poles for the over-all gain function. Then, assuming that the stages are unilateral (or very nearly so), we can build up the total gain function by cascading the individual stages.

When an over-all gain function is obtained in this way, the individual

---

[1] S. Butterworth, On the theory of filter amplifiers, *Wireless Engineering*, vol. 7, pp. 536–541, October, 1930.

[2] Maximally flat functions need not be of the all-pole type considered here.

[3] The mathematical proof of this result is given in most texts on modern network theory or amplifier theory. See, for example, Pettit and McWhorter, "Electronic Amplifier Circuits," McGraw-Hill Book Company, New York, 1961, pp. 204–206.

stages are said to be *stagger-tuned*, since they will exhibit $A(\text{db})$ vs. $\log f$ characteristics which "peak-up" at various points within the pass band. An example of a three-pole maximally flat gain function will illustrate this effect and the relative merits of using a complicated gain function in preference to that obtained from the resistively loaded cascade.

EXAMPLE   Let us return to our previous example of a two-stage cascade designed to give an over-all bandwidth of 3.5 mcps and a gain of 500 from source to load. In the simple resistively-loaded case, we found that each stage should have a bandwidth of 5.5 mcps, and that $R_s = R_i = 170$ ohms, $R_i = 130$ ohms were required.

However, we can also build a two-stage cascade in the form shown in Fig. 18·17, where we attempt to choose L_s so that the first stage provides a pair of complex conjugate poles at $(-0.5 \pm 0.866) \times 3.5$ mcps, and the second stage provides a simple pole at 3.5 mcps (instead of 5.5 mcps, as before). The result will be a maximally flat amplifier with higher midband gain for the same -3-db frequency.

To design the circuit, we will arbitrarily assume for the moment that $R_s = R_i = 170$ ohms, as before, and see whether we can choose L_s and R_L for the desired performance.

To choose L_s, we note from Eq. (18·33b) that

$$b = \frac{1}{\omega_0{}^2} = \frac{L_s C_T r_\pi}{R_s + r_b + r_\pi}$$

Using $r_\pi = 500$ ohms, $R_s = 170$ ohms, $r_b = 60$ ohms, $C_{T,1} = 180$ pf and $\omega_0 = (2\pi \times 3.5 \times 10^6)$, we find

$$L_s = \frac{730}{180 \times 500 \times 4\pi^2 \times (3.5)^2} = 16.7 \ \mu\text{henries}$$

This is the value of L_s required for the poles to lie on the required semicircle. However, we also want them to be at the specific points $-0.5 \pm j0.866$ (on the normalized circle). Since we have arbitrarily chosen R_s, this condition may not be satisfied. To see whether it is, we can simply calculate $a/2b = a\omega_0{}^2/2$ from

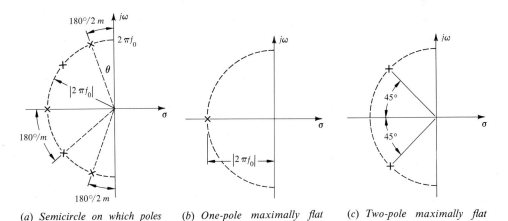

(a) Semicircle on which poles are located for an all-pole maximally flat function

(b) One-pole maximally flat function ($m = 1$)

(c) Two-pole maximally flat function ($m = 2$)

FIG. 18·16   POLE POSITIONS FOR A MAXIMALLY FLAT ALL-POLE FUNCTION

High-frequency circuits and transistor limitations   *801*

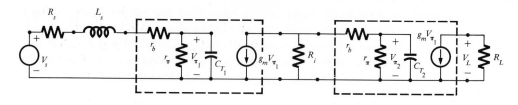

$$R_s = R_i = 170\ \Omega;\ R_L = 130\ \Omega$$

$$A_0 = \frac{V_L}{V_s} = 500;\ f(-3\ \text{db}) = 3.5\ \text{mcps}$$

$$R_s = R_i = 170\ \Omega;\ L_s = 16.3\ \mu\text{h};\ R_L = 296\ \Omega$$

$$A_0 \simeq \frac{V_L}{V_s} = 1{,}150;\ f(-3\ \text{db}) \simeq 3.5\ \text{mcps}$$

(a) *Resistive cascade*　　　　　　　　　　　　　　　　(b) *Maximally flat cascade*

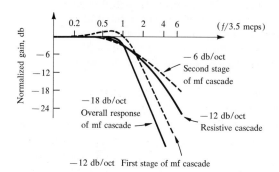

(c) *Comparative gain characteristics*

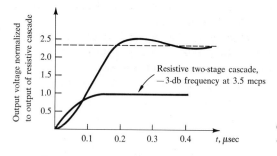

(d) *Step response of two-stage maximally flat cascade and two-stage resistive cascade*

FIG. **18·17**　CASCADE OF SERIES-PEAKED STAGE AND RESISTIVELY LOADED STAGE TO PRODUCE AN APPROXIMATION TO A THREE-POLE MAXIMALLY FLAT GAIN FUNCTION

Eqs. (18·33a) and (18·33b). Equation (18·33a) gives

$$a = \frac{(16.7 + 20.6) \times 10^{-6}}{710} = 0.52 \times 10^{-7}$$

and thus

$$\frac{a}{2}\,\omega_0 = 0.26 \times 10^{-7} \times 2\pi \times 3.5 \times 10^6 = 0.57$$

or the poles of the input stage are located at $-0.57 \pm j\,0.82$ instead of $-0.5 \pm j\,0.866$. We can change this if necessary by adjusting R_s, though the models we are using are sufficiently imprecise that this extra effort is not warranted, and the adjustment would be more profitably made experimentally in the final amplifier.

Thus, if we choose $L_s = 16.7$ μhenries, we obtain (approximately) two of the poles required in the three-pole maximally flat function. To obtain the third pole, we can now design a simple resistively-loaded second stage to give a pole at $s = -2\pi \times 3.5 \times 10^6/\text{sec}$. To do this, we simply choose R_L so that $C_{T,2}$, the input capacitance of the second stage, is a factor of $5.5/3.5$ larger than before; that is, we want $C_{T,2} = 180 \times 1.57 = 282$ pfarads. The value of R_L required is found from the equation

$$C_T = C_\pi + C_\mu(1 + g_m R_L)$$

to be 296 ohms. Since the previous R_L was 130 ohms, the gain of the maximally flat amplifier will be a factor of 2.3 greater than the resistive cascade.

Comparative gain characteristics for the two amplifiers are shown in Fig. 18·17c. The individual characteristics of two stages of the maximally flat (mf) cascade are shown dotted (the first-stage gain characteristic is obtained from Fig. 18·12 with $\sigma/\omega_0 = 1$). Note that the peaking up of the first stage partially compensates for the falling characteristic of the second stage in the mf cascade.

The price we pay for this increased gain is some overshoot, a greater delay, and a slightly *increased* rise time, as suggested in Fig. 18·17d.

Linear phase functions. If we are dissatisfied with the transient character-istics which we have obtained, we can consider the possibility of trading the increased gain for some increase in bandwidth. In doing this, it is useful to point out that overshoot and rise time are *increased* for a given cutoff frequency as the sharpness of the cutoff of the amplitude response is increased. In physical terms, the sharp cutoff of the maximally flat gain function described earlier makes it a good filter, but, like all good filters, it rings when it is driven with a transient excitation.

The ringing and overshoot can be avoided by designing an amplifier to have a *linear phase shift with frequency* in so far as possible. Again, a linear-phase-shift function can be described by an appropriate pole-zero pattern, so the amplifier design can proceed as before.

A particularly interesting set of all-pole linear-phase functions has been described by Storch.[1] The three-pole linear-phase function is of the form

$$A(s) = \frac{1}{1 + s + 0.4s^2 + 0.067s^3} \tag{18·38}$$

and has its poles at the relative positions $s_1 = +\sigma = -1.0$, s_2, $s_3 = -0.78 \pm j0.73$ (see Fig. 18·18). We can choose L_s and R_L to obtain this set of poles in our previous amplifier, and in so doing compare the three-pole linear-phase amplifier to the two-stage resistive cascade.

EXAMPLE We first choose L_s to obtain the complex poles at the correct angular positions. Again, assuming $R_s = R_i = 170$ ohms, $C_{T,1} = 180$ pf, $r_b = 60$ ohms, $r_\pi = 500$ ohms, we find that $L_s = 5$ μhenries gives us a pair of poles at the positions $s_{2,3} = -2\pi \times 6.3$ mcps $\times (0.715 \pm j0.7)$, which is very close to the angular location required for the complex poles.

[1] L. Storch, Synthesis of constant time delay ladder networks using Bessel polynomials, *Proc. IRE*, vol. 42, pp. 1666–1676, November, 1954.

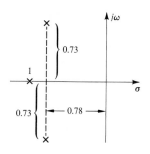

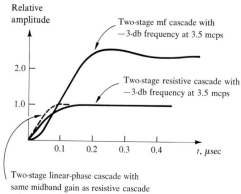

Two-stage linear-phase cascade with
same midband gain as resistive cascade

FIG. 18·18 POLE CONFIGURATIONS FOR A THREE-POLE APPROXIMATION TO A LINEAR PHASE CHARACTERISTIC, AND STEP RESPONSE OF RESULTING AMPLIFIER

The corresponding real pole should be at the position

$$\sigma = \frac{2.3}{1.8} \, (2\pi \times 6.3) \times 0.715 = 5.7 \text{ mcps}$$

Now, our two-stage resistively loaded amplifier, with $R_s = 130$ ohms, gave us a pole at 5.5 mcps, which is actually close enough to the required value, considering the expected tolerances in circuit and device parameters. If we leave R_L at 130 ohms to simplify the comparison, then basically a three-pole linear-phase (approximation) amplifier is obtained by simply using $L_s = 5$ μhenries in the first stage of the resistive cascade.

The step response of the corresponding amplifier is shown in Fig. 18·18. By comparing this result with the maximally flat three-pole amplifier, we can see that gain has been lost, but a considerable improvement in rise time and overshoot results. By comparison with the resistive cascade, the midband gain is the same, but the rise time is somewhat more than a factor of two better than the resistive cascade.

18·2 BAND-PASS AMPLIFIERS

In designing video amplifiers, the principal problem is to select the load impedance and bias point so as to get the maximum gain-bandwidth product per stage. The fundamental limitation which one faces in this task is that the transistor shunt capacitance limits the bandwidth. The nature of this restriction is such that low-impedance levels are necessary, and a relatively low-stage gain results. Furthermore, since the amplifier is a low-pass amplifier, the load impedances will be predominantly resistive, with inductance and capacitance being added only to "trim" the amplifier characteristic at the edge of the passband.

These facts tend to minimize the importance of coupling between the input and output circuits of a given stage. As a result, a simplified hybrid-π model can be used to give a reasonably accurate basis for the initial design. The input-output coupling effects of C_c are included by simply adding a magnified capacitance $C_c(1 + g_m R_L)$ in parallel with C_π at the transistor input and then treating the transistor as a unilateral element.

The situation is very different in typical bandpass (or IF) amplifiers, however. Here, the band of frequencies to be amplified is very small compared to f_T, so the gain which can be obtained in a stage is very high (assuming the center frequency is not close to f_T). Furthermore, the band of frequencies to be amplified is centered at a frequency that is considerably higher than the bandwidth. In order to achieve this condition it is necessary to "short out" the response at both low and high frequencies. This can be done by using a parallel-tuned interstage network, as shown in Fig. 18 · 19a; or it may be accomplished with transformers as in Fig. 18 · 19b, where the transformer magnetizing inductance is tuned by a capacitor to the appropriate center frequency. In either case, the variation of load susceptance with frequency determines the shape of the passband, and it is therefore necessary to adjust the L and C values in the final amplifier to

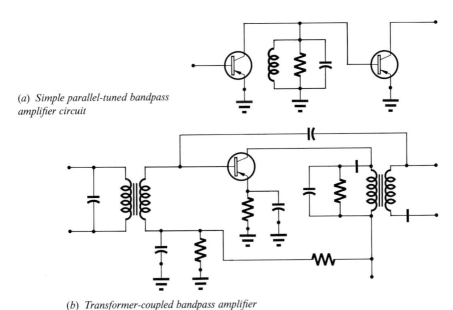

(a) *Simple parallel-tuned bandpass amplifier circuit*

(b) *Transformer-coupled bandpass amplifier*

FIG. 18 · 19   SOME TYPICAL FORMS FOR *IF* AMPLIFIERS

produce the desired gain characteristic. Unfortunately, the adjustments are rather critical, and it is quite easy to make an oscillator out of an IF amplifier. It is therefore useful to study the conditions for oscillation in an IF amplifier briefly before developing the design theory so that we can be reasonably sure that our amplifier design will be free of oscillation difficulties.

18·2·1 Oscillations in band-pass amplifiers. The conditions which produce oscillation in an IF (or any other) amplifier can be stated in several ways, though they can all be reduced to this: if the real part of the small-signal input admittance (or impedance) of the transistor is negative at some frequency, then oscillations *may* occur. *May* is emphasized because oscillations need not occur just because the input impedance has a negative real part.

The ease with which a negative small-signal input conductance can be obtained in an IF amplifier can be appreciated quite simply by referring to the phasor diagram shown in Fig. 18·20. In this figure we use V_π as a reference voltage, and assume g_m is real (that is, no excess phase shift), so that the current $g_m V_\pi$ is in phase with V_π. We also assume that the operating frequency is below the resonant frequency of the tuned load circuit, so that Z_L can be represented by a series combination of resistive and inductive components.

Under these conditions, the load voltage V_L leads V_π (note the reference direction for V_L on Fig. 18·20), and the voltage across C_c, which is $V_\pi + V_L$, also leads V_π. Now the current I_{Cc} must lead $V_\pi + V_L$ by 90°, so it leads V_π by more than 90°. The input admittance seen at the terminals AA of the active structure therefore has a *negative* real part of magnitude

$$G_{AA} = \left| \frac{I_{C_r}}{V_\pi} \right|$$

If this is sufficient to cancel g_π, plus any equivalent shunt conductance due to losses in r_b and the input circuit, then the possibility of oscillation exists.

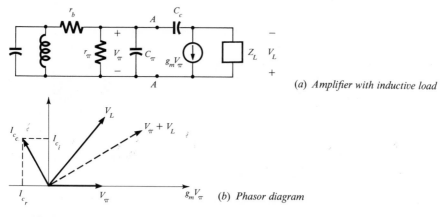

(a) Amplifier with inductive load

(b) Phasor diagram

FIG. **18·20**   CONDITIONS FOR POSSIBLE OSCILLATION IN A BANDPASS AMPLIFIER

Notice that the input admittance at the terminals AA also has an equivalent capacitive susceptance of

$$B_{AA} = \left| \frac{I_{C_i}}{V_\pi} \right|$$

If V_L were in phase with V_π (at band center where $Z_L = R_L$), then

$$B_{AA} = \frac{\omega C_c V_\pi (1 + g_m R_L)}{V_\pi}$$

giving an equivalent capacitance which is exactly the same as that added in the video amplifier studies.

18·2·2   Neutralization and mismatching.   The oscillation tendencies in an IF amplifier lead to several changes in design philosophy and realization of a practical circuit. From a theoretical standpoint, it is necessary to account for *all* the input-output coupling mechanisms which exist in a real transistor and all the small-phase shifts which are neglected in the hybrid-π model. This is most conveniently done by basing the design of an IF amplifier directly on measured two-port parameters for the transistor rather than the hybrid-π model. The parameters (usually y parameters) are either given by the manufacturer or measured by the designer.

From a practical standpoint, oscillation problems are minimized by choosing suitable structures to eliminate or substantially reduce the major sources of input-output coupling (primarily C_c and the header capacitance C_{CB}). There are two common ways of doing this. One popular scheme which can readily be used when transformers (or tapped inductances) are used in the interstages is to return a capacitor from the secondary of the output transformer to the transistor base, as shown in Fig. 18·19b. Since a phase reversal of the output voltage can be obtained by appropriate connection of the transformer, this external feedback is opposite in phase to that provided by C_c (to the extent that r_b is negligible). When the external and internal feedback exactly cancel, the stage is said to be *neutralized.*

When transformers are not used in the interstages, it is customary to reduce the importance of input-output coupling by "loading"—that is, we simply reduce the magnitude of the load impedance to the point where the total input conductance cannot be negative.[1] Thus oscillations cannot exist. While gain must be sacrificed to do this, the sacrifice is not very large, particularly when the frequency passband is wide, and this technique finds application in these cases.

18·2·3   Band-pass amplifier design.   Any transistor amplifier can be designed on the basis of two-port parameters that are measured over the frequency range of interest.[2] However, this procedure is most useful for band-pass

---

[1] This procedure is developed in detail in J. G. Linvill and J. F. Gibbons, "Transistors and Active Circuits," McGraw-Hill Book Company, New York, chaps. 18 and 19.
[2] Ibid.

amplifiers, and particularly narrow-band amplifiers, where the two-port parameters can be considered to be constant over the passband. In this section, we will develop design relations for unneutralized band-pass amplifiers and apply the results to obtain a two-stage common-emitter amplifier having a bandwidth of 10 kcps (± 5 kcps) centered at 500 kcps. We will assume that simple interstages of the form shown in Fig. $18 \cdot 19a$ are to be used, all tuned to the same center frequency. The transistors to be used have the parameters

$$y_{11} = y_{ie} = (1.52 + j0.19)\ 10^{-3}\ \text{mho}$$
$$y_{12} = y_{re} = -j1.1 \times 10^{-6}\ \text{mho}$$
$$y_{21} = y_{fe} = (24 - j1.5) \times 10^{-3}\ \text{mho}$$
$$y_{22} = y_{oe} = (8.0 + j5.0) \times 10^{-6}\ \text{mho}$$

which can be considered to be constant over the passband. (These parameters are typical of transistors intended for 455-kcps IF amplifier service.)

Choice of load admittance. To begin the design considerations, we show how a load admittance can be chosen to minimize oscillation difficulties. This is done by selecting a load admittance Y_L which is such that variations in Y_L (with frequency or in the process of alignment) do not produce significant variations in the input admittance of the transistor. To select such a load admittance, we first write out the general formula for the input admittance of a two-port structure in terms of the known y parameters and Y_L (see Fig. $18 \cdot 21$).

$$y_{in} = y_{11} - \frac{y_{12}y_{21}}{y_{22} + Y_L} \qquad (18 \cdot 39)$$

We then evaluate the *fractional* change which we can expect in input admittance for a given *fractional* change in load as follows:

$$\delta \equiv \frac{(dy_{in}/y_{in})}{(dY_L/Y_L)} \qquad (18 \cdot 40)$$

$$y_{in} = y_{11} - \frac{y_{12}y_{21}}{y_{22} + Y_L}$$

$$\delta \equiv \left| \frac{dy_{in}/y_{in}}{dY_L/Y_L} \right| \simeq \left| \frac{y_{12}\,y_{21}}{y_{11}\,Y_L} \right|$$

FIG. $18 \cdot 21$ TWO-PORT STRUCTURE WITH LOAD ADMITTANCE Y_L SELECTED TO MAKE $(dy_{in}/y_{in})/(dY_L/Y_L)$ SMALL

Using Eq. (18·39), we can find δ to be

$$\delta = \frac{y_{12}y_{21}Y_L}{(y_{22} + Y_L)[y_{11}(y_{22} + Y_L) - y_{12}y_{21}]} \qquad (18·41)$$

In a practical amplifier we want to keep δ small. This will insure that variations in Y_L have a negligible effect on y_{in}. Furthermore, this must usually be accomplished with large values of Y_L (that is, low-load impedances) on account of the impedance limitations provided by the transistor. We therefore consider values of Y_L which are large compared to y_{22}, and large enough to make $(y_{12}y_{21})/(y_{22} + Y_L) \ll y_{11}$. For these values of Y_L, Eq. (18·41) becomes

$$\delta \cong \frac{y_{12}y_{21}}{y_{11}Y_L} \qquad (18·42)$$

For the transistor parameters given earlier, this equation reduces to

$$|\delta| = \frac{1.74 \times 10^{-5}}{Y_L}$$

If we assume that a $|\delta|$ of 0.05 will give an adequate margin of protection ($|\delta| \lesssim 0.1$ is usually quite adequate), then we find

$$|Y_L| \geq 3.5 \times 10^{-4}$$

That is, the load impedance which a transistor faces should have a maximum value of 2,860 ohms.

Design of the final stage. With this estimate of $|Z_L|_{max}$ in hand, we now proceed to design the final stage of the amplifier. A circuit model for the final stage is shown in Fig. 18·22, where a load conductance $G_L = 4 \times 10^{-4}$ mho ($R_L = 2500$ ohms) has been used in accordance with the estimates made above. The voltage gain of the stage shown in Fig. 18·22 is

$$\frac{V_2}{V_1} = -y_{fe}Z_L = \frac{-y_{fe}}{y_{oe} + G_L + j\omega C_1 + \dfrac{1}{j\omega L_1}} \qquad (18·43)$$

The gain at band center, where the reactances cancel and $Z_L = R_L = 2500$ ohms, is $|V_2/V_1| = 60$; or $A_v(db) = 35$ db, which is typical for a stable unneutralized IF stage.

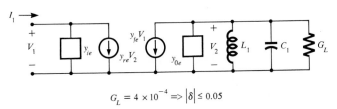

$$G_L = 4 \times 10^{-4} \Rightarrow |\delta| \leq 0.05$$

FIG. 18·22   CIRCUIT MODEL OF FINAL STAGE OF *IF* AMPLIFIER, USING *y*-PARAMETER CHARACTERIZATION OF THE TRANSISTOR

In order to obtain a two-stage passband of $\pm\Delta\omega$ at a center frequency ω_0, we want to choose C_1 and L_1 to ensure that the stage gain has dropped by 1.5 db at $\omega = \omega_0 \pm \Delta\omega$. We do this as follows.

First, the denominator of Eq. $(18\cdot43)$ can be written as

$$D(j\omega) = G_L + j\omega C_1 + \frac{1}{j\omega L_1} \qquad (18\cdot44)$$

to a very good approximation because $Y_L \gg y_{oe}$. And since y_{fe} is assumed constant over the passband (and beyond), the frequency variation of $D(j\omega)$ will determine the frequency characteristics of the amplifier.

These variations of $D(j\omega)$ with frequency can be obtained in several ways, though perhaps the simplest is as follows. We write

$$D(j\omega) = G_L + j\omega C_1 + jB(\omega) \qquad (18\cdot45a)$$

where

$$B(\omega) = \omega C - \frac{1}{\omega L} \qquad (18\cdot45b)$$

Now, at band center we want

$$B(\omega_0) = 0 = \omega_0 C - \frac{1}{\omega_0 L} \qquad (18\cdot46)$$

which gives us one relation between C and L, knowing ω_0 ($=2\pi \times 500$ kcps).

In addition, the rate of change of B with ω for a narrow band about ω_0 can be evaluated as

$$\frac{dB}{d\omega}\bigg|_{\omega_0} = C + \frac{1}{\omega_0{}^2 L} \qquad (18\cdot47)$$

Thus, G_L and B_L can be plotted as functions of ω in the manner shown in Fig. $18\cdot23$. [The dotted extension of $B(\omega)$ suggests the behavior of B, an LC reactance function, over a broader range of frequency.]

The -3-db frequencies for the voltage gain of the final stage will occur at the two points where $|B(\omega)| = G_L$ (in general). If we use the approximation that $(dB/d\omega)$ is constant over the band of interest, we have

$$\left(\frac{dB}{d\omega}\bigg|_{\omega_0} \Delta\omega\right) = G_L \qquad (18\cdot48)$$

Using Eq. $(18\cdot47)$ in Eq. $(18\cdot48)$ then gives

$$\left(C_1 + \frac{1}{\omega_0{}^2 L_1}\right)\Delta\omega = G_L \qquad (18\cdot49)$$

This equation, together with Eq. $(18\cdot46)$, provides the two relations needed to find C_1 and L_1 when G_L, ω_0, and $\Delta\omega$ are known. Formally, we obtain

$$C_1 = \frac{G_L}{2\Delta\omega} \text{ (narrow band)} \qquad (18\cdot50a)$$

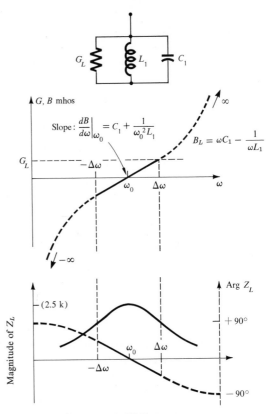

FIG. 18·23   ADMITTANCE CHARACTERISTICS OF A PARALLEL GLC CIRCUIT

$$L_1 = \frac{1}{\omega_0^2 C_1} = \frac{2\Delta\omega}{\omega_0^2 G_L} \text{ (narrow band)} \qquad (18 \cdot 50b)$$

EXAMPLE   To obtain a 3-db passband at ± 5 kcps in a cascade of two identical stages, we want each stage to have a 3-db band edge at $(\pm 5 \text{ kcps}/0.64) = \pm 7.8$ kcps. In terms of Fig. 18·23, we want the $B(\omega)$ curve to intersect the G_L line at $\Delta\omega = 2\pi(7.8 \times 10^3)$. Using $G_L = 4 \times 10^{-3}$, this requires

$$C_1 = \frac{4 \times 10^{-3}}{4\pi \times 7.8 \times 10^3} \simeq 4{,}100 \text{ pfarads}$$

$$L_1 \cong \frac{1}{\omega_0^2 C_1} = 25 \ \mu\text{henries}$$

Impedance transformation in band-pass amplifiers. While the values of L_1 and C_1 just calculated are quite reasonable, low-loss tuned circuits are usually made with somewhat larger values of L and somewhat smaller values of C. As a result, some form of impedance transformation is frequently employed to yield more desirable values of L and C. One convenient procedure is the use of an autotransformer, or tapped coil. The coil is wound on a ferrite rod or annulus, so that the coupling coefficient between turns is essentially unity. As a result, the voltage step-up of the transformer shown in Fig. 18·24 is

$$\frac{V_3}{V_2} = n \qquad\qquad (18·51)$$

The equivalent inductance appearing in parallel with G_L is then

$$L = \frac{L_T}{n^2} \qquad\qquad (18·52a)$$

where L_T is the total inductance of the coil. Similarly, the capacitance appearing in parallel with G_L is

$$C = n^2 C_T \qquad\qquad (18·52b)$$

EXAMPLE   If in the preceding example we want to use $C_T = 500$ pf (to obtain a stable, low-loss capacitor), then we need

$$n^2 = \frac{4,100}{500} = 8.2 \qquad n = 2.86$$

The corresponding total inductance would be

$$L_T = 25 \ \mu\text{henries} \times 8.2 = 205 \ \mu\text{henries}$$

The resulting stage is shown in Fig. 18·24 (with biasing elements neglected). The load impedance can also be transformed to a different value if this is required.

Design of intermediate stages.   To design the intermediate stages of a band-pass amplifier (or the first stage in a two-stage cascade), we need to evaluate the input admittance to the final stage, since this is a part of the load admittance of the preceding stage. In general, this may be done from the formula

$$y_{in} = y_{11} - \frac{y_{12}y_{21}}{y_{22} + Y_L} = y_{ie} - \frac{y_{re}y_{fe}}{y_{oe} + Y_L} \qquad\qquad (18·53)$$

Noting that the voltage gain of the last stage is

$$\frac{V_2}{V_1} = \frac{-y_{fe}}{y_{oe} + Y_L}$$

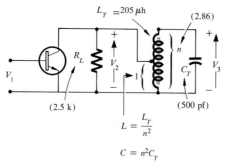

FIG. 18·24   TAPPED COIL ARRANGEMENT FOR TRANSFORMING VALUES OF L AND C

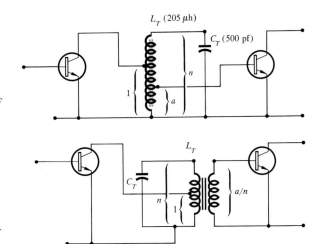

(a) Use of a tapped coil in an IF interstage ($a:1:n = 0.5:1:2.86$)

(b) Combination tapped-primary interstage transformer

FIG. 18·25   IMPEDANCE TRANSFORMATION IN THE INTERSTAGE OF AN *IF* **AMPLIFIER**

we can recast Eq. (18·53) in the form

$$y_{in} = y_{ie} + y_{re} \frac{V_2}{V_1}(j\omega) \tag{18·54}$$

EXAMPLE   For the transistor parameters given earlier, we have

$$y_{in} = (1.52 + j0.19)10^{-3} - j1.1 \times 10^{-6} \frac{V_2}{V_1}(j\omega)$$

Since $|V_2/V_1|$ has a maximum value of 60, it cannot have any significant effect on the real part of y_{in}. At band center, it reduces the imaginary part of y_{in} to $j0.13 \times 10^{-3}$, and this value is then essentially constant across the -1.5-db frequency range (or -3-db bandwidth of the two-stage amplifier). The input impedance of the final stage thus consists of a parallel *RC* network with $R = 660$ ohms, $C = 41.5$ pfarads.

If we want to increase this impedance to 2,500 ohms, so that the stage can be designed as before, then we can again use a tapped coil arrangement of the type just described, this time providing an additional tap at the point shown in Fig. 18·25a. The 660-ohm, 41.5-pfarad combination is tapped in at point *a*, so that the collector will face a resistive load of 2,500-ohms at band center. To achieve this condition we use

$$\frac{1}{a^2} = \frac{2,500}{660} = 3.8 \qquad a \simeq 0.5$$

The 41.5-pfarad capacitor is transformed to $(41.5/3.8)$ pfarads $\simeq 11$ pfarads, which is negligible in comparison to the 4,000 pfarads required at this point.

An alternate scheme which is more popular in practice is to use a transformer with a tapped primary, as shown in Fig. 18·25b. Such a transformer is usually specified only as a 455-kcps IF transformer with a tapped primary and a given total primary/secondary impedance-transformation ratio. The transformer required for our previous example would be specified as providing a $(2.5 \text{ kohm} \times 8.2)/660 =$

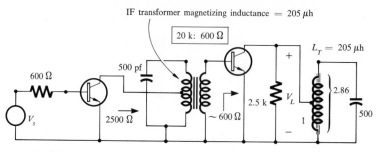

IF transformer magnetizing inductance = 205 μh

20 k: 600 Ω

$L_T = 205\ \mu h$

500 pf

600 Ω

V_s

2500 Ω

~600 Ω

2.5 k

V_L

2.86

500

1

FIG. 18·26 TWO-STAGE *IF* AMPLIFIER PROVIDING A TRANSDUCER GAIN OF **46** DB FOR TRANSISTORS WITH $y_{ie} = (1.52 + j0.19)10^{-3}$ MHO; $y_{re} = -j1.1 \times 10^{-6}$ MHO; $y_{fe} = (24 - j1.5)10^{-3}$ MHO; $y_{0e} = (8.0 + j5.0)10^{-6}$ MHO; THE LOAD ADMITTANCES ARE SELECTED TO GIVE $|\delta| \simeq 0.05$

20.5 kohm/660 ohm impedance-transformation ratio, with a primary tap at 2,500 ohms (20 kohm/600 IF transformers are standard items).

In either case, an appropriate impedance transformation can be effected which will make the load impedance and gain function for the first stage have the same form as the gain function for the final stage. Of course, the fact that the emitter is tapped down from the collector to the point $a = 0.5$ means that the voltage gain from the input terminals of the first transistor to the input terminals of the second transistor will be only 30 at band center.

> *Input termination.* In view of the essentially constant 660-ohm resistive input impedance which each stage provides at 500 kcps, the input termi-nation would usually be selected to match this impedance directly. If the source impedance is not already 600 ohms, a transformer or tapped coil can be used to transform the source resistance to this value. The amplifier will then provide maximum transducer power gain from source to load.
>
> On the assumption that the source resistance is already 600 ohms, the complete small-signal amplifier circuit is shown in Fig. 18·26. The voltage gain V_L/V_s provided by this amplifier is 900 or about 50 db. The imped-ance transformation from 600 to 2,500 ohms means that the transducer power gain is $10 \log_{10} (2,500/600) \simeq 6$ db less than the voltage gain in db, or 44 db. If more gain than this is required, an intermediate stage can be added to give about 30 db more gain.

18·3 THE MAXIMUM FREQUENCY OF OSCILLATION OF A TRANSISTOR

In Chap. 14 we studied some oscillator circuits in which the frequency of operation was low enough that the transistor acts primarily as an amplifying element. In such cases the oscillation frequency is determined primarily by circuit elements external to the transistor. As one increases the frequency of oscillation, however, the transistor parameters begin to play a dominant role, and finally they set an upper limit on the oscil-lation frequency. This limit is called the maximum frequency of oscillation and denoted by f_{max}. f_{max} is a very useful figure of merit of the transistor, since its maximum power gain will be unity at this frequency. f_{max} is thus the maximum frequency at which the transistor can be considered to be an active device.

18·3·1 *A simple circuit for analysis.* Fortunately, there is a very simple and extremely practical oscillator circuit which can be made to oscillate up to the maximum frequency of oscillation. This circuit is shown in Fig. 18·27, together with a circuit model for it which utilizes the T model for the transistor. We wish to analyze this circuit to find its oscillation frequency and condition for oscillation.

To do this, we shall attempt to find the impedance which the transistor presents at the points AA. In order for the circuit to oscillate at a given frequency, this impedance must have a negative real part equal to the resistance associated with L_2 (assuming C_2 without loss); and at the same frequency the reactance of the L_2 and C_2 combination must be the negative of that presented at the terminals AA by the transistor.

To calculate the impedance at the terminals AA, we first make some simplifying assumptions to get at the core of the matter. First of all, the r_e and C_e combination might as well be considered to be part of the external circuit. It will provide an equivalent positive series resistance to be added to the resistance of L_2. Thus, if negative resistance is going to be produced at all, it must be produced by the structure shown in Fig. 18·26c. In the right-hand part of this figure, Z' stands for the complex impedance of C_c in parallel with $r_b + Z$.

The impedance at the terminals BB is simply

$$Z_{BB} = \frac{V_{BB}}{I_e} = \frac{[(1-\alpha)I_e]Z'}{I_e} = (1-\alpha)Z' \qquad (18\cdot55)$$

Now, if oscillations are to be possible, the argument of Z_{BB} must satisfy the inequality

$$|\arg Z_{BB}| > 90° \qquad (18\cdot56)$$

In fact, we can define the maximum frequency of oscillation to be a frequency at which $|\arg Z_{BB}|$ is just equal to $90°$.

From Eq. (18·56), it is apparent that

$$\arg Z_{BB} = \arg(1-\alpha) + \arg Z' \qquad (18\cdot57)$$

so the oscillation condition becomes

$$|\arg(1-\alpha) + \arg Z'| > 90° \qquad (18\cdot58)$$

Now one fact is immediately obvious from Eq. (18·58): Since $\arg(1-\alpha)$ and $\arg Z'$ are each less than $90°$, they must both have the same sign if the inequality is to be met at all.

Let us study $(1-\alpha)$ vs. frequency for the moment, using a single-pole

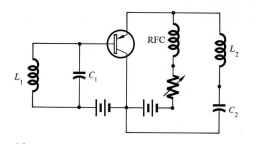

(a)

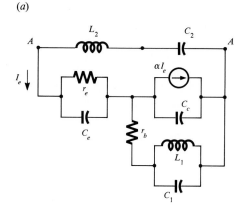

(b)

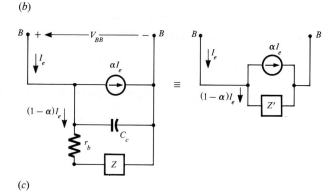

(c)

FIG. 18·27 A SIMPLE HIGH-FREQUENCY OSCILLATOR CIRCUIT

expression for α:

$$\alpha = \frac{\alpha_0}{1 + s/\omega_\alpha} \tag{18·59}$$

From Eq. (18·59), we find

$$1 - \alpha = \frac{(1 - \alpha_0) + s/\omega_\alpha}{1 + s/\omega_\alpha} \simeq \frac{s/\omega_\alpha}{1 + s/\omega_\alpha} \tag{18·60}$$

A phasor diagram (or polar plot) of $(1 - \alpha)$ vs. $s = j\omega$ is a semicircle, as shown in Fig. 18·28. The magnitude of $(1 - \alpha)$ is the length of the radius vector from the origin to the curve, and the argument of $(1 - \alpha)$ is the angle which this radius vector makes with the horizontal axis.

It is apparent that, to the extent that α may be represented by a single-pole expression, arg $(1 - \alpha)$ is always positive. Hence, the oscillation condition can only be satisfied if arg Z' is positive. This condition is shown in Fig. $18 \cdot 28b$ where the impedance

$$Z_{BB} = (1 - \alpha)Z' \qquad (18 \cdot 61)$$

is plotted for a typical case. As is indicated there, Z_{BB_r} is negative and, hopefully, large enough to compensate losses in r_e and r_{L_2}. If it is, and the reactance of L_2 and C_2 in series is the negative of Z_{BBi}, the circuit will oscillate at the frequency ω_0 (see Fig. $18 \cdot 28b$).

The question is, how high can ω_0 be made? It is quite apparent that as ω increases, arg $(1 - \alpha)$ decreases. Furthermore, as ω increases, arg Z' also decreases since Z' has a shunt capacitance C_c as the lead element. Hence, a frequency will exist at which

$$\text{arg } (1 - \alpha) + \text{arg } Z' = 90° \qquad (18 \cdot 62)$$

and this will be the maximum frequency of oscillation.

Now, it is possible to show (see problems) that the maximum positive angle of Z' at a given frequency ω_0 is

$$(\text{arg } Z')_{max} = \phi_{max} = \tan^{-1}\left(\frac{1}{4\omega_0 C_c r_b} - \omega_0 C_c r_b\right) \qquad (18 \cdot 63)$$

Furthermore, the argument of $(1 - \alpha)$ is

$$\text{arg } (1 - \alpha) = 90° - \tan^{-1}\frac{\omega}{\omega_\alpha} \qquad (18 \cdot 64)$$

Using $(18 \cdot 63)$ and $(18 \cdot 64)$ in $(18 \cdot 62)$, we find

$$\tan^{-1}\left(\frac{1}{4\omega_{max} C_c r_b} - \omega_{max} C_c r_b\right) - \tan^{-1}\frac{\omega_{max}}{\omega_\alpha} = 0 \qquad (18 \cdot 65)$$

or

$$\frac{1}{4\omega_{max} C_c r_b} - \omega_{max} C_c r_b = \frac{\omega_{max}}{\omega_\alpha} \qquad (18 \cdot 66)$$

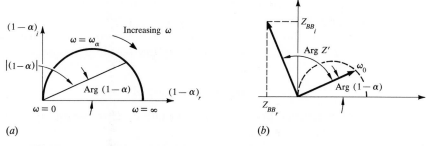

(a) (b)

FIG. $18 \cdot 28$ THE ANALYSIS OF THE HIGH-FREQUENCY OSCILLATOR GIVEN IN FIG. $18 \cdot 27$ USING A SINGLE-POLE EXPRESSION FOR α

as the equation which ω_{max} must satisfy. The maximum frequency of oscillation is thus

$$f_{max} = \left[\frac{f_\alpha}{8\pi C_c r_b (1 + \omega_\alpha C_c r_b)} \right]^{1/2} \qquad (18 \cdot 67a)$$

For many types of transistors, $\omega_\alpha C_c r_b \ll 1$, so

$$f_{max} \simeq \sqrt{\frac{f_\alpha}{8\pi r_b C_c}} \qquad (18 \cdot 67b)$$

is essentially the maximum frequency of oscillation.

EXAMPLE As a numerical example, we assume $f_\alpha = 400$ mcps, $r_b = 40$ ohms, $C_c = 1$ pfarad. Using these numbers in Eq. ($18 \cdot 67b$) gives $f_{max} = 650$ mcps. Notice that f_{max} is higher than f_α. Of course, the calculated f_{max} is somewhat higher than the real f_{max}, because we have found (approximately) the frequency at which the real part of Z_{BB} goes to zero. However, the estimate provided by Eq. ($18 \cdot 66$) is sufficiently close for most purposes that it is frequently used as a figure of merit for the transistor.[1]

Actually, Eq. ($18 \cdot 67$) may be predicted on quite general grounds from energy considerations. It therefore follows that the simple circuit shown in Fig. $18 \cdot 27$ is capable of oscillating at any frequency up to the maximum allowable under the present approximations.

The transit-time mode. So far we have been working with the assumption that α could be represented reasonably well with an expression of the form

$$\alpha = \frac{\alpha_0}{1 + s/\omega_\alpha}$$

For transistors with a uniform base, this approximation is sufficiently correct that the previous analysis gives a good estimate of f_{max} [primarily because f_{max} depends on $(f_\alpha)^{1/2}$ or $(f_T)^{1/2}$]. However, we saw in Chap. 17 that the α of a drift transistor exhibits very significant excess phase shift. As a result, $|\alpha|$ can have an appreciable magnitude when $|\arg \alpha| > 180°$, as shown in Fig. $18 \cdot 29$. The interesting point about this plot is that $\arg (1 - \alpha)$ can have *negative* values. Thus the possibility exists that the basic oscillation condition given in Eq. ($18 \cdot 56$) can be met by making $\arg Z'$ negative.

Now, it is quite easy to make $\arg Z' = -90°$. All we need to do is adjust L_1 and C_1 to be resonant at the required frequency. Then the inequality

$$\arg (1 - \alpha) + \arg Z' > 90°$$

is met whenever $\arg (1 - \alpha)$ is negative. This condition is shown in Fig. $18 \cdot 29c$.

---

[1] In a diffused-base transistor, a better estimate is obtained by replacing f_α by f_T.

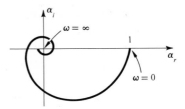

(a) Sketch of α for a transistor with drift field in the base

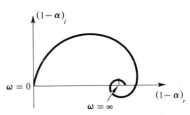

(b) Sketch of $(1 - α)$ for the α given in (a)

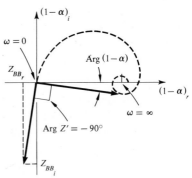

(c) Dynamic negative resistance in the transit time region

FIG. **18 · 29** **THE POSSIBILITY OF OSCILLATION FOR FREQUENCIES WHERE ARG** $(1 - α)$ **IS NEGATIVE**

It is interesting to note that if L_1 and C_1 are in resonance, then no rf current flows in r_b. Thus there is no rf loss in r_b, and r_b cannot have anything to do with the determination of the oscillation frequency. For the case presently being considered, the transistor is behaving like a two-terminal negative resistance. The negative resistance is produced by the fact that injected carriers take time to traverse the base region, and if the frequency is high enough, the ac emitter-to-collector voltage may have changed sign before the injected carriers are collected. This oscillation mode is thus sometimes called the *transit-time mode,* distinct from the *three-terminal mode* discussed in the previous section.

Thus the simple oscillation circuit shown in Fig. 18 · 27 *may* be capable of oscillating in two modes. For transistors with uniform doping density in the base, the transit-time possibility is quite remote because $(1 - α)$ does not have enough phase shift in the fourth quadrant of Fig. 18 · 29c. For a diffused-base transistor the possibility of oscillation in both the three-terminal and transit-time modes is less remote, though transit-time oscillations have not been reproducibly observed with commercially available transistors.

18 · 4 SUMMARY

In this concluding chapter, we have described video amplifier and band-pass amplifier design and considered briefly the maximum frequency of oscillation of the transistor. Each of these topics brings us up against some of the fundamental limitations of transistors as active circuit elements, and as a result each topic is a field of specialization in its own right. The reader should therefore consider this chapter as only an introduction to a field of considerable literature.

REFERENCES

Linvill, J. G., and J. F. Gibbons: "Transistors and Active Circuits," McGraw-Hill Book Company, New York, 1961, chaps. 18 and 19.

Pettit, J. M., and M. M. McWhorter: "Electronic Amplifier Circuits," McGraw-Hill Book Company, New York, 1961, chaps. 4, 5, 7 and 9.

Hunter, L. B.: "Handbook of Semiconductor Electronics," McGraw-Hill Book Company, New York, 1964, 2nd ed., chap. 14.

Searle, C. D., et al., "Elementary Circuit Properties of Transistors," SEEC Notes, vol. 3, John Wiley & Sons, Inc., New York, 1965, chap. 8.

Joyce, M. V., and K. K. Clarke: "Transistor Circuit Analysis," Addison-Wesley Publishing Company, Inc., Reading, Mass., 1961, chaps. 8 and 9.

DEMONSTRATION

STEP RESPONSE OF A LOW-PASS AMPLIFIER. An interesting verification of the results of this chapter can be obtained by making step-response measurements on the experimental amplifier shown in Fig. 13 · 28 and described in Fig. D13 · 1. The results are (1) a rise time of 9 μs, (2) a delay time of 3 μs, and (3) a sag of 2% in 80 μs. These results can be related to the amplifier frequency response given in Chap. 13.

PROBLEMS

18 · 1 A 2N916 (relevant parameters given on p. 789) is to be used in the amplifier shown in Fig. 18 · 2. Calculate T_R, T_D, and the sag for this stage.

18 · 2 To minimize sag, a decoupling network of the form shown in Fig. 18 · 6 is to be used in the amplifier of Prob. 18 · 1. What value of C_d is required to give zero initial slope on $v_2(t)$?

18 · 3 Calculate the upper cutoff frequency of the amplifier of Prob. 18 · 1 and verify numerically that $T_R B = 0.35$.

18 · 4 The gain-bandwidth product of a stage can be slightly improved by the judicious use of emitter degeneration. To do this, we insert in the emitter lead of Fig. 18 · 2 an emitter resistor R_e and a capacitor C_e in parallel with it.

Two $R \| C$ networks are now in the emitter lead in series with each other. The values of R_e and C_e will be selected to affect only the high-frequency response of the stage; R_E and C_E control the bias point and low-frequency cutoff.

1. Show that, if $R_e C_e = 1/(2\pi f_T)$, the hybrid-π model is transformed into a new hybrid-π model with $r_\pi{}^* = r_\pi(1 + g_m R_e)$, $C_\pi{}^* = C_\pi/(1 + g_m R_e)$, $g_m{}^* = g_m(1 + g_m R_e)$.

2. Using the parameters of the 2N916 given on p. 789, compute the gain-bandwidth product and the C_e required for $R_e = 2, 5, 10, 20$ ohms.

3. Compute the performance you expect in the amplifier shown in Fig. $18 \cdot 2$ when these values of R_e are used (i.e., T_R versus R_e).

4. Discuss your results by showing that the ratio of r_b to $r_\pi{}^*$ is decreased by using R_e.

5. For $C_E = 100$ μfd, what effect will the values of R_e given in part 2 have on the low-frequency cutoff (and hence on the sag)? (Cf. Fig. $13 \cdot 15$ and relevant discussion.)

18·5 Verify the results shown in Fig. $18 \cdot 13$.

18·6 The amplifier shown in Fig. $13 \cdot 28$ has the following step response: (1) rise time $\simeq 9$ μs; (2) delay $\simeq 3$ μs; (3) sag $\simeq 2\%$ in 80 μs. The transistors are 2N324 with $f_T = 4$ mcps, $C_c = 18$ pf, and $\beta \simeq 130$. Using the methods of this chapter, compare these measurements with the values theoretically expected for T_R, T_D and sag.

18·7 It is desired to build a video amplifier having a gain of 1,000, a rise time of 0.1 μs, and a sag of no more than 10% in 50 μs. The transistors to be used are 2N916 (relevant parameters given on p. 789). Design the amplifier. Note that the parameters for the 2N916 are at a bias point $I_C = 5$ ma, $V_{CE} = 5$ volts. (You should design the biasing network assuming a supply voltage of 22.5 volts.)

18·8 The y parameters for a 2N918 are given in Fig. $17 \cdot 29$. Design a two-stage IF amplifier using this transistor, using $\delta = 0.05$. The center frequency should be 100 mcps and the 3-db points should be at ± 1 mcps. The biasing network can be of the type given in Fig. $18 \cdot 19b$. How much transducer gain does this amplifier provide when a matched source is used?

18·9 Using the parameters for a 2N916, as given on p. 789, estimate the maximum frequency of oscillation of this transistor. Compare this frequency with f_T. The transistor can give power amplification (via impedance transformation) in the frequency range $f_T < f < f_{max}$.

18·10 Using the formula for $\alpha(s)$ given in Eq. $(17 \cdot 61)$ with the m and f_α values given in Fig. $17 \cdot 25$, estimate the negative resistance that can be obtained from this transistor in the frequency range where arg $(1 - \alpha)$ is negative.

APPENDIX: NATURAL FREQUENCIES AND THEIR RELATION TO AMPLIFIER FREQUENCY RESPONSE

THE SMALL-SIGNAL amplifiers considered in Chaps. 13 and 14 are *linear systems*. This means that when they are excited by sources with an exponential time dependence of waveform $e^{j\omega t}$ (or more generally, e^{st}), they will respond with waveforms at the same frequency. However, in addition to the response at the frequency of the excitation, a linear system will also exhibit responses at its *natural frequencies*. The natural frequencies are the frequencies at which signals can circulate in a system without applied sources.

Generally, the natural frequencies of a system can lie anywhere in the complex-frequency plane, and it is the designer's job to place them in desirable positions for the application at hand. In a sine-wave oscillator, for example, the natural frequencies should lie on the $j\omega$ axis, or perhaps slightly into the right half of the complex-frequency plane (see Sec. $14 \cdot 3$). In an amplifier, on the other hand, they should lie in the left half-plane, at positions which depend on the application. There are several ways of finding the natural frequencies of a system, which complement each other in evaluating system performance. In developing each method, we shall always start from the basic definition of a natural frequency: a frequency at which signals can circulate in a system without applied sources. We shall also use two-port (h) parameters to characterize the network under consideration to maintain generality in the development.

Loop-impedance criterion. We shall first use the basic criterion to show that *the loop impedance around any loop must be zero at a natural frequency* (assuming that the network does not have isolated parts). To do this, consider the h-parameter system shown in Fig. $13 \cdot 6$, which is driven by a voltage source V_s. The source impedance is Z_s. The input impedance of the two-port network is, from Eq. $(13 \cdot 10c)$,

$$Z_{in} = h_{11} - \frac{h_{12}h_{21}}{h_{22} + Y_L} \qquad (A \cdot 1)$$

The current flowing in the input loop of the network must satisfy the equation

$$V_s = I_1(Z_s + Z_{in}) \qquad (A \cdot 2)$$

If we want a current I_1 to circulate, with V_s equal to zero, we must evidently have

$$Z_{loop} = Z_s + Z_{in} = 0 \qquad (A \cdot 3a)$$

which implies

$$(h_{11} + Z_s)(h_{22} + Y_L) - h_{12}h_{21} = 0 \qquad (A \cdot 3b)$$

Equation $(A \cdot 3b)$ is called the *characteristic equation* of the system. When each term in this equation is expressed as a function of s (including the h parameters), the left-hand side will be a function of s whose zeros will be the natural frequencies of the system. Of course, we have proved the criterion only for the input loop, but we could choose other ports in the system, redefine the h parameters Z_s and Y_L appropriately, and still establish the same criterion.

Nodal-admittance criterion. We can also show that *the nodal admittance across any node pair is zero at a natural frequency*. This condition will lead to the same characteristic equation as Eq. $(A \cdot 3b)$. To begin the development we replace the voltage source V_s and series source impedance Z_s of Fig. $13 \cdot 6$ with a current source I_s and parallel source admittance Y_s. The input admittance of the two-port is, from Eq. $(13 \cdot 10e)$,

$$Y_{in} = \frac{1}{Z_{in}} = \frac{h_{22} + Y_L}{h_{11}(h_{22} + Y_L) - h_{12}h_{21}}$$

Now, the voltage across the input port, with I_s as a driving source, must satisfy

$$I_s = V_1(Y_s + Y_{in})$$

and if we want V_1 to exist without external drive ($I_s = 0$), then we require

$$Y_{node} = Y_s + Y_{in} = 0 \qquad (A \cdot 4a)$$

Since $Y_s = 1/Z_s$, Eq. $(A \cdot 4a)$ can be written in the equivalent form

$$(h_{11} + Z_s)(h_{22} + Y_L) - h_{12}h_{21} = 0 \qquad (A \cdot 4b)$$

which is the same as Eq. $(A \cdot 3b)$.

Alternately, if we use Eq. $(13 \cdot 12e)$ for Y_{out}, the total nodal admittance at the output port of the two-port, including Y_L, is

$$Y_{node} = Y_L + Y_{out} = Y_L + h_{22} - \frac{h_{12}h_{21}}{h_{11} + Z_s}$$

which can also be manipulated to yield the characteristic equation.

Transfer-function criterion. As our final criterion, we can show that *the natural frequencies of a network are the poles of its transfer function*. In the *h*-parameter case, we simply note that, from Eq. $(13 \cdot 15e)$, we have

$$A_v = \frac{V_2}{V_s} = \frac{h_{21}}{(h_{11} + Z_s)(h_{22} + Y_L) - h_{12}h_{21}}$$

where it is assumed that all the h parameters Z_s and Y_L are functions of frequency. In order to obtain an output voltage V_2 with no external drive V_s, we want $A_v \to \infty$, which requires that the denominator of Eq. $(A \cdot 5) \to 0$. But the denominator is just the characteristic equation obtained earlier [Eqs. $(A \cdot 3b)$ and $(A \cdot 4b)$].

The fact that the characteristic equation can be found in several ways sometimes simplifies analysis. Some examples of oscillator analysis simplified in this manner are given in Sec. $14 \cdot 3$. Following is an example showing how the concept of natural frequencies is useful in determining amplifier gain functions.

EXAMPLE   To calculate the low frequency response of the amplifier shown in Fig. $(13 \cdot 19a)$, we begin by simply writing the amplifier gain function in the general form

$$A_v(s) = K \frac{s + s_0}{s + s_p} \qquad (A \cdot 6)$$

We can evaluate $A_v(s)$ by finding the constants K, s_0, and s_p. Since we know that the poles of the system (s_p) are also its natural frequencies, we can choose any criterion to evaluate s_p. In this case the loop-impedance criterion is convenient:

$$Z_{loop} = \frac{1}{s_p C_s} + R_s + r_b + r_\pi = 0$$

$$\Rightarrow s_p = \frac{-1}{C_s(R_s + r_b + r_\pi)} \qquad (A \cdot 7)$$

This then determines the denominator of the low-frequency gain function.

To determine K and s_0, we use physical reasoning to ensure that $A_v(s)$ has the proper behavior as $s \to 0$ and $s \to \infty$. To be specific,

1. As $s \to \infty$ (that is, $s \gg s_p$ or s_0), $A_v(s) \to A_{v0}$. This requires $K = A_{v0}$, the midband gain.

2. As $s \to 0$, the response must tend to zero, because C_s will not pass dc. Hence we must set $s_0 = 0$.

The gain function is therefore

$$A_v(s) = A_{v0} \frac{s}{s + s_p}$$

for the amplifier shown in Fig. (13·19), where s_p is given by Eq. (A·7).

It is frequently true that amplifier gain functions can be most readily obtained by using a loop-impedance or nodal-admittance criterion to find the natural frequencies and then physical reasoning to determine the form of the numerator. For further examples, see C. L. Searle, et al., "Elementary Circuit Properties of Transistors," SEEC Notes, vol. 3, John Wiley & Sons, Inc., 1964, pp. 187–193.

INDEX